# Pathogenesis of Pregnancy-Related Complication

# Pathogenesis of Pregnancy-Related Complication

Editor

**Ilona Hromadnikova**

MDPI • Basel • Beijing • Wuhan • Barcelona • Belgrade • Manchester • Tokyo • Cluj • Tianjin

*Editor*
Ilona Hromadnikova
Department of Molecular
Biology and Cell Pathology
Third Faculty of Medicine
Charles University
Prague
Czech Republic

*Editorial Office*
MDPI
St. Alban-Anlage 66
4052 Basel, Switzerland

This is a reprint of articles from the Special Issue published online in the open access journal *International Journal of Molecular Sciences* (ISSN 1422-0067) (available at: www.mdpi.com/journal/ijms/special_issues/Pathogenesis_Pregnancy).

For citation purposes, cite each article independently as indicated on the article page online and as indicated below:

LastName, A.A.; LastName, B.B.; LastName, C.C. Article Title. *Journal Name* **Year**, *Volume Number*, Page Range.

**ISBN 978-3-0365-7327-4 (Hbk)**
**ISBN 978-3-0365-7326-7 (PDF)**

# Contents

# About the Editor

**Ilona Hromadnikova**

Prof. Ilona Hromadnikova, PhD., is the Head of the Department of Molecular Biology and Cell Pathology, Third Faculty of Medicine, Charles University, located in the Institute for the Care of Mother and Child in Prague, Czech Republic. Prof. Ilona Hromadnikova, PhD., is the laureate of the Czech Brains National Award 2022 for outstanding research in the field of Health (LOREM).

Prof. Ilona Hromadnikova, PhD., developed a novel method to predict pregnancy-related complications in the first trimester of gestation. These pregnancy-related complications involve gestational hypertension, preeclampsia, fetal growth restriction, small for gestational age fetuses, HELLP syndrome, preterm delivery, gestational diabetes mellitus, and pregnancy loss. The early identification of pregnancies at risk enables clinical teams to start efficient prevention if available, to implement more frequent monitoring, and to start therapy if necessary. The method is based on the detection of microRNA biomarkers that function in the cardiovascular system and play roles in the pathogenesis of cardiovascular diseases using real-time PCR. The method is applicable in every routine medical genetics laboratory.

Furthermore, Prof. Ilona Hromadnikova, PhD., developed a novel method to predict cardiovascular risk in mothers after complicated pregnancies and in children born from complicated pregnancies. This method is again based on screening of microRNA biomarkers that function in the cardiovascular system and play roles in the pathogenesis of cardiovascular diseases using real-time PCR. This approach enables an early implementation of preventive programmes in mothers and children at risk to decrease the incidence of later development of cardiovascular diseases. Prof. Ilona Hromadnikova, PhD., has been awarded two national patents, has filed three national patent applications, and has created three PCT applications.

# Preface to "Pathogenesis of Pregnancy-Related Complication"

Preeclampsia, HELLP syndrome, fetal growth restriction, gestational diabetes mellitus, preterm birth (preterm prelabor rupture of membranes and spontaneous preterm labor), infection/inflammation, recurrent pregnancy loss, and pregnancy-induced chronic venous disease are major complications responsible for maternal and perinatal morbidity and mortality.

Elucidation of the pathogenetic mechanisms related to human reproduction, the implantation process, and the initiation and onset of severe pregnancy-related complications is crucial, since it enables the identification of potential biomarkers for early stratification of patients at risk, early right diagnosis, and the development of novel therapeutical options. All studies performed using either in vitro or in vivo experimental models or samples derived from patients have significantly contributed to the broadening of our understanding, since altogether, they give comprehensive insight into these particular mechanisms.

Additionally, pregnancy-related complications have induced long-term metabolic and vascular abnormalities that might increase the overall risk of metabolic, cardiovascular, cerebrovascular, kidney, and other diseases later in life in mothers and their offspring. In addition, children from pregnancies with an estimated first trimester risk of complications based on the fetal nuchal translucency thickness and abnormal maternal serum levels of first trimester routine biomarkers have a higher likelihood of adverse outcomes. Therefore, they should be monitored long term.

This Special Issue provides an overview of the latest research in the field and may be of interest to scientists, gynaecologists, and obstetricians working in this highly demanding and quickly expanding area.

**Ilona Hromadnikova**
*Editor*

International Journal of
*Molecular Sciences*

MDPI

*Editorial*

# Pathogenesis of Pregnancy-Related Complications

**Ilona Hromadnikova**

Department of Molecular Biology and Cell Pathology, Third Faculty of Medicine, Charles University, 100 00 Prague, Czech Republic; ilona.hromadnikova@lf3.cuni.cz; Tel.: +42-029-651-1336

In this special edition (closed on 31 October 2022), 4 reviews, 13 original papers, 1 communication, and 1 case report are published. These papers were published by scientific teams from 24 countries, including China, Tunisia, Canada, France, Serbia, Portugal, Spain, Chile, Singapore, Australia, USA, Mexico, Czech Republic, Germany, Sweden, Finland, United Kingdom, Norway, Poland, Greece, Japan, Italy, Switzerland, and Israel, which is a really great achievement.

In the review of Chen et al. [1], the authors summarize the role of galectins, multifunctional regulators of cellular biological processes involving innate and adaptive immune responses, in human reproduction, pregnancies with normal course of gestation and pregnancy-related disorders such as preeclampsia, gestational diabetes mellitus, fetal growth restriction and preterm birth.

The review of Benkhalifa et al. [2] was dedicated to the endometrium immunomodulation to prevent recurrent implantation failures and repeated pregnancy loss in women undergoing an in vitro fertilization programme. Various possible therapeutical options are reviewed together with a profiling of predictive biomarkers of implantation before embryo transfer.

The review of Vilotić et al. [3] provided a comprehensive overview on the role of IL-6 and IL-8, inflammatory cytokines, in cycling endometrium, the feto–maternal interface, establishment of pregnancy, parturition, and in the pathogenesis of pregnancy-related complications such as pregnancy loss, preeclampsia, gestational diabetes mellitus and infection/inflammation.

Ângelo-Dias et al. [4] performed a systematic review and meta-analysis assessing the association of B cells with idiopathic recurrent pregnancy loss. They highlighted a potential association between increased levels of endometrial B cells and idiopathic recurrent pregnancy loss.

Ortega et al. [5] demonstrated that pregnancy-induced chronic venous disease is associated with a proinflammatory environment characterized by altered serum levels of multiple inflammatory cytokines and chemokines in affected women and their newborns, which might have serious consequences for both.

Peñailillo et al. [6] described the potential communication between maternal mesenchymal stem cells derived from menstrual fluid and invading trophoblast cells during the implantation process. Reduced mesenchymal stem cell-induced trophoblast invasion was observed in patients with a history of preeclampsia.

Martinez-Fierro et al. [7] present interesting findings on the pathogenesis of preeclampsia, evaluating the effect of the administration of fibroblast growth factor type 2 on the placental expression of various genes involved in angiogenesis and apoptosis in an experimental murine model of preeclampsia. The administration of fibroblast growth factor type 2 reduced the effects generated by proteinuria and hypertension and also impacted the expression of studied genes.

Our research group [8] introduced a set of cardiovascular disease-associated microRNAs as potential biomarkers for the early identification of pregnancies at risk of later development of gestational diabetes mellitus. A first trimester screening of particular microRNAs

**Citation:** Hromadnikova, I. Pathogenesis of Pregnancy-Related Complications. *Int. J. Mol. Sci.* **2023**, *24*, 5584. https://doi.org/10.3390/ijms24065584

Received: 8 March 2023
Accepted: 13 March 2023
Published: 15 March 2023

allowed the authors to predict the later occurrence of gestational diabetes mellitus both irrespective of or with respect to the severity of the disease (group of all patients, group of patients on diet only, and group of patients on the combination of diet and therapy).

Czamara et al. [9] showed that children from pregnancies with an estimated first trimester risk based on fetal nuchal translucency thickness and abnormal maternal serum levels of first trimester routine biomarkers have a higher likelihood of adverse outcomes, even if initial testing of known genetic conditions is negative. In these children, congenital malformations of the circulatory system are more frequent. These children also have more copy number duplications. These cases should therefore be followed-up during pregnancy and after the birth.

Huang et al. [10] demonstrated that PEG2-induced pyroptosis affects the progression of endometriosis by changing the migration ability of pyroptotic cells and through the upregulation of HMGB1, E-cadherin, and vimentin. These findings might support the usage of anti-inflammatory drugs in patients with endometriosis.

Lynch et al. [11] reported that a deficiency in SLC2A3, a glucose transporter located on the maternal-facing apical trophoblast membrane, results in fetal hypoglycemia, reduced fetal development, and altered metabolic hormone concentrations in sheep.

Misan et al. [12] described the destabilization of the blood–brain barrier in pregnancies complicated by fetal growth restriction. Neurological disorders in newborns, including intraventricular hemorrhage, were associated with higher serum levels of NME1, nucleoside diphosphate kinase A, and the decreased placental expression of CLN4. Both biomarkers may be predictive of the appearance of intraventricular hemorrhage in newborns in FGR pregnancies.

Mavreli et al. [13] introduced miR-125a-3p as a promising early biomarker for prediction of spontaneous preterm birth. Mir-125a-3p was identified as a potential biomarker for the early prediction of spontaneous preterm birth using small RNA-seq and confirmed by qRT-PCR.

Szala-Poździej et al. [14] reported that some FCN2 gene promoter region polymorphisms that are associated with relatively low serum levels of ficolin-2 significantly increase the risk of very low birthweight in preterm neonates.

Kedziora et al. [15] studied gene expression signatures related to the diabetic placental pathology and concluded that fetal sex has a prominent effect on the placental transcriptome in diabetic pregnancies.

Kadife et al. [16] demonstrated that the expression of SLC38A4, a system A transporter controlling non-essential amino acid uptake and supply, is persistently low in placentas derived from pregnancies complicated with early preterm intrauterine growth restriction regardless of disease etiology.

Dymara-Konopka et al. [17] evaluated the serum levels of anti- and pro-angiogenic factors in pregnancies with preeclampsia and/or fetal growth restriction. They concluded that the angiogenic imbalance reflects placental disease regardless of its clinical manifestation in the mother.

Chighine et al. [18] presented a very rare case of a fatal respiratory failure two weeks after the delivery of a healthy newborn at home in a woman with normal course of gestation who suffered from a primary hyperparathyroidism secondary to a parathyroid carcinoma.

Finally, Sammar et al. [19] showed on in vitro models of preeclampsia that the overexpression of STOX1-A/B transcription gene, discovered in familial forms of preeclampsia, leads to the decreased expression of CD24, a mucin-like immunosuppressing glycoprotein, which was observed in syncytiothrophoblasts and cytotrophoblasts in early and preterm preeclampsia.

**Conflicts of Interest:** The author declares no conflict of interest.

## References


1. Chen, M.; Shi, J.L.; Zheng, Z.M.; Lin, Z.; Li, M.Q.; Shao, J. Galectins: Important Regulators in Normal and Pathologic Pregnancies. *Int. J. Mol. Sci.* **2022**, *23*, 10110. [CrossRef] [PubMed]
2. Benkhalifa, M.; Joao, F.; Duval, C.; Montjean, D.; Bouricha, M.; Cabry, R.; Bélanger, M.C.; Bahri, H.; Miron, P.; Benkhalifa, M. Endometrium Immunomodulation to Prevent Recurrent Implantation Failure in Assisted Reproductive Technology. *Int. J. Mol. Sci.* **2022**, *23*, 12787. [CrossRef] [PubMed]
3. Vilotić, A.; Nacka-Aleksić, M.; Pirković, A.; Bojić-Trbojević, Ž.; Dekanski, D.; Jovanović Krivokuća, M. IL-6 and IL-8: An Overview of Their Roles in Healthy and Pathological Pregnancies. *Int. J. Mol. Sci.* **2022**, *23*, 14574. [CrossRef] [PubMed]
4. Ângelo-Dias, M.; Martins, C.; Dias, S.S.; Borrego, L.M.; Lima, J. Association of B Cells with Idiopathic Recurrent Pregnancy Loss: A Systematic Review and Meta-Analysis. *Int. J. Mol. Sci.* **2022**, *23*, 15200. [CrossRef] [PubMed]
5. Ortega, M.A.; Gómez-Lahoz, A.M.; Sánchez-Trujillo, L.; Fraile-Martinez, O.; García-Montero, C.; Guijarro, L.G.; Bravo, C.; De Leon-Luis, J.A.; Saz, J.V.; Bujan, J.; et al. Chronic Venous Disease during Pregnancy Causes a Systematic Increase in Maternal and Fetal Proinflammatory Markers. *Int. J. Mol. Sci.* **2022**, *23*, 8976. [CrossRef] [PubMed]
6. Peñailillo, R.; Acuña-Gallardo, S.; García, F.; Monteiro, L.J.; Nardocci, G.; Choolani, M.A.; Kemp, M.W.; Romero, R.; Illanes, S.E. Mesenchymal Stem Cells-Induced Trophoblast Invasion Is Reduced in Patients with a Previous History of Preeclampsia. *Int. J. Mol. Sci.* **2022**, *23*, 9071. [CrossRef] [PubMed]
7. Martinez-Fierro, M.L.; Garza-Veloz, I.; Castañeda-Lopez, M.E.; Wasike, D.; Castruita-De la Rosa, C.; Rodriguez-Sanchez, I.P.; Delgado-Enciso, I.; Flores-Mendoza, J. Evaluation of the Effect of the Fibroblast Growth Factor Type 2 (FGF-2) Administration on Placental Gene Expression in a Murine Model of Preeclampsia Induced by L-NAME. *Int. J. Mol. Sci.* **2022**, *23*, 10129. [CrossRef] [PubMed]
8. Hromadnikova, I.; Kotlabova, K.; Krofta, L. Cardiovascular Disease-Associated MicroRNAs as Novel Biomarkers of First-Trimester Screening for Gestational Diabetes Mellitus in the Absence of Other Pregnancy-Related Complications. *Int. J. Mol. Sci.* **2022**, *23*, 10635. [CrossRef] [PubMed]
9. Czamara, D.; Cruceanu, C.; Lahti-Pulkkinen, M.; Dieckmann, L.; Ködel, M.; Sauer, S.; Rex-Haffner, M.; Sammallahti, S.; Kajantie, E.; Laivuori, H.; et al. Genome-Wide Copy Number Variant and High-Throughput Transcriptomics Analyses of Placental Tissues Underscore Persisting Child Susceptibility in At-Risk Pregnancies Cleared in Standard Genetic Testing. *Int. J. Mol. Sci.* **2022**, *23*, 11448. [CrossRef] [PubMed]
10. Huang, Y.; Li, R.; Hu, R.; Yao, J.; Yang, Y. PEG2-Induced Pyroptosis Regulates the Expression of HMGB1 and Promotes hEM15A Migration in Endometriosis. *Int. J. Mol. Sci.* **2022**, *23*, 11707. [CrossRef] [PubMed]
11. Lynch, C.S.; Kennedy, V.C.; Tanner, A.R.; Ali, A.; Winger, Q.A.; Rozance, P.J.; Anthony, R.V. Impact of Placental SLC2A3 Deficiency during the First-Half of Gestation. *Int. J. Mol. Sci.* **2022**, *23*, 12530. [CrossRef] [PubMed]
12. Misan, N.; Michalak, S.; Rzymski, P.; Poniedziałek, B.; Kapska, K.; Osztynowicz, K.; Ropacka-Lesiak, M. Molecular Indicators of Blood-Brain Barrier Breakdown and Neuronal Injury in Pregnancy Complicated by Fetal Growth Restriction. *Int. J. Mol. Sci.* **2022**, *23*, 13798. [CrossRef]
13. Mavreli, D.; Theodora, M.; Avgeris, M.; Papantoniou, N.; Antsaklis, P.; Daskalakis, G.; Kolialexi, A. First Trimester Maternal Plasma Aberrant miRNA Expression Associated with Spontaneous Preterm Birth. *Int. J. Mol. Sci.* **2022**, *23*, 14972. [CrossRef]
14. Szala-Poździej, A.; Świerzko, A.S.; Gajek, G.; Kufelnicka-Babout, M.; Chojnacka, K.; Kobiela, P.; Jarych, D.; Sobczuk, K.; Mazela, J.; Domżalska-Popadiuk, I.; et al. Association of the FCN2 Gene Promoter Region Polymorphisms with Very Low Birthweight in Preterm Neonates. *Int. J. Mol. Sci.* **2022**, *23*, 15336. [CrossRef] [PubMed]
15. Kedziora, S.M.; Obermayer, B.; Sugulle, M.; Herse, F.; Kräker, K.; Haase, N.; Langmia, I.M.; Müller, D.N.; Staff, A.C.; Beule, D.; et al. Placental Transcriptome Profiling in Subtypes of Diabetic Pregnancies Is Strongly Confounded by Fetal Sex. *Int. J. Mol. Sci.* **2022**, *23*, 15388. [CrossRef] [PubMed]
16. Kadife, E.; Harper, A.; De Alwis, N.; Chien, K.; Hannan, N.; Brownfoot, F.C. SLC38A4 Amino Acid Transporter Expression Is Significantly Lower in Early Preterm Intrauterine Growth Restriction Complicated Placentas. *Int. J. Mol. Sci.* **2022**, *24*, 403. [CrossRef] [PubMed]
17. Dymara-Konopka, W.; Laskowska, M.; Grywalska, E.; Hymos, A.; Błażewicz, A.; Leszczyńska-Gorzelak, B. Similar Pro- and Antiangiogenic Profiles Close to Delivery in Different Clinical Presentations of Two Pregnancy Syndromes: Preeclampsia and Fetal Growth Restriction. *Int. J. Mol. Sci.* **2023**, *24*, 972. [CrossRef] [PubMed]
18. Chighine, A.; Corona, A.; Catani, G.; Conte, C.; Demontis, R.; Nioi, M. A Fatal Case of Metastatic Pulmonary Calcification during the Puerperium. *Int. J. Mol. Sci.* **2022**, *23*, 15131. [CrossRef] [PubMed]
19. Sammar, M.; Apicella, C.; Altevogt, P.; Meiri, H.; Vaiman, D. Modeling Preeclampsia In Vitro: Polymorphic Variants of STOX1-A/B Genes Can Downregulate CD24 in Trophoblast Cell Lines. *Int. J. Mol. Sci.* **2022**, *23*, 15927. [CrossRef] [PubMed]

International Journal of
*Molecular Sciences*

MDPI

*Review*

# Galectins: Important Regulators in Normal and Pathologic Pregnancies

**Min Chen [1], Jia-Lu Shi [1], Zi-Meng Zheng [1], Zhi Lin [1], Ming-Qing Li [1,2,3,*] and Jun Shao [1,4,*]**

[1] Laboratory for Reproductive Immunology, Hospital of Obstetrics and Gynecology, Shanghai Medical School, Fudan University, Shanghai 200080, China
[2] NHC Key Lab of Reproduction Regulation, Shanghai Institute for Biomedical and Pharmaceutical Technologies, Fudan University, Shanghai 201203, China
[3] Shanghai Key Laboratory of Female Reproductive Endocrine Related Diseases, Hospital of Obstetrics and Gynecology, Fudan University, Shanghai 200080, China
[4] Department of Obstetrics, Hospital of Obstetrics and Gynecology, Shanghai Medical School, Fudan University, Shanghai 200011, China
* Correspondence: mqli@fudan.edu.cn (M.-Q.L.); junshao@fudan.edu.cn (J.S.)

**Abstract:** Galectins (Gal) are characterized by their affinity for galactoside structures on glycoconjugates. This relationship is mediated by carbohydrate recognition domains, which are multifunctional regulators of basic cellular biological processes with high structural similarity among family members. They participate in both innate and adaptive immune responses, as well as in reproductive immunology. Recently, the discovery that galectins are highly expressed at the maternal–fetal interface has garnerd the interest of experts in human reproduction. Galectins are involved in a variety of functions such as maternal–fetal immune tolerance, angiogenesis, trophoblast invasion and placental development and are considered to be important mediators of successful embryo implantation and during pregnancy. Dysregulation of these galectins is associated with abnormal and pathological pregnancies (e.g., preeclampsia, gestational diabetes mellitus, fetal growth restriction, preterm birth). Our work reviews the regulatory mechanisms of galectins in normal and pathological pregnancies and has implications for clinicians in the prevention, diagnosis and treatment of pregnancy-related diseases.

**Keywords:** galectin; maternal–fetal interface; pathologic pregnancy; preeclampsia; gestational diabetes mellitus; fetal growth restriction; preterm birth

**Citation:** Chen, M.; Shi, J.-L.; Zheng, Z.-M.; Lin, Z.; Li, M.-Q.; Shao, J. Galectins: Important Regulators in Normal and Pathologic Pregnancies. *Int. J. Mol. Sci.* **2022**, *23*, 10110. https://doi.org/10.3390/ijms231710110

Academic Editor: Ilona Hromadnikova

Received: 5 August 2022
Accepted: 29 August 2022
Published: 3 September 2022



## 1. Introduction

The maternal–fetal interface features a number of complex and tightly regulated mechanisms. These include physical defenses through intercellular junctions, the secretion of cytokines, and protection against microbial infections through the innate immune response. In addition, the interaction between fetal trophoblast and maternal decidual immune cells can also promote immune tolerance [1,2]. Studies suggest that a successful pregnancy is the result of multiple steps, including maternal immunological adaptation, normal blastocyst development, functional placental development and endometrial receptivity formation [3]. During early gestation, trophoblasts in the human placenta have two main differentiation pathways: they differentiate into chorionic villous trophoblasts and extrachorionic villous trophoblasts. Extravillous trophoblasts (EVT) have both migratory and invasive phenotypes for remodeling spiral arteries and can invade the uterus [4].

Galectins are believed to play a crucial role in reproductive processes, such as maternal–fetal immune tolerance, embryo implantation and angiogenesis [5]. Evidence points out that galectins are involved in the establishment and maintenance of normal pregnancy. Gal-1, gal-3 and gal-9 are the main participants in these processes, with other family members also contributing significantly to immune–endocrine interactions and maternal–fetal immunological responses [5,6]. Galectins show diverse intra and extracellular localization and biological functions [7]. Through a number of clinical trials, scientists have found that their most considerable role in pregnancy appears to be the modulation of the maternal–fetal immune response, with some galectins suppressing the mother's immune response to the fetus, thus maintaining a normal pregnancy [8]. Therefore, we reviewed the role of galectins in normal and pathological pregnancies.

## 2. The Galectins (Gal) Family

Galectins are a family of galactoside-binding proteins found in animals, bacteria and fungi [9]. They consist of a core sequence of 130 amino acids and a highly conserved carbohydrate recognition domain (CRD) [10]. Two defining characteristics of the lectin family including: significant similarity in the amino acid sequence and high affinity for galactoside [11]. Currently, researchers have found 15 galectins in mammals, including gal-1-15. On the basis of their structural differences, these galectins have been categorized into three types (Figure 1): prototype galectin, chimeric galectin and tandem repeat galectin. Prototype galectins contain a single carbohydrate recognition domain, including gal-1, gal-2, gal-5, gal-7, gal-10, gal-11, gal-13, gal-14 and gal-15. Chimeric galectins are self-associated with a c-terminal CRD and a non-carbohydrate bound n-terminal structural domain, while gal-3 is the only member of this group. Tandem repeat galectins are dimers consisting of a linker peptide joining two CRDs, and these include gal-4, gal-6, gal-8, gal-9 and gal-12 [12,13].

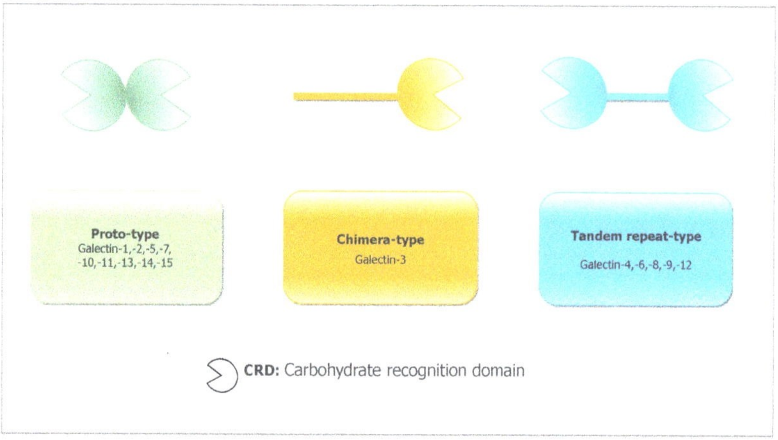


**Figure 1.** The classification and structure of galectins. The proto-type galectins are in green, which are non-covalently connected. The chimera-type galectins are in yellow, which are self-associated with a c-terminal CRD and a non-carbohydrate bound n-terminal structural domain. The tandem repeat-type galectins are in blue, which are dimers consisting of a linker peptide joining two CRDs.

In a large number of different cell types, galectins are found both intracellularly, such as in the nucleus, cytoplasm and cell membrane, and extracellularly, especially in the extracellular matrix [14]. Galectins are synthesized in the cytosol on free ribosomes [15,16]. Because they lack a signal sequence for secretion, which is probably related to their ability to bind glycans, they utilize nonclassical secretion pathways [14]. As a result, vesicle-disruption-induced accumulation and binding to luminal glycans may serve as a molecular link between autophagy and galectin secretion [17]. However, in another study, the maturation of N-linked glycoprotein was not necessary for galectin-3 transport from the cytosol to the extracellular space, and the majority of the released galectin-3 was not packaged into the extracellular vesicles [18]. The results of a study on galectin-3 showed that oligomerization may be required for galectins secretion [19–21].

Galectin-1 may secrete similarly to fibroblast growth factor-2 (FGF-2), the most researched unconventionally secreted protein that follows the direct translocation pathway [22]. Galecitins are released from cells in the extracellular vesicles (EVs), either in the microvesicles created by membrane blebbing or in the exosomes derived from multivesicular bodies (MVBs) [23]. Extracellular matrix components and some inflammatory factors can affect the secretion of galectins [13,24]. Extracellular galectins are capable of binding to various cell surface receptors to form carbohydrates [25]. Tyrosine kinase receptors and T cell glycoproteins (TCR, CD3, CD43, CD45 and CD7) are just two of the many receptors that galectins interact with to control cytokine signaling and receptor activation. Integrins and galectins collaborate to control cell adhesion and death. Additionally, contact with Gal3 encourages lamellipodia generation, actin and focal adhesion turnover and fibronectin modification all increase cell adhesion and migration [26]. Additionally, recombinant galectins can exert multiple in vitro activities by binding cell surface glycans and the extracellular matrix [27].

## 3. The Expression of Galectins at Maternal-Fetal Interface

The maternal–fetal interface connects maternal tissue and fetal components in direct contact, and its local immune response plays a role in protecting the fetus during the establishment and maintenance of pregnancy and the onset of labor [28]. Gal1-4, gal-9, and gal-12 are expressed in the endometrium, with gal-1 and gal-3 expressing the most. Immunohistochemistry detected gal-1 in the endometrial stroma, as well as the decidua and gal-3 in the endometrial glandular epithelium [29]. This was also demonstrated by immunohistochemistry, in post-pregnancy mice, where the gal-3 was located mainly in the luminal epithelium and glandular epithelium and reached a maximum at 2–4 and 6–8 days, respectively [30]. Gal-3 was rare in non-pregnant animals or during preimplantation. Later in pregnancy, gal-3 was found in the decidual basement, uterine glands and placental trophoblast cells. Decidualized and non-decidualized endometrium lacked gal-3 [31]. Finally, gal-3 was shown to be associated with endometrial receptivity and implantation [30,32].

Galectins are abundantly expressed in the reproductive system, and their expression in the human reproductive system is summarized in Table 1. During the first trimester, gal-1 is mainly expressed in human placental cytotrophoblasts (CTB) and syncytiotrophoblast (STB) [33,34]. Gal-1 also was detected in decidual cells, suggesting that this galectin promotes inter-trophoblast and trophoblast–stromal cell interactions during placental formation interactions [35]. Additionally, the double immunofluorescence confirmed that the expression of gal-2 was mainly in syncytiotrophoblast and maternal decidua [36] but also in extravillous trophoblast and fetal endothelial cells [37]. Gal-2 is primarily immunomodulatory through its anti-inflammatory effects [38].

Gal-3 was found by immunohistochemistry, which is expressed in all trophoblastic lineages but decreased in the villous trophoblast (VT) and trophoblast columns in first and third trimester [35,37], suggesting that gal-3 expression is associated with a shift in cell phenotype, namely, the change from a proliferative to a migratory phenotype [39]. As pregnancy progresses, circulating levels of gal-3 gradually increase, indicating that its expression is regulated throughout development [39,40]. Through in vitro experiments, investigators found that exogenous gal-3 positively regulates trophoblast function and induces cell invasion, tube formation and fusion [41]. Gal-7 was immunolocalized to syncytiotrophoblast, extravillous cytotrophoblast and glandular epithelial cells of the placenta in early pregnancy and to syncytiotrophoblast and endothelial cells in term placentas, but no endothelial cell staining was seen in pre-eclamptic placentas [42]. Results by immunolocalization showed gal-8 expression in both villi and EVT [43].

Gal-9 mRNA was expressed in the human endometrium, especially in endometrial epithelial cells, and was significantly increased in the mid-and to late-secretory stages, in the window of implantation and in the decidua [44]. Ultrastructural immunocytochemistry confrimed the localization of gal-9 in the endometrium was, where it was mainly located in the apical protrusions of the endometrial epithelium, a type of protrusion also known as the uterodome [44]. It plays a role in endometrial receptivity; however, there was no gal-9 observed between the uterodomes [45]. Importantly, compared with other kinds of T cells, gal-1 and -10 have much higher expression levels in CD4+ CD25+ Treg cells, where they perform an essential role in the suppression of immune responses [46,47].

Gal-12 is minimally expressed in adipocytes and is required for in vitro adipocyte differentiation. This protein can regulate in vivo lipolysis, total body adiposity, and glucose homeostasis [48]. Gal-13 (placental protein 13, PP13) was found in STB in chorionic villi and sometimes in multinucleated luminal trophoblasts within converted decidual spiral arterioles. However, gal-13 was not detected in the cytotrophoblast, and the anchoring trophoblast and invasive trophoblast [49,50]. In early gestation, gal-14 is predominantly expressed in the STB and its placental expression is decreased in women with miscarriage [7]. LGALS15 (a gene of gal-15) is only expressed in Caprinae endometrium and serves as a peri-implantation attachment factor [51].

**Table 1.** The expression of galectins in maternal–fetal interface.

| Galectins | Expression | Reference |
|---|---|---|
| Galectin-1 | Decidua; Endometrium; Extravillous trophoblast Immune cells (i.e., dNK; CD4+CD25+Treg cells); Syncytiotrophoblast | [29,47,52,53] |
| Galectin-2 | Decidua; Endometrium; Extravillous trophoblast; Syncytiotrophoblast | [29,38] |
| Galectin-3 | Cytotrophoblast; Endometrium; Villous trophoblast | [29,39,40] |
| Galectin-4 | Endometrium | [29] |
| Galectin-7 | Decidua; Endometrium; Extravillous trophoblast; Glandular epithelial cells; Syncytiotrophoblast | [29,42,54] |
| Galectin-8 | Endometrium; Syncytiotrophoblast; Extravillous trophoblast; Villous trophoblast | [29,43] |
| Galectin-9 | Cytotrophoblast; Decidua; Endometrium | [29,44] |
| Galectin-10 | Immune cells (i.e., CD4+CD25+Treg cells) | [47] |
| Galectin-12 | Endometrium | [29] |
| Galectin-13 | Syncytiotrophoblast | [49,50] |
| Galectin-14 | Syncytiotrophoblast | [7] |
| Galectin-15 | Caprinae endometrium | [51] |

## 4. The Role of Galectins in the Maternal–Fetal Interface

### *4.1. Immune Regulation*

Successful pregnancy is a complex physiological process and a major immune challenge [55]. It requires the synchronization of endometrial and embryonic development and maintenance of a delicate equilibrium between inflammation and immune tolerance [56]. The endometrium undergoes decidualization in response to the regulation of pregnancy-related hormones, accompanied by the enrichment and reissue of immune cell subpopulations [57,58]. A normal pregnancy is similar to a successful pure heterozygous semi-allogeneic transfer in which the mother does not reject the embryo carrying the father's antigens, but rather establishes a unique immune tolerance mechanism through a subtle immune dialogue [28,59,60]. The maternal–fetal interface consists mainly of trophoblast cells, decidua stromal cells and decidua immune cells, which, under hormonal regulation, produce a variety of cytokines that create a specific immune tolerance environment between the mother and the fetus, facilitating a successful pregnancy [3,61].

Additionally, the regulation of multiple galectins is involved in regulating maternal–fetal immune tolerance [62,63]. Almost all immune cells can express galectins, which are upregulated in activated B cells, T cells, macrophages and decidual natural killer (dNK) cells [63,64]. Thus, galectins make a difference in maternal–fetal immunotolerance (Figure 2).

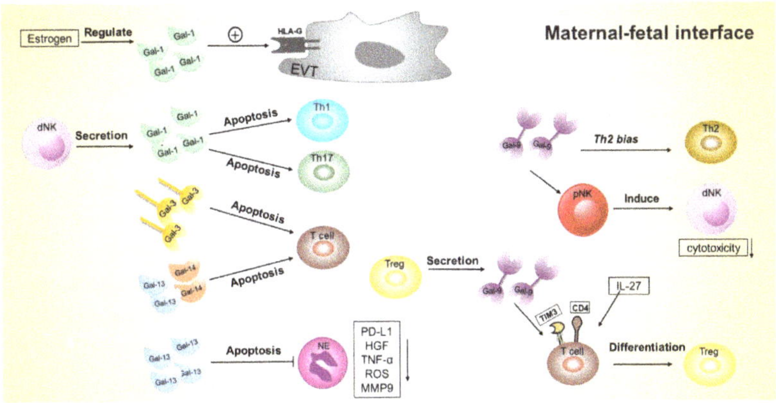


**Figure 2.** Galectin affects the function of immune cells at the maternal–fetal interface, maintaining maternal–fetal immunological tolerance. (1) Gal-1, regulated by estrogen, modulates HLA-G expression on EVT. (2) dNK cells produce gal-1, which can induce apoptosis of Th1 and Th17 cells. (3) Gal-3 induces the apoptosis of T cells. (4) Gal-13 and gal-14 can induce the apoptosis of T cells. (5) Gal-13 reduces the rate of apoptosis in neutrophils and increases the expression of PD-L1 and the production of HGF, TNF-$\alpha$, ROS and MMP-9 in neutrophils. (6) Gal-9 signal is important for the regulation of PBMC function toward a Th2 bias. (7) Gal-9 induce peripheral NK cells to a dNK-like phenotype. (8) Treg cells have high levels of gal-9, and gal-9 interacts with Tim-3 to promote the differentiation of decidual Tim-3+ CD4+ T cells into Treg cells.

During pregnancy, gal-1 modulates the inflammatory response, promotes immune tolerance and prevents maternal rejection of the fetus. Therefore, it is a key regulator of maternal–fetal immune tolerance and may have therapeutic implications for pathologic pregnancies [63]. This is because recombinant gal-1 induces tolerogenic dendritic cells, promotes the secretion of interleukin-10 (IL-10) and regulates the expansion of T cells. In addition, gal-1 and progesterone have a synergistic effect in maintaining pregnancy [63].

Gal-1 has pro-apoptotic activity on activated CD4+ T cells of the subtypes Th1, Th17 and CD8+ T cells. dNK cells can produce gal-1 [53], and the supernatant of cultured dNK cells can induce apoptosis in T cells, which can be blocked by anti-gal-1 antibodies. Instead of peripheral T cells, decidual T cells can bind gal-1. This suggests that decidual immune cells form a privileged maternal–fetal immune microenvironment by secreting gal-1 and binding gal-1 [65]. Molvarec et al. found that gal-1 is possibly related to innate and acquired immune cell activation [66].

Gal-1 is regulated by estrogen, which may be one of the regulatory mechanisms involved in maternal–fetal immune tolerance [67]. The estrogen-responsive element in the promoter of LGALS1 (gene of gal-1) is conserved human cis motifs, present in the placenta and involved in how steroid hormones influence the level of gal-1 expression. Amino acid substitutions occur at key residues in early mammalian evolution, including the acquisition of cysteine residues, which regulate immune function through redox-state-mediated conformational changes that disable sugar binding and dimerization [68]. In addition, gal-1 regulates the expression of human leukocyte antigen G (HLA-G) on EVT cells, thereby promoting maternal–fetal immune tolerance [69]. These are the possible mechanisms by which hormonal and redox responses regulate the involvement of the gal-1 in immune responses.

Gal-3 is essential during gestation, and in mice it is mainly located at the embryo implantation site. The binding partner of gal-3, cubilin, was isolated in the uteroplacental complex. In the last week of pregnancy in mice, cubilin co-localized with gal-3 in the yolk sac epithelium and was found in uterine NK cells [70]. Furthermore, cytoplasmic gal-3 protects T cells from apoptosis while increasing cell proliferation [71]. In contrast, gal-3 induced T cell apoptosis [72] and inhibited CD-66a expression [73].

Gal-9 expression is increased during pregnancy [74], which was expressed in the placental spongiotrophoblast, where decidual immune cells displayed lower toxicity and higher PD-L1 expression levels relative to peripheral immune cells. T-cell immunoglobulin and mucin domain 3 (Tim-3) plays a function in immune control by attaching to its ligand, gal-9, which triggers effector T cell exhaustion or apoptosis [75]. Local Tim-3 expression was higher than in the periphery and decidua, which reduced lysis activity. In preeclampsia, maternal immune cells (T cells, cytotoxic T cells, NK cells, CD56(dim) NK cells) express less Tim-3 [76,77]. Tim-3/Gal-9 decreases NK cell toxicity to trophoblast cells by converting NK cells into dNK [78]. Regulatory T cells (Tregs) produce gal-9 at increasing levels to maintain maternal–fetal immunological tolerance as pregnancy progresses [79]. He et al. suggest that gal-9 can modify PBMC function to Th2 bias, maintaining pregnancy [80]. IL-27 and gal-9 can synergistically induce Tim-3+Treg cells in vitro [81]. Tim-3(+) pNK cells release anti-inflammatory cytokines and activate regulatory T cells in a TGF-1-dependent manner, causing immunosuppression. Tim-3(+) pNK cells decreased miscarriage risk in NK-deficient mice. Moreover, Tim-3/Gal-9 signaling regulates immunological control by pNK cells, a critical player in maternal–fetal immune tolerance [82].

The Tim-3/Gal-9 pathway stimulation causes midterm M2 macrophage conversion. Tim-3 and gal-9 expression were elevated in dysfunctional decidual macrophages at embryonic day 9(E9), showing that this pathway is engaged in early pregnancy and embryo development [83]. The LGALS9 D5 isoform inhibits interferon production by decidual natural killer cells. In human spontaneous abortion patients, researchers detected six LGALS9 splice variants [84]. Tim-3/Gal-9 modulates the cellular activity of dNK to maintain normal pregnancy and a result has been confirmed in human aborted decidua and in a mice miscarriage model with a reduced percentage of Tim(+) dNK cells [85].

Gal-10, also known as Charcot–Leyden crystal protein, is the most abundant protein in eosinophils and can form characteristic crystals in tissues and secretions at sites of eosinophil-associated inflammation. Gal-10 is a characteristic expression feature of suppressor T cells, eosinophils and basophils [46,86,87].

Than et al. demonstrated that placenta-specific galectins (such as gal-13 and gal-14) are predominantly expressed by STB and induce apoptosis in T lymphocytes [64]. Gal-13 and gal-14 can induce apoptosis in Th and Tc cells. Examining activation markers revealed that gal-13 increased CD25 expression and gal-14 decreased CD71 expression on the cell surface, while both galectins increased CD95 expression on T cells [7]. In the presence of gal-13 and gal-14, inactivated T cells were capable of producing substantial amounts of IL-8. These are the mechanisms involved in the regulation of maternal–fetal immunity by gal-13 and gal-14 [7]. By decreasing the rate of apoptosis, gal-13 appears to make a significant contribution to the control of placental neutrophils by raising the production of PD-L1, HGF, TNF-$\alpha$, ROS, and MMP-9, therefore polarizing them toward a placental-growth-permissive phenotype [88]. In addition, gal-13 aggregates in the decidua may act as decoys to induce apoptosis and promote maternal immune tolerance to pregnancy [89].

### 4.2. Embryo Implantation

Successful embryo implantation is a complicated procedure, which requires the embryo and the endometrium to work together in order to coordinate a sequence of events that occur during the procedure [90,91]. An intricate chemical chain reaction, which is controlled by endocrine, paracrine and autocrine regulators present in both the embryo and the mother, is essential to the progression of embryo implantation [92]. The results of in vitro and in vivo studies have demonstrated that galectins are crucial mediators in the implantation process [93]. During embryo implantation, the enhanced expression of gal-1, -3 and -9 in endometrial epithelial cells reflect the galectins' most crucial involvement in endometrial receptivity. This is due to the increased expression of these genes in endometrial epithelial cells [29,45].

Gal-1 is an important downstream target of the P(4)-FKBP52-PR signaling pathway in the uterus, which enhances P(4) responsiveness during pregnancy, and the activation of this pathway reduced the rate of mid-gestation resorption in mice and rescued implantation failure [94]. In the absence of embryos, gal-1 expression decreases. In delayed implantation mice, gal-1 mRNA levels increase with the termination of the implantation delay [67]. Progesterone and estrogen oppose each other in uterine gal-1 mRNA levels. RU486 (progesterone receptor antagonist) reduced progesterone-induced gal-1 mRNA in ovariectomized mice uterine tissues. This regulation corresponded with the implantation procedure [67].

The discovery of complex connections between gal-3 and integrin $\beta 3$ in the control of endometrial cell proliferation and adhesion provides an in vitro model applicable to embryo implantation and endometrial receptivity [95]. BeWo cells are stimulated to produce and secrete gal-3 by 17-estradiol (E2), progesterone and human chorionic gonadotropin (hCG). Through activation of integrin $\beta 1$, recombinant gal-3 inhibited endometrial cell (RL95-2 cells) proliferation and induced apoptosis. In vitro tests confirmed the pro-apoptotic action of trophoblast-secreted gal-3 on endometrial cells [96]. Gal-3 has an anti-apoptotic impact on endometrial cells, and estrogen and progesterone can modify gal-3 synthesis [97]. In animal trials, when the gal-3 gene was knocked out in mice, considerably fewer embryos were implanted in the mice endometrium. In conclusion, embryo implantation requires an increase in gal-3 expression after pregnancy [30]. In contrast, researchers demonstrated that mice can undergo embryo implantation even without gal-1 and gal-3. Additionally, gal-5 is present during the window of implantation in the blastocyst [98].

Gal-7 is a potentially useful blood biomarker for preeclampsia and may play a significant part in the implantation of the placenta [42]. In addition, gal-7 expression is elevated in the endometrial epithelium and stroma of women with a history of miscarriage. The findings imply that gal-7 promotes the embryo's adhesion to the endometrium and that higher gal-7 levels may cause pathological pregnancy [54]. It is expressed at low levels throughout the proliferative and early secretory phases and sharply increases during the mid- and late-secretory phases, the window of implantation and in the decidua [44], indicating that gal-9 may have been involved in embryo implantation.

### 4.3. *Angiogenesis*

A number of vascular processes that need to be coordinated in a spatial and temporal manner at the interface between the mother and the fetus for pregnancy to be successful [99,100]. In the early stages, the embryo is able to successfully implant in a vascularized and receptive uterus thanks to the hormone-mediated modification of the endometrial vascular system [101,102]. In order to guarantee that the embryo is provided with oxygen and nutrients prior to the formation of a definitive placenta, this is accompanied by vasodilation and the formation of neonatal structures during the decidualization process [103]. The vascular system of the placenta continues to be remodeled as the pregnancy progresses reach needs of the fetus [103]. Disturbances in these processes are frequently associated with unfavorable pregnancy outcomes such as preeclampsia, intrauterine growth restriction (IUGR) or preterm birth [104].

The result of oncology studies have found that gal-1 can regulate tumor angiogenesis and can be a potential therapeutic target to reduce angiogenesis [105]. Researchers discovered that gal-1 demonstrates a pro-angiogenic function in early gestation, promoting vasodilation through vascular endothelial-derived growth factor (VEGF) receptor 2 signaling [106]. Gal-1 may be involved in mechanisms related to placental and maternal spiral artery remodeling. Gal-1 deficiency manifests as spontaneous PE-like disease in mice [106]. Through its mRNA-binding function, gal-1 is able to regulate angiogenesis by binding to the mRNAs of genes that are linked with angiogenesis [107]. OTX008, an inhibitor of gal-1, was found to inhibit tumor proliferation, invasion and angiogenesis [108].

The result of endometriosis studies have confirmed the involvement of gal-1 in angiogenesis. Using experimental endometriosis models induced in wild-type and gal-1-deficient (LGALS1(−/−)) mice, researchers demonstrated that gal-1 regulates the formation of vascular networks in endometriotic lesions and contributes to the growth of their ectopic foci, independently of VEGF and plasmacytoid-derived CXC-motif (CXC-KC) chemokines [109]. An increase in angiogenesis can be attributed to the synergistic impact of gal-1 and gal-3, which works by activating VEGFR-1. This activation of VEGFR (vascular endothelial-derived growth factor receptor)-1 may be related to a decrease in receptor endocytosis [110,111]. The absence of gal-3 or its inhibition led to a significant reduction in the implantation and size of endometriotic lesions, as well as the expression of VEGF and VEGFR-2, and the vascular density [112].

Gal-2 expression was significantly lower in decidua and extravillous cytotrophoblast, possibly due to its role in angiogenesis [113]. In tumor patients, an increased circulation of gal-2, -4 and -8 interacts with the vascular endothelium and significantly promotes the increased circulation of granulocyte colony-stimulating factor (G-CSF), IL-6 and monocyte chemoattractant protein-1 (MCP-1). In turn, these cytokines and chemokines boost the activity of endothelial cells during angiogenesis and metastasis [114]. Only gal-2 expression was dramatically reduced in spontaneous abortion and recurrent abortion (RA) placental trophoblast cells [113]. In studies of patients with coronary artery disease, research has confirmed that gal-2 is a novel inhibitor of arteriogenesis. The modulation of gal-2 may become a new therapeutic strategy for stimulating arteriogenesis in patients with coronary artery disease [115].

Troncoso et al. describe a unique role for gal-8 in the regulation of vascular and lymphatic angiogenesis and give evidence of its critical impacts on tumor progression [116]. Gal-8 may synergize with VEGF to promote pro-angiogenesis [117]. Gal-9 stimulates monocyte migration in vitro and produces acute inflammatory arthritis in mice, suggesting a unique role for gal-9 in angiogenesis, joint inflammation and other inflammatory diseases [118,119]. Reduced gal-9 and VEGF-A concentrations in women with previous miscarriages may be associated with angiogenic regulation [120]. Gal-12 has been shown to have an angiogenic effect in adipose tissue [121].

During pregnancy, gal-13 stimulates the dilatation of uterine arteries and veins via endothelium-dependent endothelial NO synthase (eNOS) and prostaglandin signaling pathways [89]. The carbohydrate recognition domain of gal-13 causes the structural stability of vasodilation by binding to the sugar residues of extracellular and connective tissue components [89].

## 5. Galectins in Pregnancy Disorders

The function of galectin dysregulation in abnormal pregnancies is becoming the focal point of investigation for an increasing number of researchers, with some issues linked to faulty placental development, abnormal angiogenesis and others with inflammatory responses and inappropriate maternal–fetal immunological tolerance. Figure 3 summarizes the expression of galectins in these four categories of pregnancy-related disorders and the possible pathogenic processes.

### 5.1. *Preeclampsia*

Preeclampsia is a multisystemic disorder specific to pregnancy, mainly associated with systemic small vessel spasms [122]. In early pregnancy, immune, genetic and endothelial cell dysfunction factors can lead to the spasm of the small spiral arteries of the uterus and a reduced invasiveness of trophoblast cells due to ischemia [123]. It has also been suggested that PE is associated with trophoblast immaturity, which has poor trophoblast differentiation in pathologies [124]. In mid to late pregnancy, due to local oxidative stress in the placenta from ischemia and hypoxia, endothelial cell damage is induced, resulting in the release of a large number of inflammatory factors, thus causing various clinical symptoms such as preeclampsia and eclampsia. The main clinical manifestations are hypertension and kidney damage [123].

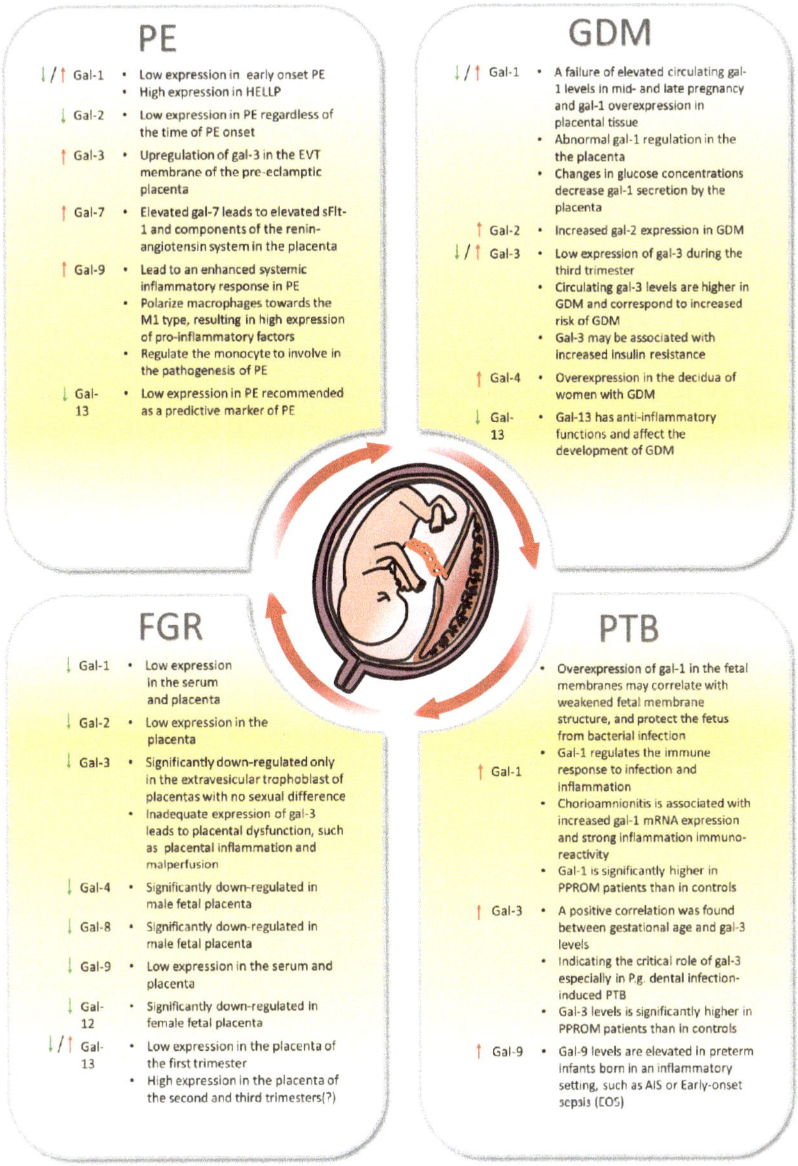


**Figure 3.** Expression of galectins in four types of pregnancy-related diseases and their possible pathological mechanisms. The four pregnancy-related diseases are preeclampsia (PE), gestational diabetes mellitus (GDM), fetal growth restriction (FGR) and preterm birth (PTB). The figure describes the dysregulation of galectins in these diseases. Red up-arrow means up regulation; Green down-arrow means down regulation.

PE is classified as early onset (before 34 weeks of gestation) or late onset based on its onset timing (after 34 weeks of gestation) [125]. Compared with the normal group, researchers found that gal-1 was elevated in early-onset and late-onset HELLP patients [126]. The term HELLP is used to describe a clinical disease that produces hemolysis, elevated liver enzymes and low platelet count. As a serious complication of hypertension during pregnancy, it can be fatal for both mother and fetus [127]. Low gal-1 levels at 18–24 weeks, but not 27–31 weeks, predicted early-onset and late-onset PE, as well as gestational hypertension (GH). After adjusting for the effects of high blood pressure and an elevated soluble fms-like tyrosine kinase-1 (sFlt-1)/placental growth factor (PlGF) ratio at 18–24 weeks, decreased gal-1 expression is considered to be a risk factor for the development of GH and PE in pregnancy [128]. Because it is expressed at low levels in late pregnancy and at high levels after the onset of the condition, serum gal-1 levels have been proposed as an independent risk factor for gestational hypertension and preeclampsia [128]. This is due to the fact that it is expressed at low levels in late pregnancy and at high levels following the onset of the condition [128].

According to the findings of Freitag et al., patients who presented with early-onset PE showed a reduced expression of gal-1 [106]. Galectin-1 suppresses trophoblast cell growth and promotes the development of syncytium. Its suppression in the syncytiotrophoblast has been linked to early pregnancy loss [129]. Recombinant Gal-1 also promotes differentiation and invasion of trophoblast stem cells [130]. These findings provide credence to the hypothesis that gal-1 is necessary for a healthy pregnancy and highlight gal-1 as a biomarker that has the potential to be helpful in the early diagnosis of PE [106].

Gal-2 is thought to be an inhibitor of atherogenesis by regulating monocyte/macrophage numbers in a mice model [115]. In PE, both spiral artery formation and macrophage inward flow were dramatically altered. As a result, abnormal spiral artery transformation in PE may be connected to trophoblast downregulation of gal-2 expression [131]. In preeclampsia, both mRNA and protein levels of gal-2 are decreased in EVT, independent of the time of PE onset. This discovery was made possible by the fact that gal-2 was shown to be one of the molecules with reduced expression [131]. A recent study found that downregulated gal-2 seems to prevent the apoptosis of Tregs in PE [132]. In trophoblasts, gal-2 has been linked to the histone modifications H3K4me3 (trimethylated lysine 4 of the histone H3) and H3K9ac (acetylated lysine 9 of the histone H3). In addition, an increase in syncytialization was seen following incubation with Gal-2 [133]. According to Charkiewicz et al. gal-2 participates in the immunological pathogenic process of PE [134]. The reduced gal-2 may be associated with autoantibodies against this protein along a potential immunological pathway. Individuals with antiphospholipid syndrome produce autoantibodies against gal-2 [135]. Antiphospholipid syndrome is prevalent in individuals with PE, which may help to explain why women who have mild forms of PE have lower levels of gal-2 in their blood [136].

During mouse pregnancy, gal-3 deficiency leads to placental dysfunction, which is manifested by inflammation and poor perfusion. This may be associated with an elevated expression of gal-3 in the EVT, leading to the development of PE [137,138]. Furthermore, the gal-3 levels of patients with late-onset PE were considerably lower than those of patients with early-onset PE [139].

Gal-7 increased anti-angiogenic sFlt-1 splice variants in the placenta and reduced the production and secretion of ADAM12 (a catabolite and metalloenzyme 12) and angiotensinogen, which may lead to the development of PE [140]. Tim-3/Gal-9 causes a pro-inflammatory response by regulating the polarization of decidual macrophages, causing macrophages to polarize towards M1, resulting in an increased expression of pro-inflammatory factors, such as TNF-$\alpha$ and IL-1$\beta$, and a decreased expression of anti-inflammatory factors such as TGF-$\beta$ and IL-10. It can also cause placental dysfunction impairment [83]. After the administration of the recombinant gal-9 (rGal-9) protein, researchers found that liver and kidney damage as well as maternal placental dysfunction were reversed [83]. The monocyte may be implicated in the pathogenesis of PE via the Tim-3/Gal-9 pathway, and reducing Tim-3 may lower monocyte inhibitory activity [141].

The elevated expression of the Tim-3/Gal-9 pathway in PE may be involved in its systemic inflammatory response in PE, which shows that gal-9 may serve as a marker for PE [142].

Gal-13 is a lectin expressed by syncytiotrophoblast. The actin cytoskeleton, likely in conjunction with lipid rafts, regulates "nonclassical" PP13 exports from trophoblasts [143]. According to the results of studies, pregnant women with PE had lower levels of gal-13 than normal groups. However, there is no difference in gal-13 expression in the placenta or serum between pregnant women with PE and those who deliver at term [144]. In contrast, immune responses in STB microvilli were stronger in both preeclampsia and HELLP syndrome patients with term and preterm births than in controls [145]. In addition, gal-13 is highly expressed in syncytial cytoplasm protrusions, membrane vesicles and shed particles in PE and HELLP syndrome patients [143,145]. This mechanism may be due to the secretion of gal-13 into the intervillous space and perivenous area through the basal vein of the decidua, which forms gal-13 aggregates that attract and activate maternal immune cells and promote the transformation of trophoblast cells and small maternal spiral arteries [49]. PP13 induced uterine veins dilatation via the SKca–NO–BKca pathway in late pregnancy and became endothelium-dependent [146]. Researchers have proposed a gal-13 complementary therapy for the treatment of PE [89]. Screening serum gal-13 levels in the first trimester is a potential diagnostic method with high sensitivity and specificity for predicting preterm birth [147,148].

### 5.2. Fetal Growth Restriction

Fetal growth restriction (FGR), which can be diagnosed by ultrasound and maternal abdominal circumference measurements, is a disease in which the fetus does not grow to its proper potential in the uterus, usually due to placental hypoplasia and dysfunction [149,150]. A fetus with FGR is more susceptible to perinatal morbidity and mortality, as well as long-term health problems. Immediate complications include neonatal asphyxia, hypothermia and hypoglycemia, while long-term complications include intellectual disability, behavioral abnormalities and an increased incidence of hypertension, coronary heart disease and diabetes in adulthood [150,151]. Due to the importance of fetal growth for subsequent development and health and the absence of indicators for early identification and appropriate care, there is a need for in-depth study on this pregnancy issue and its prognosis [152].

Gal-1 expression is low in the serum and placenta of FGR-affected pregnant women. In addition, gal-1 may contribute to the development of FGR and serve as a possible diagnostic marker for the disease [153]. By immunohistochemistry, Jeschke et al. discovered a considerable elevation of gal-1 and gal-3 expression on EVT in placentas from patients with preeclampsia or HELLP, but no significant change was seen in FGR placentas compared to normal controls [137].

In the FGR, gal-2 and gal-13 expression was decreased in placental villous and extravillous trophoblasts; furthermore, this decrease was more pronounced in male placentas, which demonstrates significant sex differences [154]. In contrast, gal-3, the only chimeric lectin, showed no sex differences and was only significantly downregulated in extravillous cytotrophoblast of the placenta [154]. Decreased gal-3 expression during pregnancy in mice without preeclampsia syndrome leads to placental dysfunction, as well as FGR. This suggests that gal-3 deficiency leads to placental inflammation and malperfusion. Gal-3 dysregulation leads to abnormal uterine natural killer cell activation and infiltration, further contributing to the development of FGR [138]. These same findings were also observed in human pregnancies, where reduced maternal serum gal-3 levels were associated with the development of FGR in mid and late pregnancy [138].

Decreased gal-13 levels in early pregnancy are associated with a number of pathological pregnancy disorders, in particular fetal growth restriction and early onset preeclampsia. In the second and third trimesters, pregnant women with PE and FGR had concentrations that were greater than normal [155], mostlikely due to its strong association with placental growth before the transition of uterine spiral arteries begins [156]. One study, however, revealed no link between decreasing gal-13 levels and FGR. Further research is required to determine whether gal-13 assays have any relevance for assessing early pregnancies [157].

The expression of certain galectins is believed to correlate with the fetal gender. Gal-4, gal-8 and gal-9 expression in male fetal FGR trophoblast cells is dramatically reduced. In contrast, gal-9 and gal-12 expression increased in the EVT and endothelial cells of female fetuses with FGR [158]. Thus, researchers argue that tandem repeat galectins in FGR placentas have a fetal sex-specific role [158]. Using double immunofluorescence with trophoblast-specific markers, cells expressing galectin at the maternal–fetal interface in the decidua were identified. Only the extravillous trophoblast was significantly downregulated for gal-3 in FGR placentas. In contrast, both the villous and extravillous trophoblast cells of FGR placentas revealed lower gal-2 and gal-13 expressions [145,154].

### 5.3. Gestational Diabetes Mellitus

Gestational diabetes mellitus (GDM) is a serious pregnancy complication that includes two conditions: a pregnancy in a patient with pre-existing diabetes, known as pregestational diabetes mellitus (PGDM), and a first occurrence of diabetes after pregnancy, also known as gestational diabetes mellitus [159]. GDM is observed in about 90% of diabetic pregnant women. Changes to the diagnostic criteria for GDM have resulted in a considerable increase in the prevalence of GDM to above 15% [160]. Most patients with GDM recover from abnormal glucose metabolism after delivery; however, 60% will develop diabetes in the future [161]. GDM can lead to an increased incidence of spontaneous abortion in early pregnancy, complications of hypertension during pregnancy and a decrease in maternal resistance, which leads to infections. Effects on the fetus include fetal malformations and fetal growth restriction [162]. There is evidence that galectins contribute to the pathophysiology of GDM.

Elevated circulating gal-1 expression contributes to the development of GDM in mid- to late-term pregnancy, in addition to placental tissue gal-1 overexpression. Researchers detected aberrant gal-1 regulation in the local and peripheral circulation of the placenta in pregnancies with combined GDM. In addition, the relationship between LGALS1 polymorphism and GDM may suggest that genetic factors play a role in this unfavorable pregnancy outcome [163]. In GDM patients, there was an inverse relationship between glucose and gal-1, confirming these results [163].

A recent study revealed elevated gal-2 expression in STB and GDM placental decidua. These data suggest two possible conclusions about the function of gal-2 dysregulation in the pathophysiology of GDM: that elevated gal-2 expression is a reaction to the inflammatory state of GDM or whether it promotes the development of GDM [36]. Consequently, its significance in GDM and potential treatment implications must be further studied.

Gal-3 levels increased when mothers presented with gestational diabetes [164–167]. Women with late-onset GDM exhibited raised gal-3 levels in the first trimester, suggesting that gal-3 in the first trimester can serve as a predictor of GDM and may be related with increased insulin resistance [166,168]. Other researchers found that GDM patients' circulating gal-3 levels were significantly lower than those of normal pregnant women [41]. As mid- to late-term placental development correlates with increased maternal peripheral gal-3 levels, it has been postulated that trophoblast cells may be a source of circulating lectins. In the same study, it was revealed that patients with GDM had reduced serum gal-3 levels, which were only detectable in late pregnancy. This may suggest that gal-3 is vulnerable to GDM-specific hormonal and metabolic changes [41].

Gal-4 overexpression in the decidua of women with type 2 diabetes, with significantly increased nuclear and cytoplasmic levels [169].

The gal-13 serum levels in the blood of GDM patients were considerably lower than those of healthy controls, as tested by ELISA. Since gal-13 has anti-inflammatory functions and regulates the maternal immune system, a deficiency of gal-13 may lead to an imbalance in the inflammatory process of the placenta during pregnancy, thereby influencing the development of GDM [170].

### *5.4. Preterm Birth*

Preterm birth is defined as less than 37 weeks of gestation [171]. The shorter the weeks of gestation at delivery, the lower the birth weight, the worse the prognosis for the perinatal baby and the more complications after birth, accounting for 5% to 18% of pregnancy complications [172]. The causes of the majority of PTBs, which are spontaneous and idiopathic, are largely unknown [172].

Preterm premature rupture of the membranes (PPROM) refers to the rupture of the fetal membranes prior to the 37th week of pregnancy. This problem is estimated to affect 4% of pregnant women and accounts for nearly half of all preterm births [173]. Despite significant improvements in perinatal care over the past few decades, the number of preterm births caused by PPROM continues to increase [174]. It is hypothesized that galectins are related to premature membrane rupture and that premature membrane rupture is associated with up to 75% bacterial membrane and/or amniotic fluid infections [175]. Compared with healthy controls, PPROM patients showed significantly higher levels of gal-1 and gal-3. Gal-3 concentrations in maternal serum are significantly and adversely linked with gestational age and birth weight. The data show that gal-1 and gal-3 regulate critical biological processes and may be initiators of PPROM pathogenesis, predictive indicators, and targets for preventative treatments [176]. Chorioamnionitis is an acute inflammatory reaction associated with early rupture of the amniotic sac. Inflammatory factors cause chorioamnionitis by disrupting membrane junctions [177]. Gal-1 is associated with an inflammatory response [178]. IL-1, a key cytokine in chorioamnionitis, can upregulate gal-1 expression [179]. Overexpression of gal-1 in fetal membranes may correlate with weakened membrane structure, increased susceptibility to infection and eventual membrane rupture, and it is believed that gal-1 protects the fetus against bacterial infection [180]. At 21–24 weeks of gestation, the majority of spontaneous births are accompanied by histological chorioamnionitis, whereas at 35–36 weeks, this rate is approximately 10% [181]. Than et al. revealed that gal-1 modulates the immunological response to infection and inflammation and that chorioamnionitis is related with high gal-1 mRNA expression and robust inflammatory immunoreactivity. Therefore, gal-1 may be implicated in the modulation of the inflammatory response of chorioamnionitis [180].

The immunological regulator Gal-3 was highly increased in placenta, amniotic fluid and serum. Co-stimulation with gal-3 and Porphyromonas gingivalis (*P.g.*)-LPS raised cytokine levels; however, co-stimulation with gal-3 and Aggregatibacter actinomycetemcomitans (*A.a.*)- or Escherichia coli (*E. coli*)-LPS lowered cytokine levels, showing the essential involvement of gal-3 in P.g.-induced PTB. Infection with P.g.-dental produced PTB, which was related with gal-3-dependent cytokine production, prompting the development of gal-3 targeted therapy or diagnostic systems for PTB treatment [73]. A link between gestational age and gal-3 levels was identified by analyzing gal-3 levels in the cord blood of term and preterm newborns. Expression of gal-3 was induced by invasive but not definitive streptococcus lactis strains, indicating a role for gal-3 in innate immunity [182]. Nevertheless, studies found the gal-3 binding protein levels in cord blood have revealed that gal-3BP levels are elevated in cases of preterm birth, which may reflect the inflammatory status of the infant and mother, and this warrants further investigation [183].

A single facility conducted a prospective analysis of 170 preterm infants delivered at fewer than 35 weeks gestation. During the neonatal period, peripheral venous blood was collected, and gal-1, gal-3 and gal-9 were measured by ELISA [184]. Researchers found that negative correlation existed between the levels of gal-1 and gal-3 at delivery and gestational age. Gal-1 and gal-9 concentrations were significantly higher in neonates with Amniotic Infection Syndrome (AIS). Neonates with early-onset sepsis exhibited greater gal-3 concentrations than healthy infants. Observational research revealed that preterm children born in an inflammatory milieu, such as AIS or early-onset sepsis (EOS), have higher levels of galectin-1, -3 and -9. Future studies must identify if galectins mediate inflammation-induced preterm birth; if so, galectins might be the subject of treatment studies [184].

## 6. Conclusions

Galectins are a family of conserved, soluble proteins distinguished by their affinity for the galactoside structures present on a variety of glycoconjugates. Over the past several decades, galectins have been recognized as crucial for implantation and pregnancy maintenance. The results of several studies have revealed their role in trophoblast cell function and placental development. In addition, evidence suggest that they play key roles in the control of fetal–maternal immunological tolerance and angiogenesis. Additionally, an increasing number of studies reported pregnancy-related diseases with altered or dysregulated galectin expression.

This review seeks to elucidate the expression of the galectin family in the reproductive system and its function in normal and pathological pregnancy. An increasing number of researchers haves found galectins, and a substantial body of this literature indicates that specific galactose lectins may be recommended to predict pregnancy-related diseases. However, additional research is required to confirm this.

**Author Contributions:** M.C., J.-L.S., Z.-M.Z., Z.L., M.-Q.L. and J.S. drafted and revised the manuscript. M.C. generated the tables and figures; J.S. and M.-Q.L. conceived and designed the review. All authors have read and agreed to the published version of the manuscript.

**Funding:** This study was supported by the Major Research Program of National Natural Science Foundation of China (NSFC) (No. 81971362, 92057119, 31970798), the Program for Zhuoxue of Fudan University (JIF157602), the Support Project for Original Personalized Research of Fudan University (IDF157014/002).

**Institutional Review Board Statement:** Not applicable.

**Informed Consent Statement:** Not applicable.

**Data Availability Statement:** Not applicable.

**Conflicts of Interest:** The authors declare that they have no competing interest.

## Abbreviations

The following abbreviations are used in this manuscript:

| | |
|---|---|
| A.a. | Aggregatibacter actinomycetemcomitans |
| ADAM12 | A catabolite and metalloenzyme 12 |
| AIS | Amniotic Infection Syndrome |
| CKC-KC | Keratinocyte-derived CXC-motif chemokine |
| CRDs | Carbohydrate recognition domains CTB Cytotrophoblasts |
| DMs | Dysfunction of decidual macrophages |
| dNK cells | Decidual natural killer cells |
| dpc 5 | Fifth day post-coitum |
| E.coli | Escherichia coli |
| E2 | 17-estradiol |
| EVs | Extracellular Vesicles |
| E9 | Embryonic day 9 |
| eNOS | endothelial NO synthase |
| EOS | Early-onset sepsis |
| EVT | Extravillous trophoblast |
| FGF-2 | Fibroblast growth factor-2 |
| FGR | Fetal growth restriction |
| GDM | Gestational diabetes mellitus |
| GH | Gestational hypertension |
| GTD | Gestational trophoblastic disease |
| hCG | human chorionic gonadotropin |
| HELLP | Hemolysis, elevated liver enzymes, low platelet count |

| | |
|---|---|
| HLA-G | Human leukocyte antigen G |
| H3K4me3 | trimethylated lysine 4 of the histone H3 |
| H3K9ac | acetylated lysine 9 of the histone H3 |
| IUGR | intrauterine growth restriction |
| LPS | Lipopolysaccharide |
| MVBs | Multivesicular Bodies |
| MUC1 | Mucin1 |
| P.g. | Porphyromonas gingivalis |
| P4 | Progesterone |
| PBMC | Peripheral blood mononuclear cells |
| PE | Preeclampsia |
| PGDM | Pregestational diabetes mellitus |
| PIGF | Placental growth factor |
| pNK cells | peripheral natural killer cells |
| PP13 | Placental protein 13 |
| PPROM | Preterm premature rupture of the membranes |
| PTB | Preterm birth |
| RA | Recurrent abortion |
| rGal-9 | Recombinant Gal-9 |
| sFlt-1 | Soluble fms-like tyrosine kinase-1 |
| SA | Spontaneous abortion |
| STB | Syncytiotrophoblast |
| TF | Thomsen–Friedenreich |
| Tim-3 | T-cell immunoglobulin and mucin domain 3 |
| VEGF | Vascular endothelial-derived growth factor |
| VEGFR | Vascular endothelial-derived growth factor receptor |
| VT | Villous trophoblast |

## References


1. Megli, C.J.; Coyne, C.B. Infections at the maternal-fetal interface: An overview of pathogenesis and defence. *Nat. Rev. Microbiol.* **2022**, *20*, 67–82. [CrossRef]
2. Prabhudas, M.; Bonney, E.; Caron, K.; Dey, S.; Erlebacher, A.; Fazleabas, A.; Fisher, S.; Golos, T.; Matzuk, M.; Mccune, J.M. Immune mechanisms at the maternal-fetal interface: Perspectives and challenges. *Nat. Immunol.* **2015**, *16*, 328–334. [CrossRef]
3. Cha, J.; Sun, X.; Dey, S.K. Mechanisms of implantation: Strategies for successful pregnancy. *Nat. Med.* **2012**, *18*, 1754–1767. [CrossRef] [PubMed]
4. Red-Horse, K.; Zhou, Y.; Genbacev, O.; Prakobphol, A.; Foulk, R.; Mcmaster, M.; Fisher, S.J. Trophoblast differentiation during embryo implantation and formation of the maternal-fetal interface. *J. Clin. Investig.* **2004**, *114*, 744–754. [CrossRef] [PubMed]
5. Than, N.G.; Romero, R.; Chong, J.K.; Mcgowen, M.R.; Papp, Z.; Wildman, D.E. Galectins: Guardians of eutherian pregnancy at the maternal-fetal interface. *Trends Endocrinol. Metab.* **2012**, *23*, 23–31. [CrossRef]
6. Blois, S.M.; Conrad, M.L.; Freitag, N.; Barrientos, G. Galectins in angiogenesis: Consequences for gestation. *J. Reprod. Immunol.* **2014**, *108*, 33–41. [CrossRef] [PubMed]
7. Balogh, A.; Toth, E.; Romero, R.; Parej, K.; Than, N.G. Placental Galectins Are Key Players in Regulating the Maternal Adaptive Immune Response. *Front. Immunol.* **2019**, *10*, 1240. [CrossRef]
8. Blidner, A.G.; Rabinovich, G.A. 'Sweetening' pregnancy: Galectins at the fetomaternal interface. *Am. J. Reprod. Immunol.* **2013**, *69*, 369–382. [CrossRef]
9. Cooper, D. Galectinomics: Finding themes in complexity. *Biochim. Biophys. Acta (BBA)-Gen. Subj.* **2002**, *1572*, 209–231. [CrossRef]
10. Barondes, S.H.; Castronovo, V.; Cooper, D.N.; Cummings, R.D.; Kasai, K. Galectins: A family of animal beta-galactoside-binding lectins. *Cell* **1994**, *76*, 597–598. [CrossRef]
11. Liu, F.T.; Rabinovich, G.A. Galectins as novel modulators of tumor progression. *Nat. Rev. Cancer* **2005**, *5*, 29–41. [CrossRef] [PubMed]
12. Yang, R.Y.; Rabinovich, G.A.; Liu, F.T. Galectins: Structure, function and therapeutic potential. *Expert Rev. Mol. Med.* **2008**, *10*, e17. [CrossRef] [PubMed]
13. Leffler, H. Galectins Structure and Function—A Synopsis. *Results Probl. Cell Differ.* **2001**, *33*, 57–83.
14. Seelenmeyer, C.; Wegehingel, S.; Tews, I.; Künzler, M.; Nickel, A.W. Cell Surface Counter Receptors Are Essential Components of the Unconventional Export Machinery of Galectin-1. *J. Cell Biol.* **2005**, *171*, 373–381. [CrossRef] [PubMed]
15. Wilson, T.J.G.; Firth, M.N.; Powell, J.T.; Harrison, F.L. The sequence of the mouse 14 kDa $\beta$-galactoside-binding lectin and evidence for its synthesis on free cytoplasmic ribosomes. *Biochem. J.* **1989**, *261*, 847–852. [CrossRef]
16. Rabinovich, G.A.; Rubinstein, N.; Fainboim, L. Unlocking the secrets of galectins: A challenge at the frontier of glyco-immunology. *J. Leukoc. Biol.* **2002**, *71*, 741–752.

17. Hughes, R.C. Secretion of the galectin family of mammalian carbohydrate-binding proteins. *Biochim. Biophys. Acta (BBA)-Gen. Subj.* **1999**, *1473*, 172–185. [CrossRef]
18. Stewart, S.E.; Menzies, S.A.; Popa, S.J.; Savinykh, N.; Petrunkina Harrison, A.; Lehner, P.J.; Moreau, K. A genome-wide CRISPR screen reconciles the role of N-linked glycosylation in galectin-3 transport to the cell surface. *J. Cell Sci.* **2017**, *130*, 3234–3247. [CrossRef]
19. Delacour, D.; Koch, A.; Jacob, R. The role of galectins in protein trafficking. *Traffic* **2009**, *10*, 1405–1413. [CrossRef]
20. Lannoo, N.; Van Damme, E.J. Nucleocytoplasmic plant lectins. *Biochim. Biophys. Acta (BBA)-Gen. Subj.* **2010**, *1800*, 190–201. [CrossRef]
21. Ermakova, E.; Miller, M.C.; Nesmelova, I.V.; López-Merino, L.; Berbís, M.A.; Nesmelov, Y.; Tkachev, Y.V.; Lagartera, L.; Daragan, V.A.; André, S.; et al. Lactose binding to human galectin-7 (p53-induced gene 1) induces long-range effects through the protein resulting in increased dimer stability and evidence for positive cooperativity. *Glycobiology* **2013**, *23*, 508–523. [CrossRef] [PubMed]
22. Schafer, T.; Zentgraf, H.; Zehe, C.; Brugger, B.; Bernhagen, J.; Nickel, W. Unconventional secretion of fibroblast growth factor 2 is mediated by direct translocation across the plasma membrane of mammalian cells. *J. Biol. Chem.* **2004**, *279*, 6244–6251. [CrossRef]
23. Popa, S.J.; Stewart, S.E.; Moreau, K. Unconventional secretion of annexins and galectins. *Semin. Cell Dev. Biol.* **2018**, *83*, 42–50. [CrossRef] [PubMed]
24. Griffioen, A.W.; Thijssen, V.L. Galectins in tumor angiogenesis. *Ann. Transl. Med.* **2014**, *2*, 90. [PubMed]
25. Haudek, K.C.; Spronk, K.J.; Voss, P.G.; Patterson, R.J.; Wang, J.L.; Arnoys, E.J. Dynamics of galectin-3 in the nucleus and cytoplasm. *Biochim. Biophys. Acta (BBA)-Gen. Subj.* **2010**, *1800*, 181–189. [CrossRef] [PubMed]
26. Boscher, C.; Dennis, J.W.; Nabi, I.R. Glycosylation, galectins and cellular signaling. *Curr. Opin. Cell Biol.* **2011**, *23*, 383–392. [CrossRef]
27. Elola, M.; Wolfenstein-Todel, C.; Troncoso, M.; Vasta, G.; Rabinovich, G. Galectins: Matricellular glycan-binding proteins linking cell adhesion, migration, and survival. *Cell. Mol. Life Sci.* **2007**, *64*, 1679–1700. [CrossRef]
28. Erlebacher, A. Immunology of the maternal-fetal interface. *Annu. Rev. Immunol.* **2013**, *31*, 387–411. [CrossRef]
29. Von, W.M.; Wang, X.; H.-J., G.; Strowitzki, T. Galectin fingerprinting in human endometrium and decidua during the menstrual cycle and in early gestation. *Mol. Hum. Reprod.* **2005**, *11*, 189–194.
30. Yang, H.; Lei, C.; Wei, Z. Expression of galectin-3 in mouse endometrium and its effect during embryo implantation. *Reprod. Biomed. Online* **2012**, *24*, 116–122. [CrossRef] [PubMed]
31. Phillips, B. Differential expression of two beta-galactoside-binding lectins in the reproductive tracts of pregnant mice. *Biol. Reprod.* **1996**, *55*, 548–558. [CrossRef] [PubMed]
32. Orazizadeh, M.; Khorsundi, L.; Saki, G. Immunohistochemical assessment of galectin-3 during pre-implantation in mouse endometrium. *Iran. J. Reprod. Med.* **2013**, *11*, 119–126. [PubMed]
33. Viovac, L.; Jankovi, M.; Cuperlovi, M. Galectin-1 and -3 in cells of the first trimester placental bed. *Hum. Reprod.* **1998**, *13*, 730–735. [CrossRef] [PubMed]
34. Kolundi, N.; Boji-Trbojevi, A.; Kovaevi, T.; Stefanoska, I.; Viovac, L. Galectin-1 Is Part of Human Trophoblast Invasion Machinery—A Functional Study In Vitro. *PLoS ONE* **2011**, *6*, e28514.
35. Maquoi, E.; Van den Brule, F.; Castronovo, V.; Foidart, J.M. Changes in the distribution pattern of galectin-1 and galectin-3 in human placenta correlates with the differentiation pathways of trophoblasts. *Placenta* **1997**, *18*, 433–439. [CrossRef]
36. Hepp, P.; Unverdorben, L.; Hutter, S.; Kuhn, C.; Heidegger, H.H. Placental Galectin-2 Expression in Gestational Diabetes: A Systematic, Histological Analysis. *Int. J. Mol. Sci.* **2020**, *21*, 2404. [CrossRef]
37. Jeschke, U.; Hutter, S.; Heublein, S.; Vrekoussis, T.; Andergassen, U.; Unverdorben, L.; Papadakis, G.; Makrigiannakis, A. Expression and function of galectins in the endometrium and at the human feto-maternal interface. *Placenta* **2013**, *34*, 863–872. [CrossRef]
38. Tian, J.; Hu, S.; Wang, F.; Yang, X.; Li, Y.; Huang, C. PPARG, AGTR1, CXCL16 and LGALS2 polymorphisms are correlated with the risk for coronary heart disease. *Int. J. Clin. Exp. Pathol.* **2015**, *8*, 3138–3143.
39. Bojić-Trbojević, Ž.; Jovanović Krivokuća, M.; Vilotić, A.; Kolundžić, N.; Stefanoska, I.; Zetterberg, F.; Nilsson, U.; Leffler, H.; Vićovac, L. Human trophoblast requires galectin-3 for cell migration and invasion. *Sci. Rep.* **2019**, *9*, 2136. [CrossRef]
40. Iglesias, M.M.; Rabinovich, G.A.; Ambrosio, A.L.; Castagna, L.F.; Sotomayor, C.E.; Wolfenstein-Todel, C. Purification of galectin-3 from ovine placenta: Developmentally regulated expression and immunological relevance. *Glycobiology* **1998**, *8*, 59–65. [CrossRef]
41. Freitag, N.; Tirado-González, I.; Barrientos, G.; Cohen, M.; Daher, S.; Goldman-Wohl, D.; Mincheva-Nilsson, L.; John, C.M.; Jeschke, U.; Blois, S.M. The chimera-type galectin-3 is a positive modulator of trophoblast functions with dysregulated expression in gestational diabetes mellitus. *Am. J. Reprod. Immunol.* **2020**, *84*, e13311. [CrossRef] [PubMed]
42. Menkhorst, E.; Koga, K.; Van Sinderen, M.; Dimitriadis, E. Galectin-7 serum levels are altered prior to the onset of pre-eclampsia. *Placenta* **2014**, *35*, 281–285. [CrossRef]
43. Kolundžić, N.; Bojić-Trbojević, Ž.; Radojčić, L.; Petronijević, M.; Vićovac, L. Galectin-8 is expressed by villous and extravillous trophoblast of the human placenta. *Placenta* **2011**, *32*, 909–911. [CrossRef] [PubMed]
44. Popovici, R.M.; Krause, M.S.; Germeyer, A.; Strowitzki, T.; von Wolff, M. Galectin-9: A new endometrial epithelial marker for the mid-and late-secretory and decidual phases in humans. *J. Clin. Endocrinol. Metab.* **2005**, *90*, 6170–6176. [CrossRef] [PubMed]
45. Shimizu, Y.; Kabir-Salmani, M.; Azadbakht, M.; Sugihara, K.; Sakai, K.; Iwashita, M. Expression and localization of galectin-9 in the human uterodome. *Endocr. J.* **2008**, *55*, 879–887. [CrossRef]

46. Kubach, J.; Lutter, P.; Bopp, T.; Stoll, S.; Becker, C.; Huter, E.; Richter, C.; Weingarten, P.; Warger, T.; Knop, J.; et al. Human CD4+ CD25+ regulatory T cells: Proteome analysis identifies galectin-10 as a novel marker essential for their anergy and suppressive function. *Blood J. Am. Soc. Hematol.* **2007**, *110*, 1550–1558. [CrossRef]
47. Garín, M.I.; Chu, C.C.; Golshayan, D.; Cernuda-Morollón, E.; Wait, R.; Lechler, R.I. Galectin-1: A key effector of regulation mediated by CD4+ CD25+ T cells. *Blood* **2007**, *109*, 2058–2065. [CrossRef]
48. Yang, R.Y.; Xue, H.; Yu, L.; Velayos-Baeza, A.; Monaco, A.P.; Liu, F.T. Identification of VPS13C as a galectin-12-binding protein that regulates galectin-12 protein stability and adipogenesis. *PLoS ONE* **2016**, *11*, e0153534. [CrossRef]
49. Kliman, H.J.; Sammar, M.; Grimpel, Y.; Lynch, S.K.; Milano, K.; Pick, E.; Bejar, J.; Arad, A.; Lee, J.; Meiri, H.; et al. Placental protein 13 and decidual zones of necrosis: An immunologic diversion that may be linked to preeclampsia. *Reprod. Sci.* **2012**, *19*, 16–30. [CrossRef]
50. Than, N.G.; Pick, E.; Bellyei, S.; Szigeti, A.; Burger, O.; Berente, Z.; Janaky, T.; Boronkai, A.; Kliman, H.; Meiri, H.; et al. Functional analyses of placental protein 13/galectin-13. *Eur. J. Biochem.* **2004**, *271*, 1065–1078. [CrossRef]
51. Lewis, S.K.; Farmer, J.L.; Burghardt, R.C.; Newton, G.R.; Johnson, G.A.; Adelson, D.L.; Bazer, F.W.; Spencer, T.E. Galectin 15 (LGALS15): A gene uniquely expressed in the uteri of sheep and goats that functions in trophoblast attachment. *Biol. Reprod.* **2007**, *77*, 1027–1036. [CrossRef] [PubMed]
52. Barrientos, G.; Freitag, N.; Tirado-Gonzalez, I.; Unverdorben, L.; Jeschke, U.; Thijssen, V.L.; Blois, S.M. Involvement of galectin-1 in reproduction: Past, present and future. *Hum. Reprod. Update* **2014**, *20*, 175–193. [CrossRef]
53. Koopman, L.A.; Kopcow, H.D.; Rybalov, B.; Boyson, J.E.; Orange, J.S.; Schatz, F.; Masch, R.; Lockwood, C.J.; Schachter, A.D.; Park, P.J.; et al. Human decidual natural killer cells are a unique NK cell subset with immunomodulatory potential. *J. Exp. Med.* **2003**, *198*, 1201–1212. [CrossRef]
54. Menkhorst, E.; Gamage, T.; Cuman, C.; Kaitu'u-Lino, T.; Tong, S.; Dimitriadis, E. Galectin-7 acts as an adhesion molecule during implantation and increased expression is associated with miscarriage. *Placenta* **2014**, *35*, 195–201. [CrossRef] [PubMed]
55. Medawar, P.B. Some immunological and endocrinological problems raised by the evolution of viviparity in vertebrates. *Symp. Soc. Exp. Biol.* **1953**, *7*, 320–338.
56. Deshmukh, H.; Way, S.S. Immunological basis for recurrent fetal loss and pregnancy complications. *Annu. Rev. Pathol.* **2019**, *14*, 185. [CrossRef] [PubMed]
57. Dey, S.; Lim, H.; Das, S.K.; Reese, J.; Paria, B.; Daikoku, T.; Wang, H. Molecular cues to implantation. *Endocr. Rev.* **2004**, *25*, 341–373. [CrossRef]
58. Wang, H.; Dey, S.K. Roadmap to embryo implantation: Clues from mouse models. *Nat. Rev. Genet.* **2006**, *7*, 185–199. [CrossRef]
59. Billington, W. The normal fetomaternal immune relationship. *Baillière's Clin. Obstet. Gynaecol.* **1992**, *6*, 417–438. [CrossRef]
60. Ferreira, L.M.; Meissner, T.B.; Tilburgs, T.; Strominger, J.L. HLA-G: At the interface of maternal–fetal tolerance. *Trends Immunol.* **2017**, *38*, 272–286. [CrossRef]
61. Ander, S.E.; Diamond, M.S.; Coyne, C.B. Immune responses at the maternal-fetal interface. *Sci. Immunol.* **2019**, *4*, eaat6114. [CrossRef] [PubMed]
62. Wang, L.; Zhao, Y.; Wang, Y.; Wu, X. The role of galectins in cervical cancer biology and progression. *BioMed Res. Int.* **2018**, *2018*, 2175927. [CrossRef] [PubMed]
63. Blois, S.M.; Ilarregui, J.M.; Tometten, M.; Garcia, M.; Orsal, A.S.; Cordo-Russo, R.; Toscano, M.A.; Bianco, G.A.; Kobelt, P.; Handjiski, B.; et al. A pivotal role for galectin-1 in fetomaternal tolerance. *Nat. Med.* **2007**, *13*, 1450–1457. [CrossRef] [PubMed]
64. Than, N.G.; Romero, R.; Goodman, M.; Weckle, A.; Xing, J.; Dong, Z.; Xu, Y.; Tarquini, F.; Szilagyi, A.; Gal, P.; et al. A primate subfamily of galectins expressed at the maternal–fetal interface that promote immune cell death. *Proc. Natl. Acad. Sci. USA* **2009**, *106*, 9731–9736. [CrossRef] [PubMed]
65. Kopcow, H.D.; Rosetti, F.; Leung, Y.; Allan, D.S.; Kutok, J.L.; Strominger, J.L. T cell apoptosis at the maternal–fetal interface in early human pregnancy, involvement of galectin-1. *Proc. Natl. Acad. Sci. USA* **2008**, *105*, 18472–18477. [CrossRef] [PubMed]
66. Molvarec, A.; Blois, S.M.; Stenczer, B.; Toldi, G.; Tirado-Gonzalez, I.; Ito, M.; Shima, T.; Yoneda, S.; Vásárhelyi, B.; Rigó Jr, J.; et al. Peripheral blood galectin-1-expressing T and natural killer cells in normal pregnancy and preeclampsia. *Clin. Immunol.* **2011**, *139*, 48–56. [CrossRef]
67. Choe, Y.S.; Shim, C.; Choi, D.; Lee, C.S.; Lee, K.K.; Kim, K. Expression of galectin-1 mRNA in the mouse uterus is under the control of ovarian steroids during blastocyst implantation. *Mol. Reprod. Dev. Inc. Gamete Res.* **1997**, *48*, 261–266. [CrossRef]
68. Than, N.G.; Romero, R.; Erez, O.; Weckle, A.; Tarca, A.L.; Hotra, J.; Abbas, A.; Han, Y.M.; Kim, S.S.; Kusanovic, J.P.; et al. Emergence of hormonal and redox regulation of galectin-1 in placental mammals: Implication in maternal–fetal immune tolerance. *Proc. Natl. Acad. Sci. USA* **2008**, *105*, 15819–15824. [CrossRef]
69. Tirado-Gonzalez, I.; Freitag, N.; Barrientos, G.; Shaikly, V.; Nagaeva, O.; Strand, M.; Kjellberg, L.; Klapp, B.F.; Mincheva-Nilsson, L.; Cohen, M.; et al. Galectin-1 influences trophoblast immune evasion and emerges as a predictive factor for the outcome of pregnancy. *MHR Basic Sci. Reprod. Med.* **2012**, *19*, 43–53. [CrossRef]
70. Crider-Pirkle, S.; Billingsley, P.; Faust, C.; Hardy, D.M.; Lee, V.; Weitlauf, H. Cubilin, a binding partner for galectin-3 in the murine utero-placental complex. *J. Biol. Chem.* **2002**, *277*, 15904–15912. [CrossRef]
71. Yang, R.Y.; Hsu, D.K.; LIu, F.T. Expression of galectin-3 modulates T-cell growth and apoptosis. *Proc. Natl. Acad. Sci. USA* **1996**, *93*, 6737–6742. [CrossRef] [PubMed]

72. Fukumori, T.; Takenaka, Y.; Yoshii, T.; Kim, H.R.C.; Hogan, V.; Inohara, H.; Kagawa, S.; Raz, A. CD29 and CD7 mediate galectin-3-induced type II T-cell apoptosis. *Cancer Res.* **2003**, *63*, 8302–8311. [PubMed]
73. Miyauchi, M.; Ao, M.; Furusho, H.; Chea, C.; Nagasaki, A.; Sakamoto, S.; Ando, T.; Inubushi, T.; Kozai, K.; Takata, T. Galectin-3 plays an important role in preterm birth caused by dental infection of Porphyromonas gingivalis. *Sci. Rep.* **2018**, *8*, 2867. [CrossRef] [PubMed]
74. Enninga, E.A.L.; Harrington, S.M.; Creedon, D.J.; Ruano, R.; Markovic, S.N.; Dong, H.; Dronca, R.S. Immune checkpoint molecules soluble program death ligand 1 and galectin-9 are increased in pregnancy. *Am. J. Reprod. Immunol.* **2018**, *79*, e12795. [CrossRef] [PubMed]
75. Hu, X.H.; Tang, M.X.; Mor, G.; Liao, A.H. Tim-3: Expression on immune cells and roles at the maternal-fetal interface. *J. Reprod. Immunol.* **2016**, *118*, 92–99. [CrossRef]
76. Meggyes, M.; Lajko, A.; Palkovics, T.; Totsimon, A.; Illes, Z.; Szereday, L.; Miko, E. Feto-maternal immune regulation by TIM-3/galectin-9 pathway and PD-1 molecule in mice at day 14.5 of pregnancy. *Placenta* **2015**, *36*, 1153–1160. [CrossRef]
77. Miko, E.; Meggyes, M.; Bogar, B.; Schmitz, N.; Barakonyi, A.; Varnagy, A.; Farkas, B.; Tamas, P.; Bodis, J.; Szekeres-Bartho, J.; et al. Involvement of Galectin-9/TIM-3 pathway in the systemic inflammatory response in early-onset preeclampsia. *PLoS ONE* **2013**, *8*, e71811. [CrossRef]
78. Sun, J.; Yang, M.; Ban, Y.; Gao, W.; Song, B.; Wang, Y.; Zhang, Y.; Shao, Q.; Kong, B.; Qu, X. Tim-3 is upregulated in NK cells during early pregnancy and inhibits NK cytotoxicity toward trophoblast in galectin-9 dependent pathway. *PLoS ONE* **2016**, *11*, e0147186. [CrossRef]
79. Meggyes, M.; Miko, E.; Polgar, B.; Bogar, B.; Farkas, B.; Illes, Z.; Szereday, L. Peripheral blood TIM-3 positive NK and CD8+ T cells throughout pregnancy: TIM-3/galectin-9 interaction and its possible role during pregnancy. *PLoS ONE* **2014**, *9*, e92371.
80. He, M.; Jiang, M.; Zhou, Y.; Li, F.; Yang, M.; Fan, Y.; Xie, Y.; Beejadhursing, R.; Feng, L.; Deng, D. Impaired Gal-9 dysregulates the PBMC-induced Th1/Th2 imbalance in abortion-prone matings. *J. Immunol. Res.* **2018**, *2018*, 9517842. [CrossRef]
81. Hu, X.; Zhu, Q.; Wang, Y.; Wang, L.; Li, Z.; Mor, G.; Liao, A. Newly characterized decidual Tim-3+ Treg cells are abundant during early pregnancy and driven by IL-27 coordinately with Gal-9 from trophoblasts. *Hum. Reprod.* **2020**, *35*, 2454–2466. [CrossRef] [PubMed]
82. Li, Y.; Zhang, J.; Zhang, D.; Hong, X.; Tao, Y.; Wang, S.; Xu, Y.; Piao, H.; Yin, W.; Yu, M.; et al. Tim-3 signaling in peripheral NK cells promotes maternal-fetal immune tolerance and alleviates pregnancy loss. *Sci. Signal.* **2017**, *10*, eaah4323. [CrossRef] [PubMed]
83. Li, Z.H.; Wang, L.L.; Liu, H.; Muyayalo, K.P.; Huang, X.B.; Mor, G.; Liao, A.H. Galectin-9 alleviates LPS-induced preeclampsia-like impairment in rats via switching decidual macrophage polarization to M2 subtype. *Front. Immunol.* **2019**, *9*, 3142. [CrossRef] [PubMed]
84. Heusschen, R.; Freitag, N.; Tirado-González, I.; Barrientos, G.; Moschansky, P.; Muñoz-Fernández, R.; Leno-Durán, E.; Klapp, B.F.; Thijssen, V.L.; Blois, S.M. Profiling Lgals9 splice variant expression at the fetal-maternal interface: Implications in normal and pathological human pregnancy. *Biol. Reprod.* **2013**, *88*, 22. [CrossRef]
85. Li, Y.H.; Zhou, W.H.; Tao, Y.; Wang, S.C.; Jiang, Y.L.; Zhang, D.; Piao, H.L.; Fu, Q.; Li, D.J.; Du, M.R. The Galectin-9/Tim-3 pathway is involved in the regulation of NK cell function at the maternal–fetal interface in early pregnancy. *Cell. Mol. Immunol.* **2016**, *13*, 73–81. [CrossRef]
86. Devouassoux, G.; Pachot, A.; Laforest, L.; Diasparra, J.; Freymond, N.; Van Ganse, E.; Mougin, B.; Pacheco, Y. Galectin-10 mRNA is overexpressed in peripheral blood of aspirin-induced asthma. *Allergy* **2008**, *63*, 125–131.
87. Swaminathan, G.J.; Leonidas, D.D.; Savage, M.P.; Ackerman, S.J.; Acharya, K.R. Selective recognition of mannose by the human eosinophil Charcot-Leyden crystal protein (galectin-10): A crystallographic study at 1.8 Å resolution. *Biochemistry* **1999**, *38*, 13837–13843. [CrossRef]
88. Vokalova, L.; Balogh, A.; Toth, E.; Van Breda, S.V.; Schäfer, G.; Hoesli, I.; Lapaire, O.; Hahn, S.; Than, N.G.; Rossi, S.W. Placental protein 13 (Galectin-13) polarizes neutrophils toward an immune regulatory phenotype. *Front. Immunol.* **2020**, *11*, 145. [CrossRef]
89. Sammar, M.; Drobnjak, T.; Mandala, M.; Gizurarson, S.; Huppertz, B.; Meiri, H. Galectin 13 (PP13) facilitates remodeling and structural stabilization of maternal vessels during pregnancy. *Int. J. Mol. Sci.* **2019**, *20*, 3192. [CrossRef]
90. Wilcox, A.J.; Weinberg, C.R.; O'Connor, J.F.; Baird, D.D.; Schlatterer, J.P.; Canfield, R.E.; Armstrong, E.G.; Nisula, B.C. Incidence of early loss of pregnancy. *N. Engl. J. Med.* **1988**, *319*, 189–194. [CrossRef]
91. Zhang, S.; Lin, H.; Kong, S.; Wang, S.; Wang, H.; Wang, H.; Armant, D.R. Physiological and molecular determinants of embryo implantation. *Mol. Asp. Med.* **2013**, *34*, 939–980. [CrossRef] [PubMed]
92. Hernández-Vargas, P.; Muñoz, M.; Domínguez, F. Identifying biomarkers for predicting successful embryo implantation: Applying single to multi-OMICs to improve reproductive outcomes. *Hum. Reprod. Update* **2020**, *26*, 264–301. [CrossRef] [PubMed]
93. Jovanović Krivokuća, M.; Vilotić, A.; Nacka-Aleksić, M.; Pirković, A.; Ćujić, D.; Legner, J.; Dekanski, D.; Bojić-Trbojević, Ž. Galectins in Early Pregnancy and Pregnancy-Associated Pathologies. *Int. J. Mol. Sci.* **2021**, *23*, 69. [CrossRef] [PubMed]
94. Hirota, Y.; Burnum, K.E.; Acar, N.; Rabinovich, G.A.; Daikoku, T.; Dey, S.K. Galectin-1 markedly reduces the incidence of resorptions in mice missing immunophilin FKBP52. *Endocrinology* **2012**, *153*, 2486–2493. [CrossRef]
95. Lei, C.X.; Zhang, W.; Zhou, J.P.; Liu, Y.K. Interactions between galectin-3 and integrin$\beta$3 in regulating endometrial cell proliferation and adhesion. *Hum. Reprod.* **2009**, *24*, 2879–2889. [CrossRef]

96. Yang, H.; Taylor, H.S.; Lei, C.; Cheng, C.; Zhang, W. Hormonal regulation of galectin 3 in trophoblasts and its effects on endometrium. *Reprod. Sci.* **2011**, *18*, 1118–1127. [CrossRef]
97. Yang, H.; Lei, C.; Cheng, C.; Feng, Y.; Zhang, W.; Petracco, R.G.; Sak, S. The antiapoptotic effect of galectin-3 in human endometrial cells under the regulation of estrogen and progesterone. *Biol. Reprod.* **2012**, *87*, 39. [CrossRef]
98. Colnot, C.; Fowlis, D.; Ripoche, M.A.; Bouchaert, I.; Poirier, F. Embryonic implantation in galectin 1/galectin 3 double mutant mice. *Dev. Dyn. Off. Publ. Am. Assoc. Anat.* **1998**, *211*, 306–313. [CrossRef]
99. Rizov, M.; Andreeva, P.; Dimova, I. Molecular regulation and role of angiogenesis in reproduction. *Taiwan. J. Obstet. Gynecol.* **2017**, *56*, 127–132. [CrossRef]
100. Ferrara, N.; Chen, H.; Davis-Smyth, T.; Gerber, H.P.; Nguyen, T.N.; Peers, D.; Chisholm, V.; Hillan, K.J.; Schwall, R.H. Vascular endothelial growth factor is essential for corpus luteum angiogenesis. *Nat. Med.* **1998**, *4*, 336–340. [CrossRef]
101. Gargett, C.E.; Rogers, P.A. Human endometrial angiogenesis. *Reproduction* **2001**, *121*, 181–186. [CrossRef] [PubMed]
102. Gambino, L.S.; Wreford, N.G.; Bertram, J.F.; Dockery, P.; Lederman, F.; Rogers, P.A. Angiogenesis occurs by vessel elongation in proliferative phase human endometrium. *Hum. Reprod.* **2002**, *17*, 1199–1206. [CrossRef] [PubMed]
103. Ballard, K.; Seaman, H.; de Vries, C.; Wright, J. BJOG: An international journal of obstetrics and gynaecology. *BJOG* **2008**, *115*, 1382–1391. [CrossRef] [PubMed]
104. Reynolds, L.P.; Caton, J.S.; Redmer, D.A.; Grazul-Bilska, A.T.; Vonnahme, K.A.; Borowicz, P.P.; Luther, J.S.; Wallace, J.M.; Wu, G.; Spencer, T.E. Evidence for altered placental blood flow and vascularity in compromised pregnancies. *J. Physiol.* **2006**, *572*, 51–58. [CrossRef] [PubMed]
105. Thijssen, V.L.; Postel, R.; Brandwijk, R.J.; Dings, R.P.; Nesmelova, I.; Satijn, S.; Verhofstad, N.; Nakabeppu, Y.; Baum, L.G.; Bakkers, J.; et al. Galectin-1 is essential in tumor angiogenesis and is a target for antiangiogenesis therapy. *Proc. Natl. Acad. Sci. USA* **2006**, *103*, 15975–15980. [CrossRef]
106. Freitag, N.; Tirado-González, I.; Barrientos, G.; Herse, F.; Thijssen, V.L.; Weedon-Fekjær, S.M.; Schulz, H.; Wallukat, G.; Klapp, B.F.; Nevers, T.; et al. Interfering with Gal-1–mediated angiogenesis contributes to the pathogenesis of preeclampsia. *Proc. Natl. Acad. Sci. USA* **2013**, *110*, 11451–11456. [CrossRef]
107. Wei, J.; Li, D.K.; Hu, X.; Cheng, C.; Zhang, Y. Galectin-1–RNA interaction map reveals potential regulatory roles in angiogenesis. *FEBS Lett.* **2021**, *595*, 623–636. [CrossRef]
108. Astorgues-Xerri, L.; Riveiro, M.E.; Tijeras-Raballand, A.; Serova, M.; Rabinovich, G.A.; Bieche, I.; Vidaud, M.; de Gramont, A.; Martinet, M.; Cvitkovic, E.; et al. OTX008, a selective small-molecule inhibitor of galectin-1, downregulates cancer cell proliferation, invasion and tumour angiogenesis. *Eur. J. Cancer* **2014**, *50*, 2463–2477. [CrossRef]
109. Bastón, J.I.; Barañao, R.I.; Ricci, A.G.; Bilotas, M.A.; Olivares, C.N.; Singla, J.J.; Gonzalez, A.M.; Stupirski, J.C.; Croci, D.O.; Rabinovich, G.A.; et al. Targeting galectin-1-induced angiogenesis mitigates the severity of endometriosis. *J. Pathol.* **2014**, *234*, 329–337. [CrossRef]
110. D'Haene, N.; Sauvage, S.; Maris, C.; Adanja, I.; Le Mercier, M.; Decaestecker, C.; Baum, L.; Salmon, I. VEGFR1 and VEGFR2 involvement in extracellular galectin-1-and galectin-3-induced angiogenesis. *PLoS ONE* **2013**, *8*, e67029.
111. Eguchi, R.; Wakabayashi, I. HDGF enhances VEGF-dependent angiogenesis and FGF-2 is a VEGF-independent angiogenic factor in non-small cell lung cancer. *Oncol. Rep.* **2020**, *44*, 14–28. [CrossRef] [PubMed]
112. de Mattos, R.M.; Machado, D.E.; Perini, J.A.; Alessandra-Perini, J.; da Costa, N.d.O.M.; de Oliveira, A.F.d.R.; dos Santos Cabral, K.M.; Takiya, C.M.; Carvalho, R.S.; Nasciutti, L.E.; et al. Galectin-3 plays an important role in endometriosis development and is a target to endometriosis treatment. *Mol. Cell. Endocrinol.* **2019**, *486*, 1–10. [CrossRef] [PubMed]
113. Unverdorben, L.; Haufe, T.; Santoso, L.; Hofmann, S.; Jeschke, U.; Hutter, S. Prototype and chimera-type galectins in placentas with spontaneous and recurrent miscarriages. *Int. J. Mol. Sci.* **2016**, *17*, 644. [CrossRef] [PubMed]
114. Chen, C.; Duckworth, C.; Fu, B.; Pritchard, D.M.; Rhodes, J.; Yu, L. Circulating galectins-2,-4 and-8 in cancer patients make important contributions to the increased circulation of several cytokines and chemokines that promote angiogenesis and metastasis. *Br. J. Cancer* **2014**, *110*, 741–752. [CrossRef]
115. van der Laan, A.M.; Schirmer, S.H.; de Vries, M.R.; Koning, J.J.; Volger, O.L.; Fledderus, J.O.; Bastiaansen, A.J.; Hollander, M.R.; Baggen, J.M.; Koch, K.T.; et al. Galectin-2 expression is dependent on the rs7291467 polymorphism and acts as an inhibitor of arteriogenesis. *Eur. Heart J.* **2012**, *33*, 1076–1084. [CrossRef] [PubMed]
116. Troncoso, M.F.; Ferragut, F.; Bacigalupo, M.L.; Cárdenas Delgado, V.M.; Nugnes, L.G.; Gentilini, L.; Laderach, D.; Wolfenstein-Todel, C.; Compagno, D.; Rabinovich, G.A.; et al. Galectin-8: A matricellular lectin with key roles in angiogenesis. *Glycobiology* **2014**, *24*, 907–914. [CrossRef] [PubMed]
117. Varinska, L.; Faber, L.; Petrovova, E.; Balážová, L.; Ivančová, E.; Kolář, M.; Gal, P. Galectin-8 favors VEGF-induced angiogenesis: In vitro study in human umbilical vein endothelial cells and in vivo study in chick chorioallantoic membrane. *Anticancer. Res.* **2020**, *40*, 3191–3201. [CrossRef] [PubMed]
118. O'Brien, M.J.; Shu, Q.; Stinson, W.A.; Tsou, P.S.; Ruth, J.H.; Isozaki, T.; Campbell, P.L.; Ohara, R.A.; Koch, A.E.; Fox, D.A.; et al. A unique role for galectin-9 in angiogenesis and inflammatory arthritis. *Arthritis Res. Ther.* **2018**, *20*, 1–8. [CrossRef]
119. Wang, Y.; Song, L.; Sun, J.; Sui, Y.; Li, D.; Li, G.; Liu, J.; Shu, Q. Expression of Galectin-9 and correlation with disease activity and vascular endothelial growth factor in rheumatoid arthritis. *Clin. Exp. Rheumatol.* **2019**, *38*, 654–661.

120. Wyatt, M.A.; Baumgarten, S.C.; Weaver, A.L.; Van Oort, C.C.; Fedyshyn, B.; Ruano, R.; Shenoy, C.C.; Enninga, E.A.L. Evaluating markers of immune tolerance and angiogenesis in maternal blood for an association with risk of pregnancy loss. *J. Clin. Med.* **2021**, *10*, 3579. [CrossRef]
121. Maller, S.M.; Cagnoni, A.J.; Bannoud, N.; Sigaut, L.; Pérez Sáez, J.M.; Pietrasanta, L.I.; Yang, R.Y.; Liu, F.T.; Croci, D.O.; Di Lella, S.; et al. An adipose tissue galectin controls endothelial cell function via preferential recognition of 3-fucosylated glycans. *FASEB J.* **2020**, *34*, 735–753. [CrossRef] [PubMed]
122. Bartsch, E.; Medcalf, K.E.; Park, A.L.; Ray, J.G. Clinical risk factors for pre-eclampsia determined in early pregnancy: Systematic review and meta-analysis of large cohort studies. *BMJ* **2016**, *353*, i1753. [CrossRef] [PubMed]
123. Burton, G.J.; Redman, C.W.; Roberts, J.M.; Moffett, A. Pre-eclampsia: Pathophysiology and clinical implications. *Bmj* **2019**, *366*, il2381. [CrossRef] [PubMed]
124. Fantone, S.; Mazzucchelli, R.; Giannubilo, S.R.; Ciavattini, A.; Marzioni, D.; Tossetta, G. AT-rich interactive domain 1A protein expression in normal and pathological pregnancies complicated by preeclampsia. *Histochem. Cell Biol.* **2020**, *154*, 339–346. [CrossRef]
125. Robillard, P.Y.; Dekker, G.; Scioscia, M.; Bonsante, F.; Iacobelli, S.; Boukerrou, M.; Hulsey, T.C. Validation of the 34-week gestation as definition of late onset preeclampsia: Testing different cutoffs from 30 to 37 weeks on a population-based cohort of 1700 preeclamptics. *Acta Obstet. Gynecol. Scand.* **2020**, *99*, 1181–1190. [CrossRef]
126. Schnabel, A.; Blois, S.M.; Meint, P.; Freitag, N.; Ernst, W.; Barrientos, G.; Conrad, M.L.; Rose, M.; Seelbach-Göbel, B. Elevated systemic galectin-1 levels characterize HELLP syndrome. *J. Reprod. Immunol.* **2016**, *114*, 38–43. [CrossRef]
127. Dusse, L.M.; Alpoim, P.N.; Silva, J.T.; Rios, D.R.A.; Brandão, A.H.; Cabral, A.C.V. Revisiting HELLP syndrome. *Clin. Chim. Acta* **2015**, *451*, 117–120. [CrossRef]
128. Hirashima, C.; Ohkuchi, A.; Nagayama, S.; Suzuki, H.; Takahashi, K.; Ogoyama, M.; Takahashi, H.; Shirasuna, K.; Matsubara, S. Galectin-1 as a novel risk factor for both gestational hypertension and preeclampsia, specifially its expression at a low level in the second trimester and a high level after onset. *Hypertens. Res.* **2018**, *41*, 45–52. [CrossRef]
129. Fischer, I.; Jeschke, U.; Friese, K.; Daher, S.; Betz, A.G. The role of galectin-1 in trophoblast differentiation and signal transduction. *J. Reprod. Immunol.* **2011**, *90*, 35–40. [CrossRef]
130. Tang, M.; You, J.; Wang, W.; Lu, Y.; Hu, X.; Wang, C.; Liu, A.; Zhu, Y. Impact of galectin-1 on trophoblast stem cell differentiation and invasion in in vitro implantation model. *Reprod. Sci.* **2018**, *25*, 700–711. [CrossRef]
131. Hutter, S.; Martin, N.; Von Schönfeldt, V.; Messner, J.; Kuhn, C.; Hofmann, S.; Andergassen, U.; Knabl, J.; Jeschke, U. Galectin 2 (gal-2) expression is downregulated on protein and mRNA level in placentas of preeclamptic (PE) patients. *Placenta* **2015**, *36*, 438–445. [CrossRef] [PubMed]
132. Meister, S.; Hahn, L.; Beyer, S.; Mannewitz, M.; Perleberg, C.; Schnell, K.; Anz, D.; Corradini, S.; Schmoeckel, E.; Mayr, D.; et al. Regulatory T cell apoptosis during preeclampsia may be prevented by Gal-2. *Int. J. Mol. Sci.* **2022**, *23*, 1880. [CrossRef] [PubMed]
133. Hahn, L.; Meister, S.; Mannewitz, M.; Beyer, S.; Corradini, S.; Hasbargen, U.; Mahner, S.; Jeschke, U.; Kolben, T.; Burges, A. Gal-2 Increases H3K4me3 and H3K9ac in Trophoblasts and Preeclampsia. *Biomolecules* **2022**, *12*, 707. [CrossRef] [PubMed]
134. Charkiewicz, K.; Goscik, J.; Raba, G.; Laudanski, P. Syndecan 4, galectin 2, and death receptor 3 (DR3) as novel proteins in pathophysiology of preeclampsia. *J. Matern.-Fetal Neonatal Med.* **2021**, *34*, 2965–2970. [CrossRef] [PubMed]
135. Janko, C.; André, S.; Munoz, L.E.; Briand, J.P.; Schorn, C.; Winkler, S.; Schiller, M.; Andreoli, L.; Manfredi, A.A.; Isenberg, D.; et al. Autoantibodies against galectin-2 peptides as biomarkers for the antiphospholipid syndrome. *Lupus* **2012**, *21*, 781–783. [CrossRef]
136. Ruiz-Irastorza, G.; Crowther, M.; Branch, W.; Khamashta, M.A. Antiphospholipid syndrome. *Lancet* **2010**, *376*, 1498–1509. [CrossRef]
137. Jeschke, U.; Mayr, D.; Schiessl, B.; Mylonas, I.; Schulze, S.; Kuhn, C.; Friese, K.; Walzel, H. Expression of galectin-1,-3 (gal-1, gal-3) and the Thomsen–Friedenreich (TF) antigen in normal, IUGR, preeclamptic and HELLP placentas. *Placenta* **2007**, *28*, 1165–1173. [CrossRef]
138. Freitag, N.; Tirado-Gonzalez, I.; Barrientos, G.; Powell, K.L.; Boehm-Sturm, P.; Koch, S.P.; Hecher, K.; Staff, A.C.; Arck, P.C.; Diemert, A.; et al. Galectin-3 deficiency in pregnancy increases the risk of fetal growth restriction (FGR) via placental insufficiency. *Cell Death Dis.* **2020**, *11*, 1–9. [CrossRef]
139. Atakul, N.; Atamer, Y.; Selek, Ş.; Kılıç, B.; Koktasoglu, F. ST2 and galectin-3 as novel biomarkers for the prediction of future cardiovascular disease risk in preeclampsia. *J. Obstet. Gynaecol.* **2022**, 42, 1023–1029. [CrossRef]
140. Menkhorst, E.; Zhou, W.; Santos, L.L.; Delforce, S.; So, T.; Rainczuk, K.; Loke, H.; Syngelaki, A.; Varshney, S.; Williamson, N.; et al. Galectin-7 impairs placentation and causes preeclampsia features in mice. *Hypertension* **2020**, *76*, 1185–1194. [CrossRef]
141. Dong, S.; Shah, N.K.; He, J.; Han, S.; Xie, M.; Wang, Y.; Cheng, T.; Liu, Z.; Shu, C. The abnormal expression of Tim-3 is involved in the regulation of myeloid-derived suppressor cells and its correlation with preeclampsia. *Placenta* **2021**, *114*, 108–114. [CrossRef] [PubMed]
142. Hao, H.; He, M.; Li, J.; Zhou, Y.; Dang, J.; Li, F.; Yang, M.; Deng, D. Upregulation of the Tim-3/Gal-9 pathway and correlation with the development of preeclampsia. *Eur. J. Obstet. Gynecol. Reprod. Biol.* **2015**, *194*, 85–91. [CrossRef] [PubMed]
143. Balogh, A.; Pozsgay, J.; Matkó, J.; Dong, Z.; Kim, C.J.; Várkonyi, T.; Sammar, M.; Rigó Jr, J.; Meiri, H.; Romero, R.; et al. Placental protein 13 (PP13/galectin-13) undergoes lipid raft-associated subcellular redistribution in the syncytiotrophoblast in preterm preeclampsia and HELLP syndrome. *Am. J. Obstet. Gynecol.* **2011**, *205*, 156.e1–156.e4. [CrossRef]

144. Sammar, M.; Nisemblat, S.; Fleischfarb, Z.; Golan, A.; Sadan, O.; Meiri, H.; Huppertz, B.; Gonen, R. Placenta-bound and body fluid PP13 and its mRNA in normal pregnancy compared to preeclampsia, HELLP and preterm delivery. *Placenta* **2011**, *32*, S30–S36. [CrossRef] [PubMed]
145. Than, N.G.; Abdul Rahman, O.; Magenheim, R.; Nagy, B.; Fule, T.; Hargitai, B.; Sammar, M.; Hupuczi, P.; Tarca, A.L.; Szabo, G.; et al. Placental protein 13 (galectin-13) has decreased placental expression but increased shedding and maternal serum concentrations in patients presenting with preterm pre-eclampsia and HELLP syndrome. *Virchows Arch.* **2008**, *453*, 387–400. [CrossRef] [PubMed]
146. Mariacarmela, G.; Milena, E.; Sveinbjorn, G.; Daniel, H.; Maurizio, M. Placental protein 13 dilation of pregnant rat uterine vein is endothelium dependent and involves nitric oxide/calcium activated potassium channels signals. *Placenta* **2022**, *126*, 233–238. [CrossRef]
147. Chafetz, I.; Kuhnreich, I.; Sammar, M.; Tal, Y.; Gibor, Y.; Meiri, H.; Cuckle, H.; Wolf, M. First-trimester placental protein 13 screening for preeclampsia and intrauterine growth restriction. *Am. J. Obstet. Gynecol.* **2007**, *197*, 35-e1. [CrossRef]
148. Gonen, R.; Shahar, R.; Grimpel, Y.; Chefetz, I.; Sammar, M.; Meiri, H.; Gibor, Y. Placental protein 13 as an early marker for pre-eclampsia: A prospective longitudinal study. *BJOG Int. J. Obstet. Gynaecol.* **2008**, *115*, 1465–1472. [CrossRef]
149. American College of Obstetricians and Gynecologists. ACOG Practice Bulletin No. 204: Fetal growth restriction. *Obstet. Gynecol.* **2019**, *133*, e97–e109. [CrossRef]
150. Frøen, J.F.; Gardosi, J.O.; Thurmann, A.; Francis, A.; Stray-Pedersen, B. Restricted fetal growth in sudden intrauterine unexplained death. *Acta Obstet. Gynecol. Scand.* **2004**, *83*, 801–807.
151. Nardozza, L.M.M.; Caetano, A.C.R.; Zamarian, A.C.P.; Mazzola, J.B.; Silva, C.P.; Marçal, V.M.G.; Lobo, T.F.; Peixoto, A.B.; Araujo Júnior, E. Fetal growth restriction: Current knowledge. *Arch. Gynecol. Obstet.* **2017**, *295*, 1061–1077. [CrossRef] [PubMed]
152. Albu, A.; Anca, A.; Horhoianu, V.; Horhoianu, I. Predictive factors for intrauterine growth restriction. *J. Med. Life* **2014**, *7*, 165. [PubMed]
153. Jin, X.X.; Ying, X.; Dong, M.Y. Galectin-1 expression in the serum and placenta of pregnant women with fetal growth restriction and its significance. *BMC Pregnancy Childbirth* **2021**, *21*, 14. [CrossRef]
154. Hutter, S.; Knabl, J.; Andergassen, U.; Hofmann, S.; Kuhn, C.; Mahner, S.; Arck, P.; Jeschke, U. Placental expression patterns of galectin-1, galectin-2, galectin-3 and galectin-13 in cases of intrauterine growth restriction (IUGR). *Int. J. Mol. Sci.* **2016**, *17*, 523. [CrossRef]
155. Burger, O.; Pick, E.; Zwickel, J.; Klayman, M.; Meiri, H.; Slotky, R.; Mandel, S.; Rabinovitch, L.; Paltieli, Y.; Admon, A.; et al. Placental protein 13 (PP-13): Effects on cultured trophoblasts, and its detection in human body fluids in normal and pathological pregnancies. *Placenta* **2004**, *25*, 608–622. [CrossRef] [PubMed]
156. Bujold, E. Prediction of preeclampsia and fetal growth restriction using placental protein 13. *Placenta* **2012**, *33*, 955. [CrossRef] [PubMed]
157. Cowans, N.J.; Spencer, K.; Meiri, H. First-trimester maternal placental protein 13 levels in pregnancies resulting in adverse outcomes. *Prenat. Diagn.* **2008**, *28*, 121–125. [CrossRef]
158. Hutter, S.; Knabl, J.; Andergassen, U.; Mayr, D.; Hofmann, S.; Kuhn, C.; Mahner, S.; Arck, P.; Jeschke, U. Fetal gender specific expression of tandem-repeat galectins in placental tissue from normally progressed human pregnancies and intrauterine growth restriction (IUGR). *Placenta* **2015**, *36*, 1352–1361. [CrossRef]
159. Ornoy, A.; Becker, M.; Weinstein-Fudim, L.; Ergaz, Z. Diabetes during pregnancy: A maternal disease complicating the course of pregnancy with long-term deleterious effects on the offspring. A clinical review. *Int. J. Mol. Sci.* **2021**, *22*, 2965. [CrossRef]
160. Chiefari, E.; Arcidiacono, B.; Foti, D.; Brunetti, A. Gestational diabetes mellitus: An updated overview. *J. Endocrinol. Investig.* **2017**, *40*, 899–909. [CrossRef]
161. Peters, R.K.; Xiang, A.; Kjos, S.; Buchanan, T.A. Long-term diabetogenic effect of single pregnancy in women with previous gestational diabetes mellitus. *Lancet* **1996**, *347*, 227–230. [CrossRef]
162. Plows, J.F.; Stanley, J.L.; Baker, P.N.; Reynolds, C.M.; Vickers, M.H. The pathophysiology of gestational diabetes mellitus. *Int. J. Mol. Sci.* **2018**, *19*, 3342. [CrossRef] [PubMed]
163. Blois, S.M.; Gueuvoghlanian-Silva, B.Y.; Tirado-Gonzalez, I.; Torloni, M.R.; Freitag, N.; Mattar, R.; Conrad, M.L.; Unverdorben, L.; Barrientos, G.; Knabl, J.; et al. Getting too sweet: Galectin-1 dysregulation in gestational diabetes mellitus. *Mol. Hum. Reprod.* **2014**, *20*, 644–649. [CrossRef] [PubMed]
164. Briana, D.D.; Malamitsi-Puchner, A. Galectin-3: An early marker of gestational diabetes, subclinical atherosclerosis, and tumor progression. *Angiology* **2020**, *71*, 474. [CrossRef]
165. Boutsikou, T.; Giotaki, M.; Boutsikou, M.; Briana, D.D.; Baka, S.; Piatopoulou, D.; Hassiakos, D.; Gourgiotis, D.; Malamitsi-Puchner, A. Cord blood galectin-1 and-3 concentrations in term pregnancies with normal restricted and increased fetal growth. *J. Perinat. Med.* **2015**, *43*, 305–309. [CrossRef] [PubMed]
166. Talmor-Barkan, Y.; Chezar-Azerrad, C.; Kruchin, B.; Leshem-Lev, D.; Levi, A.; Hadar, E.; Kornowski, R.; Tenenbaum-Gavish, K.; Porter, A. Elevated galectin-3 in women with gestational diabetes mellitus, a new surrogate for cardiovascular disease in women. *PLoS ONE* **2020**, *15*, e0234732. [CrossRef]
167. Heusler, I.; Biron-Shental, T.; Farladansky-Gershnabel, S.; Pasternak, Y.; Kidron, D.; Vulih-Shuitsman, I.; Einbinder, Y.; Cohen-Hagai, K.; Benchetrit, S.; Zitman-Gal, T. Enhanced expression of Galectin-3 in gestational diabetes. *Nutr. Metab. Cardiovasc. Dis.* **2021**, *31*, 1791–1797. [CrossRef]

168. Zhang, Z.; Kang, X.; Guo, Y.; Zhang, J.; Xie, J.; Shao, S.; Xiang, Y.; Chen, G.; Yu, X. Association of circulating galectin-3 with gestational diabetes mellitus, progesterone, and insulin resistance. *J. Diabetes* **2021**, *13*, 54–62. [CrossRef]
169. Schrader, S.; Unverdorben, L.; Hutter, S.; Knabl, J.; Schmoeckel, E.; Meister, S.; Beyer, S.; Vilsmaier, T.; Mahner, S.; Jeschke, U.; et al. Overexpression of galectin-4 in placentas of women with gestational diabetes. *J. Reprod. Immunol.* **2022**, *151*, 103629. [CrossRef]
170. Unverdorben, L.; Hüttenbrenner, R.; Knabl, J.; Jeschke, U.; Hutter, S. Galectin-13/PP-13 expression in term placentas of gestational diabetes mellitus pregnancies. *Placenta* **2015**, *36*, 191–198. [CrossRef]
171. Dbstet, A. WHO: Recommended definitions, terminology and format for statistical tables related to the perinatal period and use of a new certificate for cause of perinatal deaths. *Acta Obstet. Gynecol. Scand.* **1977**, *56*, 247–253.
172. Green, E.S.; Arck, P.C. Pathogenesis of preterm birth: Bidirectional inflammation in mother and fetus. *Semin. Immunopathol.* **2020**, *42*, 413–429. [CrossRef] [PubMed]
173. Mercer, B.M.; Crouse, D.T.; Goldenberg, R.L.; Miodovnik, M.; Mapp, D.C.; Meis, P.J.; Dombrowski, M.P.; The Eunice Kennedy Shriver National Institute of Child Health; Human Development Maternal-Fetal Medicine Units Network. The antibiotic treatment of PPROM study: Systemic maternal and fetal markers and perinatal outcomes. *Am. J. Obstet. Gynecol.* **2012**, *206*, 145-e1. [CrossRef] [PubMed]
174. Montenegro, D.; Romero, R.; Kim, S.S.; Tarca, A.L.; Draghici, S.; Kusanovic, J.P.; Kim, J.S.; Lee, D.C.; Erez, O.; Gotsch, F.; et al. Expression patterns of microRNAs in the chorioamniotic membranes: A role for microRNAs in human pregnancy and parturition. *J. Pathol.* **2009**, *217*, 113–121. [CrossRef] [PubMed]
175. DiGiulio, D.B.; Romero, R.; Amogan, H.P.; Kusanovic, J.P.; Bik, E.M.; Gotsch, F.; Kim, C.J.; Erez, O.; Edwin, S.; Relman, D.A. Microbial prevalence, diversity and abundance in amniotic fluid during preterm labor: A molecular and culture-based investigation. *PLoS ONE* **2008**, *3*, e3056. [CrossRef] [PubMed]
176. Kaya, B.; Turhan, U.; Sezer, S.; Kaya, S.; Dağ, İ.; Tayyar, A. Maternal serum galectin-1 and galectin-3 levels in pregnancies complicated with preterm prelabor rupture of membranes. *J. Matern.-Fetal Neonatal Med.* **2020**, *33*, 861–868. [CrossRef]
177. Licini, C.; Tossetta, G.; Avellini, C.; Ciarmela, P.; Lorenzi, T.; Toti, P.; Gesuita, R.; Voltolini, C.; Petraglia, F.; Castellucci, M.; et al. Analysis of cell-cell junctions in human amnion and chorionic plate affected by chorioamnionitis. *Histol. Histopathol.* **2016**, *31*, 759–767.
178. Liu, F.T.; Yang, R.Y.; Hsu, D.K. Galectins in acute and chronic inflammation. *Ann. N. Y. Acad. Sci.* **2012**, *1253*, 80–91. [CrossRef]
179. Tarassishin, L.; Lim, J.; Weatherly, D.B.; Angeletti, R.H.; Lee, S.C. Interleukin-1-induced changes in the glioblastoma secretome suggest its role in tumor progression. *J. Proteom.* **2014**, *99*, 152–168. [CrossRef]
180. Than, N.G.; Kim, S.S.; Abbas, A.; Han, Y.M.; Hotra, J.; Tarca, A.L.; Erez, O.; Wildman, D.E.; Kusanovic, J.P.; Pineles, B.; et al. Chorioamnionitis and increased galectin-1 expression in PPROM—An anti-inflammatory response in the fetal membranes? *Am. J. Reprod. Immunol.* **2008**, *60*, 298–311. [CrossRef]
181. Mueller-Heubach, E.; Rubinstein, D.N.; Schwarz, S.S. Histologic chorioamnionitis and preterm delivery in different patient populations. *Obstet. Gynecol.* **1990**, *75*, 622–626. [PubMed]
182. Demmert, M.; Faust, K.; Bohlmann, M.; Tröger, B.; Göpel, W.; Herting, E.; Härtel, C. Galectin-3 in cord blood of term and preterm infants. *Clin. Exp. Immunol.* **2012**, *167*, 246–251. [CrossRef] [PubMed]
183. Chan, C.; Bode, L.; Kim, J. Galectin-3 binding protein in human preterm infant umbilical cord plasma. *J. Neonatal-Perinat. Med.* **2015**, *8*, 99–104. [CrossRef] [PubMed]
184. Faust, K.; Freitag, N.; Barrientos, G.; Hartel, C.; Blois, S.M. Galectin-levels are elevated in infants born preterm due to amniotic infection and rapidly decline in the neonatal period. *Front. Immunol.* **2021**, *11*, 599104. [CrossRef]

International Journal of
*Molecular Sciences*

MDPI

*Review*

# Endometrium Immunomodulation to Prevent Recurrent Implantation Failure in Assisted Reproductive Technology

**Mustapha Benkhalifa [1,2], Fabien Joao [3], Cynthia Duval [3], Debbie Montjean [3], Molka Bouricha [4], Rosalie Cabry [4], Marie-Claire Bélanger [3], Hatem Bahri [1], Pierre Miron [3] and Moncef Benkhalifa [4,5,*]**

1 HB Laboratory, Tunis TN 1007, Tunisia
2 Faculty of Sciences of Bizerte, University of Carthage, Bizerte TN 7021, Tunisia
3 Fertilys Reproductive Center, Laval, QC H7S 1Z5, Canada
4 Department of Reproductive Medicine, Reproductive Biology & Genetics, University Hospital and School of Medicine Picardie University Jules Verne, 80054 Amiens, France
5 PeriTox Laboratory, CURS, Amiens Sud, 80480 Salouël, France
* Correspondence: benkhalifamoncef78@gmail.com; Tel.: +33-677-867-390

**Abstract:** After more than four decades of assisted reproductive technology (ART) practice worldwide, today more than 60% of women undergoing in vitro fertilization (IVF) treatments fail to become pregnant after the first embryo transfer and nearly 20% of patients are suffering from unexplained recurrent implantation failures (RIFs) and repeated pregnancy loss (RPL). The literature reported different causes of RIF–RPL, mainly multifactorial, endometrial and idiopathic. RIF remains a black box because of the complicated categorization and causes of this physio-pathological dysregulation of implantation and pregnancy process after ovarian stimulation. Many options were suggested as solutions to treat RIF–RPL with controversial results on their usefulness. In this article, we reviewed different possible therapeutic options to improve implantation rates and clinical outcomes. Based on our experience we believe that endometrium immunomodulation after intrauterine insemination of activated autologous peripheral blood mononuclear cells (PBMCs) or platelet-rich plasma (PRP) can be a promising therapeutic solution. On the other hand, peripheral lymphocyte balance typing, specific cytokines and interleukins profiling can be proposed as predictive biomarkers of implantation before embryo transfer.

**Keywords:** infertility; assisted reproductive technology; implantation failure; endometrium immunomodulation

**Citation:** Benkhalifa, M.; Joao, F.; Duval, C.; Montjean, D.; Bouricha, M.; Cabry, R.; Bélanger, M.-C.; Bahri, H.; Miron, P.; Benkhalifa, M. Endometrium Immunomodulation to Prevent Recurrent Implantation Failure in Assisted Reproductive Technology. *Int. J. Mol. Sci.* **2022**, *23*, 12787. https://doi.org/10.3390/ijms232112787

Academic Editor: Ilona Hromadnikova

Received: 2 October 2022
Accepted: 21 October 2022
Published: 24 October 2022



## 1. Introduction

In assisted reproductive technology (ART) programs, 60–70% of women fail to become pregnant after embryo transfer. Repeated implantation failure (RIF) remains a black box in daily practice due to the complicated categorization and causes of this physio-pathological dysregulation [1]. Different causes of RIF were reported, mainly multifactorial, endometrial and idiopathic. Multifactorial RIF can be caused by maternal and paternal factors, gamete and embryo quality, infections and lifestyle changes in combination with psychological status and oxidative stress [1,2]. Impaired endometrium function such as abnormal growth or loss of vascularization can account for endometrial RIF, but idiopathic RIF, caused mainly by abnormal cross-talk between the embryo and endometrium, remains the principal question and needs to be elucidated [1].

RIF may be defined as a failure to obtain a pregnancy after multiple viable embryo transfers during IVF treatment [3], but its definition is inconsistent between studies. The most common definition was portrayed by Bashiri and colleagues [4] who describe RIF as three or more pregnancy failures following the transfer of at least three good-quality embryos [4]. However, other authors such as Coughlan and colleagues [5] suggest including maternal age, number of embryos transferred and number of previous cycles to the definition of RIF [5]. Interestingly, a consensus is emerging thanks to a recent extensive survey. It was proposed to define RIF as the failure to achieve a clinical pregnancy after 2–3 IVF cycles with 1–4 good-quality embryos [6]. RIF is a challenge for clinicians as its etiology includes various possible causes [2].

The causes of RIF can be divided into two categories: maternal (uterine anatomic abnormalities, chronic endometritis, non-receptive endometrium, antiphospholipid antibody syndrome and immunological factors) and embryonic (genetic defects and other factors specific to embryonic development) causes [3]. In the absence of male factors, oxidative stress, bad-quality embryos and anatomical abnormalities such as hydro-salpinx and thrombophilia, RIF seems to be caused by impaired endometrial function such as abnormal endometrial growth or loss of vascularization [4]. However, RIF caused by immunological factors could be manageable using several innovative therapeutic options. Among them, intrauterine administration of human chorionic gonadotropin (HCG), granulocyte colony-stimulating factor (G-CSF) or autologous peripheral blood mononuclear cells (PBMCs) has been suggested as a treatment for patients suffering from RIF [4,7–16].

Intrauterine administration of autologous PBMC prior to embryo transfer was proposed to regulate the immune environment of the endometrial tissue [4]. In 2006, Yoshioka and colleagues were the first to propose this immunotherapy to help RIF patients [7]. Since then, this therapeutic option was recommended as an effective treatment for RIF according to numerous studies [4,7–14]. The present study is a review aiming at summarizing studies that used this immunotherapy to evaluate its benefit regarding RIF patients.

## 2. Endometrium Immunomodulation via Intrauterine Insemination of Activated Autologous Peripheral Blood Mononuclear Cells (PBMCs)

PBMCs from patients with RIF are usually isolated during the ovulation period using a lymphocyte separation medium composed of an iso-osmotic poly-sucrose and sodium diatrizoate solution to separate mononuclear cells (including B-lymphocytes, T-lymphocytes and monocytes) from the other blood cells. After separation, PBMCs are generally activated with hCG or corticotropin-releasing hormone (CRH) and cultured in vitro for 24–72 h in a humidified incubator with 5% $CO_2$ at 37 °C (Figure 1).

After culture, PBMCs are administered in utero using a catheter [4,7–15]. However, the number of cells administered in utero is not homogeneous among all studies investigating the use of PBMC in the treatment of RIF (Table 1). Although there were some methodological variations between studies in terms of the number of previous cycles, cycle type, and number and quality of transferred embryos, patients were generally administered with 10 to 30 million PBMCs [7–16]. Madkour and colleagues showed a significant increase in clinical pregnancy rate (CPR) with only 1 million cells [10]. Furthermore, in a recent meta-analysis, Qin and colleagues have demonstrated that CPR was higher when less than 100 million PBMCs/mL were administered in utero, suggesting that although the quantity of cells inseminated is not homogeneous, intrauterine administration of PBMC does appear to be an effective treatment for patients suffering from RIF [17].

**Table 1.** Main studies using PBMCs to treat RIF.

| Study | Number of Previous Failed IVF Cycles | Sample Size | | Day of Blood Collection | PBMCs Co-cultured with | Duration of PBMC Culture | Number of PBMCs Administered In Utero | Transfer Type | Stage of Embryo | Implantation Rate (Control vs. Case) | Clinical Pregnancy Rate (Control vs. Case) | Miscarriage Rate (Control vs. Case) | Live Birth Rate (Control vs. Case) |
|---|---|---|---|---|---|---|---|---|---|---|---|---|---|
| | | Control | Case | | | | | | | | | | |
| Yoshioka et al., 2006 [7] | ≥4 | 18 | 17 | On the day of oocyte retrieval | hCG: 5 IU/mL | 48 h | $20 \times 10^6$ | Fresh | 1, 2 or 3 blastocysts | 4.1% vs. 23.4% ($p = 0.0034$) | 11.1% vs. 41.2% ($p = 0.042$) | Not specified | 7.6% vs. 55.6% ($p = 0.013$) |
| Okitsu et al., 2011 [16] | ≥1 | 170 | 83 | On the day following ovulation or the day after | Not activated | No culture | $30 \times 10^6$ | Frozen/thawed | early cleavage embryo or blastocyst | ≥1 RIF: 21.1% vs. 21.6% (ns); 3 RIF: 9.38% vs. 25.0% ($p = 0.041$) | ≥1 RIF: 32.9% vs. 34.9% (ns); ≥3 RIF: 16.7% vs. 42.1% ($p = 0.039$) | Not specified | ≥ 1 RIF: 21.8% vs. 21.7% (ns); ≥3 RIF: 11.1% vs. 21.2% (ns) |
| Makrigiannakis et al., 2015 [9] | ≥3 | 45 | 45 | On the day of oocyte retrieval | CRH: $10^7$ M/$1.10^6$ cells/mL | 48 h | $20 \times 10^6 + 10^7$ CRH | Fresh | 2 or 3 blastocysts (grade 3BB and above) | Not specified | 0% vs. 44.44% ($p < 0.001$) | Not specified | Not specified |
| Madkour et al., 2016 [10] | ≥2 | 27 | 27 | On the day of ovulation induction | Complete culture medium + 75 IU of hMG | 72 h | $1 \times 10^6$ | Fresh | 1, 2 or 3 early cleavage embryos | ≥2 RIF: 9% vs. 22% ($p = 0.02$); 2 RIF vs. ≥ 3 RIF: 15% vs. 35% ($p = 0.09$) | ≥2 RIF: 15% vs. 44% ($p = 0.045$); 2 RIF vs. ≥ 3 RIF: 29% vs. 70% ($p = 0.04$) | ≥2 RIF: 17% vs. 75% ($p = 0.08$) 2 RIF vs. ≥ 3 RIF: 20% vs. 14% ($p = 0.8$) | Not specified |
| Yu et al., 2016 [11] | ≥3 | 105 | 93 | On the day following ovulation | hCG: 10 IU/mL | 24 h | $10–20 \times 10^6$ | Frozen/thawed | early cleavage embryo | 11.43% vs. 23.66% ($p < 0.05$) | 20.95% vs. 46.24% ($p < 0.05$) | 31.8% vs. 20.9% (ns) | 14.28% vs. 34.41% ($p < 0.05$) |
| Li et al., 2017 [12] | ≥1 | 339 | 294 | Two days before embryo transfer | hCG: 10 IU/mL | 24 h | $10–20 \times 10^6$ | Fresh and frozen/thawed | 2 or 3 early cleavage embryos or 2 or 3 grade 2 blastocysts at day 5 and 3BB and above at day 6 | 1 RIF: 32.33% vs. 29.35% (ns); 2 RIF: 27.74% vs. 35.98% ($p = 0.048$); 3 RIF: 26.23% vs. 23.20% (ns); ≥4 RIF: 4.88% vs. 22.00% ($p = 0.014$) | 1 RIF: 41.23% vs. 43.75% (ns); 2 RIF: 42.18% vs. 48.15% ($p = 0.016$); 3 RIF: 36.84% vs. 42.22% (ns); ≥4 RIF: 14.29% vs. 39.58% ($p = 0.038$) | Not specified | 1 RIF: 36.84% vs. 37.5% (ns); 2 RIF: 33.33% vs. 34.26% (ns); 3 RIF: 24.56% vs. 28.89% (ns); ≥4 RIF: 9.58% vs. 33.33% ($p = 0.038$) |
| Makrigiannakis et al., 2019 [13] | ≥3 | 26 | 26 | On the day of oocyte retrieval | CRH: $10^7$ M/$1.10^6$ cells/mL | 48 h | $20 \times 10^6 + 10^7$ M CRH | Fresh | 2 or 3 grade 1 or 2 early cleavage embryos | Not specified | 0% vs. 57,69% ($p < 0.01$) | Not specified | Not specified |
| Nobijari et al., 2019 [15] | ≥1 | 128 | 122 | 5 days before the frozen/thawed embryo transfer | CRH (concentration not specified) | 48–72 h | $20 \times 10^6 + 10^7$ M CRH | Frozen/thawed | early cleavage embryo or blastocyst | Not specified | <3 RIF: 30.4% vs. 30.8% ($p = 0.91$); ≥3 RIF: 19,7% vs. 38,6% ($p = 0.01$) | Not specified | Not specified |
| Pourmoghadam et al., 2020 [14] | ≥3 | 50 | 50 | On the day of ovulation induction | hCG: 10 IU/mL daily | 48 h | $15–20 \times 10^6$ | Frozen/thawed | early cleavage embryo or blastocyst | Not specified | 22% vs. 42% ($p = 0.032$) | 24% vs. 8% ($p = 0.029$) | 20% vs. 38% ($p = 0.047$) |

CRH: corticotropin-releasing hormone; hCG: human chorionic gonadotropin; RIF: recurrent implantation failure.

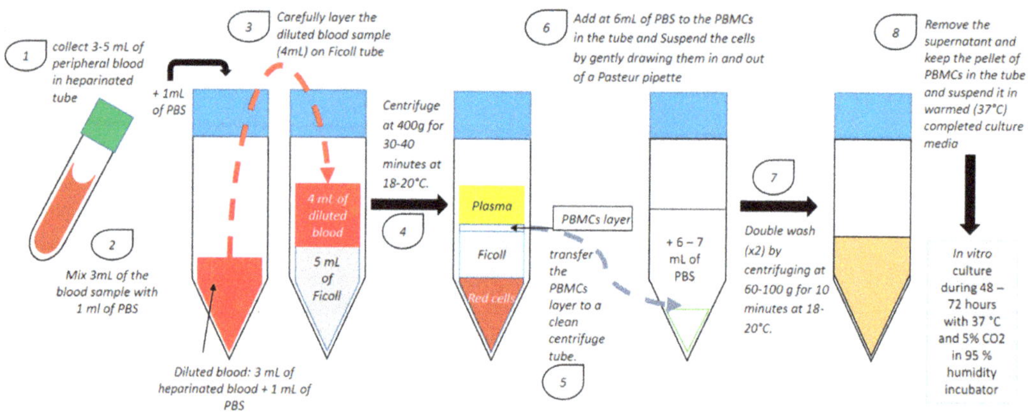


**Figure 1.** PBMC isolation technique and in vitro culture (PBS: phosphate-buffered saline; PBMC: peripheral blood mononuclear cell).

## 3. Immunoregulation of the Endometrium during Embryo Implantation: Biological Function and Molecular Pathway

To achieve successful embryo implantation and pregnancy, an appropriate dialogue between the embryo and the endometrium must take place [18].

In the uterine environment, a particular form of natural killer (NK) cells with a unique transcriptional profile, the uterine NK (uNK) cells, represents the most abundant lymphocyte population, especially in the endometrium [19–21]. In fact, most of the immune cells present in the uterus usually display a unique phenotype [18]. Peripheral blood NK cells express $CD56^{+}CD16^{+}$ at their membrane surface and are characterized by a highly cytotoxic profile [22]. However, uNK cells are less toxic since they do not express CD16 on their membrane surface [23]. During the menstrual cycle, levels of uNK cells start to increase in the mid-secretory phase, which could explain their importance in embryo implantation [24–27].

Dendritic cells (DCs), another type of innate immune cells, have a crucial role in the site of embryo implantation and maternal–fetal interface. DCs act as antigen-presenting cells to T cells and have the unique ability to induce a primary immune response, a phenomenon crucial for successful pregnancy [28]. In addition, DCs can influence trophoblast invasion by regulating the secretion of cytokines and the production of endometrial cell-surface proteins. Through the regulation of immune cell functions and actions, DCs have a major role in the establishment of a special local immune environment essential for embryo implantation and placental development [29]. Human decidual DCs, however, seem to have an immature phenotype characterized by a low expression of CD40, CD80, CD86 and CD205 [30,31]. DCs seem to be involved in the immune tolerance of the implantation site through the regulation of T-cell proliferation and the elimination of antigen-specific T cells. In the decidua, uterine dendritic cells (uDCs) are also crucial in maintaining pregnancy [32]. Since the 1990s, it has been known that maternal T cells are essential to the complex mechanisms of immune tolerance, a phenomenon critical to the invasion of the endometrium by the blastocyst [33].

T-cell interactions can be performed directly by cell–cell contact or indirectly through the secretion of pro-inflammatory or anti-inflammatory cytokines [34]. Pro-inflammatory cytokines such as interleukin (IL)-1β, -6, -12, -2 and -18; tumor necrosis factor alpha (TNF-α) and interferon gamma (IFN-γ) are mainly produced by T helper (Th) 1 cells, while anti-inflammatory cytokines such as IL-4, IL-10, IL-13 and TGF-β1 are mostly secreted by Th2 cells [35]. The pro-inflammatory Th1 profile was shown to be associated with successful and normal pregnancy at early and late pregnancy stages. In the midgestation stage, however, a shift to an anti-inflammatory Th2 profile must take place to establish tolerance to the foreign fetal antigens [36]. An imbalance in these cytokine profiles has been associated with spontaneous abortion and common complications of pregnancy [37–39]. Moreover, it has been shown that levels of pro-inflammatory cytokines (such as IL-2 and IFN-γ) decreased while levels of anti-inflammatory cytokines (such as IL-4 and IL-10) increased in the induction of immune tolerance to allografts [40,41]. The implication of T cells, especially CD4+ CD25+ Foxp3+ Treg cells, in the initial stages of pregnancy is therefore needed for the prevention of an alloreactivity action by the endometrium against the fetus through cascades of immunoregulation actions [42,43].

Treg, Th1 and Th2 cells are, however, not the only T-cell subtypes known to be crucial for successful embryo implantation. Th17 cells, a subset of T cells showing remarkable plasticity, are also indispensable in the immunoregulation of embryo implantation as well as in maintaining normal pregnancy [44].

Monocytes and macrophages also play an important role during the menstrual cycle and pregnancy [14,45,46]. Macrophages regulate trophoblast activity by promoting endometrial tissue remodeling and angiogenesis [47]. Pregnancy hormones directly and indirectly modulate the recruitment of monocytes in the uterus and participate in their differentiation and stimulation into functional macrophages [48]. Intrauterine administration of PBMCs could also be a source of hCG-activated macrophages and regulate the uterine environment at the embryo implantation site [14].

## 4. Endometrium Immunomodulation with Activated PBMCs and Embryo Implantation

Intrauterine administration of PBMCs in patients suffering from RIF aims to improve endometrial receptivity by regulating the Th1/Th2 cytokine ratio and growth factors to stimulate many cascades of cytokines and matrix metalloproteinase actions [1,7,10,15]. Increased peripheral blood Th1/Th2 ratio was shown to be detrimental to embryo implantation [39]. However, PBMCs produce many cytokines that can regulate Th1/Th2 imbalance in women suffering from RIF [39]. Furthermore, PBMCs are known to increase the secretion of growth factors and Th1 pro-inflammatory and anti-inflammatory cytokines at the time of embryo implantation to boost endometrial receptivity [4,9–11]. This immunotherapy was shown to improve progesterone (P4) production in cultured human granulosa luteal cells [49]. Ovarian steroids such as P4 and β-hCG are among the most crucial factors needed in the immunoregulation of embryo implantation [50]. Luteinizing hormone (LH) and hCG have also an important role in establishing the immune tolerance mechanisms of embryo implantation. These two gonadotropins were shown to affect immune cells by binding to the LH/hCG receptors present at the surface of several immune cell types [50]. Furthermore, it has been shown that hCG has the capacity to downregulate pro-inflammatory immune responses during pregnancy [51]. During the embryo implantation window, β-hCG seems to play a role in the immunoregulation of the endometrium in increasing Fas ligand expression (APO-1, CD95) in the endometrial cells to facilitate trophoblast invasion [52]. Increased peripheral blood Treg cell levels have also been shown to be positively associated with higher pregnancy rates in IVF treatment [53]. These cells being attracted to trophoblasts by hCG [51] supports the fact that the administration of hCG could be an effective treatment for some infertile women. Moreover, it has been shown by Mansour and colleagues that intrauterine hCG injection before embryo transfer could significantly improve implantation and pregnancy rates [54].

Intrauterine administration of PBMCs for patients suffering from RIF has been shown to be specifically efficient for increasing implantation and pregnancy rates in women with three or more previous implantation failures [17,55]. Recently, Nobijari and colleagues and Pourmoghadam and colleagues presented a different strategy to administer PBMCs in RIF patients using frozen–thawed embryo transfers [14,15]. Nobijari and colleagues confirmed the effectiveness of this immunotherapy by showing an increase in CPR in patients with three or more implantation failures undergoing frozen–thawed embryo transfer [15]. Pourmoghadam and colleagues only administrated PBMCs in utero in RIF patients with a low Th-17/Treg cell ratio [14]. Furthermore, in the study of Pourmoghadam and colleagues, PBMCs were activated in vitro with 10 IU/mL hCG for 48 h before the intrauterine administration, while Nobijari and colleagues activated the PBMCs in vitro with CRH for 48–72 h [14,15]. In RIF patients, it has also been shown that levels of IL-1β, TNF-α and IFN-γ, three pro-inflammatory cytokines, were increased in the PBMC culture medium, suggesting that PBMCs secrete these Th1 cytokines when treated with hCG [14]. Moreover, Pourmoghadam and colleagues have shown that CPR and live birth rates increased significantly and miscarriage rates decreased significantly in RIF patients treated with PBMCs compared to control [14]. In addition, Makrigiannakis and colleagues have shown that the insemination of autologous PBMCs treated with CRH before blastocyst or early cleaved embryo transfer presented better results than PBMCs without CRH treatment in terms of CPR in women with RIF [13].

Therefore, these three studies supported the effectiveness of this immunotherapy for patients suffering from RIF undergoing fresh or frozen–thawed embryo transfer, especially when PBMCs are treated with CRH [13–15]. However, these findings are still limited because, in the study of Pourmoghadam and colleagues for example, the authors measured only three pro-inflammatory cytokines, and they did not show anti-inflammatory cytokine levels with PBMC administration for RIF women or in a control group [14]. The increase in these cytokine levels should be compared to a control, not treated cells, but the authors did not perform this comparison [14].

## 5. Other Endometrium Immunomodulation Options

Immunological therapy approaches other than intrauterine administration of PBMCs for the management of RIF patients were reported in the literature. These immunotherapies focus on elevated Th1/Th2 ratio, abnormal TNF-α/IL-10 ratio, elevated NK cells and auto-antibodies. One of these immunomodulatory agents that have been described for RIF patients is intravenous immunoglobulin IgG (IVIg). Patients receiving this treatment have shown significantly higher implantation and clinical pregnancy rates compared to non-treated patients [56]. This treatment has been extensively used, but the results are heterogeneous [57–60]. According to many studies, the application of IVIg has shown positive effects on RIF patient pregnancy rates and in patients with increased immunological risk factors [24,61–64].

Granulocyte colony-stimulating factor (G-CSF) was also shown to have positive effects on embryo implantation in women suffering from RIF, especially when endometrial thickness was insufficient [65]. Furthermore, a recent meta-analysis showed that G-CSF was an effective treatment for women with thin endometrium or with recurrent IVF failures [66]. G-CSF was originally used as a treatment for thin endometrium to thicken it. Increased implantation rates were shown after G-CSF treatment in patients with an endometrium thickness $\geq$7 mm on the day of embryo transfer [67]. These results were confirmed by another study conducted by Xu and colleagues in 2015 in which they showed a higher implantation rate in women treated with G-CSF compared to controls [68]. Furthermore, Kalem and colleagues have shown that the administration of G-CSF into the uterine cavity in RIF patients with normal endometrium did not alter the endometrial thickness, clinical pregnancy rates or live birth rates in comparison with a control group [69].

Vitamin E, which has been shown to improve capillary blood flow in different organs [70,71], and sildenafil citrate (Viagra), which improves uterine artery blood flow [72,73], were also proposed as a treatment for thin endometrium [72–74]. In the study of Miwa and colleagues, 23 out of 25 patients showed improved radial artery, 17 patients had increased endometrial thickness and 13 patients developed an endometrium thickness of more than 8 mm [74]. Sher and Fisch were the first to suggest the use of sildenafil during the follicular phase and until ovulation trigger as a treatment for thin endometrium of women undergoing IVF with fresh embryo transfer [72]. They reported an improvement in uterine blood flow and in endometrial thickness [72]. These results were confirmed in a larger cohort study showing a 45% pregnancy rate [73]. However, a randomized controlled trial study reported no significant difference in endometrial thickness and pregnancy rate after administration of sildenafil and valerate estradiol during the luteal phase following fresh embryo transfer [75]. Another randomized controlled trial study did not show any improvement in uterine blood flow or in endometrial thickness [76]. Recently, a randomized placebo-controlled trial study reported that the use of vaginal sildenafil on the hCG injection day did not present a statistically significant improvement in endometrium thickness; however, the implantation (chemical pregnancy) was significantly higher in women who received sildenafil with placebo compared to women who received only sildenafil or only placebo [77]. More trials are needed to confirm the effectiveness of these treatments on endometrium thickness and/or RIF.

In 2015, Nakagawa and colleagues proposed a treatment using immunosuppressive drugs such as tacrolimus, one of the major immune-suppressive agents that have been used after allogeneic organ transplantation to reduce the alloreactivity of a recipient's immune system and to decrease the risk of the rejection [78,79]. This treatment has shown positive results on successful implantation and pregnancy outcome in RIF patients with elevated Th1/Th2 ratios, suggesting that this immunological imbalance plays a crucial role in causing RIF [78]. However, the posology of this drug must be determined more accurately to maintain the levels of the essential Th1 cytokines necessary for embryo implantation [39].

Another treatment using atosiban administration was proposed for RIF women. In fact, atosiban is a receptor of oxytocin and V1a vasopressin, proposed to avoid uterine contractions during embryo transfer, which could be detrimental in embryonic apposition [80]. However, according to the review of Makrigiannakis and colleagues, various randomized controlled trial studies reported a non-significant effect on reproductive outcomes [81–85], and only two non-randomized studies on RIF patients report a significant benefit after atosiban treatment [86,87]. Therefore, more randomized studies are needed to verify the efficiency of atosiban as a benefic treatment for RIF women.

In 2015, Chang and colleagues reported that autologous platelet-rich plasma (PRP) promotes endometrial growth and improves pregnancy outcomes during IVF [88]. After being collected from the peripheral vein in acid citrate dextrose solution A (ACD-A) anticoagulant tubes, PRP was prepared by separating the various components of the blood using multiple centrifugations [89]. This PRP, within 10 min after clotting, can activate cytokines and growth factors which become bioactive and increasingly secreted. These factors include vascular endothelial growth factor (VEGF), transforming growth factor (TGF), platelet-derived growth factor (PDGF) and epidermal growth factor (EGF), which can regulate cell migration, attachment, proliferation and differentiation, while promoting extracellular matrix accumulation [90]. This could lead to ameliorated implantation conditions and improved pregnancy, as was revealed by Chang and collaborators [88]. Other studies could confirm these results; for example, in 2019, Kim et al. showed that autologous PRP treatment increases the activity of cytokines and growth factors compared to that observed without the use of PRP, especially when combined with frozen–thawed embryo transfer [91]. These studies support the suggestion of PRP as a useful treatment for RIF. However, in a recent study that used PRP treatment in patients with a history of failed implantation before frozen–thawed embryo transfer, the authors did not find significant differences in the pregnancy results in comparison with controls [92]. A recent study by Ibañez-Perez and colleagues suggested a non-invasive method of microRNA-based signatures obtained from very small volumes of endometrial fluid collected just before day 5 frozen embryo transfers to identify the competence of the endometrium in implantation [93]. This technique could help physicians to avoid RIF by changing the embryo transfer strategy when the results show an unfavorable implantation pattern by using immunomodulation techniques from the first IVF cycle [93].

## 6. Conclusions

There is no scientific consensus about the best immunological treatment for RIF patients presenting an imbalanced Th1/Th2 ratio or immune dysregulation. However, recent studies have shown the potential of the intrauterine administration of hCG-activated PBMCs and activated PRP as a good way to modulate endometrial receptivity. The immunotherapy field strategy needs to be further elucidated for a better understanding of maternal immunotolerance to embryo implantation. Proteomic investigations of biomarkers produced by immunological cells and their pathways should be continued to identify the exact combination of immunological factors needed for successful implantation. Correcting immunological dysregulations in embryo implantation by intrauterine administration of PBMCs or treatment with activated PRP seems to be a promising solution in RIF. It is clear that we need to know much more about maternal immune tolerance and the exact role of each biomarker involved in embryo–endometrium cross-talk to improve implantation and reduce repeated implantation failure and pregnancy loss.

**Author Contributions:** Conceptualization, M.B. (Mustapha Benkhalifa) and M.B. (Moncef Benkhalifa); methodology, M.B. (Mustapha Benkhalifa); software, M.B. (Mustapha Benkhalifa); validation, M.B. (Moncef Benkhalifa), D.M., F.J., H.B., P.M., M.-C.B., R.C. and C.D.; formal analysis, D.M.; investigation, M.B. (Mustapha Benkhalifa) and M.B. (Molka Bouricha); resources, M.B. (Mustapha Benkhalifa), M.B. (Molka Bouricha) and M.B. (Moncef Benkhalifa); data curation, Mustapha Benkhaifa and D.M.; writing—original draft preparation, M.B. (Mustapha Benkhalifa); writing—review and editing, M.B. (Mustapha Benkhalifa); visualization, D.M.; supervision, M.B. (Moncef Benkhalifa) and D.M.; project administration, M.B. (Moncef Benkhalifa); funding acquisition, not applicable. All authors have read and agreed to the published version of the manuscript.

**Funding:** This research received no external funding.

**Institutional Review Board Statement:** Not applicable.

**Conflicts of Interest:** The authors declare no conflict of interest.

## References


1. Benkhalifa, M.; Zayani, Y.; Bach, V.; Copin, H.; Feki, M.; Benkhalifa, M.; Allal-Elasmi, M. Does the dysregulation of matrix metalloproteinases contribute to recurrent implantation failure? *Expert Rev. Proteom.* **2018**, *15*, 311–323. [CrossRef]
2. Busnelli, A.; Somigliana, E.; Cirillo, F.; Baggiani, A.; Levi-Setti, P.E. Efficacy of therapies and interventions for repeated embryo implantation failure: A systematic review and meta-analysis. *Sci. Rep.* **2021**, *11*, 1747. [CrossRef]
3. Simon, A.; Laufer, N. Repeated implantation failure: Clinical approach. *Fertil. Steril.* **2012**, *97*, 1039–1043. [CrossRef] [PubMed]
4. Bashiri, A.; Halper, K.I.; Orvieto, R. Recurrent Implantation Failure-update overview on etiology, diagnosis, treatment and future directions. *Reprod. Biol. Endocrinol.* **2018**, *16*, 121. [CrossRef] [PubMed]
5. Coughlan, C.; Ledger, W.; Wang, Q.; Liu, F.; Demirol, A.; Gurgan, T.; Cutting, R.; Ong, K.; Sallam, H.; Li, T.C. Recurrent implantation failure: Definition and management. *Reprod. Biomed. Online* **2014**, *28*, 14–38. [CrossRef] [PubMed]
6. Cimadomo, D.; Craciunas, L.; Vermeulen, N.; Vomstein, K.; Toth, B. Definition, diagnostic and therapeutic options in recurrent implantation failure: An international survey of clinicians and embryologists. *Hum. Reprod.* **2021**, *36*, 305–317. [CrossRef] [PubMed]
7. Yoshioka, S.; Fujiwara, H.; Nakayama, T.; Kosaka, K.; Mori, T.; Fujii, S. Intrauterine administration of autologous peripheral blood mononuclear cells promotes implantation rates in patients with repeated failure of IVF-embryo transfer. *Hum. Reprod.* **2006**, *21*, 3290–3294. [CrossRef]
8. Bielfeld, A.P.; Pour, S.J.; Poschmann, G.; Stühler, K.; Krüssel, J.S.; Baston-Büst, D.M. A Proteome Approach Reveals Differences between Fertile Women and Patients with Repeated Implantation Failure on Endometrial Level Does hCG Render the Endometrium of RIF Patients? *Int J. Mol. Sci.* **2019**, *2*, 425. [CrossRef]
9. Makrigiannakis, A.; BenKhalifa, M.; Vrekoussis, T.; Mahjub, S.; Kalantaridou, S.N.; Gurgan, T. Repeated implantation failure: A new potential treatment option. *Eur. J. Clin. Investig.* **2015**, *45*, 380–384. [CrossRef]
10. Madkour, A.; Bouamoud, N.; Louanjli, N.; Kaarouch, I.; Copin, H.; Benkhalifa, M.; Sefrioui, O.; Madkour, L. Intrauterine insemination of cultured peripheral blood mononuclear cells prior to embryo transfer improves clinical outcome for patients with repeated implantation failures. *Zygote* **2016**, *24*, 58–69. [CrossRef]
11. Yu, N.; Zhang, B.; Xu, M.; Wang, S.; Liu, R.; Wu, J.; Yang, J.; Feng, L. Intrauterine administration of autologous peripheral blood mononuclear cells (PBMCs) activated by HCG improves the implantation and pregnancy rates in patients with repeated implantation failure: A prospective randomized study. *Am. J. Reprod. Immunol.* **2016**, *76*, 212–216. [CrossRef]
12. Li, S.; Wang, J.; Cheng, Y.; Zhou, D.; Yin, T.; Xu, W.; Yu, N.; Yang, J. Intrauterine administration of hCG-activated autologous human peripheral blood mononuclear cells (PBMC) promotes live birth rates in frozen/thawed embryo transfer cycles of patients with repeated implantation failure. *J. Reprod. Immunol.* **2017**, *119*, 15–22. [CrossRef]

13. Makrigiannakis, A.; Vrekoussis, T.; Makrygiannakis, F.; Ruso, H.; Kalantaridou, S.N.; Gurgan, T. Intrauterine CRH-treated PBMC in repeated implantation failure. *Eur. J. Clin. Investig.* **2019**, *49*, e13084. [CrossRef] [PubMed]
14. Pourmoghadam, Z.; Soltani-Zangbar, M.S.; Sheikhansari, G.; Azizi, R.; Eghbal-Fard, S.; Mohammadi, H.; Siahmansouri, H.; Aghebati-Maleki, L.; Danaii, S.; Mehdizadeh, A.; et al. Intrauterine administration of autologous hCG- activated peripheral blood mononuclear cells improves pregnancy outcomes in patients with recurrent implantation failure; A double-blind, randomized control trial study. *J. Reprod. Immunol.* **2020**, *142*, 103182. [CrossRef]
15. Nobijari, F.F.; Arefi, S.S.; Moini, A.; Taheripanah, R.; Fazeli, E.; Kharazi, H.; Hosseini, S.Z.; Hosseini, A.; Valojerdi, M.R.; Copin, H. Endometrium immunomodulation by intrauterine insemination administration of treated peripheral blood mononuclear cell prior frozen/thawed embryos in patients with repeated implantation failure. *Zygote* **2019**, *27*, 214–218. [CrossRef] [PubMed]
16. Okitsu, O.; Kiyokawa, M.; Oda, T.; Miyake, K. Intrauterine administration of autologous blood mononuclear cells increases clinical pregnancy rates in frozen/thawed embryo transfer cycles of patients with repeated implantation failure. *J. Reprod. Immunol.* **2011**, *92*, 82–87. [CrossRef]
17. Qin, Q.; Chang, H.; Zhou, S.; Zhang, S.; Yuan, D.; Yu, L.L.; Qu, T. Intrauterine administration of peripheral blood mononuclear cells activated by human chorionic gonadotropin in patients with repeated implantation failure: A meta-analysis. *J. Reprod. Immunol.* **2021**, *145*, 103323. [CrossRef]
18. Zenclussen, A.C.; Hämmerling, G.J. Cellular regulation of the uterine microenvironment that enables embryo implantation. *Front. Immunol.* **2015**, *6*, 321. [CrossRef] [PubMed]
19. King, A.; Balendran, N.; Wooding, P.; Carter, N.P.; Loke, Y.W. CD3- leukocytes present in the human uterus during early placentation: Phenotypic and morphologic characterization of the CD56++ population. *Dev. Immunol.* **1991**, *1*, 169–190. [CrossRef] [PubMed]
20. Koopman, L.A.; Kopcow, H.D.; Rybalov, B.; Boyson, J.E.; Orange, J.S.; Schatz, F.; Masch, R.; Lockwood, C.J.; Schachter, A.D.; Park, P.J.; et al. Human decidual natural killer cells are a unique NK cell subset with immunomodulatory potential. *J. Exp. Med.* **2003**, *198*, 1201–1212. [CrossRef]
21. Mori, M.; Bogdan, A.; Balassa, T.; Csabai, T.; Szekeres-Bartho, J. The deciduas-the maternal bed embracing the embryo-maintains the pregnancy. *Semin. Immunopathol.* **2016**, *38*, 635–649. [CrossRef] [PubMed]
22. Poli, A.; Michel, T.; Thérésine, M.; Andrès, E.; Hentges, F.; Zimmer, J. CD56bright natural killer (NK) cells: An important NK cell subset. *Immunology* **2009**, *126*, 458–465. [CrossRef]
23. Mandal, A.; Viswanathan, C. Natural killer cells: In health and disease. *Hematol. Oncol. Stem Cell Ther.* **2015**, *8*, 47–55. [CrossRef]
24. Timeva, T.; Shterev, A.; Kyurkchiev, S. Recurrent Implantation Failure: The Role of the Endometrium. *J. Reprod. Infertil.* **2014**, *15*, 173–183.
25. Helige, C.; Ahammer, H.; Hammer, A.; Huppertz, B.; Frank, H.G.; Dohr, G. Trophoblastic invasion in vitro and in vivo: Similarities and differences. *Hum. Reprod.* **2008**, *23*, 2282–2291. [CrossRef]
26. Wallace, A.E.; Fraser, R.; Cartwright, J.E. Extravillous trophoblast and decidual natural killer cells: A remodelling partnership. *Hum. Reprod. Update* **2012**, *18*, 458–471. [CrossRef]
27. Lee, J.Y.; Lee, M.; Lee, S.K. Role of endometrial immune cells in implantation. *Clin. Exp. Reprod. Med.* **2011**, *38*, 119–125. [CrossRef]
28. Huang, X.; Venet, F.; Chung, C.S.; Lomas-Neira, J.; Ayala, A. Changes in dendritic cell function in the immune response to sepsis: Cell- & tissue-based therapy. *Expert Opin. Biol. Ther.* **2007**, *7*, 929–938.
29. Gellersen, B.; Brosens, J.J. Cyclic decidualization of the human endometrium in reproductive health and failure. *Endocr. Rev.* **2014**, *35*, 851–905. [CrossRef] [PubMed]
30. Péguet-Navarro, J.; Dalbiez-Gauthier, C.; Rattis, F.M.; Van Kooten, C.; Banchereau, J.; Schmitt, D. Functional expression of CD40 antigen on human epidermal Langerhans cells. *J. Immunol.* **1995**, *155*, 4241–4247. [PubMed]
31. Rattis, F.M.; Péguet-Navarro, J.; Staquet, M.J.; Dezutter-Dambuyant, C.; Courtellemont, P.; Redziniak, G.; Schmitt, D. Expression and function of B7-1 (CD80) and B7-2 (CD86) on human epidermal Langerhans cells. *Eur. J. Immunol.* **1996**, *26*, 449–453. [CrossRef] [PubMed]
32. Tagliani, E.; Erlebacher, A. Dendritic cell function at the maternal–fetal interface. *Expert Rev. Clin. Immunol.* **2011**, *7*, 593–602. [CrossRef] [PubMed]
33. Tafuri, A.; Alferink, J.; Möller, P.; Hämmerling, G.J.; Arnold, B. T cell awareness of paternal alloantigensduring pregnancy. *Science* **1995**, *270*, 630–633. [CrossRef]
34. Hudic, I.; Fatušic, Z. Progesterone-induced blocking factor (PIBF) and Th1/Th2 cytokine in women with threatened spontaneous abortion. *J. Perinat. Med.* **2009**, *37*, 338–342. [CrossRef]
35. Kaiko, G.E.; Horvat, J.C.; Beagley, K.W.; Hansbro, P.M. Immunological decision-making: How does the immune system decide to mount a helper T-cell response? *Immunology* **2008**, *123*, 326–338. [CrossRef] [PubMed]
36. Mor, G.; Cardenas, I.; Abrahams, V.; Guller, S. Inflammation and pregnancy: The role of the immune system at the implantation site. *Ann. N. Y. Acad. Sci.* **2011**, *1221*, 80–87. [CrossRef]
37. Jena, M.K.; Nayak, N.; Chen, K.; Nayak, N.R. Role of Macrophages in Pregnancy and Related Complications. *Arch. Immunol. Ther. Exp.* **2019**, *67*, 295–309. [CrossRef] [PubMed]
38. Deshmukh, H.; Way, S.S. Immunological Basis for Recurrent Fetal Loss and Pregnancy Complications. *Annu. Rev. Pathol.* **2019**, *14*, 185–210. [CrossRef]

39. Liang, P.Y.; Diao, L.H.; Huang, C.Y.; Lian, R.C.; Chen, X.; Li, G.G.; Zhao, J.; Li, Y.Y.; He, X.B.; Zeng, Y. The pro-inflammatory and anti-inflammatory cytokine profile in peripheral blood of women with recurrent implantation failure. *Reprod. Biomed. Online* **2015**, *31*, 823–826. [CrossRef]
40. Robertson, S.A.; Care, A.S.; Moldenhauer, L.M. Regulatory T cells in embryo implantation and the immune response to pregnancy. *J. Clin. Investig.* **2018**, *128*, 4224–4235. [CrossRef]
41. Von Wolff, M.; Thaler, C.J.; Strowitzki, T.; Broome, J.; Stolz, W.; Tabibzadeh, S. Regulated expression of cytokines in human endometrium throughout the menstrual cycle: Dysregulation in habitual abortion. *Mol. Hum. Reprod.* **2000**, *6*, 627–634. [CrossRef] [PubMed]
42. Ahmadi, M.; Abdolmohammadi-Vahid, S.; Ghaebi, M.; Aghebati-Maleki, L.; Dolati, S.; Farzadi, L.; Ghasemzadeh, A.; Hamdi, K.; Younesi, V.; Nouri, M.; et al. Regulatory T cells improve pregnancy rate in RIF patients after additional IVIG treatment. *Syst. Biol. Reprod. Med.* **2017**, *63*, 350–359. [CrossRef] [PubMed]
43. Teles, A.; Schumacher, A.; Kühnle, M.C.; Linzke, N.; Thuere, C.; Reichardt, P.; Tadokoro, C.E.; Hämmerling, G.J.; Zenclussen, A.C. Control of uterine microenvironment by foxp3 (+) cells facilitates embryo implantation. *Front. Immunol.* **2013**, *4*, 158. [CrossRef] [PubMed]
44. Saito, S.; Nakashima, A.; Shima, T.; Ito, M. Th1/Th/Th17 and regulatory T-cell paradigm in pregnancy. *Am. J. Reprod. Immunol.* **2010**, *63*, 601–610. [CrossRef] [PubMed]
45. Tedesco, S.; Adorni, M.P.; Ronda, N.; Cappellari, R.; Mioni, R.; Barbot, M.; Pinelli, S.; Plebani, M.; Bolego, C.; Scaroni, C.; et al. Activation profiles of monocyte-macrophages and HDL function in healthy women in relation to menstrual cycle and in polycystic ovary syndrome patients. *Endocrine* **2019**, *66*, 360–369. [CrossRef] [PubMed]
46. Faas, M.M.; Spaans, F.; De Vos, P. Monocytes and macrophages in pregnancy and pre-eclampsia. *Front. Immunol.* **2014**, *5*, 298. [CrossRef]
47. Cervar, M.; Blaschitz, A.; Dohr, G.; Desoye, G. Paracrine regulation of distinct trophoblast functions in vitro by placental macrophages. *Cell Tissue Res.* **1999**, *295*, 297–305. [CrossRef]
48. Tonello, A.; Poli, G. Tubal ectopic pregnancy: Macrophages under the microscope. *Hum. Reprod.* **2007**, *22*, 2577–2584. [CrossRef]
49. Emi, N.; Kanzaki, H.; Yoshida, M.; Takakura, K.; Kariya, M.; Okamoto, N.; Imai, K.; Mori, T. Lymphocytes stimulate progesterone production by cultured human granulosa luteal cells. *Am. J. Obstet. Gynecol.* **1991**, *165 Pt 1*, 1469–1474. [CrossRef]
50. Schumacher, A.; Costa, S.D.; Zenclussen, A.C. Endocrine factors modulating immune responses in pregnancy. *Front. Immunol.* **2014**, *5*, 196. [CrossRef]
51. Schumacher, A.; Brachwitz, N.; Sohr, S.; Engeland, K.; Langwisch, S.; Dolaptchieva, M.; Alexander, T.; Taran, A.; Malfertheiner, S.F.; Costa, S.-D.; et al. Human chorionic gonadotropin attracts regulatory T cells into the fetal-maternal interface during early human pregnancy. *J. Immunol.* **2009**, *182*, 5488–5497. [CrossRef] [PubMed]
52. Kayisli, U.A.; Selam, B.; Guzeloglu-Kayisli, O.; Demir, R.; Arici, A. Human chorionic gonadotropin contributes to maternal immunotolerance and endometrial apoptosis by regulating Fas-Fas ligand system. *J. Immunol.* **2003**, *171*, 2305–2313. [CrossRef] [PubMed]
53. Zhou, J.; Wang, Z.; Zhao, X.; Wang, J.; Sun, H.; Hu, Y. An increase of Treg cells in the peripheral blood is associated with a better in vitro fertilization treatment outcome. *Am. J. Reprod. Immunol.* **2012**, *68*, 100–106. [CrossRef] [PubMed]
54. Mansour, R.; Tawab, N.; Kamal, O.; El-Faissal, Y.; Serour, A.; Aboulghar, M.; Serour, G. Intra uterine injection of human chorionic gonadotropin before embryo transfer significantly improves the implantation and pregnancy rates in in vitro fertilization/intra cytoplasmic sperm injection: A prospective randomized study. *Fertil. Steril.* **2011**, *96*, 1370–1374. [CrossRef] [PubMed]
55. Yakin, K.; Oktem, O.; Urman, B. Intrauterine administration of peripheral mononuclear cells in recurrent implantation failure: A systematic review and meta-analysis. *Sci. Rep.* **2019**, *9*, 3897. [CrossRef]
56. Li, J.; Chen, Y.; Liu, C.; Hu, Y.; Li, L. Intravenous immunoglobulin treatment for repeated IVF/ICSI failure and unexplained infertility: A systematic review and a meta-analysis. *Am. J. Reprod. Immunol.* **2013**, *70*, 434–447. [CrossRef]
57. Ho, Y.-K.; Chen, H.-H.; Huang, C.-C.; Lee, C.-I.; Lin, P.-Y.; Lee, M.-S.; Lee, T.-H. Peripheral CD56+CD16+ NK Cell Populations in the Early Follicular Phase Are Associated with Successful Clinical Outcomes of Intravenous Immunoglobulin Treatment in Women With Repeated Implantation Failure. *Front. Endocrinol.* **2020**, *10*, 937. [CrossRef]
58. Sung, N.; Han, A.R.; Park, C.W.; Park, D.W.; Park, J.C.; Kim, N.Y.; Lim, K.S.; Shin, J.E.; Joo, C.W.; Lee, S.E.; et al. Intravenous immunoglobulin G in women with reproductive failure: The Korean Society for Reproductive Immunology practice guidelines. *Clin. Exp. Reprod. Med.* **2017**, *44*, 1–7. [CrossRef]
59. Stephenson, M.D.; Fluker, M.R. Treatment of repeated unexplained in vitro fertilization failure with intravenous immunoglobulin: A randomized, placebo controlled Canadian trial. *Fertil. Steril.* **2000**, *74*, 1108–1113. [CrossRef]
60. Stephenson, M.D.; Kutteh, W.H.; Purkiss, S.; Librach, C.; Schultz, P.; Houlihan, E.; Liao, C. Intravenous immunoglobulin and idiopathic secondary recurrent miscarriage: A multicentered randomized placebo-controlled trial. *Hum. Reprod.* **2010**, *25*, 2203–2209. [CrossRef]
61. Balasch, J.; Creus, M.; Fabregues, F.; Font, J.; Martorell, J.; Vanrell, J.A. Intravenous immunoglobulin preceding in vitro fertilization-embryo transfer for patients with repeated failure of embryo transfer. *Fertil. Steril.* **1996**, *65*, 655–658. [CrossRef]
62. Sher, G.; Zouves, C.; Feinman, M.; Maassarani, G.; Matzner, W.; Chong, P.; Ching, W. A rational basis for the use of combined heparin/aspirin and IVIG immunotherapy in the treatment of recurrent IVF failure associated with antiphospholipid antibodies. *Am. J. Reprod. Immunol.* **1998**, *39*, 391–394. [CrossRef] [PubMed]

63. Christiansen, O.B.; Pedersen, B.; Rosgaard, A.; Husth, M. A randomized, double-blind, placebo controlled trial of intravenous immunoglobulin in the prevention of recurrent miscarriage: Evidence for a therapeutic effect in women with secondary recurrent miscarriage. *Hum. Reprod.* **2002**, *17*, 809–816. [CrossRef] [PubMed]
64. Coulam, C.B.; Acacio, B. Does immunotherapy for treatment of reproductive failure enhance live births? *Am. J. Reprod. Immunol.* **2012**, *67*, 296–304. [CrossRef]
65. Gleicher, N.; Kim, A.; Michaeli, T.; Lee, H.J.; Shohat-Tal, A.; Lazzaroni, E.; Barad, D.H. A pilot cohort study of granulocyte colony-stimulating factor in the treatment of unresponsive thin endometrium resistant to standard therapies. *Hum. Reprod.* **2013**, *28*, 172–177. [CrossRef]
66. Li, J.; Mo, S.; Chen, Y. The effect of G-CSF on infertile women undergoing IVF treatment: A meta-analysis. *Syst. Biol. Reprod. Med.* **2017**, *63*, 239–247. [CrossRef]
67. Gleicher, N.; Vidali, A.; Barad, D.H. Successful treatment of unresponsive thin endometrium. *Fertil. Steril.* **2011**, *95*, 2123.e13–2123.e17. [CrossRef]
68. Xu, B.; Zhang, Q.; Hao, J.; Xu, D.; Li, Y. Two protocols to treat thin endometrium with granulocyte colony-stimulating factor during frozen embryo transfer cycles. *Reprod. Biomed. Online* **2015**, *30*, 349–358. [CrossRef]
69. Kalem, Z.; Namli Kalem, M.; Bakirarar, B.; Kent, E.; Makrigiannakis, A.; Gurgan, T. Intrauterine G-CSF Administration in Recurrent Implantation Failure (RIF): An Rct. *Sci. Rep.* **2020**, *10*, 5139. [CrossRef]
70. Chung, T.W.; Chen, T.Z.; Yu, J.J.; Lin, S.Y.; Chen, S.C. Effects of a-tocopherol nicotinate on hemorheology and retinal capillary blood flow in female NIDDM with retinopathy. *Clin. Hemorheol.* **1995**, *15*, 775–782. [CrossRef]
71. Chung, T.W.; Yu, J.J.; Liu, D.Z. Reducing lipid peroxidation stress of erythrocyte membrane by a-tocopherol nicotinate plays an important role in improving blood rheological properties in type 2 diabetic patients with retinopathy. *Diabet. Med.* **1998**, *15*, 380–385. [CrossRef]
72. Sher, G.; Fisch, J.D. Vaginal sildenafil (Viagra): A preliminary report of a novel method to improve uterine artery blood flow and endometrial development in patients undergoing IVF. *Hum. Reprod.* **2000**, *15*, 806–809. [CrossRef] [PubMed]
73. Sher, G.; Fisch, J.D. Effect of vaginal sildenafil on the outcome of in vitro fertilization (IVF) after multiple IVF failures attributed to poor endometrial development. *Fertil. Steril.* **2002**, *78*, 1073–1076. [CrossRef]
74. Miwa, I.; Tamura, H.; Takasaki, A.; Yamagata, Y.; Shimamura, K.; Sugiro, N. Pathophysiologic features of "thin" endometrium. *Fertil. Steril.* **2009**, *91*, 998–1004. [CrossRef]
75. Kim, K.R.; Lee, H.S.; Ryu, H.E.; Park, C.Y.; Min, S.H.; Park, C.; Jee, B.C. Efficacy of luteal supplementation of vaginal sildenafil and oral estrogen on pregnancy rate following IVF-ET in women with a history of thin endometria: A pilot study. *J. Womens Med.* **2010**, *3*, 155–158. [CrossRef]
76. Check, J.H.; Graziano, V.; Lee, G.; Nazari, A.; Choe, J.K.; Dietterich, C. Neither sildenafil nor vaginal estradiol improves endometrial thickness in women with thin endometria after taking oral estradiol in graduating dosages. *Clin. Exp. Obstet. Gynecol.* **2004**, *31*, 99–102.
77. Moini, A.; Zafarani, F.; Jahangiri, N.; Sadatmahalleh, S.H.J.; Sadeghi, M.; Chehrazi, M.; Ahmadi, F. The Effect of Vaginal Sildenafil on The Outcome of Assisted Reproductive Technology Cycles in Patients with Repeated Implantation Failures: A Randomized Placebo-Controlled Trial. *Int. J. Fertil. Steril.* **2020**, *13*, 289–295.
78. Nakagawa, K.; Kwak-Kim, J.; Ota, K.; Kuroda, K.; Hisano, M.; Sugiyama, R.; Yamaguchi, K. Immunosuppression with tacrolimus improved reproductive outcome of women with repeated implantation failure and elevated peripheral blood TH1/TH2 cell ratios. *Am. J. Reprod. Immunol.* **2015**, *73*, 353–361. [CrossRef]
79. Uchida, K. Long-term Prograf multicenter retrospective study in kidney transplantation: Seven-year follow-up. *Transpl. Now* **2006**, *19*, 380–389.
80. Pierzynski, P.; Reinheimer, T.M.; Kuczynski, W. Oxytocin antagonists may improve infertility treatment. *Fertil. Steril.* **2007**, *88*, e19–e22. [CrossRef]
81. Makrigiannakis, A.; Makrygiannakis, F.; Vrekoussis, T. Approaches to Improve Endometrial Receptivity in Case of Repeated Implantation Failures. *Front. Cell Dev. Biol.* **2021**, *9*, 613277. [CrossRef] [PubMed]
82. Moraloglu, O.; Tonguc, E.; Var, T.; Zeyrek, T.; Batioglu, S. Treatment with oxytocin antagonists before embryo transfer may increase implantation rates after IVF. *Reprod. Biomed. Online* **2010**, *21*, 338–343. [CrossRef] [PubMed]
83. Ng, E.H.; Li, R.H.; Chen, L.; Lan, V.T.; Tuong, H.M.; Quan, S. A randomized double blind comparison of atosiban in patients undergoing IVF treatment. *Hum. Reprod.* **2014**, *29*, 2687–2694. [CrossRef]
84. He, Y.; Wu, H.; He, X.; Xing, Q.; Zhou, P.; Cao, Y.; Wei, Z. Administration of atosiban in patients with endometriosis undergoing frozen-thawed embryo transfer: A prospective, randomized study. *Fertil. Steril.* **2016**, *106*, 416–422. [CrossRef]
85. Yuan, C.; Song, H.; Fan, L.; Su, S.; Dong, B. The Effect of Atosiban on Patients with Difficult Embryo Transfers Undergoing In Vitro Fertilization-Embryo Transfer. *Reprod. Sci.* **2019**, *26*, 1613–1617. [CrossRef]
86. Chou, P.Y.; Wu, M.H.; Pan, H.A.; Hung, K.H.; Chang, F.M. Use of an oxytocin antagonist in in vitro fertilization-embryo transfer for women with repeated implantation failure: A retrospective study. *Taiwan J. Obstet. Gynecol.* **2011**, *50*, 136–140. [CrossRef] [PubMed]
87. Lan, V.T.; Khang, V.N.; Nhu, G.H.; Tuong, H.M. Atosiban improves implantation and pregnancy rates in patients with repeated implantation failure. *Reprod. Biomed. Online* **2012**, *25*, 254–260. [CrossRef] [PubMed]

88. Chang, Y.; Li, J.; Chen, Y.; Wei, L.; Yang, X.; Shi, Y.; Liang, X. Autologous platelet-rich plasma promotes endometrial growth and improves pregnancy outcome during in vitro fertilization. *Int. J. Clin. Exp. Med.* **2015**, *8*, 1286–1290.
89. Amable, P.R.; Carias, R.B.; Teixeira, M.V.; da Cruz Pacheco, I.; Correa do Amaral, R.J.; Granjeiro, J.M.; Borojevic, R. Platelet-rich plasma preparation for regenerative medicine: Optimization and quantification of cytokines and growth factors. *Stem Cell Res. Ther.* **2013**, *4*, 67. [CrossRef]
90. Lee, J.W.; Kwon, O.H.; Kim, T.K.; Cho, Y.K.; Choi, K.Y.; Chung, H.Y.; Cho, B.C.; Yang, J.D.; Shin, J.H. Platelet-rich plasma: Quantitative assessment of growth factor levels and comparative analysis of activated and inactivated groups. *Arch. Plast. Surg.* **2013**, *40*, 530–535. [CrossRef]
91. Kim, H.; Shin, J.E.; Koo, H.S.; Kwon, H.; Choi, D.H.; Kim, J.H. Effect of Autologous Platelet-Rich Plasma Treatment on Refractory Thin Endometrium During the Frozen Embryo Transfer Cycle: A Pilot Study. *Front. Endocrinol.* **2019**, *10*, 61. [CrossRef] [PubMed]
92. Allahveisi, A.; Seyedoshohadaei, F.; Rezaei, M.; Bazrafshan, N.; Rahimi, K. The effect of platelet-rich plasma on the achievement of pregnancy during frozen embryo transfer in women with a history of failed implantation. *Heliyon* **2020**, *6*, e03577. [CrossRef] [PubMed]
93. Ibañez-Perez, J.; Díaz-Nuñez, M.; Clos-García, M.; Lainz, L.; Iglesias, M.; Díez-Zapirain, M.; Rabanal, A.; Bárcena, L.; González, M.; Lozano, J.J.; et al. microRNA-based signatures obtained from endometrial fluid identify implantative endometrium. *Hum. Reprod.* **2022**, *37*, 2375–2391. [CrossRef] [PubMed]

International Journal of
*Molecular Sciences*

MDPI

*Review*

# IL-6 and IL-8: An Overview of Their Roles in Healthy and Pathological Pregnancies

**Aleksandra Vilotić †, Mirjana Nacka-Aleksić *,†, Andrea Pirković, Žanka Bojić-Trbojević, Dragana Dekanski and Milica Jovanović Krivokuća ***

Department for Biology of Reproduction, Institute for Application of Nuclear Energy (INEP), University of Belgrade, Banatska 31b, 11080 Belgrade, Serbia
* Correspondence: mnacka@inep.co.rs (M.N.-A.); milicaj@inep.co.rs (M.J.K.)
† These authors contributed equally to this work.

**Abstract:** Interleukin-6 (IL-6) is an acknowledged inflammatory cytokine with a pleiotropic action, mediating innate and adaptive immunity and multiple physiological processes, including protective and regenerative ones. IL-8 is a pro-inflammatory CXC chemokine with a primary function in attracting and activating neutrophils, but also implicated in a variety of other cellular processes. These two ILs are abundantly expressed at the feto-maternal interface over the course of a pregnancy and have been shown to participate in numerous pregnancy-related events. In this review, we summarize the literature data regarding their role in healthy and pathological pregnancies. The general information related to IL-6 and IL-8 functions is followed by an overview of their overall expression in cycling endometrium and at the feto-maternal interface. Further, we provide an overview of their involvement in pregnancy establishment and parturition. Finally, the implication of IL-6 and IL-8 in pregnancy-associated pathological conditions, such as pregnancy loss, preeclampsia, gestational diabetes mellitus and infection/inflammation is discussed.

**Keywords:** IL-6; IL-8; inflammation; pregnancy; pregnancy complications

**Citation:** Vilotić, A.; Nacka-Aleksić, M.; Pirković, A.; Bojić-Trbojević, Ž.; Dekanski, D.; Jovanović Krivokuća, M. IL-6 and IL-8: An Overview of Their Roles in Healthy and Pathological Pregnancies. *Int. J. Mol. Sci.* **2022**, *23*, 14574. https://doi.org/10.3390/ijms232314574

Academic Editor: Ilona Hromadnikova

Received: 21 October 2022
Accepted: 18 November 2022
Published: 23 November 2022



## 1. Introduction

The initial step in the establishment of a pregnancy is the implantation of an dembryo at the blastocyst stage into the receptive endometrium (decidua). This is followed by the formation of the placenta, a transient organ essential for fetal development and growth [1]. The development of the placenta depends on the differentiation of the cells from the outer layer of the blastocyst, i.e., trophectoderm, into specialized trophoblast cell subpopulations, with specific roles in the process of placentation and in placental function. Cytotrophoblast cells (CTBs) are progenitor cells which constantly proliferate and differentiate into other trophoblast populations. By fusion, CTBs give rise of multinucleated syncytiotrophoblast (STB) and through epithelial-mesenchymal transition, CTBs differentiate to invasive extravillous trophoblast cells (EVTs). STB covers chorionic villi and is in direct contact with the maternal circulation, facilitating the exchange of nutrients, gases and excretory materials between the mother and fetus. EVTs invade the decidual stroma and part of the myometrium anchoring the placenta to the mother's uterus while a portion of the EVTs, representing endovascular EVT cells (eEVTs), reaches the decidual spiral arteries, and transforms their walls by degrading and replacing the endothelial and smooth muscle cell layers. Inadequate implantation and impaired trophoblast differentiation and function lead to placental dysfunction and the development of different pregnancy complications which affect the maternal health and fetal development, with a possible lifelong impact on the offspring health (all in [1]).

The highly regulated and coordinated feto-maternal communication is essential for a successful pregnancy. Cytokines, small multifunctional molecules abundantly expressed at the feto-maternal interface, are one of the key players involved in this finely orchestrated interaction [2]. Data presented in the literature, together with our previous research, show that interleukins (IL)-6 and IL-8, among other cytokines, are one of the decisive drivers of the physiological pregnancy-related processes and pathological conditions [2].

IL-6 is a pleiotropic cytokine, a member of the IL-6 family of cytokines, implicated in a wide range of physiological processes, such as organ development, acute-phase response, inflammation, immune responses, metabolic regulation and others [3]. IL-6 exerts its effects upon binding to its receptors and subsequently activating the Janus kinase-signal transducer and activator of the transcription (JAK/STAT) pathway [4]. The activated STAT3 downstream induces the expression of the suppressor of the cytokine signaling 3 (SOCS3), a potent negative regulator of the JAK/STAT signaling that prevents excessive inflammation [5]. The classical signal transduction of IL-6 is induced by the binding of IL-6 to its specific membrane IL-6α-receptor (IL-6R), which is followed by the formation of a signaling complex with the signal-transducing receptor β-subunit, transmembrane glycoprotein 130 (gp130). This pathway is thought to be limited to a few tissues only, due to the restricted expression of IL-6R [6]. The signaling receptor β-subunit is shared by several other cytokines that comprise the IL-6 cytokine family, enabling redundant effects upon engaging with different ligands [6]. By contrast, IL-6 trans-signaling denotes the IL-6-binding to a soluble form of IL-6Rα (sIL-6R), which enables increased IL-6 bioavailability and broadening of its target cell repertoire, due to the ubiquitous expression of gp130 [6]. IL-6 and other members of this cytokine family may exert both pro- and anti-inflammatory functions. It is generally accepted that the pro-inflammatory effects of IL-6 are exerted by IL-6 trans-signaling, via the sIL-6R [6]. Moreover, the anti-inflammatory effects of IL-6 are mainly conducted via the classic membrane-bound IL-6R-mediated signaling [6]. The IL-6 signaling pathways and functions are described in more detail elsewhere [6,7].

IL-8, also known as C-X-C motif chemokine ligand 8 (CXCL8), is a pro-inflammatory chemokine, a member of the CXC family of chemokines, produced under inflammatory conditions by immune and other cell types [8]. The most prominent role of IL-8 is the attraction of neutrophils to the sites of inflammation, but also the promotion of monocyte-macrophage growth and differentiation [9], endothelial cell survival, proliferation and angiogenesis [10]. IL-8 also enhances the oxidative metabolism and generation of reactive oxygen species, possibly leading to oxidative stress [11]. The biological effects of IL-8 are induced upon engaging with its transmembrane, G protein-coupled receptors CXCR1 and CXCR2 [12], and the activation of the inflammatory Akt/protein kinase B (PKB), mitogen-activated protein kinase (MAPK) and protein kinase C (PKC) pathways [13].

This review discusses the role of IL-6 and IL-8 in processes important for the establishment of a healthy pregnancy and for parturition, as well as the involvement of these cytokines in the development of selected common pregnancy-related pathologies (Figure 1).

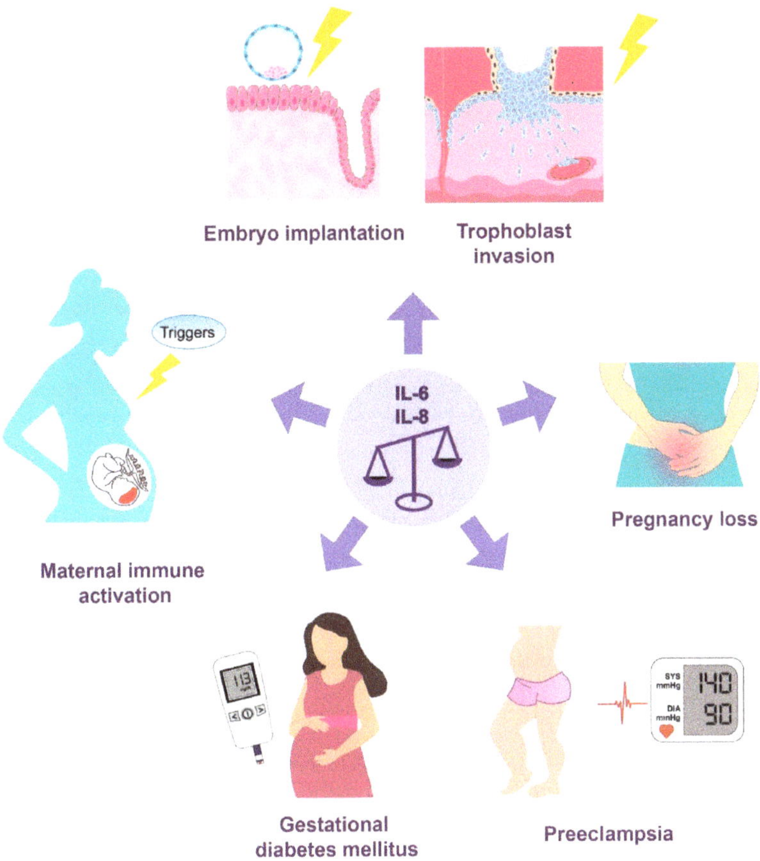


**Figure 1.** Implication of the IL-6 and IL-8 disbalance in the embryo implantation and trophoblast invasion, and the pathological pregnancy conditions – pregnancy loss, preeclampsia, gestational diabetes mellitus and maternal immune activation, based on the literature data.

## 2. IL-6 and IL-8 in a Healthy Pregnancy

### *2.1. Expression of IL-6 and IL-8 and Their Receptors at the Feto-Maternal Interface*

IL-6, IL-8 and their respective receptors are expressed in the human endometrium throughout the menstrual cycle [14–16]. IL-6 and IL-8 are predominantly localized to the endometrial epithelial and glandular cells [14,15]. IL-6R and gp130 are expressed in the endometrial glands [14], while CXCR1 and CXCR2 are immunolocalized into the surface of the endometrial epithelium, endometrial glandular cells, and, to a lesser extent, on stromal cells [16]. Both IL-6 and IL-8 show a menstrual cycle-dependent expression pattern, suggesting their role in endometrial physiology. IL-6 is weakly expressed during the proliferative phase, but its expression progressively increases after ovulation and peaks during the mid- to late-secretory phase [14]. This expression pattern temporally corresponds to the window of implantation, suggesting a role for IL-6 in the endometrial receptivity. Moreover, across the menstrual cycle, the IL-8 mRNA expression peaks at the early- to mid-proliferative phase and once again at the late secretory phase [15]. The first peak-expression suggests a role for IL-8 in the neovascularization of the growing endometrium, whereas the second peak could be related with the neutrophil recruitment into the endometrium, right before the onset of menstruation [15].

In the first trimester of pregnancy, IL-6 expression is detected in different cell populations from the uteroplacental tissues. In the decidua, decidual stromal cells (DSCs) and different populations of immune cells were immunostained for IL-6 [17–19]. In the placenta, IL-6 expression was found in CTBs, STB and EVTs [17,19,20]. In line with immunohistochemical analyses, IL-6 expression at the mRNA and protein level was also shown in isolated first-trimester CTBs and EVTs [20,21], as in primary DSCs, decidual natural killer cells (dNKs), $CD8^+$ T cells and macrophages [18,20]. Moreover, IL-6 production by DSCs, CTBs and EVTs, is shown to increase with gestational age [20]. In first-trimester decidual sections and isolated cells, IL-8 is shown to be expressed in DSCs and the glandular epithelium, as well as in dNKs, decidual $CD8^+$ T cells and macrophages [18,22–24]. In the first-trimester placenta, this protein is detected in different trophoblast subpopulations, as STB, CTBs and EVTs [22,24–26]. Study on explants from the first and the second trimesters, and the at term placenta, show that IL-8 secretion increases during gestation, with the maximal production in the second trimester and at term [25]. A widespread expression of IL-6R, gp130, CXCR1 and CXCR2 is found in first-trimester placental bed sections [18]. Among the trophoblast subpopulations, all of the named receptors were immunolocalized to EVTs [20,24].

### 2.2. Pregnancy Establishment

Considerable data now indicate that uterine epithelium-derived factors control embryo-implantation processes. In turn, the developing embryos have been shown to produce a variety of cytokines that may act in an autocrine fashion or on the endometrium to influence its receptivity [27]. The IL-6 immunoreactivity in the human endometrium is strong during the putative window of implantation, suggesting its role in this process [14]. A microarray analysis of the mid-secretory phase endometrium of patients suffering from recurrent implantation failure show a lower expression of IL-6, compared to healthy controls [28], further supporting the role of IL-6 in endometrial receptivity. Blastocysts also express and secrete IL-6 [29–32], and IL-6 in vitro treatment increases the embryos' blastulation and hatching rates, compared to untreated embryos [33]. Furthermore, a higher IL-6 level in the follicular fluid also correlates with implantation success in patients undergoing in vitro fertilization (IVF) [34]. However, studies on animal models regarding the role of IL-6 in pregnancy establishment and maintenance are inconsistent. Reduced fertility, a decreased number of viable implantation sites and the increased rate of miscarriage in mid-gestation [35], but also no changes in fertility, implantation or early embryonic development [36], have both been observed in IL-6 knockout mice, compared to control mice. These discrepancies are suggested to reflect the differences in the lines of the IL-6 knockouts used in the studies, housing conditions, and other factors that influence the immune system development [36].

The results from our study and those by other groups also highlight IL-6 as one of the major regulators of multiple cellular processes at the feto-maternal interface. Namely, we have shown that IL-6 stimulates trophoblast invasion and migration of both primary first trimester CTBs and the EVT cell line HTR-8/SVneo [21]. This was partly mediated by the upregulated trophoblast expression of integrin $\alpha5$, $\alpha1$ and $\beta1$ [21], and the activation of MMP-2 and MMP-9 [37,38], molecules, particularly relevant for trophoblast invasion [39–42]. Our findings were also confirmed on other trophoblast cell models, such as ACH-3P and JEG3 cell lines [43,44]. However, Champion and collaborators did not observe a stimulatory effect of HTR-8/SVneocell invasion by IL-6 [20]. This discrepancy could be due to the different experimental conditions, compared to the ones found effective in our study [21].

Further, a role for IL-6 in the remodeling of spiral arteries has also been indicated. IL-6, along with IL-8, has been proposed as a key EVT-derived factor that activates endothelial cells to release chemoattractants for the dNKs and macrophages from the surrounding tissue into the upper segments of the spiral arteries, to initiate the remodeling process [45]. Along with mediating the trophoblast invasion and spiral artery remodeling, IL-6 has also been shown to mediate the immune-endocrine crosstalk in pregnancy. Namely, IL-6 was found to regulate the synthesis of the β-subunit of the human chorionic gonadotropin and human placental lactogen, two major placental hormones essential for pregnancy [46,47].

IL-8 is also proposed to contribute to endometrial receptivity and to participate in the dialogue between the embryo and the human endometrium during implantation. Namely, the endometrial IL-8 mRNA expression starts to increase at the receptive phase of the menstrual cycle [15,48]. Furthermore, in vitro findings demonstrate an upregulation of the IL-8 and CXCR1 expression in endometrial epithelial cells in the presence of an embryo [48,49]. Additionally, in vitro decidualization of endometrial stromal fibroblasts is shown to increase their IL-8 secretion in respect to the non-differentiated cells [50]. In patients undergoing IVF, IL-8 is suggested to be a predictor of the embryo developmental potential in the pre-transfer assessment of embryos. The pregnancy and the implantation rates, as well as the number of live births per IVF or intracytoplasmic sperm injection, were higher when the pre-transfer embryo-conditioned medium contained IL-8 [51].

Moreover, IL-8 is shown to stimulate progesterone secretion from the BeWo trophoblast cell line, suggesting a role for IL-8 in the maintenance of pregnancy [52]. dNKs are shown to regulate human trophoblast invasion in vitro as well as in in vivo mouse models by the production of IL-8 [23]. Recombinant IL-8 is shown to stimulate both first-trimester CTBs and HTR-8/SVneo-cell invasion [24,26], while silencing of IL-8 mRNA had the opposite effect [53]. The levels of the secreted MMP-2 and MMP-9 were significantly elevated by IL-8 treatment, as was the trophoblast expression of integrin α5 and β1, compared to the non-treated cells [24,26]. Furthermore, the viability and proliferation of HTR-8/SVneo cells increased following IL-8 treatment [26,54], as well as HTR-8/SVneo- and JEG-3-cell migration [26,55]. IL-8 has also been designated as one of the key EVT-secreted factors involved in spiral artery remodeling [45]. This chemokine is also shown to stimulate the endothelial-like phenotype in HTR-8/SVneo cells, reflected in an increased tube formation on Matrigel [54]. The presented literature data about the IL-6 and IL-8 roles in pregnancy establishment are summarized in Table 1.

In conclusion, the spatial and temporal distribution of IL-6, IL-8 and their corresponding receptors, clearly indicate a role for IL-6 and IL-8 signaling in the early events of pregnancy establishment—from the initial crosstalk between the embryo and the endometrium, to the subsequent formation of the functional placenta. Their disturbed expression in a range of clinical situations discussed further herein, suggests that manipulation with IL-6, IL-8, their receptors and/or components of their effector pathways, may be a plausible strategy for achieving a successful pregnancy in selected patients.

**Table 1.** Involvement of IL-6 and IL-8 in a healthy pregnancy.

| Process | Role | Cytokine | Reference |
|---|---|---|---|
| **Pregnancy establishment** | Endometrial receptivity and implantation | **IL-6**<br>**IL-8** | [14,28]<br>[48,49] |
| | Stimulation of human embryo blastulation and blastocyst hatching in vitro | **IL-6** | [33] |
| | Potential indicator of human embryo developmental potential in ARTs * | **IL-8** | [51] |
| | Regulation of trophoblast invasion and migration in vitro | **IL-6**<br>**IL-8** | [21,43,44]<br>[23,24,26,53,55] |
| | Spiral artery remodeling | **IL-6**<br>**IL-8** | [45]<br>[45,54] |

**Table 1.** *Cont.*

| Process | Role | Cytokine | Reference |
| --- | --- | --- | --- |
| Parturition | Key gene triggering specific mechanisms in different gestational tissues leading to labor onset | *IL6* | [56] |
| | Most upregulated gene in the myometrium and cervix with the onset of labor | *CXCL8* | [57] |
| | Stimulation of prostaglandin synthesis by decidual cells and chorioamniotic membranes | IL-6 | [58,59] |
| | Stimulation of oxytocin secretion and oxytocin receptors' expression on MSMCs * | IL-6 | [60,61] |
| | Cervical remodeling and rupture of the gestational membranes | IL-8 | [62,63] |
| | **Circulating concentrations** | | |
| Pregnancy | Increased over the course of pregnancy | IL-6 | [64–67] |
| | No difference between trimesters | IL-6 | [68–72] |
| | Decreased over the course of pregnancy | IL-6 | [73,74] |
| | | IL-8 | [75] |
| | U shape (decreased between the 1st and 2nd trimesters and increased between the 2nd and 3rd) | IL-8 | [68–70,76] |

* ARTs—artificial reproductive technologies; MSMCs—myometrial smooth muscle cells.

### 2.3. Parturition

Human labor is a physiological process that involves a coordinated activation and transformation of several reproductive tissues – myometrium, cervix and choriamniotic membranes. The timely development of the inflammatory/immune response in all main reproductive tissues involved in parturition, is shown to be indispensable for initiating the uterine contractions, cervical ripening and dilatation, as for the activation and rupture of the gestational membranes [77]. Recent concepts suggest that human parturition occurs when the upregulated pro-inflammatory mediators are amplified above a threshold level that stimulates the uterine transition to its activated state for labor [78,79]. This paracrine inflammatory amplification is based on multiple positive feedback loops between ligands, cells, and tissues [77]. Due to cellular stress, necrosis and senescence, more and more damage-associated molecular patterns (DAMPs) are released from the cells of the increasingly stressed uterus, maturing fetus and ageing placenta, as parturition nears [80]. Engaging with toll-like receptors (TLRs), DAMPs cause immune cell activation and inflammasome assembly, a release of pro-inflammatory cytokines and chemokines, and leukocyte chemoattraction and activation [81]. The activated leucocytes, along with the resident cells, jointly release inflammatory mediators, such as cytokines, chemoattractants, prostaglandins and other molecules. This enables the transition from a quiescent to an active uterine phenotype [78,79,82]. Consistently, a number of studies report increased mRNA expression and protein levels of IL-6 and IL-8 in the myometrium, cervix, decidua, amnion and choriodecidua, during labor [57,83–87]. This is supported by transcriptome studies showing upregulation of the relevant signaling pathways in the laboring gestational tissues [57,88–92].

According to the gene-gene association analysis, *IL6* might be the key gene to trigger specific mechanisms in the gestational tissues that eventually lead to the onset of labor [56]. This study showed that *IL6* was linked to genes which stimulate proliferation, maturation, chemoattraction and activation of neutrophils, including *CXCL8* [56]. Furthermore, IL-6 participates in the initiation and progression of parturition by stimulating the production of prostaglandins and oxytocin, which facilitate cervical ripening and induce uterine contractions. IL-6 is shown to stimulate prostaglandin synthesis by decidual cells and chorioamniotic membranes, the major sources of intrauterine prostaglandins [58,59]. Additionally, IL-6 treatment increases oxytocin secretion by myometrial smooth muscle cells (MSMCs) [60] and the expression of oxytocin receptors on them [61], thus establishing a positive feedback loop for amplifying the oxytocin-induced effects. Vice versa, oxytocin activates the master inflammatory transcription factor NFκB, followed by an upregulation of gene expression for IL-6, IL-8 and other inflammatory molecules in MSMCs [93], further amplifying the inflammatory process. In addition, the mechanical stretching of the uterus significantly elevates the expression of pro-inflammatory cytokines, such as IL-6 and IL-8 [94]. Nonetheless, according to rodent models, IL-6 alone is incapable of stimulating uterine contractions [95], or to induce preterm labor (PTL) [96].

An analysis of the inflammatory signaling pathways has shown that *CXCL8* is the most upregulated gene in both the myometrium and cervix with labor onset [57]. Consistently, the cells of the cervix and the lower segments of the uterus produce great amounts of IL-8 with the progression of the cervical dilatation and parturition [86,97,98]. In cervical stroma, expression of CXCR1 and CXCR2 was seen only after vaginal delivery [99], possibly providing a feedforward mechanism for amplification of IL-8 signaling during human vaginal parturition. Upregulation of IL-8 expression in the laboring intrauterine tissues is associated with an increased leukocyte infiltration [85,100]. In addition to neutrophil chemoattraction, IL-8 also stimulates neutrophil degranulation and release of MMPs and elastase, leading to extracellular matrix degradation [101,102]. This contributes to cervical remodeling and rupture of the gestational membranes—processes that promote spontaneous labor [62,63]. In line with that, a correlation between IL-8 expression and the neutrophil-derived MMPs in laboring tissues was detected [97,103].

Considering all the aforementioned (summarized in Table 1), it may be hypothesized that human parturition is driven by multiple local interactions between the pro-inflammatory and pro-contractile mechanisms. This perpetuates the inflammation towards a point-of-no-return value which sets in motion parturition cascades [77]. Although IL-6 and IL-8 are repeatedly implicated in various steps of labor, the exact role of each of them is still unclear. Further research on the specific inflammatory interactions associated with the onset and progression of parturition could lead to a better risk assessment and treatment of PTL. For instance, tracking IL-8 serum levels in pregnant women is suggested as a main marker to determine the time of parturition [104]. In addition, the maternal IL-6 and IL-8 serum levels are also considered as appropriate markers for monitoring the effects of tocolytics in PTL [105]. Being central to the initiation and propagation of the inflammatory signaling cascade in parturition, IL-6 and IL-8 may also serve as potent therapeutic targets for PTL and other adverse birth outcomes. As suggested, targeting the inflammatory cascade at an earlier stage, could be a promising tocolytic strategy [82,106].

### 2.4. Circulating IL-6 and IL-8 Levels in a Healthy Pregnancy

Measurements of the circulating cytokine levels show that IL-6 levels are generally found to be elevated in pregnant vs. non-pregnant women [67], especially in the second and third trimester [66,67]. Longitudinal assessments of IL-6 concentrations in the maternal circulation over the course of pregnancy, adjusted for maternal body mass index (BMI) and other confounders, yield rather conflicting findings (Table 1). While several studies report a progressive increase of IL-6 with advancing gestational age [64–67,69], no significant differences between trimesters [68,70–72], or a decrease in the circulating IL-6 levels during gestation [73,74], have also been observed.

The maternal serum levels of IL-8 have been shown to decrease with gestational age during the first half of a non-complicated pregnancy [69,76], but to increase between the second and third trimester [68,70] (Table 1). This pattern of circulatory IL-8 levels might suggest a Th1/Th2 cytokine shift towards a pro-inflammatory profile, as the term for parturition approaches, following the predominantly immunotolerant state that protects the feto-placental unit from the maternal immune system [107]. Still, a progressive decline in serum IL-8 over the course of pregnancy has also been observed [75].

Although the influence of the fetoplacental sex on the maternal immune milieu is growingly recognized [108–110], only a few existing studies investigated gestational cytokine levels in respect to the fetoplacental sex. Although data indicate a generally more pro-inflammatory milieu in women carrying male vs. female fetuses [108,110], no significant sex-specific differences regarding the IL-6/IL-8 levels under steady state conditions are reported [108–111]. However, upon lipopolysaccharide (LPS) stimulation, a more robust inflammatory response, reflected in greater production of cytokines, including IL-6, was shown in cultures of peripheral blood mononuclear cells (PBMCs) from mothers carrying female vs. male fetuses, at all tested time-points across pregnancies [109]. In summary, it may be concluded that the circulating IL-6 and IL-8 levels fluctuate over the course of pregnancy. This may reflect the continuous immune modulation across gestation, consistent with the aim of pregnancy maintenance. Identifying typical patterns in immune parameter trajectories over the course of pregnancy should enable recognizing relevant deviations and predict adverse perinatal outcomes [69]. However, comparisons of absolute cytokine levels between studies are often inconclusive in practice. The abundance of inconsistent data probably reflects the methodological inconsistencies between the studies and inter-assay differences (i.e., the detection of varying amounts of free and/or bound cytokines, using plasma vs. serum, differences in population size—statistical power, population characteristics), and other factors which cannot be corrected [67,70]. Conflicting results may also be related to a number of factors that influence cytokine levels, such as age, ethnicity, genetic polymorphisms and epigenetic marks, fetoplacental sex, pre-gestational BMI, HbA1c, diet, smoking, intestinal microbiota, and other [112–115]. Therefore, more standardized fundamental mechanistic research and longitudinal study designs accounting for confounding variables may help to clarify the role of IL-6/IL-8 in a healthy pregnancy and related disorders. Nevertheless, a panel of cytokines/chemokines and other related parameters, adjusted for the multicollinearity among them, is more likely to describe the inflammatory milieu in pregnancy and predict the outcomes than the individual measures.

## 3. IL-6 and IL-8 in Selected Pregnancy Pathologies

### *3.1. Pregnancy Loss*

Pregnancy loss (PL) is the most common pregnancy complication. The pooled risk of PL is 15.3% of all clinically recognized pregnancies, usually before the 12th week of gestation (wg) [116]. Most PLs stay undetected since they happen soon after implantation. With these cases included, the incidence of PLs rises to 30% [117,118]. Spontaneous PLs are usually sporadic (SPL), but 1% to 5% of women experience recurrent PLs (RPL). RPL is defined as more than two or three consecutive PLs, depending on the definition [119]. Genetic abnormalities of the conceptus represent a major cause of early PL [119]. Other common risk factors and causes of PL, especially recurrent, include maternal anatomical malformations, infections, endocrine, thrombophilic and immune disorders [119]. Nevertheless, the etiology of more than 50% of PLs remains unexplained [120].

An inadequate expression/secretion of IL-6 and IL-8 at the feto-maternal interface has been indicated in unexplained early PLs in several studies (Table 2). In isolated SPL, decidual macrophages and dNKs are found to produce less IL-6 and IL-8, compared to the corresponding cells from a normal pregnancy [18]. Considering that: (i) IL-6 and IL-8 are involved in the regulation of trophoblast invasion and spiral artery remodeling [18,21,26], and (ii) dNKs stimulate EVT invasion, at least partly, through IL-8 signaling [24], suboptimal decidual IL-6 and IL-8 levels could lead to an inadequate trophoblast invasion and spiral artery remodeling, and eventually to an early PL. However, recent studies report increased IL-6 and IL-8 expression in decidual tissue [121–123], and increased IL-8 expression in decidual macrophages and dNKs [124,125], in RPL, compared to normal pregnancy. Furthermore, increased IL-8 levels in products of conception containing tissue of maternal and fetal origin, were reported in RPL patients, compared to healthy controls [126]. Increased IL-6 and IL-8 levels in decidual tissue indicate an enhanced pro-inflammatory state at the feto-maternal interface that could be detrimental to the implanted embryo and compromise the pregnancy. The previous findings indicate that both insufficient and excessive levels of IL-6/IL-8 disturb the inflammatory network at the feto-maternal interface, which may compromise the pregnancy. Furthermore, the differences in the expression profile of IL-6 and IL-8 in reproductive tissues between SPL and RPL, support the hypothesis that these complications may have a substantially different etiopathogenetic background [127].

Patients experiencing PL may exhibit altered systemic cytokine levels, compared to women having uneventful pregnancies, although the results are varying significantly. An increased IL-6 concentration in plasma [128] and serum [123,129,130], as well as greater expression in PBMCs [123,128], is found in SPL and RPL patients vs. controls. Unaltered [131,132] or decreased IL-6 levels in serum [133–136] of SPL and RPL patients are also shown. Similarly, unaltered [131], increased [133,136] and decreased [134] levels of IL-8 in PL patients are all reported. These inconsistent findings could be explained, at least in part, by the methodological differences between the studies (inclusion criteria for the participants, gestational age, inclusion of both SPL and RPL patients in the study, assigned control groups, etc.).

Women experiencing RPL exhibit a heightened immune activity, both locally and systemically, regardless of the gestational status. Thus, in non-pregnant RPL patients, higher IL-6 plasma levels, compared with women without reproductive problems [137], along with increased circulating levels of sIL-6R [138] are found. This, in addition to the lower levels of soluble gp130 (sgp 130) — a selective antagonist of the IL-6/sIL-R trans-signaling pathway [139], points to increased pro-inflammatory IL-6 trans-signaling in RPL patients [138]. Moreover, the PBMCs of RPL patients isolated at the mid-luteal phase of the cycle, which corresponds to the window of implantation, were shown to more readily respond to stimulation, expressing a greater amount of IL-6 mRNA, compared to healthy controls [140]. Increased IL-8 mRNA and protein levels in peripheral blood samples of non-pregnant RPL patients vs. controls are also reported [141,142].

Locally, the peri-implantation endometrial tissue of RPL-prone patients is shown to express lower IL-6 and IL-8 levels, compared to controls [143–146]. The impaired expression of IL-6, IL-8 and other cytokines in the mid-secretory endometrium could affect endometrial receptivity and thus compromise the establishment of a pregnancy. Or, it may impair the decidual selection of high-quality embryos, which could lead to a subsequent abortion.

Taken together, the reported data suggest that IL-6 and IL-8 may be closely associated with PL, possibly through different underlying mechanisms, consistent with the multifactorial nature of PL (Table 2). The variety of findings regarding the local or systemic IL-6 and IL-8 expression between PL-prone women and controls, suggests that neither IL-6 nor IL-8 could be a sole underlying factor of PL. Hence, neither IL-6 nor IL-8 levels alone represent a reliable diagnostic test for PL. Instead, a panel of several cytokines and/or other parameters in combination may provide a greater predictive value than any single factor utilized alone. Furthermore, a better understanding of the specific underlying mechanisms of PL is also needed, to explore the therapeutic potential of the IL-6 and IL-8 pathway regulation in PL-prone patients. The previous implies a more personalized approach towards the diagnostics and the therapeutic treatments—ideally, adjusted to the individual patient.

**Table 2.** Pathology-related changes of the IL-6/IL-8 in pregnancy pathologies.

| Pathology | Sample | Pathology-Related change | Cytokine | Reference |
|---|---|---|---|---|
| Pregnancy loss | Decidual macrophages and dNKs * | Decreased expression in SPL * | IL-6, IL-8 | [18] |
| | | Increased expression in RPL * | IL-8 | [124,125] |
| | Decidua | Increased expression in RPL * | IL-6, IL-8 | [121–123] |
| | Serum | Increased concentration | IL-6 | [123,129,130] |
| | | | IL-8 | [133,136] |
| | | No change | IL-6 | [131,132] |
| | | | IL-8 | [131] |
| | | Decreased concentration | IL-6 | [133–136] |
| | | | IL-8 | [134] |
| Preeclampsia | Placenta | Increased expression | IL-6 | [147–151] |
| | | | IL-8 | [149,152,153] |
| | Serum | Increased concentration | IL-6 | [150,151,154–158] |
| | | | IL-8 | [151,153,156,159–162] |
| Gestational diabetes mellitus | Placenta | Increased expression | IL-6 | [115,163–166] |
| | | | IL-8 | [115,167] |
| | | No change | IL-6 | [168,169] |
| | | | IL-8 | [169,170] |
| | | Increased levels in extravillous and decreased in villi tissue | IL-6 | [167] |
| | | Sex-specific expression in STB * and EVTs * | IL-8 | [171] |
| | Serum | Increased concentration | IL-6 | [113,115,172–179] |
| | | | IL-8 | [115,176,179,180] |
| | | No change | IL-6 | [180–183] |
| | | | IL-8 | [167,184,185] |
| | | Decreased concentration in early pregnancy | IL-8 | [184] |
| Maternal immune activation | Placenta/trophoblast and fetal membranes | Increased expression in response to inflammatory stimuli | IL-6 | [186–191] |
| | | | IL-8 | [189–191] |
| | Amniotic fluid and cervicovaginal lavage | Increased concentration in API * | IL-6 | [192–198] |
| | | | IL-8 | [192–197,199,200] |

* API—acute placental inflammation; dNKs—decidual natural killer cells; EVTs—extravillous trophoblast cells; RPL—recurrent pregnancy loss; SPL—sporadic pregnancy loss; STB—syncytiotrophoblast.

### 3.2. *Preeclampsia*

According to the 2018 recommendations from The International Society for the Study of Hypertension in Pregnancy (ISSHP), preeclampsia (PE) is defined as de novo hypertension after the 20th wg, accompanied by one or more of the following features: proteinuria, maternal organ dysfunction (including hepatic, renal, neurological), or hematological involvement, such as thrombocytopenia, and/or uteroplacental dysfunction, such as fetal growth restriction (FGR) and/or abnormal Doppler ultrasound findings of the uteroplacental blood flow [201]. PE affects approx. 2% to 8% of pregnancies worldwide [202]. Maternal and perinatal outcomes in PE are predicted based on the gestational age at the onset. Accordingly, PE is classified as early-onset PE (EOPE), occurring before the 34th wg, and late-onset PE (LOPE), which manifests at or after the 34th wg [201]. LOPE comprises around 80% to 95% of all PE cases, while EOPE, although less common, is associated with a higher maternal morbidity and FGR or neonatal mortality rates [203,204]. Although the exact mechanisms are not fully understood, there are indications that EOPE is related to abnormal placental development and consequent placental inflammation/dysfunction, whereas LOPE seems to be related to inherent maternal cardiovascular dysfunction and systemic inflammation [204,205].

Common features of the both PE phenotypes are placental ischemia and increased oxidative stress, with excessive systemic inflammation and endothelial dysfunction, which may be deleterious to the fetal and maternal health [206]. A reduced trophoblast invasion into the decidua and defective spiral artery remodeling are thought to be the earliest pathophysiological events in PE [207]. The shallow trophoblast invasion and development of placental hypoxia, induced tissue injury and increased release of inflammatory mediators from the placental cells [207]. The increased levels of pro-inflammatory cytokines, reactive oxygen and nitrogen species, lytic enzymes and other aggressive molecules damage the endothelial cells, causing their dysfunction and increased endothelial production of vasoconstrictors over vasodilators, leading to maternal hypertension and uteroplacental dysfunction [208]. An analysis of the dynamic connections within the pro-inflammatory cytokine network in PE cases identified a positive correlation between IL-6 and IL-8, suggesting these cytokines are implicated in the pathophysiology of PE [156] (Table 2). Consistent with the previous notion, IL-6 is recognized as a circulating marker of endothelial dysfunction and increased levels have been observed in the sera of women suffering from PE [150,151,154–158]. Interestingly, besides IL-6, increased level of sgp130 in the maternal circulation [209] and lower release of sIL-6R from the maternal neutrophils [210] are also shown in PE patients, compared to healthy pregnant women. As previously suggested, these findings possibly indicate a compensatory mechanism to control IL-6 signaling and prevent an overactivation of the IL-6/sIL-6R pathway [211,212].

An increased IL-6 expression is found in decidual cells and placentas of PE patients, associated with elevated levels of plasma IL-6 in PE [147–151]. The local excess of IL-6 could increase the trophoblast shedding, as shown in vitro, possibly contributing to the development of PE [213]. Furthermore, IL-6 aids to the recruitment and activation of the decidual macrophages that could lead to disturbed EVT invasion and spiral artery transformation [147,214,215]. A shift in the macrophage differentiation from the anti-inflammatory M2 to the pro-inflammatory M1 phenotype is observed in PE deciduas [151,214,216], consistent with a greater production of pro-inflammatory cytokines and decreased levels of anti-inflammatory cytokines in PE placentas [150,151,214,216]. In addition, an excess of IL-6 favors the differentiation of naïve $CD4^+$ T cells towards Th17 and cytotoxic T cell phenotype, whereas it inhibits the differentiation of Th2 and regulatory T (Treg) cells [217,218]. This contributes to the immune maladaptation and sustained systemic inflammation observed in PE [219]. Therefore, applying IL-6-reduction strategies as anti-IL-6 monoclonal Abs (mAbs) or TLR inhibitors to the treatment of PE, could shift the differentiation of naïve $CD4^+$ T cells towards the anti-inflammatory Treg and Th2 phenotype, rather than the pro-inflammatory Th1 and Th17 one [219]. Of note, although the current data do not indicate a substantially increased malformation risk of using anti-IL-6 mAbs in pregnancy, they are insufficient to prove safety [220,221].

IL-8 is considered to contribute to the PE pathogenesis by attracting more neutrophils into the endothelium [222]. Neutrophils infiltrate the vessel tunica intima and release reactive oxygen species, myeloperoxidase, MMP-8 and thromboxane, causing cell injury/death, endothelial inflammation and vasoconstriction [156]. Neutrophil extracellular traps (NET) have been found in placental intervillous spaces in PE patients [223]. The presence of NETs in the maternal circulation during pregnancy can contribute to thrombotic events, inflammation, and ultimately, to fetal death [224]. Consistently, higher IL-8 serum levels [151,153,156,159–162], and placental tissue expression of IL-8 [149,152,153] are reported in PE patients, compared to healthy controls.

According to some studies, sex-specific susceptibility is noticed for different pregnancy complications, including PE [225]. Hence, pregnant non-Asian women bearing male fetuses were found to have an increased risk for developing PE [226]. Moreover, the placental inflammatory response in PE was found to be significantly influenced by the fetal sex [149]. For instance, the IL-6 and IL-8 expression was found to be more pronounced in male, compared to female PE placentas [149]. Of note, healthy placentas did not exhibit any sexual dimorphism in the expression of IL-6/IL-8 [149]. The underlying mechanisms of the reported sexual dimorphism in PE remain unclear, but they could be associated with sex-specific gene expression in early to mid-gestation placenta [227]. In that vein, placental transcriptome profiling revealed that genes upregulated in male placentas are the ones involved in the regulation of the immune response [227].

Conflicting results are also shown for the serum levels of IL-6 and IL-8, when comparing severe vs. mild PE, or the time of PE onset. A number of studies reported elevated levels of both cytokines in severe, compared to mild PE [162,228–232]. However, an absence of an association between the maternal serum IL-6 levels and the severity of PE is also reported [155,233,234]. Comparing maternal blood concentrations of IL-6 between EOPE and LOPE on the one hand, and a healthy pregnancy on the other, also yielded inconsistent findings [205,232,235,236]. These inconsistencies could reflect individual differences in age, hormonal status, lipid concentration, chronic inflammation and other factors which are shown to affect individual circulating IL-6 levels [237–239], and whether these confounders were factored into the analyses or not.

Collectively, it can be concluded that the altered trophoblast invasion and spiral artery remodeling, as well as the endothelial dysfunction in PE, are interrelated with the immune maladaptation and disturbed homeostasis of IL-6 and IL-8. However, what remains unresolved is to what extent the levels of these cytokines relate to the severity of PE and its phenotypes, due to a large number of studies with conflicting results. The latter probably reflects not only the inconsistencies between the study methods, but also the multifaceted nature of the PE syndrome and the heterogeneity of risk factors and mechanisms leading to its development. In that context, PE, or at least EOPE, is seen as just one in a spectrum of complications of pregnancy that share a common pathophysiology rooted in aberrant placentation. In general, an elevated level of pro-inflammatory cytokines in the maternal circulation, with a shift in the "IL-8 $\times$ IL-6" axis towards the pro-inflammatory Th1 response is thought to drive the cytokine network in PE women towards an excessive systemic inflammatory state [156]. Thus, while the mechanistic relevance of IL-6 and IL-8 in the pathogenesis of PE is to some extent obvious, the mechanisms influencing their dysregulation are noteworthy objectives of additional investigations.

### 3.3. *Gestational Diabetes Mellitus*

Gestational diabetes mellitus (GDM) is a common gestational complication, discernable by de novo spontaneous hyperglycemia that develops during the course of pregnancy [240]. It is formally defined as "diabetes first diagnosed in the second or third trimester of pregnancy that is not clearly either preexisting type 1 or type 2 diabetes mellitus" (DM) [240]. GDM usually resolves following delivery, however, it may have long-lasting health consequences for the mother and the fetus, including an increased risk for type 2 DM and cardiovascular diseases [241,242]. It may be associated with premature delivery and PE [243,244], and quite often with neonate hypoglycemia, macrosomia and obstructed labor, due to the endogenous production of fetal insulin and insulin-like growth factor 1, in response to maternal metabolic alterations [245].

Analyses of the risk factors for GDM, despite methodological inconsistencies, point to an advanced maternal age, overweight/obesity, excessive gestational weight gain, ethnicity, genetic polymorphisms, low or high birth weight, family or past history of GDM, and other insulin-resistant states, such as polycystic ovarian syndrome, as common risk factors for developing GDM [246–248]. As in type 2 DM, insulin resistance (IR) and β-cell dysfunction play a central role in the pathophysiology of GDM [249]. In normal pregnancy, during early gestation, insulin sensitivity increases, promoting adipose storage of glucose in preparation for the increased energy demands in gestation [250]. As pregnancy advances, the surge of adipokines and diabetogenic placental hormones (progesterone, cortisol, prolactin and human placental lactogen) promotes a state of decreased insulin sensitivity and hyperglycemia [251]. Thus, a physiological pregnancy is considered an insulin-resistant state, with a 50% reduction in the insulin-mediated glucose clearance, and a more than two-fold increase in insulin production to maintain maternal euglycemia [252]. Chronic hyperglycemia and hyperinsulinemia, along with increased inflammatory and oxidative stress, is detrimental for the maternal pancreatic β-cells, leading to their functional exhaustion and injury/death [253]. When β-cell function fails to compensate the additional metabolic stress imposed by the diabetogenic state of pregnancy, the glucose metabolism becomes further dysregulated, leading to the development of GDM [254].

Evidence points that GDM is associated with changes in the maternal, fetal and placental inflammatory profile [254]. Systemic, chronic, subclinical inflammation that involves unbalanced cytokine production, is a key feature of GDM. Such metabolically induced inflammation, appropriately termed "metainflammation" [255], also accompanies obesity, IR, metabolic syndrome, type 2 DM and other related metabolic disorders. Although a lack of a significant association between the circulating IL-6 levels and GDM has been observed in some studies [180–183,256], an elevated concentration of IL-6 has been frequently reported in GDM patients, even regardless of obesity [113,115,172–179]. The results of the most recent systematic review indicate that serum IL-6 levels seem to be significantly higher in the majority of GDM patients, compared to euglycemic pregnant women [257]. Thus, as the authors suggest, assessing the serum IL-6 level could be a feasible diagnostic criterion for GDM [257]. As observed in PE and other chronic low grade inflammations [258,259], the increase in the systemic IL-6 in GDM patients may be accompanied by an increase in the sgp130 concentration [211]. This finding could be indicative of a compensatory anti-inflammatory mechanism to prevent overt inflammation induced by IL-6 trans-signaling [211,212]. Moreover, animal studies reveal that the blockade of peripheral IL-6 trans-signaling by recombinant sgp130, induces mature-onset obesity, glucose intolerance and IR [260].

The first (and, so far, the only) meta-analysis and systematic review of chemokines and their cognate receptors, suggests a role for IL-8 in the shaping of the complex immune microenvironment in GDM [261]. This is supported by case-control studies reporting increased circulating IL-8 levels in GDM patients, compared to healthy pregnancies [115,176,179,180]. However, comparable concentrations of IL-8 in the plasma of GDM patients vs. healthy pregnant women are also reported [167,184,185], as well as a lower IL-8 level in early pregnancy [184]. These inter-study inconsistencies probably emerge from the differences in the applied protocols and assays.

Current data point that both innate and adaptive immune system components respond to hyperglycemic and IR conditions, participating in the development of metainflammation [262]. In an obese state, the immune cells maintaining an anti-inflammatory environment in the adipose tissue are replaced with a pro-inflammatory immune-cell infiltrate [262]. This is accompanied by an increased secretion of pro-inflammatory cytokines and chemokines that act in an autocrine, paracrine, and endocrine manner, to promote inflammation and IR in the adipose and other target tissues [262]. Additionally, evidence shows that activated adipocytes are one of the main sources of the IL-6 and IL-8 production in obesity-associated IR and also in GDM [263,264]. In line with this, Kleiblova and coauthors indicated an upregulated IL-6 mRNA expression in subcutaneous adipose tissue of pregnant women with GDM [169]. Unlike TNF-α, which is suggested to act locally (in the adipose tissue) in an autocrine/paracrine manner contributing to the local IR and inducing IL-6 secretion, IL-6 rather appears to be released systemically by the adipose tissue, acting more as an endocrine signal that induces the hepatic acute-phase response and IR [265]. In fact, under basal conditions, up to 35% of systemic IL-6 is shown to originate from visceral adipose tissue in obese states, secreted by adipocytes and resident/infiltrated immune cells [265]. IL-6 contributes to IR primarily by impairing the phosphorylation of the insulin receptor and insulin receptor substrate-1, and inducing the expression of SOCS3–which impairs insulin signaling [266]. IL-6 is also known to promote lipolysis and secretion of free fatty acids from the adipose tissue into the circulation, which contributes to IR and to the increased gluconeogenesis in hepatocytes [267].

In obesity-related IR, visceral white adipose tissue (WAT) is considered to be the main source of IL-8, along with subcutaneous WAT and the infiltrated macrophages [268]. Considering that IL-8 attracts not only neutrophils and other immune cells, but also adipocytes, IL-8 secreted from hypertrophic adipocytes may contribute to the further accumulation of excess intra-abdominal fat in obesity [112]. Moreover, IL-8 itself enhances

IL-8 mRNA expression in human adipocytes, thus providing an autoamplifying loop via CXCR1 and CXCR2 expressed on the human adipocytes [269]. Data suggest that persistent inflammatory stimuli may perpetuate this vicious circle of IL-8 production in human adipocytes over the p38 MAPK pathway, which is also implicated in promoting IR in human adipocytes [269]. Moreover, IL-8 in obesity may downregulate adiponectin in adipocytes [169,261]. A decreased adiponectin level is a common finding in IR, DM and GDM [270]. Given that adiponectin stimulates insulin secretion, enhances its signaling and inhibits gluconeogenesis [271], by decreasing the adiponectin production, IL-8 may play a crucial role in obesity-linked IR and GDM. Consistently, increased levels of IL-8 are detected in visceral adipose tissue of women with GDM [169,272,273].

The placental common repertoire of cytokines also becomes overexpressed in a diabetic environment [274–276]. Data from the literature show an increased IL-6 mRNA expression in GDM placentas [115,163–166], possibly associated with the enhanced macrophage infiltration in GDM placentas, compared to a physiological pregnancy [163,168]. However, no significant differences in the IL-6 mRNA expression in placentas of women with GDM, compared with healthy controls were also reported [168,169]. Results on the IL-8 placental expression in GDM patients are also conflicting. No significant difference [169,170], as well as increased IL-8 levels in GDM placentas [115,167], compared to healthy controls were both detected. Interestingly, one recent study revealed that the expression of IL-8 in GDM placentas might be sex-specific [171]. Namely, the male GDM placentas exhibited a lower IL-8 expression in EVTs and STB, compared to sex-matched controls [171]. On the other hand, the female GDM placentas expressed comparable and higher levels of IL-8 in EVTs and STB, respectively, than the sex-matched controls [171]. However, the implications of this sex-specific expression of IL-8 in GDM placentas remain to be elucidated. Data regarding the IL-6 and IL-8 expression/circulating level changes in GDM are summarized in Table 2.

Collectively, it may be concluded that a certain level of metabolically induced inflammation, reflected in the more pronounced maternal cellular and biochemical inflammatory profile than in non-diabetic pregnancies, accompanies maternal IR and GDM. IL-6 appears to be decisively involved in the development of IR and GDM, primarily by the impaired phosphorylation of the insulin receptors and the induced expression of SOCS3, which inhibits the insulin signaling in peripheral tissues, adipocytes and hepatocytes [266]. Moreover, IL-8 seems to participate in promoting obesity, inflammation and IR, by attracting additional adipocytes and immune cells in the adipose tissue, and by interfering with insulin signaling by downregulation of adiponectin and/or activation of the p38 MAPK pathway [112,261,269]. However, studies investigating inflammatory mediators in the maternal and placental compartments in GDM are not always consistent, whereas data regarding the fetal inflammatory state in GDM are lacking. A better understanding of the inflammatory process in GDM is urgently needed, considering that the placental inflammation in GDM may have a central role in shaping the in utero environment that "programs" the offspring development [169,277]. As the prevalence of pre-gestational DM and GDM continues to rise worldwide [278], such an understanding will be critical to optimize long-term health outcomes for both the mother and the offspring. In the meantime, it is necessary to once again stress the importance of correcting the modifiable risk factors and applying inflammation-reducing life-style changes during pregnancy, especially in the early stages of pregnancy, to reduce the risk of developing GDM and the consequences associated with it [279].

### 3.4. Maternal Immune Activation

Maternal immune activation (MIA) in pregnancy is usually associated with acute infections, or with a sterile, low-grade, persistent inflammation, accompanying a number of systemic conditions, such as metabolic syndrome, type 2 DM, autoimmune diseases, cardiovascular disease, anxiety, depression, socio-economic adversity, micronutrient deficiencies, microbiome alterations, exposure to cigarette smoke or Δ-9-tetrahydrocannabinol, air pollution and other factors [280–287]. This myriad of exogenous and endogenous environmental exposures during pregnancy may cause tissue injury and trigger maternal inflammatory/immune responses, leading to a release of a plethora of effector molecules, with IL-6 and IL-8 having one of the key roles [288].

Maternal systemic infections in pregnancy are recognized as the principal non-genetic risk for neurodevelopmental and neuropsychiatric disorders in the child [289–291]. The variety of infectious agents associated with a shared neurodevelopmental risk points to the maternal immune response rather than a particular pathogen as a common denominator of the dysregulated offspring development [292,293]. This concept is particularly plausible considering that an increased risk of developmental abnormalities was also associated with pathogens that typically do not cross the placental barrier (influenza, agents causing upper respiratory infections) [280,294–296] or with serologic evidence of the maternal pre-gestational exposure to pathogens in the absence of an active infection [297–299]. It has been suggested that MIA during the sensitive window of in utero brain development, may alter neurodevelopmental trajectories, following the proposed sequence of events: (i) infection or other noxious stimuli trigger maternal inflammatory/immune responses, (ii) released cytokines and other inflammatory mediators cross the blood-placental barrier and (iii) activate the fetal immune system enabling the establishment of a self-propagating, low-grade inflammatory cascade [300]. Inflammatory cytokines reaching the fetal brain may trigger microglial activation and upregulation of pro-inflammatory transcription factors. Consequently, aberrant gene and protein expression may in long term give rise to neurological, immunological and behavioral disturbances in a predisposed offspring [300]. Supported by a fair amount of epidemiological data and animal research, IL-6 has been identified as the critical mediator in this unfortunate series of events [301–304].

Apart from systemic infection, acute placental inflammation (API), the microscopic equivalent to the clinical diagnosis of chorioamnionitis [305], is another common MIA-associated event in pregnancy, even in a clinically inapparent disease [306,307]. Low-stage API is observed in up to 50% of uncomplicated vaginal deliveries following uncomplicated pregnancies [308]. It is even more frequent in the absence of infectious agents, than due to intra-amnionic infection [192]. Whether the microbial invasion of the chorioamniotic membranes or the release of DAMPs during the course of cellular injury/death, an increase in the expression of IL-6 and IL-8 follows, along with the release of other acute phase mediators [25,193,309]. This is supported by findings of an increased expression of IL-6 and IL-8 in the trophoblast in response to LPS stimulation [188,190,191]. Consistently, elevated concentrations of IL-6 and IL-8 in amniotic fluid (AF) or cervicovaginal lavage, as indicators of API in PTL, are reported in a vast number of studies [192–198]. Additionally, a high concentration of IL-6 in AF was significantly associated with an increased risk of perinatal morbidity and mortality [197]. The AF level of IL-8 sampled in the third trimester of pregnancy was shown to correlate with the severity of API, irrespective of the presence/severity of funisitis [200], suggesting that the high AF IL-8 levels in API are mainly a product of the maternal immune response. Moreover, some studies indicate that human IL-8 does not seem to cross placentas obtained from pregnancies at term, implying that IL-8 in AF and in fetal blood is solely of fetal origin [310,311]. This is consistent with the data showing human fetal amnion and chorion as an important source of IL-8 in the setting of an ongoing inflammation [312–314]. The source of IL-6 in MIA has also been debated. Through a rodent MIA model, Hsiao and Patterson proposed that elevated level of IL-6 in the placenta was of maternal origin only, including both circulating IL-6 and the one secreted by the placental resident cells [189]. Nevertheless, the fetus itself can also mount an inflammatory response to maternal infection, especially in terms of IL-6 production [315,316]. Fetuses with fetal inflammatory response syndrome (FIRS), characterized by high levels of IL-6 in fetal plasma, had a higher rate of severe neonatal morbidity and a shorter procedure-to-delivery interval [317]. Considering these risks, rapid tests for the timely identification of inflammatory processes in the amniotic cavity that would circumvent amniocentesis as an invasive technique are highly needed in clinical practice. To this end, the determination of IL-8 [318] in the maternal serum and IL-8 [319] or IL-6 [318] in the cervical secretion for non-invasive screening for chorioamnionitis was suggested. However, there are data indicating that high levels of cervical IL-6 and IL-8 are only moderately predictive of intrauterine infection/inflammation and preterm delivery [194]. The reported data are summarized in Table 2.

The role of IL-6 and IL-8 in the pathogenesis of the neurodevelopmental and neuropsychiatric endophenotypes as MIA sequelae is widely supported by epidemiological studies indicating that in utero exposure to elevated concentrations of both, IL-6 [320–322], and IL-8 [323–326], may partially account for an increased risk of neurodevelopmental/neuropsychiatric disorders. However, remarkably little is known about the mechanistic pathways that connect these molecules with neurodevelopmental disorders. Emerging data suggest that both IL-6 and IL-8 inhibit synaptic long-term potentiation and induce changes in the hippocampal-dependent learning and memory tasks [327]. IL-6 may also influence the fate switching and cell differentiation in development, acting directly on the progenitor cells to regulate fetal neurogenesis and gliogenesis [328,329], or altering many parameters that influence neuronal migration, axonal pathfinding and synapse formation [330], or fetal growth in general, including nutrient transfer, anoxia and vascular permeability at the feto-maternal interface [331–333]. IL-8 dysregulation has also been found to have a role in atypical white matter development in preterm infants [334] and brain dysmaturation [326]. In addition, the activation of the JAK/STAT3 signaling axis by the maternal IL-6 in murine placenta, indirectly influenced fetal neurodevelopment through the diminished production of both the placental growth hormone and the insulin-growth factor 1 [189] indispensable for the proper fetal development [335]. Moreover, IL-6 can disrupt the immunological homeostasis of the placenta and the maintenance of the maternal tolerance by altering the Th1/Th2 ratio and by activation of the uterine immune cells [329,330].

One murine study employing a trophoblast IL-6R knockout model, reported no sexual dimorphism in the MIA-related behavioral abnormalities [303]. Noteworthy, it is generally recognized that MIA affects the offspring neurodevelopment in a sexually dimorphic manner and those sex-specific effects persists across the lifespan [320,336]. Sex differences in the placental responses to MIA, fetal brain structure/function and immune response could account for these sexually dimorphic effects of MIA [337–339]. Both human [322,340,341] and animal [336,342,343] studies indicate that male offsprings are more frequently affected with MIA-induced neurodevelopmental outcomes than females. Consistent with this notion, a general trend toward the heightened acute inflammation and elevated cytokine levels was shown in murine male vs. female placentas, especially for the abundant cytokines, such as IL-6, upon maternal stimulation with LPS [344].

Considering all of the aforementioned, it may be concluded that the maternal immune activity in pregnancy may potentially affect the offspring development, cognition, and behavior, through mechanisms including, at least partly, IL-6 and IL-8 signaling. As therapeutic interventions that significantly alter the prenatal environment and systemic non-specific immunomodulatory agents are not likely candidates in pregnant humans, focusing on eliminating the risk of maternal infection remains the main prevention strategy to reduce the incidence of neurodevelopmental abnormalities.

Last, but not least important, it must be taken into consideration that in humans, most gestational infections do not lead to overt neurological/psychiatric disease in the offspring, despite the strong evidence that the stimulation of the maternal immune response during gestation has a potential for profound effects on the offspring neurodevelopment [326,345]. In fact, it may be assumed that MIA acts more as a "disease primer", by establishing a susceptible neuroanatomical/neurophysiological setting which, coupled with a permissive genetic background, may increase sensitivity to the disrupting effects of postnatal stressors and ultimately result in pathological behaviors and functions later in life [346]. This is reasonable, considering the multitude of highly diverse factors that contri-bute to neurodevelopment, and the multifactorial etiology and complex pathogenesis of neuropsychiatric and developmental diseases [347].

## 4. Conclusions

According to the reviewed literature, IL-6 and IL-8 clearly play multiple functional roles in pregnancy physiology. They appear to contribute to the establishment and maintenance of pregnancy by mediating uterine receptivity, trophoblast function at the implantation site and parturition, the immune-endocrine interactions at the feto-maternal interface and other processes. The hereby presented body of evidence also indicates that a dysregulated IL-6/IL-8 expression, either at the feto-maternal interface or systemically, may contribute to the development of various gestational complications. Therefore, it appears that targeting the IL-6/IL-8 pathways may rescue some pregnancy trajectories and prevent or ameliorate sequelae. Animal models and empirical data suggest several preventive/therapeutic strategies which, directly or indirectly, affect the IL-6/IL-8 production/function. Classic anti-inflammatory drugs, both steroid [348–350] and non-steroid [350,351], the application of anti-IL-6 or anti-IL-6R mAbs [352–355], or anti-inflammatory cytokines [356,357], dietary interventions [358–360], the use of probiotics [361,362] or vitamin D [363], microbiota transplants [281,364], and other immunomodulatory interventions have been examined in the context of gestational complications with some success. However, considering that (i) interference with the prenatal inflammatory/immune environment may lead to devastating consequences [365], and (ii) due to complex ethical issues pregnant women are traditionally excluded from clinical trials [366], novel immunomodulatory treatments require an extensive evaluation on both a scientific and ethical basis before being routinely implemented in a clinical setting. Thus, increasing awareness and optimizing prevention by correcting modifiable risk factors for gestational complications associated with dysregulated inflammatory/immune responses, should remain one of the main strategies in prenatal care.

Finally, one of the limitations of the present review for a successful translation to the clinical level, is its focus on IL-6 and IL-8 only. This reductionist approach enables a better overview of the role of IL-6 and IL-8 in pregnancy-related processes, but is insufficient to draw accurate conclusions about the inflammatory status. This is reasonable considering the complex cytokine networks that underlie these processes and the dynamic relationship between the pro- and anti-inflammatory factors over the course of inflammation.

The emerging advances in biomedical research that enable the computer modeling of data and the new insights into the fields of genomics, epigenetics, proteomics, metagenomics (the microbiome), and metabolomics, and will hopefully improve our understanding of the molecular mechanisms of pregnancy and its possible complications. Such an understanding could be employed to tailor the diagnostic/therapeutic strategies for a more personalized healthcare. Bridging the gaps in knowledge identified herein, could contribute to optimizing the current practices to improve pregnancy outcomes.

**Author Contributions:** Conceptualization, M.J.K., M.N.-A. and A.V.; writing—original draft preparation, A.V., M.N.-A., A.P., Ž.B.-T., D.D. and M.J.K.; writing—review and editing, M.N.-A. and A.V.; visualization, M.J.K. All authors have read and agreed to the published version of the manuscript.

**Funding:** This work was supported by the Ministry of Education, Science and Technological Development, Republic of Serbia (contract no. 451-03-68/2022 14/200019).

**Institutional Review Board Statement:** Not applicable.

**Informed Consent Statement:** Not applicable.

**Data Availability Statement:** Not applicable.

**Conflicts of Interest:** The authors declare no conflict of interest.

## References


1. Benirschke, K.; Burton, G.J.; Baergen, R.N. *Pathology of the Human Placenta*; Springer: Berlin/Heidelberg, Germany, 2012; ISBN 978-3-642-23940-3.
2. Saito, S. Cytokine Network at the Feto-Maternal Interface. *J. Reprod. Immunol.* **2000**, *47*, 87–103. [CrossRef] [PubMed]
3. Matsuda, T.; Hirano, T. *IL-6. Cytokine Reference*; Oppenheim, J., Feldmann, M., Eds.; Academic Press: San Diego, CA, USA, 2000; Volume 1, pp. 537–563.
4. Heinrich, P.C.; Behrmann, I.; Haan, S.; Hermanns, H.M.; Müller-Newen, G.; Schaper, F. Principles of Interleukin (IL)-6-Type Cytokine Signalling and Its Regulation. *Biochem. J.* **2003**, *374*, 1–20. [CrossRef]
5. Naka, T.; Narazaki, M.; Hirata, M.; Matsumoto, T.; Minamoto, S.; Aono, A.; Nishimoto, N.; Kajita, T.; Taga, T.; Yoshizaki, K.; et al. Structure and Function of a New STAT-Induced STAT Inhibitor. *Nature* **1997**, *387*, 924–929. [CrossRef] [PubMed]
6. Jones, S.A.; Jenkins, B.J. Recent Insights into Targeting the IL-6 Cytokine Family in Inflammatory Diseases and Cancer. *Nat. Rev. Immunol.* **2018**, *18*, 773–789. [CrossRef] [PubMed]
7. Rose-John, S. Interleukin-6 Signalling in Health and Disease. *F1000Research* **2020**, *9*, 1013. [CrossRef]
8. Matsushima, K.; Yang, D.; Oppenheim, J.J. Interleukin-8: An Evolving Chemokine. *Cytokine* **2022**, *153*, 155828. [CrossRef]
9. Corre, I.; Pineau, D.; Hermouet, S. Interleukin-8: An Autocrine/Paracrine Growth Factor for Human Hematopoietic Progenitors Acting in Synergy with Colony Stimulating Factor-1 to Promote Monocyte-Macrophage Growth and Differentiation. *Exp. Hematol.* **1999**, *27*, 28–36. [CrossRef]
10. Li, A.; Dubey, S.; Varney, M.L.; Dave, B.J.; Singh, R.K. IL-8 Directly Enhanced Endothelial Cell Survival, Proliferation, and Matrix Metalloproteinases Production and Regulated Angiogenesis. *J. Immunol.* **2003**, *170*, 3369–3376. [CrossRef]
11. Bréchard, S.; Bueb, J.-L.; Tschirhart, E.J. Interleukin-8 Primes Oxidative Burst in Neutrophil-like HL-60 through Changes in Cytosolic Calcium. *Cell Calcium* **2005**, *37*, 531–540. [CrossRef]
12. Iizasa, H.; Matsushima, K. *IL-8. Cytokine Reference*; Oppenheim, J., Feldmann, M., Eds.; Academic Press: San Diego, CA, USA, 2000; Volume 1, pp. 1061–1067.
13. Chan, L.-P.; Liu, C.; Chiang, F.-Y.; Wang, L.-F.; Lee, K.-W.; Chen, W.-T.; Kuo, P.-L.; Liang, C.-H. IL-8 Promotes Inflammatory Mediators and Stimulates Activation of P38 MAPK/ERK-NF-KB Pathway and Reduction of JNK in HNSCC. *Oncotarget* **2017**, *8*, 56375–56388. [CrossRef]
14. Tabibzadeh, S.; Kong, Q.F.; Babaknia, A.; May, L.T. Progressive Rise in the Expression of Interleukin-6 in Human Endometrium during Menstrual Cycle Is Initiated during the Implantation Window. *Hum. Reprod.* **1995**, *10*, 2793–2799. [CrossRef] [PubMed]
15. Arici, A.; Seli, E.; Senturk, L.M.; Gutierrez, L.S.; Oral, E.; Taylor, H.S. Interleukin-8 in the Human Endometrium. *J. Clin. Endocrinol. Metab.* **1998**, *83*, 1783–1787. [CrossRef] [PubMed]
16. Mulayim, N.; Palter, S.F.; Kayisli, U.A.; Senturk, L.; Arici, A. Chemokine Receptor Expression in Human Endometrium. *Biol. Reprod.* **2003**, *68*, 1491–1495. [CrossRef] [PubMed]
17. Jauniaux, E.; Gulbis, B.; Schandene, L.; Collette, J.; Hustin, J. Molecular Interactions during Pregnancy. *MHR Basic Sci. Reprod. Med.* **1996**, *2*, 239–243. [CrossRef]
18. Pitman, H.; Innes, B.A.; Robson, S.C.; Bulmer, J.N.; Lash, G.E. Altered Expression of Interleukin-6, Interleukin-8 and Their Receptors in Decidua of Women with Sporadic Miscarriage. *Hum. Reprod.* **2013**, *28*, 2075–2086. [CrossRef]
19. Pietro, L.; Bottcher-Luiz, F.; Velloso, L.A.; Morari, J.; Nomura, M.; Lucci De Angelo Andrade, L.A. Expression of Interleukin-6 (IL-6), Signal Transducer and Activator of Transcription-3 (STAT-3) and Telomerase in Choriocarcinomas. *Surg. Exp. Pathol.* **2020**, *3*, 28. [CrossRef]
20. Champion, H.; Innes, B.A.; Robson, S.C.; Lash, G.E.; Bulmer, J.N. Effects of Interleukin-6 on Extravillous Trophoblast Invasion in Early Human Pregnancy. *MHR Basic Sci. Reprod. Med.* **2012**, *18*, 391–400. [CrossRef]
21. Jovanović, M.; Vićovac, L. Interleukin-6 Stimulates Cell Migration, Invasion and Integrin Expression in HTR-8/SVneo Cell Line. *Placenta* **2009**, *30*, 320–328. [CrossRef]
22. Saito, S.; Kasahara, T.; Sakakura, S.; Umekage, H.; Harada, N.; Ichijo, M. Detection and Localization of Interleukin-8 MRNA and Protein in Human Placenta and Decidual Tissues. *J. Reprod. Immunol.* **1994**, *27*, 161–172. [CrossRef]
23. Hanna, J.; Goldman-Wohl, D.; Hamani, Y.; Avraham, I.; Greenfield, C.; Natanson-Yaron, S.; Prus, D.; Cohen-Daniel, L.; Arnon, T.I.; Manaster, I.; et al. Decidual NK Cells Regulate Key Developmental Processes at the Human Fetal-Maternal Interface. *Nat. Med.* **2006**, *12*, 1065–1074. [CrossRef]
24. De Oliveira, L.G.; Lash, G.E.; Murray-Dunning, C.; Bulmer, J.N.; Innes, B.A.; Searle, R.F.; Sass, N.; Robson, S.C. Role of Interleukin 8 in Uterine Natural Killer Cell Regulation of Extravillous Trophoblast Cell Invasion. *Placenta* **2010**, *31*, 595–601. [CrossRef] [PubMed]
25. Shimoya, K.; Matsuzaki, N.; Taniguchi, T.; Kameda, T.; Koyama, M.; Neki, R.; Saji, F.; Tanizawa, O. Human Placenta Constitutively Produces Interleukin-8 during Pregnancy and Enhances Its Production in Intrauterine Infection. *Biol. Reprod.* **1992**, *47*, 220–226. [CrossRef] [PubMed]
26. Jovanović, M.; Stefanoska, I.; Radojcić, L.; Vićovac, L. Interleukin-8 (CXCL8) Stimulates Trophoblast Cell Migration and Invasion by Increasing Levels of Matrix Metalloproteinase (MMP)2 and MMP9 and Integrins Alpha5 and Beta1. *Reproduction* **2010**, *139*, 789–798. [CrossRef]
27. Makrigiannakis, A.; Minas, V. Mechanisms of Implantation. *Reprod. Biomed. Online* **2007**, *14*, 75–82. [CrossRef] [PubMed]

28. Zhou, M.; Xu, H.; Zhang, D.; Si, C.; Zhou, X.; Zhao, H.; Liu, Q.; Xu, B.; Zhang, A. Decreased PIBF1/IL6/p-STAT3 during the Mid-Secretory Phase Inhibits Human Endometrial Stromal Cell Proliferation and Decidualization. *J. Adv. Res.* **2021**, *30*, 15–25. [CrossRef] [PubMed]
29. Sharkey, A.M.; Dellow, K.; Blayney, M.; Macnamee, M.; Charnock-Jones, S.; Smith, S.K. Stage-Specific Expression of Cytokine and Receptor Messenger Ribonucleic Acids in Human Preimplantation Embryos. *Biol. Reprod.* **1995**, *53*, 974–981. [CrossRef]
30. Zolti, M.; Ben-Rafael, Z.; Meirom, R.; Shemesh, M.; Bider, D.; Mashiach, S.; Apte, R.N. Cytokine Involvement in Oocytes and Early Embryos. *Fertil. Steril.* **1991**, *56*, 265–272. [CrossRef]
31. Zhong, H.; Sun, Q.; Chen, P.; Xiong, F.; Li, G.; Wan, C.; Yao, Z.; Zeng, Y. Detection of IL-6, IL-10, and TNF-α Level in Human Single-Blastocyst Conditioned Medium Using Ultrasensitive Single Molecule Array Platform and Its Relationship with Embryo Quality and Implantation: A Pilot Study. *J. Assist. Reprod. Genet.* **2020**, *37*, 1695–1702. [CrossRef]
32. Plana-Carmona, M.; Stik, G.; Bulteau, R.; Segura-Morales, C.; Alcázar, N.; Wyatt, C.D.R.; Klonizakis, A.; de Andrés-Aguayo, L.; Gasnier, M.; Tian, T.V.; et al. The Trophectoderm Acts as a Niche for the Inner Cell Mass through C/EBPα-Regulated IL-6 Signaling. *Stem Cell Rep.* **2022**, *17*, 1991–2004. [CrossRef]
33. Desai, N.; Scarrow, M.; Lawson, J.; Kinzer, D.; Goldfarb, J. Evaluation of the Effect of Interleukin-6 and Human Extracellullar Matrix on Embryonic Development. *Hum. Reprod.* **1999**, *14*, 1588–1592. [CrossRef]
34. Yang, J.; Yang, X.; Yang, H.; Bai, Y.; Zha, H.; Jiang, F.; Meng, Y. Interleukin 6 in Follicular Fluid Reduces Embryo Fragmentation and Improves the Clinical Pregnancy Rate. *J. Assist. Reprod. Genet.* **2020**, *37*, 1171–1176. [CrossRef] [PubMed]
35. Robertson, S.; O'Connell, A.; Ramsey, A. The Effect of Interleukin-6 Deficiency on Implantation, Fetal Development and Parturition in Mice. *Proc. Aust. Soc. Reprod. Biol.* **2000**, *31*, 97.
36. Sakurai, T.; Takai, R.; Bürgin, H.; Ishihara, K.; Sakamoto, Y.; Amano, J.; Higuchi, Y.; Chiba, S.; Singer, T.; Kawamura, A.; et al. The Effects of Interleukin-6 Signal Blockade on Fertility, Embryo-Fetal Development, and Immunization In Vivo. *Birth Defects Res. Part B-Dev. Reprod. Toxicol.* **2012**, *95*, 304–317. [CrossRef] [PubMed]
37. Meisser, A.; Cameo, P.; Islami, D.; Campana, A.; Bischof, P. Effects of Interleukin-6 (IL-6) on Cytotrophoblastic Cells. *Mol. Hum. Reprod.* **1999**, *5*, 1055–1058. [CrossRef]
38. Jovanovic, M.; Kovacevic, T.; Stefanoska, I.; Vicovac, L. The Effect of IL-6 on the Trophoblast Cell Line HTR-8/SVneo. *Arch. Biol. Sci.* **2010**, *62*, 531–538. [CrossRef]
39. Damsky, C.H.; Fitzgerald, M.L.; Fisher, S.J. Distribution Patterns of Extracellular Matrix Components and Adhesion Receptors Are Intricately Modulated during First Trimester Cytotrophoblast Differentiation along the Invasive Pathway, in Vivo. *J. Clin. Investig.* **1992**, *89*, 210–222. [CrossRef]
40. Vićovac, L.; Jones, C.J.; Aplin, J.D. Trophoblast Differentiation during Formation of Anchoring Villi in a Model of the Early Human Placenta in Vitro. *Placenta* **1995**, *16*, 41–56. [CrossRef]
41. Lala, P.K.; Graham, C.H. Mechanisms of Trophoblast Invasiveness and Their Control: The Role of Proteases and Protease Inhibitors. *Cancer Metastasis Rev.* **1990**, *9*, 369–379. [CrossRef]
42. Librach, C.L.; Werb, Z.; Fitzgerald, M.L.; Chiu, K.; Corwin, N.M.; Esteves, R.A.; Grobelny, D.; Galardy, R.; Damsky, C.H.; Fisher, S.J. 92-KD Type IV Collagenase Mediates Invasion of Human Cytotrophoblasts. *J. Cell Biol.* **1991**, *113*, 437–449. [CrossRef]
43. Godbole, G.; Suman, P.; Malik, A.; Galvankar, M.; Joshi, N.; Fazleabas, A.; Gupta, S.K.; Modi, D. Decrease in Expression of HOXA10 in the Decidua After Embryo Implantation Promotes Trophoblast Invasion. *Endocrinology* **2017**, *158*, 2618–2633. [CrossRef]
44. Dubinsky, V.; Poehlmann, T.G.; Suman, P.; Gentile, T.; Markert, U.R.; Gutierrez, G. Role of Regulatory and Angiogenic Cytokines in Invasion of Trophoblastic Cells. *Am. J. Reprod. Immunol.* **2010**, *63*, 193–199. [CrossRef] [PubMed]
45. Choudhury, R.H.; Dunk, C.E.; Lye, S.J.; Aplin, J.D.; Harris, L.K.; Jones, R.L. Extravillous Trophoblast and Endothelial Cell Crosstalk Mediates Leukocyte Infiltration to the Early Remodeling Decidual Spiral Arteriole Wall. *J. Immunol.* **2017**, *198*, 4115–4128. [CrossRef] [PubMed]
46. Nishino, E.; Matsuzaki, N.; Masuhiro, K.; Kameda, T.; Taniguchi, T.; Takagi, T.; Saji, F.; Tanizawa, O. Trophoblast-Derived Interleukin-6 (IL-6) Regulates Human Chorionic Gonadotropin Release through IL-6 Receptor on Human Trophoblasts. *J. Clin. Endocrinol. Metab.* **1990**, *71*, 436–441. [CrossRef] [PubMed]
47. Stephanou, A.; Handwerger, S. Interleukin-6 Stimulates Placental Lactogen Expression by Human Trophoblast Cells. *Endocrinology* **1994**, *135*, 719–723. [CrossRef] [PubMed]
48. Caballero-Campo, P. Hormonal and Embryonic Regulation of Chemokines IL-8, MCP-1 and RANTES in the Human Endometrium during the Window of Implantation. *Mol. Hum. Reprod.* **2002**, *8*, 375–384. [CrossRef] [PubMed]
49. Dominguez, F.; Galan, A.; Martin, J.J.L.; Remohi, J.; Pellicer, A.; Simón, C. Hormonal and Embryonic Regulation of Chemokine Receptors CXCR1, CXCR4, CCR5 and CCR2B in the Human Endometrium and the Human Blastocyst. *Mol. Hum. Reprod.* **2003**, *9*, 189–198. [CrossRef]
50. Sharma, S.; Godbole, G.; Modi, D. Decidual Control of Trophoblast Invasion. *Am. J. Reprod. Immunol.* **2016**, *75*, 341–350. [CrossRef]
51. Huang, G.; Zhou, C.; Wei, C.-J.; Zhao, S.; Sun, F.; Zhou, H.; Xu, W.; Liu, J.; Yang, C.; Wu, L.; et al. Evaluation of In Vitro Fertilization Outcomes Using Interleukin-8 in Culture Medium of Human Preimplantation Embryos. *Fertil. Steril.* **2017**, *107*, 649–656. [CrossRef]
52. Tsui, K.-H.; Chen, L.-Y.; Shieh, M.-L.; Chang, S.-P.; Yuan, C.-C.; Li, H.-Y. Interleukin-8 Can Stimulate Progesterone Secretion from a Human Trophoblast Cell Line, BeWo. *In Vitro Cell. Dev. Biol. Anim.* **2004**, *40*, 331–336. [CrossRef]

53. Banerjee, P.; Malik, A.; Malhotra, S.S.; Gupta, S.K. Role of STAT Signaling and Autocrine Action of Chemokines during H2 O 2 Induced HTR-8/SVneo Trophoblastic Cells Invasion. *J. Cell. Physiol.* **2019**, *234*, 1380–1397. [CrossRef]
54. Das, M.K.; Basak, S.; Ahmed, M.S.; Attramadal, H.; Duttaroy, A.K. Connective Tissue Growth Factor Induces Tube Formation and IL-8 Production in First Trimester Human Placental Trophoblast Cells. *Eur. J. Obstet. Gynecol. Reprod. Biol.* **2014**, *181*, 183–188. [CrossRef] [PubMed]
55. Sokolov, D.I.; Furaeva, K.N.; Stepanova, O.I.; Sel'kov, S.A. Proliferative and Migration Activity of JEG-3 Trophoblast Cell Line in the Presence of Cytokines. *Bull. Exp. Biol. Med.* **2015**, *159*, 550–556. [CrossRef] [PubMed]
56. Ding, W.; Chim, S.S.C.; Wang, C.C.; Lau, C.S.L.; Leung, T.Y. Molecular Mechanism and Pathways of Normal Human Parturition in Different Gestational Tissues: A Systematic Review of Transcriptome Studies. *Front. Physiol.* **2021**, *12*, 730030. [CrossRef] [PubMed]
57. Bollopragada, S.; Youssef, R.; Jordan, F.; Greer, I.; Norman, J.; Nelson, S. Term Labor Is Associated with a Core Inflammatory Response in Human Fetal Membranes, Myometrium, and Cervix. *Am. J. Obstet. Gynecol.* **2009**, *200*, 104.e1–104.e.11. [CrossRef] [PubMed]
58. Kent, A.S.H.; Sullivan, M.H.F.; Sun, M.-Y.; Zosmer, A.; Elder, M.G. Effects of Interleukin-6 and Tumor Necrosis Factor-α on Prostaglandin Production by Cultured Human Fetal Membranes. *Prostaglandins* **1993**, *46*, 351–359. [CrossRef]
59. Mitchell, M.D.; Dudley, D.J.; Edwin, S.S.; Schiller, S.L. Interleukin-6 Stimulates Prostaglandin Production by Human Amnion and Decidual Cells. *Eur. J. Pharmacol.* **1991**, *192*, 189–191. [CrossRef]
60. Friebe-Hoffmann, U.; Chiao, J.P.; Rauk, P.N. Effect of IL-1beta and IL-6 on Oxytocin Secretion in Human Uterine Smooth Muscle Cells. *Am. J. Reprod. Immunol.* **2001**, *46*, 226–231. [CrossRef]
61. Rauk, P.N.; Friebe-Hoffmann, U.; Winebrenner, L.D.; Chiao, J.P. Interleukin-6 up-Regulates the Oxytocin Receptor in Cultured Uterine Smooth Muscle Cells. *Am. J. Reprod. Immunol.* **2001**, *45*, 148–153. [CrossRef]
62. Timmons, B.; Akins, M.; Mahendroo, M. Cervical Remodeling during Pregnancy and Parturition. *Trends Endocrinol. Metab.* **2010**, *21*, 353–361. [CrossRef]
63. Strauss, J.F. Extracellular Matrix Dynamics and Fetal Membrane Rupture. *Reprod. Sci.* **2013**, *20*, 140–153. [CrossRef]
64. Curry, A.E.; Vogel, I.; Skogstrand, K.; Drews, C.; Schendel, D.E.; Flanders, W.D.; Hougaard, D.M.; Thorsen, P. Maternal Plasma Cytokines in Early- and Mid-Gestation of Normal Human Pregnancy and Their Association with Maternal Factors. *J. Reprod. Immunol.* **2008**, *77*, 152–160. [CrossRef] [PubMed]
65. Blackmore, E.R.; Moynihan, J.A.; Rubinow, D.R.; Pressman, E.K.; Gilchrist, M.; O'Connor, T.G. Psychiatric Symptoms and Proinflammatory Cytokines in Pregnancy. *Psychosom. Med.* **2011**, *73*, 656–663. [CrossRef] [PubMed]
66. Björkander, S.; Bremme, K.; Persson, J.-O.; van Vollenhoven, R.F.; Sverremark-Ekström, E.; Holmlund, U. Pregnancy-Associated Inflammatory Markers Are Elevated in Pregnant Women with Systemic Lupus Erythematosus. *Cytokine* **2012**, *59*, 392–399. [CrossRef] [PubMed]
67. Fu, Y.; Tang, L.; Hu, M.; Xiang, Z.; Hu, Y. Changes of Serum Interleukin-6 in Healthy Pregnant Women and Establishment of Relevant Reference Intervals. *Clin. Chim. Acta* **2020**, *502*, 116–119. [CrossRef]
68. Farah, N.; Hogan, A.E.; O'Connor, N.; Kennelly, M.M.; O'Shea, D.; Turner, M.J. Correlation between Maternal Inflammatory Markers and Fetomaternal Adiposity. *Cytokine* **2012**, *60*, 96–99. [CrossRef]
69. Christian, L.M.; Porter, K. Longitudinal Changes in Serum Proinflammatory Markers across Pregnancy and Postpartum: Effects of Maternal Body Mass Index. *Cytokine* **2014**, *70*, 134–140. [CrossRef]
70. Ross, K.M.; Miller, G.; Culhane, J.; Grobman, W.; Simhan, H.N.; Wadhwa, P.D.; Williamson, D.; McDade, T.; Buss, C.; Entringer, S.; et al. Patterns of Peripheral Cytokine Expression during Pregnancy in Two Cohorts and Associations with Inflammatory Markers in Cord Blood. *Am. J. Reprod. Immunol.* **2016**, *76*, 406–414. [CrossRef]
71. Atta, D.S.; Girbash, E.F.; Abdelwahab, S.M.; Abdeldayem, H.M.; Tharwat, I.; Ghonaim, R. Maternal Cytokines and Disease Severity Influence Pregnancy Outcomes in Women with Rheumatoid Arthritis. *J. Matern. Fetal. Neonatal Med.* **2016**, *29*, 3358–3363. [CrossRef]
72. Mitchell, A.M.; Porter, K.; Christian, L.M. Examination of the Role of Obesity in the Association between Childhood Trauma and Inflammation during Pregnancy. *Health Psychol.* **2018**, *37*, 114–124. [CrossRef]
73. Doria, A.; Cutolo, M.; Ghirardello, A.; Zen, M.; Villalta, D.; Tincani, A.; Punzi, L.; Iaccarino, L.; Petri, M. Effect of Pregnancy on Serum Cytokines in SLE Patients. *Arthritis Res. Ther.* **2012**, *14*, R66. [CrossRef]
74. Iaccarino, L.; Ghirardello, A.; Zen, M.; Villalta, D.; Tincani, A.; Punzi, L.; Doria, A. Polarization of TH2 Response Is Decreased during Pregnancy in Systemic Lupus Erythematosus. *Reumatismo* **2012**, *64*, 314–320. [CrossRef] [PubMed]
75. Jarmund, A.H.; Giskeødegård, G.F.; Ryssdal, M.; Steinkjer, B.; Stokkeland, L.M.T.; Madssen, T.S.; Stafne, S.N.; Stridsklev, S.; Moholdt, T.; Heimstad, R.; et al. Cytokine Patterns in Maternal Serum From First Trimester to Term and Beyond. *Front. Immunol.* **2021**, *12*, 752660. [CrossRef] [PubMed]
76. Stokkeland, L.M.T.; Giskeødegård, G.F.; Stridsklev, S.; Ryan, L.; Steinkjer, B.; Tangerås, L.H.; Vanky, E.; Iversen, A.-C. Serum Cytokine Patterns in First Half of Pregnancy. *Cytokine* **2019**, *119*, 188–196. [CrossRef]
77. Leimert, K.B.; Xu, W.; Princ, M.M.; Chemtob, S.; Olson, D.M. Inflammatory Amplification: A Central Tenet of Uterine Transition for Labor. *Front. Cell. Infect. Microbiol.* **2021**, *11*, 660983. [CrossRef] [PubMed]
78. Talati, A.N.; Hackney, D.N.; Mesiano, S. Pathophysiology of Preterm Labor with Intact Membranes. *Semin. Perinatol.* **2017**, *41*, 420–426. [CrossRef]

79. Keelan, J.A. Intrauterine Inflammatory Activation, Functional Progesterone Withdrawal, and the Timing of Term and Preterm Birth. *J. Reprod. Immunol.* **2018**, *125*, 89–99. [CrossRef] [PubMed]
80. Nadeau-Vallée, M.; Obari, D.; Palacios, J.; Brien, M.-È.; Duval, C.; Chemtob, S.; Girard, S. Sterile Inflammation and Pregnancy Complications: A Review. *Reproduction* **2016**, *152*, R277–R292. [CrossRef]
81. Menon, R.; Bonney, E.A.; Condon, J.; Mesiano, S.; Taylor, R.N. Novel Concepts on Pregnancy Clocks and Alarms: Redundancy and Synergy in Human Parturition. *Hum. Reprod. Update* **2016**, *22*, 535–560. [CrossRef]
82. Shynlova, O.; Nadeem, L.; Zhang, J.; Dunk, C.; Lye, S. Myometrial Activation: Novel Concepts Underlying Labor. *Placenta* **2020**, *92*, 28–36. [CrossRef] [PubMed]
83. Keelan, J.A.; Marvin, K.W.; Sato, T.A.; Coleman, M.; McCowan, L.M.E.; Mitchell, M.D. Cytokine Abundance in Placental Tissues: Evidence of Inflammatory Activation in Gestational Membranes with Term and Preterm Parturition. *Am. J. Obstet. Gynecol.* **1999**, *181*, 1530–1536. [CrossRef]
84. Kemp, B.; Menon, R.; Fortunato, S.J.; Winkler, M.; Maul, H.; Rath, W. Quantitation and Localization of Inflammatory Cytokines Interleukin-6 and Interleukin-8 in the Lower Uterine Segment During Cervical Dilatation. *J. Assist. Reprod. Genet.* **2002**, *19*, 215–219. [CrossRef] [PubMed]
85. Osman, I.; Young, A.; Ledingham, M.A.; Thomson, A.J.; Jordan, F.; Greer, I.A.; Norman, J.E. Leukocyte Density and Pro-Inflammatory Cytokine Expression in Human Fetal Membranes, Decidua, Cervix and Myometrium before and during Labour at Term. *Mol. Hum. Reprod.* **2003**, *9*, 41–45. [CrossRef]
86. Sennström, M.B.; Ekman, G.; Westergren-Thorsson, G.; Malmström, A.; Byström, B.; Endrésen, U.; Mlambo, N.; Norman, M.; Ståbi, B.; Brauner, A. Human Cervical Ripening, an Inflammatory Process Mediated by Cytokines. *Mol. Hum. Reprod.* **2000**, *6*, 375–381. [CrossRef]
87. Singh, N.; Herbert, B.; Sooranna, G.; Shah, N.M.; Das, A.; Sooranna, S.R.; Johnson, M.R. Is There an Inflammatory Stimulus to Human Term Labour? *PLoS ONE* **2021**, *16*, e0256545. [CrossRef]
88. Haddad, R.; Tromp, G.; Kuivaniemi, H.; Chaiworapongsa, T.; Kim, Y.M.; Mazor, M.; Romero, R. Human Spontaneous Labor without Histologic Chorioamnionitis Is Characterized by an Acute Inflammation Gene Expression Signature. *Am. J. Obstet. Gynecol.* **2006**, *195*, 394–405.e12. [CrossRef] [PubMed]
89. Lee, K.J.; Shim, S.H.; Kang, K.M.; Kang, J.H.; Park, D.Y.; Kim, S.H.; Farina, A.; Shim, S.S.; Cha, D.H. Global Gene Expression Changes Induced in the Human Placenta during Labor. *Placenta* **2010**, *31*, 698–704. [CrossRef] [PubMed]
90. Mittal, P.; Romero, R.; Tarca, A.L.; Gonzalez, J.; Draghici, S.; Xu, Y.; Dong, Z.; Nhan-Chang, C.-L.; Chaiworapongsa, T.; Lye, S.; et al. Characterization of the Myometrial Transcriptome and Biological Pathways of Spontaneous Human Labor at Term. *J. Perinat. Med.* **2010**, *38*, 617–643. [CrossRef]
91. Rinaldi, S.F.; Makieva, S.; Saunders, P.T.; Rossi, A.G.; Norman, J.E. Immune Cell and Transcriptomic Analysis of the Human Decidua in Term and Preterm Parturition. *MHR Basic Sci. Reprod. Med.* **2017**, *23*, 708–724. [CrossRef]
92. Stephen, G.L.; Lui, S.; Hamilton, S.A.; Tower, C.L.; Harris, L.K.; Stevens, A.; Jones, R.L. Transcriptomic Profiling of Human Choriodecidua During Term Labor: Inflammation as a Key Driver of Labor. *Am. J. Reprod. Immunol.* **2015**, *73*, 36–55. [CrossRef] [PubMed]
93. Kim, S.H.; MacIntyre, D.A.; Firmino Da Silva, M.; Blanks, A.M.; Lee, Y.S.; Thornton, S.; Bennett, P.R.; Terzidou, V. Oxytocin Activates NF-KB-Mediated Inflammatory Pathways in Human Gestational Tissues. *Mol. Cell. Endocrinol.* **2015**, *403*, 64–77. [CrossRef]
94. Adams Waldorf, K.M.; Singh, N.; Mohan, A.R.; Young, R.C.; Ngo, L.; Das, A.; Tsai, J.; Bansal, A.; Paolella, L.; Herbert, B.R.; et al. Uterine Overdistention Induces Preterm Labor Mediated by Inflammation: Observations in Pregnant Women and Nonhuman Primates. *Am. J. Obstet. Gynecol.* **2015**, *213*, 830.e1–830.e19. [CrossRef] [PubMed]
95. Dajani, N.; Idriss, E.; Collins, P.L. Interleukin-6 Does Not Stimulate Rat Myometrial Contractions in an In Vitro Model. *Am. J. Reprod. Immunol.* **1994**, *32*, 248–254. [CrossRef] [PubMed]
96. Yoshimura, K., Hirsch, E. Interleukin-6 Is Neither Necessary Nor Sufficient for Preterm Labor in a Murine Infection Model. *J. Soc. Gynecol. Investig.* **2003**, *10*, 423–427. [CrossRef] [PubMed]
97. Osmers, R. Interleukin-8 Synthesis and the Onset of Labor. *Obstet. Gynecol.* **1995**, *86*, 223–229. [CrossRef]
98. Winkler, M.; Fischer, D.C.; Hlubek, M.; van de Leur, E.; Haubeck, H.D.; Rath, W. Interleukin-1beta and Interleukin-8 Concentrations in the Lower Uterine Segment during Parturition at Term. *Obstet. Gynecol.* **1998**, *91*, 945–949. [CrossRef]
99. el Maradny, E.; Kanayama, N.; Maehara, K.; Kobayashi, T.; Terao, T. Expression of Interleukin-8 Receptors in the Gestational Tissues before and after Initiation of Labor: Immunohistochemical Study. *Acta Obstet. Gynecol. Scand.* **1996**, *75*, 790–796. [CrossRef]
100. Hamilton, S.A.; Tower, C.L.; Jones, R.L. Identification of Chemokines Associated with the Recruitment of Decidual Leukocytes in Human Labour: Potential Novel Targets for Preterm Labour. *PLoS ONE* **2013**, *8*, e56946. [CrossRef]
101. Willems, J.; Joniau, M.; Cinque, S.; van Damme, J. Human Granulocyte Chemotactic Peptide (IL-8) as a Specific Neutrophil Degranulator: Comparison with Other Monokines. *Immunology* **1989**, *67*, 540–542.
102. Pugin, J.; Widmer, M.C.; Kossodo, S.; Liang, C.M.; Preas, H.L., 2nd; Suffredini, A.F. Human Neutrophils Secrete Gelatinase B in Vitro and in Vivo in Response to Endotoxin and Proinflammatory Mediators. *Am. J. Respir. Cell Mol. Biol.* **1999**, *20*, 458–464. [CrossRef]

103. Winkler, M.; Fischer, D.C.; Ruck, P.; Marx, T.; Kaiserling, E.; Oberpichler, A.; Tschesche, H.; Rath, W. Parturition at Term: Parallel Increases in Interleukin-8 and Proteinase Concentrations and Neutrophil Count in the Lower Uterine Segment. *Hum. Reprod.* **1999**, *14*, 1096–1100. [CrossRef]
104. Ehsani, V.; Mortazavi, M.; Ghorban, K.; Dadmanesh, M.; Bahramabadi, R.; Rezayati, M.-T.; Javadi-Moghadam, E.; Rezaei, Z.; Sabzali, Z.; Fatemi, I.; et al. Role of Maternal Interleukin-8 (IL-8) in Normal-Term Birth in the Human. *Reprod. Fertil. Dev.* **2019**, *31*, 1049. [CrossRef] [PubMed]
105. Shahshahan, Z.; Hashemi, L.; Rasouli, O. Maternal Serum Interleukin 6 and 8 and C-Reactive Protein in Predicting the Tocolytic Therapy in Preterm Labor. *J. Res. Med. Sci.* **2014**, *19*, 537–541. [PubMed]
106. Rinaldi, S.F.; Hutchinson, J.L.; Rossi, A.G.; Norman, J.E. Anti-Inflammatory Mediators as Physiological and Pharmacological Regulators of Parturition. *Expert Rev. Clin. Immunol.* **2011**, *7*, 675–696. [CrossRef] [PubMed]
107. Spence, T.; Allsopp, P.J.; Yeates, A.J.; Mulhern, M.S.; Strain, J.J.; McSorley, E.M. Maternal Serum Cytokine Concentrations in Healthy Pregnancy and Preeclampsia. *J. Pregnancy* **2021**, 2021. [CrossRef]
108. Enninga, E.A.L.; Nevala, W.K.; Creedon, D.J.; Markovic, S.N.; Holtan, S.G. Fetal Sex-Based Differences in Maternal Hormones, Angiogenic Factors, and Immune Mediators during Pregnancy and the Postpartum Period. *Am. J. Reprod. Immunol.* **2015**, *73*, 251–262. [CrossRef]
109. Mitchell, A.M.; Palettas, M.; Christian, L.M. Fetal Sex Is Associated with Maternal Stimulated Cytokine Production, but Not Serum Cytokine Levels, in Human Pregnancy. *Brain. Behav. Immun.* **2017**, *60*, 32–37. [CrossRef]
110. Ramiro-Cortijo, D.; de la Calle, M.; Böger, R.; Hannemann, J.; Lüneburg, N.; López-Giménez, M.R.; Rodríguez-Rodríguez, P.; Martín-Cabrejas, M.Á.; Benítez, V.; de Pablo, Á.L.L.; et al. Male Fetal Sex Is Associated with Low Maternal Plasma Anti-Inflammatory Cytokine Profile in the First Trimester of Healthy Pregnancies. *Cytokine* **2020**, *136*, 155290. [CrossRef] [PubMed]
111. Burns, C.; Hall, S.T.; Smith, R.; Blackwell, C. Cytokine Levels in Late Pregnancy: Are Female Infants Better Protected Against Inflammation? *Front. Immunol.* **2015**, *6*, 318. [CrossRef]
112. Straczkowski, M.; Dzienis-Straczkowska, S.; Stêpieñ, A.; Kowalska, I.; Szelachowska, M.; Kinalska, I. Plasma Interleukin-8 Concentrations Are Increased in Obese Subjects and Related to Fat Mass and Tumor Necrosis Factor-Alpha System. *J. Clin. Endocrinol. Metab.* **2002**, *87*, 4602–4606. [CrossRef]
113. Yu, H.; Liu, Z.; Dong, S. Changes in Intestinal Flora, TNF-α, L-17, and IL-6 Levels in Patients with Gestational Diabetes Mellitus. *Eur. J. Inflamm.* **2018**, *16*, 205873921879355. [CrossRef]
114. Zhao, X.; Liu, J.; Shen, L.; Wang, A.; Wang, R. Correlation between Inflammatory Markers (Hs-CRP, TNF-α, IL-1β, IL-6, IL-18), Glucose Intolerance, and Gestational Diabetes Mellitus in Pregnant Women. *Int. J. Clin. Exp. Med.* **2018**, *11*, 8310–8316.
115. Zhang, J.; Chi, H.; Xiao, H.; Tian, X.; Wang, Y.; Yun, X.; Xu, Y. Interleukin 6 (IL-6) and Tumor Necrosis Factor α (TNF-α) Single Nucleotide Polymorphisms (SNPs), Inflammation and Metabolism in Gestational Diabetes Mellitus in Inner Mongolia. *Med. Sci. Monit.* **2017**, *23*, 4149–4157. [CrossRef] [PubMed]
116. Quenby, S.; Gallos, I.D.; Dhillon-Smith, R.K.; Podesek, M.; Stephenson, M.D.; Fisher, J.; Brosens, J.J.; Brewin, J.; Ramhorst, R.; Lucas, E.S.; et al. Miscarriage Matters: The Epidemiological, Physical, Psychological, and Economic Costs of Early Pregnancy Loss. *Lancet* **2021**, *397*, 1658–1667. [CrossRef] [PubMed]
117. Wang, X.; Chen, C.; Wang, L.; Chen, D.; Guang, W.; French, J. Conception, Early Pregnancy Loss, and Time to Clinical Pregnancy: A Population-Based Prospective Study. *Fertil. Steril.* **2003**, *79*, 577–584. [CrossRef] [PubMed]
118. Wilcox, A.J.; Weinberg, C.R.; O'Connor, J.F.; Baird, D.D.; Schlatterer, J.P.; Canfield, R.E.; Armstrong, E.G.; Nisula, B.C. Incidence of Early Loss of Pregnancy. *N. Engl. J. Med.* **1988**, *319*, 189–194. [CrossRef] [PubMed]
119. Rai, R.; Regan, L. Recurrent Miscarriage. *Lancet* **2006**, *368*, 601–611. [CrossRef]
120. The Practice Committee of the American Society for Reproductive Medicine. Evaluation and Treatment of Recurrent Pregnancy Loss: A Committee Opinion. *Fertil. Steril.* **2012**, *98*, 1103–1111. [CrossRef]
121. Krieg, S.A.; Fan, X.; Hong, Y.; Sang, Q.-X.; Giaccia, A.; Westphal, L.M.; Lathi, R.B.; Krieg, A.J.; Nayak, N.R. Global Alteration in Gene Expression Profiles of Deciduas from Women with Idiopathic Recurrent Pregnancy Loss. *MHR Basic Sci. Reprod. Med.* **2012**, *18*, 442–450. [CrossRef] [PubMed]
122. Qian, J.; Zhang, N.; Lin, J.; Wang, C.; Pan, X.; Chen, L.; Li, D.; Wang, L. Distinct Pattern of Th17/Treg Cells in Pregnant Women with a History of Unexplained Recurrent Spontaneous Abortion. *Biosci. Trends* **2018**, *12*, 157–167. [CrossRef]
123. Zhao, L.; Han, L.; Hei, G.; Wei, R.; Zhang, Z.; Zhu, X.; Guo, Q.; Chu, C.; Fu, X.; Xu, K.; et al. Diminished MiR-374c-5p Negatively Regulates IL (Interleukin)-6 in Unexplained Recurrent Spontaneous Abortion. *J. Mol. Med.* **2022**, *100*, 1043–1056. [CrossRef]
124. Chen, P.; Zhou, L.; Chen, J.; Lu, Y.; Cao, C.; Lv, S.; Wei, Z.; Wang, L.; Chen, J.; Hu, X.; et al. The Immune Atlas of Human Deciduas With Unexplained Recurrent Pregnancy Loss. *Front. Immunol.* **2021**, *12*, 689019. [CrossRef] [PubMed]
125. Wang, F.; Jia, W.; Fan, M.; Shao, X.; Li, Z.; Liu, Y.; Ma, Y.; Li, Y.-X.; Li, R.; Tu, Q.; et al. Single-Cell Immune Landscape of Human Recurrent Miscarriage. *Genom. Proteom. Bioinform.* **2021**, *19*, 208–222. [CrossRef] [PubMed]
126. Madhappan, B.; Kempuraj, D.; Christodoulou, S.; Tsapikidis, S.; Boucher, W.; Karagiannis, V.; Athanassiou, A.; Theoharides, T.C. High Levels of Intrauterine Corticotropin-Releasing Hormone, Urocortin, Tryptase, and Interleukin-8 in Spontaneous Abortions. *Endocrinology* **2003**, *144*, 2285–2290. [CrossRef] [PubMed]
127. Laisk, T.; Soares, A.L.G.; Ferreira, T.; Painter, J.N.; Censin, J.C.; Laber, S.; Bacelis, J.; Chen, C.-Y.; Lepamets, M.; Lin, K.; et al. The Genetic Architecture of Sporadic and Multiple Consecutive Miscarriage. *Nat. Commun.* **2020**, *11*, 5980. [CrossRef] [PubMed]

128. Calleja-Agius, J.; Jauniaux, E.; Pizzey, A.R.; Muttukrishna, S. Investigation of Systemic Inflammatory Response in First Trimester Pregnancy Failure. *Hum. Reprod.* **2012**, *27*, 349–357. [CrossRef] [PubMed]
129. Drozdzik, M.; Szlarb, N.; Kurzawski, M. Interleukin-6 Level and Gene Polymorphism in Spontaneous Miscarriage. *Tissue Antigens* **2013**, *82*, 171–176. [CrossRef] [PubMed]
130. Thaker, R.; Oza, H.; Verma, V.; Gor, M.; Kumar, S. The Association of Circulatory Cytokines (IL-6 and IL-10) Level with Spontaneous Abortion—A Preliminary Observation. *Reprod. Sci.* **2021**, *28*, 857–864. [CrossRef]
131. Hattori, Y.; Nakanishi, T.; Ozaki, Y.; Nozawa, K.; Sato, T.; Sugiura-Ogasawara, M. Uterine Cervical Inflammatory Cytokines, Interleukin-6 and -8, as Predictors of Miscarriage in Recurrent Cases. *Am. J. Reprod. Immunol.* **2007**, *58*, 350–357. [CrossRef]
132. Luo, M.; Xiao, H.; Wang, L.; Zhao, J.; Gao, J.; Ma, W. The Expression and Clinical Significance of Three LncRNAs in Patients with a Missed Abortion. *Exp. Ther. Med.* **2020**, *21*, 8. [CrossRef]
133. AlJameil, N.; Tabassum, H.; AlMayouf, H.; Alshenefy, A.; Almohizea, M.M.; Ali, M.N. Identification of Serum Cytokines as Markers in Women with Recurrent Pregnancy Loss or Miscarriage Using MILLIPLEX Analysis. *Biomed. Res.* **2018**, *29*, 3512–3517. [CrossRef]
134. Koumantaki, Y.; Matalliotakis, I.; Sifakis, S.; Kyriakou, D.; Neonaki, M.; Goymenou, A.; Koumantakis, E. Detection of Interleukin-6, Interleukin-8, and Interleukin-11 in Plasma from Women with Spontaneous Abortion. *Eur. J. Obstet. Gynecol. Reprod. Biol.* **2001**, *98*, 66–71. [CrossRef] [PubMed]
135. Kwiatek, M.; Gęca, T.; Kwaśniewska, A. Pro- and Anti-Inflammatory Cytokines in the First Trimester—Comparison of Missed Miscarriage and Normal Pregnancy. *Int. J. Environ. Res. Public Health* **2021**, *18*, 8538. [CrossRef]
136. Tyagi, P.; Alharthi, N. Evaluation of Pro-Inflammatory Cytokine Level in Cases of Idiopathic Recurrent Spontaneous Miscarriage in Saudi Arabia. *Biomed. Biotechnol. Res. J.* **2020**, *4*, 225. [CrossRef]
137. Ozkan, Z.S.; Devecı, D.; Sımsek, M.; Ilhan, F.; Rısvanlı, A.; Sapmaz, E. What Is the Impact of SOCS3, IL-35 and IL17 in Immune Pathogenesis of Recurrent Pregnancy Loss? *J. Matern. Neonatal Med.* **2015**, *28*, 324–328. [CrossRef]
138. Arruvito, L.; Billordo, A.; Capucchio, M.; Prada, M.E.; Fainboim, L. IL-6 Trans-Signaling and the Frequency of CD4+FOXP3+ Cells in Women with Reproductive Failure. *J. Reprod. Immunol.* **2009**, *82*, 158–165. [CrossRef] [PubMed]
139. Wolf, J.; Waetzig, G.H.; Chalaris, A.; Reinheimer, T.M.; Wege, H.; Rose-John, S.; Garbers, C. Different Soluble Forms of the Interleukin-6 Family Signal Transducer Gp130 Fine-Tune the Blockade of Interleukin-6 Trans-Signaling. *J. Biol. Chem.* **2016**, *291*, 16186–16196. [CrossRef]
140. Saifi, B.; Rezaee, S.A.; Tajik, N.; Ahmadpour, M.E.; Ashrafi, M.; Vakili, R.; SoleimaniAsl, S.; Aflatoonian, R.; Mehdizadeh, M. Th17 Cells and Related Cytokines in Unexplained Recurrent Spontaneous Miscarriage at the Implantation Window. *Reprod. Biomed. Online* **2014**, *29*, 481–489. [CrossRef]
141. Ota, K.; Yamagishi, S.; Kim, M.; Dambaeva, S.; Gilman-Sachs, A.; Beaman, K.; Kwak-Kim, J. Elevation of Soluble Form of Receptor for Advanced Glycation End Products (SRAGE) in Recurrent Pregnancy Losses (RPL): Possible Participation of RAGE in RPL. *Fertil. Steril.* **2014**, *102*, 782–789. [CrossRef] [PubMed]
142. Gu, H.; Li, L.; Du, M.; Xu, H.; Gao, M.; Liu, X.; Wei, X.; Zhong, X. Key Gene and Functional Pathways Identified in Unexplained Recurrent Spontaneous Abortion Using Targeted RNA Sequencing and Clinical Analysis. *Front. Immunol.* **2021**, *12*, 717832. [CrossRef]
143. Jasper, M.J.; Tremellen, K.P.; Robertson, S.A. Reduced Expression of IL-6 and IL-1α MRNAs in Secretory Phase Endometrium of Women with Recurrent Miscarriage. *J. Reprod. Immunol.* **2007**, *73*, 74–84. [CrossRef] [PubMed]
144. Lim, K.J.; Odukoya, O.A.; Ajjan, R.A.; Li, T.-C.; Weetman, A.P.; Cooke, I.D. The Role of T-Helper Cytokines in Human Reproduction. *Fertil. Steril.* **2000**, *73*, 136–142. [CrossRef] [PubMed]
145. von Wolff, M. Regulated Expression of Cytokines in Human Endometrium throughout the Menstrual Cycle: Dysregulation in Habitual Abortion. *Mol. Hum. Reprod.* **2000**, *6*, 627–634. [CrossRef] [PubMed]
146. Banerjee, P.; Jana, S.K.; Pasricha, P.; Ghosh, S.; Chakravarty, B.; Chaudhury, K. Proinflammatory Cytokines Induced Altered Expression of Cyclooxygenase-2 Gene Results in Unreceptive Endometrium in Women with Idiopathic Recurrent Spontaneous Miscarriage. *Fertil. Steril.* **2013**, *99*, 179–187.e2. [CrossRef] [PubMed]
147. Lockwood, C.J.; Yen, C.-F.; Basar, M.; Kayisli, U.A.; Martel, M.; Buhimschi, I.; Buhimschi, C.; Huang, S.J.; Krikun, G.; Schatz, F. Preeclampsia-Related Inflammatory Cytokines Regulate Interleukin-6 Expression in Human Decidual Cells. *Am. J. Pathol.* **2008**, *172*, 1571–1579. [CrossRef] [PubMed]
148. Bernardi, F.C.B.; Felisberto, F.; Vuolo, F.; Petronilho, F.; Souza, D.R.; Luciano, T.F.; de Souza, C.T.; Ritter, C.; Dal-Pizzol, F. Oxidative Damage, Inflammation, and Toll-like Receptor 4 Pathway Are Increased in Preeclamptic Patients: A Case-Control Study. *Oxid. Med. Cell. Longev.* **2012**, *2012*, 636419. [CrossRef] [PubMed]
149. Muralimanoharan, S.; Maloyan, A.; Myatt, L. Evidence of Sexual Dimorphism in the Placental Function with Severe Preeclampsia. *Placenta* **2013**, *34*, 1183–1189. [CrossRef] [PubMed]
150. Aggarwal, R.; Jain, A.K.; Mittal, P.; Kohli, M.; Jawanjal, P.; Rath, G. Association of Pro- and Anti-Inflammatory Cytokines in Preeclampsia. *J. Clin. Lab. Anal.* **2019**, *33*, e22834. [CrossRef]
151. Ma, Y.; Ye, Y.; Zhang, J.; Ruan, C.-C.; Gao, P.-J. Immune Imbalance Is Associated with the Development of Preeclampsia. *Medicine* **2019**, *98*, e15080. [CrossRef]
152. Pang, Z.-J.; Xing, F.-Q. Comparative Study on the Expression of Cytokine—Receptor Genes in Normal and Preeclamptic Human Placentas Using DNA Microarrays. *J. Perinat. Med.* **2003**, *31*, 153–162. [CrossRef]

153. Sun, L.; Mao, D.; Cai, Y.; Tan, W.; Hao, Y.; Li, L.; Liu, W. Association between Higher Expression of Interleukin-8 (IL-8) and Haplotype −353A/−251A/+678T of IL-8 Gene with Preeclampsia. *Medicine* **2016**, *95*, e5537. [CrossRef]
154. Casart, Y.C.; Tarrazzi, K.; Camejo, M.I. Serum Levels of Interleukin-6, Interleukin-1β and Human Chorionic Gonadotropin in Pre-Eclamptic and Normal Pregnancy. *Gynecol. Endocrinol.* **2007**, *23*, 300–303. [CrossRef]
155. Lau, S.Y.; Guild, S.-J.; Barrett, C.J.; Chen, Q.; McCowan, L.; Jordan, V.; Chamley, L.W. Tumor Necrosis Factor-Alpha, Interleukin-6, and Interleukin-10 Levels Are Altered in Preeclampsia: A Systematic Review and Meta-Analysis. *Am. J. Reprod. Immunol.* **2013**, *70*, 412–427. [CrossRef] [PubMed]
156. Pinheiro, M.B.; Martins-Filho, O.A.; Mota, A.P.L.; Alpoim, P.N.; Godoi, L.C.; Silveira, A.C.O.; Teixeira-Carvalho, A.; Gomes, K.B.; Dusse, L.M. Severe Preeclampsia Goes along with a Cytokine Network Disturbance towards a Systemic Inflammatory State. *Cytokine* **2013**, *62*, 165–173. [CrossRef] [PubMed]
157. Krasnyi, A.M.; Gracheva, M.I.; Sadekova, A.A.; Vtorushina, V.V.; Balashov, I.S.; Kan, N.E.; Borovikov, P.I.; Krechetova, L.V.; Tyutyunnik, V.L. Complex Analysis of Total and Fetal DNA and Cytokines in Blood Plasma of Pregnant Women with Preeclampsia. *Bull. Exp. Biol. Med.* **2018**, *164*, 721–725. [CrossRef] [PubMed]
158. Negi, R.; Haritha, V.; Aziz, N.; Siddiqui, A.H. Biochemical Markers in the Pathogenesis of Preeclampsia: Novel Link between Placental Growth Factor and Interleukin-6. *Mol. Cell. Biochem.* **2022**, *477*, 1765–1774. [CrossRef] [PubMed]
159. Kauma, S.; Takacs, P.; Scordalakes, C.; Walsh, S.; Green, K.; Peng, T. Increased Endothelial Monocyte Chemoattractant Protein-1 and Interleukin-8 in Preeclampsia. *Obstet. Gynecol.* **2002**, *100*, 706–714. [CrossRef]
160. Molvarec, A.; Szarka, A.; Walentin, S.; Bekő, G.; Karádi, I.; Prohászka, Z.; Rigó, J. Serum Leptin Levels in Relation to Circulating Cytokines, Chemokines, Adhesion Molecules and Angiogenic Factors in Normal Pregnancy and Preeclampsia. *Reprod. Biol. Endocrinol.* **2011**, *9*, 124. [CrossRef]
161. Moreno-Eutimio, M.A.; Tovar-Rodríguez, J.M.; Vargas-Avila, K.; Nieto-Velázquez, N.G.; Frías-De-León, M.G.; Sierra-Martinez, M.; Acosta-Altamirano, G. Increased Serum Levels of Inflammatory Mediators and Low Frequency of Regulatory T Cells in the Peripheral Blood of Preeclamptic Mexican Women. *Biomed Res. Int.* **2014**, *2014*, 413249. [CrossRef] [PubMed]
162. Sahin, H.; Gunel, T.; Benian, A.; Onay Ucar, E.; Guralp, O.; Kilic, A. Genomic and Proteomic Investigation of Preeclampsia. *Exp. Ther. Med.* **2015**, *10*, 711–716. [CrossRef] [PubMed]
163. Yu, J.; Zhou, Y.; Gui, J.; Li, A.-Z.; Su, X.-L.; Feng, L. Assessment of the Number and Function of Macrophages in the Placenta of Gestational Diabetes Mellitus Patients. *J. Huazhong Univ. Sci. Technol. Med. Sci.* **2013**, *33*, 725–729. [CrossRef]
164. Bari, M.F.; Weickert, M.O.; Sivakumar, K.; James, S.G.; Snead, D.R.J.; Tan, B.K.; Randeva, H.S.; Bastie, C.C.; Vatish, M. Elevated Soluble CD163 in Gestational Diabetes Mellitus: Secretion from Human Placenta and Adipose Tissue. *PLoS ONE* **2014**, *9*, e101327. [CrossRef] [PubMed]
165. Stirm, L.; Kovářová, M.; Perschbacher, S.; Michlmaier, R.; Fritsche, L.; Siegel-Axel, D.; Schleicher, E.; Peter, A.; Pauluschke-Fröhlich, J.; Brucker, S.; et al. BMI-Independent Effects of Gestational Diabetes on Human Placenta. *J. Clin. Endocrinol. Metab.* **2018**, *103*, 3299–3309. [CrossRef] [PubMed]
166. Lacroix, M.; Lizotte, F.; Hivert, M.-F.; Geraldes, P.; Perron, P. Calcifediol Decreases Interleukin-6 Secretion by Cultured Human Trophoblasts From GDM Pregnancies. *J. Endocr. Soc.* **2019**, *3*, 2165–2178. [CrossRef] [PubMed]
167. Hara, C.d.C.P.; França, E.L.; Fagundes, D.L.G.; de Queiroz, A.A.; Rudge, M.V.C.; Honorio-França, A.C.; Calderon, I.d.M.P. Characterization of Natural Killer Cells and Cytokines in Maternal Placenta and Fetus of Diabetic Mothers. *J. Immunol. Res.* **2016**, *2016*, 7154524. [CrossRef]
168. Mrizak, I.; Grissa, O.; Henault, B.; Fekih, M.; Bouslema, A.; Boumaiza, I.; Zaouali, M.; Tabka, Z.; Khan, N.A. Placental Infiltration of Inflammatory Markers in Gestational Diabetic Women. *Gen. Physiol. Biophys.* **2014**, *33*, 169–176. [CrossRef]
169. Kleiblova, P.; Dostalova, I.; Bartlova, M.; Lacinova, Z.; Ticha, I.; Krejci, V.; Springer, D.; Kleibl, Z.; Haluzik, M. Expression of Adipokines and Estrogen Receptors in Adipose Tissue and Placenta of Patients with Gestational Diabetes Mellitus. *Mol. Cell. Endocrinol.* **2010**, *314*, 150–156. [CrossRef]
170. Lappas, M.; Permezel, M.; Rice, G.E. Release of Proinflammatory Cytokines and 8-Isoprostane from Placenta, Adipose Tissue, and Skeletal Muscle from Normal Pregnant Women and Women with Gestational Diabetes Mellitus. *J. Clin. Endocrinol. Metab.* **2004**, *89*, 5627–5633. [CrossRef]
171. Keckstein, S.; Pritz, S.; Amann, N.; Meister, S.; Beyer, S.; Jegen, M.; Kuhn, C.; Hutter, S.; Knabl, J.; Mahner, S.; et al. Sex Specific Expression of Interleukin 7, 8 and 15 in Placentas of Women with Gestational Diabetes. *Int. J. Mol. Sci.* **2020**, *21*, 8026. [CrossRef]
172. Kuzmicki, M.; Telejko, B.; Szamatowicz, J.; Zonenberg, A.; Nikolajuk, A.; Kretowski, A.; Gorska, M. High Resistin and Interleukin-6 Levels Are Associated with Gestational Diabetes Mellitus. *Gynecol. Endocrinol.* **2009**, *25*, 258–263. [CrossRef]
173. Hassiakos, D.; Eleftheriades, M.; Papastefanou, I.; Lambrinoudaki, I.; Kappou, D.; Lavranos, D.; Akalestos, A.; Aravantinos, L.; Pervanidou, P.; Chrousos, G. Increased Maternal Serum Interleukin-6 Concentrations at 11 to 14 Weeks of Gestation in Low Risk Pregnancies Complicated with Gestational Diabetes Mellitus: Development of a Prediction Model. *Horm. Metab. Res.* **2016**, *48*, 35–41. [CrossRef]
174. Morisset, A.-S.; Dubé, M.-C.; Côté, J.A.; Robitaille, J.; Weisnagel, S.J.; Tchernof, A. Circulating Interleukin-6 Concentrations during and after Gestational Diabetes Mellitus. *Acta Obstet. Gynecol. Scand.* **2011**, *90*, 524–530. [CrossRef]
175. Siddiqui, S.; Waghdhare, S.; Goel, C.; Panda, M.; Soneja, H.; Sundar, J.; Banerjee, M.; Jha, S.; Dubey, S. Augmentation of IL-6 Production Contributes to Development of Gestational Diabetes Mellitus: An Indian Study. *Diabetes Metab. Syndr. Clin. Res. Rev.* **2019**, *13*, 895–899. [CrossRef] [PubMed]

176. Sudharshana Murthy, K.; Bhandiwada, A.; Chandan, S.; Gowda, S.; Sindhusree, G. Evaluation of Oxidative Stress and Proinflammatory Cytokines in Gestational Diabetes Mellitus and Their Correlation with Pregnancy Outcome. *Indian J. Endocrinol. Metab.* **2018**, *22*, 79. [CrossRef]
177. Yang, Y.; Liu, L.; Liu, B.; Li, Q.; Wang, Z.; Fan, S.; Wang, H.; Wang, L. Functional Defects of Regulatory T Cell Through Interleukin 10 Mediated Mechanism in the Induction of Gestational Diabetes Mellitus. *DNA Cell Biol.* **2018**, *37*, 278–285. [CrossRef] [PubMed]
178. Winzer, C.; Wagner, O.; Festa, A.; Schneider, B.; Roden, M.; Bancher-Todesca, D.; Pacini, G.; Funahashi, T.; Kautzky-Willer, A. Plasma Adiponectin, Insulin Sensitivity, and Subclinical Inflammation in Women with Prior Gestational Diabetes Mellitus. *Diabetes Care* **2004**, *27*, 1721–1727. [CrossRef] [PubMed]
179. Li, Y.-X.; Long, D.-L.; Liu, J.; Qiu, D.; Wang, J.; Cheng, X.; Yang, X.; Li, R.-M.; Wang, G. Gestational Diabetes Mellitus in Women Increased the Risk of Neonatal Infection via Inflammation and Autophagy in the Placenta. *Medicine* **2020**, *99*, e22152. [CrossRef]
180. Gümüş, P.; Özçaka, Ö.; Ceyhan-Öztürk, B.; Akcali, A.; Lappin, D.F.; Buduneli, N. Evaluation of Biochemical Parameters and Local and Systemic Levels of Osteoactive and B-Cell Stimulatory Factors in Gestational Diabetes in the Presence or Absence of Gingivitis. *J. Periodontol.* **2015**, *86*, 387–397. [CrossRef]
181. Abell, S.K.; Shorakae, S.; Harrison, C.L.; Hiam, D.; Moreno-Asso, A.; Stepto, N.K.; De Courten, B.; Teede, H.J. The Association between Dysregulated Adipocytokines in Early Pregnancy and Development of Gestational Diabetes. *Diabetes. Metab. Res. Rev.* **2017**, *33*, e2926. [CrossRef]
182. Braga, F.O.; Negrato, C.A.; Matta, M. de F.B. da; Carneiro, J.R.I.; Gomes, M.B. Relationship between Inflammatory Markers, Glycated Hemoglobin and Placental Weight on Fetal Outcomes in Women with Gestational Diabetes. *Arch. Endocrinol. Metab.* **2019**, *63*, 22–29. [CrossRef]
183. Özyer, Ş.; Engin-Üstün, Y.; Uzunlar, Ö.; Katar, C.; Danışman, N. Inflammation and Glycemic Tolerance Status in Pregnancy: The Role of Maternal Adiposity. *Gynecol. Obstet. Investig.* **2014**, *78*, 53–58. [CrossRef]
184. Tang, M.; Luo, M.; Lu, W.; Zhang, R.; Liang, W.; Gu, J.; Yu, X.; Zhang, X.; Hu, C. Nerve Growth Factor Is Closely Related to Glucose Metabolism, Insulin Sensitivity and Insulin Secretion in the Second Trimester: A Case-Control Study in Chinese. *Nutr. Metab.* **2020**, *17*, 98. [CrossRef]
185. Tagoma, A.; Haller-Kikkatalo, K.; Oras, A.; Roos, K.; Kirss, A.; Uibo, R. Plasma Cytokines during Pregnancy Provide Insight into the Risk of Diabetes in the Gestational Diabetes Risk Group. *J. Diabetes Investig.* **2022**, *13*, 1596–1606. [CrossRef]
186. Benyo, D.F.; Miles, T.M.; Conrad, K.P. Hypoxia Stimulates Cytokine Production by Villous Explants from the Human Placenta. *J. Clin. Endocrinol. Metab.* **1997**, *82*, 1582–1588. [CrossRef] [PubMed]
187. Fortunato, S.J.; Menon, R.P.; Swan, K.F.; Menon, R. Inflammatory Cytokine (Interleukins 1, 6, and 8 and Tumor Necrosis Factor-α) Release from Cultured Human Fetal Membranes in Response to Endotoxic Lipopolysaccharide Mirrors Amniotic Fluid Concentrations. *Am. J. Obstet. Gynecol.* **1996**, *174*, 1855–1862. [CrossRef]
188. Anton, L.; Brown, A.G.; Parry, S.; Elovitz, M.A. Lipopolysaccharide Induces Cytokine Production and Decreases Extravillous Trophoblast Invasion through a Mitogen-Activated Protein Kinase-Mediated Pathway: Possible Mechanisms of First Trimester Placental Dysfunction. *Hum. Reprod.* **2012**, *27*, 61–72. [CrossRef] [PubMed]
189. Hsiao, E.Y.; Patterson, P.H. Activation of the Maternal Immune System Induces Endocrine Changes in the Placenta via IL-6. *Brain. Behav. Immun.* **2011**, *25*, 604–615. [CrossRef] [PubMed]
190. Jovanovic-Krivokuca, M.; Stefanoska, I.; Rabi, A.; Vilotic, A.; Petronijevic, M.; Vrzic-Petronijevic, S.; Radojcic, L.; Vicovac, L. MIF Is among the Proinflammatory Cytokines Increased by LPS in the Human Trophoblast Line. *Arch. Biol. Sci.* **2016**, *68*, 715–722. [CrossRef]
191. Riewe, S.D.; Mans, J.J.; Hirano, T.; Katz, J.; Shiverick, K.T.; Brown, T.A.; Lamont, R.J. Human Trophoblast Responses to Porphyromonas Gingivalis Infection. *Mol. Oral Microbiol.* **2010**, *25*, 252–259. [CrossRef]
192. Romero, R.; Miranda, J.; Kusanovic, J.P.; Chaiworapongsa, T.; Chaemsaithong, P.; Martinez, A.; Gotsch, F.; Dong, Z.; Ahmed, A.I.; Shaman, M.; et al. Clinical Chorioamnionitis at Term I: Microbiology of the Amniotic Cavity Using Cultivation and Molecular Techniques. *J. Perinat. Med.* **2015**, *43*, 19–36. [CrossRef]
193. Cherouny, P.H.; Pankuch, G.A.; Romero, R.; Botti, J.J.; Kuhn, D.C.; Demers, L.M.; Appelbaum, P.C. Neutrophil Attractant/Activating Peptide-1/Interleukin-8: Association with Histologic Chorioamnionitis, Preterm Delivery, and Bioactive Amniotic Fluid Leukoattractants. *Am. J. Obstet. Gynecol.* **1993**, *169*, 1299–1303. [CrossRef]
194. Holst, R.-M.; Mattsby-Baltzer, I.; Wennerholm, U.-B.; Hagberg, H.; Jacobsson, B. Interleukin-6 and Interleukin-8 in Cervical Fluid in a Population of Swedish Women in Preterm Labor: Relationship to Microbial Invasion of the Amniotic Fluid, Intra-Amniotic Inflammation, and Preterm Delivery. *Acta Obstet. Gynecol. Scand.* **2005**, *84*, 551–557. [CrossRef]
195. Gomez-Lopez, N.; Hernandez-Santiago, S.; Lobb, A.P.; Olson, D.M.; Vadillo-Ortega, F. Normal and Premature Rupture of Fetal Membranes at Term Delivery Differ in Regional Chemotactic Activity and Related Chemokine/Cytokine Production. *Reprod. Sci.* **2013**, *20*, 276–284. [CrossRef] [PubMed]
196. Agrawal, V.; Hirsch, E. Intrauterine Infection and Preterm Labor. *Semin. Fetal Neonatal Med.* **2012**, *17*, 12–19. [CrossRef]
197. Leaños-Miranda, A.; Nolasco-Leaños, A.G.; Carrillo-Juárez, R.I.; Molina-Pérez, C.J.; Isordia-Salas, I.; Ramírez-Valenzuela, K.L. Interleukin-6 in Amniotic Fluid: A Reliable Marker for Adverse Outcomes in Women in Preterm Labor and Intact Membranes. *Fetal Diagn. Ther.* **2021**, *48*, 313–320. [CrossRef] [PubMed]

198. Kacerovsky, M.; Drahosova, M.; Hornychova, H.; Pliskova, L.; Bolehovska, R.; Forstl, M.; Tosner, J.; Andrys, C. Value of Amniotic Fluid Interleukin-8 for the Prediction of Histological Chorioamnionitis in Preterm Premature Rupture of Membranes. *Neuro Endocrinol. Lett.* **2009**, *30*, 733–738.
199. Yoon, B.H.; Romero, R.; Moon, J.B.; Shim, S.S.; Kim, M.; Kim, G.; Jun, J.K. Clinical Significance of Intra-Amniotic Inflammation in Patients with Preterm Labor and Intact Membranes. *Am. J. Obstet. Gynecol.* **2001**, *185*, 1130–1136. [CrossRef] [PubMed]
200. Yoneda, S.; Shiozaki, A.; Ito, M.; Yoneda, N.; Inada, K.; Yonezawa, R.; Kigawa, M.; Saito, S. Accurate Prediction of the Stage of Histological Chorioamnionitis before Delivery by Amniotic Fluid IL-8 Level. *Am. J. Reprod. Immunol.* **2015**, *73*, 568–576. [CrossRef]
201. Brown, M.A.; Magee, L.A.; Kenny, L.C.; Karumanchi, S.A.; McCarthy, F.P.; Saito, S.; Hall, D.R.; Warren, C.E.; Adoyi, G.; Ishaku, S. Hypertensive Disorders of Pregnancy. *Hypertension* **2018**, *72*, 24–43. [CrossRef]
202. Karrar, S.A.; Hong, P.L. *Preeclampsia;* StatPearls Publishing: Tampa, FL, USA, 2022.
203. Aneman, I.; Pienaar, D.; Suvakov, S.; Simic, T.P.; Garovic, V.D.; McClements, L. Mechanisms of Key Innate Immune Cells in Early- and Late-Onset Preeclampsia. *Front. Immunol.* **2020**, *11*, 01864. [CrossRef]
204. Masini, G.; Foo, L.F.; Tay, J.; Wilkinson, I.B.; Valensise, H.; Gyselaers, W.; Lees, C.C. Preeclampsia Has Two Phenotypes Which Require Different Treatment Strategies. *Am. J. Obstet. Gynecol.* **2022**, *226*, S1006–S1018. [CrossRef]
205. Valencia-Ortega, J.; Zárate, A.; Saucedo, R.; Hernández-Valencia, M.; Cruz, J.G.; Puello, E. Placental Proinflammatory State and Maternal Endothelial Dysfunction in Preeclampsia. *Gynecol. Obstet. Investig.* **2019**, *84*, 12–19. [CrossRef] [PubMed]
206. Geldenhuys, J.; Rossouw, T.M.; Lombaard, H.A.; Ehlers, M.M.; Kock, M.M. Disruption in the Regulation of Immune Responses in the Placental Subtype of Preeclampsia. *Front. Immunol.* **2018**, *9*, 01659. [CrossRef] [PubMed]
207. Rana, S.; Lemoine, E.; Granger, J.P.; Karumanchi, S.A. Preeclampsia: Pathophysiology, Challenges, and Perspectives. *Circ. Res.* **2019**, *124*, 1094–1112. [CrossRef]
208. Opichka, M.A.; Rappelt, M.W.; Gutterman, D.D.; Grobe, J.L.; McIntosh, J.J. Vascular Dysfunction in Preeclampsia. *Cells* **2021**, *10*, 3055. [CrossRef]
209. Wang, Y.; Lewis, D.F.; Gu, Y.; Zhao, S.; Groome, L.J. Elevated Maternal Soluble Gp130 and IL-6 Levels and Reduced Gp130 and SOCS-3 Expressions in Women Complicated with Preeclampsia. *Hypertension* **2011**, *57*, 336–342. [CrossRef]
210. Wang, Y.; Gu, Y.; Alexander, J.S.; Lewis, D.F. Preeclampsia Status Controls Interleukin-6 and Soluble IL-6 Receptor Release from Neutrophils and Endothelial Cells: Relevance to Increased Inflammatory Responses. *Pathophysiology* **2021**, *28*, 202–211. [CrossRef]
211. Kuźmicki, M.; Telejko, B.; Lipińska, D.; Pliszka, J.; Wilk, J.; Wawrusiewicz-Kurylonek, N.; Zielińska, A.; Sobota, A.; Krętowski, A.; Górska, M.; et al. Stężenie Interleukiny-6, Receptora Dla Interleukiny-6 i Glikoproteiny 130 Oraz Cytokin Zależnych Od Limfocytów Th17 u Pacjentek z Cukrzycą Ciążową. *Endokrynol. Pol.* **2014**, *65*, 169–175. [CrossRef]
212. Rebouissou, C.; Wijdenes, J.; Autissier, P.; Tarte, K.; Costes, V.; Liautard, J.; Rossi, J.F.; Brochier, J.; Klein, B. A Gp130 Interleukin-6 Transducer-Dependent SCID Model of Human Multiple Myeloma. *Blood* **1998**, *91*, 4727–4737. [CrossRef] [PubMed]
213. Chen, L.M.; Liu, B.; Zhao, H.B.; Stone, P.; Chen, Q.; Chamley, L. IL-6, TNFalpha and TGFbeta Promote Nonapoptotic Trophoblast Deportation and Subsequently Causes Endothelial Cell Activation. *Placenta* **2010**, *31*, 75–80. [CrossRef]
214. Faas, M.M.; Spaans, F.; De Vos, P. Monocytes and Macrophages in Pregnancy and Pre-Eclampsia. *Front. Immunol.* **2014**, *5*, 298. [CrossRef]
215. Ning, F.; Liu, H.; Lash, G.E. The Role of Decidual Macrophages During Normal and Pathological Pregnancy. *Am. J. Reprod. Immunol.* **2016**, *75*, 298–309. [CrossRef]
216. Michalczyk, M.; Celewicz, A.; Celewicz, M.; Woźniakowska-Gondek, P.; Rzepka, R. The Role of Inflammation in the Pathogenesis of Preeclampsia. *Mediat. Inflamm.* **2020**, *2020*, 3864941. [CrossRef]
217. Kimura, A.; Kishimoto, T. IL-6: Regulator of Treg/Th17 Balance. *Eur. J. Immunol.* **2010**, *40*, 1830–1835. [CrossRef]
218. Vargas-Rojas, M.I.; Solleiro-Villavicencio, H.; Soto-Vega, E. Th1, Th2, Th17 and Treg Levels in Umbilical Cord Blood in Preeclampsia. *J. Matern. Fetal. Neonatal Med.* **2016**, *29*, 1642–1645. [CrossRef] [PubMed]
219. Hosseini, A.; Dolati, S.; Hashemi, V.; Abdollahpour-Alitappeh, M.; Yousefi, M. Regulatory T and T Helper 17 Cells: Their Roles in Preeclampsia. *J. Cell. Physiol.* **2018**, *233*, 6561–6573. [CrossRef] [PubMed]
220. Hoeltzenbein, M.; Beck, E.; Rajwanshi, R.; Gøtestam Skorpen, C.; Berber, E.; Schaefer, C.; Østensen, M. Tocilizumab Use in Pregnancy: Analysis of a Global Safety Database Including Data from Clinical Trials and Post-Marketing Data. *Semin. Arthritis Rheum.* **2016**, *46*, 238–245. [CrossRef]
221. Jorgensen, S.C.J.; Lapinsky, S.E. Tocilizumab for Coronavirus Disease 2019 in Pregnancy and Lactation: A Narrative Review. *Clin. Microbiol. Infect.* **2022**, *28*, 51–57. [CrossRef] [PubMed]
222. Bellos, I.; Karageorgiou, V.; Kapnias, D.; Karamanli, K.-E.; Siristatidis, C. The Role of Interleukins in Preeclampsia: A Comprehensive Review. *Am. J. Reprod. Immunol.* **2018**, *80*, e13055. [CrossRef] [PubMed]
223. Gardiner, C.; Vatish, M. Impact of Haemostatic Mechanisms on Pathophysiology of Preeclampsia. *Thromb. Res.* **2017**, *151*, S48–S52. [CrossRef] [PubMed]
224. Giaglis, S.; Stoikou, M.; Grimolizzi, F.; Subramanian, B.Y.; van Breda, S.V.; Hoesli, I.; Lapaire, O.; Hasler, P.; Than, N.G.; Hahn, S. Neutrophil Migration into the Placenta: Good, Bad or Deadly? *Cell Adh. Migr.* **2016**, *10*, 208–225. [CrossRef]
225. Broere-Brown, Z.A.; Schalekamp-Timmermans, S.; Hofman, A.; Jaddoe, V.; Steegers, E. Fetal Sex Dependency of Maternal Vascular Adaptation to Pregnancy: A Prospective Population-Based Cohort Study. *BJOG* **2016**, *123*, 1087–1095. [CrossRef]

226. Jaskolka, D.; Retnakaran, R.; Zinman, B.; Kramer, C.K. Fetal Sex and Maternal Risk of Pre-Eclampsia/Eclampsia: A Systematic Review and Meta-Analysis. *BJOG* **2017**, *124*, 553–560. [CrossRef] [PubMed]
227. Braun, A.E.; Muench, K.L.; Robinson, B.G.; Wang, A.; Palmer, T.D.; Winn, V.D. Examining Sex Differences in the Human Placental Transcriptome During the First Fetal Androgen Peak. *Reprod. Sci.* **2021**, *28*, 801–818. [CrossRef]
228. Ellis, J.; Wennerholm, U.B.; Bengtsson, A.; Lilja, H.; Pettersson, A.; Sultan, B.; Wennergren, M.; Hagberg, H. Levels of Dimethylarginines and Cytokines in Mild and Severe Preeclampsia. *Acta Obstet. Gynecol. Scand.* **2001**, *80*, 602–608. [PubMed]
229. Ouyang, Y.-Q.; Li, S.-J.; Zhang, Q.; Cai, H.-B.; Chen, H.-P. Interactions between Inflammatory and Oxidative Stress in Preeclampsia. *Hypertens. Pregnancy* **2009**, *28*, 56–62. [CrossRef]
230. Guven, M.A.; Coskun, A.; Ertas, I.E.; Aral, M.; Zencırcı, B.; Oksuz, H. Association of Maternal Serum CRP, IL-6, TNF-α, Homocysteine, Folic Acid and Vitamin B12 Levels with the Severity of Preeclampsia and Fetal Birth Weight. *Hypertens. Pregnancy* **2009**, *28*, 190–200. [CrossRef]
231. Tosun, M.; Celik, H.; Avci, B.; Yavuz, E.; Alper, T.; Malatyalioğlu, E. Maternal and Umbilical Serum Levels of Interleukin-6, Interleukin-8, and Tumor Necrosis Factor-α in Normal Pregnancies and in Pregnancies Complicated by Preeclampsia. *J. Matern. Neonatal Med.* **2010**, *23*, 880–886. [CrossRef] [PubMed]
232. Xiao, J.P.; Yin, Y.X.; Gao, Y.F.; Lau, S.; Shen, F.; Zhao, M.; Chen, Q. The Increased Maternal Serum Levels of IL-6 Are Associated with the Severity and Onset of Preeclampsia. *Cytokine* **2012**, *60*, 856–860. [CrossRef]
233. Ozler, A.; Turgut, A.; Sak, M.E.; Evsen, M.S.; Soydinc, H.E.; Evliyaoglu, O.; Gul, T. Serum Levels of Neopterin, Tumor Necrosis Factor-Alpha and Interleukin-6 in Preeclampsia: Relationship with Disease Severity. *Eur. Rev. Med. Pharmacol. Sci.* **2012**, *16*, 1707–1712.
234. Ovayolu, A.; Turksoy, V.A.; Ovayolu, G.; Ozek, M.A.; Dogan, I.; Karaman, E. Analyses of Interleukin-6, Presepsin and Pentraxin-3 in the Diagnosis and Severity of Late-Onset Preeclampsia. *J. Matern. Neonatal Med.* **2022**, *35*, 299–307. [CrossRef]
235. Küçük, M.; Sezer, S.D.; Yenisey, Ç.; Yüksel, H.; Odabaşı, A.R. Comparison of Interleukin-6 Levels in Maternal and Umbilical Cord Blood in Early- and Late-Onset Preeclampsia. *Gynecol. Endocrinol.* **2012**, *28*, 640–643. [CrossRef]
236. Wang, Y.; Shi, D.; Chen, L. Lipid Profile and Cytokines in Hypertension of Pregnancy: A Comparison of Preeclampsia Therapies. *J. Clin. Hypertens.* **2018**, *20*, 394–399. [CrossRef]
237. El-Mikkawy, D.M.E.; EL-Sadek, M.A.; EL-Badawy, M.A.; Samaha, D. Circulating Level of Interleukin-6 in Relation to Body Mass Indices and Lipid Profile in Egyptian Adults with Overweight and Obesity. *Egypt. Rheumatol. Rehabil.* **2020**, *47*, 7. [CrossRef]
238. Kim, O.Y.; Chae, J.S.; Paik, J.K.; Seo, H.S.; Jang, Y.; Cavaillon, J.-M.; Lee, J.H. Effects of Aging and Menopause on Serum Interleukin-6 Levels and Peripheral Blood Mononuclear Cell Cytokine Production in Healthy Nonobese Women. *Age* **2012**, *34*, 415–425. [CrossRef] [PubMed]
239. Naugler, W.E.; Karin, M. The Wolf in Sheep's Clothing: The Role of Interleukin-6 in Immunity, Inflammation and Cancer. *Trends Mol. Med.* **2008**, *14*, 109–119. [CrossRef] [PubMed]
240. American Diabetes Association 2. Classification and Diagnosis of Diabetes: Standards of Medical Care in Diabetes-2021. *Diabetes Care* **2021**, *44*, S15–S33. [CrossRef] [PubMed]
241. Angueira, A.R.; Ludvik, A.E.; Reddy, T.E.; Wicksteed, B.; Lowe, W.L.; Layden, B.T. New Insights Into Gestational Glucose Metabolism: Lessons Learned From 21st Century Approaches. *Diabetes* **2015**, *64*, 327–334. [CrossRef]
242. Goldstein, R.F.; Abell, S.K.; Ranasinha, S.; Misso, M.; Boyle, J.A.; Black, M.H.; Li, N.; Hu, G.; Corrado, F.; Rode, L.; et al. Association of Gestational Weight Gain with Maternal and Infant Outcomes. *JAMA* **2017**, *317*, 2207. [CrossRef]
243. Immanuel, J.; Simmons, D. Screening and Treatment for Early-Onset Gestational Diabetes Mellitus: A Systematic Review and Meta-Analysis. *Curr. Diabetes Rep.* **2017**, *17*, 115. [CrossRef]
244. Moon, J.H.; Jang, H.C. Gestational Diabetes Mellitus: Diagnostic Approaches and Maternal-Offspring Complications. *Diabetes Metab. J.* **2022**, *46*, 3–14. [CrossRef]
245. Kc, K.; Shakya, S.; Zhang, H. Gestational Diabetes Mellitus and Macrosomia: A Literature Review. *Ann. Nutr. Metab.* **2015**, *66* (Suppl. 2), 14–20. [CrossRef] [PubMed]
246. Durnwald, C. Gestational Diabetes: Linking Epidemiology, Excessive Gestational Weight Gain, Adverse Pregnancy Outcomes, and Future Metabolic Syndrome. *Semin. Perinatol.* **2015**, *39*, 254–258. [CrossRef] [PubMed]
247. Jenum, A.K.; Mørkrid, K.; Sletner, L.; Vangen, S.; Vange, S.; Torper, J.L.; Nakstad, B.; Voldner, N.; Rognerud-Jensen, O.H.; Berntsen, S.; et al. Impact of Ethnicity on Gestational Diabetes Identified with the WHO and the Modified International Association of Diabetes and Pregnancy Study Groups Criteria: A Population-Based Cohort Study. *Eur. J. Endocrinol.* **2012**, *166*, 317–324. [CrossRef]
248. Anghebem-Oliveira, M.I.; Martins, B.R.; Alberton, D.; Ramos, E.A. de S.; Picheth, G.; Rego, F.G. de M. Type 2 Diabetes-Associated Genetic Variants of FTO, LEPR, PPARg, and TCF7L2 in Gestational Diabetes in a Brazilian Population. *Arch. Endocrinol. Metab.* **2017**, *61*, 238–248. [CrossRef]
249. Plows, J.F.; Stanley, J.L.; Baker, P.N.; Reynolds, C.M.; Vickers, M.H. The Pathophysiology of Gestational Diabetes Mellitus. *Int. J. Mol. Sci.* **2018**, *19*, 3342. [CrossRef] [PubMed]
250. Di Cianni, G.; Miccoli, R.; Volpe, L.; Lencioni, C.; Del Prato, S. Intermediate Metabolism in Normal Pregnancy and in Gestational Diabetes. *Diabetes. Metab. Res. Rev.* **2003**, *19*, 259–270. [CrossRef] [PubMed]
251. Mirghani Dirar, A.; Doupis, J. Gestational Diabetes from A to Z. *World J. Diabetes* **2017**, *8*, 489–511. [CrossRef]

252. Buchanan, T.A. Pancreatic B-Cell Defects in Gestational Diabetes: Implications for the Pathogenesis and Prevention of Type 2 Diabetes. *J. Clin. Endocrinol. Metab.* **2001**, *86*, 989–993. [CrossRef]
253. Cerf, M.E. Beta Cell Dysfunction and Insulin Resistance. *Front. Endocrinol.* **2013**, *4*, 37. [CrossRef]
254. Pantham, P.; Aye, I.L.M.H.; Powell, T.L. Inflammation in Maternal Obesity and Gestational Diabetes Mellitus. *Placenta* **2015**, *36*, 709–715. [CrossRef]
255. Gregor, M.F.; Hotamisligil, G.S. Inflammatory Mechanisms in Obesity. *Annu. Rev. Immunol.* **2011**, *29*, 415–445. [CrossRef]
256. Kim, J.-H.; Bachmann, R.A.; Chen, J. Interleukin-6 and Insulin Resistance. *Vitam. Horm.* **2009**, *80*, 613–633. [CrossRef] [PubMed]
257. Amirian, A.; Mahani, M.B.; Abdi, F. Role of Interleukin-6 (IL-6) in Predicting Gestational Diabetes Mellitus. *Obstet. Gynecol. Sci.* **2020**, *63*, 407–416. [CrossRef] [PubMed]
258. Nikolajuk, A.; Kowalska, I.; Karczewska-Kupczewska, M.; Adamska, A.; Otziomek, E.; Wolczynski, S.; Kinalska, I.; Gorska, M.; Straczkowski, M. Serum Soluble Glycoprotein 130 Concentration Is Inversely Related to Insulin Sensitivity in Women with Polycystic Ovary Syndrome. *Diabetes* **2010**, *59*, 1026–1029. [CrossRef] [PubMed]
259. Zuliani, G.; Galvani, M.; Maggio, M.; Volpato, S.; Bandinelli, S.; Corsi, A.M.; Lauretani, F.; Cherubini, A.; Guralnik, J.M.; Fellin, R.; et al. Plasma Soluble Gp130 Levels Are Increased in Older Subjects with Metabolic Syndrome. The Role of Insulin Resistance. *Atherosclerosis* **2010**, *213*, 319–324. [CrossRef]
260. Lanton, T.; Levkovitch-Siany, O.; Udi, S.; Tam, J.; Abramovitch, R.; Perles, S.; Williams, E.; Rachmilewitz, J.; Mor, U.; Elinav, E.; et al. Peripheral Sgp130-Mediated Trans-Signaling Blockade Induces Obesity and Insulin Resistance in Mice via PPARα Suppression. *bioRxiv* **2020**. [CrossRef]
261. Liu, H.; Liu, A.; Kaminga, A.C.; McDonald, J.; Wen, S.W.; Pan, X. Chemokines in Gestational Diabetes Mellitus. *Front. Immunol.* **2022**, *13*, 705852. [CrossRef]
262. Michailidou, Z.; Gomez-Salazar, M.; Alexaki, V.I. Innate Immune Cells in the Adipose Tissue in Health and Metabolic Disease. *J. Innate Immun.* **2022**, *14*, 4–30. [CrossRef]
263. Richardson, A.C.; Carpenter, M.W. Inflammatory Mediators in Gestational Diabetes Mellitus. *Obstet. Gynecol. Clin. N. Am.* **2007**, *34*, 213–224. [CrossRef]
264. Waki, H.; Tontonoz, P. Endocrine Functions of Adipose Tissue. *Annu. Rev. Pathol.* **2007**, *2*, 31–56. [CrossRef]
265. Mohamed-Ali, V.; Goodrick, S.; Rawesh, A.; Katz, D.R.; Miles, J.M.; Yudkin, J.S.; Klein, S.; Coppack, S.W. Subcutaneous Adipose Tissue Releases Interleukin-6, but Not Tumor Necrosis Factor-Alpha, in Vivo. *J. Clin. Endocrinol. Metab.* **1997**, *82*, 4196–4200. [CrossRef]
266. Rehman, K.; Akash, M.S.H.; Liaqat, A.; Kamal, S.; Qadir, M.I.; Rasul, A. Role of Interleukin-6 in Development of Insulin Resistance and Type 2 Diabetes Mellitus. *Crit. Rev. Eukaryot. Gene Expr.* **2017**, *27*, 229–236. [CrossRef] [PubMed]
267. Blumberg, D.; Hochwald, S.; Brennan, M.F.; Burt, M. Interleukin-6 Stimulates Gluconeogenesis in Primary Cultures of Rat Hepatocytes. *Metabolism* **1995**, *44*, 145–146. [CrossRef]
268. Piya, M.K.; McTernan, P.G.; Kumar, S. Adipokine Inflammation and Insulin Resistance: The Role of Glucose, Lipids and Endotoxin. *J. Endocrinol.* **2013**, *216*, T1–T15. [CrossRef]
269. Kobashi, C.; Asamizu, S.; Ishiki, M.; Iwata, M.; Usui, I.; Yamazaki, K.; Tobe, K.; Kobayashi, M.; Urakaze, M. Inhibitory Effect of IL-8 on Insulin Action in Human Adipocytes via MAP Kinase Pathway. *J. Inflamm.* **2009**, *6*, 25. [CrossRef] [PubMed]
270. Williams, M.A.; Qiu, C.; Muy-Rivera, M.; Vadachkoria, S.; Song, T.; Luthy, D.A. Plasma Adiponectin Concentrations in Early Pregnancy and Subsequent Risk of Gestational Diabetes Mellitus. *J. Clin. Endocrinol. Metab.* **2004**, *89*, 2306–2311. [CrossRef] [PubMed]
271. Tretjakovs, P.; Jurka, A.; Bormane, I.; Mackevics, V.; Mikelsone, I.; Balode, L.; Reihmane, D.; Stukena, I.; Bahs, G.; Aivars, J.I.; et al. Relation of Inflammatory Chemokines to Insulin Resistance and Hypoadiponectinemia in Coronary Artery Disease Patients. *Eur. J. Intern. Med.* **2009**, *20*, 712–717. [CrossRef]
272. Urakaze, M.; Temaru, R.; Satou, A.; Yamazaki, K.; Hamazaki, T.; Kobayashi, M. The IL-8 Production in Endothelial Cells Is Stimulated by High Glucose. *Horm. Metab. Res.* **1996**, *28*, 400–401. [CrossRef]
273. Kuzmicki, M.; Telejko, B.; Wawrusiewicz-Kurylonek, N.; Citko, A.; Lipinska, D.; Pliszka, J.; Wilk, J.; Kalejta, K.; Lemancewicz, A.; Grabiec, M.; et al. The Expression of Suppressor of Cytokine Signaling 1 and 3 in Fat and Placental Tissue from Women with Gestational Diabetes. *Gynecol. Endocrinol.* **2012**, *28*, 841–844. [CrossRef]
274. Bowen, J.M.; Chamley, L.; Mitchell, M.D.; Keelan, J.A. Cytokines of the Placenta and Extra-Placental Membranes: Biosynthesis, Secretion and Roles in Establishment of Pregnancy in Women. *Placenta* **2002**, *23*, 239–256. [CrossRef]
275. Radaelli, T.; Uvena-Celebrezze, J.; Minium, J.; Huston-Presley, L.; Catalano, P.; Hauguel-de Mouzon, S. Maternal Interleukin-6: Marker of Fetal Growth and Adiposity. *J. Soc. Gynecol. Investig.* **2006**, *13*, 53–57. [CrossRef] [PubMed]
276. Desoye, G.; Hauguel-de Mouzon, S. The Human Placenta in Gestational Diabetes Mellitus. *Diabetes Care* **2007**, *30*, S120–S126. [CrossRef] [PubMed]
277. Fetita, L.-S.; Sobngwi, E.; Serradas, P.; Calvo, F.; Gautier, J.-F. Consequences of Fetal Exposure to Maternal Diabetes in Offspring. *J. Clin. Endocrinol. Metab.* **2006**, *91*, 3718–3724. [CrossRef]
278. Egan, A.M.; Dunne, F.P. *Epidemiology of Gestational and Pregestational Diabetes Mellitus*; Karger Publishers: Basel, Switzerland, 2020; pp. 1–10.
279. Song, C.; Li, J.; Leng, J.; Ma, R.C.; Yang, X. Lifestyle Intervention Can Reduce the Risk of Gestational Diabetes: A Meta-Analysis of Randomized Controlled Trials. *Obes. Rev.* **2016**, *17*, 960–969. [CrossRef]

280. Brown, A.S. The Environment and Susceptibility to Schizophrenia. *Prog. Neurobiol.* **2011**, *93*, 23–58. [CrossRef]
281. Hsiao, E.Y.; McBride, S.W.; Hsien, S.; Sharon, G.; Hyde, E.R.; McCue, T.; Codelli, J.A.; Chow, J.; Reisman, S.E.; Petrosino, J.F.; et al. Microbiota Modulate Behavioral and Physiological Abnormalities Associated with Neurodevelopmental Disorders. *Cell* **2013**, *155*, 1451–1463. [CrossRef]
282. Moretti, S.; Franchi, S.; Castelli, M.; Amodeo, G.; Somaini, L.; Panerai, A.; Sacerdote, P. Exposure of Adolescent Mice to Delta-9-Tetrahydrocannabinol Induces Long-Lasting Modulation of Pro- and Anti-Inflammatory Cytokines in Hypothalamus and Hippocampus Similar to That Observed for Peripheral Macrophages. *J. Neuroimmune Pharmacol.* **2015**, *10*, 371–379. [CrossRef] [PubMed]
283. Debost, J.-C.P.G.; Larsen, J.T.; Munk-Olsen, T.; Mortensen, P.B.; Meyer, U.; Petersen, L. Joint Effects of Exposure to Prenatal Infection and Peripubertal Psychological Trauma in Schizophrenia. *Schizophr. Bull.* **2017**, *43*, 171–179. [CrossRef]
284. Money, K.M.; Barke, T.L.; Serezani, A.; Gannon, M.; Garbett, K.A.; Aronoff, D.M.; Mirnics, K. Gestational Diabetes Exacerbates Maternal Immune Activation Effects in the Developing Brain. *Mol. Psychiatry* **2018**, *23*, 1920–1928. [CrossRef]
285. Furman, D.; Campisi, J.; Verdin, E.; Carrera-Bastos, P.; Targ, S.; Franceschi, C.; Ferrucci, L.; Gilroy, D.W.; Fasano, A.; Miller, G.W.; et al. Chronic Inflammation in the Etiology of Disease across the Life Span. *Nat. Med.* **2019**, *25*, 1822–1832. [CrossRef]
286. Mattei, D.; Pietrobelli, A. Micronutrients and Brain Development. *Curr. Nutr. Rep.* **2019**, *8*, 99–107. [CrossRef] [PubMed]
287. Han, V.X.; Patel, S.; Jones, H.F.; Nielsen, T.C.; Mohammad, S.S.; Hofer, M.J.; Gold, W.; Brilot, F.; Lain, S.J.; Nassar, N.; et al. Maternal Acute and Chronic Inflammation in Pregnancy Is Associated with Common Neurodevelopmental Disorders: A Systematic Review. *Transl. Psychiatry* **2021**, *11*, 71. [CrossRef] [PubMed]
288. Gruys, E.; Toussaint, M.J.M.; Niewold, T.A.; Koopmans, S.J. Acute Phase Reaction and Acute Phase Proteins. *J. Zhejiang Univ. Sci.* **2005**, *6*, 1045–1056. [CrossRef]
289. Patterson, P.H. Maternal Infection and Immune Involvement in Autism. *Trends Mol. Med.* **2011**, *17*, 389–394. [CrossRef]
290. Atladóttir, H.Ó.; Thorsen, P.; Østergaard, L.; Schendel, D.E.; Lemcke, S.; Abdallah, M.; Parner, E.T. Maternal Infection Requiring Hospitalization During Pregnancy and Autism Spectrum Disorders. *J. Autism Dev. Disord.* **2010**, *40*, 1423–1430. [CrossRef] [PubMed]
291. Brown, A.S.; Derkits, E.J. Prenatal Infection and Schizophrenia: A Review of Epidemiologic and Translational Studies. *Am. J. Psychiatry* **2010**, *167*, 261–280. [CrossRef]
292. Brown, A.S.; Patterson, P.H. Maternal Infection and Schizophrenia: Implications for Prevention. *Schizophr. Bull.* **2011**, *37*, 284–290. [CrossRef]
293. Lins, B. Maternal Immune Activation as a Risk Factor for Psychiatric Illness in the Context of the SARS-CoV-2 Pandemic. *Brain, Behav. Immun.-Health* **2021**, *16*, 100297. [CrossRef]
294. Brown, A.S.; Schaefer, C.A.; Wyatt, R.J.; Goetz, R.; Begg, M.D.; Gorman, J.M.; Susser, E.S. Maternal Exposure to Respiratory Infections and Adult Schizophrenia Spectrum Disorders: A Prospective Birth Cohort Study. *Schizophr. Bull.* **2000**, *26*, 287–295. [CrossRef]
295. McGrath, J.J.; Murray, R.M. Risk Factors for Schizophrenia: From Conception to Birth. In *Schizophrenia*; John Wiley & Sons, Ltd.: Hoboken, NJ, USA, 2003; pp. 232–250. ISBN 9780470987353.
296. Brown, A.S.; Begg, M.D.; Gravenstein, S.; Schaefer, C.A.; Wyatt, R.J.; Bresnahan, M.; Babulas, V.P.; Susser, E.S. Serologic Evidence of Prenatal Influenza in the Etiology of Schizophrenia. *Arch. Gen. Psychiatry* **2004**, *61*, 774. [CrossRef]
297. Brown, A.S.; Schaefer, C.A.; Quesenberry, C.P.; Liu, L.; Babulas, V.P.; Susser, E.S. Maternal Exposure to Toxoplasmosis and Risk of Schizophrenia in Adult Offspring. *Am. J. Psychiatry* **2005**, *162*, 767–773. [CrossRef] [PubMed]
298. Mortensen, P.B.; Norgaard-Pedersen, B.; Waltoft, B.L.; Sorensen, T.L.; Hougaard, D.; Yolken, R.H. Early Infections of Toxoplasma Gondii and the Later Development of Schizophrenia. *Schizophr. Bull.* **2007**, *33*, 741–744. [CrossRef] [PubMed]
299. Ellman, L.M.; Yolken, R.H.; Buka, S.L.; Torrey, E.F.; Cannon, T.D. Cognitive Functioning Prior to the Onset of Psychosis: The Role of Fetal Exposure to Serologically Determined Influenza Infection. *Biol. Psychiatry* **2009**, *65*, 1040–1047. [CrossRef] [PubMed]
300. Zawadzka, A.; Cieślik, M.; Adamczyk, A. The Role of Maternal Immune Activation in the Pathogenesis of Autism: A Review of the Evidence, Proposed Mechanisms and Implications for Treatment. *Int. J. Mol. Sci.* **2021**, *22*, 11516. [CrossRef]
301. Samuelsson, A.-M.; Jennische, E.; Hansson, H.-A.; Holmäng, A. Prenatal Exposure to Interleukin-6 Results in Inflammatory Neurodegeneration in Hippocampus with NMDA/GABA $_A$ Dysregulation and Impaired Spatial Learning. *Am. J. Physiol. Integr. Comp. Physiol.* **2006**, *290*, R1345–R1356. [CrossRef]
302. Prins, J.R.; Gomez-Lopez, N.; Robertson, S.A. Interleukin-6 in Pregnancy and Gestational Disorders. *J. Reprod. Immunol.* **2012**, *95*, 1–14. [CrossRef]
303. Wu, W.-L.; Hsiao, E.Y.; Yan, Z.; Mazmanian, S.K.; Patterson, P.H. The Placental Interleukin-6 Signaling Controls Fetal Brain Development and Behavior. *Brain. Behav. Immun.* **2017**, *62*, 11–23. [CrossRef]
304. Mirabella, F.; Desiato, G.; Mancinelli, S.; Fossati, G.; Rasile, M.; Morini, R.; Markicevic, M.; Grimm, C.; Amegandjin, C.; Termanini, A.; et al. Prenatal Interleukin 6 Elevation Increases Glutamatergic Synapse Density and Disrupts Hippocampal Connectivity in Offspring. *Immunity* **2021**, *54*, 2611–2631.e8. [CrossRef] [PubMed]
305. Khong, T.Y.; Mooney, E.E.; Ariel, I.; Balmus, N.C.M.; Boyd, T.K.; Brundler, M.-A.; Derricott, H.; Evans, M.J.; Faye-Petersen, O.M.; Gillan, J.E.; et al. Sampling and Definitions of Placental Lesions: Amsterdam Placental Workshop Group Consensus Statement. *Arch. Pathol. Lab. Med.* **2016**, *140*, 698–713. [CrossRef]

306. Smulian, J.C.; Vintzileos, A.M.; Lai, Y.L.; Santiago, J.; Shen-Schwarz, S.; Campbell, W.A. Maternal Chorioamnionitis and Umbilical Vein Interleukin-6 Levels for Identifying Early Neonatal Sepsis. *J Matern Fetal Med* **1999**, *3*, 88–94.
307. Roberts, D.J.; Celi, A.C.; Riley, L.E.; Onderdonk, A.B.; Boyd, T.K.; Johnson, L.C.; Lieberman, E. Acute Histologic Chorioamnionitis at Term: Nearly Always Noninfectious. *PLoS ONE* **2012**, *7*, e31819. [CrossRef] [PubMed]
308. Romero, R.; Kim, Y.M.; Pacora, P.; Kim, C.J.; Benshalom-Tirosh, N.; Jaiman, S.; Bhatti, G.; Kim, J.-S.; Qureshi, F.; Jacques, S.M.; et al. The Frequency and Type of Placental Histologic Lesions in Term Pregnancies with Normal Outcome. *J. Perinat. Med.* **2018**, *46*, 613–630. [CrossRef] [PubMed]
309. Kamity, R.; Patel, H.; Younis, S.; Nasim, M.; Miller, E.; Ahmed, M. Inhibition of Cxcr 1 and 2 Delays Preterm Delivery and Reduces Neonatal Mortality in a Mouse Model of Chorioamnionitis. *Eur. J. Inflamm.* **2014**, *12*, 447–457. [CrossRef]
310. Reisenberger, K.; Egarter, C.; Vogl, S.; Sternberger, B.; Kiss, H.; Husslein, P. The Transfer of Interleukin-8 across the Human Placenta Perfused in Vitro. *Obstet. Gynecol.* **1996**, *87*, 613–616. [CrossRef]
311. Aaltonen, R.; Heikkinen, T.; Hakala, K.; Laine, K.; Alanen, A. Transfer of Proinflammatory Cytokines Across Term Placenta. *Obstet. Gynecol.* **2005**, *106*, 802–807. [CrossRef]
312. Ito, A.; Nakamura, T.; Uchiyama, T.; Hirose, K.; Hirakawa, S.; Sasaguri, Y.; Mori, Y. Stimulation of the Biosynthesis of Interleukin 8 by Interleukin 1 and Tumor Necrosis Factor Alpha in Cultured Human Chorionic Cells. *Biol. Pharm. Bull.* **1994**, *17*, 1463–1467. [CrossRef]
313. Fortunato, S.J.; Menon, R.; Swan, K.F. Amniochorion: A Source of Interleukin-8. *Am. J. Reprod. Immunol.* **1995**, *34*, 156–162. [CrossRef]
314. Hsu, C.-D.; Meaddough, E.; Aversa, K.; Hong, S.-F.; Lu, L.-C.; Jones, D.C.; Copel, J.A. Elevated Amniotic Fluid Levels of Leukemia Inhibitory Factor, Interleukin 6, and Interleukin 8 in Intra-Amniotic Infection. *Am. J. Obstet. Gynecol.* **1998**, *179*, 1267–1270. [CrossRef]
315. Gomez, R.; Romero, R.; Ghezzi, F.; Yoon, B.H.; Mazor, M.; Berry, S.M. The Fetal Inflammatory Response Syndrome. *Am. J. Obstet. Gynecol.* **1998**, *179*, 194–202. [CrossRef]
316. Yoon, B.H.; Romero, R.; Moon, J.; Chaiworapongsa, T.; Espinoza, J.; Kim, Y.M.; Edwin, S.; Kim, J.C.; Camacho, N.; Bujold, E.; et al. Differences in the Fetal Interleukin-6 Response to Microbial Invasion of the Amniotic Cavity between Term and Preterm Gestation. *J. Matern. Neonatal Med.* **2003**, *13*, 32–38. [CrossRef]
317. Romero, R.; Gomez, R.; Ghezzi, F.; Yoon, B.H.; Mazor, M.; Edwin, S.S.; Berry, S.M. A Fetal Systemic Inflammatory Response Is Followed by the Spontaneous Onset of Preterm Parturition. *Am. J. Obstet. Gynecol.* **1998**, *179*, 186–193. [CrossRef]
318. Saji, F.; Samejima, Y.; Kamiura, S.; Sawai, K.; Shimoya, K.; Kimura, T. Cytokine Production in Chorioamnionitis. *J. Reprod. Immunol.* **2000**, *47*, 185–196. [CrossRef]
319. Oh, K.J.; Lee, J.; Romero, R.; Park, H.S.; Hong, J.-S.; Yoon, B.H. A New Rapid Bedside Test to Diagnose and Monitor Intraamniotic Inflammation in Preterm PROM Using Transcervically Collected Fluid. *Am. J. Obstet. Gynecol.* **2020**, *223*, 423.e1–423.e15. [CrossRef]
320. Goldstein, J.M.; Cohen, J.E.; Mareckova, K.; Holsen, L.; Whitfield-Gabrieli, S.; Gilman, S.E.; Buka, S.L.; Hornig, M. Impact of Prenatal Maternal Cytokine Exposure on Sex Differences in Brain Circuitry Regulating Stress in Offspring 45 Years Later. *Proc. Natl. Acad. Sci. USA* **2021**, *118*, e2014464118. [CrossRef]
321. Allswede, D.M.; Buka, S.L.; Yolken, R.H.; Torrey, E.F.; Cannon, T.D. Elevated Maternal Cytokine Levels at Birth and Risk for Psychosis in Adult Offspring. *Schizophr. Res.* **2016**, *172*, 41–45. [CrossRef]
322. Mac Giollabhui, N.; Breen, E.C.; Murphy, S.K.; Maxwell, S.D.; Cohn, B.A.; Krigbaum, N.Y.; Cirillo, P.M.; Perez, C.; Alloy, L.B.; Drabick, D.A.G.; et al. Maternal Inflammation during Pregnancy and Offspring Psychiatric Symptoms in Childhood: Timing and Sex Matter. *J. Psychiatr. Res.* **2019**, *111*, 96–103. [CrossRef]
323. Brown, A.S.; Hooton, J.; Schaefer, C.A.; Zhang, H.; Petkova, E.; Babulas, V.; Perrin, M.; Gorman, J.M.; Susser, E.S. Elevated Maternal Interleukin-8 Levels and Risk of Schizophrenia in Adult Offspring. *Am. J. Psychiatry* **2004**, *161*, 889–895. [CrossRef]
324. Ellman, L.M.; Deicken, R.F.; Vinogradov, S.; Kremen, W.S.; Poole, J.H.; Kern, D.M.; Tsai, W.Y.; Schaefer, C.A.; Brown, A.S. Structural Brain Alterations in Schizophrenia Following Fetal Exposure to the Inflammatory Cytokine Interleukin-8. *Schizophr. Res.* **2010**, *121*, 46–54. [CrossRef]
325. Aguilar-Valles, A.; Rodrigue, B.; Matta-Camacho, E. Maternal Immune Activation and the Development of Dopaminergic Neurotransmission of the Offspring: Relevance for Schizophrenia and Other Psychoses. *Front. Psychiatry* **2020**, *11*, 852. [CrossRef]
326. Nist, M.D.; Pickler, R.H. An Integrative Review of Cytokine/Chemokine Predictors of Neurodevelopment in Preterm Infants. *Biol. Res. Nurs.* **2019**, *21*, 366–376. [CrossRef]
327. Nisticò, R.; Salter, E.; Nicolas, C.; Feligioni, M.; Mango, D.; Bortolotto, Z.A.; Gressens, P.; Collingridge, G.L.; Peineau, S. Synaptoimmunology—Roles in Health and Disease. *Mol. Brain* **2017**, *10*, 26. [CrossRef] [PubMed]
328. Gilmore, J.H.; Fredrik Jarskog, L.; Vadlamudi, S.; Lauder, J.M. Prenatal Infection and Risk for Schizophrenia: IL-1β, IL-6, and TNFα Inhibit Cortical Neuron Dendrite Development. *Neuropsychopharmacology* **2004**, *29*, 1221–1229. [CrossRef] [PubMed]
329. Deverman, B.E.; Patterson, P.H. Cytokines and CNS Development. *Neuron* **2009**, *64*, 61–78. [CrossRef] [PubMed]
330. Parker-Athill, E.C.; Tan, J. Maternal Immune Activation and Autism Spectrum Disorder: Interleukin-6 Signaling as a Key Mechanistic Pathway. *Neurosignals* **2010**, *18*, 113–128. [CrossRef] [PubMed]
331. Desai, T.R.; Leeper, N.J.; Hynes, K.L.; Gewertz, B.L. Interleukin-6 Causes Endothelial Barrier Dysfunction via the Protein Kinase C Pathway. *J. Surg. Res.* **2002**, *104*, 118–123. [CrossRef] [PubMed]

332. Kendall, G.; Peebles, D. Acute Fetal Hypoxia: The Modulating Effect of Infection. *Early Hum. Dev.* **2005**, *81*, 27–34. [CrossRef] [PubMed]
333. Jones, H.N.; Jansson, T.; Powell, T.L. IL-6 Stimulates System A Amino Acid Transporter Activity in Trophoblast Cells through STAT3 and Increased Expression of SNAT2. *Am. J. Physiol. Physiol.* **2009**, *297*, C1228–C1235. [CrossRef]
334. Sullivan, G.; Galdi, P.; Cabez, M.B.; Borbye-Lorenzen, N.; Stoye, D.Q.; Lamb, G.J.; Evans, M.J.; Quigley, A.J.; Thrippleton, M.J.; Skogstrand, K.; et al. Interleukin-8 Dysregulation Is Implicated in Brain Dysmaturation Following Preterm Birth. *Brain. Behav. Immun.* **2020**, *90*, 311–318. [CrossRef]
335. Ravaccia, D.; Ghafourian, T. Critical Role of the Maternal Immune System in the Pathogenesis of Autism Spectrum Disorder. *Biomedicines* **2020**, *8*, 557. [CrossRef]
336. Makinson, R.; Lloyd, K.; Rayasam, A.; McKee, S.; Brown, A.; Barila, G.; Grissom, N.; George, R.; Marini, M.; Fabry, Z.; et al. Intrauterine Inflammation Induces Sex-Specific Effects on Neuroinflammation, White Matter, and Behavior. *Brain. Behav. Immun.* **2017**, *66*, 277–288. [CrossRef]
337. Barke, T.L.; Money, K.M.; Du, L.; Serezani, A.; Gannon, M.; Mirnics, K.; Aronoff, D.M. Sex Modifies Placental Gene Expression in Response to Metabolic and Inflammatory Stress. *Placenta* **2019**, *78*, 1–9. [CrossRef] [PubMed]
338. Bale, T.L. The Placenta and Neurodevelopment: Sex Differences in Prenatal Vulnerability. *Dialogues Clin. Neurosci.* **2016**, *18*, 459–464. [CrossRef] [PubMed]
339. Kalish, B.T.; Kim, E.; Finander, B.; Duffy, E.E.; Kim, H.; Gilman, C.K.; Yim, Y.S.; Tong, L.; Kaufman, R.J.; Griffith, E.C.; et al. Maternal Immune Activation in Mice Disrupts Proteostasis in the Fetal Brain. *Nat. Neurosci.* **2021**, *24*, 204–213. [CrossRef]
340. Werling, D.M.; Geschwind, D.H. Sex Differences in Autism Spectrum Disorders. *Curr. Opin. Neurol.* **2013**, *26*, 146–153. [CrossRef] [PubMed]
341. Rana, S.A.; Aavani, T.; Pittman, Q.J. Sex Effects on Neurodevelopmental Outcomes of Innate Immune Activation during Prenatal and Neonatal Life. *Horm. Behav.* **2012**, *62*, 228–236. [CrossRef] [PubMed]
342. Carpentier, P.A.; Haditsch, U.; Braun, A.E.; Cantu, A.V.; Moon, H.M.; Price, R.O.; Anderson, M.P.; Saravanapandian, V.; Ismail, K.; Rivera, M.; et al. Stereotypical Alterations in Cortical Patterning Are Associated with Maternal Illness-Induced Placental Dysfunction. *J. Neurosci.* **2013**, *33*, 16874–16888. [CrossRef] [PubMed]
343. Xuan, I.C.Y.; Hampson, D.R. Gender-Dependent Effects of Maternal Immune Activation on the Behavior of Mouse Offspring. *PLoS ONE* **2014**, *9*, e104433. [CrossRef] [PubMed]
344. Braun, A.E.; Carpentier, P.A.; Babineau, B.A.; Narayan, A.R.; Kielhold, M.L.; Moon, H.M.; Shankar, A.; Su, J.; Saravanapandian, V.; Haditsch, U.; et al. "Females Are Not Just 'Protected' Males": Sex-Specific Vulnerabilities in Placenta and Brain after Prenatal Immune Disruption. *eNeuro 6*. [CrossRef]
345. Selten, J.-P.; Morgan, V.A. Prenatal Exposure to Influenza and Major Affective Disorder. *Bipolar Disord.* **2010**, *12*, 753–754. [CrossRef]
346. Estes, M.L.; McAllister, A.K. Maternal Immune Activation: Implications for Neuropsychiatric Disorders. *Science* **2016**, *353*, 772–777. [CrossRef]
347. Kentner, A.C.; Bilbo, S.D.; Brown, A.S.; Hsiao, E.Y.; McAllister, A.K.; Meyer, U.; Pearce, B.D.; Pletnikov, M.V.; Yolken, R.H.; Bauman, M.D. Maternal Immune Activation: Reporting Guidelines to Improve the Rigor, Reproducibility, and Transparency of the Model. *Neuropsychopharmacology* **2019**, *44*, 245–258. [CrossRef]
348. Stefanatos, G.A.; Grover, W.; Geller, E. Case Study: Corticosteroid Treatment of Language Regression in Pervasive Developmental Disorder. *J. Am. Acad. Child Adolesc. Psychiatry* **1995**, *34*, 1107–1111. [CrossRef] [PubMed]
349. Shenoy, S.; Arnold, S.; Chatila, T. Response to Steroid Therapy in Autism Secondary to Autoimmune Lymphoproliferative Syndrome. *J. Pediatr.* **2000**, *136*, 682–687. [CrossRef] [PubMed]
350. Sun, F.; Zhang, Y.; Li, Q. Therapeutic Mechanisms of Ibuprofen, Prednisone and Betamethasone in Osteoarthritis. *Mol. Med. Rep.* **2017**, *15*, 981–987. [CrossRef] [PubMed]
351. Asadabadi, M.; Mohammadi, M.-R.; Ghanizadeh, A.; Modabbernia, A.; Ashrafi, M.; Hassanzadeh, E.; Forghani, S.; Akhondzadeh, S. Celecoxib as Adjunctive Treatment to Risperidone in Children with Autistic Disorder: A Randomized, Double-Blind, Placebo-Controlled Trial. *Psychopharmacology* **2013**, *225*, 51–59. [CrossRef]
352. Katagiri, R.; Ishihara-Hattori, K.; Frings, W.; Amano, J.; Fuchs, A.; Chiba, S. Effects of SA237, a Humanized Anti-Interleukin-6 Receptor Monoclonal Antibody, on Pre- and Postnatal Development in Cynomolgus Monkey. *Birth Defects Res.* **2017**, *109*, 843–856. [CrossRef]
353. Smith, S.E.P.; Li, J.; Garbett, K.; Mirnics, K.; Patterson, P.H. Maternal Immune Activation Alters Fetal Brain Development through Interleukin-6. *J. Neurosci.* **2007**, *27*, 10695–10702. [CrossRef]
354. Pineda, E.; Shin, D.; You, S.J.; Auvin, S.; Sankar, R.; Mazarati, A. Maternal Immune Activation Promotes Hippocampal Kindling Epileptogenesis in Mice. *Ann. Neurol.* **2013**, *74*, 11–19. [CrossRef]
355. Tanaka, T.; Ogata, A.; Narazaki, M. Tocilizumab: An Updated Review of Its Use in the Treatment of Rheumatoid Arthritis and Its Application for Other Immune-Mediated Diseases. *Clin. Med. Insights Ther.* **2013**, *5*, CMT.S9282. [CrossRef]
356. Dördelmann, M.; Kerk, J.; Dressler, F.; Brinkhaus, M.-J.; Bartels, D.; Dammann, C.; Dörk, T.; Dammann, O. Interleukin-10 High Producer Allele and Ultrasound-Defined Periventricular White Matter Abnormalities in Preterm Infants: A Preliminary Study. *Neuropediatrics* **2006**, *37*, 130–136. [CrossRef]

357. Zanno, A.E.; Romer, M.A.; Fox, L.; Golden, T.; Jaeckle-Santos, L.; Simmons, R.A.; Grinspan, J.B. Reducing Th2 Inflammation through Neutralizing IL-4 Antibody Rescues Myelination in IUGR Rat Brain. *J. Neurodev. Disord.* **2019**, *11*, 34. [CrossRef]
358. Basil, P.; Li, Q.; Gui, H.; Hui, T.C.K.; Ling, V.H.M.; Wong, C.C.Y.; Mill, J.; McAlonan, G.M.; Sham, P.-C. Prenatal Immune Activation Alters the Adult Neural Epigenome but Can Be Partly Stabilised by a N-3 Polyunsaturated Fatty Acid Diet. *Transl. Psychiatry* **2018**, *8*, 125. [CrossRef] [PubMed]
359. Hennigar, S.R.; McClung, J.P.; Pasiakos, S.M. Nutritional Interventions and the IL-6 Response to Exercise. *FASEB J.* **2017**, *31*, 3719–3728. [CrossRef] [PubMed]
360. Wang, Y.; Song, Z.; Guo, Y.; Xie, H.; Zhang, Z.; Sun, D.; Li, H.; Chen, L. Diterpenoids from the Seeds of Euphorbia Lathyris and Their Anti-Inflammatory Activity. *Bioorg. Chem.* **2021**, *112*, 104944. [CrossRef] [PubMed]
361. Wang, X.; Yang, J.; Zhang, H.; Yu, J.; Yao, Z. Oral Probiotic Administration during Pregnancy Prevents Autism-related Behaviors in Offspring Induced by Maternal Immune Activation via Anti-inflammation in Mice. *Autism Res.* **2019**, *12*, 576–588. [CrossRef]
362. Custodero, C.; Mankowski, R.T.; Lee, S.A.; Chen, Z.; Wu, S.; Manini, T.M.; Hincapie Echeverri, J.; Sabbà, C.; Beavers, D.P.; Cauley, J.A.; et al. Evidence-Based Nutritional and Pharmacological Interventions Targeting Chronic Low-Grade Inflammation in Middle-Age and Older Adults: A Systematic Review and Meta-Analysis. *Ageing Res. Rev.* **2018**, *46*, 42–59. [CrossRef] [PubMed]
363. Luan, W.; Hammond, L.A.; Vuillermot, S.; Meyer, U.; Eyles, D.W. Maternal Vitamin D Prevents Abnormal Dopaminergic Development and Function in a Mouse Model of Prenatal Immune Activation. *Sci. Rep.* **2018**, *8*, 9741. [CrossRef] [PubMed]
364. Kang, D.-W.; Adams, J.B.; Coleman, D.M.; Pollard, E.L.; Maldonado, J.; McDonough-Means, S.; Caporaso, J.G.; Krajmalnik-Brown, R. Long-Term Benefit of Microbiota Transfer Therapy on Autism Symptoms and Gut Microbiota. *Sci. Rep.* **2019**, *9*, 5821. [CrossRef]
365. Janssen, N.M.; Genta, M.S. The Effects of Immunosuppressive and Anti-Inflammatory Medications on Fertility, Pregnancy, and Lactation. *Arch. Intern. Med.* **2000**, *160*, 610–619. [CrossRef]
366. Morse, J.E.; Calvert, S.B.; Jurkowski, C.; Tassinari, M.; Sewell, C.A.; Myers, E.R. Evidence-Based Pregnancy Testing in Clinical Trials: Recommendations from a Multi-Stakeholder Development Process. *PLoS ONE* **2018**, *13*, e0202474. [CrossRef]

International Journal of
*Molecular Sciences*

MDPI

*Review*

# Association of B Cells with Idiopathic Recurrent Pregnancy Loss: A Systematic Review and Meta-Analysis

**Miguel Ângelo-Dias [1,2], Catarina Martins [1,2], Sara Simões Dias [1,3], Luís Miguel Borrego [1,2,4] and Jorge Lima [1,2,5,*]**

1 CHRC, NOVA Medical School, Faculdade de Ciências Médicas, NMS, FCM, Universidade NOVA de Lisboa, 1169-056 Lisboa, Portugal
2 Immunology Department, NOVA Medical School, Faculdade de Ciências Médicas, NMS, FCM, Universidade NOVA de Lisboa, 1169-056 Lisboa, Portugal
3 ciTechCare—Center for Innovative Care and Health Technology, Polytechnic of Leiria, 2411-901 Leiria, Portugal
4 Department of Imunoallergy, LUZ SAÚDE, Hospital da Luz Lisboa, 1500-650 Lisboa, Portugal
5 Department of Obstetrics and Gynecology, LUZ SAÚDE, Hospital da Luz Lisboa, 1500-650 Lisboa, Portugal
* Correspondence: jorgeramoslima@sapo.pt

**Abstract:** Recurrent pregnancy loss (RPL) affects 1–2% of women and is defined as having experienced two or more failed pregnancies. In almost 50% of cases, the causes are idiopathic (IRPL), but increasing evidence has suggested an immunological cause. B cells are known to provide crucial support for a successful pregnancy outcome. However, their involvement in the mechanisms underlying IRPL is still unclear. This systematic review and meta-analysis aimed to comprehensively summarise the existing evidence regarding the levels and profiles of B cells in IRPL. An extensive computerized search in PubMed/Medline, Embase, Scopus, and Web of Science databases was performed with no imposed limits. Two reviewers independently screened all retrieved studies, extracted all the data, and assessed the methodological quality. Disagreements were resolved by a third reviewer. From a total of 1125 retrieved studies, 19 studies were included in the systematic review, and 8 studies were quantitatively analysed. We highlight a potential association between women with IRPL and increased levels of endometrial B cells. In addition, the flow cytometry technique seems to be preferred over immunohistochemistry for identifying those differences, while further studies are necessary to clarify the role of B cells as an immunological risk factor for RPL.

**Keywords:** B cells; idiopathic recurrent pregnancy loss; reproductive immunology; meta-analysis; MeSH

**Citation:** Ângelo-Dias, M.; Martins, C.; Dias, S.S.; Borrego, L.M.; Lima, J. Association of B Cells with Idiopathic Recurrent Pregnancy Loss: A Systematic Review and Meta-Analysis. *Int. J. Mol. Sci.* **2022**, *23*, 15200. https://doi.org/10.3390/ijms232315200

Academic Editor: Ilona Hromadnikova

Received: 14 November 2022
Accepted: 26 November 2022
Published: 2 December 2022



## 1. Introduction

Recurrent pregnancy loss (RPL) is defined as experiencing two or more failed pregnancies prior to the 24th week of gestation and affects 1–2% of women [1,2]. Several causes and risk factors have been considered, including advanced maternal age, parental chromosomal abnormalities, uterine anatomical disorders, antiphospholipid syndrome, inherited thrombophilia, thyroid disorders, and even environmental factors. However, in nearly 50% of cases, the specific aetiology cannot be determined, and such cases are commonly referred to as unexplained or idiopathic RPL (IRPL) [3–5], with a significant psychological impact for the couples involved [6]. Many IRPL cases are treated empirically using several therapeutic strategies, including acetylsalicylic acid, progesterone, corticosteroids, low-molecular-weight heparin, intravenous immunoglobulin, lipid emulsion, leukocyte immune therapy, pre-implantation genetic screening, and tender loving care, but there is a paucity of high-quality evidence for the medical treatment of women with IRPL, with the exception of the use of progesterone [3,7].

Increasing experimental and clinical evidence suggests that immunological causes, such as immunological rejection or the presence of an unbalanced intrauterine immune homeostasis that is adverse for the embryo and pregnancy, could be significantly implicated in IRPL [8]. In fact, the maternal immune system plays a fundamental and challenging role during pregnancy. It ensures a state of tolerance for genetically foreign content while maintaining important protections against pathogens for both the mother and the developing foetus [9,10]. Looking at the maternal–foetal interface, we observed a maternal immune system acquiring distinctive features and articulating new functions and particular cell phenotypes, which suggest its commitment to assure the necessary processes taking place at these sites [11]. In this way, endometrial immune cells may contribute to the proper mechanisms for embryo implantation, survival and development. Therefore, is it of utmost importance to clarify the molecular mechanisms and specific cell types and cellular pathways involved in mediating endometrial receptivity.

B cells are a major component of the immune system. Recent studies have proven that during pregnancy, these cells undergo important adaptations, with physiological circulating B cell lymphopenia observed from mid-gestation onwards and a decreased presence of the more differentiated B cell subsets in the peripheral blood [12,13]. Moreover, altered B cell proportions and changes in their activation states were reported in different obstetric complications, including RPL, preterm birth, and pre-eclampsia [14–16]. In fact, differences in the peripheral B cell compartment have been previously observed in women with recurrent pregnancy losses. In particular, Kwak et al. [17] reported increased percentages of CD19$^+$ B cells in pregnant women with RPL compared to multiparous pregnant normal controls. Later, Jablonowska et al. [18] also obtained similar results, with increased percentages and levels of peripheral B cells in first-trimester RPL pregnant women. In contrast, Darmochwal et al. [19] reported decreased percentages in CD19$^+$ B cells when considering non-pregnant women with RPL compared to normal non-pregnant controls., while Ghafourian et al. [20] observed similar proportions of CD20$^+$ B cells in both non-pregnant RPL and normal control women. Regarding the specific subsets of B cell compartments, it has been demonstrated that IL-10-producing regulatory B cells were decreased in a murine model of pregnancy loss compared to the normal pregnancy model, which present elevated levels in the first pregnancy trimester. Furthermore, it was shown that IL-10 administration and the transfer of IL-10-producing regulatory B cells in aborting animals could prevent foetal rejection [21]. In humans, increased regulatory B cell counts observed in the first trimester of pregnancy may also indicate the higher necessity to suppress possible unwanted immune maternal responses, thus protecting against pregnancy loss [22].

Interestingly, given the different leucocyte compositions of tissue, such as the decidua and the endometrium, which are rich in T cells, uterine NK cells (uNK), macrophages, and dendritic cells in comparison to peripheral blood stress, characterising these tissue types as only addressing circulating cells can be a limitation and is sometimes a biased perspective [23–25]. Thus, despite being considered a rare population with undetermined functions at these locations, B cells are present in the endometrium and they should be further characterised, particularly in pathological reproductive processes [23,26,27]. The initial characterisation of endometrial immune populations by Lachapelle and colleagues reported the presence of 6% of lymphocytic cells expressing the CD20 B cell marker in the endometrium of normal non-pregnant women [28]. Recently, similar proportions of CD19$^+$ B cells using menstrual blood and term decidua parietalis samples of healthy non-pregnant women were also reported [29], while other studies presented different results [30,31]. Importantly, differences in endometrial B cells have also been demonstrated, with some studies reporting increased proportions of B cells in the endometrium of women with recurrent pregnancy losses [30,32].

The increasing number of studies characterising uterine B cells confirms the growing interest in the role of B cells in reproductive immunology. However, information on their potential involvement in the mechanisms underlying IRPL is still scarce and unclear. Hence, studying the role of B cells during IRPL deserves proper consideration. To congregate information and discuss what is known so far, we conducted a systematic review of the literature by addressing the association between idiopathic recurrent pregnancy loss and B cells, whether local or systemic.

## 2. Materials and Methods

This systematic review and meta-analysis aimed to identify and analyse human studies that assessed the role of B cell levels and profiles in non-pregnant women with recurrent pregnancy loss of unknown aetiology compared to non-pregnant healthy women.

This work was conducted and reported in line with the criteria of Preferred Reporting Items for Systematic Reviews and Meta-analyses (PRISMA) guidelines (Table S1) [33,34]. The updated methodology used in this systematic review is in accordance with the Cochrane Handbook of Systematic Reviews of Interventions [35] and is registered in the PROSPERO database (ID: CRD42020181418).

### 2.1. Search Strategy

An extensive computerised literature search was performed to retrieve studies that were included in this systematic review. The searches were performed in PubMed/Medline, Embase, Scopus, and Web of Science databases using database-specific subject heading terms and all variants in free-text words according to the specificities of each database (Table S2). Additionally, a supplementary search of the grey literature and of the reference sections of the selected studies and reviews was performed to identify any additional relevant missing publications that were not retrieved in the electronic search. No date or language limits were imposed on the search.

### 2.2. Eligibility Criteria and Study Selection

Study selection was independently performed by two individuals (MAD and JL) who screened the title and abstracts of all yielded articles from the queries according to the eligibility criteria displayed in Table 1.

**Table 1.** Eligibility criteria for study selection.

| Inclusion Criteria | Exclusion Criteria |
|---|---|
| Studies on living humans | Animal studies |
| Women of reproductive age (18–45 years) | Women with current pregnancy |
| Recurrent pregnancy loss of unknown aetiology | Studies not reporting B cell levels |
| IRPL group with at least two consecutives miscarriages | Genetic studies |
| B cell compartment evaluation | No primary research |
| | Case-report studies |

IRPL, idiopathic recurrent pregnancy loss.

Idiopathic RPL is defined as experiencing more than 2 or 3 failed pregnancies before 24 weeks of gestation in the absence of risk factors or commonly known causes of RPL, namely, uterine anatomic disorders, thyroid dysfunctions, inherited and acquired thrombophilia, and/or parental chromosomal disorders. Studies clearly not matching the eligibility criteria were excluded. No restrictions on geographical location, language of publication, or year of publication were applied, and all the non-primary literature was excluded, such as literature reviews, dissertations, theses, editorials, protocol studies, clinical guidelines, and abstracts or reports from meetings.

The full text of the remaining studies was analysed, and studies were included or discarded according to the eligibility criteria. Any disagreements were referred to a third reviewer (CM) to reach consensus. All decisions, including reasons for exclusion and the number of selected articles in each step, are recorded and depicted in a flow chart following the PRISMA 2020 guidelines. None of the review authors was blinded to journal titles or the study's authors or institutions.

EndNote version 20 (bibliographic software) was used to store, organise, and manage all references arising from the literature search, including the management and removal of duplicates and scanning the titles/abstracts of all records.

### 2.3. Data Extraction and Quality Assessment

All relevant data were extracted from each selected study by two independent reviewers following a standardised piloting form methodology to minimise the risk of bias and to ensure full data extraction. Only information relevant to this review from studies assessing multiple outcomes and variables was collected. If applicable, the original study's authors were contacted to clarify missing or unreported data. Any disagreements were referred to a third reviewer (CM) to reach consensus. The following data were extracted:

Identification of the study: title, authors, year of publication, journal title, country of origin, study design, and number of participants;

Participant characteristics: sample size, age, race, IRPL definition, clinical data of IRPL, and control groups;

Methodological features: sample characteristics, phase of sample collection, methodology used for B cell characterisation, and B cell markers;

Outcomes: B cell levels, B cell profiles, and number of miscarriages.

Methodological quality of each individual study was assessed independently by two reviewers (MAD and JL), using the NHLBI quality assessment tool for case–control studies. This scale evaluates 12 components of a study to determine its methodological quality. Therefore, evaluation criterion was answered either by "Yes", "No", "CD" (cannot determine), "NA" (not applicable), or "NR" (not reported). Subsequently, studies were graded as "Poor", "Fair", or "Good". The level of bias will influence the evidence and results obtained in the systematic review; thus, studies deemed to be of "poor quality" were not included in the meta-analysis. Any disagreement between the two reviewers was referred to a third reviewer to reach consensus.

### 2.4. Data Synthesis

Meta-analyses were conducted using standardised mean differences (SMDs) and 95% confidence intervals (CIs) to allow a comparison of data from different instruments [36]. A random-effect model was used in the meta-analysis as it combines sampling errors and between-study variances to estimate the effect size [37]. To interpret the effect sizes, the following thresholds were used: <0.2 = trivial effect; 0.2–0.5 = small effect; 0.5–0.8 = moderate effect; >0.8 denoting large effects [38]. The statistical heterogeneity among studies was assessed using the I-squared ($I^2$) value, which represents the percentage of variation across studies that is attributable to heterogeneities rather than chance [39]. We adopted the following thresholds: $I^2$ = 25%: low heterogeneity; $I^2$ = 50%: moderate heterogeneity; $I^2$ = 75%: high heterogeneity [39]. Evident heterogeneities were investigated via subgroup analyses. Studies that did not report data as means $\pm$ SD were not suitable for inclusion in the meta-analysis.

All statistical analyses were conducted using statistical software R and using the package "meta" to perform the meta-analysis [40,41]. Statistical significance was defined as a $p$-value < 0.05.

## 3. Results

### 3.1. Study Selection

Using the aforementioned methodology, the database search yielded a total of 1125 studies, of which 452 were removed due to duplication. The abstract and title screening of 673 records revealed 576 studies that clearly did not meet the eligibility criteria, resulting in their exclusion. From the 97 records sought for retrieval, 94 full-text articles were obtained and critically analysed, leading to the exclusion of 75 studies due to ineligibility-related reasons. Finally, a total of 19 studies [17,19,20,30–32,42–54] were included in the qualitative analysis and 8 [20,32,42,43,46,48,50,54] were included in the quantitative analysis. Eleven studies were excluded from the meta-analysis because one had a high risk of bias [47] and ten [17,19,30,31,44,45,49,51–53] did not have the data required for performing the analysis. Figure 1 presents an outline of the study's selection process.

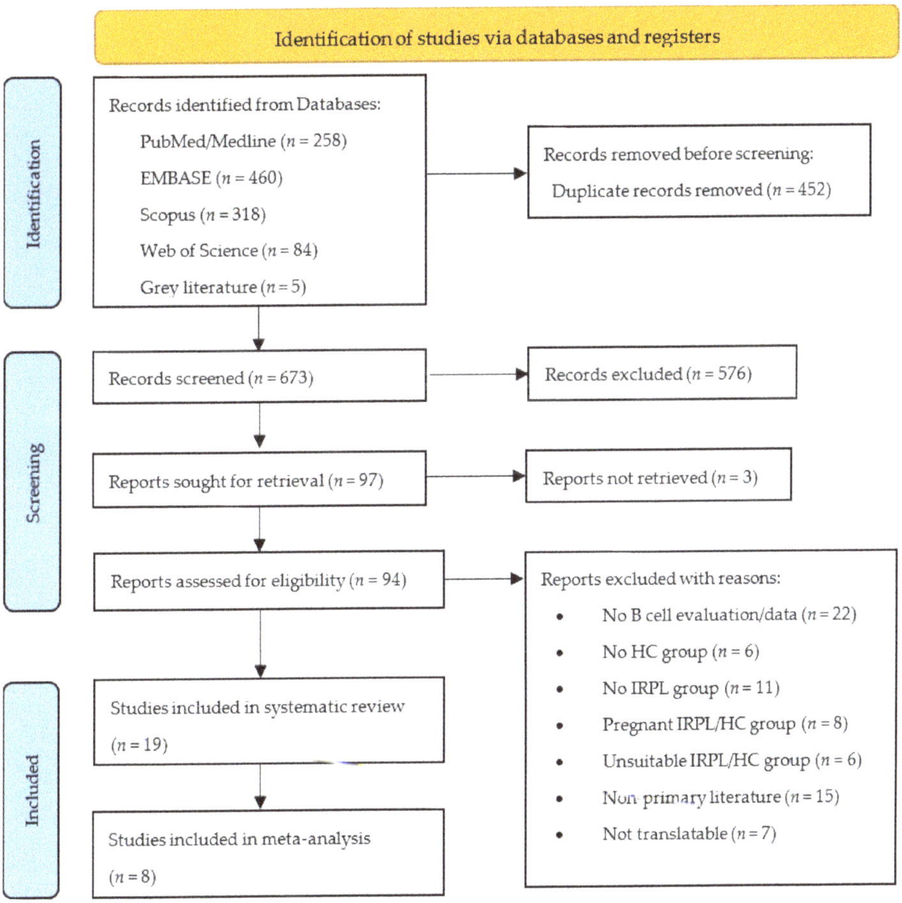


**Figure 1.** PRISMA 2020 flow diagram. PRISMA, Preferred Reporting Items for Systematic Reviews and Meta-analyses; HC, healthy control; IRPL, idiopathic recurrent pregnancy loss.

### 3.2. Characteristics of Included Studies

All relevant data collected from each selected study are summarised in Table 2.

All studies were case–control observational studies conducted in 12 countries: China ($n$ = 4) [45,46,53,54], Ireland ($n$ = 2) [30,49], Spain ($n$ = 2) [43,44], the US ($n$ = 3) [17,47,48], and one study each from Egypt [42], Germany [31], Greece [50], Iran [20], Poland [19], the UK [51], Brazil [52], and Canada [32]. The RPL definition criteria used were not consistent among the studies, with 9 studies defining RPL as two or more pregnancy losses [30,31,43,46,47,49,50,52,54] and 10 studies using three or more pregnancy losses as the criterion for RPL [17,19,20,32,42,44,45,48,51,53]. Six studies did not report information regarding the number of miscarriages in the IRPL group [19,20,44,48,50,52].

As part of the inclusion criteria, all studies included and compared an IRPL group with a healthy control group. In six studies, the control group was composed of women with at least one live birth and no history of miscarriages [17,20,43,44,52,53], three studies included women with at least one live birth but sporadic cases of miscarriages in some [32,51,54], four studies included women with at least one successful pregnancy with no information regarding the previous history of miscarriages [19,42,45,46], two studies included a control group with no previous history of miscarriage but with no information regarding parity history [31,48], two studies included a control group with some miscarriages and some live births [30,49], and two studies did not report clinical information for the control group [47,50].

A total of 1386 IRPL women and 581 control women were included, with a mean sample size amongst studies of 73 (SD = 101; 9–411 range) in the IRPL group and 31 women (SD = 38; 8–179 range) in the control group. The studies used two different types of samples to characterise the immune compartment: peripheral blood ($n$ = 12) [17,19,20,43–48,50,52,54] and endometrial biopsies ($n$ = 7) [30–32,42,49,51,53]. Nine studies performed sample collection during the luteal phase of the menstrual cycle [30–32,42,46,49,51–53], while only one study performed sample collection during the follicular phase [44]. The remaining studies did not specify the phase of the menstrual cycle in which the samples were collected [17,19,20,43,45,47,48,50,54]. Flow cytometry was the preferred methodology used to assess the B cell population in studies ($n$ = 16) [17,19,20,30,32,42–50,52,54], while immunohistochemistry was used in three studies [31,51,53]. B cells were identified mostly through CD19 lineage markers ($n$ = 14) [17,19,30,31,43–50,52,54], followed by CD20 ($n$ = 4) [20,32,42,53] and CD22 ($n$ = 1) [51]. Additional markers were used in seven studies to identify specific B cell subsets, namely, CD5, IgD, CD27, and CD40 [17,19,43,44,47,48,50].

### 3.3. Methodological Quality

The methodological quality (risk of bias) of the included studies was assessed independently by MAD and JL, while any disagreement was discussed with a third author (CM) to reach consensus. The NHLBI Assessment Tool for case–control studies was used, which assesses the quality of a study through 12 questions that can be answered as yes, no, not applicable, not reported, or cannot determine, as shown in Table S3.

All studies explicitly defined the research question. With the exception of two studies [47,48], the study's population was clearly specified and well defined, whereas only one study included a sample size justification [44]. Most of the studies had a low risk of bias when considering the group's population recruitment, the specification and application of the inclusion and exclusion criteria, and the definition of cases and their differentiation from the controls. None of the studies performed a random selection of study participants, although this was not considered to be a fatal flaw, while most of the studies did not use concurrent controls. In none of the studies did the exposure (B cell evaluation) precede the outcome (miscarriages). Nevertheless, we consider that this question is not applicable in this context, since immune profile evaluation is usually recommended when women have already had two or more miscarriages with an unknown aetiology where immune dysregulation might be occurring. Finally, all studies measured the exposure in a consistent and valid manner, and most applied an adjustment for potential confounding variables in the statistical analysis. It was not possible to determine whether the exposure assessors were blinded to the case or control status of the participants.

**Table 2.** Characteristics of included studies.

| Reference | Country | $n$ (Mean ± SD Age in Years) | iRPL Criteria | N of Miscarriages * | $n$ (Mean ± SD Age in Years) | Obstetric History * | $n$ of Studied Groups | Sample & Method | Sample Collection Phase | Markers | Results * |
|---|---|---|---|---|---|---|---|---|---|---|---|
| Alosh et al., 1998 [42] | Egypt | $n = 20$ (31.6 ± 4.2 yrs; 24–36 range) | 3+ | 4.3 miscarriages (3–6 range) | $n = 12$ (33.2 ± 5.5; 28–38 range) | ≥2 successful pregnancies. | 2 | EB I FC | Lutheal | CD20 | ↑ proportion of endometrial B cells in IRPL (17 ± 7%) compared to HC (5.5 ± 6%)—$p < 0.05$ |
| Bohlmann et al., 2010 [31] | Germany | $n = 25$ (32.8 ± 5.6 yrs; 21–41 range) | 2+ | 3.3 ± 1.17 miscarriages (2–6 range) | $n = 10$ (33.5 ± 4.3; 23–37 range) | No history of miscarriage. | 2 | EB I IHC | Lutheal | CD19 | Similar CD19 staining score in IRPL (0.66 ± 0.64) and HC (0.38 ± 0.72). |
| Carbone et al., 2009 [43] | Spain | $n = 36$ (37 yrs, 30–43 range) | 2+ | 2.89 miscarriages (2–7 range) 0.44 live-born babies (0–3 range) | $n = 37$ (37.0; 22–48 range) | ≥1 live-born babies (1–3 range). No history of miscarriage. | 5 | PB I FC | Not specified | CD19, IgD, CD27 CD40, CD5 | Similar % of B cells in IRPL (8 ± 3%, 7–8 95% CI) and HC (8 ± 4%, 7–10 95% CI). Similar B cell counts in IRPL (155 ± 76 cells/µL) and HC (159 ± 95 cells/µL). No significant differences regarding other B cell subsets. |
| Carbone et al., 2016 [44] | Spain | $n = 24$ (37 yrs; 32–43 range) | 3+ | NA | $n = 37$ (37,0) | History of a live child. No history of miscarriage. | 3 | PB I FC | Follicular | CD19, CD27, IgD | Similar total, naive (85 ± 52 vs. 99 ± 72 cells/µL) and class-switched memory B cell levels in IRPL and HC. ↑ levels of unswitched memory B cells in the IRPL group |
| Darmochwal-Kolarz et al., 2002 [19] | Poland | $n = 14$ (28.92 yrs; 25–34 range) | 3+ | NA | $n = 18$ (27.42; 26–35 range) | History of successful pregnancies. | 2 | PB I FC | Not specified | $CD19^+$ $CD19^+$ $CD5^+$ | ↓ total B cell% in IRPL [8.5 (3.2–15.9)] # compared to HC [14.45 (10.9–20.7)] #—$p < 0.005$. ↑ % of $CD5^+$ B cell in IRPL [2.0 (0.7–5.9)] # compared to HC [0.9 (0.5–2.5)] #—$p < 0.05$ |
| Gao et al., 2014 [46] | China | $n = 67$ (30.28 ± 4.12 yrs) | 2+ | Total of 182 miscarriages | $n = 22$ (29.67 ± 3.29) | ≥1 live birth. | 3 | PB I FC | Lutheal | CD19 | Similar % of B cells in IRPL (13.19 ± 4.31) and HC (12.56 ± 3.36)—$p = 0.232$ |
| Gao et al., 2021 [45] | China | $n = 411$ (30.22 ± 4.10 yrs) | 3+ | 3.39 ± 0.66 miscarriages | $n = 179$ (30.82 ± 3.70) | ≥1 live birth. | 3 | PB I FC | Not specified | CD19 | Similar % of B cells in IRPL [11.0 (8.8–13.9)] $ and HC [11.8 (10.4–13.0)] $. |
| Ghafourian et al., 2014 [20] | Iran | $n = 25$ (20–35 yrs) | 3+ | NA | $n = 25$ | History of a live child. No history of miscarriage. | 2 | PB I FC | Not specified | CD20 | Similar % of B cells in IRPL (9.45 ± 0.71) and HC (11.34 ± 0.76). |
| Kwak et al., 1995 [17] | USA | $n = 81$ (33.6 ± 4.8 yrs) | 3+ | 4.1 ± 1.4 miscarriages | $n = 17$ (36.5 ±7.0) | ≥2 successful pregnancies. No history of miscarriage. | 4 | PB I FC | Not specified | $CD19^+$ $CD19^+$ $CD5^+$ | Similar % of B cells and $CD5^+$ B cells in IRPL and HC. |
| Kwak et al., 1998 [47] | USA | $n = 33$ (34.0 yrs; 25–43 range) | 2+ | 3.0 miscarriages (2–8 range) | $n = 8$ | NA | 2 | PB I FC | Not specified | $CD19^+$ $CD19^+$ $CD5^+$ | The % of B cells was 12.8 ± 0.8 in the IRPL and 10.4 ± 1.4 in HC. The % of $CD5^+$ B cells within total B cells was 43.7 ± 4.4 in the IRPL and 55.2 ± 13.0 in HC. |

**Table 2.** *Cont.*

| Reference | Country | $n$ (Mean ± SD Age in Years) | iRPL Criteria | N of Miscarriages * | $n$ (Mean ± SD Age in Years) | Obstetric History * | $n$ of Studied Groups | Sample & Method | Sample Collection Phase | Markers | Results * |
|---|---|---|---|---|---|---|---|---|---|---|---|
| Lachapelle et al., 1996 [32] | Canada | 1 ary IRPL: $n$ = 11 (30 ± 4 yrs; 22–37 range) 2 ary IRPL: $n$ = 9 (33 ± 2 yrs; 26–39 range) | 3+ | 1 ary IRPL: 4 ± 1 miscarriages (3–6 range) 2 ary IRPL: 4 ± 1 miscarriages (3–5 range) | $n$ = 15 (35.0 ± 4.0; 27–40 range) | ≥1 live birth. 0.3 ± 0.5 miscarriages (0–1 range). | 3 | EB \| FC | Lutheal | CD20 | ↑ % of endometrial B cells in IRPL (16.0 ± 8.0%) compared to HC (5.0 ± 6.0%)—$p < 0.05$. ↓ % of endometrial B cells in the IRPL group who had maintained an intact conceptus for extended periods compared to patients with continued miscarriages. |
| Mahmoud et al., 2001 [48] | USA | $n$ = 10 (31.4 ± 2.2 yrs; 22–42 range) | 3+ | NA | $n$ = 20 (29.5 ± 1.8, 20–43 range) | No history of RPL. | 3 | PB \| FC | Not specified | $CD19^+$ $CD19^+$ $CD5^+$ | ↓ % of B cells in RPL subjects (9.9 ± 1.1) compared to HC (13.9 ± 1.0)—$p < 0.05$. |
| Marron et al., 2019 [49] | Ireland | $n$ = 121 (37.9 ± 4.0 yrs) | 2+ | Total of 320 miscarriages. Total of 47 live births. | $n$ = 29 (35.2 ± 3.1) | Total of 4 miscarriages. Total of 10 live births. | 5 | EB \| FC | Lutheal | CD19 | ↑ concentration endometrial of B cells in IRPL (79.6 cells/mg) compared to HC (48.8 cells/mg)—$p = 0.002$; |
| Marron et al., 2019 [30] | Ireland | $n$ = 155 (38.0 ± 4.0 yrs) | 2+ | Total of 442 miscarriages. Total of 61 live births. | $n$ = 35 (35.1 ± 2.9) | Total of 6 miscarriages. Total of 10 live births. | 5 | EB \| FC | Lutheal | CD19 | ↑ % of endometrial B cells (within total CD45+ endometrial lymphocytes) in IRPL (0.77%) compared to HC (0.43%)—$p < 0.001$. |
| Psarra et al., 2001 [50] | Greece | $n$ = 244 (26–39 yrs) | 2+ | NA | $n$ = 44 (23–42 range) | NA | 2 | PB \| FC | Not specified | $CD19^+$ $CD19^+$ $CD5^+$ | Similar % of B cells in IRPL (10.6 ± 3.8) and HC (11.4 ± 6.0). ↓ % of CD19+CD5+B cells within total lymphocytes in IRPL (0.4 ± 0.6) compared to HC (1.4 ± 0.8). |
| Quenby et al., 1999 [51] | UK | $n$ = 22 (33.9 yrs; 20–41 range) | 3+ | 4.4 miscarriages (3–17 range) 0.3 live births (0–2 range) | $n$ = 9 (33.1; 24–41 range) | ≥2 live births (2–4 range). 0.4 miscarriages (0–1 range). | 2 | EB \| IHC | Lutheal | CD22 | Similar median % of endometrial B cells within total cells in IRPL (0.18, 0–4 range) and HC (0, 0–0.8 range). |
| Souza et al., 2002 [52] | Brazil | $n$ = 9 | 2+ | NA | $n$ = 9 (<40) | ≥2 term pregnancies. No history of miscarriage. | 2 | PB \| FC | Lutheal | CD19 | ↑ B cell counts in IRPL [215 (188–236) cells/$mm^3$] $ than in HC [182 (151–185) cells/$mm^3$] $—$p = 0.05$ |
| Zhao et al., 2020 [53] | China | $n$ = 30 (35.40 ± 0.62 yrs) | 3+ | 3.0 miscarriages (3–5 range) | $n$ = 30 (29.47 ± 0.66) | ≥1 live births. No history of spontaneous miscarriages. | 2 | EB \| IHC | Lutheal | CD20 | Similar B cell density in IRPL (0.5%, 0.2–2.5% range) and HC (0.4%, 0.1–2.2% range)—$p = 0.0693$ |
| Zhu et al., 2015 [54] | China | $n$ = 39 (28.3 ± 3.22 yrs) | 2+ | 2.8 ± 0.6 miscarriages | $n$ = 25 (26.8 ± 2.34) | Normal pregnancy history. 0.7 ± 0.34 miscarriages. | 4 | PB \| FC | Not specified | CD19 | Similar % of CD19+ B cells in IRPL (11.7 ± 3.31) and HC (11.7 ± 2.45) |

IRPL, idiopathic recurrent pregnancy loss; PB, peripheral blood; FC, flow cytometry; EB, endometrial biopsy; IHC, immunohistochemistry; NA, not available; ↑ (increased); ↓ (decreased). * Mean ± SD, unless otherwise indicated; # median (min–max); $ median (25th–75th percentile).

Overall, 10 studies were considered to be at a low risk of bias and had good methodological quality [19,20,31,32,43–46,52,54]. Eight were considered to have fair methodological quality [17,30,42,48–51,53]. One study was deemed to be of poor quality [47]; thus, it was not included in the meta-analysis due to its potential high risk of bias.

### 3.4. *Results of Individual Studies and Meta-Analyses*

Considering all included studies, a majority ($n$ = 11) reported no significant differences between the proportion or concentration of total B cells in women with IRPL compared to HC women [17,20,31,43–46,50,51,53,54]. Seven studies reported statistically significant differences between groups [19,28,30,42,48,49,52], and one study did not perform statistical analyses on B cell data [47].

We pooled data from eight studies ($n$ = 652 women) and observed a non-significant tendency towards lower proportions of total B cells in the IRPL group (SMD = −0.36 [95% CI, −1.63–0.92]), with a high heterogeneity among the included studies ($I^2$ = 93%) (Figure 2).

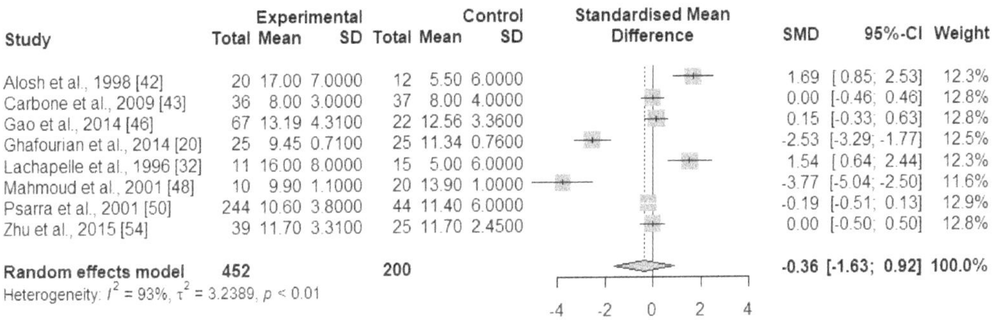


**Figure 2.** Random-effect meta-analysis for the levels of B cells in IRPL and HC women.

Subgroup analyses were carried out to explore possible causes of heterogeneity according to the type of sample used in the studies: peripheral blood or endometrial biopsies.

From the seven studies using endometrial biopsies, all those ($n$ = 4) characterising the B cell compartment with flow cytometry reported significant differences towards increased percentages or concentrations of total B cells in IRPL compared to the HC group [30,32,42,49]. The remaining three studies used immunohistochemistry to evaluate B cell compartment and did not find significant differences between groups [31,51,53]. We pooled data from two studies ($n$ = 58 women) that reported B cell data in endometrial biopsies. Overall, there were higher proportions of total B cells in the IRPL group (SMD = 1.62 [95% CI, 1.00–2.23]; $p < 0.001$, $I^2$ = 0%) (Figure 3).

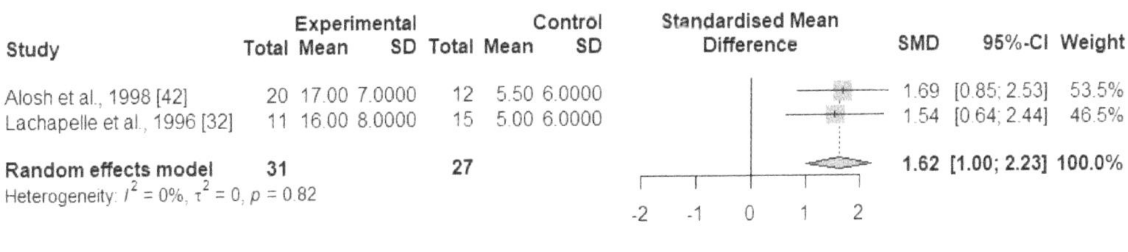


**Figure 3.** Random-effect meta-analysis for the levels of endometrial B cells in IRPL and HC women.

From the 12 studies using peripheral blood samples, a majority ($n$ = 8) reported similar percentages of B cells between groups [17,20,43–46,50,54], while two reported significantly lower percentages of total B cells in IRPL women compared to HC [19,48]. These observations were not shared by Souza et al., 2002 [52], who reported increased B cell counts in IRPL compared to HC women. We pooled data from six studies ($n$ = 594 women) that evaluated B cells in peripheral blood. Overall, the analyses revealed no statistically significant associations between peripheral B cell levels and IRPL (SMD = −0.99 [95% CI, −2.29–0.32]; $p > 0.05$, $I^2 = 93\%$) (Figure 4).

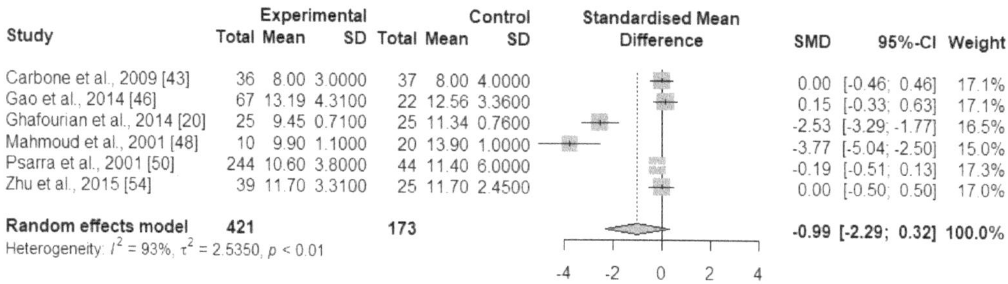


**Figure 4.** Random-effect meta-analysis for the levels of circulating B cells in IRPL and HC women.

Regarding the evaluation of specific B cell subsets in the included studies, one study reported increased levels of circulating unswitched memory B cells in IRPL compared to the HC group [44], one study reported increased percentages of circulating $CD5^+$ B cells within total lymphocytes in IRPL compared to the HC group [19], and one study reported decreased percentages of circulating $CD5^+$ B cells within total lymphocytes in IRPL compared to the HC group [50]. Due to the small number of studies addressing specific B cell subsets, a meta-analysis was not feasible for this subgroup.

## 4. Discussion

Reproductive failure is a pregnancy-related complication that represents a significant concern for human reproduction. In this context, immunological abnormalities have been implicated in many female reproductive pathologies, including recurrent pregnancy loss [8,55]. Unlike other immune compartments [56–58], the current scarce, sparse, and methodologically unstandardised available data on the involvement of B cells (their levels and profiles) as a risk factor in the aetiology of recurrent pregnancy loss represent and point to the challenge in obtaining consistent conclusions on this matter. Therefore, it is of utmost interest to summarise, in a transparent, structured, and organised manner, the existing literature evaluating B cell compartments in women with IRPL. To the best of our knowledge, this is the first systematic review with a meta-analysis that addresses studies evaluating the involvement of B cells in non-pregnant women with IRPL compared to non-pregnant healthy controls. In this review, a total of 19 studies were selected in the qualitative analysis, and from these, 8 were included in the meta-analysis.

Here, we highlight the potential association between women with IRPL and increased levels of endometrial B cells (compared to HC women). In contrast, no such associations were found when considering peripheral B cells, which is not surprising, since distributions and functions of immune populations, including B cells, are not equal when considering different tissue types [59,60]. Unfortunately, insufficient studies evaluating specific B cell subsets were retrieved. Interestingly, although only two studies characterising endometrial B cells were quantitatively analysed, all retrieved studies evaluating endometrial B cells with flow cytometry ($n$ = 4) reported significantly increased levels of endometrial B cells in IRPL compared to the HC group, whereas studies using immunohistochemistry did not report differences between groups. This might indicate that the endometrial characterisation of B cells is preferred over peripheral characterisation to identify differences between these subsets in IRPL women. In addition, the use of flow cytometry as the methodological tool to characterise those differences seems to be advantageous compared to immunohistochemistry.

Our systematic review has several strengths. The predefined methodology used in this review was based on the Cochrane Handbook for Systematic Reviews of Interventions [35], and we conducted our review using a prospectively registered protocol. In addition, we employed a comprehensive and extensive search strategy using the most representative electronic bibliographic databases for biomedical research. No restrictions were applied to the search so that all possibly relevant studies were retrieved, and the risk of bias was evaluated independently by two individuals. Finally, a meta-analysis was performed using the random-effect model to pool the data as much as possible.

However, we also acknowledge some limitations of this review. First, although we were able to identify a large number of studies reporting the levels of B cells in IRPL women, there were significant heterogeneities among them, namely, in the criteria for the definition of RPL, the lack of or limited clinical reported information from the included groups, particularly for the control group, the large number of countries in which the selected studies were conducted, and the distinct sample types and methodologies used to characterise B cell data. Importantly, our significant findings should be interpreted with caution, as some of our meta-analysis included data from a reduced number of studies, which—although valid—might limit the value of pooling data. In fact, although it is a limitation of the included studies, only a small portion of the selected studies ($n$ = 8) reported B cell data suitable for inclusion in the meta-analysis. Other conclusions or more robust findings might have been identified if all studies were included in the meta-analysis. Due to the small number of studies, we were unable to adjust for some important confounding variables, such as age, ethnicity, country of origin and number of previous miscarriages, which could affect our findings. In addition, we did not consider formal tests for publication bias, since fewer than 10 studies were included in the meta-analysis [35].

Finally, although this review emphasises that increased levels of endometrial B cells might play a role during the processes of miscarriage, it is still unclear whether these higher levels of B cells are causes or consequences of the RPL. To further explore this causality, it would be more accurate to compare women with a known cause for the RPL as a control group with women with IRPL. Thus, if B cells are equally increased in this control group, then B cells may likely represent a consequence of miscarriages rather than the cause.

We have identified a systematic review published in HROpen journal that is somewhat similar to our present work [61]. However, in this review, the authors did not specifically analyse the RPL of unknown origins or analyse both circulating and local B cell compartments in the IRPL. In addition, this review includes data from both pregnant and non-pregnant RPL women; thus, the results obtained in those studies might reflect pregnancy-induced changes in B cell populations, rather than changes associated with the pathology of interest. Importantly, in this systematic review, the authors were unable to perform a quantitative synthesis, so the conclusiveness of the presented evidence in that review is unclear. Nevertheless, we recognise that the existing heterogeneity among the available studies (either due to differences in the methodology, study design, or selection criteria) on this matter represents a problem in the attempt to provide solid conclusions regarding the role of B cells in IRPL.

Overall, although there is an apparent association between increased endometrial B cells and IRPL, their role and levels in the development of this condition are not well understood. The use of flow cytometry could be a valuable tool to evaluate different endometrial B cell phenotypes in IRPL and to further explore this association. Nevertheless, further studies are necessary to clarify the role of B cells as an immunological risk factor for RPL, and we expect that this review will provide clues and important data to stimulate further research on this matter.

**Supplementary Materials:** The following supporting information can be downloaded at https://www.mdpi.com/article/10.3390/ijms232315200/s1. Table S1: Preferred Reporting Items for Systematic Reviews and Meta-analyses checklist; Table S2: Retrieval search strategy; Table S3: Methodological quality assessment of selected studies.

**Author Contributions:** Conceptualisation, J.L.; methodology, M.Â.-D., J.L. and S.S.D.; software and formal analysis, S.S.D. and M.Â.-D.; validation, C.M.; investigation, M.Â.-D. and J.L.; resources, J.L. and S.S.D.; data curation, M.Â.-D.; writing—original draft preparation, M.Â.-D., S.S.D. and J.L.; writing—review and editing, M.Â.-D., S.S.D., C.M. and L.M.B.; visualisation, M.Â.-D.; supervision, C.M., L.M.B. and J.L.; project administration, M.Â.-D. and J.L.; funding acquisition, J.L. All authors have read and agreed to the published version of the manuscript.

**Funding:** This work in the context of the "Women's Health and Maternal & Fetal Research Group" was partially co-financed by Hospital da Luz Lisboa under the initiative "Luz Investigação".

**Institutional Review Board Statement:** Not applicable.

**Informed Consent Statement:** Not applicable.

**Data Availability Statement:** Not applicable.

**Conflicts of Interest:** The authors declare no conflict of interest.

## References

1. The Practice Committee of the American Society for Reproductive, Medicine. Evaluation and Treatment of Recurrent Pregnancy Loss: A Committee Opinion. *Fertil. Steril.* **2012**, *98*, 1103–1111.
2. The ESHRE Guideline Group on RPL; Bender Atik, R.; Christiansen, O.B.; Elson, J.; Kolte, A.M.; Lewis, S.; Middeldorp, S.; Nelen, W.; Peramo, B.; Quenby, S.; et al. ESHRE guideline: Recurrent pregnancy loss. *Hum. Reprod. Open* **2018**, *2018*, hoy004. [PubMed]
3. El Hachem, H.; Crepaux, V.; May-Panloup, P.; Descamps, P.; Legendre, G.; Bouet, P. Recurrent pregnancy loss: Current perspectives. *Int. J. Women's Health* **2017**, *9*, 331–345. [CrossRef] [PubMed]
4. Stephenson, M.D. Frequency of factors associated with habitual abortion in 197 couples. *Fertil. Steril.* **1996**, *66*, 25–29.
5. Jaslow, C.R.; Carney, J.L.; Kutteh, W.H. Diagnostic factors identified in 1020 women with two versus three or more recurrent pregnancy losses. *Fertil. Steril.* **2010**, *93*, 1234–1243. [CrossRef]
6. Kolte, A.M.; Olsen, L.R.; Mikkelsen, E.M.; Christiansen, O.B.; Nielsen, H.S. Depression and emotional stress is highly prevalent among women with recurrent pregnancy loss. *Hum. Reprod.* **2015**, *30*, 777–782. [CrossRef]
7. Rasmark Roepke, E.; Hellgren, M.; Hjertberg, R.; Blomqvist, L.; Matthiesen, L.; Henic, E.; Lalitkumar, S.; Strandell, A. Treatment efficacy for idiopathic recurrent pregnancy loss—A systematic review and meta-analyses. *Acta Obstet. Gynecol. Scand.* **2018**, *97*, 921–941. [CrossRef]
8. Wang, N.F.; Kolte, A.M.; Larsen, E.C.; Nielsen, H.S.; Christiansen, O.B. Immunologic Abnormalities, Treatments, and Recurrent Pregnancy Loss: What Is Real and What Is Not? *Clin. Obstet. Gynecol.* **2016**, *59*, 509–523. [CrossRef]
9. Guzman-Genuino, R.M.; Diener, K.R. Regulatory B Cells in Pregnancy: Lessons from Autoimmunity, Graft Tolerance, and Cancer. *Front. Immunol.* **2017**, *8*, 172. [CrossRef]
10. Fettke, F.; Schumacher, A.; Costa, S.D.; Zenclussen, A.C. B cells: The old new players in reproductive immunology. *Front. Immunol.* **2014**, *5*, 285. [CrossRef]
11. Dutta, S.; Sengupta, P. Defining pregnancy phases with cytokine shift. *J. Pregnancy* **2017**, *1*, 1–3. [CrossRef]
12. Muzzio, D.; Zenclussen, A.C.; Jensen, F. The role of B cells in pregnancy: The good and the bad. *Am. J. Reprod. Immunol.* **2013**, *69*, 408–412. [CrossRef] [PubMed]
13. Lima, J.; Martins, C.; Leandro, M.J.; Nunes, G.; Sousa, M.J.; Branco, J.C.; Borrego, L.M. Characterization of B cells in healthy pregnant women from late pregnancy to post-partum: A prospective observational study. *BMC Pregnancy Childbirth* **2016**, *16*, 139. [CrossRef] [PubMed]
14. Jensen, F.; Wallukat, G.; Herse, F.; Budner, O.; El-Mousleh, T.; Costa, S.-D.; Dechend, R.; Zenclussen, A.C. $CD19^+$ $CD5^+$ Cells as Indicators of Preeclampsia. *Hypertension* **2012**, *59*, 861–868. [CrossRef] [PubMed]
15. Nguyen, T.G.; Ward, C.M.; Morris, J.M. To B or not to B cells-mediate a healthy start to life. *Clin. Exp. Immunol.* **2013**, *171*, 124–134. [CrossRef]
16. Saccone, G.; Berghella, V.; Maruotti, G.M.; Ghi, T.; Rizzo, G.; Simonazzi, G.; Rizzo, N.; Facchinetti, F.; Dall'Asta, A.; Visentin, S.; et al. Antiphospholipid antibody profile based obstetric outcomes of primary antiphospholipid syndrome: The PREGNANTS study. *Am. J. Obstet. Gynecol.* **2017**, *216*, 525.e1–525.e12. [CrossRef]
17. Kwak, J.Y.; Beaman, K.D.; Gilman-Sachs, A.; Ruiz, J.E.; Schewitz, D.; Beer, A.E. Up-regulated expression of $CD56^+$, $CD56^+/CD16^+$, and $CD19^+$ cells in peripheral blood lymphocytes in pregnant women with recurrent pregnancy losses. *Am. J. Reprod. Immunol.* **1995**, *34*, 93–99. [CrossRef]

18. Jablonowska, B.; Palfi, M.; Matthiesen, L.; Selbing, A.; Kjellberg, S.; Ernerudh, J. T and B lymphocyte subsets in patients with unexplained recurrent spontaneous abortion: IVIG versus placebo treatment. *Am. J. Reprod. Immunol.* **2002**, *48*, 312–318. [CrossRef]
19. Darmochwal-Kolarz, D.; Leszczynska-Gorzelak, B.; Rolinski, J.; Oleszczuk, J. The immunophenotype of patients with recurrent pregnancy loss. *Eur. J. Obstet. Gynecol. Reprod. Biol.* **2002**, *103*, 53–57. [CrossRef]
20. Ghafourian, M.; Band, N.; Pour, A.; Kooti, W.; Foroutan-rad, M.; Badiee, M. The role of $CD16^+$, $CD56^+$, NK ($CD16^+/CD56^+$) and B $CD20^+$ cells in the outcome of pregnancy in women with recurrent spontaneous abortion. *Int. J. Women's Health Reprod. Sci.* **2015**, *3*, 61–66. [CrossRef]
21. Jensen, F.; Muzzio, D.; Soldati, R.; Fest, S.; Zenclussen, A.C. Regulatory B10 cells restore pregnancy tolerance in a mouse model. *Biol. Reprod.* **2013**, *89*, 90. [CrossRef] [PubMed]
22. Rolle, L.; Tehran, M.M.; Morell, A.; Raeva, Y.; Schumacher, A.; Hartig, R.; Costa, S.-D.; Jensen, F.; Zenclussen, A.C. Cutting edge: IL-10-producing regulatory B cells in early human pregnancy. *Am. J. Reprod. Immunol.* **2013**, *70*, 448–453. [CrossRef]
23. Mor, G.; Aldo, P.; Alvero, A.B. The unique immunological and microbial aspects of pregnancy. *Nat. Rev. Immunol.* **2017**, *17*, 469–482. [CrossRef] [PubMed]
24. Trundley, A.; Moffett, A. Human uterine leukocytes and pregnancy. *Tissue Antigens* **2004**, *63*, 1–12. [CrossRef] [PubMed]
25. Bulmer, J.N.; Williams, P.J.; Lash, G.E. Immune cells in the placental bed. *Int. J. Dev. Biol.* **2010**, *54*, 281–294. [CrossRef] [PubMed]
26. Zenclussen, A.C.; Hammerling, G.J. Cellular Regulation of the Uterine Microenvironment That Enables Embryo Implantation. *Front. Immunol.* **2015**, *6*, 321. [CrossRef]
27. Harrity, C.; Bereir, M.M.; Walsh, D.J.; Marron, K.D. Moving from peripheral blood to local uterine immunophenotype analysis in patients with poor reproductive history: Pilot study of a novel technique. *Ir. J. Med Sci.* **2018**, *188*, 893–901. [CrossRef]
28. Lachapelle, M.H.; Miron, P.; Hemmings, R.; Baron, C.; Roy, D.C. Flow-cytometric characterization of hematopoietic cells in non-pregnant human endometrium. *Am. J. Reprod. Immunol.* **1996**, *35*, 5–13. [CrossRef]
29. Feyaerts, D.; Benner, M.; van Cranenbroek, B.; van der Heijden, O.W.H.; Joosten, I.; van der Molen, R.G. Human uterine lymphocytes acquire a more experienced and tolerogenic phenotype during pregnancy. *Sci. Rep.* **2017**, *7*, 2884. [CrossRef]
30. Marron, K.; Walsh, D.; Harrity, C. Detailed endometrial immune assessment of both normal and adverse reproductive outcome populations. *J. Assist. Reprod. Genet.* **2019**, *36*, 199–210. [CrossRef]
31. Bohlmann, M.K.; Luedders, D.W.; Strowitzki, T.; von Wolff, M. Specific secretory phase endometrial leukocytes of women with two and more consecutive idiopathic abortions are not significantly different from healthy controls. *Arch. Gynecol. Obstet.* **2010**, *281*, 983–990. [CrossRef] [PubMed]
32. Lachapelle, M.H.; Miron, P.; Hemmings, R.; Roy, D.C.; Endometrial, T.B. NK cells in patients with recurrent spontaneous abortion. Altered profile and pregnancy outcome. *J. Immunol.* **1996**, *156*, 4027–4034. [PubMed]
33. Page, M.J.; McKenzie, J.E.; Bossuyt, P.M.; Boutron, I.; Hoffmann, T.C.; Mulrow, C.D.; Shamseer, L.; Tetzlaff, J.M.; Akl, E.A.; Brennan, S.E.; et al. The PRISMA 2020 statement: An updated guideline for reporting systematic reviews. *Int. J. Surg.* **2021**, *88*, 89. [CrossRef]
34. Page, M.J.; Moher, D.; Bossuyt, P.M.; Boutron, I.; Hoffmann, T.C.; Mulrow, C.D.; Shamseer, L.; Tetzlaff, J.M.; Akl, E.A.; Brennan, S.E.; et al. PRISMA 2020 explanation and elaboration: Updated guidance and exemplars for reporting systematic reviews. *BMJ* **2021**, *372*, n160. [CrossRef]
35. Higgins, J.P.T.; Thomas, J.; Chandler, J.; Cumpston, M.; Li, T.; Page, M.J.; Welch, V.A. (Eds.) Cochrane Handbook for Systematic Reviews of Interventions Version 6.3 (Updated February 2022). Cochrane, 2022. Available online: www.training.cochrane.org/handbook (accessed on 8 August 2022).
36. Takeshima, N.; Sozu, T.; Tajika, A.; Ogawa, Y.; Hayasaka, Y.; Furukawa, T.A. Which is more generalizable, powerful and interpretable in meta-analyses, mean difference or standardized mean difference? *BMC Med. Res. Methodol.* **2014**, *14*, 30. [CrossRef] [PubMed]
37. Borenstein, M.; Hedges, L.V.; Higgins, J.P.; Rothstein, H.R. A basic introduction to fixed-effect and random-effects models for meta-analysis. *Res. Synth. Methods* **2010**, *1*, 97–111. [CrossRef] [PubMed]
38. Cohen, J. *Statistical Power Analysis for the Behavioral Sciences*, 2nd ed.; Lawrence Erlbaum Associates Publishers: Hillsdale, NJ, USA, 1988.
39. Higgins, J.P.; Thompson, S.G.; Deeks, J.J.; Altman, D.G. Measuring inconsistency in meta-analyses. *BMJ* **2003**, *327*, 557–560. [CrossRef]
40. R Core Team. *R: A Language and Environment for Statistical Computing*; R Foundation for Statistical Computing: Vienna, Austria, 2020.
41. Balduzzi, S.; Rucker, G.; Schwarzer, G. How to perform a meta-analysis with R: A practical tutorial. *Evid. Based Ment. Health* **2019**, *22*, 153–160. [CrossRef]
42. Alosh, M.; Badawy, A.; Shafik, A.O.; El-Sabour, M.A.; Hissin, N.A. Endometrial differentiation in patients with recurrent abortion: A morphometric and immunophenotypic study. *ZUMJ* **1998**, *4*, 78–87.
43. Carbone, J.; Gallego, A.; Lanio, N.; Navarro, J.; Orera, M.; Aguaron, A.; Fernandez-Cruz, E.; Sarmiento, E. Quantitative abnormalities of peripheral blood distinct T, B, and natural killer cell subsets and clinical findings in obstetric antiphospholipid syndrome. *J. Rheumatol.* **2009**, *36*, 1217–1225. [CrossRef]

44. Carbone, J.; Sarmiento, E.; Gallego, A.; Lanio, N.; Navarro, J.; García, S.; Fernandez-Cruz, E. Peripheral blood T- and B-cell immunophenotypic abnormalities in selected women with unexplained recurrent miscarriage. *J. Reprod. Immunol.* **2016**, *113*, 50–53. [CrossRef] [PubMed]
45. Gao, J.; Guo, H.; Zhu, L.; Yang, B.; Zhang, J.; Xu, K.; Hao, C. Analysis of blocking antibodies and lymphocyte subsets in patients with recurrent reproductive failure. *Hum. Fertil.* **2021**, *13*, 1169–1178. [CrossRef] [PubMed]
46. Gao, L.; Zhang, J.; Chen, H.; Zhang, S.; Chen, L.; Tan, J.; Liu, M.; Meng, L.; Wang, Y.; Zhang, R.; et al. Characteristics of immune cell changes before and after immunotherapy and their clinical significance in patients with unexplained recurrent spontaneous abortion. *Genet. Mol. Res.* **2014**, *13*, 1169–1178. [CrossRef] [PubMed]
47. Kwak, J.Y.H.; Gilman-Sachs, A.; Moretti, M.; Beaman, K.D.; Beer, A.E. Natural killer cell cytotoxicity and paternal lymphocyte immunization in women with recurrent spontaneous abortions. *Am. J. Reprod. Immunol.* **1998**, *40*, 352–358. [CrossRef] [PubMed]
48. Mahmoud, F.; Diejomaoh, M.; Omu, A.E.; Abul, H.; Haines, D. Lymphocyte subpopulation frequency and presence of anti-cardiolipin and anti-nuclear antibodies in peripheral blood of Kuwaiti women experiencing recurrent pregnancy loss. *J. Obstet. Gynaecol.* **2001**, *21*, 587–590. [CrossRef] [PubMed]
49. Marron, K.; Harrity, C. Endometrial lymphocyte concentrations in adverse reproductive outcome populations. *J. Assist. Reprod. Genet.* **2019**, *36*, 837–846. [CrossRef]
50. Psarra, K.; Kapsimali, V.; Tarassi, K.; Dendrinos, S.; Athanasiadis, T.; Botsis, D.; Kreatsas, G.; Papasteriades, C. TCR$\gamma\delta^+$ T lymphocytes in unexplained recurrent spontaneous abortions. *Am. J. Reprod. Immunol.* **2001**, *45*, 6–11. [CrossRef]
51. Quenby, S.; Bates, M.; Doig, T.; Brewster, J.; Lewis-Jonesn, D.I.; Johnson, P.M.; Vince, G. Pre-implantation endometrial leukocytes in women with recurrent miscarriage. *Hum. Reprod.* **1999**, *14*, 2386–2391. [CrossRef]
52. Souza, S.S.; Ferriani, R.A.; Santos, C.M.P.; Voltarelli, J.C. Immunological evaluation of patients with recurrent abortion. *J. Reprod. Immunol.* **2002**, *56*, 111–121. [CrossRef]
53. Zhao, Y.; Chen, X.; Zhang, T.; Chan, L.K.Y.; Liu, Y.; Chung, J.P.W.; Kwong, J.; Li, T.C. The use of multiplex staining to measure the density and clustering of four endometrial immune cells around the implantation period in women with recurrent miscarriage: Comparison with fertile controls. *J. Mol. Histol.* **2020**, *51*, 593–603. [CrossRef]
54. Zhu, L.Y.; Chen, X.; Xu, Z.Z.; Xu, L.; Mao, T.; Zhang, H. Changes and clinical significance of peripheral blood helper T lymphocyte and natural killer (NK) cells in unexplained recurrent spontaneous abortion (URSA) patients after abortion and successful pregnancy. *Clin. Exp. Obstet. Gynecol.* **2015**, *42*, 62–66. [CrossRef] [PubMed]
55. Li, D.; Zheng, L.; Zhao, D.; Xu, Y.; Wang, Y. The Role of Immune Cells in Recurrent Spontaneous Abortion. *Reprod. Sci.* **2021**, *28*, 3303–3315. [CrossRef] [PubMed]
56. Guerrero, B.; Hassouneh, F.; Delgado, E.; Casado, J.G.; Tarazona, R. Natural killer cells in recurrent miscarriage: An overview. *J. Reprod. Immunol.* **2020**, *142*, 103209. [CrossRef] [PubMed]
57. Sfakianoudis, K.; Rapani, A.; Grigoriadis, S.; Pantou, A.; Maziotis, E.; Kokkini, G.; Tsirligkani, C.; Bolaris, S.; Nikolettos, K.; Chronopoulou, M.; et al. The Role of Uterine Natural Killer Cells on Recurrent Miscarriage and Recurrent Implantation Failure: From Pathophysiology to Treatment. *Biomedicines* **2021**, *9*, 1425. [CrossRef] [PubMed]
58. Keller, C.C.; Eikmans, M.; van der Hoorn, M.P.; Lashley, L. Recurrent miscarriages and the association with regulatory T cells; A systematic review. *J. Reprod. Immunol.* **2020**, *139*, 103105. [CrossRef] [PubMed]
59. Lee, J.Y.; Lee, M.; Lee, S.K. Role of endometrial immune cells in implantation. *Clin. Exp. Reprod. Med.* **2011**, *38*, 119–125. [CrossRef]
60. Schumacher, A.; Sharkey, D.J.; Robertson, S.A.; Zenclussen, A.C. Immune Cells at the Fetomaternal Interface: How the Microenvironment Modulates Immune Cells To Foster Fetal Development. *J. Immunol.* **2018**, *201*, 325–334. [CrossRef]
61. Shen, M.; O'Donnell, E.; Leon, G.; Kisovar, A.; Melo, P.; Zondervan, K.; Granne, I.; Southcombe, J. The role of endometrial B cells in normal endometrium and benign female reproductive pathologies: A systematic review. *Hum. Reprod. Open* **2022**, *2022*, hoab043. [CrossRef]

International Journal of
*Molecular Sciences*

MDPI

*Article*

# Chronic Venous Disease during Pregnancy Causes a Systematic Increase in Maternal and Fetal Proinflammatory Markers

**Miguel A. Ortega [1,2,*], Ana M. Gómez-Lahoz [1,2], Lara Sánchez-Trujillo [1,3], Oscar Fraile-Martinez [1], Cielo García-Montero [1], Luis G. Guijarro [2,4], Coral Bravo [5,6,7], Juan A. De Leon-Luis [5,6,7], Jose V. Saz [7], Julia Bujan [1,2], Natalio García-Honduvilla [1,2], Jorge Monserrat [1,2,†] and Melchor Alvarez-Mon [1,2,8,9,†]**

1 Department of Medicine and Medical Specialities, Faculty of Medicine and Health Sciences, University of Alcalá, 28801 Alcalá de Henares, Spain
2 Ramón y Cajal Institute of Sanitary Research (IRYCIS), 28034 Madrid, Spain
3 Service of Pediatric, Hospital Universitario Principe de Asturias, 28801 Alcalá de Henares, Spain
4 Department of Systems Biology, Faculty of Medicine and Health Sciences (Networking Research Center on for Liver and Digestive Diseases (CIBEREHD)), University of Alcalá, 28801 Alcalá de Henares, Spain
5 Department of Public and Maternal and Child Health, School of Medicine, Complutense University of Madrid, 28040 Madrid, Spain
6 Department of Obstetrics and Gynecology, University Hospital Gregorio Marañón, 28009 Madrid, Spain
7 Health Research Institute Gregorio Marañón, 28009 Madrid, Spain
8 Department of Biomedicine and Biotechnology, Faculty of Medicine and Health Sciences, University of Alcalá, 28801 Alcalá de Henares, Spain
9 Immune System Diseases-Rheumatology and Internal Medicine Service, University Hospital Príncipe de Asturias, CIBEREHD, 28806 Alcalá de Henares, Spain
* Correspondence: miguel.angel.ortega92@gmail.com
† These authors contributed equally to this work.

**Citation:** Ortega, M.A.; Gómez-Lahoz, A.M.; Sánchez-Trujillo, L.; Fraile-Martinez, O.; García-Montero, C.; Guijarro, L.G.; Bravo, C.; De Leon-Luis, J.A.; Saz, J.V.; Bujan, J.; et al. Chronic Venous Disease during Pregnancy Causes a Systematic Increase in Maternal and Fetal Proinflammatory Markers. *Int. J. Mol. Sci.* **2022**, *23*, 8976. https://doi.org/10.3390/ijms23168976

Academic Editor: Ilona Hromadnikova

Received: 15 July 2022
Accepted: 10 August 2022
Published: 11 August 2022



**Abstract:** Chronic venous disease (CVD) is a common vascular disorder characterized by increased venous hypertension and insufficient venous return from the lower limbs. Pregnancy is a high-risk situation for developing CVD. Approximately a third of the women will develop this condition during pregnancy, and similarly to arterial hypertensive disorders, previous evidence has described a plethora of alterations in placental structure and function in women with pregnancy-induced CVD. It is widely known that arterial-induced placenta dysfunction is accompanied by an important immune system alteration along with increased inflammatory markers, which may provide detrimental consequences for the women and their offspring. However, to our knowledge, there are still no data collected regarding cytokine profiling in women with pregnancy-induced CVD. Thus, the aim of the present work was to examine cytokine signatures in the serum of pregnant women (PW) with CVD and their newborns (NB). This study was conducted through a multiplex technique in 62 PW with pregnancy-induced CVD in comparison to 52 PW without CVD (HC) as well as their NB. Our results show significant alterations in a broad spectrum of inflammatory cytokines (IL-6, IL-12, TNF-α, IL-10, IL-13, IL-2, IL-7, IFN-γ, IL-4, IL-5, IL-21, IL-23, GM-CSF, chemokines (fractalkine), MIP-3α, and MIP-1β). Overall, we demonstrate that pregnancy-induced CVD is associated with a proinflammatory environment, therefore highlighting the potentially alarming consequences of this condition for maternal and fetal wellbeing.

**Keywords:** pregnancy-induced CVD; chronic venous disease; proinflammatory cytokines; hypertensive vascular disorder; MeSH

## 1. Introduction

Chronic venous disease (CVD) is a common vascular disorder characterized by insufficient venous return from the lower extremities and an increase in venous pressure, known as venous hypertension [1,2]. Clinical manifestations of CVD can range from mild to severe, such as telangiectasia, venous ulceration, lipodermatosclerosis, and, most notably, varicose veins. Risk factors for developing CVD include advanced age, obesity, genetics, and female sex [3–5]. Pregnancy also appears to be a major risk factor for developing CVD. Indeed, it is estimated that approximately 40% of women will suffer from this condition during pregnancy, and this increased risk is even higher with the number of pregnancies [6,7]. During pregnancy, there are many changes in the hemodynamics of the cardiovascular system accompanied by hormonal and mechanical variations. Some of these changes include vasodilation, iliac vein compression, stasis, decreased venous flow velocity, and venous valve incompetence, hence leading to the development of CVD [8–12]. Previous studies have demonstrated the impact of CVD-induced venous hypertension on placental integrity. Specifically, evidence of cellular damage, hypoxia, increased calcification, oxidative stress, and increased vascularization was observed in the placentas of CVD patients [13–16]. These pathological manifestations are also observed in pre-eclampsia, an analogous condition also characterized by vascular hypertension [2,17,18]. However, the effects of CVD on maternal and fetal wellbeing have not been as widely studied as pre-eclampsia.

The pathogenesis of CVD in pregnancy remains poorly understood. Cytokines are central players in immune system modulation and also show modulatory effects on different tissues and organs [19]. Alterations in cytokine production and circulating levels have been involved in the pathogenesis of organ and systemic damage [20]. Different patterns of variation in cytokine serum levels have been to be associated with different inflammatory diseases [21]. They may provide either beneficial effects, collaborating with host defense, or they can be related to adverse outcomes under pathological conditions when they are dysregulated [19]. Pregnancy is, in part, inflammatory status, and a broad range of studies have demonstrated the relevance of analyzing cytokine signatures in both physiological and pathological conditions [22–27]. Conversely, elevated levels of inflammatory cytokines, such as TNF-$\alpha$ and IL-6, have been implicated in the pathogenesis of vascular hypertension during pregnancy [28,29]. Inflammation often mediates the deterioration of healthy tissue, and proinflammatory cytokines instigate apoptotic pathways responsible for many of the clinical presentations of CVD [30]. It has been shown that proinflammatory cytokines are significantly elevated in patients with CVD compared to healthy controls. On aggregate, these previous data on proinflammatory cytokine levels in both CVD and pregnancy-induced vascular hypertension suggest that proinflammatory cytokines play an important role in pregnancy-induced CVD. This study aims to identify the systemic effects of pregnancy-induced CVD by measuring proinflammatory cytokines in the peripheral blood of mothers as well as of their newborns.

## 2. Results

### *2.1. Women with CVD during Pregnancy Show an Increase in Different Proinflammatory Cytokines*

Analysis of plasma samples has shown a significant increase in many proinflammatory cytokines. For instance, we observed a significant increase in IL-6 levels in the plasma of PW-CVD patients (PW-HC = 3.168 $\pm$ 4.897 pg/mL vs. PW-CVD = 4.791 $\pm$ 314.900 pg/mL, $p$ ** = 0.029, Figure 1A). This trend of a significant increase is similarly observed in the plasma of NB-CVD without being statistically significant (NB-HC = 1.988 $\pm$ 1.555 pg/mL vs. NB-CVD = 10.685 $\pm$ 19.350 pg/mL, $p$ = 0.3167, Figure 1A). In contrast, no significant differences were observed in the plasma levels of IL-1B in the study patients (PW-HC = 5861 $\pm$ 11,486 pg/mL vs. PW-CVD = 2.692 $\pm$ 3.598 pg/mL, $p$ = 0.9302, NB-HC = 1.231 $\pm$ 1.220 pg/mL vs. NB-CVD = 2.625 $\pm$ 4.536 pg/mL, $p$ = 0.8366, Figure 1B).

Simultaneously, an increase in TNF-$\alpha$ was reported in pregnant women with CVD (PW-CVD) compared to PW-HC (PW-HC = 14.002 $\pm$ 23.096 pg/mL vs. PW-CVD = 12,295 $\pm$ 5477 pg/mL, $p$ * = 0.0167, Figure 2A). Likewise, NB-CVD display a notable increase in this cytokine in comparison to NB-HC (NB-HC = 6225 $\pm$ 2360 pg/mL vs. NB-CVD = 12,076 $\pm$ 3079 pg/mL, $p$ *** < 0.0001). Likewise, our results reported a significant increase in the proinflammatory cytokine IL-12 in pregnant women with CVD (PW-CVD) compared to PW-HC, as well as in NB-CVD (PW-HC = 0.698 $\pm$ 0.320 pg/mL vs. PW-CVD = 3.569 $\pm$ 1.617 pg/mL,

$p$ *** $< 0.0001$, NB-HC = 0.965 ± 0.469 pg/mL vs. NB-CVD = 3917 ± 1335 pg/mL, $p$ *** $< 0.0001$, Figure 2B). Subsequently, a significant increase in the proinflammatory cytokine IL-2 was observed in PW-CVD with respect to PW-HC, as well as in NB-CVD (PW-HC = 1112 ± 3343 pg/mL vs. PW-CVD = 1583 ± 0.987 pg/mL, $p$ *** $< 0.0001$, NB-HC = 1948 ± 6268 pg/mL vs. NB-CVD = 2309 ± 3108 pg/mL, $p$ *** = 0.0002, Figure 2C).

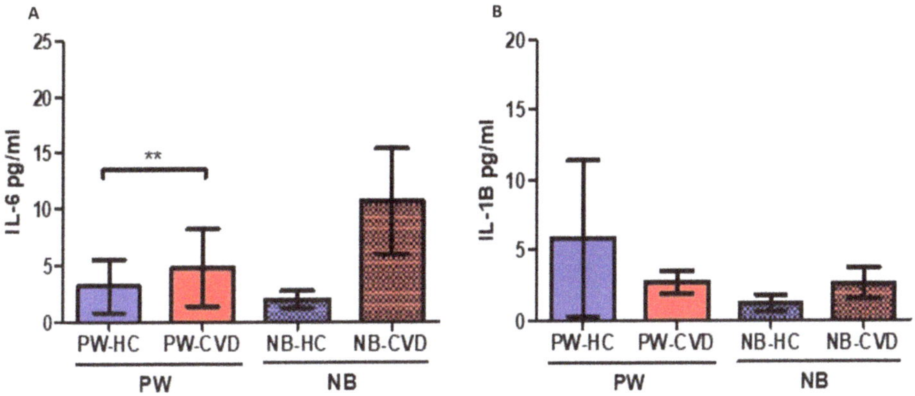


**Figure 1.** Histogram showing the significant increase in IL-6 in PW-CVD but not in their NB (**A**). IL-1B did not show any variation neither in the plasma of PW-CVD or NB-CVD (**B**). PW-HC = pregnant women without vascular pathology, PW-CVD = pregnant women with chronic venous disease during gestation, NB-HC = newborns of mothers without vascular pathology, NB-CVD = newborns of mothers with chronic venous disease during gestation. $p < 0.01$ (**).

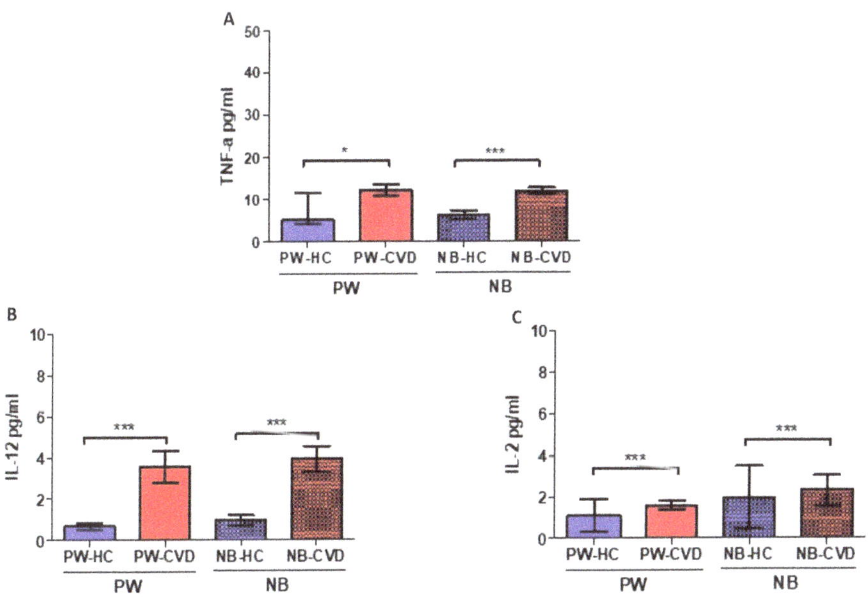


**Figure 2.** Histogram showing the significant increase in TNF-a (**A**), IL-12 (**B**), and IL-2 (**C**) in PW-CVD plasma and in NB-CVD. PW-HC = pregnant women without vascular pathology, PW-CVD = pregnant women with chronic venous disease during gestation, NB-HC = newborns of mothers without vascular pathology, NB-CVD = newborns of mothers with chronic venous disease during gestation. $p < 0.05$ (*), $p < 0.001$ (***).

Similarly, IL-17A levels were significantly higher in PW-CVD compared to PW-HC, as well as in NB-CVD (PW-HC = 1555 ± 1055 pg/mL vs. PW-CVD = 6119 ± 3244 pg/mL, *p* *** < 0.0001, NB-HC = 2384 ± 1711 pg/mL vs. NB-CVD = 7245 ± 2381 pg/mL, *p* *** < 0.0001, Figure 3A). No significant differences were observed in plasma levels of IL-21 in PW-CVD with respect to PW-HC (PW-HC = 1142 ± 0.843 pg/mL vs. PW-CVD = 2197 ± 1991 pg/mL, *p* = 0.0871, Figure 3B). However, plasma IL-21 levels were significantly higher in NB-CVD compared to NB-HC (NB-HC = 2036 ± 1501 pg/mL vs. NB-CVD = 5.124 ± 5.428 pg/mL, *p* * = 0.0174, Figure 3B). Moreover, we found a significant increase in the proinflammatory cytokine IL-23 in pregnant women with CVD (PW-CVD) compared to PW-HC, as well as in NB-CVD (PW-HC = 54,831 ± 51,632 pg/mL vs. PW-CVD = 208.095 ± 144.753 pg/mL, *p* *** < 0.0001, NB-HC = 93.715 ± 78.808 pg/mL vs. NB-CVD = 273,872 ± 196,395 pg/mL, *p* *** = 0.0008, Figure 3C). Similarly, the levels of the proinflammatory cytokine IL-7 were significantly higher in PW-CVD, being similar in NB-CVD (PW-HC = 8778 ± 4162 pg/mL vs. PW-CVD = 31,549 ± 42,609 pg/mL, *p* *** < 0.0001, NB-HC = 10,281 ± 4534 pg/mL vs. NB-CVD = 23,503 ± 6550 pg/mL, *p* *** < 0.0001, Figure 3D).

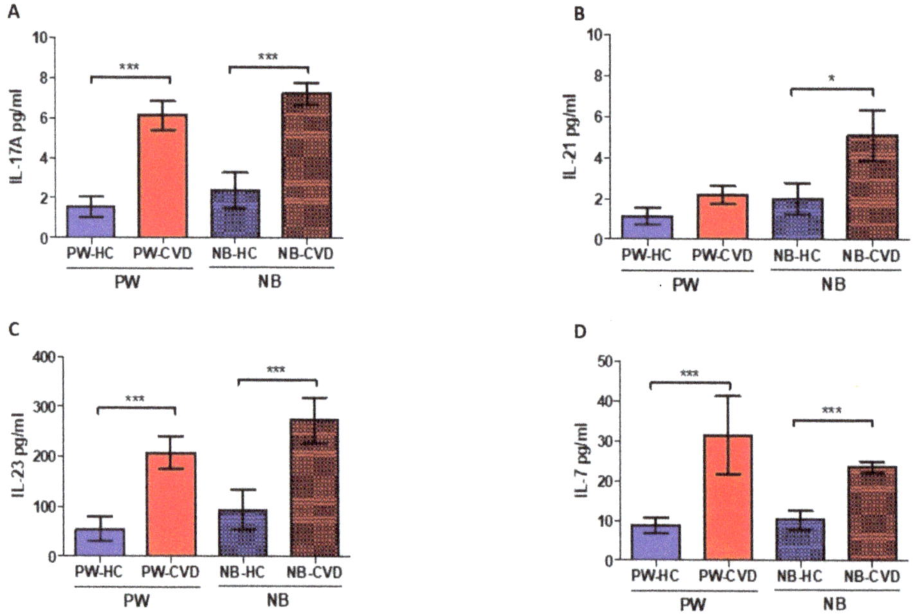


**Figure 3.** Histogram showing a significant increase in IL-17A (**A**), IL-23 (**C**), and IL-7 (**D**) in PW-CVD plasma and in NB-CVD. Likewise, an increase in IL-21 (**B**) in NB-CVD is also observed. PW-HC = pregnant women without vascular pathology, PW-CVD = pregnant women with chronic venous disease during gestation, NB-HC = newborns of mothers without vascular pathology, NB-CVD = newborns of mothers with chronic venous disease during gestation. $p < 0.05$ (*), $p < 0.001$ (***).

### *2.2. Women with CVD during Pregnancy Show a Decrease in Anti-Inflammatory Cytokines*

Our results showed a significant decrease in IL-4 levels in PW with CVD (PW-CVD) with respect to PW-HC (PW-HC = 23,642 ± 19,978 pg/mL vs. PW-CVD= 19,149 ± 66,704 pg/mL, *p* *** < 0.0001, Figure 4A). In parallel, we have observed a decrease in IL-4 in the umbilical cord plasma of NB-CVD with respect to NB-HC (NB-HC = 22,469 ± 13,756 pg/mL vs. NB-CVD = 37,471 ± 88,639 pg/mL, *p* * = 0.0265, Figure 4A). Furthermore, our analyses have shown a significant decrease in the anti-inflammatory cytokine IL-10 in PW with CVD (PW-CVD) with respect to PW-HC (PW-CVD = 6958 ± 3949 pg/mL vs. PW-HC = 4.354 ± 4.596 pg/mL, *p* * = 0.0102, Figure 4B). In parallel, we have observed a decrease in IL-10 in the umbilical cord plasma of NB-CVD with respect to NB-

HC (NB-CVD = 8611 ± 5201 pg/mL vs. NB-HC = 8.307 ± 16.487 pg/mL, $p$ * = 0.0127, Figure 4B). Our analysis has reported a significant decrease in the anti-inflammatory cytokine IL-13 in PW-CVD with respect to to PW-HC (PW-CVD = 2094 ± 5902 pg/mL vs. PW-HC = 5973 ± 2975 pg/mL, $p$ *** < 0.0001, Figure 4C). Simultaneously, we have observed a decrease in IL-13 in the umbilical cord plasma of NB-CVD with respect to NB-HC (NB-CVD = 3453 ± 7995 pg/mL vs. NB-HC = 7.121 ± 3.439 pg/mL, $p$ *** = 0.0002, Figure 4C).

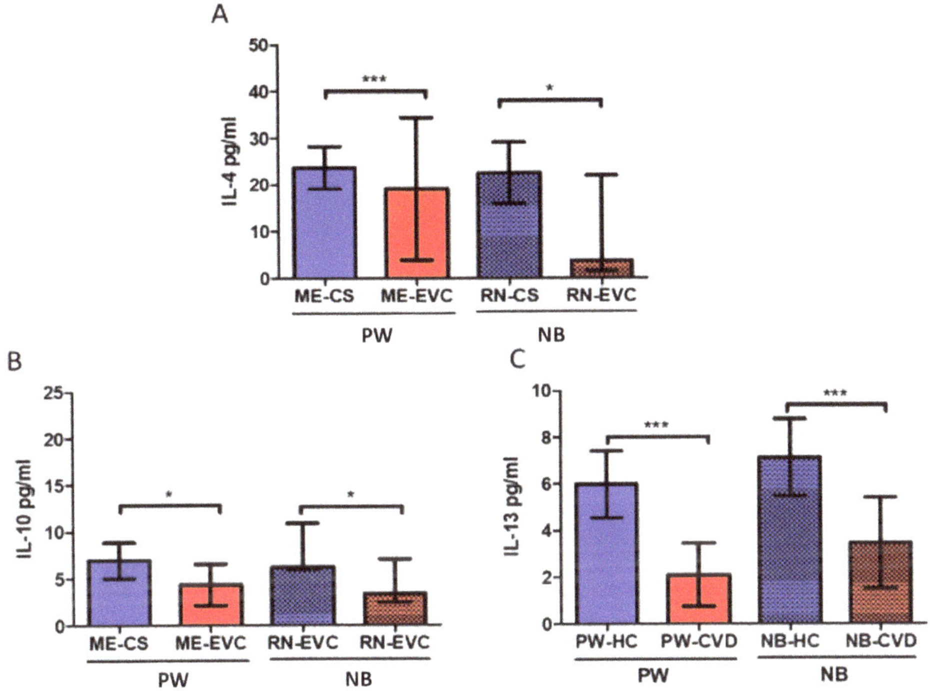


**Figure 4.** Histogram showing the significant decrease in the anti-inflammatory cytokines (**A**) IL-4, (**B**) IL-10, and (**C**) IL-13 in PW-CVD plasma and in NB-CVD. PW-HC = pregnant women without vascular pathology, PW-CVD = pregnant women with chronic venous disease during gestation, NB-HC = newborns of mothers without vascular pathology, NB-CVD = newborns of mothers with chronic venous disease during gestation. $p < 0.05$ (*), $p < 0.001$ (***).

### 2.3. *Women with CVD during Pregnancy Showed a Decrease in IFN-γ*

A significant decrease in INF-γ levels was observed in pregnant women with CVD (PW-CVD) with respect to PW-HC (PW-HC = 35,221 ± 15,473 pg/mL vs. PW-CVD = 9390 ± 9130 pg/mL, $p$ *** < 0.0001, Figure 5. In parallel, we have observed a decrease in INF-γ in the umbilical cord plasma of NB-CVD with respect to NB-HC (NB-HC = 40,815 ± 15,181 pg/mL vs. NB-CVD = 15,969 ± 16,285 pg/mL, $p$ *** < 0.0001, Figure 5).

### 2.4. *Women with CVD during Pregnancy Show an Increase in the Eosinopoietins GM-CSF and IL-5*

Analysis of plasma samples has shown a significant increase in GM-CSF in pregnant women with CVD (PW-CVD) compared to PW-HC (PW-HC = 5900 ± 3276 pg/mL vs. PW-CVD = 12,359 ± 10,980 pg/mL, $p$ * = 0.0104, Figure 2). In parallel, we have observed an increase in GM-CSF in the umbilical cord plasma of NB-CVD with respect to NB-HC (NB-HC = 6567 ± 5234 pg/mL vs. NB-CVD = 18.108 ± 17.329 pg/mL, $p$ ** = 0.0084, Figure 6A).

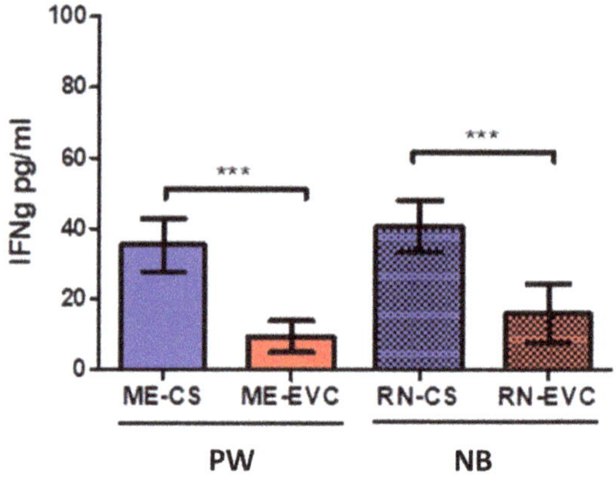


**Figure 5.** Histogram showing a significant decrease in IFN-γ in PW-CVD plasma and in NB-CVD. PW-CVD = pregnant women with chronic venous disease during gestation, NB-CVD = newborns of mothers with chronic venous disease during gestation. $p < 0.001$ (***).

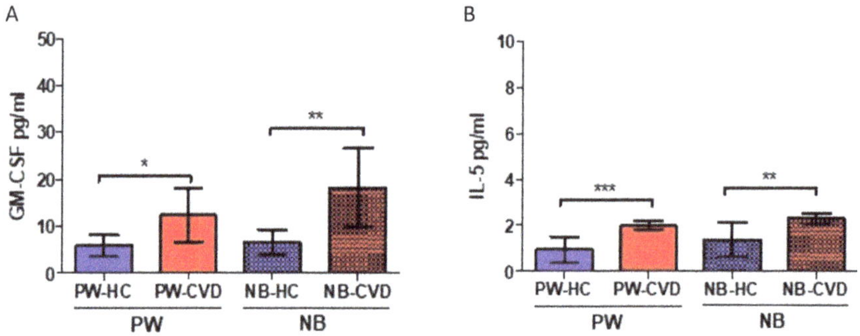


**Figure 6.** Histogram representing the significant increase in GM-CSF (**A**) and IL-5 (**B**) in plasma from PW-CVD and NB-CVD. PW-HC = pregnant women without vascular pathology; PW-CVD = pregnant women with chronic venous disease during pregnancy; NB-HC = newborn from mothers without vascular pathology; NB-CVD = newborn from mothers with CVD during pregnancy. $p < 0.05$ (*), $p < 0.01$ (**), $p < 0.001$ (***).

Similarly, IL-5 levels were significantly higher in PW-CVD compared to PW-HC, as well as in NB-CVD (PW-HC = 0.936 ± 1.196 pg/mL vs. PW-CVD = 1.987 ± 0.884 pg/mL, *p* *** = 0.0002, NB-HC = 1.387 ± 1.447 pg/mL vs. NB-CVD = 2.316 ± 0.957 pg/mL, *p* *** = 0.0019, Figure 6B).

*2.5. Women with CVD during Pregnancy Show a Significant Increase in Plasmatic Chemokines*

The study of plasma samples did not show significant differences in MIP-1a levels in PW-CVD compared to PW-HC, just as in NB (PW-HC = 11.153 ± 19.986 pg/mL vs. PW-CVD = 75.099 ± 176.796 pg/mL, $p = 0.8674$, NB-HC = 9.165 ± 7.764 pg/mL vs. NB-CVD = 5.077 ± 3.881 pg/mL, $p = 0.1322$, Figure 7A). On the contrary, a significant increase in MIP-1b was observed in PW-CVD compared to PW-HC, just as in NB (PW-HC = 34.131 ± 47.936 pg/mL vs. PW-CVD = 48.097 ± 19.065 pg/mL, *p* = *** 0.007, NB-HC = 16.585 ± 11.144 pg/mL vs. NB-CVD = 550.822 ± 17.412 pg/mL, *p* *** < 0.0001, Figure 7B). In this line, a significant increase in MIP-3a was evinced in PW-CVD compared to PW-HC, just as in NB (PW-HC = 12.096 ± 6.086 pg/mL vs. PW-CVD = 30.241 ± 21.189 pg/mL, *p* = *** 0.0003, NB-HC = 12.759 ± 3.407 pg/mL vs. NB-CVD = 26.338 ± 13.532 pg/mL, *p* *** < 0.0001, Figure 7C).

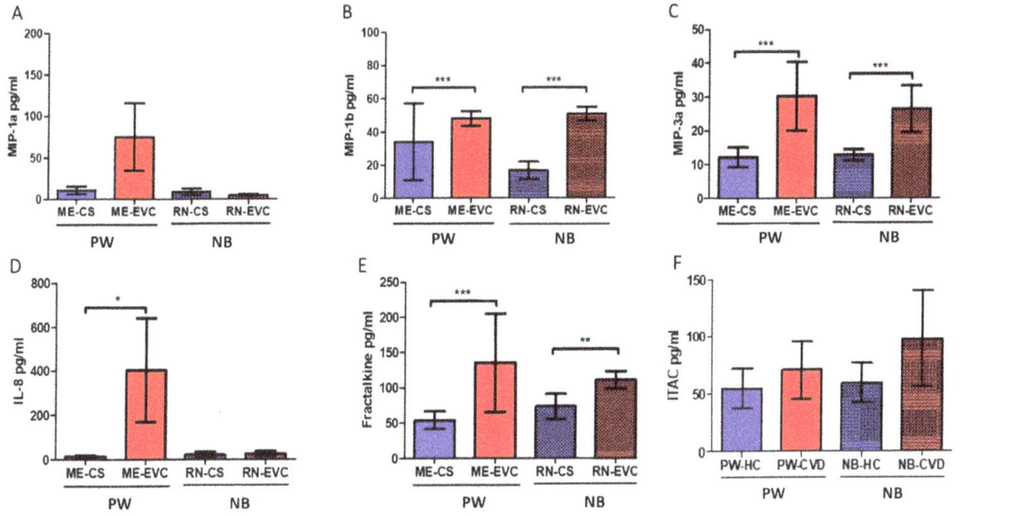


**Figure 7.** Histogram representing a significantly increased level of plasmatic chemokines (**A**) MIP-1a, (**B**) MIP-1b, (**C**) MIP-3a, (**D**) IL-8, (**E**) Fractalkine, and (**F**) ITAC in PW-CVD and in NB-CVD. PW-CVD = pregnant women with chronic venous disease during pregnancy; NB-HC = newborn from mothers without vascular pathology; NB-CVD = newborn from mothers with CVD during pregnancy. $p < 0.05$ (*), $p < 0.01$ (**), $p < 0.001$ (***).

Our results have demonstrated a significant increase in IL-8 plasmatic levels in PW-CVD; however, an upward trend was only observed in NB-CVD (PW-HC = 14.050 ± 19.501 pg/mL vs. PW-CVD = 405.486 ± 915.893 pg/mL, $p$ = * 0.0209, NB-HC = 20.878 ± 25.607 pg/mL vs. NB-CVD = 25.127 ± 48.059 pg/mL, $p$ = 0.3581, Figure 7D). Moreover, a significant increase in fractalkine was observed in PW-CVD compared to PW-HC (PW-HC = 54.148 ± 26.064 pg/mL vs. PW-CVD = 135.082 ± 143.891 pg/mL, $p$ *** < 0.0001, Figure 7E). Moreover, significant increased levels were observed in NB-CVD's umbilical cord compared to NB-HC (NB-HC = 73.367 ± 34.607 pg/mL vs. NB-CVD = 110.285 ± 26.028 pg/mL, $p$ ** = 0.0022, Figure 7E).

Finally, the study of plasma samples has not shown significant differences either in ITAC levels in PW-CVD with respect to PW-HC or NB-CVD's umbilical cord compared to NB-HC (PW-HC = 54,823 ± 35,371 pg/mL vs. PW-CVD = 70,630 ± 51,407 pg/mL, $p$ = 0.3209, NB-HC = 59,398 ± 35,982 pg/mL vs. NB-CVD = 97.948 ± 814.592 pg/mL, $p$ = 0.2540, Figure 7F).

## 3. Discussion

In the present work, we have demonstrated that CVD leads to an altered cytokine signature in the PW and their NB in comparison to those without this condition. More detailly, we have observed an increased serum level of several proinflammatory cytokines but reduced levels of anti-inflammatory cytokines and INF-γ. Simultaneously, we have observed raised serum chemokines and GCSF, measurable in the PW with CVD and their NB.

CVD is a multifactorial disease with complex pathophysiological mechanisms involved, associated with an important inflammatory response [31]. CVD involves a powerful alteration in the immune inflammatory system, with a significant increase in plasmatic innate and adaptive cytokines. In fact, it has been demonstrated that CVD itself causes noteworthy changes in cytokine production by the immune cells, hence inducing proinflammatory profiling [32]. CVD has also been associated with placental, umbilical cord, and systemic oxidative stress [14,33]. Moreover, an altered local detection of some cytokines has also been observed in the placenta of women with CVD, which is closely related to abnormal cell and molecular behavior [34–36]. This inflammatory, hypoxic, and also oxidative environment could be part of fetal programming, as some previous studies suggest [37–39]. In this sense, we show the possible role of a group of cytokines in PW undergoing CVD that could simultaneously affect NB, showing a proinflammatory state. To our knowledge, our study is the first to evidence a unique cytokine signature in this group of patients, therefore showing that CVD may be a deteriorating condition with important consequences for both PW and NB.

Cytokines can be classified according to different criteria such as molecular composition, interaction receptor, main cell producers, and targets [40]. However, from a pathogenic point of view, cytokines are defined by their effect on the immunoinflammatory response as proinflammatory and anti-inflammatory. They are produced by leukocytes and other cells, being essential to orchestrate immune cells growth, differentiation, and activation [41]. Moreover, many of these cytokines have provided their clinical relevance in a wide variety of conditions, including during normal pregnancy or its complications [26,29]. This is mainly due to the fact that many of these cytokines can cross the placental barrier, although it is in dispute to what extent this fact occurs [42]. In our study, we found significant alterations in diverse interleukins, including IL-6, IL-2, IL-12, IL-7, IL-21, IL-23, IL-10, IL-13, IL-4, and IL-5.

Our data clearly show a marked increase in serum levels of proinflammatory cytokines in PW-CVD. Unfortunately, this maternal immunoinflammatory disturbance is also found in NB-CVD. We have found increased IL-6 and TNF-α serum levels in both populations. In agreement with this fact, we also observed increased IL-6 levels in the placenta of women with CVD [34]. IL-6 is mostly a proinflammatory but also anti-inflammatory cytokine with pleiotropic effects in the organism [43]. For instance, IL-6 participates in B-cell differentiation and stimulation of acute phase proteins [41]. Increased maternal IL-6 levels have been related to the development and severity of different pregnancy-associated complications such as pre-eclampsia or chorioamnionitis [44,45]. IL-6 with TNF-α exert synergistic proinflammatory effects [46]. It seems that high levels of both cytokines promote trophoblasts cell death, hence impairing placental function [47]. In addition, high levels of IL-6 in the umbilical cord have been associated with the requirement of oxygen at 36 weeks of post-menstrual age in small for gestational age newborns [48]. Furthermore, increased maternal serum levels of IL-6 and TNF-α have been associated with hypertensive disorders during pregnancy [49–51]. In this line, our study might indicate an important correlation between CVD with high levels of IL-6 and TNF-α. Because of that, TNF-α has been proposed as a potential target for preventing placental and fetal complications of pregnancy [52]. Further studies could be designed to evaluate therapeutical approaches of TNF-α inhibitors in pregnant women with CVD as well as to avoid possible fetus or newborn complications. Moreover, we cannot disregard the fact that IL-6 has also been described as an elevated inflammatory mediator during labor onset [50], but this condition was shared by both groups of women.

In contrast with the increased serum levels of IL-6 with TNF-α, our results show normal concentrations of IL-1β. Different patterns of alterations in the levels of these cytokines have been observed in inflammatory diseases, as well as different clinical responses to specific anti-cytokine treatments [53]. These findings suggest that the cellular mechanisms involved in the pathogenesis of the proinflammatory status of PW-CVD are specific.

Associated with the systemic proinflammatory environment observed in PW-CVD, we have found a marked disbalance of the circulating cytokines secreted by the different Th subsets. An increase in Th1 cytokines has been observed in these women and their NB. IL-2, IL-12, and TNF-α are critical cytokines involved in Th1 responses, while IL-4 and IL-10 inhibit this polarization [54]. IL-2, also called T-cell growth factor, is a central cytokine involved in the proliferation and differentiation of both adaptative and innate immune cells [55]. IL-2 is produced by polarized Th1 cells, and it has central effects on the activation of B cells, monocytes, natural killers (NKs), innate lymphoid cells (ILCs), as well as modulating effector T cells and T reg activity [56]. However, it is hypothesized that IL-2 proinflammatory/anti-inflammatory effects might be determined by the amount and kinetics of IL-2. Thus, a high but transient level of IL-2 appears to be associated with effector cell development, while low-grade IL-2 presence could be related to T reg induction [57]. T reg populations are essential for gestational success, and a correct IL-2—STAT5 signaling with adequate levels of T reg has been associated with the prevention of autoimmunity and human recurrent abortions [58]. Oppositely, increased levels of IL-2 have been related to higher NK cytotoxicity, which has been proposed as a risk factor for human recurrent abortions [59]. Increased IL-2 levels described in PW and NB related to CVD may indicate a likely imbalance of Th1/T regs and NK cytotoxicity, therefore supporting a proinflammatory status affecting both individuals. Moreover, we have also reported increased levels of further Th1 cytokines, including the proper TNF-α and IL-12, along with a reduction in IL-4 and IL-10. IL-12 is a crucial cytokine involved in IFN-γ production [60], also related to pathogenic Th1 differentiation [61]. Simultaneously, IL-12 is also associated with an imbalance in Th1/Th2 cells, which has been associated with pregnancy complications such as recurrent spontaneous abortion, obstetric complications, and poor pregnancy outcomes [62]. Despite the elevated IL-12 levels found, we report a significant IFN-γ decrease in both PW with CVD and NB. The role of IFN-γ in pregnancy has already been well-described and substantial alterations of this cytokine appear to be related to different pregnancy complications such as preterm labor [63]. In this line, Scott et al. [64] also reported high levels of IL-12 without IFN-γ induction by immune cells extracted from cord blood. More recently, a reduction in IFN-γ levels was observed in PW with pre-eclampsia [65]. It is probable that IFN-γ diminishment could be associated with pathological conditions such as CVD, although further works should clarify the mechanisms involved in its dysregulation.

IL-4, IL-10, and IL-13 are three anti-inflammatory cytokines significantly decreased in our study in both PW and NB. IL-10 was first discovered as a product secreted by Th2 cells, although this cytokine is secreted by many types of immune cells, being capable of reducing proinflammatory cytokines release and Th1 responses [66,67]. It causes inhibition of IL-2 and interferon gamma [41]. One of the most important roles of IL-10 is to provide contrary effects to TNF-α. Thus, an adequate balance between IL-10 and TNF-α is crucial during pregnancy, and reductions in IL-10 levels with augmented TNF-α might be related to pathological inflammation during this period [68]. In addition, deficiencies in IL-4 and IL-10 cytokines have been associated with a plethora of pregnancy-related disorders, including infertility, spontaneous miscarriage, preterm birth, fetal growth restriction, pre-eclampsia, gestational hypertension [69], and as we have just demonstrated with CVD. IL-4 is synthesized by CD4+T cells, and it is a major inductor of Th2 differentiation while inhibiting Th1 phenotyping, acting co-ordinately with IL-13 in the alternative macrophage polarization (M2 responses), among other effects [70]. Animal models show that the absence of IL-4 is sufficient to induce pregnancy hypertension accompanied by excessive inflammation in IL-4-deficient mice [71]. Low levels of IL-4 and IL-10 have been described in pregnancies with severe pre-eclampsia [72]. Similarly, low maternal levels of IL-4 and IL-13 have been positively correlated with an increased risk of NB for developing overweight during childhood [73]. Therefore, our results might indicate the pathological role of IL-4, IL-10, and IL-13 reduction due to CVD, furthermore promoting a proinflammatory status and a Th1/Th2 imbalance, which has also been implicated with preterm labor [74].

Additionally, we have observed increased Th17 cytokine levels in PW with CVD and their NB. Accordingly, elevated levels of circulating IL-23 and IL-17A are found in PW-CVD and NB-CVD. IL-23 is a member of the IL-12 family, and it is key to inducing the Th17 pathogenic phenotype through the stabilization of IL-17 (Also known as IL-17A) [60]. Our results show increased IL-17A and IL-23 levels, denoting an abnormal Th17 polarization associated with CVD in PW and NB. IL-17A dysregulation is associated with the development and progression of different inflammatory diseases [75]. A study conducted Eghbal-Fard et al. [76] in 50 women with pre-eclampsia also reported the contribution of higher serum levels of IL-17A and IL-23 in the pathogenesis of the disease, with impaired Th17/Treg ratio. Conversely, other studies only detected significant differences in IL-17 but not in IL-23 [77,78]. In this line, we show that CVD is responsible for the induction of both IL-17 and IL-23 production, which may be implicated in the proper pathogenesis of the disease. In the same line, we report a significant increase in IL-7 in both maternal and fetal serum. IL-7 is crucial for B-cell proliferation, T-cell development, and homeostasis [79]. Additionally, it has been associated with pregnancy complications such as recurrent pregnancy losses due to its ability to induce aberrant Th17 responses and reductions in Treg cells in animal models [80]. In addition, it has been proposed that IL-7 crosses the placental barrier and triggers IL-17R, and could affect fetal neurons producing cortical and behavioral abnormalities [24]. Furthermore, we reported an increased IL-21 in the cord blood obtained from NB. IL-21 is another cytokine produced by T cells and NKT cells, inducing Th17 phenotyping while stimulating NKT, NK, and T cytotoxic subsets proliferation and cytotoxicity [81]. However, in immune cells derived from cord blood, IL-21 seems to stimulate the expression of immunosuppressive IL-10 to diminish Th1 responses [82]. It is probable that increased levels of IL-21 in the NB could emerge as a protective mechanism to diminish the global proinflammatory status.

IL-5 causes B-cell growth factor and differentiation and IgA selection. IL-5, together with granulocyte-macrophage colony-stimulating factor (GM-CSF), plays a key role in eosinophilic function and development, being frequently designed as "eosinopoietins" [83]. Previous research has established the synergic action of GM-CSF and IL-5 on eosinophil activation under inflammatory diseases [84]. Although, classically, the eosinophils were associated with anti-parasite responses, nowadays, it is widely accepted their importance in maintaining tissue homeostasis [85]. Furthermore, eosinophils are involved in the secretion of many immunomodulatory cytokines, integrating different signals and directing inflammatory responses [86]. An altered eosinophilic activity might be a clinical risk of note during mild to late gestation of preterm labor related to type I hypersensitivity reaction [87]. Recently, Lebold et al. [88] have demonstrated that intra-utero exposition to IL-5 results in fetal eosinophilia and as a developmental origin of airway hyperreactivity in the adult offspring. Regarding GM-CSF, this cytokine is produced by different cells from the innate and adaptative immune system with major effects in bone marrow, where stem cells are provoked to mature not only into eosinophils but also into monocytes and macrophages [89]. GM-CSF is importantly produced both by Th17 and Th1 cells after IL-23 and IL-12, respectively [90,91]. In addition, the IL-7 axis is involved in GM-CSF production by Th subsets that could lead to autoimmune diseases such as type 1 diabetes mellitus [92]. Besides its role in cell growth, it may act as a proinflammatory cytokine in infections and activates the following pathways: JAK/STAT, PI3K, MAPK, and NFκB [93]. This factor has an important role in fertility and in embryo implantation and is crucial for placental development [94]. In fact, reduced levels of this cytokine during pregnancy were related to recurrent miscarriages [95], placental dysfunction, and abnormal fetal growth [96]. Conversely, Huang et al. [97] described the crucial role of aberrant GM-CSF expression in the pathogenesis of pre-eclampsia, acting as a powerful inductor of inflammatory cells. Increased levels of IL-5 and GM-CSF may indicate an abnormal activation of eosinophils in pregnancy-associated CVD. Future studies should be conducted to unravel the possible role of eosinophils in the pathogenesis of the disease and its consequences in newborns.

Chemotactic cytokines are produced by mast cells and stimulate the migration of several cells, mostly WBCs, not only being involved in all kinds of immune responses but also in many other biological processes such as angiogenesis, embryonic development, phagocytosis, survival, and apoptosis [98]. By following gradients of several kinds of chemokines, cells are usually guided to the site of interest in homeostasis and inflammation and linking innate and adaptative responses [99]. In this study, there were four chemokines found from two different subfamilies, CC and CXC (classification is based upon cysteine residues position): CCL4, CCL20, IL-8, and CX3CL1, all of them with significance in the PW and the NB. CXC chemokines convey chemotactic activity for neutrophils and CC for monocytes and Th subsets, although there are exceptions [100]. Chemokines are critical regulators for trophoblasts invasion. The rising evidence alleges that chemokines are considered regulatory molecules that, due to their selective trafficking of immune cells, settle a normal or a pathological placental status and delivery [101]. The expression of chemokines in the endometrium orchestrates the appropriate infiltration of immune cells and invasion of trophoblasts in the maternal vasculature. It is known that trophoblast cells express countless membrane receptors for these chemokines as well, contributing to fetal immunity besides placental development.

IL-8 (CXCL8) is released by NK cells implying the migration of trophoblast cells for endovascular invasion and maternal vascular remodeling [102]. Elevated levels of cord blood IL-8 have been associated with pre-eclampsia [103] and moderate-severe bronchopulmonary dysplasia in NBs [48]. The chemokine network at the fetal-maternal interface also looks decisive in the future adult's health. Our results denote a decrease in the inhibitors of IL-8, which are anti-inflammatory cytokines IL-4 and IL-13. When blocking IL-8, and hence neutrophil migration, by these two, the polarization of Th subsets tends to Th2 type [104]. As in this case, there is no impedance for IL-8 activity together with other chemokines and interleukins, and Th2 response is decreased. We also found increased TNF-α, which upregulates IL-8 [105], agreeing with the high IL-8 obtained. At the same time, at normal term, choriodecidua and amnion also produce IL-8 [50], boosting the activity of MMPs and other compounds; meanwhile, IL-6 and TNF-α also stimulate these components leading to collagenolysis [101]. In previous studies, we found increased levels of MMP-9 and COL-III, affecting the structure of the placentas of women with venous insufficiency during pregnancy [106]. Then, if certain chemokines may upregulate the expression of collagenolytic components, CVD may increase these even more.

Furthermore, we found significant plasma levels of fractalkine (CX3CL1) and CCL4 (MIP-1β). Some studies have found an association between fractalkine and later pre-eclampsia, concretely decidual cell secreted CX3CL1 but not circulating [107]. Other trials have denoted that pregnant women with pre-eclampsia present an overexpression of fractalkine, coinciding with poor pregnancy outcomes [108]. Recently, elevated levels of fractalkine in maternal serum in pre-eclampsia have been described [109]. Moreover, CX3CL1 chemoattractant and adhesive properties breeding inflammation and angiogenesis processes are especially upregulated by inflammatory conditions such as diabetic placenta. The evidence also demonstrates a robust upregulation by hypoxia conditions [110]; hence, we could see CVD women also have overexpression of fractalkine compared to the control group, which could be promoted by hypoxia pathways such as HIF-1α, as we previously found in placenta from women with venous insufficiency [18]. CCL4 has been observed with enhanced expression related to implantation competence, serving as a predictor of pregnancy labor [111]. Increased detection of serum CCL4 has been associated with the presence of active infections during pregnancy [112]. Trophoblast migration also reacts to CCL4 and CX3CL1, being also key for maternal-fetal communication [113].

Finally, CCL20 (MIP-3α) was significantly increased in PW affected by CVD and their NB as well. CCL20 is known to be chemotactic and antimicrobial [114], and evidence says that it is a Th-17 response associated with chemokine, inducing inflammation [115]. The presence of this cytokine within amniotic fluid has been associated with microbial invasion and amniotic inflammation in preterm labors [111]. In the absence of infection, the bioavailability of CCL20 in amniotic fluid was associated with the partum process. However, it remains elusive if maternal serum concentrations of CCL20 might be indicative of intra-amniotic infection or inflammation [116].

Taken together, our results demonstrate a severe disturbance of cytokines and chemokines in PW with CVD and their NB. The interactions and possible implications of the abnormal pool of the cytokines mentioned before and the NB remains to be explored. In Table 1, the main findings and discussion about the different cytokines detected in our study are summarized.

**Table 1.** Cytokines found in multiplex analysis. Results from PW with CVD and their NB, description, and possible implications. ↑ (increased), ↓ (decreased), - (no change), $p < 0.05$ (*), $p < 0.01$ (**), and $p < 0.001$ (***).

| Significantly Altered Cytokines | | | | | | |
|---|---|---|---|---|---|---|
| **Cytokines** | **Original Designation** | **Abbreviatures** | **Targets and Functions** | **Pregnancy-Induced CVD** | **Previous Studies and Possible Implications** | **References** |
| **Interleukins** | Interleukin-6 | IL-6 | A major proinflammatory cytokine. Synergic effects with TNF-α. B-cell differentiation and stimulation of acute phase proteins. | PW: ↑ ** NB: - | Increased maternal IL-6 levels have been related to the development and severity of different pregnancy-associated complications. High levels of IL-6 in the umbilical cord have been associated with the requirement of oxygen at 36 weeks of post-menstrual age in small for gestational age newborns. | [44,45,48] |
| | Interleukin-12 | IL-12 | Involved in pathogenic Th1 responses and IFN-γ production. It causes inhibition of IL-2 and interferon gamma. | PW: ↑ *** NB: ↑ *** | High IL-12 and low IFN-γ were observed in mononuclear cord blood cells. Th2-type response has been associated with pregnancy complications such as recurrent spontaneous abortion, obstetric complications, and poor pregnancy outcomes. | [41,62,64] |
| | Interleukin-10 | IL-10 | Anti-inflammatory cytokine Diminish Th1 responses and induce T reg activity. | PW: ↓ * NB: ↓ * | IL-10 and IL-4 reduction is associated with a plethora of pregnancy-related disorders, including infertility, spontaneous abortion, preterm birth, fetal growth restriction, pre-eclampsia, gestational hypertension | [68,69] |
| | Interleukin-13 | IL-13 | Anti-inflammatory effects acting synergically with IL-4 to promote Th2 responses | PW: ↓ *** NB: ↓ *** | Maternal levels of IL-4 and IL-13 were directly correlated with a decreased risk of NB for developing overweight in 1–2 years old | [73] |
| | Interleukin-2 | IL-2 | Pleiotropic effects on multiple immune populations. At high levels, it induces Th1 responses | PW: ↑ *** NB: ↑ *** | IL-2 dysregulation may negatively affect Treg expansion during pregnancy. Increased levels of IL-2 have been related to higher NK cytotoxicity, which has been proposed as a risk factor for human recurrent abortions. | [59,79] |

**Table 1.** *Cont.*

| Significantly Altered Cytokines | | | | | | |
|---|---|---|---|---|---|---|
| **Cytokines** | **Original Designation** | **Abbreviatures** | **Targets and Functions** | **Pregnancy-Induced CVD** | **Previous Studies and Possible Implications** | **References** |
| | Interleukin-7 | IL-7 | Involved in T-cell development and homeostasis. B-cell proliferation. | PW: ↑ *** NB: ↑ *** | During pregnancy, IL-7 promotes an aberrant Th17 response with Treg reductions. Also, IL-7 could affect fetal neurons producing cortical and behavioral abnormalities. | [80] |
| | Interleukin-4 | IL-4 | Anti-inflammatory effects. IL-4 is a central inductor of Th2 responses and Th1 inhibition | PW: ↓ *** NB: ↓ * | IL-10 and IL-4 reduction are associated with a plethora of pregnancy-related disorders, including infertility, spontaneous miscarriage, preterm birth, fetal growth restriction, pre-eclampsia, gestational hypertension. Low maternal levels of IL-4 have been positively correlated with an increased risk of NB for developing overweight during childhood. | [69,73] |
| | Interleukin-5 | IL-5 | Together with GM-CSF and IL-3, they are "eosinopoietins" because of their ability to induce eosinophils proliferation and activation | PW: ↑ *** NB: ↑ ** | An altered eosinophilic activity might be a clinical risk of preterm labor In utero exposition to IL-5 result in fetal eosinophilia and is a developmental origin of airway hyperreactivity | [87,88] |
| | Interleukin-17A | IL-17A | Along with IL-23, it mediates Th17 responses. Involved in the development of many inflammatory diseases | PW: ↑ *** NB: ↑ *** | Studies in women with pre-eclampsia show increased IL-17A levels alone or in combination with IL-23 | [76–78] |
| | Interleukin-21 | IL-21 | Inductor of Th17 responses | PW: - NB: ↑ * | In cord blood cells, it may induce IL-10 production | [82] |
| | Interleukin-23 | IL-23 | Along with IL-17A, it mediates Th17 responses. | PW: ↑ *** NB: ↑ *** | Studies in women with pre-eclampsia show increased IL-17A in combination with IL-23 | [76] |

**Table 1.** *Cont.*

| Significantly Altered Cytokines | | | | | | |
|---|---|---|---|---|---|---|
| **Cytokines** | **Original Designation** | **Abbreviatures** | **Targets and Functions** | **Pregnancy-Induced CVD** | **Previous Studies and Possible Implications** | **References** |
| **Tumor necrosis factor** | Tumor necrosis factor-α | TNF-α | Proinflammatory cytokine that coordinates Th1 responses | PW: ↑ *<br>NB: ↑ *** | High levels of TNF-α alone or with increased IL-6 and low IL-10 are related to pregnancy hypertensive disorders and other complications. | [24,47,51,52] |
| **Interferons** | Type II interferon gamma | IFN-γ | Proinflammatory cytokine that coordinates Th1 responses | PW: ↓ ***<br>NB: ↓ *** | Low IFN-γ levels were detected in women with pre-eclampsia and blood cord despite high IL-12 levels. | [64,65] |
| **Colony-stimulating factors** | Granulocyte-macrophage colony-stimulating factor or colony-stimulating factor 2 | GM-CSF (CSF-2) | Participates in Th1 and Th17 responses<br>Together with IL-5 and IL-3, they are "eosinopoietins" because of their ability to induce eosinophils proliferation and activation | PW: ↑ *<br>NB: ↑ ** | Reduced levels of this cytokine were related to recurrent miscarriage, placental dysfunction, and abnormal fetal growth.<br>Increased levels of this cytokine might be implicated in the pathogenesis of pre-eclampsia.<br>An altered eosinophilic activity might be a clinical risk of preterm labor | [62,87,96,97] |
| **Chemokines** | Fractalkine or chemokine (C-X3-C motif) ligand 1 | $CX_3CL1$ | Chemoatractive properties.<br>Upregulated by hypoxia | PW: ↑ ***<br>NB: ↑ ** | Overexpression of this cytokine is related to poor pregnancy outcomes such as pre-eclampsia and gestational diabetes | [108,110] |
| | Chemokine (CXC motif) ligand-8 or Interleukin-8 | CXCL8 (IL-8) | Neutrophils recruitment.<br>Involved in Th1 responses and inhibited by Th2 cytokines (IL-4 and IL-13) | PW: ↑<br>NB: ↑ | Some studies have found a positive correlation between maternal IL-8 levels and the risk of mental disorders in adulthood offspring.<br>IL-8 induces matrix remodeling in placental tissue.<br>Elevated levels of cord blood IL-8 have been associated with pre-eclampsia and moderate-severe bronchopulmonary dysplasia in newborns | [48,101,103] |

**Table 1.** *Cont.*

| Significantly Altered Cytokines | | | | | | |
|---|---|---|---|---|---|---|
| **Cytokines** | **Original Designation** | **Abbreviatures** | **Targets and Functions** | **Pregnancy-Induced CVD** | **Previous Studies and Possible Implications** | **References** |
| | Macrophage inflammatory protein-1β or Chemokine (C-C motif) ligand 4 | MIP-1β (CCL4) | Chemoattractive molecule of T lymphocites, dendritic cells, monocytes, and NKs; HIV correceptor | PW: ↑ ***<br>NB: ↑ *** | Increased levels of this molecule appear to be indicative of active infections during pregnancy.<br>Together with fractalkine, it is a central component in maternal-fetal dialogue | [112,113] |
| | Macrophage inflammatory protein-3α or chemokine (C-C motif) ligand 20 | MIP-3α (CCL20) | Chemotactic and antimicrobial activity; associated with Th17 polarization and inflammation | PW: ↑ ***<br>NB: ↑ *** | The presence of this cytokine in the amniotic fluid is a marker of infection or inflammation affecting the amniotic cavity. It remains to be elucidated is correlation with serum levels | [111,116] |

## 4. Materials and Methods

### 4.1. Experimental Design

An observational, analytical, and prospective cohort study was conducted on 114 pregnant women (PW) and their newborns (NB). A total of 62 plasma samples from PW diagnosed with CVD during pregnancy and their NB were obtained with a median age of 33 years (22–40) and a median gestational age of 40.5 weeks (39–41.5). Similarly, 52 plasma samples from PW and NB without CVD were also studied during pregnancy, with a median age of 34 years (27–41) and a median gestational age of 41 weeks (39–42).

Exclusion criteria were defined by women with endocrine diseases such as diabetes mellitus; high blood pressure (HBP); body mass index (BMI) > 25 kg/m$^2$; unhealthy habits; active infectious diseases; autoimmune diseases; venous malformations; renal insufficiency; heart failure; pulmonary insufficiency; pre-eclampsia and/or hemolysis, elevated liver enzymes and low platelet syndrome (HELLP); uterine growth restriction of unknown cause; pathological lesions such as placental infarcts, avascular villi, late maturation and chronic inflammation affecting the chorionic villi; the appearance of any of these exclusion criteria described at any time before delivery or prior evidence of CVD.

All the participants were women who had visited their obstetrician at week 32 of gestation (time of blood sample collection). Once the informed consent was signed, her medical history was obtained, and a general physical exploration along with laboratory measurements was performed. An Echo-Doppler (portable M-Turbo Echo-Doppler; SonoSite, Inc., Washington, DC, USA) examination of the lower extremity was performed at 7.5 MHz while the women were in the orthostatic position, and the leg was examined by external rotation with support on the contralateral leg. The study included the greater saphenous axis from the inguinal region to the ankle and the femoral veins. Classification of CVD in participating PW was based on CEAP (Clinical-Etiological-Anatomical-Pathophysiological) [14]. All participants had CEAP scores $\geq 1$ (C1 = 59.67% ($n$ = 37), C2 = 33.87% ($n$ = 21), C3 = 6.45% ($n$ = 4)).

The gestational period of the studied participants was routinely monitored and followed at the Hospital Central de la Defensa Gómez Ulla-UAH (Madrid, Spain), and plasma samples were obtained from the umbilical cord vein at the time of delivery.

The study was carried out in accordance with the basic ethical principles of autonomy, beneficence, non-maleficence, and distributive justice, and its development followed the statements of Good Clinical Practice, the principles contained in the most recent Declaration of Helsinki (2013), and the Convention of Oviedo (1997). The data and information collected complied with current legislation on data protection (Organic Law 3/2018 of December 5, Protection of Personal Data and Guarantee of Digital Rights and Regulation (EU) 2016/679). The project was approved by the Clinical Research Ethics Committee of the Gómez Ulla Military Hospital (37/17).

### 4.2. Determination of Inflammatory Status

Plasma levels of ITAC, IL-10, granulocyte-macrophage colony-stimulating factor (GM-CSF), fractalkine, IFN-γ, MIP-3α, IL-12p70, IL-1β, IL-2, IL-5, IL-13, IL-21, IL-17A, IL-4, IL-23, IL-6, IL-7, IL-8, MIP-1α, MIP-1β, and TNFα were determined. With this aim, we used an aliquot of serum samples that had previously been obtained from peripheral blood in a dry tube by centrifugation at 2000 rpm for 20 min and kept at −80 °C until the moment of quantification.

This study was carried out using the *Luminex* technique with a high sensitivity kit (Milliplex MAP kit, HSTCMAG-28SK-21) from the Merck laboratory (Darmstadt, Germany). For the study of cytokines, microspheres were used, each one encoded with a percentage of red and infrared depending on the element to be studied. These were incubated for 16–18 h with the antigen for binding to the capture antibody of each microsphere in 96-well plates. After incubation, the biotilinated detection antibody for every cytokine was added. Lastly, a streptavidin-phycoerythrin complex (Strep-PE) was employed, which bound the detection antibody. The plate was read on the MAGpix equipment (Merk). Using the standard curve, the Merck analysis program (Analyst) calculated the concentration of each cytokine of interest using the mean fluorescence intensity (MFI). Detection limits were established for all cytokines analyzed according to the protocol.

### *4.3. Statistical Analysis*

For the statistical analysis, the GraphPad Prism® 9.0 program (San Diego, CA, USA) was used, and the Mann–Whitney U test was applied. Data are expressed as the mean with SD. The significant results were established at $p < 0.05$ (*), $p < 0.01$ (**), and $p < 0.001$ (***).

## 5. Conclusions

Overall, our study is the first to demonstrate a proinflammatory cytokine profiling in both PW and NB associated with pregnancy-induced CVD. This could have important consequences in the maternal and fetal environment, thereby affecting different immune populations from the innate and adaptative systems. Assuming the choriodecidual interface (where mother and fetal tissues are in contact) is a complex network of signals where each component (cells, cytokines, and many molecules) is a critical regulator, it is undeniable that the formed environment could be a determinant for the future child with echo in adulthood (as summarized in Figure 8). Future research could be approached to evaluate the impact of the inflammatory environment associated with CVD in women affected by this condition and their offspring.

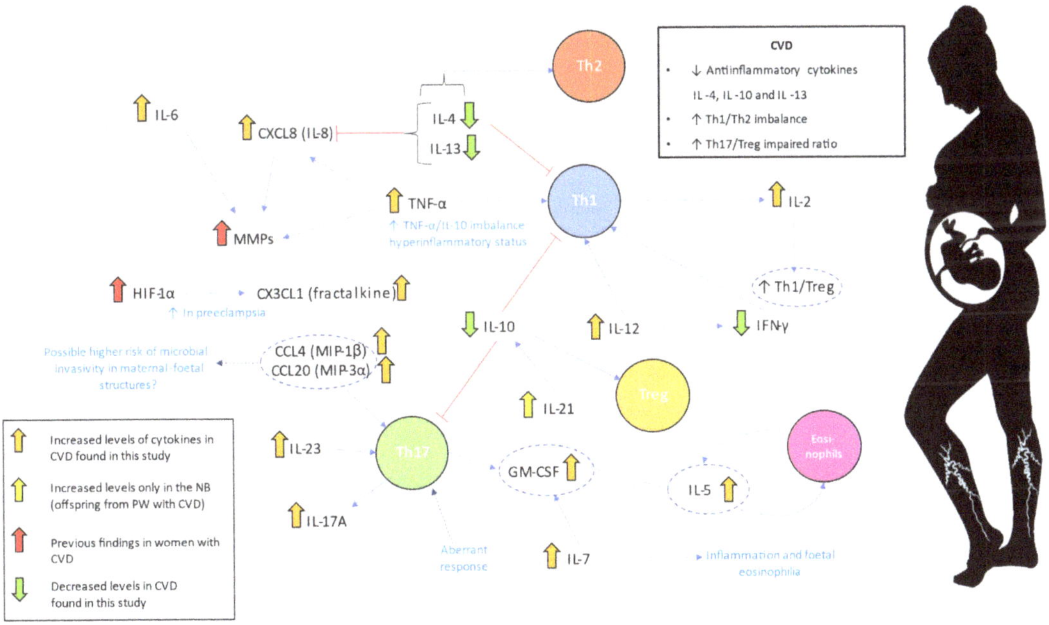


**Figure 8.** Summary of cytokines network studied in PW with CVD and their NB. The picture describes significant impaired levels of cytokines and conclusions from the study, being the Th1/Th2 imbalance a notable hallmark, besides the Th17/Treg impaired ratio. The complex network of signals could be determinant for the future child. A low level of anti-inflammatory cytokines is also associated with pre-eclampsia, gestational hypertension, spontaneous miscarriage, infertility, fetal growth restriction, and preterm birth. Low levels of IL-4 and IL-13 have been related to overweight in 1–2-year-old NB and could have their echo in health for adulthood. PW = pregnant woman; CVD = chronic venous disease; NB = newborn.

**Author Contributions:** Conceptualization, M.A.O., J.B., N.G.-H. and M.A.-M.; methodology, M.A.O., C.B., J.A.D.L.-L., J.B., N.G.-H., J.M. and M.A.-M.; validation, J.M.; formal analysis, M.A.O., A.M.G.-L., J.M. and M.A.-M.; investigation, M.A.O., A.M.G.-L., L.S.-T., O.F.-M., C.G.-M., L.G.G., C.B., J.A.D.L.-L., J.V.S., J.B., N.G.-H., J.M. and M.A.-M.; data curation, M.A.O., A.M.G.-L., J.M. and M.A.-M.; writing—original draft preparation, M.A.O., A.M.G.-L., L.S.-T., O.F.-M., C.G.-M., L.G.G., C.B., J.A.D.L.-L., J.V.S., J.B., N.G.-H., J.M. and M.A.-M.; writing—review and editing, M.A.O., A.M.G.-L., L.S.-T., O.F.-M., C.G.-M., L.G.G., C.B., J.A.D.L.-L., J.V.S., J.B., N.G.-H., J.M. and M.A.-M.; supervision, M.A.-M.; project administration, M.A.O.; funding acquisition, M.A.O., N.G.-H. and M.A.-M. All authors have read and agreed to the published version of the manuscript.

**Funding:** The study (FIS-PI21/01244) was supported by the Instituto de Salud Carlos III (grant no. Estatal de I + D + I 2020–2027) and co-financed by the European Development Regional Fund "A way to achieve Europe" and B2017/BMD-3804 MITIC-CM (Comunidad de Madrid), Halekulani S.L. and MJR.

**Institutional Review Board Statement:** The study was conducted according to the guidelines of the Declaration of Helsinki and approved by the Clinical Research Ethics Committee of the Central University Hospital of Defense-UAH (37/17).

**Informed Consent Statement:** Informed consent was obtained from all subjects involved in the study.

**Data Availability Statement:** The data used to support the findings of the present study are available from the corresponding author upon request.

**Conflicts of Interest:** The authors declare no conflict of interest.

## References

1. Youn, Y.J.; Lee, J. Chronic Venous Insufficiency and Varicose Veins of the Lower Extremities. *Korean J. Intern. Med.* **2019**, *34*, 269–283. [CrossRef]
2. Raffetto, J.D.; Mannello, F. Pathophysiology of Chronic Venous Disease. *Int. Angiol.* **2014**, *33*, 212–221. [PubMed]
3. Vlajinac, H.D.; Radak, D.J.; Marinković, J.M.; Maksimović, M.Ž. Risk Factors for Chronic Venous Disease. *Phlebology* **2012**, *27*, 416–422. [CrossRef] [PubMed]
4. Homs-Romero, E.; Romero-Collado, A.; Verdú, J.; Blanch, J.; Rascón-Hernán, C.; Martí-Lluch, R. Validity of Chronic Venous Disease Diagnoses and Epidemiology Using Validated Electronic Health Records From Primary Care: A Real-World Data Analysis. *J. Nurs. Scholarsh.* **2021**, *53*, 296–305. [CrossRef] [PubMed]
5. Zolotukhin, I.A.; Seliverstov, E.I.; Shevtsov, Y.N.; Avakiants, I.P.; Nikishkov, A.S.; Tatarintsev, A.M.; Kirienko, A.I. Prevalence and Risk Factors for Chronic Venous Disease in the General Russian Population. *Eur. J. Vasc. Endovasc. Surg.* **2017**, *54*, 752–758. [CrossRef] [PubMed]
6. Ortega, M.; Fraile-Martínez, O.; García-Montero, C.; Álvarez-Mon, M.; Chaowen, C.; Ruiz-Grande, F.; Pekarek, L.; Monserrat, J.; Asúnsolo, A.; García-Honduvilla, N.; et al. Understanding Chronic Venous Disease: A Critical Overview of Its Pathophysiology and Medical Management. *J. Clin. Med.* **2021**, *10*, 3239. [CrossRef] [PubMed]
7. Morton, A. Physiological Changes and Cardiovascular Investigations in Pregnancy. *Heart Lung Circ.* **2021**, *30*, e6–e15. [CrossRef]
8. Ropacka-Lesiak, M.; Jaroslaw, K.; Bręborowicz, G. Pregnancy-Dependent Blood Flow Velocity Changes in Lower Extremities Veins in Venous Insufficiency. *Ginekol. Pol.* **2015**, *86*, 659–665. [CrossRef]
9. Taylor, J.; Hicks, C.W.; Heller, J.A. The Hemodynamic Effects of Pregnancy on the Lower Extremity Venous System. *J. Vasc. Surg. Venous Lymphat. Disord.* **2018**, *6*, 246–255. [CrossRef]
10. Lohr, J.M.; Bush, R.L. Venous Disease in Women: Epidemiology, Manifestations, and Treatment. *J. Vasc. Surg.* **2013**, *57*, 37S–45S. [CrossRef]
11. NH, T. Physiologic and Hemodynamic Changes During Pregnancy. *AACN Adv. Crit. Care* **2018**, *29*, 273–283. [CrossRef]
12. Labropoulos, N. How Does Chronic Venous Disease Progress from the First Symptoms to the Advanced Stages? A Review. *Adv. Ther.* **2019**, *36*, 13–19. [CrossRef]
13. Ortega, M.A.; Saez, M.Á.; Asúnsolo, Á.; Romero, B.; Bravo, C.; Coca, S.; Sainz, F.; Álvarez-Mon, M.; Buján, J.; García-Honduvilla, N. Upregulation of VEGF and PEDF in Placentas of Women with Lower Extremity Venous Insufficiency during Pregnancy and Its Implication in Villous Calcification. *BioMed Res. Int.* **2019**, *2019*, 5320902. [CrossRef]
14. Ortega, M.A.; Romero, B.; Asúnsolo, Á.; Martínez-Vivero, C.; Sainz, F.; Bravo, C.; de León-Luis, J.; Álvarez-Mon, M.; Buján, J.; García-Honduvilla, N. Pregnancy-Associated Venous Insufficiency Course with Placental and Systemic Oxidative Stress. *J. Cell. Mol. Med.* **2020**, *24*, 4157–4170. [CrossRef]
15. Ortega, M.A.; Saez, M.A.; Fraile-Martínez, O.; Asúnsolo, Á.; Pekarek, L.; Bravo, C.; Coca, S.; Sainz, F.; Álvarez-Mon, M.; Buján, J.; et al. Increased Angiogenesis and Lymphangiogenesis in the Placental Villi of Women with Chronic Venous Disease during Pregnancy. *Int. J. Mol. Sci.* **2020**, *21*, 2487. [CrossRef]

16. García-Honduvilla, N.; Ortega, M.A.; Asúnsolo, Á.; Álvarez-Rocha, M.J.; Romero, B.; de León-Luis, J.; Álvarez-Mon, M.; Buján, J. Placentas from Women with Pregnancy-Associated Venous Insufficiency Show Villi Damage with Evidence of Hypoxic Cellular Stress. *Hum. Pathol.* **2018**, *77*, 45–53. [CrossRef]
17. Chaiworapongsa, T.; Chaemsaithong, P.; Yeo, L.; Romero, R. Pre-Eclampsia Part 1: Current Understanding of Its Pathophysiology. *Nat. Rev. Nephrol.* **2014**, *10*, 466–480. [CrossRef]
18. Ortega, M.A.; Chaowen, C.; Fraile-Martinez, O.; García-Montero, C.; Saez, M.A.; Cruza, I.; Pereda-Cerquella, C.; Alvarez-Mon, M.A.; Guijarro, L.G.; Fatych, Y.; et al. Chronic Venous Disease in Pregnant Women Causes an Increase in ILK in the Placental Villi Associated with a Decrease in E-Cadherin. *J. Pers. Med.* **2022**, *12*, 277. [CrossRef]
19. Ramani, T.; Auletta, C.S.; Weinstock, D.; Mounho-Zamora, B.; Ryan, P.C.; Salcedo, T.W.; Bannish, G. Cytokines: The Good, the Bad, and the Deadly. *Int. J. Toxicol.* **2015**, *34*, 355–365. [CrossRef]
20. Feldmann, M. Many Cytokines Are Very Useful Therapeutic Targets in Disease. *J. Clin. Investig.* **2008**, *118*, 3533–3536. [CrossRef]
21. Kany, S.; Vollrath, J.T.; Relja, B. Cytokines in Inflammatory Disease. *Int. J. Mol. Sci.* **2019**, *20*, 6008. [CrossRef]
22. Aggarwal, R.; Jain, A.K.; Mittal, P.; Kohli, M.; Jawanjal, P.; Rath, G. Association of Pro- and Anti-Inflammatory Cytokines in Preeclampsia. *J. Clin. Lab. Anal.* **2019**, *33*, e22834. [CrossRef]
23. Raghupathy, R.; Kalinka, J. Cytokine Imbalance in Pregnancy Complications and Its Modulation. *Front. Biosci.* **2008**, *13*, 985–994. [CrossRef]
24. Yockey, L.J.; Iwasaki, A. Interferons and Proinflammatory Cytokines in Pregnancy and Fetal Development. *Immunity* **2018**, *49*, 397–412. [CrossRef]
25. Velez, D.R.; Fortunato, S.J.; Morgan, N.; Edwards, T.L.; Lombardi, S.J.; Williams, S.M.; Menon, R. Patterns of Cytokine Profiles Differ with Pregnancy Outcome and Ethnicity. *Hum. Reprod.* **2008**, *23*, 1902–1909. [CrossRef]
26. Hernández-Trejo, M.; Montoya-Estrada, A.; Torres-Ramos, Y.; Espejel-Núñez, A.; Guzmán-Grenfell, A.; Morales-Hernández, R.; Tolentino-Dolores, M.; Laresgoiti-Servitje, E. Oxidative Stress Biomarkers and Their Relationship with Cytokine Concentrations in Overweight/Obese Pregnant Women and Their Neonates. *BMC Immunol.* **2017**, *18*, 3. [CrossRef]
27. Agarwal, S.; Karmaus, W.; Davis, S.; Gangur, V. Immune Markers in Breast Milk and Fetal and Maternal Body Fluids: A Systematic Review of Perinatal Concentrations. *J. Hum. Lact.* **2011**, *27*, 171–186. [CrossRef]
28. Conrad, K.P.; Miles, T.M.; Benyo, D.F. Circulating Levels of Immunoreactive Cytokines in Women with Preeclampsia. *Am. J. Reprod. Immunol.* **1998**, *40*, 102–111. [CrossRef]
29. Szarka, A.; Rigó, J.; Lázár, L.; Beko, G.; Molvarec, A. Circulating Cytokines, Chemokines and Adhesion Molecules in Normal Pregnancy and Preeclampsia Determined by Multiplex Suspension Array. *BMC Immunol.* **2010**, *11*, 59. [CrossRef]
30. Lattimer, C.R.; Kalodiki, E.; Geroulakos, G.; Hoppensteadt, D.; Fareed, J. Are Inflammatory Biomarkers Increased in Varicose Vein Blood? *Clin. Appl. Thromb. Hemost.* **2016**, *22*, 656–664. [CrossRef]
31. Ligi, D.; Croce, L.; Mannello, F. Chronic Venous Disorders: The Dangerous, the Good, and the Diverse. *Int. J. Mol. Sci.* **2018**, *19*, 2544. [CrossRef] [PubMed]
32. Grudzinska, E.; Lekstan, A.; Szliszka, E.; Czuba, Z.P. Cytokines Produced by Lymphocytes in the Incompetent Great Saphenous Vein. *Mediat. Inflamm.* **2018**, *2018*, 7161346. [CrossRef] [PubMed]
33. Ortega, M.A.; Sánchez-Trujillo, L.; Bravo, C.; Fraile-Martinez, O.; García-Montero, C.; Saez, M.A.; Alvarez-Mon, M.A.; Sainz, F.; Alvarez-Mon, M.; Bujan, J.; et al. Newborns of Mothers with Venous Disease during Pregnancy Show Increased Levels of Lipid Peroxidation and Markers of Oxidative Stress and Hypoxia in the Umbilical Cord. *Antioxidants* **2021**, *10*, 980. [CrossRef] [PubMed]
34. Ortega, M.; Fraile-Martínez, O.; Saez, M.; Álvarez-Mon, M.; Gómez-Lahoz, A.M.; Bravo, C.; Luis, J.A.D.L.; Sainz, F.; Coca, S.; Asúnsolo; et al. Abnormal Proinflammatory and Stressor Environmental with Increased the Regulatory Cellular IGF-1/PAPP-A/STC and Wnt-1/β-Catenin Canonical Pathway in Placenta of Women with Chronic Venous Disease during Pregnancy. *Int. J. Med. Sci.* **2021**, *18*, 2814–2827. [CrossRef]
35. Ortega, M.A.; Asúnsolo, Á.; Fraile-Martínez, O.; Sainz, F.; Saez, M.A.; Bravo, C.; De León-Luis, J.A.; Alvarez-Mon, M.A.; Coca, S.; Álvarez-Mon, M.; et al. An Increase in Elastogenic Components in the Placental Villi of Women with Chronic Venous Disease during Pregnancy Is Associated with Decreased EGFL7 Expression Level. *Mol. Med. Rep.* **2021**, *24*, 556. [CrossRef]
36. Ortega, M.A.; Saez, M.A.; Sainz, F.; Fraile-Martínez, O.; García-Gallego, S.; Pekarek, L.; Bravo, C.; Coca, S.; Álvarez-Mon, M.; Buján, J.; et al. Lipidomic Profiling of Chorionic Villi in the Placentas of Women with Chronic Venous Disease. *Int. J. Med. Sci.* **2020**, *17*, 2790–2798. [CrossRef]
37. Myatt, L. Placental Adaptive Responses and Fetal Programming. *J. Physiol.* **2006**, *572*, 25. [CrossRef]
38. Fajersztajn, L.; Veras, M.M. Hypoxia: From Placental Development to Fetal Programming. *Birth Defects Res.* **2017**, *109*, 1377–1385. [CrossRef]
39. Konkel, L. Lasting Impact of an Ephemeral Organ: The Role of the Placenta in Fetal Programming. *Environ. Health Perspect.* **2016**, *124*, A124–A129. [CrossRef]
40. Loppnow, H. Cytokines: Classification, Receptors, Mechanisms of Action. *Internist* **2001**, *42*, 13–27. [CrossRef]
41. Justiz Vaillant, A.A.; Qurie, A. *Interleukin*; StatPearls Publishing: Treasure Island, FL, USA, 2021.
42. Zaretsky, M.V.; Alexander, J.M.; Byrd, W.; Bawdon, R.E. Transfer of Inflammatory Cytokines across the Placenta. *Obstet. Gynecol.* **2004**, *103*, 546–550. [CrossRef]
43. Uciechowski, P.; Dempke, W.C.M. Interleukin-6: A Masterplayer in the Cytokine Network. *Oncology* **2020**, *98*, 131–137. [CrossRef]

44. Xiao, J.P.; Yin, Y.X.; Gao, Y.F.; Lau, S.; Shen, F.; Zhao, M.; Chen, Q. The Increased Maternal Serum Levels of IL-6 Are Associated with the Severity and Onset of Preeclampsia. *Cytokine* **2012**, *60*, 856–860. [CrossRef]
45. Martinez-Portilla, R.J.; Hawkins-Villarreal, A.; Alvarez-Ponce, P.; Chinolla-Arellano, Z.L.; Moreno-Espinosa, A.L.; Sandoval-Mejia, A.L.; Moreno-Uribe, N. Maternal Serum Interleukin-6: A Non-Invasive Predictor of Histological Chorioamnionitis in Women with Preterm-Prelabor Rupture of Membranes. *Fetal Diagn. Ther.* **2019**, *45*, 168–175. [CrossRef]
46. Deon, D.; Ahmed, S.; Tai, K.; Scaletta, N.; Herrero, C.; Lee, I.-H.; Krause, A.; Ivashkiv, L.B. Cross-Talk between IL-1 and IL-6 Signaling Pathways in Rheumatoid Arthritis Synovial Fibroblasts. *J. Immunol.* **2001**, *167*, 5395–5403. [CrossRef]
47. Chen, L.M.; Liu, B.; Zhao, H.B.; Stone, P.; Chen, Q.; Chamley, L. IL-6, TNFα and TGFβ Promote Nonapoptotic Trophoblast Deportation and Subsequently Causes Endothelial Cell Activation. *Placenta* **2010**, *31*, 75–80. [CrossRef]
48. Rocha, G.; Proença, E.; Guedes, A.; Carvalho, C.; Areias, A.; Ramos, J.P.; Rodrigues, T.; Guimarães, H. Cord Blood Levels of IL-6, IL-8 and IL-10 May Be Early Predictors of Bronchopulmonary Dysplasia in Preterm Newborns Small for Gestational Age. *Dis. Markers* **2012**, *33*, 51–60. [CrossRef]
49. Jonsson, Y.; Rubèr, M.; Matthiesen, L.; Berg, G.; Nieminen, K.; Sharma, S.; Ernerudh, J.; Ekerfelt, C. Cytokine Mapping of Sera from Women with Preeclampsia and Normal Pregnancies. *J. Reprod Immunol.* **2006**, *70*, 83–91. [CrossRef]
50. Huang, Q.; Jin, X.; Li, P.; Zheng, Z.; Jiang, Y.; Liu, H. Elevated Inflammatory Mediators from the Maternal-Fetal Interface to Fetal Circulation during Labor. *Cytokine* **2021**, *148*, 155707. [CrossRef]
51. Li, Y.; Wang, Y.; Ding, X.; Duan, B.; Li, L.; Wang, X. Serum Levels of TNF-α and IL-6 Are Associated with Pregnancy-Induced Hypertension. *Reprod. Sci.* **2016**, *23*, 1402–1408. [CrossRef]
52. Carpentier, P.A.; Dingman, A.L.; Palmer, T.D. Placental TNF-α Signaling in Illness-Induced Complications of Pregnancy. *Am. J. Pathol.* **2011**, *178*, 2802–2810. [CrossRef] [PubMed]
53. Stefańska, K.; Zieliński, M.; Jankowiak, M.; Zamkowska, D.; Sakowska, J.; Adamski, P.; Jassem-Bobowicz, J.; Piekarska, K.; Leszczyńska, K.; Świątkowska-Stodulska, R.; et al. Cytokine Imprint in Preeclampsia. *Front. Immunol.* **2021**, *12*, 667841. [CrossRef] [PubMed]
54. Hanlon, A.M.; Jang, S.; Salgame, P. Signaling from Cytokine Receptors That Affect Th1 Responses. *Front. Biosci.* **2002**, *7*, 1247–1254. [CrossRef]
55. Gaffen, S.L.; Liu, K.D. Overview of Interleukin-2 Function, Production and Clinical Applications. *Cytokine* **2004**, *28*, 109–123. [CrossRef] [PubMed]
56. Mitra, S.; Leonard, W.J. Biology of IL-2 and Its Therapeutic Modulation: Mechanisms and Strategies. *J. Leukoc. Biol.* **2018**, *103*, 643–655. [CrossRef] [PubMed]
57. Abbas, A.K.; Trotta, E.; Simeonov, D.R.; Marson, A.; Bluestone, J.A. Revisiting IL-2: Biology and Therapeutic Prospects. *Sci. Immunol.* **2018**, *3*, eaat1482. [CrossRef] [PubMed]
58. Fainboim, L.; Arruvito, L. Mechanisms Involved in the Expansion of Tregs during Pregnancy: Role of IL-2/STAT5 Signalling. *J. Reprod. Immunol.* **2011**, *88*, 93–98. [CrossRef] [PubMed]
59. Hadinedoushan, H.; Mirahmadian, M.; Aflatounian, A. Increased Natural Killer Cell Cytotoxicity and IL-2 Production in Recurrent Spontaneous Abortion. *Am. J. Reprod. Immunol.* **2007**, *58*, 409–414. [CrossRef]
60. Vignali, D.A.A.; Kuchroo, V.K. IL-12 Family Cytokines: Immunological Playmakers. *Nat. Immunol.* **2012**, *13*, 722–728. [CrossRef]
61. Becker, C.; Wirtz, S.; Neurath, M.F. Stepwise Regulation of TH1 Responses in Autoimmunity: IL-12-Related Cytokines and Their Receptors. *Inflamm. Bowel Dis.* **2005**, *11*, 755–764. [CrossRef]
62. Perricone, C.; de Carolis, C.; Perricone, R. Pregnancy and Autoimmunity: A Common Problem. *Best Pract. Res. Clin. Rheumatol.* **2012**, *26*, 47–60. [CrossRef]
63. Murphy, S.P.; Tayade, C.; Ashkar, A.A.; Hatta, K.; Zhang, J.; Croy, B.A. Interferon Gamma in Successful Pregnancies. *Biol. Reprod.* **2009**, *80*, 848–859. [CrossRef]
64. Scott, M.E.; Kubin, M.; Kohl, S. High Level Interleukin-12 Production, but Diminished Interferon-γ Production, by Cord Blood Mononuclear Cells. *Pediatric Res.* **1997**, *41*, 547–553. [CrossRef]
65. Giurgescu, C.; Sanguanklin, N.; Engeland, C.G.; White-Traut, R.C.; Park, C.; Mathews, H.L.; Janusek, L.W. Relationships among Psychosocial Factors, Biomarkers, Preeclampsia, and Preterm Birth in African American Women: A Pilot. *Appl. Nurs. Res.* **2015**, *28*, e1–e6. [CrossRef]
66. Saraiva, M.; O'Garra, A. The Regulation of IL-10 Production by Immune Cells. *Nat. Rev. Immunol.* **2010**, *10*, 170–181. [CrossRef]
67. Wang, X.; Wong, K.; Ouyang, W.; Rutz, S. Targeting IL-10 Family Cytokines for the Treatment of Human Diseases. *Cold Spring Harb. Perspect. Biol.* **2019**, *11*, a028548. [CrossRef]
68. Brogin Moreli, J.; Cirino Ruocco, A.M.; Vernini, J.M.; Rudge, M.V.C.; Calderon, I.M.P. Interleukin 10 and Tumor Necrosis Factor-Alpha in Pregnancy: Aspects of Interest in Clinical Obstetrics. *ISRN Obstet. Gynecol.* **2012**, *2012*, 230742. [CrossRef]
69. Chatterjee, P.; Chiasson, V.L.; Bounds, K.R.; Mitchell, B.M. Regulation of the Anti-Inflammatory Cytokines Interleukin-4 and Interleukin-10 during Pregnancy. *Front. Immunol.* **2014**, *5*, 253. [CrossRef]
70. Junttila, I.S. Tuning the Cytokine Responses: An Update on Interleukin (IL)-4 and IL-13 Receptor Complexes. *Front. Immunol.* **2018**, *9*, 888. [CrossRef]
71. Chatterjee, P.; Kopriva, S.E.; Chiasson, V.L.; Young, K.J.; Tobin, R.P.; Newell-Rogers, K.; Mitchell, B.M. Interleukin-4 Deficiency Inducesmild Preeclampsia Inmice. *J. Hypertens.* **2013**, *31*, 1414–1423. [CrossRef]

72. Daneva, A.; Hadži-Lega, M.; Stefanovic, M. Correlation of the System of Cytokines in Moderate and Severe Preeclampsia. *Clin. Exp. Obs. Gynecol.* **2016**, *43*, 220–224. [CrossRef]
73. Englich, B.; Herberth, G.; Rolle-Kampczyk, U.; Trump, S.; Röder, S.; Borte, M.; Stangl, G.I.; von Bergen, M.; Lehmann, I.; Junge, K.M. Maternal Cytokine Status May Prime the Metabolic Profile and Increase Risk of Obesity in Children. *Int. J. Obes.* **2017**, *41*, 1440–1446. [CrossRef]
74. Sykes, L.; MacIntyre, D.A.; Yap, X.J.; Teoh, T.G.; Bennett, P.R. The Th1:Th2 Dichotomy of Pregnancy and Preterm Labour. *Mediat. Inflamm.* **2012**, *2012*, 12. [CrossRef]
75. Zenobia, C.; Hajishengallis, G. Basic Biology and Role of Interleukin-17 in Immunity and Inflammation. *Periodontol. 2000* **2015**, *69*, 142–159. [CrossRef]
76. Eghbal-Fard, S.; Yousefi, M.; Heydarlou, H.; Ahmadi, M.; Taghavi, S.; Movasaghpour, A.; Jadidi-Niaragh, F.; Yousefi, B.; Dolati, S.; Hojjat-Farsangi, M.; et al. The Imbalance of Th17/Treg Axis Involved in the Pathogenesis of Preeclampsia. *J. Cell. Physiol.* **2019**, *234*, 5106–5116. [CrossRef]
77. Poordast, T.; Najib, F.; Baharlou, R.; Bijani, A.; Alamdarloo, S.; Poordast, A. Assessment of T Helper 17-Associated Cytokines in Third Trimester of Pregnancy. *Iran. J. Immunol.* **2017**, *14*, 172–179.
78. Darmochwal-Kolarz, D.; Michalak, M.; Kolarz, B.; Przegalinska-Kalamucka, M.; Bojarska-Junak, A.; Sliwa, D.; Oleszczuk, J. The Role of Interleukin-17, Interleukin-23, and Transforming Growth Factor- β in Pregnancy Complicated by Placental Insufficiency. *BioMed Res. Int.* **2017**, *2017*, 6904325. [CrossRef]
79. Elkassar, N.; Gress, R.E. An Overview of IL-7 Biology and Its Use in Immunotherapy. *J. Immunotoxicol.* **2010**, *7*, 1–7. [CrossRef]
80. Wu, L.; Li, J.; Xu, H.L.; Xu, B.; Tong, X.H.; Kwak-Kim, J.; Liu, Y.S. IL-7/IL-7R Signaling Pathway Might Play a Role in Recurrent Pregnancy Losses by Increasing Inflammatory Th17 Cells and Decreasing Treg Cells. *Am. J. Reprod. Immunol.* **2016**, *76*, 454–464. [CrossRef]
81. Leonard, W.J.; Wan, C.K. IL-21 Signaling in Immunity. *F1000Research* **2016**, *5*, 224. [CrossRef]
82. Doganci, A.; Birkholz, J.; Gehring, S.; Puhl, A.G.; Zepp, F.; Meyer, C.U. In the Presence of IL-21 Human Cord Blood T Cells Differentiate to IL-10-Producing Th1 but Not Th17 or Th2 Cells. *Int. Immunol.* **2013**, *25*, 157–169. [CrossRef] [PubMed]
83. Radonjic-Hoesli, S.; Valent, P.; Klion, A.D.; Wechsler, M.E.; Simon, H.U. Novel Targeted Therapies for Eosinophil-Associated Diseases and Allergy. *Annu. Rev. Pharmacol. Toxicol.* **2015**, *55*, 633–656. [CrossRef] [PubMed]
84. Griseri, T.; Arnold, I.C.; Pearson, C.; Krausgruber, T.; Schiering, C.; Franchini, F.; Schulthess, J.; McKenzie, B.S.; Crocker, P.R.; Powrie, F. Granulocyte Macrophage Colony-Stimulating Factor-Activated Eosinophils Promote Interleukin-23 Driven Chronic Colitis. *Immunity* **2015**, *43*, 187–199. [CrossRef] [PubMed]
85. Ramirez, G.A.; Yacoub, M.R.; Ripa, M.; Mannina, D.; Cariddi, A.; Saporiti, N.; Ciceri, F.; Castagna, A.; Colombo, G.; Dagna, L. Eosinophils from Physiology to Disease: A Comprehensive Review. *Biomed. Res. Int* **2018**, *2018*, 9095275. [CrossRef]
86. Spencer, L.A.; Szela, C.T.; Perez, S.A.C.; Kirchhoffer, C.L.; Neves, J.S.; Radke, A.L.; Weller, P.F. Human Eosinophils Constitutively Express Multiple Th1, Th2, and Immunoregulatory Cytokines That Are Secreted Rapidly and Differentially. *J. Leukoc. Biol.* **2008**, *85*, 117–123. [CrossRef]
87. Romero, R.; Kusanovic, J.P.; Gomez, R.; Lamont, R.; Bytautiene, E.; Garfield, R.E.; Mittal, P.; Hassan, S.S.; Yeo, L. The Clinical Significance of Eosinophils in the Amniotic Fluid in Preterm Labor. *J. Matern. Fetal Neonatal Med.* **2010**, *23*, 320–329. [CrossRef]
88. Lebold, K.M.; Drake, M.G.; Hales-Beck, L.B.; Fryer, A.D.; Jacoby, D.B. IL-5 Exposure in Utero Increases Lung Nerve Density and Airway Reactivity in Adult Offspring. *Am. J. Respir. Cell Mol. Biol.* **2020**, *62*, 493–502. [CrossRef]
89. Stone, W.L.; Leavitt, L.; Varacallo, M. Excerpt. In *Physiology, Growth Factor*; StatPearls: Treasure Island, FL, USA, 2021.
90. Duvallet, E.; Semerano, L.; Assier, E.; Falgarone, G.; Boissier, M.C. Interleukin-23: A Key Cytokine in Inflammatory Diseases. *Ann. Med.* **2011**, *43*, 503–511. [CrossRef]
91. Grifka-Walk, H.M.; Giles, D.A.; Segal, B.M. IL-12-Polarized Th1 Cells Produce GM-CSF and Induce EAE Independent of IL-23. *Eur. J. Immunol.* **2015**, *45*, 2780–2786. [CrossRef]
92. Herndler-Brandstetter, D.; Flavell, R.A. Producing GM-CSF: A Unique T Helper Subset? *Cell Res.* **2014**, *24*, 1379–1380. [CrossRef]
93. Petrina, M.; Martin, J.; Basta, S. Granulocyte Macrophage Colony-Stimulating Factor Has Come of Age: From a Vaccine Adjuvant to Antiviral Immunotherapy. *Cytokine Growth Factor Rev.* **2021**, *59*, 101–110. [CrossRef]
94. Robertson, S.A. GM-CSF Regulation of Embryo Development and Pregnancy. *Cytokine Growth Factor Rev.* **2007**, *18*, 287–298. [CrossRef]
95. Perricone, R.; De Carolis, C.; Giacomelli, R.; Guarino, M.D.; De Sanctis, G.; Fontana, L. GM-CSF and Pregnancy: Evidence of Significantly Reduced Blood Concentrations in Unexplained Recurrent Abortion Efficiently Reverted by Intravenous Immunoglobulin Treatment. *Am. J. Reprod. Immunol.* **2003**, *50*, 232–237. [CrossRef]
96. Sjöblom, C.; Roberts, C.T.; Wikland, M.; Robertson, S.A. Granulocyte-Macrophage Colony-Stimulating Factor Alleviates Adverse Consequences of Embryo Culture on Fetal Growth Trajectory and Placental Morphogenesis. *Endocrinology* **2005**, *146*, 2142–2153. [CrossRef]
97. Huang, S.J.; Zenclussen, A.C.; Chen, C.P.; Basar, M.; Yang, H.; Arcuri, F.; Li, M.; Kocamaz, E.; Buchwalder, L.; Rahman, M.; et al. The Implication of Aberrant GM-CSF Expression in Decidual Cells in the Pathogenesis of Preeclampsia. *Am. J. Pathol.* **2010**, *177*, 2472–2482. [CrossRef]
98. Hughes, C.E.; Nibbs, R.J.B. A Guide to Chemokines and Their Receptors. *FEBS J.* **2018**, *285*, 2944–2971. [CrossRef]
99. Sokol, C.L.; Luster, A.D. The Chemokine System in Innate Immunity. *Cold Spring Harb. Perspect. Biol.* **2015**, *7*, a016303. [CrossRef]

100. Palomino, D.C.; Arolina, T.; Marti, L.C. Avalheiro Chemokines and Immunity. *Einstein* **2015**, *13*, 469–473. [CrossRef]
101. Gomez-Lopez, N.; Laresgoiti-Servitje, E.; Olson, D.M.; Estrada-Gutiérrez, G.; Vadillo-Ortega, F. The Role of Chemokines in Term and Premature Rupture of the Fetal Membranes: A Review1. *Biol. Reprod.* **2010**, *82*, 809–814. [CrossRef]
102. Hannan, N.J.; Salamonsen, L.A. Role of Chemokines in the Endometrium and in Embryo Implantation. *Curr. Opin. Obstet. Gynecol.* **2007**, *19*, 266–272. [CrossRef]
103. Spence, T.; Allsopp, P.J.; Yeates, A.J.; Mulhern, M.S.; Strain, J.J.; McSorley, E.M. Maternal Serum Cytokine Concentrations in Healthy Pregnancy and Preeclampsia. *J. Pregnancy* **2021**, *2021*, 6649608. [CrossRef]
104. Bonecchi, R.; Facchetti, F.; Dusi, S.; Luini, W.; Lissandrini, D.; Simmelink, M.; Locati, M.; Bernasconi, S.; Allavena, P.; Brandt, E.; et al. Induction of Functional IL-8 Receptors by IL-4 and IL-13 in Human Monocytes. *J. Immunol.* **2000**, *164*, 3862–3869. [CrossRef]
105. Osawa, Y.; Nagaki, M.; Banno, Y.; Brenner, D.A.; Asano, T.; Nozawa, Y.; Moriwaki, H.; Nakashima, S. Tumor Necrosis Factor Alpha-Induced Interleukin-8 Production via NF-KB and Phosphatidylinositol 3-Kinase/Akt Pathways Inhibits Cell Apoptosis in Human Hepatocytes. *Infect. Immun.* **2002**, *70*, 6294–6301. [CrossRef]
106. Ortega, M.A.; Asúnsolo, Á.; Álvarez-Rocha, M.J.; Romero, B.; de León-Luis, J.; Álvarez-Mon, M.; Buján, J.; García-Honduvilla, N. Remodelling of Collagen Fibres in the Placentas of Women with Venous Insufficiency during Pregnancy. *Histol. Histopathol.* **2018**, *33*, 567–576. [CrossRef]
107. Huang, S.J.; Chen, C.P.; Buchwalder, L.; Yu, Y.C.; Piao, L.; Huang, C.Y.; Schatz, F.; Lockwood, C.J. Regulation of CX3CL1 Expression in Human First-Trimester Decidual Cells: Implications for Preeclampsia. *Reprod. Sci.* **2019**, *26*, 1256–1265. [CrossRef] [PubMed]
108. Usta, A.; Turan, G.; Sancakli Usta, C.; Avci, E.; Adali, E. Placental Fractalkine Immunoreactivity in Preeclampsia and Its Correlation with Histopathological Changes in the Placenta and Adverse Pregnancy Outcomes. *J. Matern. Fetal Neonatal Med.* **2020**, *33*, 806–815. [CrossRef] [PubMed]
109. Szewczyk, G.; Pyzlak, M.; Pankiewicz, K.; Szczerba, E.; Stangret, A.; Szukiewicz, D.; Skoda, M.; Bierła, J.; Cukrowska, B.; Fijałkowska, A. The Potential Association between a New Angiogenic Marker Fractalkine and a Placental Vascularization in Preeclampsia. *Arch. Gynecol. Obs.* **2021**, *304*, 365–376. [CrossRef] [PubMed]
110. Szukiewicz, D.; Kochanowski, J.; Pyzlak, M.; Szewczyk, G.; Stangret, A.; Mittal, T.K. Fractalkine (CX3CL1) and Its Receptor CX3CR1 May Contribute to Increased Angiogenesis in Diabetic Placenta. *Mediat. Inflamm.* **2013**, *2013*, 437576. [CrossRef] [PubMed]
111. Du, M.R.; Wang, S.C.; Li, D.J. The Integrative Roles of Chemokines at the Maternal-Fetal Interface in Early Pregnancy. *Cell. Mol. Immunol.* **2014**, *11*, 438–448. [CrossRef]
112. Chaisavaneeyakorn, S.; Moore, J.M.; Mirel, L.; Othoro, C.; Otieno, J.; Chaiyaroj, S.C.; Shi, Y.P.; Nahlen, B.L.; Lal, A.A.; Udhayakumar, V. Levels of Macrophage Inflammatory Protein 1$\alpha$-(MIP-1$\alpha$) and MIP-1$\beta$ in Intervillous Blood Plasma Samples from Women with Placental Malaria and Human Immunodeficiency Virus Infection. *Clin. Diagn. Lab. Immunol.* **2003**, *10*, 631–636. [CrossRef]
113. Hannan, N.J.; Jones, R.L.; White, C.A.; Salamonsen, L.A. The Chemokines, CX3CL1, CCL14, and CCL4, Promote Human Trophoblast Migration at the Feto-Maternal Interface. *Biol. Reprod.* **2006**, *74*, 896–904. [CrossRef]
114. Schutyser, E.; Struyf, S.; Van Damme, J. The CC Chemokine CCL20 and Its Receptor CCR6. *Cytokine Growth Factor Rev.* **2003**, *14*, 409–426. [CrossRef]
115. Li, Q.; Laumonnier, Y.; Syrovets, T.; Simmet, T. Recruitment of CCR6-Expressing Th17 Cells by CCL20 Secreted from Plasmin-Stimulated Macrophages. *Acta Biochim. Biophys. Sin.* **2013**, *45*, 593–600. [CrossRef]
116. Hamill, N.; Romero, R.; Gotsch, F.; Pedro Kusanovic, J.; Edwin, S.; Erez, O.; Gabor Than, N.; Mittal, P.; Espinoza, J.; Friel, L.A.; et al. Exodus-1 (CCL20): Evidence for the Participation of This Chemokine in Spontaneous Labor at Term, Preterm Labor, and Intrauterine Infection. *J. Perinat. Med.* **2008**, *36*, 217–227. [CrossRef]

International Journal of *Molecular Sciences*

MDPI

*Article*

# Mesenchymal Stem Cells-Induced Trophoblast Invasion Is Reduced in Patients with a Previous History of Preeclampsia

**Reyna Peñailillo [1,†], Stephanie Acuña-Gallardo [1,2,†], Felipe García [1], Lara J. Monteiro [1,2,3], Gino Nardocci [2,3,4], Mahesh A. Choolani [5], Matthew W. Kemp [5,6], Roberto Romero [7,8,9,10,11] and Sebastián E. Illanes [1,2,3,*]**

1 Laboratory of Reproductive Biology, Center for Biomedical Research and Innovation (CIIB), Universidad de los Andes, Santiago 7620001, Chile
2 Faculty of Medicine, Universidad de los Andes, Santiago 7620001, Chile
3 IMPACT, Center of Interventional Medicine for Precision and Advanced Cellular Therapy, Santiago 7620001, Chile
4 Molecular Biology and Bioinformatics Lab, Program in Molecular Biology and Bioinformatics, Center for Biomedical Research and Innovation (CIIB), Universidad de los Andes, Santiago 7620001, Chile
5 Department of Obstetrics and Gynaecology, Yong Loo Lin School of Medicine, National University of Singapore, Singapore 119228, Singapore
6 Division of Obstetrics and Gynaecology, The University of Western Australia, Crawley, WA 6009, Australia
7 Perinatology Research Branch, Division of Obstetrics and Maternal-Fetal Medicine, Division of Intramural Research, *Eunice Kennedy Shriver* National Institute of Child Health and Human Development, National Institutes of Health, United States Department of Health and Human Services, Bethesda, MD 20892, and Detroit, MI 48201, USA
8 Department of Obstetrics and Gynecology, University of Michigan, Ann Arbor, MI 48109, USA
9 Department of Epidemiology and Biostatistics, Michigan State University, East Lansing, MI 48824, USA
10 Center for Molecular Medicine and Genetics, Wayne State University, Detroit, MI 48201, USA
11 Detroit Medical Center, Detroit, MI 48201, USA
* Correspondence: sillanes@uandes.cl
† These authors contributed equally to this work.

**Citation:** Peñailillo, R.; Acuña-Gallardo, S.; García, F.; Monteiro, L.J.; Nardocci, G.; Choolani, M.A.; Kemp, M.W.; Romero, R.; Illanes, S.E. Mesenchymal Stem Cells-Induced Trophoblast Invasion Is Reduced in Patients with a Previous History of Preeclampsia. *Int. J. Mol. Sci.* **2022**, *23*, 9071. https://doi.org/10.3390/ijms23169071

Academic Editor: Ilona Hromadnikova

Received: 15 June 2022
Accepted: 7 August 2022
Published: 13 August 2022



**Abstract:** Endometrial stromal cells play an important role in reproductive success, especially in implantation and placentation. Although Mesenchymal stem cells (MSCs) have been studied to assess decidualization disorders in preeclampsia (PE), their role during trophoblast invasion remains unclear. This study aims to determine: (i) whether MSCs isolated from menstrual fluid (MenSCs) from nulliparous, multiparous, and women with a previous history of preeclampsia exhibited different patterns of proliferation and migration and (ii) whether reproductive history (i.e., prior pregnancy or prior history of PE) was able to produce changes in MenSCs, thus altering trophoblast invasion capacity. MenSCs were collected from nulliparous and multiparous women without a history of PE and from non-pregnant women with a history of PE. Proliferation and migration assays were performed on MenSCs with sulforhodamine B and transwell assays, respectively. Trophoblast invasion was analyzed by culturing HTR-8/SVneo trophospheres on a matrigel overlying MenSCs for 72 h at 5% $O_2$, simulating a 3D implantation model. A previous history of pregnancy or PE did not impact the proliferative capacity or migratory behavior of MenSCs. Following exposure to physiological endometrial conditions, MenSCs demonstrated upregulated expression of *IGFBP-1* and *LIF* mRNA, decidualization and window of implantation markers, respectively. The mRNA expression of *VIM*, *NANOG*, and *SOX2* was upregulated upon trophosphere formation. Relative to co-culture with multiparous MenSCs, co-culture with PE-MenSCs was associated with reduced trophoblast invasion. The findings of this study suggest a potential role for communication between maternal MenSCs and invading trophoblast cells during the implantation process that could be implicated in the etiology of PE.

**Keywords:** preeclampsia; MenSCs; trophoblast invasion

## 1. Introduction

Preeclampsia (PE) is a pregnancy-specific disorder characterized by new-onset hypertension and proteinuria after 20 weeks of gestation [1]. It affects 2–8% of all pregnancies and is associated with an increased risk of maternal and fetal morbidity and mortality [2]. Although the precise etiology of PE remains unclear, it is now widely accepted that its pathophysiological process involves deficient trophoblast invasion of the maternal decidua and impaired remodeling of the maternal spiral arteries during the first trimester of pregnancy [3,4]. For successful implantation and placentation, the interaction between decidual stromal cells and extravillous trophoblast (EVT) cells is crucial. Specifically, trophoblast invasion relies on communication between the blastocyst and the maternal decidua [5].

The endometrium is highly dynamic and undergoes cyclical regeneration, differentiation, and shedding during the menstrual cycle. In humans, decidualization occurs during the mid-secretory phase of the cycle, begins around the spiral arteries, and is independent of the presence of the conceptus [6]. During this process, regulated by estradiol and progesterone, cells acquire a secretory phenotype that discharges specific products such as prolactin and insulin-like growth factor binding protein 1 (IGFBP1) [7]. Decidualization helps regulate embryo implantation and, subsequently, cytotrophoblast interaction with the uterus, making this process an essential component of establishing the maternal–fetal interface during normal pregnancy [8]. Failed decidualization has been an important contributor of altered cytotrophoblast invasion in human endometrial stromal cells from women with a previous pregnancy complicated by severe PE [9].

Placentation requires the invasion of fetal-derived EVT cells into the maternal uterine spiral arteries [10]. EVTs, that differentiate from cytotrophoblast cells, lose some epithelial phenotypes at the villous tips and acquire additional mesenchymal phenotypes, improving their migration and invasion capacity. EVT invasion involves the degradation and remodeling of the extracellular matrix, which is achieved mainly by matrix metalloproteinases (MMPs), such as MMP-2 and MMP-9 [11], and alters the abundances of adhesion-associated molecules such as E-cadherin and vimentin [12,13]. Alteration in MMP expression and activity could cause uterine and vascular dysfunction, contributing to the pathogenesis of PE [14].

Endometrial cells have been demonstrated to play a central role in normal and abnormal early pregnancy development. Mesenchymal stem cells (MSCs) are pluripotent progenitor cells with a self-renewing capacity and potential ability to differentiate. MSCs of endometrial origin have been identified and characterized from human endometrial tissue and menstrual fluid [15,16]. The properties of MSCs isolated from menstrual fluid (MenSCs) have demonstrated improved angiogenic properties, including vascular endothelial growth factor (VEGF) secretion, in comparison to MSCs isolated from the bone marrow [17]. We previously studied the angiogenic properties of MenSCs obtained from patients with a history of PE (PE-MenSCs), further demonstrating less endoglin and VEGF expression as well as less VEGF secretion but higher expression of interleukin (IL)-6 compared to MenSCs obtained from women with a previous normal pregnancy [18]. These results suggested that PE-MenSCs had reduced angiogenic capacity and were more proinflammatory than those of MenSCs from women with a previous normal pregnancy [18].

In this study, we aim to: (i) characterize the proliferative capacity and migratory behavior of MenSCs isolated from women with or without a history of PE (nulliparous and multiparous women); and to (ii) determine the capacity of MenSCs to induce trophoblast invasion by utilizing a blastocyst-like structure in a 3D in vitro model that simulated communication between endometrial and trophoblast cells.

## 2. Results

### 2.1. Characteristics of the Donors

The demographic characteristics of the nulliparous, multiparous, and PE donors are presented in Table 1. Nulliparous donors were younger than multiparous and PE women (adjusted $p$ values = 0.021 and $p$ = 0.008, respectively); however, there was no difference between multiparous and PE women (adjusted $p$ value > 0.99). Body mass index and gravidity did not differ between groups. Preeclamptic women presented with significantly lower gestational ages and newborn weights in their last pregnancy compared to multiparous women (control group).

**Table 1.** Characteristics of controls and preeclampsia donors.

| Characteristics | Nulliparous (n = 10) | Multiparous (n = 10) | Preeclampsia (n = 9) | $p$ Value |
|---|---|---|---|---|
| Age (years) | 27.6 ± 4.5 | 34 ± 4.6 | 35.0 ± 5.3 | 0.005 |
| BMI ($Kg/m^2$) | 23.2 ± 3.4 | 22.2 ± 3.2 | 24.8 ± 4.7 | 0.319 |
| Gravidity | 0 | 3.1 ± 1.3 | 2.1 ± 1.1 | 0.086 |
| Gestational age at last delivery | - | 37.7 ± 1.1 | 33.3 ± 5.9 | 0.045 |
| Newborn weight | - | 3240 ± 388.2 | 2100 ± 1014.8 | 0.004 |

BMI, body mass index. Statistical analyses were performed with one-way ANOVA with Bonferroni post hoc test for the variables age and BMI (between the 3 groups), meanwhile for gravidity, gestational age at last delivery and newborn weight, Student's $t$-test was performed (MUL and PE groups).

## 2.2. *Proliferation and Migration of MenSCs*

To assess functional properties of MenSCs, the proliferative potential and migratory behavior of cells obtained from the 3 different groups were compared. The proliferation rate was similar in nulliparous (NUL), multiparous (MUL), and preeclamptic (PE) women (Figure 1A) from day 1 to day 9. The migratory behavior of MenSCs isolated from multiparous women did not differ from nulliparous MenSCs ($p$ = 0.393, Figure 1B) or from MenSCs isolated from women with a history of PE ($p$ > 0.999, Figure 1C).

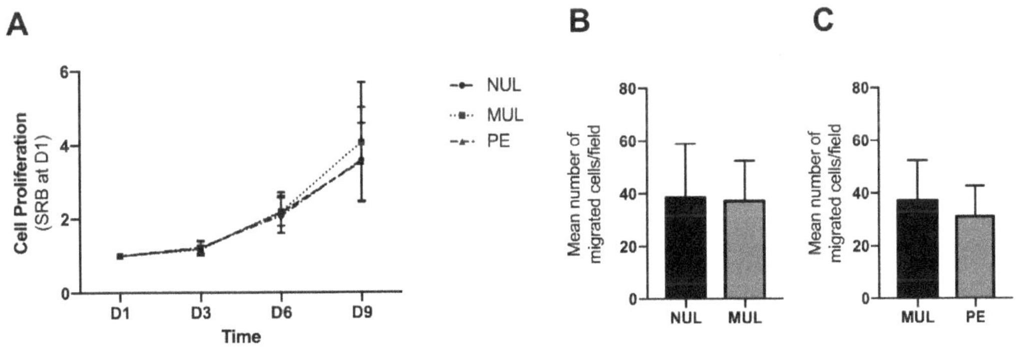


**Figure 1.** Proliferation and migration of MenSCs derived from nulliparous (NUL), multiparous (MUL) and PE women. (**A**) MenSCs were seeded for SRB assays to quantify cell proliferation at day 3, day 6, and day 9. Results are presented as mean values and SDs of eight MenSCs per group. (**B**) NUL and MUL MenSCs migration assays. (**C**) MUL and PE MenSCs trans-migration assays. Results are presented as mean values and SDs of six MenSCs per group. Statistical analyses were performed with Mann–Whitney tests. MenSCs, mesenchymal stem cells derived from the menstrual fluid; PE, preeclampsia; SD, standard deviation; SRB, sulforhodamine B.

## 2.3. *Estradiol and Progesterone Treatment Increase the Expression of LIF and IGFBP1 in MenSCs*

To characterize the effect of estradiol and progesterone treatment on MenSCs cultured in 5% $O_2$ (endometrial physiological conditions, defined as Mimic), the gene expression of decidualization and the window of implantation markers were evaluated in all MenSCs. We determined the expression of insulin-like growth factor binding protein 1 (IGFBP1) and prolactin (PRL) as decidualization markers. *IGFBP1* demonstrated a significant increase ($p$ = 0.02, Figure 2B) compared to cells without hormonal treatment (control). MenSCs also demonstrated an increase in leukemia inhibitory factor (LIF) ($p$ = 0.04, Figure 2C), the window of implantation marker after treatment with the endometrial hormonal mimic, suggesting that the exposure of MSCs of menstrual origin to estrogen for 24 h and then to estrogen plus progesterone for another 24 h at levels of oxygen present in the endometrium can simulate endometrial conditions that precede implantation.

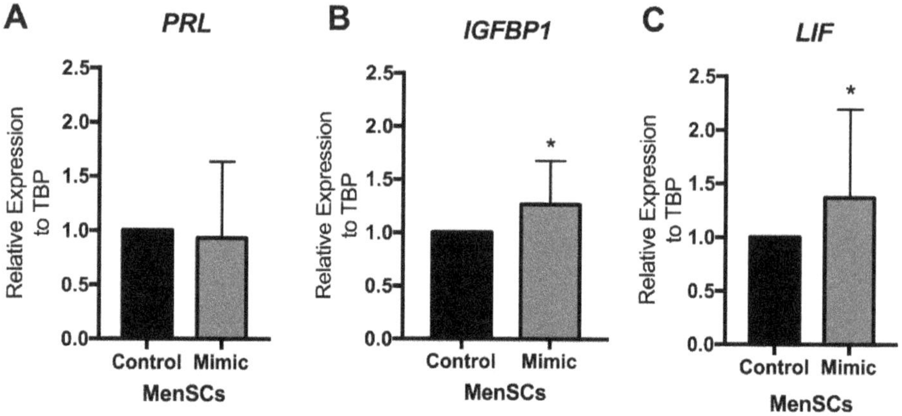


**Figure 2.** Characterization of MenSCs after endometrial physiological conditions (Mimic). Gene expression of decidualization markers: (**A**) Prolactin (PRL) and (**B**) insulin-like growth factor binding protein 1 (IGFBP1) were detected in MenSCs. Window of implantation marker (**C**) leukemia inhibitory factor (LIF) was detected in MenSCs after hormonal treatment. Results are expressed as mean and SD of nine MenSCs per group. * $p < 0.05$. Statistical analyses were performed by using the Wilcoxon matched-pairs signed-rank test. MenSCs, mesenchymal stem cells derived from the menstrual fluid; SD, standard deviation.

### 2.4. Characterization of Trophospheres from HTR-8/SVneo Cells Origin

Relative to adherent 2D cultures, trophospheres had a significant decrease in mRNA expression of the epithelial marker *CDH1* ($p = 0.046$, Figure 3A), an increase in mRNA expression of the mesenchymal gene *VIM* ($p = 0.03$, Figure 3C) and the stemness markers *NANOG* and *SOX2* ($p = 0.015$ both, Figure 3D,E) after 72 h in culture. There was no difference in the expression of mesenchymal *SNAIL* transcript levels in the trophospheres compared to the adherent cells ($p = 0.21$) (Figure 3B). These findings suggest spheroid formation and culture are associated with the induction of epithelial-mesenchymal transition (EMT) and that the formed spheres are able to express higher levels of stemness genes.

### 2.5. The Invasive Potential of Trophospheres Is Reduced When Co-Cultured with PE-MenSCs

Trophoblast invasion involves an active process of migration through various layers of endometrial tissues, including the extracellular matrix and stroma. To elucidate the communication between endometrial stem cells and trophoblasts during the invasion process, we developed a 3D in vitro invasion model by using MenSCs, Matrigel, and trophospheres (Figure 4A). After 72 h of co-culture, we observed the formation of projections that penetrated the Matrigel in a radial orientation. Using phase-contrast images captured with an inverted microscope, we measured the area of invasion and compared it to the area of invasion measured in 3D cultures lacking MenSCs (Figure 4H). First, we analyzed the effect of gravidity on trophoblast invasion. The results demonstrated that when trophospheres are co-cultured with MUL-MenSCs there was a modest but non-significant difference in the degree of invasion compared to NUL-MenSCs ($p = 0.064$, Figure 4B). These trophospheres had an increased expression of matrix metalloproteinase 2 (*MMP2*; $p = 0.036$, Figure 4C). Secondly, we compared whether the PE history of donors can influence trophoblast invasion capacity induced by MenSCs. Indeed, when trophospheres were cultured with MenSCs from women with a history of PE (PE-MenSCs), they had a reduced area of invasion compared to MUL-MenSCs ($p = 0.004$, Figure 4E). Trophospheres co-cultured with

PE-MenSCs showed an increased expression of matrix metalloproteinase 9 (*MMP9*) after invasion ($p = 0.036$, Figure 4G) but no difference in *MMP2*.

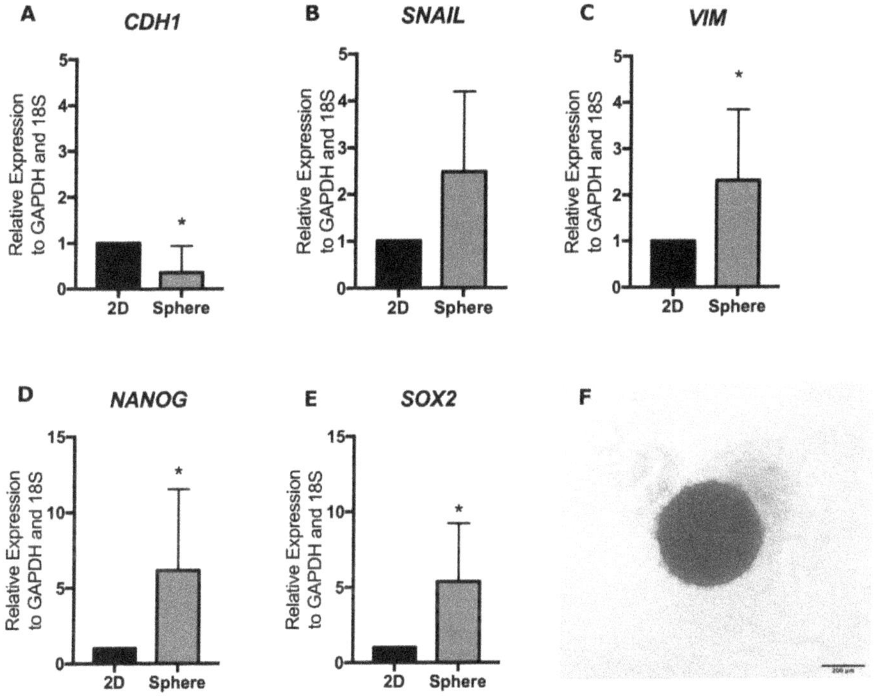


**Figure 3.** Characterization of trophospheres after 72 h culture compared to 2D adherent culture. mRNA expression of *CDH1* (**A**), *SNAIL* (**B**), *VIM* (**C**), *NANOG* (**D**) and *SOX2* (**E**) in trophospheres and 2D cultures was analyzed by qRT-PCR and normalized to *GADPH* and *18S* housekeeping genes. (**F**) Tropho-sphere following 72 h of culture. Results are expressed as mean and SD of seven trophospheres and HTR-8/SVneo monolayer cultures per group. * $p < 0.05$. Statistical analyses were performed by using the Wilcoxon matched-pairs signed-rank test. SD, standard deviation.

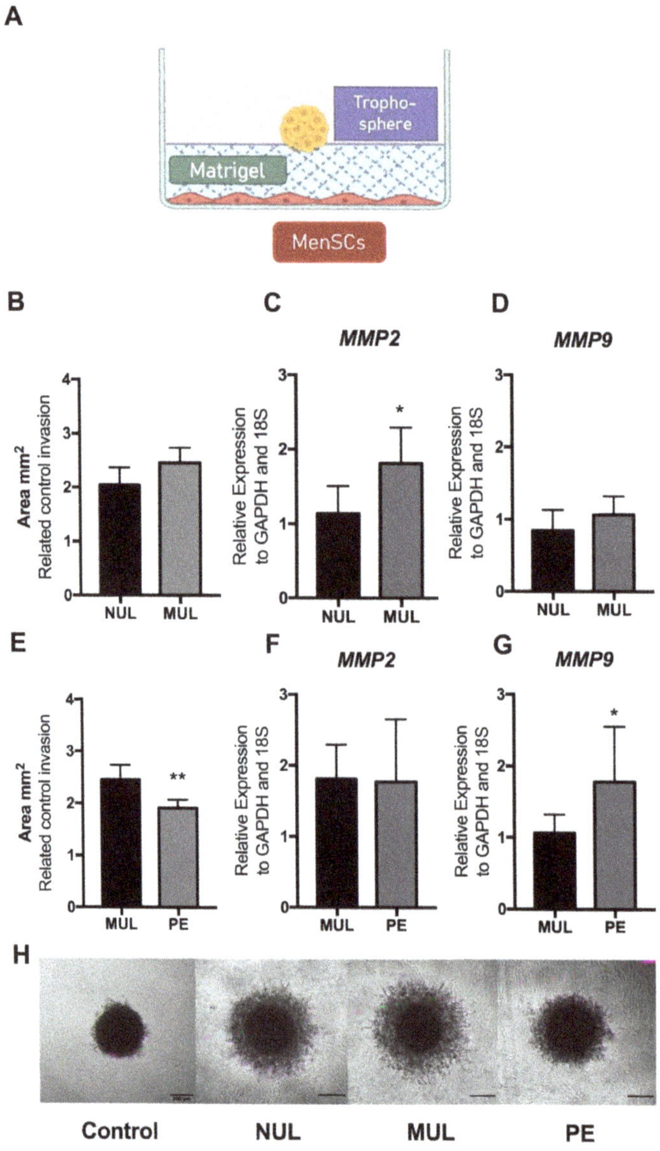


**Figure 4.** In vitro 3D model of trophoblast invasion. (**A**) Schematic depicts the components of the model: MenSCs on the bottom of the plate as a monolayer of endometrial cells. Matrigel mimics the extracellular matrix, and trophospheres mimic the trophectoderm of a blastocyst that is transferred onto the Matrigel. (**B**,**E**) Invasion area of trophospheres co-cultured with MenSCs of nulliparous, multiparous, and PE women. Each area is expressed as relative to control invasion without MenSCs. Results are the means and SD of six MenSCs per group. (**C**,**F**) *MMP2* and (**D**,**G**) *MMP9* mRNA expression in trophospheres after invasion assay compared to the control invasion trophospheres. * $p < 0.05$, ** $p < 0.01$. Statistical analysis was performed by using the Mann–Whitney U test. (**H**) Representative images of trophospheres invasion after 72 h of co-culture with MenSCs. MenSCs, mesenchymal stem cells derived from the menstrual fluid, SD, standard deviation.

## 3. Discussion

MenSCs are characterized by their capabilities of self-renewal and differentiation under standard conditions [19]. They are the focus of a growing amount of interest due to their clinical potential in the study of pregnancy complications, given the minimally invasive manner in which they may be obtained. Data published previously suggested that MenSCs retain phenotypic variation postpartum that may be associated with, and perhaps account for the abnormal implantation process observed in PE [18]. Moreover, human decidual natural killer cells also possess a trained memory after the first pregnancy, which includes expression of receptors that interact with EVTs and growth factors promoting an efficient placentation in subsequent pregnancies [20,21]. This pregnancy-imprinted memory could be a part of the reason why a history of PE could impact the capacity to induce trophoblast invasion by MenSCs. In this study, troposphere co-culture with PE-MenSCs was associated with decreased trophoblast invasion capacity compared to the control (MUL-MenSCs). This impaired invasion capacity was not associated with changes detectable in the proliferation and migration capacity of the same cells. However, deficient trophoblast invasion during placentation in combination with maternal conditions are accepted as a cause of endothelial dysfunction, inflammatory alteration, and the appearance of symptoms of PE [22].

The EMT-related molecules are linked to the migration and invasion capacity of the trophoblast. This process includes the reduction of epithelial and the increase in mesenchymal markers in trophoblast cells. Our model of trophospheres demonstrated the induction of EMT as decreasing *CDH1* expression and as increasing *VIM*, *NANOG* and *SOX2* compared to the 2D monolayer culture of the same cells. These results have been reported previously [23,24] and suggest that spheroids have the capacity to drive EVTs, as we observed in our 3D invasion model.

Trophoblast invasion is regulated by several factors, including MMPs, the proteases that degrade different components of the extracellular matrix, and are expressed by utero-placental interphases such as trophoblast, endometrial stromal cells, and natural killers cells [25]. MMPs play a major role in trophoblast invasion into the uterine wall to create an optimum environment for embryonic development. Studies have demonstrated that MMP-2 and MMP-9 may be implicated in the early and late stages of PE due to their role in vasodilation, placentation, and uterine expansion during normal pregnancy [14]. Herein, however, we demonstrated that despite the reduced trophoblast invasion capacity in the presence of PE-MenSCs, the expression of *MMP2* in trophospheres did not exhibit differences after the 72-h invasion compared to the control group (MUL-MenSCs), while *MMP9* demonstrated an increased expression. The altered expression of MMPs have also been reported in preeclamptic women. Higher levels of MMP-2 and lower levels of MMP-9 in maternal serum have been associated with early- and late-onset severe PE [26], and the urinary MMP-2 concentration at 12 and 16 weeks of gestation was reported as increased in women who developed PE later during pregnancy [27].

Trophoblast cell invasion is regulated by signaling events, autocrine and paracrine stimuli, specific protein recognition, and immunological tolerance [28]. Considering our control of invasion without MenSCs, communication between MenSCs and trophoblast cells during the 3D invasion model results appears necessary to induce trophosphere invasion. Therefore, we propose that MenSCs may stimulate trophosphere invasion through the release of exosomes [29]. Studies have demonstrated that exosomes secreted by MenSCs served as a convincing new type of cell-free treatment [30,31]. Exosomes contain microRNA/lncRNA and adhesion molecules as well as small vesicles with secreted proteins, which mediate cellular signaling pathways [32]. The content of exosomes secreted by MenSCs in relation to the invasion capacity remains poorly characterized. Analysis of the MenSCs-derived exosomes could help to explain the changes in the invasion capacity of trophoblast cells.

Given that the human uterus is exposed to varying hormonal profiles throughout the menstrual cycle [33], MenSCs were cultured with estradiol and progesterone to simulate the window of implantation environment in vitro. During the window of implantation, the blastocyst can attach to the endometrial epithelial cells and invade the endometrial stroma and vasculature [34]. The receptive endometrium is characterized by the appearance of pinopodes and the increased expression of the leukemia inhibitor factor (LIF). LIF regulates trophoblast cell adhesion, and it might be important for embryo invasion and placental development. Our results demonstrated an increase in *LIF* expression in MenSCs after treatment with estradiol and progesterone, simulating the signaling milieu present in the uterus. The extent of trophoblast invasion relies on communication between the placenta and maternal decidua. Studies have demonstrated increased decidualization markers such as IGFBP1 and prolactin in endometrial stromal cells after hormonal treatment [35,36]. Moreover, endometrial stromal cells of women with a history of PE failed to decidualize in vitro after hormonal treatment [9]. Herein, we only observed increased of *IGFBP1* expression in MenSCs after hormonal treatment, suggesting that the hormone concentration used in this study might be too modest to induce decidualization.

## 4. Materials and Methods

### 4.1. Isolation and Culture of MenSCs

Menstrual fluid was self-collected by consenting donors following informed consent according to a protocol reviewed and approved by the ethical scientific committee of Universidad de los Andes. Samples of MenSCs were obtained from three groups of study participants: nulliparous women, multiparous women, and women with a history of PE. All donors confirmed that they had not used hormonal contraceptives for at least three months, and a clinical/gynecological history was obtained for each donor.

Overnight menstrual fluid was collected in a silicone menstrual cup within the first 48 h of menstruation and transferred into a 50 mL tube containing 10 mL 1× phosphate-buffered saline (PBS) and the following supplements: 0.25 mg/mL amphotericin B, penicillin 100 IU, streptomycin 100 mg/mL, and 2 mM ethylenediaminetetraacetic acid (EDTA) (all Gibco, Thermo Fisher Scientific, Waltham, MA, USA). MenSCs were isolated from the mononuclear cell fraction with Ficoll®Paque Plus (GE Healthcare, Piscataway, NJ, USA) density gradient by centrifugation at 400× *g* for 30 min at room temperature, according to the manufacturer's instructions. Mononuclear cells were recovered from the interface between the plasma and Ficoll®Paque Plus, then washed twice with PBS to remove the platelets, Ficoll®Paque Plus, and plasma. Mononuclear cells were cultured in Dulbecco's Modified Eagle Medium (DMEM) with high glucose (Mediatech Inc, Manassas, VA, USA) supplemented with 1% penicillin/streptomycin (P/S), 2.5 μg/mL amphotericin B, 2 mM glutamine, and 15% fetal bovine serum (FBS) (all Gibco, Thermo Fisher Scientific) in a humidified environment at 37 °C and 5% $CO_2$ to obtain adherent cells. All experiments were performed by using MenSCs at early passages (P) P3 to P7.

### 4.2. Proliferation Assay: Sulforhodamine Assay

MenSCs were cultured at 1000 cells/$cm^2$ in 24-well plates (Falcon, Corning, NY, USA) in supplemented DMEM (10% FBS, 1% penicillin-streptomycin, 1% glutamine). Cell proliferation and viability were determined at days 3, 6, and 9 by using a sulforhodamine B (SRB) assay (BioVision, Milpitas, CA, USA). The SRB assay is based on the ability of the protein dye SRB to bind basic amino acid residues of fixed cells. Quantification was performed by spectrophotometric quantification (absorbance to 492 nm) on a Tecan Sunrise Reader, 96-well Microplate Reader, according to the manufacturer's instructions.

### 4.3. Migration Assay

MenSCs were suspended in 400 μL of DMEM with reduced serum (0.5% FBS) and seeded in a millicell insert (pore 8 μm, 12 mm, Millipore, Billerica, MA, USA) with 25,000 cells/insert. Inserts were placed on 500 μL of complete media (10% FBS) in 48-well plates. Migration capacity was evaluated at 18 h. Briefly, the insert was washed with 1× PBS, fixed with cold methanol for 2 min, and stained with 0.5% crystal violet (Winkler, Santiago, Chile). The cells inside the inserts were scraped with cotton swabs moistened with 1× PBS to ensure only migrated cells were analyzed. Five fields were captured for each insert at 40× objective magnification before and after scraping under an inverted microscope (Primo Vert, Zeiss, Jena, Germany), using the AxioCam ERc5s camera (Zeiss). Images were analyzed with AxioVision analysis software (Zeiss). The percentage of migrated cells was calculated as follows: number of cells after/before scraping × 100 (average of the 5 fields). The experiments were performed in duplicate.

### 4.4. Hormonal Treatment

MenSCs were cultured under an endometrial hormonal milieu, as described previously [18]. Briefly, 3000 MenSCs/well were cultured in a 96-well plate with DMEM phenol red free media (Mediatech Inc.), supplemented with 10% charcoal-stripped FBS, 1% penicillin-streptomycin for 24 h. The following day, medium was replaced with fresh media supplemented with 17β-estradiol (E2; 213 pg/mL) and cultured at 5% $O_2$ for another 24 h. The day after, MenSCs were exposed to 17β-estradiol and progesterone (E2; 146 pg/mL, P4; 11 ng/mL) and cultured at 37 °C for another 24 h in a humidified atmosphere hypoxia chamber with 5% $O_2$ and -5% $CO_2$. Hormone-treated MenSCs were used in real-time quantitative reverse transcription PCR (qRT-PCR) analysis or 3D invasion assays.

### 4.5. Trophoblast Sphere Formation

Trophoblast sphere formation was described previously in 2019 [19], which we used with some modifications. An adherent first-trimester trophoblast cell line, HTR-8/SVneo (HTR-8) was purchased from the American Type Culture Collection (CRL-3271; Lot #70016636, ATCC, Manassas, VA, USA). Cells were maintained in RPMI-1640 medium (GE Healthcare, Piscataway, NJ, USA), 10% heat-inactivated FBS (Gibco, Thermo Fisher Scientific), and 1% P/S (Gibco, Thermo Fisher Scientific) at 37 °C in a humidified incubator with 5% $CO_2$. For trophoblast sphere formation, HTR-8 cells were harvested and resuspended in complete media. The 2 × $10^4$ HTR-8 cells were suspended in 200 µL and placed into each well of an ultra-low attachment 96-well plate. Following centrifugation at 300× *g* for 5 min, cells were incubated for 72 h at 37 °C in a humidified atmosphere hypoxia chamber with 5% $O_2$ and 5% $CO_2$. The trophospheres were washed with 1× PBS and used for characterization by qRT-PCR or for 3D invasion assays.

### 4.6. 3D Invasion Assay

This model was first described in 2019 [19], which we used with some modifications. MenSCs treated under endometrial hormonal milieu in 96-well plates were used for 3D invasion assays. Matrigel Growth Factor Reduced and Phenol Red-free (Corning Life Sciences, Union City, CA, USA) was mixed with DMEM containing 10% charcoal-stripped FBS and 1% penicillin-streptomycin 1:1 and added to the treated MenSCs in the 96-well plates. The plate was then incubated for 30 min at 37 °C to allow the Matrigel to solidify. A single sphere of trophoblast cells was subsequently placed on each well onto the Matrigel, and 150 µL of DMEM, 10% charcoal-stripped FBS, and 1% P/S were added to embed the sphere. Trophosphere invasion was evaluated after 72 h. Phase contrast images were captured by the contrast microscope Olympus CKX41 and Axiocam 208 color (Zeiss). The invasion level (area) was quantified by using ImageJ software, and trophospheres were collected for qRT-PCR assay. Eight trophospheres were required for each condition.

### 4.7. RNA Isolation and qRT-PCR

Total RNA was extracted from MenSCs and trophospheres by using TRIzol Reagent (Invitrogen Corporation, San Diego, CA, USA), according to the manufacturer's protocol. RNA amount and quality were evaluated on the NanoDrop 2000 spectrophotometer (Thermo Fisher Scientific) at 260/280 with all samples having values between 1.8 and 2.0. Before cDNA synthesis, total RNA was treated with DNase I (Invitrogen Corporation). One microgram of RNA was used for reverse transcription using SuperScript II (Invitrogen Corporation), according to the manufacturer's instructions. Determination of gene expression was carried out using Brilliant III SYBR Green qPCR Master Mix (Agilent Technologies, Santa Clara, CA, USA), according to the manufacturer's instructions, and amplified on a qRT-PCR Stratagene Mx3000P System (Agilent Technologies). GAPDH and 18S were used as housekeeping genes for the normalization of trophospheres and TBP was used for MenSCs. The real time PCR was set at 95 °C for 10 min for enzyme activation, followed by 40 cycles of denaturation, primer annealing, and extension consisting of 95 °C for 15 s, 60 °C for 15 s, and 72 °C for 15 s, respectively. All samples were run in duplicate. After the PCR runs, a dissociation curve was generated to confirm the absence of nonspecific amplification. The expression was quantified using the $2^{-\Delta\Delta CT}$ method. Primers details are provided in Table 2. Reaction specificity was confirmed using dissociation curves, no-template, and no-RT controls for each target.

**Table 2.** Primer sequence and annealing temperatures (AT) for qRT-PCR.

| Gene | Sequence | AT °C |
|---|---|---|
| 18S Fw | GCCGCTAGAGGTGAAATTCTTGGA | 60 |
| 18S Rv | ATCGCCGGTCGGCATCGTTTAT | |
| GAPDH Fw | GTCAGGGTCTCTCTCTTCCT | 60 |
| GAPDH Rv | GCTCTCCTCTGACTTGAACA | |
| TBP Fw | CAC GAA CCA CGG CAC TGA T | 62 |
| TBP Rv | GTT GGT GGG TGA GCA CAA GG | |
| SOX2 Fw | AGC TAC AGC ATG ATG CAG GA | 60 |
| SOX2 Rv | GGT CAT GGA GTT GTA ATG CA | |
| NANOG Fw | CTG ATT CTT CCA CCA GTC CC | 60 |
| NANOG Rv | AGG TCT TCA CCT GTT TGT AG | |
| SNAIL Fw | CCC CAA TCG GAA GCC TAA CT | 62 |
| SNAIL Rv | GCT GGA AGG TAA ACT CTG GAT TAG | |
| IGFBP1 Fw | GAA GGA GCC CTC CCG AAT AG | 62 |
| IGFBP1 Rv | CCA TTC CAA GGG TAG ACG CA | |
| CDH1 Fw | CTGCCAATCCCGATGAAATTG | 60 |
| CDH1 Rv | TCCTTCATAGTCAAACACGAGC | |
| VIM Fw | AGTCCACTGAGTACCGGAGAC | 62 |
| VIM Rv | CATTTCACGCATCTGGCGTTC | |
| LIF Fw | GTTTCCTCCAAGGCCCTCT | 60 |
| LIF Rv | TGTTTCCAGTGCAGAACCAA | |

### 4.8. Statistical Analysis

Graphs were made and statistical analyses were performed with GraphPad Prism Version 7.0 software (Graphpad, San Diego, CA, USA). Statistical significance was set at $p < 0.05$ for all analyses. Data normality was tested by the Shapiro–Wilk test. For parametric data (characteristics of donors) we applied one-way ANOVA with a Bonferroni post-test and a Student's *t*-test and for nonparametric data (functional assays and qRT-PCR), Kruskal–Wallis, Dunn's multiple comparison, and Mann–Whitney U tests were applied.

## 5. Conclusions

In summary, we demonstrated that PE-MenSCs exposed to trophospheres in a 3D co-culture model was associated with a decrease in the invasive capacity of trophoblasts in vitro, a characteristic that could be associated with the pathogenesis of PE.

**Author Contributions:** Conceptualization, S.E.I., L.J.M. and S.A.-G.; methodology, S.A.-G., R.P. and F.G.; formal analysis, S.A.-G., R.P. and F.G.; writing—original draft preparation, R.P. and S.A.-G.; writing—review and editing, R.P., L.J.M., G.N., R.R., M.A.C., M.W.K. and S.E.I.; visualization, R.P. and S.A.-G.; supervision, S.E.I.; funding acquisition, S.E.I., L.J.M. and G.N. All authors have read and agreed to the published version of the manuscript.

**Funding:** This research was funded by ANID/CONICYT—FONDECYT Regular 1201851 (to S.E.I.); ANID/CONICYT—FONDECYT de Iniciación 11190998 (to G.N.) and ANID/CONICYT—FONDECYT de Iniciación 11181249 (to L.J.M.); ANID-BASAL funding for Scientific and Technological Center of Excellence, IMPACT, #FB210024 (to L.J.M, G.N. and S.E.I.). This study was also supported, in part, by the Perinatology Research Branch, Division of Obstetrics and Maternal-Fetal Medicine, Division of Intramural Research, *Eunice Kennedy Shriver* National Institute of Child Health and Human Development, National Institutes of Health, United States Department of Health and Human Services (NICHD/NIH/DHHS); and, in part, with Federal funds from NICHD/NIH/DHHS under Contract No. HHSN275201300006C. R.R. has contributed to this work as part of his official duties as an employee of the United States Federal Government.

**Institutional Review Board Statement:** The study was conducted in accordance with the Declaration of Helsinki and approved by the Ethical Scientific Committees of Universidad de los Andes (date of approval 22 April 2020).

**Informed Consent Statement:** Informed consent was obtained from all subjects involved in the study.

**Data Availability Statement:** Not applicable.

**Conflicts of Interest:** The authors declare no conflict of interest.

## References

1. Brown, M.A.; Magee, L.A.; Kenny, L.C.; Karumanchi, S.A.; McCarthy, F.P.; Saito, S.; Hall, D.R.; Warren, C.E.; Adoyi, G.; Ishaku, S. Hypertensive disorders of pregnancy: ISSHP classification, diagnosis, and management recommendations for international practice. *Hypertension* **2018**, *72*, 24–43. [CrossRef] [PubMed]
2. Steegers, E.A.P.; Von Dadelszen, P.; Duvekot, J.J.; Pijnenborg, R. Pre-eclampsia. *Lancet* **2010**, *376*, 631–644. [CrossRef]
3. Albrecht, E.D.; Pepe, G.J. Regulation of Uterine Spiral Artery Remodeling: A Review. *Reprod. Sci.* **2020**, *27*, 1932–1942. [CrossRef] [PubMed]
4. Pollheimer, J.; Vondra, S.; Baltayeva, J.; Beristain, A.G.; Knöfler, M. Regulation of Placental Extravillous Trophoblasts by the Maternal Uterine Environment. *Front. Immunol.* **2018**, *9*, 2597. [CrossRef] [PubMed]
5. James-Allan, L.B.; Whitley, G.S.; Leslie, K.; Wallace, A.E.; Cartwright, J.E. Decidual cell regulation of trophoblast is altered in pregnancies at risk of pre-eclampsia. *J. Mol. Endocrinol.* **2018**, *60*, 239–246. [CrossRef]
6. Ng, S.W.; Norwitz, G.A.; Pavlicev, M.; Tilburgs, T.; Simón, C.; Norwitz, E.R. Endometrial decidualization: The primary driver of pregnancy health. *Int. J. Mol. Sci.* **2020**, *21*, 4092. [CrossRef]
7. Liu, H.; Huang, X.; Mor, G.; Liao, A. Epigenetic modifications working in the decidualization and endometrial receptivity. *Cell. Mol. Life Sci.* **2020**, *77*, 2091–2101. [CrossRef]
8. Gellersen, B.; Reimann, K.; Samalecos, A.; Aupers, S.; Bamberger, A.M. Invasiveness of human endometrial stromal cells is promoted by decidualization and by trophoblast-derived signals. *Hum. Reprod.* **2010**, *25*, 862–873. [CrossRef]
9. Garrido-Gomez, T.; Dominguez, F.; Quiñonero, A.; Diaz-Gimeno, P.; Kapidzic, M.; Gormley, M.; Ona, K.; Padilla-Iserte, P.; McMaster, M.; Genbacev, O.; et al. Defective decidualization during and after severe preeclampsia reveals a possible maternal contribution to the etiology. *Proc. Natl. Acad. Sci. USA* **2017**, *114*, E8468–E8477. [CrossRef]
10. Pijnenborg, R.; Vercruysse, L.; Hanssens, M. The Uterine Spiral Arteries In Human Pregnancy: Facts and Controversies. *Placenta* **2006**, *27*, 939–958. [CrossRef]
11. Chen, J.; Khalil, R.A. Matrix Metalloproteinases in Normal Pregnancy and Preeclampsia. In *Progress in Molecular Biology and Translational Science*; Academic Press: Cambridge, MA, USA, 2017; Volume 148, pp. 87–165.
12. Turco, M.Y.; Moffett, A. Development of the human placenta. *Development* **2019**, *146*, dev163428. [CrossRef] [PubMed]
13. Knöfler, M. Critical growth factors and signalling pathways controlling human trophoblast invasion. *Int. J. Dev. Biol.* **2010**, *54*, 269–280. [CrossRef] [PubMed]
14. Nikolov, A.; Popovski, N. Role of gelatinases mmp-2 and mmp-9 in healthy and complicated pregnancy and their future potential as preeclampsia biomarkers. *Diagnostics* **2021**, *11*, 480. [CrossRef] [PubMed]
15. Gargett, C.E.; Schwab, K.E.; Zillwood, R.M.; Nguyen, H.P.T.; Wu, D. Isolation and Culture of Epithelial Progenitors and Mesenchymal Stem Cells from Human Endometrium. *Biol. Reprod.* **2009**, *80*, 1136–1145. [CrossRef] [PubMed]
16. Meng, X.; Ichim, T.E.; Zhong, J.; Rogers, A.; Yin, Z.; Jackson, J.; Wang, H.; Ge, W.; Bogin, V.; Chan, K.W.; et al. Endometrial regenerative cells: A novel stem cell population. *J. Transl. Med.* **2007**, *5*, 57–66. [CrossRef]
17. Alcayaga-Miranda, F.; Cuenca, J.; Luz-Crawford, P.; Aguila-Díaz, C.; Fernandez, A.; Figueroa, F.E.; Khoury, M. Characterization of menstrual stem cells: Angiogenic effect, migration and hematopoietic stem cell support in comparison with bone marrow mesenchymal stem cells. *Stem Cell Res. Ther.* **2015**, *6*, 32. [CrossRef]
18. Varas-Godoy, M.; Acuña-Gallardo, S.; Venegas-Duarte, S.; Hill, C.; Caceres-Verschae, A.; Realini, O.; Monteiro, L.J.; Zavala, G.; Khoury, M.; Romero, R.; et al. Angiogenic Properties of Menstrual Stem Cells Are Impaired in Women with a History of Preeclampsia. *Stem Cells Int.* **2019**, *2019*, 1916542. [CrossRef]

19. Patel, A.N.; Park, E.; Kuzman, M.; Benetti, F.; Silva, F.J.; Allickson, J.G. Multipotent menstrual blood stromal stem cells: Isolation, characterization, and differentiation. *Cell Transplant.* **2008**, *17*, 303–311. [CrossRef]
20. Staff, A.C.; Fjeldstad, H.E.; Fosheim, I.K.; Moe, K.; Turowski, G.; Johnsen, G.M.; Alnaes-Katjavivi, P.; Sugulle, M. Failure of physiological transformation and spiral artery atherosis: Their roles in preeclampsia. *Am. J. Obstet. Gynecol.* **2022**, *226*, S895–S906. [CrossRef]
21. Gamliel, M.; Goldman-Wohl, D.; Isaacson, B.; Gur, C.; Stein, N.; Yamin, R.; Berger, M.; Grunewald, M.; Keshet, E.; Rais, Y.; et al. Trained Memory of Human Uterine NK Cells Enhances Their Function in Subsequent Pregnancies. *Immunity* **2018**, *48*, 951–962.e5. [CrossRef]
22. Martinez-Fierro, M.L.; Hernández-Delgadillo, G.P.; Flores-Morales, V.; Cardenas-Vargas, E.; Mercado-Reyes, M.; Rodriguez-Sanchez, I.P.; Delgado-Enciso, I.; Galván-Tejada, C.E.; Galván-Tejada, J.I.; Celaya-Padilla, J.M.; et al. Current model systems for the study of preeclampsia. *Exp. Biol. Med.* **2018**, *243*, 576–585. [CrossRef] [PubMed]
23. Wong, M.K.; Wahed, M.; Shawky, S.A.; Dvorkin-Gheva, A.; Raha, S. Transcriptomic and functional analyses of 3D placental extravillous trophoblast spheroids. *Sci. Rep.* **2019**, *9*, 12607. [CrossRef] [PubMed]
24. Zhu, J.-Y.; Pang, Z.-J.; Yu, Y.-H. Regulation of trophoblast invasion: The role of matrix metalloproteinases. *Rev. Obstet. Gynecol.* **2012**, *5*, e137–e143. [CrossRef] [PubMed]
25. You, Y.; Stelzl, P.; Zhang, Y.; Porter, J.; Liu, H.; Liao, A.H.; Aldo, P.B.; Mor, G. Novel 3D in vitro models to evaluate trophoblast migration and invasion. *Am. J. Reprod. Immunol.* **2019**, *81*, e13076. [CrossRef] [PubMed]
26. Laskowska, M. Altered Maternal Serum Matrix Metalloproteinases MMP-2, MMP-3, MMP-9, and MMP-13 in Severe Early- and Late-Onset Preeclampsia. *Biomed Res. Int.* **2017**, *2017*, 6432426. [CrossRef] [PubMed]
27. Martinez-Fierro, M.L.; Perez-Favila, A.; Garza-Veloz, I.; Espinoza-Juarez, M.A.; Avila-Carrasco, L.; Delgado-Enciso, I.; Ortiz-Castro, Y.; Cardenas-Vargas, E.; Cid-Baez, M.A.; Ramirez-Santoyo, R.M.; et al. Matrix metalloproteinase multiplex screening identifies increased MMP-2 urine concentrations in women predicted to develop preeclampsia. *Biomarkers* **2018**, *23*, 18–24. [CrossRef] [PubMed]
28. Staun-Ram, E.; Shalev, E. Human trophoblast function during the implantation process. *Reprod. Biol. Endocrinol.* **2005**, *3*, 56. [CrossRef]
29. Adam, S.; Elfeky, O.; Kinhal, V.; Dutta, S.; Lai, A.; Jayabalan, N.; Nuzhat, Z.; Palma, C.; Rice, G.E.; Salomon, C. Review: Fetal-maternal communication via extracellular vesicles—Implications for complications of pregnancies. *Placenta* **2017**, *54*, 83–88. [CrossRef]
30. Rosenberger, L.; Ezquer, M.; Lillo-Vera, F.; Pedraza, P.L.; Ortúzar, M.I.; González, P.L.; Figueroa-Valdés, A.I.; Cuenca, J.; Ezquer, F.; Khoury, M.; et al. Stem cell exosomes inhibit angiogenesis and tumor growth of oral squamous cell carcinoma. *Sci. Rep.* **2019**, *9*, 663. [CrossRef]
31. Dalirfardouei, R.; Jamialahmadi, K.; Jafarian, A.H.; Mahdipour, E. Promising effects of exosomes isolated from menstrual blood-derived mesenchymal stem cell on wound-healing process in diabetic mouse model. *J. Tissue Eng. Regen. Med.* **2019**, *13*, 555–568. [CrossRef]
32. Chen, L.; Qu, J.; Xiang, C. The multi-functional roles of menstrual blood-derived stem cells in regenerative medicine. *Stem Cell Res. Ther.* **2019**, *10*, 1. [CrossRef] [PubMed]
33. Liu, S.-f.; Wang, Z.-x.; Yuan, Y.-e.; Bing, S.-m.; Zhang, B.-z.; Wu, J.-z.; Wu, Y.-e.; Peng, X.-y. Hormone changes during the menstrual cycle of Chinese women. *Reproduction* **1986**, *76*, 43–52. [CrossRef] [PubMed]
34. Ochoa-Bernal, M.A.; Fazleabas, A.T. Physiologic events of embryo implantation and decidualization in human and non-human primates. *Int. J. Mol. Sci.* **2020**, *21*, 1973. [CrossRef] [PubMed]
35. Skliutė, G.; Baušytė, R.; Borutinskaitė, V.; Valiulienė, G.; Kaupinis, A.; Valius, M.; Ramašauskaitė, D.; Navakauskienė, R. Menstrual blood-derived endometrial stem cells' impact for the treatment perspective of female infertility. *Int. J. Mol. Sci.* **2021**, *22*, 6774. [CrossRef] [PubMed]
36. Logan, P.C.; Ponnampalam, A.P.; Steiner, M.; Mitchell, M.D. Effect of cyclic amp and estrogen/progesterone on the transcription of dna methyltransferases during the decidualization of human endometrial stromal cells. *Mol. Hum. Reprod.* **2013**, *19*, 302–312. [CrossRef]

International Journal of
*Molecular Sciences*

MDPI

*Article*

# Evaluation of the Effect of the Fibroblast Growth Factor Type 2 (FGF-2) Administration on Placental Gene Expression in a Murine Model of Preeclampsia Induced by L-NAME

**Margarita L Martinez-Fierro [1,*], Idalia Garza-Veloz [1], Maria Eugenia Castañeda-Lopez [1], Dorothy Wasike [2], Claudia Castruita-De la Rosa [1], Iram Pablo Rodriguez-Sanchez [3], Ivan Delgado-Enciso [4,5] and Jose Flores-Mendoza [1]**

1 Molecular Medicine Laboratory, Unidad Academica de Medicina Humana y Ciencias de la Salud, Universidad Autonoma de Zacatecas, Carretera Zacatecas-Guadalajara Km. 6, Ejido la Escondida, Zacatecas 98160, Mexico
2 Faculty of Medicine, Comenius University Bratislava, Špitálska 24, 81372 Bratislava, Slovakia
3 Laboratorio de Fisiologia Molecular y Estructural, Facultad de Ciencias Biologicas, Universidad Autonoma de Nuevo Leon, Monterrey 66450, Mexico
4 School of Medicine, University of Colima, Colima 28040, Mexico
5 Colima State Health Services, Cancerology State Institute, Colima 28040, Mexico
* Correspondence: margaritamf@uaz.edu.mx; Tel.: +52-492-9256690 (ext. 2182)

**Citation:** Martinez-Fierro, M.L.; Garza-Veloz, I.; Castañeda-Lopez, M.E.; Wasike, D.; Castruita-De la Rosa, C.; Rodriguez-Sanchez, I.P.; Delgado-Enciso, I.; Flores-Mendoza, J. Evaluation of the Effect of the Fibroblast Growth Factor Type 2 (FGF-2) Administration on Placental Gene Expression in a Murine Model of Preeclampsia Induced by L-NAME. *Int. J. Mol. Sci.* **2022**, *23*, 10129. https://doi.org/10.3390/ijms231710129

Academic Editor: Ilona Hromadnikova

Received: 27 July 2022
Accepted: 1 September 2022
Published: 4 September 2022



**Abstract:** The abnormal implantation of the trophoblast during the first trimester of pregnancy precedes the appearance of the clinical manifestations of preeclampsia (PE), which is a hypertensive disorder of pregnancy. In a previous study, which was carried out in a murine model of PE that was induced by NG-nitro-L-arginine methyl ester (L-NAME), we observed that the intravenous administration of fibroblast growth factor 2 (FGF2) had a hypotensive effect, improved the placental weight gain and attenuated the fetal growth restriction, and the morphological findings that were induced by L-NAME in the evaluated tissues were less severe. In this study, we aimed to determine the effect of FGF2 administration on the placental gene expression of the vascular endothelial growth factor (VEGFA), VEGF receptor 2 (VEGFR2), placental growth factor, endoglin (ENG), superoxide dismutase 1 (SOD1), catalase (CAT), thioredoxin (TXN), tumor protein P53 (P53), BCL2 apoptosis regulator, Fas cell surface death receptor (FAS), and caspase 3, in a Sprague Dawley rat PE model, which was induced by L-NAME. The gene expression was determined by a real-time polymerase chain reaction using SYBR green. Taking the vehicle or the L-NAME group as a reference, there was an under expression of placental VEGFA, VEGFR2, ENG, P53, FAS, SOD1, CAT, and TXN genes in the group of L-NAME + FGF2 ($p < 0.05$). The administration of FGF2 in the murine PE-like model that was induced by L-NAME reduced the effects that were generated by proteinuria and the increased BP, as well as the response of the expression of genes that participate in angiogenesis, apoptosis, and OS. These results have generated valuable information regarding the identification of molecular targets for PE and provide new insights for understanding PE pathogenesis.

**Keywords:** preeclampsia; FGF2; angiogenesis; oxidative stress; L-NAME; gene expression; placenta

## 1. Introduction

Preeclampsia (PE) is a hypertensive syndrome of pregnancy (≥140 mmHg systolic or ≥90 mmHg diastolic) that occurs prior to the 20th week of gestation, which affects approximately 2–8% of women worldwide, contributing to both maternal and perinatal mortality and morbidity [1]. While the origin of PE has not yet been fully comprehended, it has been associated with the poor remodeling of the spiral arteries, whereby there is a decrease in the mean diameter of the external myometrial spiral artery [2]. Poor spiral remodeling by extravillous trophoblasts contributes to an alteration in the blood supply to the placenta, with subsequent ischemic episodes and changes in the oxygen supply in the maternal–fetal unit, triggering the wide variety of pathophysiological mechanisms that are associated with the clinical manifestations of PE [3].

In pregnancy, one of the vital processes during normal placental development is the formation of new blood vessel networks that enable the correct delivery of nutrients and oxygen to tissues from the mother to the fetus [4]. This process, which is called angiogenesis, is a complicated and well-ordered process that involves extensive signaling networks both between and inside the endothelial cells (ECs), the mural cells (the vascular smooth muscle cells and the pericytes), and other cell types (e.g., immune cells) [5–7]. Angiogenesis is regulated by a wide range of different angiogenic stimulators and inhibitors, and the normal endothelial cell turnover is the product of the correct balance between them [7]. During pregnancy, angiogenesis is promoted by vascular endothelial growth factor (VEGF) and placental growth factor (PLGF) [4], and it is modulated by antiangiogenic growth factors, such as soluble fms-like tyrosine kinase-1 (sFlt-1), which is a splice variant of Flt-1, and soluble endoglin (sENG), which is a truncated form of endoglin, both of which act as the antagonist of VEGF and PLGF and of transforming growth factor-β (TGFβ), respectively [8]. The appearance of the clinical manifestations of PE is related to the abnormal induction of the increased synthesis and release of sFlt-1 and sENG impairing their cell signaling pathways through distinct but additive mechanisms [8,9] and triggering placental oxidative stress (OS) [3]. Despite the efforts of the antioxidants, such as superoxide dismutase (SOD) and catalase (CAT), to ensure proper vascular functioning, the presence of placental ischemia reduces the anti-oxidative ability and enhances the OS [10]. Besides ischemia, hypoxia is one of the most powerful triggers for the increased production of VEGF and nitric oxide (NO), which induces both vasodilatation and angiogenesis [11]. NO is a known paracrine mediator that acts as a placental vasodilator and functions as a modulator during the ovulation, the implantation, the maintenance of pregnancy, the placental perfusion, the labor, and the delivery [3,12]. During the course of pregnancy, both placental angiogenic and oxidative imbalances, and/or an alteration in the bioavailability of NO, hinders normal pregnancy progression by disrupting the functioning of the placenta and contributing to the pathogenesis of PE [3,8].

Previous studies have stated the importance of fibroblast growth factor type 2 (FGF2) in the enhanced expression of VEGF [13]; furthermore, a relationship between a decreased circulating FGF2 concentration and PE development has also been reported [14]. In a previous study that was carried out in a Sprague Dawley rat model of PE that was induced by NG-nitro-L-arginine methyl ester (L-NAME), which is an inhibitor of nitric oxide synthase (NOS), we observed that the intravenous administration of recombinant FGF2 had a hypotensive effect, did not increase the maternal urine protein concentrations that are typically induced by L-NAME, improved the placental weight gain and attenuated the fetal growth restriction, and, histologically, the morphological findings that were induced by L-NAME in the evaluated tissues were less severe [12]. FGF family members have previously been noted to regulate the functioning of other growth factors, such as PLGF, monocyte chemoattractant protein 1, hepatocyte growth factor, and angiopoietin-2 [15–17]. This is consistent with our previous findings, whereby FGF2 was associated with the formation of the blood vessels [12]. Therefore, we hypothesize that these effects are due to the regulatory role of this growth factor in the well-known pathways that are associated with the pathophysiology of PE. Based on this, in this study we aimed to determine the effect of FGF2 administration on the placental gene expression of the key genes that are involved in angiogenesis, OS, and apoptosis, in a murine model that was induced by L-NAME.

## 2. Results

### *2.1. Modulatory Effect of FGF2 on BP Values and Urine Protein Concentration in the Rat PE-like Model Induced by L-NAME*

The PE-like model that has been used in this study was established in Sprague Dawley rats and it was induced using L-NAME at doses of 60 mg/kg/day, which were administered daily, starting on the 10th day of gestation until the 19th day of gestation, as previously reported [12]. Hypertension and proteinuria were the first two parameters that were considered as the reference for the establishment of the model [12]. In order to evaluate the modulatory effect of rhFGF2 on the BP values in the rat PE-like model, FGF2 was also administered daily (666.6 ng/kg/day), either alone or in combination with L-NAME. Figure 1 shows the BP measurements that were obtained for the groups of the vehicle, FGF2, L-NAME, L-NAME + FGF2, and L-NAME + hydralazine, on days 10, 15, and 20, respectively. Before the treatment, there were no significant differences in the BP values between the experimental groups ($p > 0.05$).

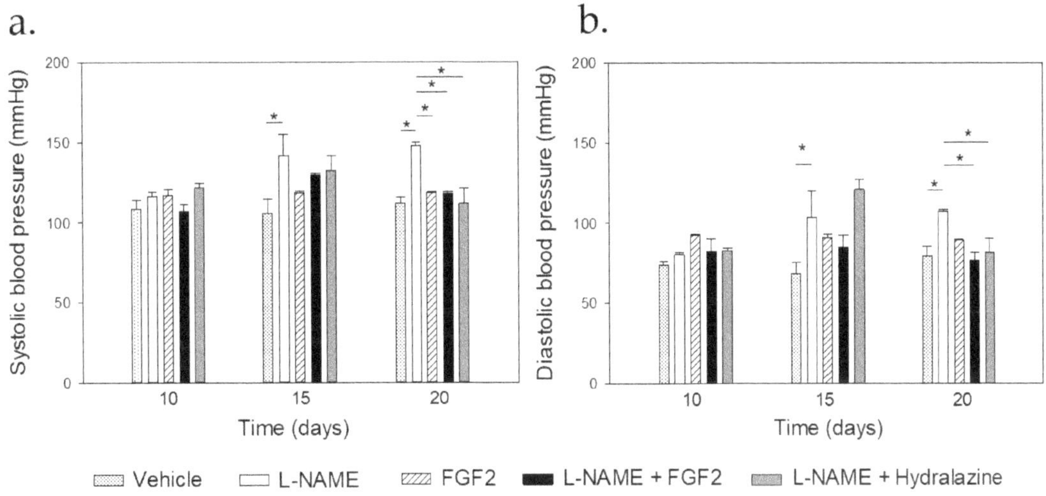


**Figure 1.** Comparisons of the blood pressure values between the treatment groups. Sprague Dawley rats ($n = 3$ for each group) were treated with vehicle, L-NAME (60 mg/kg/day), FGF-2 (666.6 ng/kg/day), L-NAME + FGF2, or L-NAME + hydralazine. Systolic (**a**) and diastolic (**b**) blood pressure values were recorded, and urine samples were collected from the 10th to the 20th day of pregnancy. * $p < 0.05$.

The mean of the SBP (Figure 1a) and the DBP (Figure 1b) before the treatment were 117.76 mmHg ± 1.9 and 74.7 mmHg ± 3.5 in the L-NAME group and were 116.5 mmHg ± 3.6 and 79.3 mmHg ± 6.7 in the vehicle group, respectively. There were differences in both the SBP and the DBP values between the L-NAME and the vehicle groups at day 15 ($p = 0.001$ for the SBP and $p = 0.023$ for the DBP) and at day 20 of pregnancy ($p < 0.001$). At days 15 and 20 of pregnancy, and compared with the vehicle group, there were no changes in the BP values in the groups that were treated with rhFGF2 alone or in the group that was treated with L-NAME + FGF2, relative to the vehicle group ($p > 0.05$). On day 20, there was a significant increase in the BP values in the L-NAME group when compared with the values that were observed in the vehicle, the FGF2, the L-NAME + hydralazine, and the L-NAME + FGF2 groups, respectively ($p < 0.05$). On day the 20th day of pregnancy, no differences in the BP values were observed between the vehicle and the L-NAME + FGF2 groups or between the vehicle and the FGF2 groups ($p > 0.05$).

Figure 2a displays the results of the urine protein concentration for each experimental group on the 20th day of pregnancy. Before the treatment (from day 10 to day 15), there were no significant changes in the urine protein concentrations between the groups ($p > 0.05$). The normal urine protein concentration before the treatment ranged from 31.4 µg/mL to 154.8 µg/mL. On day 20 of pregnancy, there were differences in the urine protein concentrations between the L-NAME group and the vehicle, the FGF2, and the L-NAME + hydralazine groups ($p < 0.05$). At this time point, the urine protein concentrations of the

L-NAME + hydralazine group decreased, reaching lower levels than those that were observed in the vehicle and the FGF2 groups. However, while there were differences in the urine protein concentrations between the vehicle and the L-NAME + FGF2 groups ($p < 0.001$), the urine protein concentrations in the FGF2 group did not differ from those that were observed in the vehicle group ($p = 0.732$).

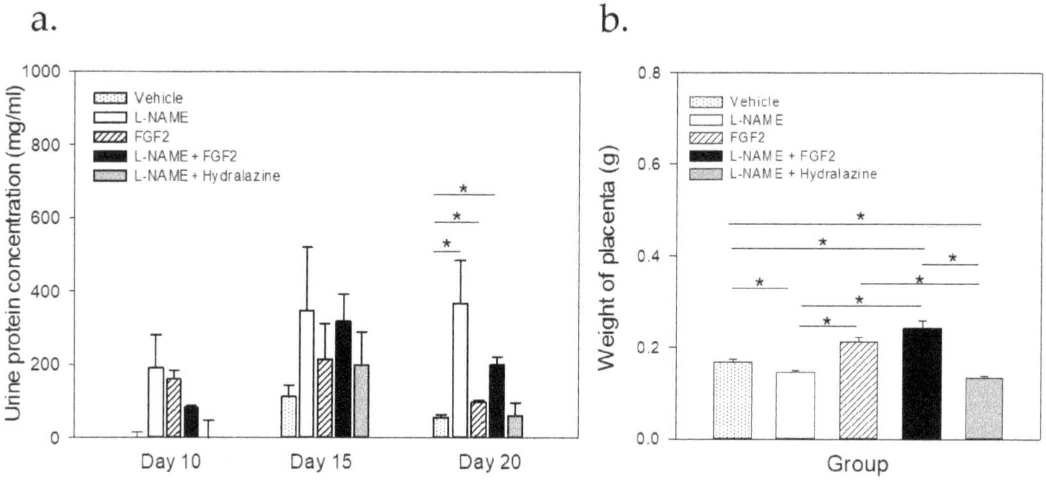


**Figure 2.** Comparisons of the urine protein concentrations and placental weight between the treatment groups. Sprague Dawley rats ($n = 3$ for each group) were treated with vehicle, L-NAME (60 mg/kg/day), FGF-2 (666.6 ng/kg/day), L-NAME + FGF2, or L-NAME + hydralazine (**a**). Urine proteins were quantified using the Bradford method. At the end of the protocol (20th day of pregnancy) placental tissue was collected ($n = 6$), weighed, and the mean of the weight was compared between groups (**b**). Data are represented as mean ± SE. * $p < 0.05$.

### 2.2. FGF2 Administration Induced Changes in Placental Weight in the Rat PE-like Model Induced by L-NAME

After the experimental protocol was completed, on the 20th day of gestation, the placentas were collected and weighed. The results of these measurements are shown in Figure 2b. The mean weights of the placentas were 0.169 g ± 0.0299, 0.146 g ± 0.0231, 0.212 g ± 0.0495, 0.242 g ± 0.0748, and 0.134 g ± 0.0227, in the vehicle, the L-NAME, the FGF2, the L-NAME + FGF2, and the L-NAME + hydralazine groups, respectively ($p < 0.05$). When considering the L-NAME group as a reference, the weight of the placentas was significantly higher in the vehicle, the FGF2, and the L-NAME + FGF2 groups ($p < 0.001$). There were also differences in the placenta weight between the vehicle group and the L-NAME + FGF2 and L-NAME + hydralazine groups ($p < 0.05$), but not between the vehicle and the FGF2 groups alone ($p > 0.05$). There were no differences between the L-NAME and the L-NAME + hydralazine groups ($p > 0.05$).

### 2.3. FGF-2 Administration Modulated the Placental Gene Expression in the Murine PE-like Model Induced by L-NAME

Figure 3 shows the results of the placental mRNA expression levels of the evaluated genes for each treatment group. Compared to the vehicle group, in the L-NAME group there was a significant under expression of TXN ($p = 0.014$). Although there was an apparent over expression of VEGF, PLGF, VEGFR-2, SOD1, p53, and FAS, and an under expression of ENG, these changes were not significant ($p > 0.05$). In the FGF2 group, only TXN showed a significant under expression ($p = 0.01$) when it was compared to the vehicle group. With the exception of PLGF, and taking the vehicle group as a reference, there was an under expression of all of the evaluated genes in the L-NAME + FGF2 group ($p < 0.05$). The

L-NAME + hydralazine group showed an under expression of TXN when compared to the vehicle group ($p = 0.017$).

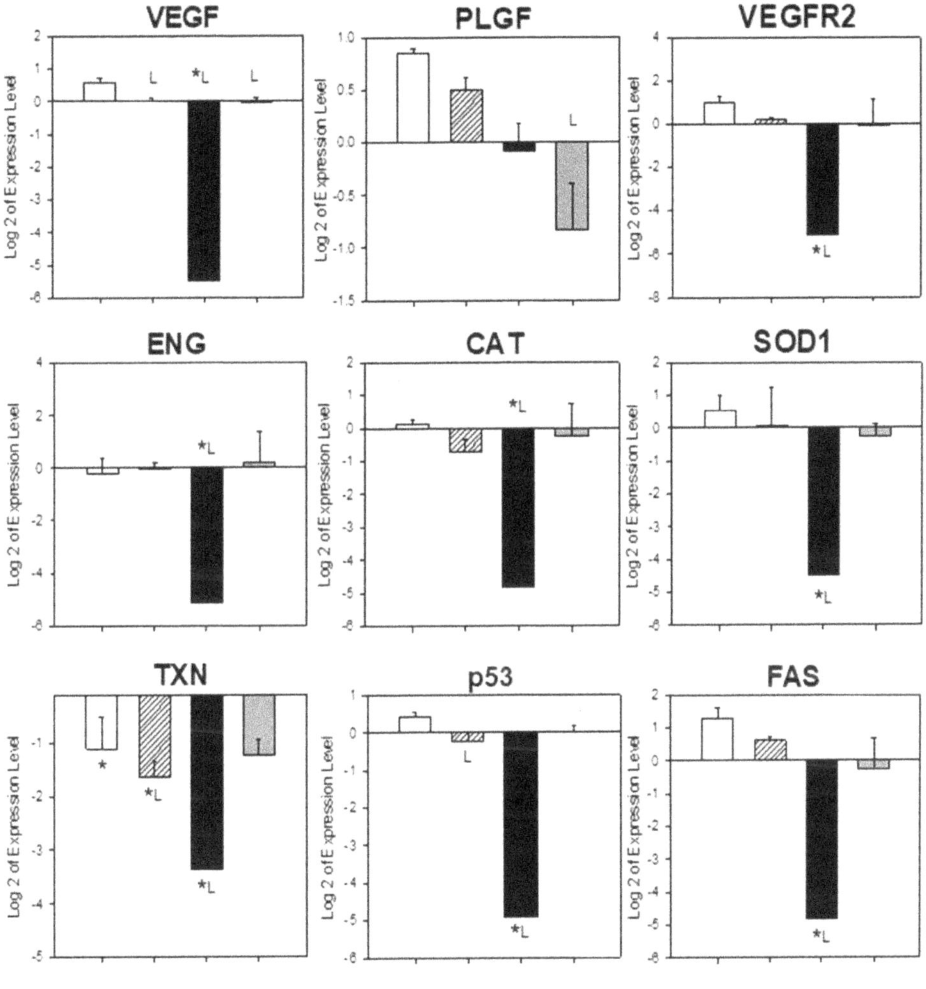


**Figure 3.** Comparison of placental gene expression between treatment groups. Sprague Dawley rats were treated with vehicle (NaCl 0.9%), L-NAME (60 mg/kg/day), FGF-2 (666.6 μg/kg/day), L-NAME + FGF2, or L-NAME + hydralazine. Placental samples ($n = 6$) were collected from three animals in each group on the 20th day of pregnancy. The expression levels of the genes involved in angiogenesis (VEGF, PLGF, VEGFR2, and ENG), oxidative stress (SOD1, CAT, and TXN) and apoptosis (p53 and FAS) were quantified by qRT-PCR, using SYBR Green, and HPRT1 as the endogenous gene. The data obtained from the vehicle group were considered as the calibrator during the gene expression calculations. One-way analysis of variance was carried out to compare each gene of interest between the treatment groups. Comparisons with $p$-values < 0.05 were subjected to multiple comparison analyses using the Holm–Sidak or Dunn's method * $p < 0.05$ versus vehicle; L $p < 0.05$ versus L-NAME.

With the exception of PLGF, all of the evaluated genes were under expressed in the placentas from the rats in the L-NAME + FGF2 group when they were compared to the

L-NAME group ($p < 0.05$). The VEGF was the only gene whose expression differed in the comparisons between the groups of the L-NAME and the FGF2, the L-NAME and the L-NAME + FGF2, and between the L-NAME and the hydralazine ($p < 0.05$). Compared with the L-NAME group, the p53 gene was also significantly under expressed in the FGF-2 group ($p = 0.047$).

Although the placental expressions of BCL2 and CASP3 were considered in this study, these genes were not included in the gene expression calculations or in the comparisons because the Cq values that were obtained for several placental tissues in the study groups were not observed in most of the animals and the amplification plots that were obtained had a Cq of > 38. Accordingly, and in order to avoid bias, these genes were excluded.

## 3. Discussion

Considering the previous evidence that the intravenous administration of FGF2 has beneficial and hypotensive effects, reducing the clinical manifestations of PE in a rat model [12], in this study we aimed to determine the effect of FGF2 administration on the placental gene expression of the key genes that are involved in angiogenesis, OS, and apoptosis, in a murine model that was induced by L-NAME.

Different working groups, including ours, have previously established the PE-like model that has been used in this study [12,18,19]. It is induced by the administration of L-NAME, which is an inhibitor of NO production, as it inhibits the NOS enzyme, prevents endothelium-dependent relaxation, and produces an increase in the BP [20]. As expected, during the establishment of the model, we observed a gradual increase in BP during the course of pregnancy and a hypotensive effect in the L-NAME group, which was derived from both the administration of the hydralazine, which is an agent that is known for its antihypertensive properties [21], and from the administration of FGF2, which may also regulate hypertension [12,22]. Similarly, besides hypertension, urinary protein excretion is a key finding during the pathogenesis of PE [23]. In this study, there was a decrease in the concentration of proteins in the urine from day 15 to day 20 of gestation in the L-NAME + FGF2 group; these results correlated with a decrease in the BP in the same group. It is well known that proteinuria causes hypoxia and the constriction of the uterine vessels, which may limit the supply of nutrients to the fetus, therefore, generating a low birth weight [24]. Similarly, previous studies have observed that the lack of endothelial synthesis of NO generates damage to the podocytes and L-NAME also causes global severe glomerular endotheliosis in rats, suggesting a beneficial decrement in the endotheliosis severity when FGF2 is administered [25]. These results suggests that FGF2 reduces the L-NAME-associated-injuries in the glomerulus, and it may also explain the decrement in the proteinuria that was observed in the L-NAME + FGF2 group in our study. Accordingly, the decrease in the BP and urine proteins after FGF2 administration represent a beneficial effect on both the mother's health and on the growth of the offspring, as previously reported [12]. Additional studies are necessary in order to investigate the molecular mechanism by which FGF2 decreases renal damage, secondary to the L-NAME treatment.

Among the main molecular markers that are associated with the development of PE are those that are associated with angiogenesis, OS, and apoptosis, which are intimately involved in adequate cell proliferation, invasion and remodeling, and placental perfusion [26]. Similarly to the molecular mechanisms that are observed during PE pathogenesis, the administration of L-NAME in animal models, as seen with other NO inhibitors, also alters the different physiological mechanisms influencing endothelial regeneration, angiogenesis, apoptosis, and cellular OS [27,28]. In our study, compared to the vehicle group, there was no apparent impact of L-NAME administration on the placental mRNA quantity of the evaluated genes (with the exception of TXN). These results were unexpected; however, they seem to indicate that, in spite of the differences in both the placenta size and weight between the vehicle and the L-NAME groups, additional mechanisms exist that may compensate for the pathological process that is induced by L-NAME, which are reflected as subtle changes in the gene expression in the placenta at term. In humans, the placenta expression profiles of women with PE differed from those of women with normal pregnancies; however, the gene expression profiles have been widely variable between studies [29–31]. Although many factors could be responsible for these differences (e.g., the lack of consensus in the criteria to select the placental tissue for the experiments and the PE classification criteria) [29–31], it is probable that the placental pathological processes in both humans and mice, may have compensatory mechanisms that allow the pregnancy to end without any obvious differences in the expression profiles in the placenta at term.

The VEGF and its receptor, VEGFR2, are two important factors that collaborate in placental cell proliferation and angiogenesis during several stages of pregnancy [32]. Interestingly, in our study, when the L-NAME group was considered as the reference, there was a lower quantity of VEGF mRNA in the placentas from the FGF2, the L-NAME + FGF2, and the L-NAME+ hydralazine groups, with the difference being the largest in the L-NAME + FGF2 group (fold change = −9.5). Lower quantities of the mRNA of VEGFR2 were also found in the L-NAME + FGF2 group when they were compared to that observed in the placentas from the L-NAME group (fold change −8.8; $p < 0.015$). The placental under expression of VEGF that was observed in our study in the L-NAME group is consistent with that observed by Abe and collaborators, who reported that the VEGF mRNA expression in both rat placentas and placental explants was temporarily decreased by L-NAME treatment [33]. However, there are no reports on the effect of FGF2 administration on these genes in any animal model that were induced with L-NAME. In order to explain the modulation of VEGF/VEGFR2 expression in our study, we must consider the following additional previously reported findings: first, the inhibition of the NO generation by L-NAME results in a decreased VEGF synthesis [27]; second, the VEGF-dependent release of NO and the angiogenic activity of VEGFA is blocked by the action of L-NAME [27,33]; third, in tumors, if VEGF-dependent angiogenesis is blocked, FGF2-driven angiogenesis takes its place [34]. Accordingly, we propose that the under expression of VEGF in our study was a consequence of the treatment with L-NAME, but when FGF2 was continuously administered, the angiogenesis that was driven by FGF2 was triggered in order to compensate for the absence of the VEGF functions. Because the inhibition of tumor angiogenesis that was mediated by the VEGF signaling blockade with bevacizumab correlated with a reversal of VEGFR1 and VEGFR2 protein levels [35], it is highly probable that the under expression of VEGFR2 that was observed in our study may reflect a VEGF modulatory effect, since FGF2 is also capable of modulating the expression of VEGF and its receptors in both an autocrine and a paracrine way [36].

Antioxidant enzymes, such as CAT, SOD1, and TXN, intervene in the normal physiological processes that help the body to counteract the negative effects of OS. These molecules are involved in the development of the fetus, they play a role in the growth and development of the fetal–placental unit [37–39], and their expression is modulated depending on the month of gestation [40]. The abnormal regulation of the placental expression of the OS genes in women with PE is controversial because some authors have demonstrated their over expression [41,42]. Other reports have shown decreased levels of these enzymes in women with PE that are associated with IUGR and lipid peroxidation [43,44], while others have reported no difference in the placental levels of antioxidants in women with PE [42]. It has been postulated that the levels of OS-related genes may be influenced by the stage and the severity of the disease, as OS can initially upregulate the antioxidant enzymes, the level of which may decrease in the presence of more severe or prolonged stress [42]. Additionally, the down-regulation of the antioxidant enzymes, such as CAT, correlate with high levels of $H_2O_2$, which is involved in the activation of the signaling pathways that induce the proliferation, the migration, and the cell invasion of cancer cells [45]. In our study, the placental gene expression of the CAT and SOD1 genes was significantly down-regulated in the L-NAME + FGF2 group, but not in the group that was treated with L-NAME or with FGF2 alone. Our results are similar to those that were reported by Tang and colleagues in alveolar epithelial cells, who observed that FGF-2 attenuated inflammation and reduced OS and apoptosis by activating the PI3K/Akt signaling pathway [46]. Accordingly, we propose that the under expression of these genes requires both the inhibition of NO synthesis (NOS inhibition) and the activation of the FGF2 signaling pathways at the same time, which may reflect a compensatory mechanism because of its proliferative and angiogenic properties (which may still be active in the placentas from this group). TXN, in turn, plays an important role in the NO pathway by reducing its production and promoting its degradation [47]. In our study, the placental expression of TXN was found to be modulated by L-NAME and FGF2 alone, or in combination, and their effect on the TXN under expression was additive. When compared with the L-NAME group, the TXN mRNA showed a fold change of $1 \times 10^{-18}$ in the vehicle group, −1.4 in the FGF2 group, and −3.0 in the L-NAME + FGF2 group. These results indicate that, in extreme conditions, such as the inhibition of the NO signaling pathway and its biological consequences, and also during the constant activation of the FGF2 route (which is associated with cell proliferation, cell migration, and angiogenesis, among others) the TXN system does not play a principal role in counteracting the OS and, therefore, is down-regulated. In agreement with this observation, in humans, the expression of the placental TXN was reduced in patients with PE [42,48].

Apoptosis is a key function in the cells that manage the induction of cell growth arrest. The FAS and p53 are involved in this pathway and are expressed in the decidua, the chorionic villi, the cytotrophoblast, and the syncytiotrophoblast, indirectly promoting the maintenance of pregnancy [49–51]. In our study, both of the genes were under expressed in the placentas from the rats that were treated with L-NAME + FGF2. The increase in the p53 expression corresponds to the hypoxia that is generated in the placenta by PE, at least in vitro [52], and has implications in the intrauterine growth restriction [53]. The FAS, on the other hand, is involved in mediating the maternal immune response, in cell remodeling, and in cell proliferation [51,54]; therefore, the increase in this protein is a direct indicator of placental apoptosis, which is a common finding in PE [55,56]. Although these findings are not completely comparable with ours, the under expression of the apoptosis-related genes may be due to the fact that the placental biopsies that were used in this study came from the rats that were at the end of their pregnancy. At this final pregnancy stage, the apoptosis is not expected to be an active process, and similarly to the OS genes, the p53 and the FAS under expression in the L-NAME +FGF2 group may be related to the chronic activation of the FGF2 signaling pathways, which is compatible with the presence of less apoptotic activity.

The study's limitations are as follows: In this study, we determined the effect of FGF2 administration on the placental gene expression of the key genes in angiogenesis, OS, and apoptosis; although the existence of regulation at the transcription level was not necessarily reflected at the protein level, a study perspective will be to validate our results at the protein level. Similarly, the placental gene expression that has been evaluated in the study was only at the end of the rat gestation and, therefore, the evaluation of the FGF2 effect during the previous stages of pregnancy at the placental levels and on the other organs or tissues should be assessed in futures studies.

## 4. Materials and Methods

### 4.1. Ethical Approval

The trial was approved by the Ethics and Biosafety Committee of the Area of Health Sciences of the Universidad Autónoma de Zacatecas Francisco García Salinas in Zacatecas, Mexico, and it was registered with the following identification number: CEB-ACS/UAZ.Ofc.002/2015. All of the experimental procedures were carried out in accordance with the recommendations of the "Technical specifications for the production, care and use of laboratory animals", Mexican guidelines (NOM-062-ZOO-1999).

### 4.2. Animal Treatment

We followed the experimental protocol that was described in detail in a previous study, carried out by Martinez-Fierro et al. 2021 [12]. Briefly, pregnant 10-week-old Sprague Dawley rats were separated into the following five treatment groups: Group 1. The vehicle group, which was administered with 0.9% of NaCl by the intragastric route, using a cannula and syringe of 4 mm diameter. Group 2. The FGF2 group, which consisted of pregnant rats treated with rhFGF2 (Sigma-Aldrich, St Louis, MO, USA), intravenously administered (666.6 ng/kg/day) using the tail vein (caudal). Group 3. The L-NAME group, which was administered with L-NAME (NG-nitro-L-arginine methyl ester; Sigma-Aldrich, St Louis, MO, United States of America) by the intragastric route at a concentration of 60 mg/kg/day. Treatments in the vehicle group, FGF2 group, and L-NAME group began on the 10th day of gestation and concluded on the 19th day of pregnancy. Group 4. The L-NAME + FGF2 group, which consisted of pregnant rats that were administered daily both L-NAME and rhFGF2 simultaneously, as described above, beginning on the day 10th up to the 19th day of pregnancy. Group 5. The L-NAME + hydralazine group, in which L-NAME was administered, as described above, along with oral hydralazine at 10 mg/mL/kg/day by the intragastric route from the 15th to 19th day of pregnancy.

### *4.3. Biological Samples and Data Collection*

The blood pressure (systolic: SBP, and diastolic: DBP) and urine protein levels were quantified and recorded on the 10th, 15th, and 19th days of pregnancy. The placental tissues were collected on the 20th day of gestation from each animal in the experimental groups. Each tissue was weighed and then all of the tissues (approximately 0.5 $cm^3$/each) were embedded in Tissue-Tek® O.C.T™ Compound (Sakura Finetek, Torrance, CA, USA). The embedded tissues were preserved by freezing at −80 °C until use.

### *4.4. Placental RNA Isolation and cDNA Synthesis*

The frozen placental tissues that were stored at −80 °C and embedded in Tissue-Tek® O.C.T™ Compound were defrosted at room temperature and were separated from the preservative. Total RNA was isolated from homogenized placental tissue, according to the RNeasy Mini Kit protocol (QIAGEN, Hilden, Germany). The RNA concentration and its purity were measured via optical density at 260 nm and the relation 260/280, respectively, using a NanoDrop 2000 spectrophotometer (Thermo Fisher Scientific, Wilmington, DE). The cDNA was synthesized from 1 mg of the total RNA using a High-Capacity cDNA Reverse Transcription Kit and random hexamers (Thermo Fisher Scientific, Wilmington, DE, USA), according to the manufacturer's instructions. The cDNA samples were stored at −20 °C until use.

### *4.5. Gene Selection and Primer Design*

The study genes were selected according to their relationship with the angiogenesis, apoptosis, and OS process. The genes selected included the following: VEGFA, PLGF, vascular endothelial growth factor receptor 2 (VEGFR2), ENG, tumor protein P53 (P53), BCL2 apoptosis regulator (BCL2), Fas cell surface death receptor (FAS), caspase 3 (CASP3), SOD1, CAT, and thioredoxin (TXN). In addition, the hypoxanthine phosphoribosyltransferase 1 (HPRT1) gene was selected as a reference gene. All gene-specific primers for the real-time qPCR assay were independently designed (Table 1) and provided by T4OLIGO® (T4OLIGO, Guanajuato, Mexico).

**Table 1.** General characteristics of the primers designed for the genes included in the study.

| Gene Symbol | Gene Bank ID | Description | * Primer Sequence (5′→3′) | Tm | Product Size (bp) |
|---|---|---|---|---|---|
| VEGFA | NM_031836.3 | Vascular Endothelial Growth Factor A | Fw: GGAGCAGAAAGCCCATGAAGTGGT<br>Rv: TCATTGCAGCAGCCCCGCACA | 65 | 168 |
| VEGFR2 | NM_013062.2 | Vascular Endothelial Growth Factor Receptor 2 | Fw: TTTGCACTGCAGGAGCGCGT<br>Rv: GGAATCGCCAGGCAAACCCACA | 65 | 171 |
| ENG | NM_001010968.3 | Endoglin | Fw: CAGGGCTTCGTACAGGTGAGCA<br>Rv: TCACACAGCTGCCCTTGGCT | 64 | 139 |
| P53 | NM_030989.3 | Tumor Protein P53 | Fw: GTTGCTCTGATGGTGACGGCCT<br>Rv: ACCACCACGCTGTGCCGAAA | 65 | 112 |

**Table 1.** *Cont.*

| Gene Symbol | Gene Bank ID | Description | * Primer Sequence (5′→3′) | Tm | Product Size (bp) |
|---|---|---|---|---|---|
| BCL2 | NM_016993.2 | BCL2 Apoptosis Regulator | Fw: TCCAGGATAACGGAGGCTGGGATGC<br>Rv: AGGCTGAGCAGCGTCTTCAGAGACA | 67 | 103 |
| FAS | NM_139194.3 | Fas Cell Surface Death Receptor | Fw: GTCAACCGTGTCAGCCTGGTGAA<br>Rv: TGGGTCCGGGTGCAGTTCGTTT | 65 | 190 |
| CASP3 | NM_012922.2 | Caspase 3 | Fw: GCGGAGCTTGGAACGCGAAGAAA<br>Rv: TCCAGAGTCCATCGACTTGCTTCCA | 65 | 120 |
| SOD1 | NM_017050.1 | Superoxide Dismutase 1 | Fw: TTCGTTTCCTGCGGCGGCTTCT<br>Rv: GGTTCACCGCTTGCCTTCTGCT | 66 | 169 |
| CAT | NM_012520.2 | Catalase | Fw: GGCACACTTTGACAGAGAGCGGA<br>Rv: TGAGCCTGACTCTCCAGCGACT | 65 | 184 |
| TXN | NM_053800.3 | Thioredoxin | Fw: TCTGCCACGTGGTGTGGACCTT<br>Rv: ACAGTCTGCAGCAACATCCTGGC | 66 | 126 |
| PLGF | NM_053595.2 | Placental Growth Factor | Fw: TGAGGAACCCCACCTGTGATGCT<br>Rv: CATTCAGCAGGGACGAGTTGGCT | 65 | 156 |
| HPRT1 | NM_012583.2 | Hypoxanthine phosphoribosyltransferase 1 | Fw: CAGTCCCAGCGTCGTGATTA<br>Rv: TGGCCTCCCATCTCCTTCAT | 60 | 168 |

* All of the primer sequences were designed based on the respective GenBank sequence for the examined gene and between exons to guarantee the specific detection of the interest gene.

### 4.6. *Quantitative Real-Time Polymerase Chain Reaction (qRT-PCR)*

Quantitative real-time PCR (qRT-PCR) was carried out using a StepOne Plus Real-Time PCR System (Applied Biosystems, Foster City, CA, USA) in 96-well PCR plates. Fifty nanograms of synthesized cDNA were used as templates for qRT-PCR amplification in a 10 μL of final reaction volume, using SYBR™ Green PCR Master Mix (Thermo Fisher Scientific, Wilmington, DE, USA), and 300 nM gene specific primers. The amplifications were performed with the following thermal cycle program: pre-denaturation for 10 min at 95 °C, amplification of 40 cycles with denaturation for 15 s at 95 °C, and annealing for 1 min at 60 °C. The cycle series were followed by a melt-curve analysis to confirm the specificity of amplification and the lack of primer dimers. All samples were analyzed in duplicate, including two non-template controls, to detect any template contamination. The $2^{-\Delta\Delta Cq}$ equation was applied to calculate the relative expression of the placenta samples [57]. The mean of quantification cycle (Cq) of the vehicle group samples was used as a calibrator. The sequences and product sizes of the forward and reverse primers for all of the evaluated genes are listed in Table 1.

### 4.7. *Statistical Analysis*

All data were expressed as mean ± standard error (SE) for three animals per group. Comparisons between two groups of data were carried out by Student's *t*-test. For multiple comparisons of data, one-way analysis of variance (ANOVA), coupled with the Holm–Sidak method, was used; for non-normally distributed variables, the Kruskal–Wallis ANOVA on ranks, and the Dunn's method as a multiple comparison procedure was applied. One-way repeated measures ANOVA, coupled with the Holm–Sidak test as a post hoc test, was used to evaluate whether there were differences in the BP values and urine protein concentrations within the same experimental group during the evaluated times. All statistical analyses were carried out using Sigma Plot® version 11 (Systat Software Inc., San Jose, CA, USA). A 95% confidence interval (CI) was used and $p < 0.05$ was considered statistically significant.

## 5. Conclusions

In conclusion, we have corroborated that the administration of FGF2 in a murine PE-like model that was induced by L-NAME reduces the effects that are generated by proteinuria and increased BP. In presence of NOS inhibition, the intravenous administration of FGF2 during pregnancy induced lower quantities of placental mRNA of the VEGFA, VEGFR2, ENG, P53, FAS, SOD1, CAT, and TXN genes, in the model that was evaluated here. These results demonstrate that the pathogenic consequences of NOS inhibition that are induced by L-NAME during pregnancy may be modulated by FGF2 and are reflected as placental under expression of genes that are related to angiogenesis, apoptosis, and OS, thus, generating valuable information for the identification of molecular targets for PE, and for understanding the complex pathogenesis of PE.

**Author Contributions:** Conceptualization, M.L.M.-F.; methodology, M.L.M.-F., J.F.-M., I.G.-V., and M.E.C.-L.; formal analysis, M.L.M.-F., C.C.-D.l.R., and D.W.; investigation, I.D.-E., I.P.R.-S., and M.E.C.-L.; resources and data curation, M.L.M.-F.; writing—original draft preparation, M.L.M.-F., D.W., and M.E.C.-L.; writing—review and editing, M.L.M.-F. and I.G.-V.; visualization, M.L.M.-F. and I.G.-V.; supervision, M.L.M.-F., and I.G.-V.; project administration, M.L.M.-F.; funding acquisition, M.L.M.-F. All authors have read and agreed to the published version of the manuscript.

**Funding:** This research was funded by Consejo Nacional de Ciencia y Tecnologia-CONACyT, Grant numbers: INFR-2014-01-225520, INFR-2015-01-254106, PDCPN-2015-01-63, SEP-CONACYT-CB-2015-258316, and SS/IMSS/ISSSTE-CONACYT-2016-01-273144. APC was funded by Molecular Medicine Laboratory from Universidad Autónoma de Zacatecas.

**Institutional Review Board Statement:** The animal study protocol was reviewed and approved by the Ethics and Biosafety Committee of the Area of Health Sciences from the Universidad Autonoma de Zacatecas (protocol ID: CEB-ACS/UAZ.Ofc.002/2015).

**Informed Consent Statement:** Not applicable.

**Data Availability Statement:** Data that support the findings of this study are available from the corresponding author, upon reasonable request.

**Acknowledgments:** The authors thank the Consejo Nacional de Ciencia y Tecnologia (CONACYT) for the project funding and the scholarships awarded to the students involved in the experimental part of the protocol.

**Conflicts of Interest:** The authors declare no conflict of interest. The funders had no role in the design of the study; in the collection, analyses, or interpretation of data; in the writing of the manuscript, or in the decision to publish the results.

## References

1. Ives, C.W.; Sinkey, R.; Rajapreyar, I.; Tita, A.T.N.; Oparil, S. Preeclampsia-Pathophysiology and Clinical Presentations: JACC State-of-the-Art Review. *J. Am. Coll. Cardiol.* **2020**, *76*, 1690–1702. [CrossRef] [PubMed]
2. Cartwright, J.E.; Fraser, R.; Leslie, K.; Wallace, A.E.; James, J.L. Remodelling at the maternal-fetal interface: Relevance to human pregnancy disorders. *Reproduction* **2010**, *140*, 803–813. [CrossRef]
3. Wu, F.; Tian, F.J.; Lin, Y. Oxidative Stress in Placenta: Health and Diseases. *BioMed Res. Int.* **2015**, *2015*, 293271. [CrossRef] [PubMed]
4. Maynard, S.E.; Karumanchi, S.A. Angiogenic factors and preeclampsia. *Semin. Nephrol.* **2011**, *31*, 33–46. [CrossRef]
5. Ucuzian, A.A.; Gassman, A.A.; East, A.T.; Greisler, H.P. Molecular mediators of angiogenesis. *J. Burn Care Res. Off. Publ. Am. Burn Assoc.* **2010**, *31*, 158–175. [CrossRef] [PubMed]
6. Wanjare, M.; Kusuma, S.; Gerecht, S. Perivascular cells in blood vessel regeneration. *Biotechnol. J.* **2013**, *8*, 434–447. [CrossRef] [PubMed]
7. Poto, R.; Cristinziano, L.; Modestino, L.; de Paulis, A.; Marone, G.; Loffredo, S.; Galdiero, M.R.; Varricchi, G. Neutrophil Extracellular Traps, Angiogenesis and Cancer. *Biomedicines* **2022**, *10*, 431. [CrossRef]
8. Pereira, R.D.; De Long, N.E.; Wang, R.C.; Yazdi, F.T.; Holloway, A.C.; Raha, S. Angiogenesis in the placenta: The role of reactive oxygen species signaling. *BioMed Res. Int.* **2015**, *2015*, 814543. [CrossRef]
9. Munaut, C.; Lorquet, S.; Pequeux, C.; Coulon, C.; Le Goarant, J.; Chantraine, F.; Noel, A.; Goffin, F.; Tsatsaris, V.; Subtil, D.; et al. Differential expression of Vegfr-2 and its soluble form in preeclampsia. *PLoS ONE* **2012**, *7*, e33475. [CrossRef]
10. Silvestro, S.; Calcaterra, V.; Pelizzo, G.; Bramanti, P.; Mazzon, E. Prenatal Hypoxia and Placental Oxidative Stress: Insights from Animal Models to Clinical Evidences. *Antioxidants* **2020**, *9*, 414. [CrossRef]
11. Krock, B.L.; Skuli, N.; Simon, M.C. Hypoxia-induced angiogenesis: Good and evil. *Genes Cancer* **2011**, *2*, 1117–1133. [CrossRef] [PubMed]
12. Martinez-Fierro, M.L.; Hernadez-Delgadillo, G.P.; Flores-Mendoza, J.F.; Alvarez-Zuniga, C.D.; Diaz-Lozano, M.L.; Delgado-Enciso, I.; Romero-Diaz, V.J.; Lopez-Saucedo, A.; Rodriguez-Sanchez, I.P.; Marino-Martinez, I.A.; et al. Fibroblast Growth Factor Type 2 (FGF2) Administration Attenuated the Clinical Manifestations of Preeclampsia in a Murine Model Induced by L-NAME. *Front. Pharmacol.* **2021**, *12*, 663044. [CrossRef] [PubMed]

13. Fujii, T.; Yonemitsu, Y.; Onimaru, M.; Inoue, M.; Hasegawa, M.; Kuwano, H.; Sueishi, K. VEGF function for upregulation of endogenous PlGF expression during FGF-2-mediated therapeutic angiogenesis. *Atherosclerosis* **2008**, *200*, 51–57. [CrossRef] [PubMed]
14. Martinez-Fierro, M.L.; Garza-Veloz, I.; Castruita-Dela Rosa, C.; Ortiz-Castro, Y.; Aceves-Medina, M.C.; Vazquez-Castro, R.; Delgado-Enciso, I.; Castaneda-Lopez, M.E. Plasma cancer biomarker multiplex screening and the risk of subsequent preeclampsia. *Int. J. Cardiol.* **2015**, *179*, 58–60. [CrossRef] [PubMed]
15. Deshmane, S.L.; Kremlev, S.; Amini, S.; Sawaya, B.E. Monocyte chemoattractant protein-1 (MCP-1): An overview. *J. Interferon Cytokine Res.* **2009**, *29*, 313–326. [CrossRef]
16. Lieu, C.; Heymach, J.; Overman, M.; Tran, H.; Kopetz, S. Beyond VEGF: Inhibition of the fibroblast growth factor pathway and antiangiogenesis. *Clin. Cancer Res.* **2011**, *17*, 6130–6139. [CrossRef]
17. Kienast, Y.; Klein, C.; Scheuer, W.; Raemsch, R.; Lorenzon, E.; Bernicke, D.; Herting, F.; Yu, S.; The, H.H.; Martarello, L.; et al. Ang-2-VEGF-A CrossMab, a novel bispecific human IgG1 antibody blocking VEGF-A and Ang-2 functions simultaneously, mediates potent antitumor, antiangiogenic, and antimetastatic efficacy. *Clin. Cancer Res.* **2013**, *19*, 6730–6740. [CrossRef]
18. Baijnath, S.; Soobryan, N.; Mackraj, I.; Gathiram, P.; Moodley, J. The optimization of a chronic nitric oxide synthase (NOS) inhibition model of pre-eclampsia by evaluating physiological changes. *Eur. J. Obstet. Gynecol. Reprod. Biol.* **2014**, *182*, 71–75. [CrossRef]
19. Soobryan, N.; Murugesan, S.; Phoswa, W.; Gathiram, P.; Moodley, J.; Mackraj, I. The effects of sildenafil citrate on uterine angiogenic status and serum inflammatory markers in an L-NAME rat model of pre-eclampsia. *Eur. J. Pharmacol.* **2017**, *795*, 101–107. [CrossRef]
20. Pfeiffer, S.; Leopold, E.; Schmidt, K.; Brunner, F.; Mayer, B. Inhibition of nitric oxide synthesis by NG-nitro-L-arginine methyl ester (L-NAME): Requirement for bioactivation to the free acid, NG-nitro-L-arginine. *Br. J. Pharmacol.* **1996**, *118*, 1433–1440. [CrossRef]
21. Alshami, A.; Romero, C.; Avila, A.; Varon, J. Management of hypertensive crises in the elderly. *J. Geriatr. Cardiol. JGC* **2018**, *15*, 504–512. [CrossRef]
22. Tassi, E.; Lai, E.Y.; Li, L.; Solis, G.; Chen, Y.; Kietzman, W.E.; Ray, P.E.; Riegel, A.T.; Welch, W.J.; Wilcox, C.S.; et al. Blood Pressure Control by a Secreted FGFBP1 (Fibroblast Growth Factor-Binding Protein). *Hypertension* **2018**, *71*, 160–167. [CrossRef] [PubMed]
23. Airoldi, J.; Weinstein, L. Clinical significance of proteinuria in pregnancy. *Obstet. Gynecol. Surv.* **2007**, *62*, 117–124. [CrossRef] [PubMed]
24. Dong, Y.J.; Gao, L.L. Effect of epidural block on 24-hour urine protein in pregnant rat models with preeclampsia. *Arch. Gynecol. Obstet.* **2012**, *286*, 609–611. [CrossRef] [PubMed]
25. Andrade, J.J.d.C.; Silva, A.C.O.; Frazão, I.d.S.; Perrelli, J.G.A.; Silva, T.T.d.M.; Cavalcanti, A.M. Family functionality and burden of family caregivers of users with mental disorders. *Rev. Bras. Enferm.* **2021**, *74*, e20201061. [CrossRef]
26. Can, M.; Guven, B.; Bektas, S.; Arikan, I. Oxidative stress and apoptosis in preeclampsia. *Tissue Cell* **2014**, *46*, 477–481. [CrossRef]
27. Dulak, J.; Jozkowicz, A.; Dembinska-Kiec, A.; Guevara, I.; Zdzienicka, A.; Zmudzinska-Grochot, D.; Florek, I.; Wojtowicz, A.; Szuba, A.; Cooke, J.P. Nitric oxide induces the synthesis of vascular endothelial growth factor by rat vascular smooth muscle cells. *Arterioscler. Thromb. Vasc. Biol.* **2000**, *20*, 659–666. [CrossRef]
28. Mortensen, K.; Skouv, J.; Hougaard, D.M.; Larsson, L.I. Endogenous endothelial cell nitric-oxide synthase modulates apoptosis in cultured breast cancer cells and is transcriptionally regulated by p53. *J. Biol. Chem.* **1999**, *274*, 37679–37684. [CrossRef]
29. Silva Carmona, A.; Mendieta Zeron, H. NF-kappaBeta and SOD expression in preeclamptic placentas. *Turk. J. Med. Sci.* **2016**, *46*, 783–788. [CrossRef]
30. Enquobahrie, D.A.; Meller, M.; Rice, K.; Psaty, B.M.; Siscovick, D.S.; Williams, M.A. Differential placental gene expression in preeclampsia. *Am. J. Obstet. Gynecol.* **2008**, *199*, 566-e1. [CrossRef]
31. Hansson, S.R.; Chen, Y.; Brodszki, J.; Chen, M.; Hernandez-Andrade, E.; Inman, J.M.; Kozhich, O.A.; Larsson, I.; Marsal, K.; Medstrand, P.; et al. Gene expression profiling of human placentas from preeclamptic and normotensive pregnancies. *Mol. Hum. Reprod.* **2006**, *12*, 169–179. [CrossRef] [PubMed]
32. Tandon, V.; Hiwale, S.; Amle, D.; Nagaria, T.; Patra, P.K. Assessment of Serum Vascular Endothelial Growth Factor Levels in Pregnancy-Induced Hypertension Patients. *J. Pregnancy* **2017**, *2017*, 3179670. [CrossRef] [PubMed]
33. Abe, H.; Ishikawa, W.; Kushima, T.; Nishimura, T.; Mori, C.; Onuki, A.; Suzuki, T.; Ishii, Y.; Kansaku, N.; Miyazaki, Y.; et al. Nitric oxide induces vascular endothelial growth factor expression in the rat placenta in vivo and in vitro. *Biosci. Biotechnol. Biochem.* **2013**, *77*, 971–976. [CrossRef] [PubMed]
34. Ichikawa, K.; Watanabe Miyano, S.; Minoshima, Y.; Matsui, J.; Funahashi, Y. Activated FGF2 signaling pathway in tumor vasculature is essential for acquired resistance to anti-VEGF therapy. *Sci. Rep.* **2020**, *10*, 2939. [CrossRef] [PubMed]
35. Zhang, Z.; Neiva, K.G.; Lingen, M.W.; Ellis, L.M.; Nor, J.E. VEGF-dependent tumor angiogenesis requires inverse and reciprocal regulation of VEGFR1 and VEGFR2. *Cell Death Differ.* **2010**, *17*, 499–512. [CrossRef]
36. Seghezzi, G.; Patel, S.; Ren, C.J.; Gualandris, A.; Pintucci, G.; Robbins, E.S.; Shapiro, R.L.; Galloway, A.C.; Rifkin, D.B.; Mignatti, P. Fibroblast growth factor-2 (FGF-2) induces vascular endothelial growth factor (VEGF) expression in the endothelial cells of forming capillaries: An autocrine mechanism contributing to angiogenesis. *J. Cell Biol.* **1998**, *141*, 1659–1673. [CrossRef] [PubMed]
37. Nakatsukasa, Y.; Tsukahara, H.; Tabuchi, K.; Tabuchi, M.; Magami, T.; Yamada, M.; Fujii, Y.; Yashiro, M.; Tsuge, M.; Morishima, T. Thioredoxin-1 and oxidative stress status in pregnant women at early third trimester of pregnancy: Relation to maternal and neonatal characteristics. *J. Clin. Biochem. Nutr.* **2013**, *52*, 27–31. [CrossRef]

38. Sahay, A.S.; Sundrani, D.P.; Wagh, G.N.; Mehendale, S.S.; Joshi, S.R. Regional differences in the placental levels of oxidative stress markers in pre-eclampsia. *Int. J. Gynaecol. Obstet. Off. Organ. Int. Fed. Gynaecol. Obstet.* **2015**, *129*, 213–218. [CrossRef]
39. Myatt, L.; Eis, A.L.; Brockman, D.E.; Kossenjans, W.; Greer, I.A.; Lyall, F. Differential localization of superoxide dismutase isoforms in placental villous tissue of normotensive, pre-eclamptic, and intrauterine growth-restricted pregnancies. *J. Histochem. Cytochem.* **1997**, *45*, 1433–1438. [CrossRef]
40. Roland, L.; Beauchemin, D.; Acteau, G.; Fradette, C.; St-Pierre, I.; Bilodeau, J.F. Effects of labor on placental expression of superoxide dismutases in preeclampsia. *Placenta* **2010**, *31*, 392–400. [CrossRef]
41. Wang, Y.; Walsh, S.W. Antioxidant activities and mRNA expression of superoxide dismutase, catalase, and glutathione peroxidase in normal and preeclamptic placentas. *J. Soc. Gynecol. Investig.* **1996**, *3*, 179–184. [CrossRef] [PubMed]
42. Perkins, A.V. Endogenous anti-oxidants in pregnancy and preeclampsia. *Aust. N. Z. J. Obstet. Gynaecol.* **2006**, *46*, 77–83. [CrossRef] [PubMed]
43. Wiktor, H.; Kankofer, M. Catalase activity in normal and preeclamptic placentas. *Ginekol. Pol.* **2001**, *72*, 1228–1232. [PubMed]
44. Krishna, T.S.; Rao, E.V.; Naidu, J.N. Alterations of antioxidant enzymes in preeclampsia. *IJRMS* **2015**, *3*, 629. [CrossRef]
45. Glorieux, C.; Zamocky, M.; Sandoval, J.M.; Verrax, J.; Calderon, P.B. Regulation of catalase expression in healthy and cancerous cells. *Free Radic. Biol. Med.* **2015**, *87*, 84–97. [CrossRef] [PubMed]
46. Tang, Q.Y.; Wei, J.X.; Xue, S.F.; Liu, G.H.; Fu, L.X. Fibrogrowth factor-2 protects against acute lung injury by activating the PI3K/Akt signaling pathway. *J. Biol. Regul. Homeost. Agents* **2020**, *34*, 1679–1688. [CrossRef]
47. Benhar, M. Nitric oxide and the thioredoxin system: A complex interplay in redox regulation. *Biochim. Biophys. Acta* **2015**, *1850*, 2476–2484. [CrossRef]
48. Sahlin, L.; Ostlund, E.; Wang, H.; Holmgren, A.; Fried, G. Decreased expression of thioredoxin and glutaredoxin in placentae from pregnancies with pre-eclampsia and intrauterine growth restriction. *Placenta* **2000**, *21*, 603–609. [CrossRef]
49. Wei, D.; Wu, Q.; Shi, H. Apoptosis and p53 expression in the placental villi of females with unexplained recurrent spontaneous abortion. *Exp. Ther. Med.* **2014**, *7*, 191–194. [CrossRef]
50. Sohr, S.; Engeland, K. The tumor suppressor p53 induces expression of the pregnancy-supporting human chorionic gonadotropin (hCG) CGB7 gene. *Cell Cycle* **2011**, *10*, 3758–3767. [CrossRef]
51. Abrahams, V.M.; Straszewski-Chavez, S.L.; Guller, S.; Mor, G. First trimester trophoblast cells secrete Fas ligand which induces immune cell apoptosis. *Mol. Hum. Reprod.* **2004**, *10*, 55–63. [CrossRef] [PubMed]
52. Levy, R. The role of apoptosis in preeclampsia. *Isr Med. Assoc. J.* **2005**, *7*, 178–181. [PubMed]
53. Baserga, M.; Hale, M.A.; Ke, X.; Wang, Z.M.; Yu, X.; Callaway, C.W.; McKnight, R.A.; Lane, R.H. Uteroplacental insufficiency increases p53 phosphorylation without triggering the p53-MDM2 functional circuit response in the IUGR rat kidney. *Am. J. Physiol. Regul. Integr. Comp. Physiol.* **2006**, *291*, R412–R418. [CrossRef] [PubMed]
54. Song, J.; Rutherford, T.; Naftolin, F.; Brown, S.; Mor, G. Hormonal regulation of apoptosis and the Fas and Fas ligand system in human endometrial cells. *Mol. Hum. Reprod.* **2002**, *8*, 447–455. [CrossRef]
55. Lan, R.; Yang, Y.; Song, J.; Wang, L.; Gong, H. Fas regulates the apoptosis and migration of trophoblast cells by targeting NF-κB. *Exp. Ther. Med.* **2021**, *22*, 1055. [CrossRef]
56. Neale, D.M.; Mor, G. The role of Fas mediated apoptosis in preeclampsia. *J. Perinat. Med.* **2005**, *33*, 471–477. [CrossRef]
57. Schmittgen, T.D.; Livak, K.J. Analyzing real-time PCR data by the comparative C(T) method. *Nat. Protoc.* **2008**, *3*, 1101–1108. [CrossRef]

International Journal of
*Molecular Sciences*

MDPI

*Article*

# Cardiovascular Disease-Associated MicroRNAs as Novel Biomarkers of First-Trimester Screening for Gestational Diabetes Mellitus in the Absence of Other Pregnancy-Related Complications

**Ilona Hromadnikova [1,*], Katerina Kotlabova [1] and Ladislav Krofta [2]**

[1] Department of Molecular Biology and Cell Pathology, Third Faculty of Medicine, Charles University, 100 00 Prague, Czech Republic
[2] Institute for the Care of the Mother and Child, Third Faculty of Medicine, Charles University, 147 00 Prague, Czech Republic
* Correspondence: ilona.hromadnikova@lf3.cuni.cz; Tel.: +420-296511336

**Abstract:** We assessed the diagnostic potential of cardiovascular disease-associated microRNAs for the early prediction of gestational diabetes mellitus (GDM) in singleton pregnancies of Caucasian descent in the absence of other pregnancy-related complications. Whole peripheral venous blood samples were collected within 10 to 13 weeks of gestation. This retrospective study involved all pregnancies diagnosed with only GDM ($n$ = 121) and 80 normal term pregnancies selected with regard to equality of sample storage time. Gene expression of 29 microRNAs was assessed using real-time RT-PCR. Upregulation of 11 microRNAs (miR-1-3p, miR-20a-5p, miR-20b-5p, miR-23a-3p, miR-100-5p, miR-125b-5p, miR-126-3p, miR-181a-5p, miR-195-5p, miR-499a-5p, and miR-574-3p) was observed in pregnancies destinated to develop GDM. Combined screening of all 11 dysregulated microRNAs showed the highest accuracy for the early identification of pregnancies destinated to develop GDM. This screening identified 47.93% of GDM pregnancies at a 10.0% false positive rate (FPR). The predictive model for GDM based on aberrant microRNA expression profile was further improved via the implementation of clinical characteristics (maternal age and BMI at early stages of gestation and an infertility treatment by assisted reproductive technology). Following this, 69.17% of GDM pregnancies were identified at a 10.0% FPR. The effective prediction model specifically for severe GDM requiring administration of therapy involved using a combination of these three clinical characteristics and three microRNA biomarkers (miR-20a-5p, miR-20b-5p, and miR-195-5p). This model identified 78.95% of cases at a 10.0% FPR. The effective prediction model for GDM managed by diet only required the involvement of these three clinical characteristics and eight microRNA biomarkers (miR-1-3p, miR-20a-5p, miR-20b-5p, miR-100-5p, miR-125b-5p, miR-195-5p, miR-499a-5p, and miR-574-3p). With this, the model identified 50.50% of GDM pregnancies managed by diet only at a 10.0% FPR. When other clinical variables such as history of miscarriage, the presence of trombophilic gene mutations, positive first-trimester screening for preeclampsia and/or fetal growth restriction by the Fetal Medicine Foundation algorithm, and family history of diabetes mellitus in first-degree relatives were included in the GDM prediction model, the predictive power was further increased at a 10.0% FPR (72.50% GDM in total, 89.47% GDM requiring therapy, and 56.44% GDM managed by diet only). Cardiovascular disease-associated microRNAs represent promising early biomarkers to be implemented into routine first-trimester screening programs with a very good predictive potential for GDM.

**Keywords:** cardiovascular microRNAs; early pregnancy; gene expression; gestational diabetes mellitus; prediction; screening; whole peripheral venous blood

**Citation:** Hromadnikova, I.; Kotlabova, K.; Krofta, L. Cardiovascular Disease-Associated MicroRNAs as Novel Biomarkers of First-Trimester Screening for Gestational Diabetes Mellitus in the Absence of Other Pregnancy-Related Complications. *Int. J. Mol. Sci.* **2022**, 23, 10635. https://doi.org/10.3390/ijms231810635

Academic Editor: Maria Luisa Balestrieri

Received: 15 August 2022
Accepted: 9 September 2022
Published: 13 September 2022

## 1. Introduction

Gestational diabetes mellitus (GDM), glucose intolerance in pregnancy [1–3], increases the risk of the onset of maternal pregnancy-related complications and neonatal morbidity. It also has long-term implications for both mother and child in form of risk of developing type 2 diabetes mellitus and cardiovascular diseases [1,4–6].

Several universal screening programs of GDM [1,2,7,8] have been implemented in the routine care of pregnant women. The first screening phase based on the monitoring of a fasting glucose is usually held at first visit during the first trimester of gestation and rules out patients with pre-existing diabetes and detects the occurrence of early GDM. The second screening phase is usually performed at 24–28 weeks of gestation in pregnancies with normal early screening with the oral glucose tolerance test (OGTT) and identifies the occurrence of GDM at the late second and early third pregnancy trimesters. If normal, the OGTT may be repeated again at 32 weeks of gestation [7].

As of now, several promising early predictive models for GDM have been established.

The initial logistic regression model based on the inclusion of maternal characteristics only (maternal age, weight, height, racial origin, family history of diabetes, use of ovulation drugs, birth weight, and previous history of GDM) showed a high accuracy for prediction of GDM at 11–13 weeks of gestation. It reached the following parameters: area under the curve (AUC) 0.823, 95% confidence interval (95% CI) 0.820–0.826, 55.0% sensitivity at a 10.0% false positive rate (FPR) [9]. A slightly older model for the prediction of GDM based on some of the above mentioned factors combined with serum concentrations of adiponectin and sex hormone binding globulin reached similar predictive results (AUC 0.842, 95% CI: 0.817–0.867, 58.6% at a 10.0% FPR) [10].

Similar data were reported by another research group which used a multivariate regression model for the early prediction of GDM. This model was also based on maternal clinical parameters such as age, body mass index (BMI), South/East Asian ethnicity, parity, family history of diabetes, and previous history of GDM (AUC 0.880, 95% CI: 0.850–0.920, 70.2% detection rate at a 10.0% FPR) [11]. Similarly, the same research group later introduced an improved first-trimester risk multivariate prediction model for GDM. This novel model incorporated family history of diabetes, previous history of GDM, South/East Asian ethnicity, parity, BMI, pregnancy-associated plasma protein A (PAPP-A), triglycerides, and lipocalin-2, and achieved a higher discrimination power (AUC 0.910, 95% CI: 0.890–0.960, 76.8% at a 10.0% FPR) [12].

Furthermore, reduced plasma levels of irisin in the first trimester of gestation were implemented into another model based on known risk factors (maternal age, BMI, gestational age at sampling, smoking, ethnicity, pre-existing hypertension or cardiovascular disease, family history of diabetes, physical activity, family history of diabetes, and blood levels of cholesterol, high-density lipoprotein cholesterol, triglycerides, insulin, fasting plasma glucose, and C-reactive protein). This improved the discrimination rate of predicting GDM in a Chinese population (AUC 0.809, 95% CI: 0.763–0.854) [13]. Another independent large-scale study performed in a Chinese population during the first trimester of pregnancy explored a total of 73 variables and also reached a high discriminative power for GDM (AUC 0.800) [14].

An additional non-invasive predictive model consisting of mean arterial blood pressure in the first trimester, age, ethnicity and previous history of GDM demonstrated relatively high predictive ability for a Singaporean population (AUC 0.820, 95% CI: 0.710–0.930), where UK NICE guidelines had poor GDM predictive outcome (AUC 0.600, 95% CI: 0.510–0.700) [15].

Additionally, metabolomics analyses performed on a Japanese population revealed novel promising metabolic biomarkers (serum glutamine, urine ethanolamine, and urine 1,3-diphosphoglycerate). Each biomarker individually demonstrated a high discrimination power for prediction of GDM during the first or early second trimesters of gestation (AUC over 0.800) [16].

First-trimester screening for GDM for an Israeli population reached very high discriminative power in both non-obese women (AUC 0.940, 95% CI: 0.850–0.990, 83.0% at a 10.0% FPR) and obese women (AUC 0.950, 95% CI: 0.880–0.990, 89% at a 10.0% FPR). These screening models were based on the combination of soluble cluster of differentiation 163 (sCD163), tumour necrosis factor alpha (TNFα), placental protein 13 (PP13), and PAPP-A or on the combination of BMI, insulin, sCD163, and TNFα [17].

The latest model was based on maternal clinical characteristics (age and pre-pregnancy BMI); maternal coagulation function (prothrombin time, international standardized ratio, activated partial thromboplastin time, fibrinogen, and thrombin time); and glycolipid metabolism indicators (fasting blood glucose, total cholesterol, triglycerides, low density lipoprotein cholesterol, small and dense low density lipoprotein cholesterol, apolipoprotein B, and apolipoprotein E). This model was applied to a Chinese population in the first trimester of gestation and reached a high clinical value for the prediction of GDM (AUC 0.892, 95% CI: 0.86–0.93) [18].

Previously, the potential usage of coagulation function examination variables such as prothrombin time and activated partial thromboplastin time as novel biomarkers for the prediction of GDM for a Chinese population at 19 weeks of gestation was demonstrated [19].

Similar results were reported for a Chinese population, when a mid-pregnancy risk prediction model for GDM was applied (AUC 0.911, 95% CI: 0.893–0.930). This model was based on maternal status in the combination with ultrasound and serological findings (age, pre-pregnancy BMI, family history of diabetes, polycystic ovary syndrome, previous history of GDM, high systolic pressure, glycosylated haemoglobin levels, triglyceride levels, total cholesterol levels, low density lipoprotein cholesterol levels, C-reactive protein levels, increased subcutaneous fat thickness, and visceral fat thickness) [20].

Similarly, a combined multivariate prediction model performed between 10 and 16 weeks of gestation in an Irish population also achieved a very high level of discrimination for the prediction of GDM (AUC 0.860, 95% CI: 0.774–0.945). This model was based on family history of diabetes, previous perinatal death, overall insulin resistant condition, ultrasound measurements of subcutaneous and visceral abdominal adipose tissue, 8-point skinfold thickness, mid-upper-arm circumference, and weight [21].

Interestingly, the latest study of Eidgahi et al. [22] presented a simplified GDM predictive model with a very good efficiency (AUC 0.83, 95% CI: 0.76–0.90) in an Irani population. This model was based on the mean values of basic indicators (haemoglobin, haematocrit, red blood cell count, and fasting blood glucose) obtained from repeated measures during the first and early second trimesters of gestation. They suggested that this GDM predictive model might be used mainly in poor and low-income countries.

Other models for the early prediction of GDM have not been as effective as the predictive models introduced above [23–40].

We focused on the exploration of gene expression profiles of selected cardiovascular disease-associated microRNAs in the whole peripheral venous blood of women during the early stages of gestation. The aim of the study was to assess the predictive potential for GDM in the absence of other pregnancy-related complications.

Previously, by searching the Medline database we identified a large number of microRNAs playing a role in pathogenesis of diabetes mellitus and cardiovascular/cerebrovascular diseases. Finally, we selected a shortlist of 29 microRNAs for the study which have been repeatedly demonstrated by numerous scientific teams to be involved in development and homeostasis of the cardiovascular system, angiogenesis, and adipogenesis. In addition, these microRNAs were reported to be associated with pathological conditions and diseases (vascular endothelial dysfunction and inflammation, hypoxia, hypertension and regulation of hypertension-related genes, obesity, dyslipidaemia, atherosclerosis and atherosclerotic plaque formation, insulin resistance, diabetes mellitus and diabetes-related complications, metabolic syndrome, cardiovascular diseases involving the blood vessels and/or the heart, chronic kidney disease, ischemia/reperfusion injury, cardiac regeneration, and cachexia) (Table 1) [41–225].

**Table 1.** The role of studied microRNAs in the pathogenesis of diabetes mellitus and cardiovascular/cerebrovascular diseases.

| miRBase ID | Gene Location on Chromosome | Role in the Pathogenesis of Diabetes Mellitus and Cardiovascular/Cerebrovascular Diseases |
|---|---|---|
| hsa-miR-1-3p | 20q13.3 [41]<br>18q11.2 | Acute myocardial infarction, heart ischemia, post-myocardial infarction complications, thoracic aortic aneurysm [43], diabetes mellitus [44,45], and vascular endothelial dysfunction [46] |
| hsa-miR-16-5p | 13q14.2 | Myocardial infarction [47,48], heart failure [49], acute coronary syndrome, cerebral ischaemic events [50], gestational diabetes mellitus [51–53], and diabetes mellitus [54–56] |
| hsa-miR-17-5p | 13q31.3 [57,58] | Cardiac development [59], ischemia/reperfusion-induced cardiac injury [60], kidney ischemia-reperfusion injury [61], diffuse myocardial fibrosis in hypertrophic cardiomyopathy [62], acute ischemic stroke [63], coronary artery disease [64], adipogenic differentiation [65], gestational diabetes mellitus [51,52], and diabetes mellitus [56,66] |
| hsa-miR-20a-5p | 13q31.3 [67] | Pulmonary hypertension [68], gestational diabetes mellitus [51,52,69], diabetic retinopathy [70], and diabetes with abdominal aortic aneurysm [71] |
| hsa-miR-20b-5p | Xq26.2 [67] | Hypertension-induced heart failure [72], insulin resistance [73], T2DM [74,75], and diabetic retinopathy [76] |
| hsa-miR-21-5p | 17q23.2 [77] | Homeostasis of the cardiovascular system [78], cardiac fibrosis and heart failure [79,80], thoracic aortic aneurysm [43], ascending aortic aneurysm [81], regulation of hypertension-related genes [82], myocardial infarction [83], insulin resistance [73], T2DM [84], T2DM with major cardiovascular events [85], T1DM [86–88], and diabetic nephropathy [89] |
| hsa-miR-23a-3p | 19p13.12 | Heart failure [90], coronary artery disease [91], cerebral ischemia-reperfusion [92], vascular endothelial dysfunction [46], small and large abdominal aortic aneurysm [93], obesity and insulin resistance [94] |
| hsa-miR-24-3p | 19p13.12 | Asymptomatic carotid stenosis [95], familial hypercholesterolemia and coronary artery disease [96], angina pectoris [97], ischemic dilated cardiomyopathy [98], small and large abdominal aortic aneurysm [93], myocardial ischemia/reperfusion [99,100], and diabetes mellitus [45,56,60,62] |
| hsa-miR-26a-5p | 3p22.2 [101]<br>12q14.1 | Heart failure, cardiac hypertrophy, myocardial infarction [83,103,104], ischemia/reperfusion injury [105], pulmonary arterial hypertension [106], T1DM [107], and diabetic nephropathy [89] |
| hsa-miR-29a-3p | 7q32.3 | Ischemia/reperfusion-induced cardiac injury [108], cardiac cachexia, heart failure [109], atrial fibrillation [110], diffuse myocardial fibrosis in hypertrophic cardiomyopathy [62], coronary artery disease [111], pulmonary arterial hypertension [106], gestational diabetes mellitus [112], and diabetes mellitus [44,55,113,114] |
| hsa-miR-92a-3p | 13q31.3<br>Xq26.2 | Mitral chordae tendineae rupture [115], children with rheumatic carditis [116], myocardial infarction [117], heart failure [118], coronary artery disease [119], and renal injury-associated atherosclerosis [120] |
| hsa-miR-100-5p | 11q24.1 | Failing human heart, idiopathic dilated cardiomyopathy, ischemic cardiomyopathy [98], regulation of hypertension-related genes [82], and T1DM [86] |
| hsa-miR-103a-3p | 5q34 [121]<br>20p13 | Hypertension, hypoxia-induced pulmonary hypertension [123], myocardial ischemia/reperfusion injury, acute myocardial infarction [124], ischemic dilated cardiomyopathy [99], obesity, and regulation of insulin sensitivity [125], T1DM [126] |
| hsa-miR-125b-5p | 11q24.1 [126]<br>21q21.1 | Acute ischemic stroke, acute myocardial infarction [128,129], ischemic dilated cardiomyopathy [98], ascending aortic aneurysm [81], gestational diabetes mellitus [130], T1DM [131,132], and T2DM [133] |

**Table 1.** *Cont.*

| miRBase ID | Gene Location on Chromosome | Role in the Pathogenesis of Diabetes Mellitus and Cardiovascular/Cerebrovascular Diseases |
|---|---|---|
| hsa-miR-126-3p | 9q34.3 [134] | Acute myocardial infarction [104], thoracic aortic aneurysm [43], T2DM [85,135], T2DM with major cardiovascular events [85], and gestational diabetes mellitus [136] |
| hsa-miR-130b-3p | 22q11.21 | Hypertriglyceridemia [137,138], intracranial aneurysms [139], hyperacute cerebral infarction [140], T2DM [84,141,142], and gestational diabetes mellitus [136] |
| hsa-miR-133a-3p | 18q11.2 [143]<br>20q13.33 | Heart failure, myocardial fibrosis in hypertrophic cardiomyopathy [62,145], arrhythmogenesis in the hypertrophic and failing hearts [146,147], coronary artery calcification [148], thoracic aortic aneurysm [43], ascending aortic aneurysm [81], and diabetes mellitus [41,45] |
| hsa-miR-143-3p | 5q33 | Intracranial aneurysms [149], coronary heart disease [150], myocardial infarction [151], myocardial hypertrophy [152], dilated cardiomyopathy [153], pulmonary arterial hypertension [154], acute ischemic stroke [127], and ascending aortic aneurysm [81], |
| hsa-miR-145-5p | 5q33 | Hypertension [155,156], dilated cardiomyopathy [157], myocardial infarction [158], stroke [159], acute cerebral ischemic/reperfusion [160], T2DM [56,161], T1DM [84], diabetic retinopathy [162], and gestational diabetes mellitus [163] |
| hsa-miR-146a-5p | 5q33.3 [164,165] | Angiogenesis [166], hypoxia, ischemia/reperfusion-induced cardiac injury [167], myocardial infarction [48], coronary atherosclerosis, coronary heart disease in patients with subclinical hypothyroidism [168], thoracic aortic aneurysm [43], acute ischemic stroke, acute cerebral ischemia [169], T2DM [56,84], T1DM [107], and diabetic nephropathy [89] |
| hsa-miR-155-5p | 21q21.3 | Thoracic aortic aneurysm [43], type 1 diabetes [125], gestational diabetes mellitus [53], adolescent obesity [170], diet-induced obesity and obesity resistance [171], atherosclerosis [172], hyperlipidemia-associated endotoxemia [173], coronary plaque rupture [174], children with cyanotic heart disease [175], chronic kidney disease and nocturnal hypertension [176], and atrial fibrillation [177] |
| hsa-miR-181a-5p | 1q32.1 [178]<br>9q33.3 | Regulation of hypertension-related genes, atherosclerosis [178], metabolic syndrome, coronary artery disease [179], non-alcoholic fatty liver disease [180], ischaemic stroke, transient ischaemic attack, acute myocardial infarction [181,182], obesity and insulin resistance [94,178,179], T1DM [84,183], and T2DM [178,182] |
| hsa-miR-195-5p | 17p13.1 [184] | Cardiac hypertrophy, heart failure [185,186], abdominal aortic aneurysms [187], aortic stenosis [188], T2DM [161], and gestational diabetes mellitus [189] |
| hsa-miR-199a-5p | 1q24.3<br>19p13.2 | T1DM, T2DM, gestational diabetes mellitus [190], diabetic retinopathy [191], cerebral ischemic injury [192], heart failure [193], hypertension [194,195], congenital heart disease [196], pulmonary artery hypertension [197], unstable angina [198], hypoxia in myocardium [196], and acute kidney injury [199] |
| hsa-miR-210-3p | 11p15.5 | Cardiac hypertrophy [200], acute kidney injury [201], myocardial infarction [202], and atherosclerosis [203] |
| hsa-miR-221-3p | Xp11.3 | Asymptomatic carotid stenosis [95], cardiac amyloidosis [204], heart failure [205], atherosclerosis [206,207], aortic stenosis [208], acute myocardial infarction [209], acute ischemic stroke [210], focal cerebral ischemia [211], pulmonary artery hypertension [212], and obesity [213] |
| hsa-miR-342-3p | 14q32.2 | Cardiac amyloidosis [204], obesity [214], T1DM [84,190,215], T2DM [190,190,216,217] and endothelial dysfunction [218] |
| hsa-miR-499a-5p | 20q11.22 | Myocardial infarction [48,219], hypoxia [220], cardiac regeneration [221], and vascular endothelial dysfunction [46] |
| hsa-miR-574-3p | 4p14 | Myocardial infarction [222], coronary artery disease [138], cardiac amyloidosis [204], stroke [223], and T2DM [142,224] |

T1DM: Diabetes mellitus type 1; T2DM: Diabetes mellitus type 2.

The epigenetic profiling of microRNAs (miR-1-3p, miR-16-5p, miR-17-5p, miR-20a-5p, miR-20b-5p, miR-21-5p, miR-23a-3p, miR-24-3p, miR-26a-5p, miR-29a-3p, miR-92a-3p, miR-100-5p, miR-103a-3p, miR-125b-5p, miR-126-3p, miR-130b-3p, miR-133a-3p, miR-143-3p, miR-145-5p, miR-146a-5p, miR-155-5p, miR-181a-5p, miR-195-5p, miR-199a-5p, miR-210-3p, miR-221-3p, miR-342-3p, miR-499a-5p, and miR-574-3p) was the subject of our interest (Table 1).

Up to now, no reports on microRNA gene profiling of the whole peripheral venous blood in early stages of gestation are at disposal in pregnancies with subsequent onset of GDM.

To our knowledge, only several studies have reported promising data on the early diagnosis of GDM during the first trimester of gestation via screening of circulating cardiovascular disease-associated microRNAs in maternal plasma or serum samples [112,130,226–228].

## 2. Results

### 2.1. Clinical Characteristics of GDM and Control Pregnancies

The clinical characteristics of GDM and control pregnancies are summarized in Table 2.

From the clinical characteristics of patients, it is obvious that maternal age (mainly advanced maternal age, ≥35 years), BMI (higher BMI values, BMI $\geq$ 30 kg/m$^2$) at early stages of gestation, the necessity to undergo an infertility treatment by assisted reproductive technology, history of miscarriage, the presence of trombophilic gene mutations, positive first-trimester screening for preeclampsia and/or FGR by FMF algorithm, and family history of diabetes mellitus in first-degree relatives represent independent significant risk factors for the subsequent onset of GDM.

**Table 2.** Clinical characteristics of the cases and controls.

| | Normal Term Pregnancies ($n$ = 80) | GDM Overall ($n$ = 121) | GDM Managed by Diet Only ($n$ = 101) | GDM Managed by Diet and Therapy ($n$ = 20) | $p$-Value [1] | $p$-Value [2] | $p$-Value [3] |
|---|---|---|---|---|---|---|---|
| *Maternal characteristics* | | | | | | | |
| Autoimmune diseases (SLE/APS/RA) | 0 (0%) | 1 (0.83%) | 1 (RA, 1.0%) | 0 (0%) | 0.672<br>OR: 2.004<br>95% CI: 0.081–49.814 | 0.593<br>OR: 2.403<br>95% CI: 0.096–59.786 | 0.497<br>OR: 3.927<br>95% CI: 0.076–203.916 |
| Other autoimmune diseases | 0 (0%) | 1 (0.83%) | 1 (vasculitis; 1.0%) | 0 (0%) | 0.672<br>OR: 2.004<br>95% CI: 0.081–49.814 | 0.593<br>OR: 2.403<br>95% CI: 0.096–59.786 | 0.497<br>OR: 3.927<br>95% CI: 0.076–203.916 |
| Any kind of autoimmune disease (SLE/APS/RA/other) | 0 (0%) | 2 (1.65%) | 2 (1.98%) | 0 (0%) | 0.435<br>OR: 3.368<br>95% CI: 0.160–71.088 | 0.369<br>OR: 4.045<br>95% CI: 0.191–85.468 | 0.497<br>OR: 3.927<br>95% CI: 0.076–203.916 |
| Trombophilic gene mutations | 0 (0%) | 11 (9.09%) | 9 (8.91%) | 2 (10.0%) | 0.052<br>OR: 16.756<br>95% CI: 0.973–288.513 | 0.055<br>OR: 16.535<br>95% CI: 0.947–288.589 | 0.050<br>OR: 21.757<br>95% CI: 1.002–472.533 |
| Family history of diabetes | | | | | | | |
| First-degree relative with DM | 10 (12.50%) | 30 (24.79%) | 26 (25.74%) | 4 (20.0%) | 0.036<br>OR: 2.308<br>95% CI: 1.057–5.037 | 0.030<br>OR: 2.427<br>95% CI: 1.092–5.394 | 0.392<br>OR: 1.750<br>95% CI: 0.486–6.297 |
| Second-degree relative with DM | 21 (26.25%) | 44 (36.36%) | 36 (35.64%) | 8 (40.0%) | 0.135<br>OR: 1.605<br>95% CI: 0.863–2.986 | 0.178<br>OR: 1.556<br>95% CI: 0.818–2.961 | 0.230<br>OR: 1.873<br>95% CI: 0.673–5.215 |
| Parity | | | | | | | |
| Nulliparous—no previous pregnancy | 40 (50.0%) | 54 (44.63%) | 46 (45.54%) | 8 (40.0%) | 0.455<br>OR: 0.806<br>95% CI: 0.458–1.419 | 0.551<br>OR: 0.836<br>95% CI: 0.465–1.505 | 0.425<br>OR: 0.667<br>95% CI: 0.246–1.805 |
| Parous—no prior GDM | 39 (48.75%) | 61 (50.41%) | 50 (49.50%) | 11 (55.0%) | 0.817<br>OR: 1.069<br>95% CI: 0.608–1.880 | 0.919<br>OR: 1.031<br>95% CI: 0.573–1.853 | 0.618<br>OR: 1.285<br>95% CI: 0.480–3.437 |
| Parous—prior GDM | 1 (1.25%) | 6 (4.96%) | 5 (4.95%) | 1 (5.0%) | | | |
| History of macrosomia (FBW > 4000 g) | 4 (5.0%) | 2 (1.65%) | 1 (0.99%) | 1 (5.0%) | 0.194<br>OR: 0.319<br>95% CI: 0.057–1.786 | 0.141<br>OR: 0.190<br>95% CI: 0.021–1.735 | 1.0<br>OR: 1.000<br>95% CI: 0.106–9.471 |

**Table 2.** *Cont.*

| | Normal Term Pregnancies ($n$ = 80) | GDM Overall ($n$ = 121) | GDM Managed by Diet Only ($n$ = 101) | GDM Managed by Diet and Therapy ($n$ = 20) | $p$-Value [1] | $p$-Value [2] | $p$-Value [3] |
|---|---|---|---|---|---|---|---|
| History of miscarriage spontaneous loss of a pregnancy before 22 weeks of gestation | 16 (20.0%) | 42 (34.71%) | 36 (35.64%) | 6 (30.0%) | 0.026<br>OR: 2.127<br>95% CI: 1.095–4.129 | 0.022<br>OR: 2.215<br>95% CI: 1.119–4.384 | 0.338<br>OR: 1.714<br>95% CI: 0.569–5.161 |
| History of perinatal death the death of a baby between 22 weeks of gestation (or weighing 500 g) and 7 days after birth | 0 (0%) | 4 (3.31%) | 3 (2.97%) | 1 (5.0%) | 0.224<br>OR: 6.166<br>95% CI:<br>0.327–116.113 | 0.251<br>OR: 5.721<br>95% CI:<br>0.291–112.387 | 0.128<br>OR: 12.385<br>95% CI:<br>0.486–315.805 |
| ART (IVF/ICSI/other) | 2 (2.5%) | 20 (16.53%) | 15 (14.85%) | 5 (25.0%) | 0.007<br>OR: 7.723<br>95% CI: 1.752–34.038 | 0.013<br>OR: 6.802<br>95% CI: 1.507–30.698 | 0.004<br>OR: 13.000<br>95% CI: 2.304–73.362 |
| Smoking during pregnancy | 2 (2.5%) | 6 (4.96%) | 4 (3.96%) | 2 (10.0%) | 0.392<br>OR: 2.035<br>95% CI: 0.108–10.343 | 0.589<br>OR: 1.608<br>95% CI: 0.287–9.012 | 0.156<br>OR: 4.333<br>95% CI: 0.572–32.859 |
| *Pregnancy details (First trimester of gestation)* | | | | | | | |
| Maternal age (years) | 32 (25–42) | 33 (21–42) | 33 (21–42) | 32 (25–42) | 0.635 | 0.572 | 0.950 |
| Advanced maternal age (≥35 years old at early stages of gestation) | 18 (22.50%) | 49 (40.49%) | 42 (41.58%) | 7 (35.0%) | 0.009<br>OR: 2.618<br>95% CI: 1.238–4.437 | 0.007<br>OR: 2.675<br>95% CI: 1.271–4.731 | 0.252<br>OR: 1.144<br>95% CI: 0.644–5.343 |
| BMI (kg/m$^2$) | 21.28 (17.16–29.76) | 24.24 (17.37–40.76) | 23.89 (17.37–40.76) | 26.55 (19.33–39.79) | <0.001 | <0.001 | <0.001 |
| BMI $\geq$ 30 kg/m$^2$ | 0 (0%) | 25 (20.66%) | 17 (16.83%) | 8 (40%) | 0.009<br>OR: 42.544<br>95% CI:<br>2.550–709.837 | 0.015<br>OR: 33.343<br>95% CI:<br>1.972–563.719 | 0.002<br>OR: 109.480<br>95% CI:<br>5.941–2017.344 |

**Table 2.** *Cont.*

| | Normal Term Pregnancies ($n$ = 80) | GDM Overall ($n$ = 121) | GDM Managed by Diet Only ($n$ = 101) | GDM Managed by Diet and Therapy ($n$ = 20) | $p$-Value [1] | $p$-Value [2] | $p$-Value [3] |
|---|---|---|---|---|---|---|---|
| Gestational age at sampling (weeks) | 10.29 (9.57–13.71) | 10.29 (9.43–13.57) | 10.29 (9.43–13.57) | 10.21 (9.43–12.71) | 0.737 | 0.548 | 0.521 |
| MAP (mmHg) | 88.75 (67.67–103.83) | 92.0 (72.83–127.58) | 91.96 (72.83–127.58) | 92.58 (82.85–101.92) | 0.051 | 0.083 | 0.022 |
| MAP (MoM) | 1.05 (0.84–1.25) | 1.05 (0.90–1.44) | 1.05 (0.90–1.44) | 1.07 (0.97–1.13) | 0.656 | 0.574 | 0.361 |
| Mean UtA-PI | 1.39 (0.56–2.43) | 1.35 (0.42–2.30) | 1.35 (0.42–2.30) | 1.25 (0.74–1.84) | 0.591 | 0.831 | 0.495 |
| Mean UtA-PI (MoM) | 0.90 (0.37–1.55) | 0.88 (0.26–1.48) | 0.89 (0.26–1.48) | 0.85 (0.52–1.26) | 0.539 | 0.710 | 0.402 |
| PIGF serum levels (pg/mL) | 27.1 (8.1–137.0) | 26.7 (9.2–71.0) | 26.8 (9.2–71.0) | 25.5 (14.5–46.0) | 0.420 | 0.377 | 0.375 |
| PIGF serum levels (MoM) | 1.04 (0.38–2.61) | 1.09 (0.44–2.0) | 1.06 (0.44–2.0) | 1.15 (0.62–1.59) | 0.934 | 0.690 | 0.065 |
| PAPP-A serum levels (IU/L) | 1.49 (0.48–15.69) | 1.28 (0.22–11.45) | 1.35 (0.22–11.45) | 1.0 (0.26–6.83) | 0.063 | 0.123 | 0.158 |
| PAPP-A serum levels (MoM) | 1.17 (0.37–3.18) | 1.05 (1.19–3.67) | 1.04 (0.28–3.02) | 1.43 (0.19–3.67) | 0.606 | 0.434 | 0.362 |
| Free b-hCG serum levels (μg/L) | 60.21 (9.9–200.6) | 50.25 (9.31–211.3) | 53.82 (9.31–211.3) | 32.62 (16.55–153.2) | 0.043 | 0.123 | 0.037 |
| Free b-hCG serum levels (MoM) | 1.02 (0.31–3.57) | 0.98 (0.18–4.54) | 1.0 (0.18–4.54) | 0.97 (0.33–2.74) | 0.317 | 0.437 | 0.446 |
| Screen positive for PE and/or FGR by FMF algorithm | 0 (0%) | 11 (9.09%) | 10 (9.90%) | 1 (5.0%) | 0.052 OR: 16.756 95% CI: 0.973–288.513 | 0.045 OR: 18.475 95% CI: 1.066–320.312 | 0.128 OR: 12.385 95% CI: 0.486–315.805 |
| Aspirin intake during pregnancy | 0 (0%) | 8 (6.61%) | 7 (6.93%) | 1 (5.0%) | 0.089 OR: 12.057 95% CI: 0.686–211.908 | 0.083 OR: 12.778 95% CI: 0.717–227.208 | 0.128 OR: 12.385 95% CI: 0.486–315.806 |
| *Pregnancy details (At delivery)* | | | | | | | |
| BMI (kg/m$^2$) | 26.66 (21.71–34.82) | 28.41 (20.11–49.31) | 28.24 (20.11–49.31) | 32.11 (23.23–44.98) | 0.004 | 0.042 | <0.001 |
| SBP (mmHg) | 122 (100–155) | 120 (90–160) | 121 (90–160) | 120 (100–140) | 0.823 | 0.950 | 0.330 |
| DBP (mmHg) | 76 (60–90) | 79 (57–109) | 79 (57–109) | 79 (60–89) | 0.898 | 0.945 | 0.816 |

**Table 2.** *Cont.*

| | Normal Term Pregnancies ($n = 80$) | GDM Overall ($n = 121$) | GDM Managed by Diet Only ($n = 101$) | GDM Managed by Diet and Therapy ($n = 20$) | *p*-Value [1] | *p*-Value [2] | *p*-Value [3] |
|---|---|---|---|---|---|---|---|
| Gestational age at delivery (weeks) | 40.07 (37.57–42.0) | 39.14 (36.14–41.29) | 39.14 (36.14–41.29) | 38.93 (36.57–41.0) | <0.001 | <0.001 | 0.009 |
| Delivery at gestational age < 37 weeks | 0 (0%) | 6 (4.96%) | 4 (3.96%)<br>1 CS for vasculitis-associated adverse obstetric history<br>3 CS for abnormal CTG | 2 (10.0%)<br>1 CS for vasculitis-associated adverse obstetric history<br>1 CS for abnormal CTG | 0.135<br>OR: 9.061<br>95% CI: 0.503–163.118 | 0.181<br>OR: 7.431<br>95% CI: 0.394–140.092 | 0.050<br>OR: 21.757<br>95% CI: 1.002–472.533 |
| Polyhydramnios | 1 (1.25%) | 28 (23.14%) | 21 (20.79%) | 7 (35.0%) | 0.002<br>OR: 23.785<br>95% CI: 3.164–178.781 | 0.003<br>OR: 20.738<br>95% CI: 2.723–157.908 | <0.001<br>OR: 42.538<br>95% CI: 4.828–374.768 |
| Fetal birth weight (grams) | 3470 (2920–4240) | 3370 (2430–4340) | 3310 (2430–4340) | 3625 (2950–4220) | 0.043 | 0.003 | 0.046 |
| LGA (FBW > 90th percentile) | 2 (2.5%) | 11 (9.09%) | 7 (6.93%) | 4 (20.0%) | 0.082<br>OR: 3.900<br>95% CI: 0.841–18.089 | 0.192<br>OR: 2.904<br>95% CI: 0.586–14.384 | 0.012<br>OR: 9.750<br>95% CI: 1.643–57.851 |
| Macrosomia (FBW > 4000g) | 5 (6.25%) | 10 (8.26%) | 8 (7.92%) | 2 (10.0%) | 0.596<br>OR: 1.351<br>95% CI: 0.444–4.112 | 0.666<br>OR: 1.290<br>95% CI: 0.405–4.108 | 0.560<br>OR: 1.667<br>95% CI: 0.299–9.295 |
| Fetal sex | | | | | | | |
| Boy | 40 (50.0%) | 60 (49.59%) | 49 (48.51%) | 11 (55.0%) | 0.954<br>OR: 0.984<br>95% CI: 0.559–1.730 | 0.843<br>OR: 0.942<br>95% CI: 0.524–1.695 | 0.689<br>OR: 1.222<br>95% CI: 0.457–3.269 |
| Girl | 40 (50.0%) | 61 (50.41%) | 52 (51.49%) | 9 (45.0%) | | | |
| Induced delivery | 8 (10.0%)<br>4 postterm pregnancy<br>1 polyhydramnios<br>1 suspicious CTG<br>2 programmed labour | 39 (32.23%) | 32 (31.68%)<br>29 term or postterm GDM pregnancy<br>2 suspicious CTG<br>1 hepatopathy | 7 (35.0%)<br>7 term or postterm GDM pregnancy | <0.001<br>OR: 4.281<br>95% CI: 1.878–9.757 | <0.001<br>OR: 4.174<br>95% CI: 1.798–9.689 | 0.008<br>OR: 4.846<br>95% CI: 1.498–15.674 |

**Table 2.** *Cont.*

| | Normal Term Pregnancies ($n$ = 80) | GDM Overall ($n$ = 121) | GDM Managed by Diet Only ($n$ = 101) | GDM Managed by Diet and Therapy ($n$ = 20) | $p$-Value [1] | $p$-Value [2] | $p$-Value [3] |
|---|---|---|---|---|---|---|---|
| Mode of delivery | | | | | | | |
| Vaginal | 69 (86.25%) | 66 (54.55%) | 58 (57.43%) | 8 (40.0%) | <0.001<br>OR: 5.227<br>95% CI: 2.519–10.848 | <0.001<br>OR: 4.651<br>95% CI: 2.199–9.832 | <0.001<br>OR: 9.409<br>95% CI: 3.139–28.205 |
| CS | 11 (13.75%) | 55 (45.45%) | 43 (42.57%) | 12 (60.0%) | | | |
| Apgar score < 7, 5 min | 0 (0%) | 0 (0%) | 0 (0%) | 0 (0%) | 0.837<br>OR: 0.663<br>95% CI: 0.013–33.732 | 0.908<br>OR: 0.793<br>95% CI: 0.015–40.411 | 0.497<br>OR: 3.927<br>95% CI:<br>0.076–203.916 |
| Apgar score < 7, 10 min | 0 (0%) | 0 (0%) | 0 (0%) | 0 (0%) | 0.837<br>OR: 0.663<br>95% CI: 0.013–33.732 | 0.908<br>OR: 0.793<br>95% CI: 0.015–40.411 | 0.497<br>OR: 3.927<br>95% CI:<br>0.076–203.916 |
| Umbilical blood pH | 7.3 (7.29–7.38) | | 7.3 (7.12–7.39) | 7.3 (7.29–7.30) | | 0.981 | 0.796 |

Continuous variables, compared using the Mann–Whitney or Kruskal–Wallis test, are presented as median (range). Categorical variables, presented as number (percent), were compared using odds ratio test. $p$-value [1,2,3]: the comparison among normal pregnancies and GDM pregnancies, the comparison among normal pregnancies and GDM pregnancies managed by diet only or GDM pregnancies managed by diet and therapy, respectively. GDM, gestational diabetes mellitus; BMI, body mass index; SBP; systolic blood pressure; DBP, diastolic blood pressure; SLE, systemic lupus erythematosus; APS, antiphospholipid syndrome; RA, rheumatoid arthritis; DM, diabetes mellitus; FBW, fetal birth weight; ART, assisted reproductive technology; IVF, in vitro fertilization; ICSI, intracytoplasmic sperm injection; MAP, mean arterial pressure; UtA-PI, uterine artery pulsatility index; PIGF, placental growth factor; PAPP-A, pregnancy-associated plasma protein-A; b-hCG, beta-subunit of human chorionic gonadotropin; PE, preeclampsia; FGR, fetal growth restriction; FMF, Fetal Medicine Foundation; LGA, large for gestational age; CS, caesarean section.

*2.2. Dysregulation of Cardiovascular Disease-Associated MicroRNAs in Early Stages of Gestation in Pregnancies Destinated to Develop GDM*

Initially, microRNA gene expression in peripheral blood leukocytes was compared in the early stages of gestation (within 10 to 13 weeks) between pregnancies destinated to develop GDM and term pregnancies with normal course of gestation (Figure 1). Afterwards, early microRNA gene expression was compared between pregnancies destinated to develop GDM and normal term pregnancies with respect to the treatment strategies (GDM pregnancies managed by diet only and GDM pregnancies requiring a combination of diet and administration of appropriate therapy).

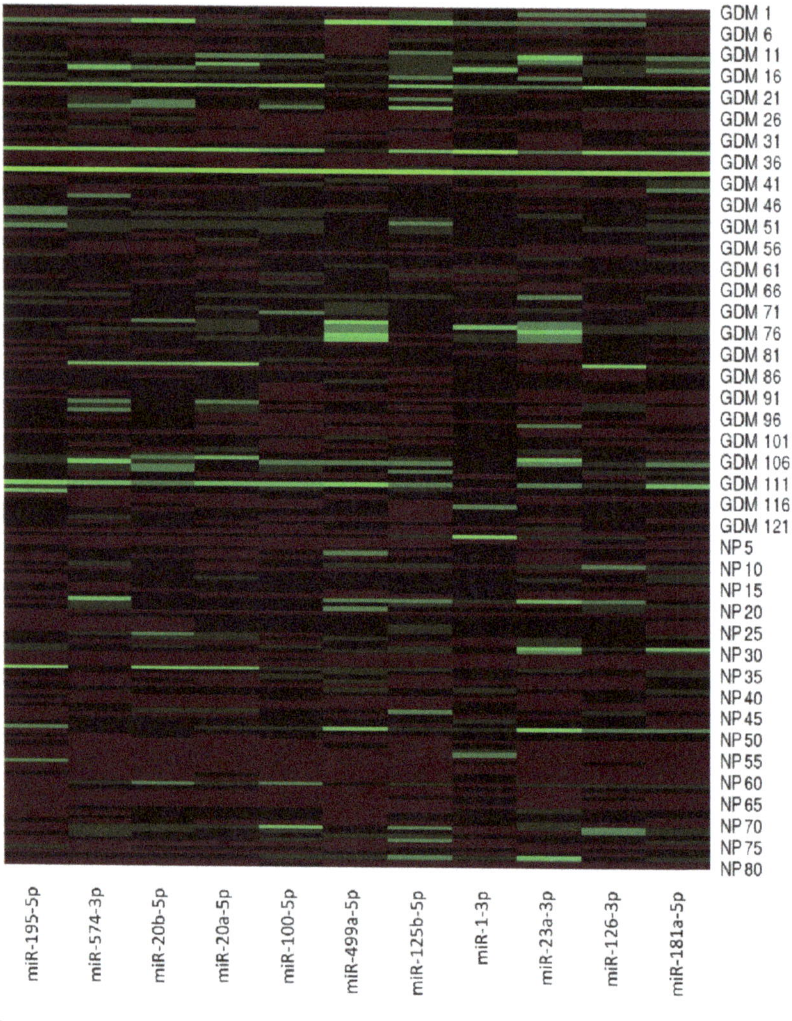


**Figure 1.** MicroRNA gene expression profile in early stages of gestation in pregnancies destinated to develop GDM and term pregnancies with normal course of gestation. MicroRNA gene expression data ($2^{-\Delta\Delta Ct}$) are visualised using the heatmap. In this setting, each row represents a sample (GDM1–GDM121, NP1–NP80) and each column represents a microRNA gene. The colour and intensity of the boxes are used to represent changes of gene expression ($2^{-\Delta\Delta Ct}$). Green colour indicates upregulation, and red colour indicates downregulation. GDM; gestational diabetes mellitus, NP; normal pregnancies.

Only the data that reached statistical significance after the application of Benjamini–Hochberg correction are discussed below (Supplementary Figures S1 and S2). To interpret the experimental data, new cutoff point *p*-values were set up. Significant results following the Benjamini–Hochberg correction are marked by asterisks for the appropriate significance levels (* for $\alpha = 0.05$, ** for $\alpha = 0.01$, and *** for $\alpha = 0.001$). The data that were statistically non-significant after the application of Benjamini–Hochberg correction (Tables 2 and 3) are also displayed (Supplementary Figures S3 and S4), but not discussed further.

Upregulation of miR-1-3p ($p = 0.0028$ **), miR-20a-5p ($p < 0.001$ ***), miR-20b-5p ($p < 0.001$ ***), miR-23a-3p ($p = 0.0065$ *), miR-100-5p ($p < 0.001$ ***), miR-125b-5p ($p = 0.0034$ **), miR-126-3p ($p = 0.0137$ *), miR-181a-5p ($p = 0.0065$ *), miR-195-5p ($p < 0.001$ ***), miR-499a-5p ($p < 0.001$ ***), and miR-574-3p ($p < 0.001$ ***) was detected during the first trimester of gestation in pregnancies destinated to develop GDM (Supplementary Figure S1, Table 3).

MiR-20a-5p (21.49%), miR-20b-5p (18.18%), miR-23a-3p (15.70%), miR-100-5p (20.66%), miR-125b-5p (14.88%), miR-126-3p (14.05%), miR-195-5p (19.83%) miR-499a-5p (14.88%), and miR-574-3p (23.14%) showed moderate sensitivities at a 10.0% FPR to distinguish between normal pregnancies and pregnancies destinated to develop GDM. In contrast, miR-1-3p (12.40%) and miR-181a-5p (10.74%) showed a low sensitivity to differentiate normal pregnancies and pregnancies with subsequent onset of GDM at a 10.0% FPR (Supplementary Figure S1). This means that the sensitivity in case of miR-1-3p and miR-181a-5p was similar to the false positive rate (10.0%) at which the expression data were assessed.

**Table 3.** MicroRNA expression profiles in peripheral blood leukocytes in early stages of gestation in pregnancies destinated to develop GDM and normal term pregnancies.

| | **Mann-Whitney Test Results<br>GDM Overall (*n* = 121) vs. Normal Term Pregnancies (*n* = 80)** | | |
|---|---|---|---|
| | **Median (IQR)** | **Mean (SD)** | ***p*-Value** |
| miR-1-3p | 0.135 (0.071–0.254) vs. 0.075 (0.033–0.198) | 0.259 (0.525) vs. 0.176 (0.303) | ***p* = 0.0028** ** |
| miR-16-5p | 1.216 (0.968–1.725) vs. 1.411 (0.890–1.980) | 1.495 (0.981) vs. 1.646 (1.129) | *p* = 0.5781 |
| miR-17-5p | 1.527 (1.181–2.311) vs. 1.384 (0.971–1.923) | 1.973 (1.473) vs. 1.748 (1.312) | *p* = 0.0538 |
| miR-20a-5p | 2.215 (1.493–3.398) vs. 1.576 (0.991–2.413) | 3.037 (3.068) vs. 1.909 (1.370) | ***p* < 0.001** *** |
| miR-20b-5p | 2.662 (1.812–3.959) vs. 1.976 (1.111–2.675) | 3.706 (3.878) vs. 2.377 (2.291) | ***p* < 0.001** *** |
| miR-21-5p | 0.344 (0.231–0.460) vs. 0.320 (0.167–0.538) | 0.433 (0.420) vs. 0.394 (0.219) | *p* = 0.2418 |
| miR-23a-3p | 0.239 (0.168–0.436) vs. 0.185 (0.103–0.376) | 0.367 (0.337) vs. 0.296 (0.329) | ***p* = 0.0065** * |
| miR-24-3p | 0.292 (0.228–0.372) vs. 0.326 (0.196–0.468) | 0.331 (0.197) vs. 0.384 (0.284) | *p* = 0.5730 |
| miR-26a-5p | 0.699 (0.500–0.926) vs. 0.633 (0.410–1.066) | 0.837 (0.670) vs. 0.776 (0.521) | *p* = 0.3022 |
| miR-29a-3p | 0.405 (0.282–0.575) vs. 0.372 (0.221–0.545) | 0.510 (0.396) vs. 0.407 (0.245) | *p* = 0.0840 |
| miR-92a-3p | 2.179 (1.604–3.084) vs. 2.327 (1.188–3.743) | 2.702 (2.226) vs. 2.807 (2.132) | *p* = 0.9812 |
| miR-100-5p | 0.0023 (0.0013–0.0036) vs. 0.0013 (0.0006–0.0027) | 0.0030 (0.0039) vs. 0.0018 (0.0016) | ***p* < 0.001** *** |
| miR-103a-3p | 1.565 (0.963–2.541) vs. 1.203 (0.815–2.425) | 2.121 (2.252) vs. 1.770 (1.466) | *p* = 0.1547 |
| miR-125b-5p | 0.0041 (0.0025–0.0057) vs. 0.0030 (0.0016–0.0054) | 0.0049 (0.0046) vs. 0.0036 (0.0027) | ***p* = 0.0034** ** |
| miR-126-3p | 0.328 (0.231–0.509) vs. 0.272 (0.140–0.432) | 0.462 (0.551) vs. 0.336 (0.270) | ***p* = 0.0137** * |
| miR-130b-3p | 0.745 (0.476–1.409) vs. 0.702 (0.407–1.157) | 1.075 (0.960) vs. 1.163 (2.425) | *p* = 0.2105 |
| miR-133a-3p | 0.109 (0.061–0.220) vs. 0.110 (0.550–0.233) | 0.193 (0.265) vs. 0.232 (0.483) | *p* = 0.8750 |
| miR-143-3p | 0.048 (0.030–0.880) vs. 0.038 (0.016–0.089) | 0.073 (0.086) vs. 0.058 (0.057) | *p* = 0.0260 |
| miR-145-5p | 0.176 (0.125–0.236) vs. 0.161 (0.980–0.243) | 0.209 (0.153) vs. 0.195 (0.143) | *p* = 0.2025 |
| miR-146a-5p | 1.224 (0.821–1.843) vs. 1.225 (0.578–1.765) | 1.658 (1.541) vs. 1.388 (1.096) | *p* = 0.1415 |
| miR-155-5p | 0.619 (0.434–0.778) vs. 0.607 (0.361–1.614) | 0.703 (0.523) vs 1.247 (1.439) | *p* = 0.2987 |

**Table 3.** *Cont.*

| | Mann-Whitney Test Results GDM Overall ($n$ = 121) vs. Normal Term Pregnancies ($n$ = 80) | | |
|---|---|---|---|
| | **Median (IQR)** | **Mean (SD)** | ***p*-Value** |
| miR-181a-5p | 0.250 (0.175–0.379) vs 0.181 (0.141–0.330) | 0.330 (0.318) vs 0.246 (0.184) | **$p$ = 0.0065** * |
| miR-195-5p | 0.267 (0.168–0.487) vs 0.106 (0.048–0.271) | 0.470 (0.690) vs 0.227 (0.364) | **$p$ < 0.001** *** |
| miR-199a-5p | 0.080 (0.037–0.159) vs 0.058 (0.023–0.111) | 0.136 (0.223) vs 0.096 (0.131) | $p$ = 0.0288 |
| miR-210-3p | 0.102 (0.074–0.154) vs 0.138 (0.075–0.224) | 0.134 (0.105) vs 0.186 (0.180) | $p$ = 0.0952 |
| miR-221-3p | 0.644 (0.448–0.969) vs 0.548 (0.293–0.906) | 0.815 (0.736) vs 0.693 (0.561) | $p$ = 0.0947 |
| miR-342-3p | 3.069 (2.122–4.110) vs 2.542 (1.551–4.206) | 3.605 (2.724) vs 3.307 (2.383) | $p$ = 0.1947 |
| miR-499a-5p | 0.460 (0.231–0.780) vs 0.269 (0.089–0.587) | 0.758 (1.070) vs 0.477 (0.566) | **$p$ < 0.001** *** |
| miR-574-3p | 0.275 (0.180–0.395) vs 0.181 (0.117–0.292) | 0.354 (0.332) vs 0.222 (0.156) | **$p$ < 0.001** *** |

MicroRNA gene expression is compared between groups using the Mann–Whitney test. Statistically significant results are marked in bold. Median (interquartile range, IQR) and mean (standard deviation, SD) fold values of relative gene expression of samples ($2^{-\Delta\Delta Ct}$) are presented. Statistical significant data after Benjamini–Hochberg correction are marked by * for $\alpha = 0.05$, ** for $\alpha = 0.01$, and *** for $\alpha = 0.001$.

### *2.3. First-Trimester Combined MicroRNA Screening Is Able to Differentiate between Pregnancies Destinated to Develop GDM and Term Pregnancies with Normal Course of Gestation*

Despite the low sensitivities of miR-1-3p (12.40%) and miR-181a-5p (10.74%), the combined screening of all 11 dysregulated microRNA biomarkers (miR-1-3p, miR-20a-5p, miR-20b-5p, miR-23a-3p, miR-100-5p, miR-125b-5p, miR-126-3p, miR-181a-5p, miR-195-5p, miR-499a-5p, and miR-574-3p) showed the highest accuracy for the early identification of pregnancies destinated to develop GDM (AUC 0.742, $p$ < 0.001, 63.64% sensitivity, 78.75% specificity, cut off >0.5850). This combined screening identified, in the early stages of gestation, 47.93% of pregnancies destinated to develop GDM at a 10.0% FPR (Figure 2).

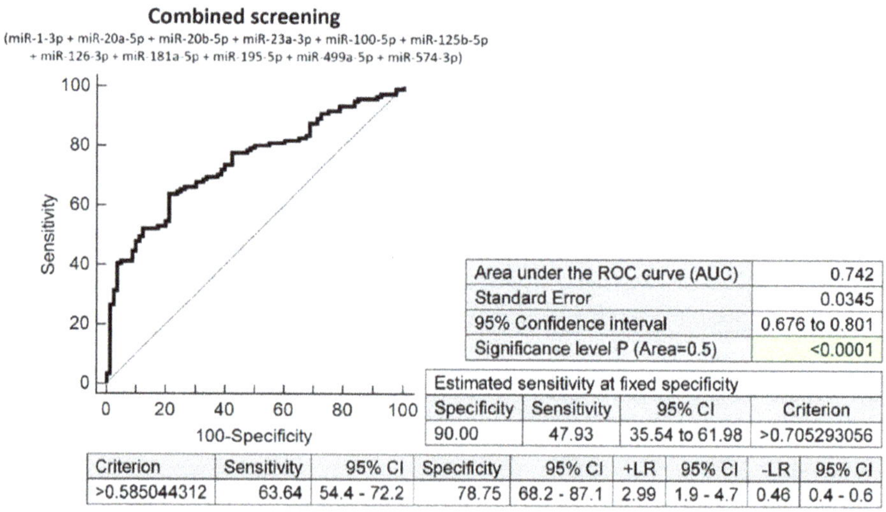


**Figure 2.** ROC analysis—the combination of microRNA biomarkers (miR-1-3p, miR-20a-5p, miR-20b-5p, miR-23a-3p, miR-100-5p, miR-125b-5p, miR-126-3p, miR-181a-5p, miR-195-5p, miR-499a-5p, and miR-574-3p). A total of 47.93% pregnancies destinated to develop GDM had an aberrant microRNA expression profile in the whole peripheral venous blood during the first trimester of gestation at a 10.0% FPR. This represents 58 out of 121 pregnancies correctly predicted to develop GDM and 8 out of 80 normal pregnancies predicted false positively to develop GDM.

### *2.4. The Very Good Accuracy of First-Trimester Combined Screening (MicroRNA Biomarkers and Selected Clinical Characteristics) to Differentiate between Pregnancies Destinated to Develop GDM and Term Pregnancies with Normal Course of Gestation*

The effective screening based on the combination of minimal number of basic clinical characteristics (maternal age and BMI at early stages of gestation and an infertility treatment by assisted reproductive technology) and 11 dysregulated microRNA biomarkers (miR-1-3p, miR-20a-5p, miR-20b-5p, miR-23a-3p, miR-100-5p, miR-125b-5p, miR-126-3p, miR-181a-5p, miR-195-5p, miR-499a-5p, and miR-574-3p) showed relatively high accuracy for the early identification of pregnancies destinated to develop GDM (AUC 0.835, $p < 0.001$, 67.50% sensitivity, 92.50% specificity, cut off >0.6929). This combined screening identified, in the early stages of gestation, 69.17% of pregnancies destinated to develop GDM at a 10.0% FPR (Figure 3).

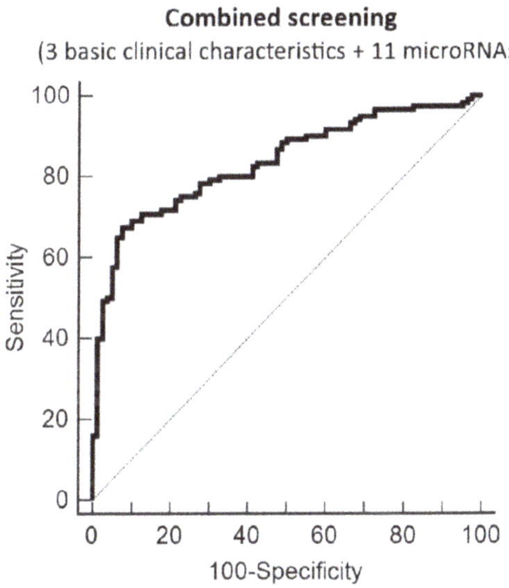


| | |
|---|---|
| Area under the ROC curve (AUC) | 0.835 |
| Standard Error | 0.0281 |
| 95% Confidence interval | 0.777 to 0.884 |
| Significance level P (Area=0.5) | <0.0001 |

| Estimated sensitivity at fixed specificity | | | |
|---|---|---|---|
| Specificity | Sensitivity | 95% CI | Criterion |
| 90.00 | 69.17 | 55.83 to 78.33 | >0.677092674 |

| Criterion | Sensitivity | 95% CI | Specificity | 95% CI | +LR | 95% CI | -LR | 95% CI |
|---|---|---|---|---|---|---|---|---|
| >0.692915478 | 67.50 | 58.3 - 75.8 | 92.50 | 84.4 - 97.2 | 9.00 | 4.1 - 19.6 | 0.35 | 0.3 - 0.5 |

**Figure 3.** ROC analysis—the combination of 3 basic clinical characteristics (maternal age and BMI values at early stages of gestation and an infertility treatment by assisted reproductive technology) and 11 dysregulated microRNA biomarkers (miR-1-3p, miR-20a-5p, miR-20b-5p, miR-23a-3p, miR-100-5p, miR-125b-5p, miR-126-3p, miR-181a-5p, miR-195-5p, miR-499a-5p, and miR-574-3p). At a 10.0% FPR, 69.17% of pregnancies destinated to develop GDM were identified during the first trimester of gestation. This represents 84 out of 121 pregnancies correctly predicted to develop GDM and 8 out of 80 normal pregnancies predicted false positively to develop GDM.

The screening based on the combination of seven clinical characteristics (maternal age and BMI at early stages of gestation, an infertility treatment by assisted reproductive technology, history of miscarriage, the presence of trombophilic gene mutations, positive first-trimester screening for preeclampsia and/or FGR by FMF algorithm, and family history of diabetes mellitus in first-degree relatives) and 11 dysregulated microRNA biomarkers (miR-1-3p, miR-20a-5p, miR-20b-5p, miR-23a-3p, miR-100-5p, miR-125b-5p, miR-126-3p, miR-181a-5p, miR-195-5p, miR-499a-5p, and miR-574-3p) showed the highest possible accuracy for the early identification of pregnancies destinated to develop GDM (AUC 0.869, $p < 0.001$, 72.50% sensitivity, 90.0% specificity, cut off >0.6572). This combined screening identified, in the early stages of gestation, 72.50% of pregnancies destinated to develop GDM at a 10.0% FPR (Figure 4).

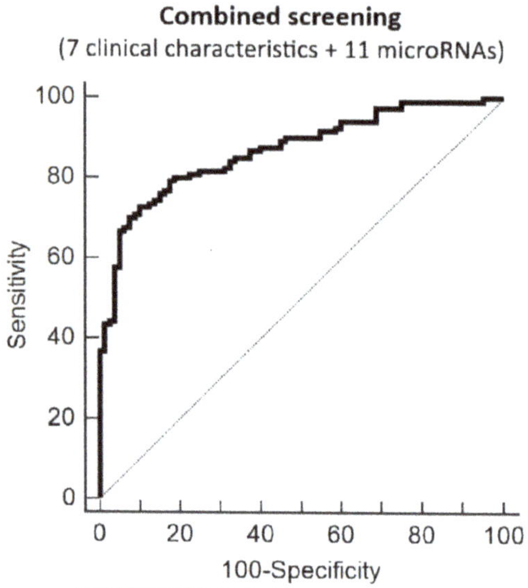


| | |
|---|---|
| Area under the ROC curve (AUC) | 0.869 |
| Standard Error | 0.0246 |
| 95% Confidence interval | 0.815 to 0.913 |
| Significance level P (Area=0.5) | <0.0001 |

| Estimated sensitivity at fixed specificity | | | |
|---|---|---|---|
| Specificity | Sensitivity | 95% CI | Criterion |
| 90.00 | 72.50 | 60.83 to 83.33 | >0.65721512 |

| Criterion | Sensitivity | 95% CI | Specificity | 95% CI | +LR | 95% CI | -LR | 95% CI |
|---|---|---|---|---|---|---|---|---|
| >0.65721512 | 72.50 | 63.6 - 80.3 | 90.00 | 81.2 - 95.6 | 7.25 | 3.7 - 14.1 | 0.31 | 0.2 - 0.4 |

**Figure 4.** ROC analysis—the combination of 7 clinical characteristics (maternal age and BMI at early stages of gestation, an infertility treatment by assisted reproductive technology, history of miscarriage, the presence of trombophilic gene mutations, positive first-trimester screening for PE and/or FGR by FMF algorithm, and family history of diabetes mellitus in first-degree relatives) and 11 dysregulated microRNA biomarkers (miR-1-3p, miR-20a-5p, miR-20b-5p, miR-23a-3p, miR-100-5p, miR-125b-5p, miR-126-3p, miR-181a-5p, miR-195-5p, miR-499a-5p, and miR-574-3p). At a 10.0% FPR, 72.50% of pregnancies destinated to develop GDM were identified during the first trimester of gestation. This represents 88 out of 121 pregnancies correctly predicted to develop GDM and 8 out of 80 normal pregnancies predicted false positively to develop GDM.

### *2.5. Dysregulation of Cardiovascular Disease-Associated MicroRNAs in Pregnancies Destinated to Develop GDM with Respect to the Treatment Strategies (Diet Only and a Combination of Diet and Administration of Appropriate Therapy)*

Concurrently, upregulation of miR-20a-5p ($p = 0.0015$ **, $p = 0.0098$ *), miR-20b-5p ($p < 0.001$ ***, $p = 0.0054$ **), and miR-195-5p ($p < 0.001$ ***, $p < 0.001$ ***) was observed in both groups of pregnancies destinated to develop GDM, irrespective of the treatment strategies (diet only or a combination of diet and therapy).

In addition, upregulation of miR-1-3p ($p = 0.0045$ *), miR-100-5p ($p = 0.0010$ **), miR-125b-5p ($p = 0.0109$ *), miR-499-5p ($p = 0.0043$ *), and miR-574-3p ($p < 0.001$ ***) was observed in only the group of pregnancies destinated to develop GDM, which was managed well by diet only (Supplementary Figure S2, Table 4).

Sensitivities at a 10.0% FPR were reported for miR-20a-5p (21.78%, 20.0%), miR-20b-5p (15.84%, 30.0%), and miR-195-5p (18.81%, 25.0%) in pregnancies destinated to develop GDM requiring management by diet only or a combination of diet and administration of appropriate therapy.

Sensitivities at a 10.0% FPR were reported for miR-1-3p (13.86%), miR-100-5p (19.80%), miR-125b-5p (14.85%), miR-499a-5p (15.84%), and miR-574-3p (21.78%) in pregnancies destinated to develop GDM requiring diet only (Supplementary Figure S2).

**Table 4.** MicroRNA expression profiles in peripheral blood leukocytes in early stages of gestation in pregnancies destinated to develop GDM with respect to the treatment strategies and normal term pregnancies.

| | **Kruskal–Wallis Test Results**<br>**GDM Managed by Diet Only (*n* = 101) vs. Normal Term Pregnancies (*n* = 80)**<br>**GDM Managed by Diet and Therapy (*n* = 20) vs. Normal Term Pregnancies (*n* = 80)** | | |
|---|---|---|---|
| | **Median (IQR)** | **Mean (SD)** | ***p*-Value** |
| miR-1-3p | 0.141 (0.075–0.274) vs. 0.075 (0.033–0.198)<br>0.099 (0.071–0.175) vs. 0.075 (0.033–0.198) | 0.278 (0.568) vs. 0.176 (0.303)<br>0.162 (0.190) vs. 0.176 (0.303) | **$p$ = 0.0045** *<br>$p$ = 1.000 |
| miR-16-5p | 1.216 (0.981–1. 785) vs. 1.411 (0.890–1.980)<br>1.268 (0.923–2.007) vs. 1.411 (0.890–1.980) | 1.469 (0.976) vs. 1.646 (1.129)<br>1.626 (1.019) vs. 1.646 (1.129) | $p$ = 1.000<br>$p$ = 1.000 |
| miR-17-5p | 1.480 (1.166–2.267) vs. 1.384 (0.971–1.923)<br>1.893 (1.346–2.362) vs. 1.384 (0.971–1.923) | 1.950 (1.553) vs. 1.748 (1.312)<br>2.085 (0.996) vs. 1.748 (1.312) | $p$ = 0.3822<br>$p$ = 0.1019 |
| miR-20a-5p | 2.144 (1.486–3.398) vs. 1.576 (0.991–2.413)<br>2.598 (1.787–3.384) vs. 1.576 (0.991–2.413) | 3.019 (3.220) vs. 1.909 (1.370)<br>3.130 (2.204) vs. 1.909 (1.370) | **$p$ = 0.0015** **<br>**$p$ = 0.0098** * |
| miR-20b-5p | 2.577 (1.784–3.719) vs. 1.976 (1.111–2.675)<br>3.072 (2.085–5.484) vs. 1.976 (1.111–2.675) | 3.678 (4.112) vs. 2.377 (2.291)<br>3.850 (2.439) vs. 2.377 (2.291) | **$p$ < 0.001** ***<br>**$p$ = 0.0054** ** |
| miR-21-5p | 0.339 (0.222–0.460) vs. 0.320 (0.167–0.538)<br>0.352 (0.260–0.464) vs. 0.320 (0.167–0.538) | 0.426 (0.436) vs. 0.394 (0.219)<br>0.472 (0.332) vs. 0.394 (0.219) | $p$ = 1.000<br>$p$ = 0.4483 |
| miR-23a-3p | 0.229 (0.160–0.444) vs. 0.185 (0.103–0.376)<br>0.299 (0.219–0.344) vs. 0.185 (0.103–0.376) | 0.364 (0.346) vs. 0.296 (0.329)<br>0.383 (0.293) vs. 0.296 (0.329) | $p$ = 0.0627<br>$p$ = 0.0371 |
| miR-24-3p | 0.292 (0.222–0.370) vs. 0.326 (0.196–0.468)<br>0.301 (0.241–0.377) vs. 0.326 (0.196–0.468) | 0.330 (0.206) vs. 0.384 (0.284)<br>0.339 (0.147) vs. 0.384 (0.284) | $p$ = 1.000<br>$p$ = 1.000 |
| miR-26a-5p | 0.729 (0.497–0.938) vs. 0.633 (0.410–1.066)<br>0.658 (0.560–0.917) vs. 0.633 (0.410–1.066) | 0.841 (0.705) vs. 0.776 (0.521)<br>0.815 (0.462) vs. 0.776 (0.521) | $p$ = 0.9599<br>$p$ = 1.000 |
| miR-29a-3p | 0.404 (0.276–0.571) vs. 0.372 (0.221–0.545)<br>0.435 (0.358–0.666) vs. 0.372 (0.221–0.545) | 0.486 (0.377) vs. 0.407 (0.245)<br>0.630 (0.471) vs. 0.407 (0.245) | $p$ = 0.5656<br>$p$ = 0.1198 |
| miR-92a-3p | 2.171 (1.604–3.036) vs. 2.327 (1.188–3.743)<br>2.258 (1.603–3.681) vs. 2.327 (1.188–3.743) | 2.647 (2.217) vs. 2.807 (2.132)<br>2.979 (2.3086) vs. 2.807 (2.132) | $p$ = 1.000<br>$p$ = 1.000 |
| miR-100-5p | 0.0024 (0.0013–0.0036) vs. 0.0013 (0.0006–0.0027)<br>0.0014 (0.0012–0.0037) vs. 0.0013 (0.0006–0.0027) | 0.0031 (0.0041) vs. 0.0018 (0.0016)<br>0.0028 (0.0025) vs. 0.0018 (0.0016) | **$p$ = 0.0010** **<br>$p$ = 0.2898 |
| miR-103a-3p | 1.531 (0.949–2.533) vs. 1.203 (0.815–2.425)<br>1.618 (1.234–2.554) vs. 1.203 (0.815–2.425) | 2.085 (2.294) vs. 1.770 (1.466)<br>2.304 (2.075) vs. 1.770 (1.466) | $p$ = 0.7368<br>$p$ = 0.4354 |
| miR-125b-5p | 0.0041 (0.0026–0.0057) vs. 0.0030 (0.0016–0.0054)<br>0.0038 (0.0021–0.0055) vs. 0.0030 (0.0016–0.0054) | 0.0050 (0.0048) vs. 0.0036 (0.0027)<br>0.0045 (0.0029) vs. 0.0036 (0.0027) | **$p$ = 0.0109** *<br>$p$ = 0.4855 |
| miR-126-3p | 0.332 (0.219–0.500) vs. 0.272 (0.140–0.432)<br>0.324 (0.280–0.546) vs. 0.272 (0.140–0.432) | 0.470 (0.595) vs. 0.336 (0.270)<br>0.418 (0.228) vs. 0.336 (0.270) | $p$ = 0.0842<br>$p$ = 0.1516 |
| miR-130b-3p | 0.707 (0.453–1.315) vs. 0.702 (0.407–1.157)<br>1.087 (0.577–1.481) vs. 0.702 (0.407–1.157) | 1.051 (0.995) vs. 1.163 (2.425)<br>1.194 (0.769) vs. 1.163 (2.425) | $p$ = 1.000<br>$p$ = 0.1983 |
| miR-133a-3p | 0.118 (0.066–0.228) vs. 0.110 (0.550–0.233)<br>0.071 (0.055–0.105) vs. 0.110 (0.550–0.233) | 0.209 (0.283) vs. 0.232 (0.483)<br>0.113 (0.109) vs. 0.232 (0.483) | $p$ = 1.000<br>$p$ = 0.4015 |
| miR-143-3p | 0.048 (0.029–0.087) vs. 0.038 (0.016–0.089)<br>0.049 (0.033–0.090) vs. 0.038 (0.016–0.089) | 0.072 (0.088) vs. 0.058 (0.057)<br>0.078 (0.077) vs. 0.058 (0.057) | $p$ = 0.1327<br>$p$ = 0.2766 |
| miR-145-5p | 0.176 (0.122–0.235) vs. 0.161 (0.980–0.243)<br>0.171 (0.131–0.242) vs. 0.161 (0.980–0.243) | 0.210 (0.162) vs. 0.195 (0.143)<br>0.200 (0.100) vs. 0.195 (0.143) | $p$ = 0.6997<br>$p$ = 1.000 |
| miR-146a-5p | 1.116 (0.800–1.798) vs. 1.225 (0.578–1.765)<br>1.451 (1.167–2.129) vs. 1.225 (0.578–1.765) | 1.634 (1.621) vs. 1.388 (1.096)<br>1.780 (1.068) vs. 1.388 (1.096) | $p$ = 0.8676<br>$p$ = 0.1619 |
| miR-155-5p | 0.624 (0.432–0.820) vs. 0.607 (0.361–1.614)<br>0.566 (0.448–0.695) vs. 0.607 (0.361–1.614) | 0.701 (0.516) vs. 1.247 (1.439)<br>0.710 (0.573) vs. 1.247 (1.439) | $p$ = 1.000<br>$p$ = 1.000 |

**Table 4.** *Cont.*

| | Kruskal–Wallis Test Results<br>GDM Managed by Diet Only ($n$ = 101) vs. Normal Term Pregnancies ($n$ = 80)<br>GDM Managed by Diet and Therapy ($n$ = 20) vs. Normal Term Pregnancies ($n$ = 80) | | |
|---|---|---|---|
| | **Median (IQR)** | **Mean (SD)** | ***p*-Value** |
| miR-181a-5p | 0.246 (0.175–0.375) vs. 0.181 (0.141–0.330)<br>0.260 (0.190–0.393) vs. 0.181 (0.141–0.330) | 0.331 (0.336) vs. 0.246 (0.184)<br>0.326 (0.208) vs. 0.246 (0.184) | $p = 0.0399$<br>$p = 0.1367$ |
| miR-195-5p | 0.269 (0.154–0.487) vs. 0.106 (0.048–0.271)<br>0.246 (0.210–0.522) vs. 0.106 (0.048–0.271) | 0.460 (0.707) vs. 0.227 (0.364)<br>0.520 (0.609) vs. 0.227 (0.364) | **$p < 0.001$** ***<br>**$p < 0.001$** *** |
| miR-199a-5p | 0.073 (0.033–0.139) vs. 0.058 (0.023–0.111)<br>0.088 (0.052–0.163) vs. 0.058 (0.023–0.111) | 0.134 (0.233) vs. 0.096 (0.131)<br>0.148 (0.165) vs. 0.096 (0.131) | $p = 0.1575$<br>$p = 0.1701$ |
| miR-210-3p | 0.102 (0.074–0.154) vs. 0.138 (0.075–0.224)<br>0.099 (0.075–0.155) vs. 0.138 (0.075–0.224) | 0.134 (0.109) vs. 0.186 (0.180)<br>0.131 (0.080) vs. 0.186 (0.180) | $p = 0.2982$<br>$p = 1.000$ |
| miR-221-3p | 0.644 (0.448–0.948) vs. 0.548 (0.293–0.906)<br>0.616 (0.459–1.032) vs. 0.548 (0.293–0.906) | 0.819 (0.776) vs. 0.693 (0.561)<br>0.796 (0.503) vs. 0.693 (0.561) | $p = 0.3698$<br>$p = 0.7241$ |
| miR-342-3p | 3.093 (2.070–3.955) vs. 2.542 (1.551–4.206)<br>2.884 (2.159–4.844) vs. 2.542 (1.551–4.206) | 3.555 (2.756) vs. 3.307 (2.383)<br>3.858 (2.610) vs. 3.307 (2.383) | $p = 0.6912$<br>$p = 1.000$ |
| miR-499a-5p | 0.459 (0.218–0.881) vs. 0.269 (0.089–0.587)<br>0.472 (0.285–0.611) vs. 0.269 (0.089–0.587) | 0.771 (1.104) vs. 0.477 (0.566)<br>0.692 (0.902) vs. 0.477 (0.566) | **$p = 0.0043$** *<br>$p = 0.1765$ |
| miR-574-3p | 0.275 (0.182–0.392) vs. 0.181 (0.117–0.292)<br>0.279 (0.178–0.485) vs. 0.181 (0.117–0.292) | 0.350 (0.339) vs. 0.222 (0.156)<br>0.375 (0.301) vs. 0.222 (0.156) | **$p < 0.001$** ***<br>$p = 0.0356$ |

MicroRNA gene expression is compared between individual groups using Kruskal–Wallis test. Statistically significant results are marked in bold. Median (interquartile range, IQR) and mean (standard deviation, SD) values of relative fold gene expression of samples ($2^{-\Delta\Delta Ct}$) are presented. Statistical significant data after Benjamini–Hochberg correction are marked by * for $\alpha = 0.05$, ** for $\alpha = 0.01$, and *** for $\alpha = 0.001$.

*2.6. First-Trimester Combined MicroRNA Screening Is Able to Differentiate between Pregnancies Destinated to Develop GDM Requiring a Combination of Diet and Administration of Appropriate Therapy and Term Pregnancies with Normal Course of Gestation*

The combined screening of three microRNA biomarkers (miR-20a-5p, miR-20b-5p and miR-195-5p) in early stages of gestation was able to detect aberrant microRNA expression profile in 30.0% pregnancies destinated to develop GDM requiring a combination of diet and administration of appropriate therapy at a 10.0% FPR (AUC 0.731, $p < 0.001$, 65.0% sensitivity, 73.75% specificity, cut off >0.1987) (Figure 5).

*2.7. The Very High Accuracy of First-Trimester Combined Screening (MicroRNA Biomarkers and Selected Clinical Characteristics) to Differentiate between Pregnancies Destinated to Develop GDM Requiring a Combination of Diet and Administration of Appropriate Therapy and Term Pregnancies with Normal Course of Gestation*

The effective screening based on the combination of minimal number of basic clinical characteristics (maternal age and BMI at early stages of gestation, and an infertility treatment by assisted reproductive technology) and three dysregulated microRNA biomarkers (miR-20a-5p, miR-20b-5p, and miR-195-5p) showed very high accuracy for the early identification of pregnancies destinated to develop GDM requiring a combination of diet and administration of appropriate therapy (AUC 0.949, $p < 0.001$, 89.47% sensitivity, 86.25% specificity, cut off >0.1912). The screening identified 78.95% of cases at a 10.0% FPR in the early stages of gestation (Figure 6).

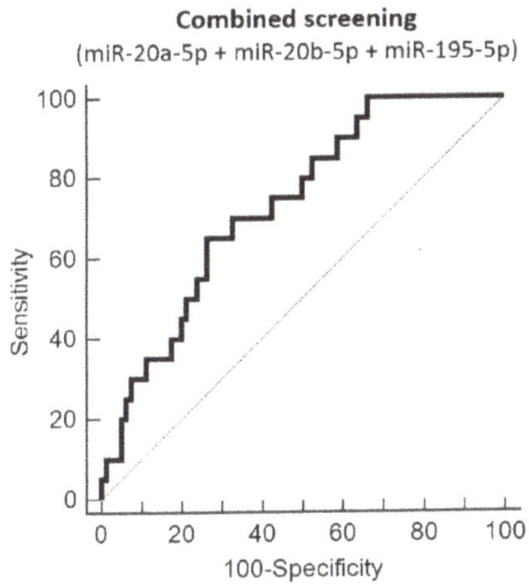


| | |
|---|---|
| Area under the ROC curve (AUC) | 0.731 |
| Standard Error | 0.0578 |
| 95% Confidence interval | 0.633 to 0.815 |
| Significance level P (Area=0.5) | 0.0001 |

| Estimated sensitivity at fixed specificity | | | |
|---|---|---|---|
| Specificity | Sensitivity | 95% CI | Criterion |
| 90.00 | 30.00 | 10.00 to 55.00 | >0.251230974 |

| Criterion | Sensitivity | 95% CI | Specificity | 95% CI | +LR | 95% CI | -LR | 95% CI |
|---|---|---|---|---|---|---|---|---|
| >0.198733282 | 65.00 | 40.8 - 84.6 | 73.75 | 62.7 - 83.0 | 2.48 | 1.5 - 4.0 | 0.47 | 0.3 - 0.9 |

**Figure 5.** ROC analysis—the combination of microRNA biomarkers (miR-20a-5p, miR-20b-5p and miR-195-5p). A total of 30.0% pregnancies destinated to develop GDM requiring a combination of diet and administration of appropriate therapy had aberrant microRNA expression profile in the whole peripheral venous blood during the first trimester of gestation at a 10.0% FPR. This represents 6 out of 20 pregnancies correctly predicted to develop GDM and 8 out of 80 normal pregnancies predicted false positively to develop GDM.

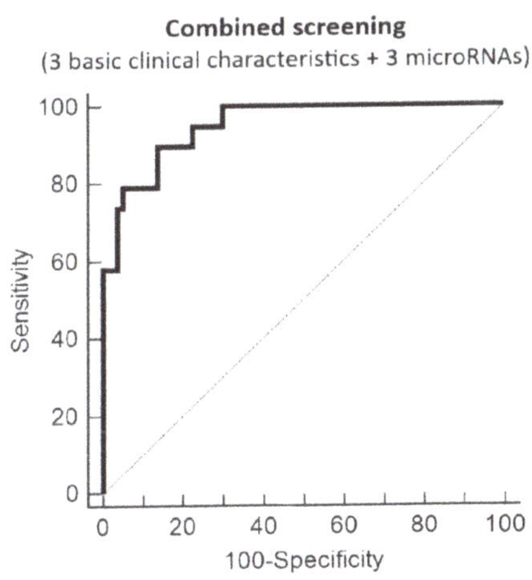


| | |
|---|---|
| Area under the ROC curve (AUC) | 0.949 |
| Standard Error | 0.0229 |
| 95% Confidence interval | 0.886 to 0.983 |
| Significance level P (Area=0.5) | <0.0001 |

| Estimated sensitivity at fixed specificity | | | |
|---|---|---|---|
| Specificity | Sensitivity | 95% CI | Criterion |
| 90.00 | 78.95 | 50.78 to 94.74 | >0.243735973 |

| Criterion | Sensitivity | 95% CI | Specificity | 95% CI | +LR | 95% CI | -LR | 95% CI |
|---|---|---|---|---|---|---|---|---|
| >0.191259321 | 89.47 | 66.9 - 98.7 | 86.25 | 76.7 - 92.9 | 6.51 | 3.7 - 11.5 | 0.12 | 0.03 - 0.5 |

**Figure 6.** ROC analysis—the combination of 3 basic clinical characteristics (maternal age and BMI values at early stages of gestation and an infertility treatment by assisted reproductive technology) and 3 dysregulated microRNA biomarkers (miR-20a-5p, miR-20b-5p, and miR-195-5p). At a 10.0% FPR, 78.95% pregnancies destinated to develop GDM requiring a combination of diet and administration of appropriate therapy were identified during the first trimester of gestation. This represents 16 out of 20 pregnancies correctly predicted to develop GDM and 8 out of 80 normal pregnancies predicted false positively to develop GDM.

The screening based on the combination of seven clinical characteristics (maternal age and BMI at early stages of gestation, an infertility treatment by assisted reproductive technology, history of miscarriage, the presence of trombophilic gene mutations, positive first-trimester screening for preeclampsia and/or FGR by FMF algorithm, family history of diabetes mellitus in first-degree relatives) and three dysregulated microRNA biomarkers (miR-20a-5p, miR-20b-5p, and miR-195-5p) showed the highest possible accuracy for the early identification of pregnancies destinated to develop GDM requiring a combination of diet and administration of appropriate therapy (AUC 0.957, $p < 0.001$, 89.47% sensitivity, 90.0% specificity, cutoff >0.2116). This screen identified 89.47% of cases in the early stages of gestation at a 10.0% FPR (Figure 7).

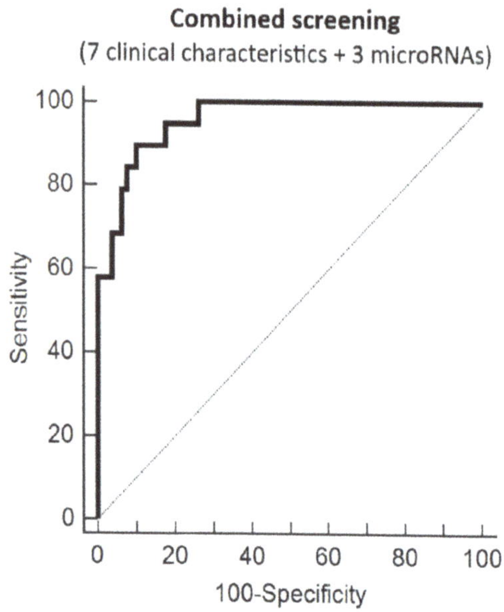


| | |
|---|---|
| Area under the ROC curve (AUC) | 0.957 |
| Standard Error | 0.0197 |
| 95% Confidence interval | 0.897 to 0.988 |
| Significance level P (Area=0.5) | <0.0001 |

| Estimated sensitivity at fixed specificity | | | |
|---|---|---|---|
| Specificity | Sensitivity | 95% CI | Criterion |
| 90.00 | 89.47 | 68.42 to 100.00 | >0.211657025 |

| Criterion | Sensitivity | 95% CI | Specificity | 95% CI | +LR | 95% CI | -LR | 95% CI |
|---|---|---|---|---|---|---|---|---|
| >0.211657025 | 89.47 | 66.9 - 98.7 | 90.00 | 81.2 - 95.6 | 8.95 | 4.6 - 17.6 | 0.12 | 0.03 - 0.4 |

**Figure 7.** ROC analysis—the combination of 7 clinical characteristics (maternal age and BMI at early stages of gestation, an infertility treatment by assisted reproductive technology, history of miscarriage, the presence of trombophilic gene mutations, positive first-trimester screening for PE and/or FGR by FMF algorithm, and a family history of diabetes mellitus in first-degree relatives) and 3 dysregulated microRNA biomarkers (miR-20a-5p, miR-20b-5p, and miR-195-5p). At a 10.0% FPR, 89.47% pregnancies destinated to develop GDM requiring a combination of diet and administration of appropriate therapy were identified during the first trimester of gestation. This represents 18 out of 20 pregnancies correctly predicted to develop GDM and 8 out of 80 normal pregnancies predicted false positively to develop GDM.

*2.8. First-Trimester Combined MicroRNA Screening Is Able to Differentiate between Pregnancies Destinated to Develop GDM Managed by Diet Only and Normal Term Pregnancies*

The combined screening of eight microRNA biomarkers (miR-1-3p, miR-20a-5p, miR-20b-5p, miR-100-5p, miR-125b-5p, miR-195-5p, miR-499a-5p, and miR-574-3p) was able to detect, in the early stages of gestation, an aberrant microRNA expression profile in 34.65% of pregnancies destinated to develop GDM managed by diet only at a 10.0% FPR (AUC 0.691, $p < 0.001$, 72.28% sensitivity, 60.0% specificity, cut off >0.4980) (Figure 8).

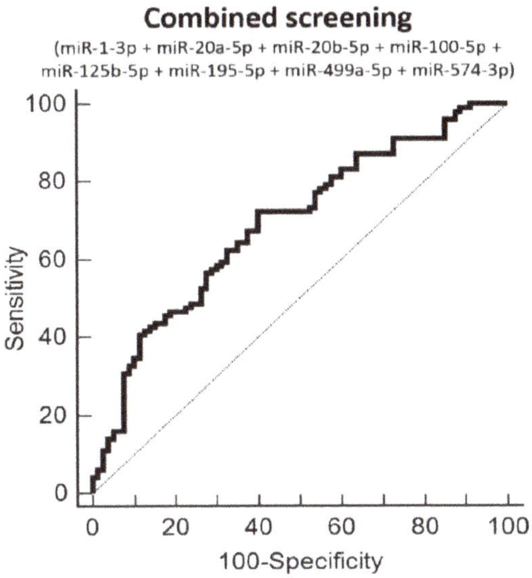


| Area under the ROC curve (AUC) | 0.691 |
|---|---|
| Standard Error | 0.0393 |
| 95% Confidence interval | 0.618 to 0.758 |
| Significance level P (Area=0.5) | <0.0001 |

| Estimated sensitivity at fixed specificity | | | |
|---|---|---|---|
| Specificity | Sensitivity | 95% CI | Criterion |
| 90.00 | 34.65 | 9.90 to 48.51 | >0.658940555 |

| Criterion | Sensitivity | 95% CI | Specificity | 95% CI | +LR | 95% CI | -LR | 95% CI |
|---|---|---|---|---|---|---|---|---|
| >0.498016311 | 72.28 | 62.5 - 80.7 | 60.00 | 48.4 - 70.8 | 1.81 | 1.3 - 2.4 | 0.46 | 0.3 - 0.7 |

**Figure 8.** ROC analysis—the combination of microRNA biomarkers (miR-1-3p, miR-20a-5p, miR-20b-5p, miR-100-5p, miR-125b-5p, miR-195-5p, miR-499a-5p, and miR-574-3p). A total of 34.65% pregnancies destinated to develop GDM on diet only had an aberrant microRNA expression profile in the whole peripheral venous blood during the first trimester of gestation at a 10.0% FPR. This represents 35 out of 101 pregnancies correctly predicted to develop GDM and 8 out of 80 normal pregnancies predicted false positively to develop GDM.

### *2.9. The Very Good Accuracy of First-Trimester Combined Screening (MicroRNA Biomarkers and Selected Clinical Characteristics) to Differentiate between Pregnancies Destinated to Develop GDM Managed by Diet Only and Term Pregnancies with Normal Course of Gestation*

The effective screening based on the combination of a minimal number of basic clinical characteristics (maternal age and BMI at early stages of gestation and an infertility treatment by assisted reproductive technology) and eight dysregulated microRNA biomarkers (miR-1-3p, miR-20a-5p, miR-20b-5p, miR-100-5p, miR-125b-5p miR-195-5p, miR-499a-5p, and miR-574-3p) showed relatively good accuracy for the early identification of pregnancies destinated to develop GDM managed by diet only (AUC 0.784, $p < 0.001$, 61.39 sensitivity, 87.50% specificity, cut off >0.6425). This screening identified 50.50% of cases during the early stages of gestation at a 10.0% FPR (Figure 9).

The screening based on the combination of seven clinical characteristics (maternal age and BMI at early stages of gestation, an infertility treatment by assisted reproductive technology, history of miscarriage, the presence of trombophilic gene mutations, positive first-trimester screening for preeclampsia and/or FGR by FMF algorithm, and family history of diabetes mellitus in first-degree relatives) and eight dysregulated microRNA biomarkers (miR-1-3p, miR-20a-5p, miR-20b-5p, miR-100-5p, miR-125b-5p, miR-195-5p, miR-499a-5p, and miR-574-3p) showed the highest possible accuracy for the early identification of pregnancies destinated to develop GDM managed by diet only (AUC 0.835, $p < 0.001$, 77.23% sensitivity, 78.75% specificity, cut off >0.5137. This combined screening identified, in the early stages of gestation, 56.44% of pregnancies destinated to develop GDM managed by diet only at a 10.0% FPR (Figure 10).

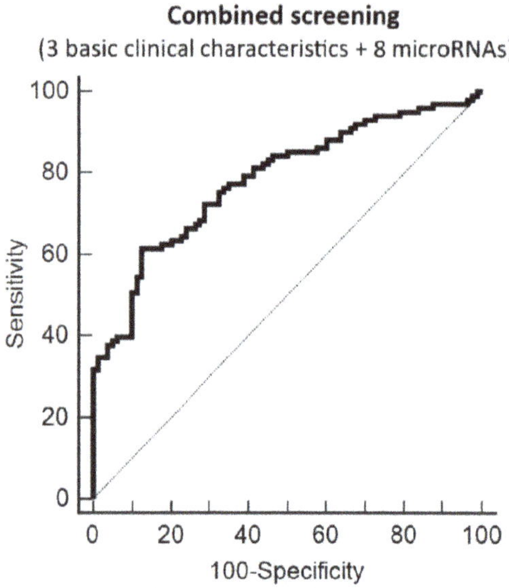


| | |
|---|---|
| Area under the ROC curve (AUC) | 0.784 |
| Standard Error | 0.0336 |
| 95% Confidence interval | 0.717 to 0.842 |
| Significance level P (Area=0.5) | <0.0001 |

| Estimated sensitivity at fixed specificity | | | |
|---|---|---|---|
| Specificity | Sensitivity | 95% CI | Criterion |
| 90.00 | 50.50 | 31.68 to 69.31 | >0.704912418 |

| Criterion | Sensitivity | 95% CI | Specificity | 95% CI | +LR | 95% CI | -LR | 95% CI |
|---|---|---|---|---|---|---|---|---|
| >0.642596346 | 61.39 | 51.2 - 70.9 | 87.50 | 78.2 - 93.8 | 4.91 | 2.7 - 8.9 | 0.44 | 0.3 - 0.6 |

**Figure 9.** ROC analysis—the combination of 3 basic clinical characteristics (maternal age and BMI values at early stages of gestation and an infertility treatment by assisted reproductive technology) and 8 dysregulated microRNA biomarkers (miR-1-3p, miR-20a-5p, miR-20b-5p, miR-100-5p, miR-125b-5p, miR-195-5p, miR-499a-5p, and miR-574-3p). At a 10.0% FPR, 50.50% pregnancies destinated to develop GDM managed by diet only were identified during the first trimester of gestation. This represents 51 out of 101 pregnancies correctly predicted to develop GDM and 8 out of 80 normal pregnancies predicted false positively to develop GDM.

### *2.10. Information on MicroRNA-Gene-Biological Pathways Interactions*

The KEGG pathway enrichment analysis of 11 microRNAs dysregulated in early stages of gestation in pregnancies destinated to develop GDM revealed a total of 62 pathways, where at least 18 (29.03%) pathways were cancer related. The cancer-related pathways with the highest $-\ln(p$-values) were proteoglycans in cancer (hsa05205; 34.738), viral carcinogenesis (hsa05203; 18.144), renal cell carcinoma (hsa05211; 12.364), glioma (hsa05214; 11.400), and pathways in cancer (hsa05200; 11.269).

Other cancer-related pathways showed slightly lower $-\ln(p$-values): transcriptional misregulation in cancer (hsa05202; 9.818), chronic myeloid leukaemia (hsa05220; 9.818), non-small cell lung cancer (hsa05223; 9.492), central carbon metabolism in cancer (hsa05230; 9.047), endometrial cancer (hsa05213; 8.698), colorectal cancer (hsa05210; 8.296), thyroid cancer (hsa05216; 7.630), bladder cancer (hsa05219; 7.099), pancreatic cancer (hsa05212; 6.996), acute myeloid leukaemia (hsa05221; 6.648), small cell lung cancer (hsa05222; 5.661), melanoma (hsa05218; 5.424), and choline metabolism in cancer (hsa05231; 4.536) (Figure 11).

The other pathways with the highest $-\ln(p$-values) have been shown to play a role in physiological processes and besides the pathogenesis of cancer. These are Hippo signalling pathway (hsa04390; 16.800), adherens junction (hsa04520; 14.198), signalling pathways regulating pluripotency of stem cells (hsa04550; 12.276), p53 signalling pathway (hsa04115; 12.276), and protein processing in endoplasmatic reticulum (hsa04141; 10.769) (Figure 12).

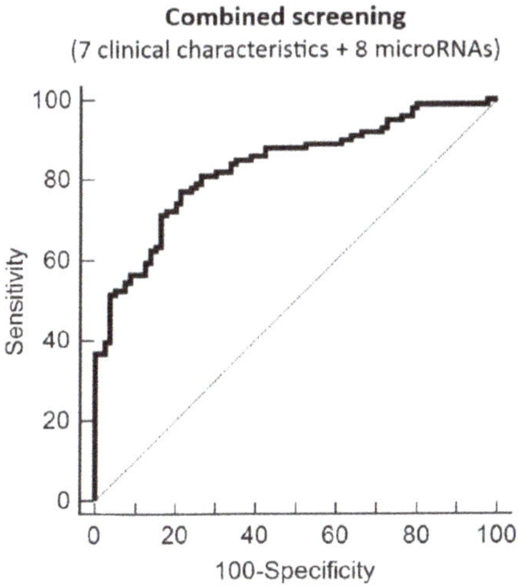


| Area under the ROC curve (AUC) | 0.835 |
|---|---|
| Standard Error | 0.0295 |
| 95% Confidence interval | 0.773 to 0.886 |
| Significance level P (Area=0.5) | <0.0001 |

Estimated sensitivity at fixed specificity

| Specificity | Sensitivity | 95% CI | Criterion |
|---|---|---|---|
| 90.00 | 56.44 | 43.56 to 73.22 | >0.679838192 |

| Criterion | Sensitivity | 95% CI | Specificity | 95% CI | +LR | 95% CI | -LR | 95% CI |
|---|---|---|---|---|---|---|---|---|
| >0.513780486 | 77.23 | 67.8 - 85.0 | 78.75 | 68.2 - 87.1 | 3.63 | 2.4 - 5.6 | 0.29 | 0.2 - 0.4 |

**Figure 10.** ROC analysis—the combination of 7 clinical characteristics (maternal age and BMI at early stages of gestation, an infertility treatment by assisted reproductive technology, history of miscarriage, the presence of trombophilic gene mutations, positive first-trimester screening for PE and/or FGR by FMF algorithm, and family history of diabetes mellitus in first-degree relatives) and 8 dysregulated microRNA biomarkers (miR-1-3p, miR-20a-5p, miR-20b-5p, miR-100-5p, miR-125b-5p, miR-195-5p, miR-499a-5p, and miR-574-3p). At a 10.0% FPR, 56.44% of pregnancies destinated to develop GDM managed by diet only were identified during the first trimester of gestation. This represents 57 out of 101 pregnancies correctly predicted to develop GDM and 8 out of 80 normal pregnancies predicted false positively to develop GDM.

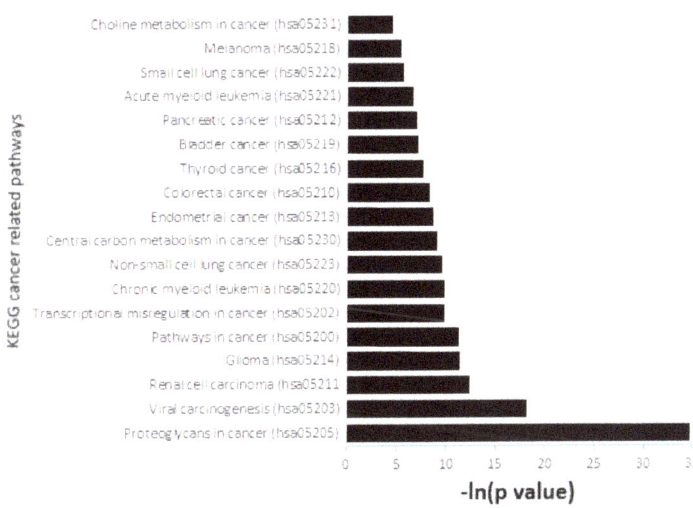


**Figure 11.** The KEGG pathway enrichment analysis of 11 microRNAs dysregulated in early pregnancies destinated to develop GDM. The analysis revealed a total of 62 pathways, where at least 18 (29.03%) pathways were cancer related. The results were expressed as –ln of the *p*-value (−ln(*p*-value)).

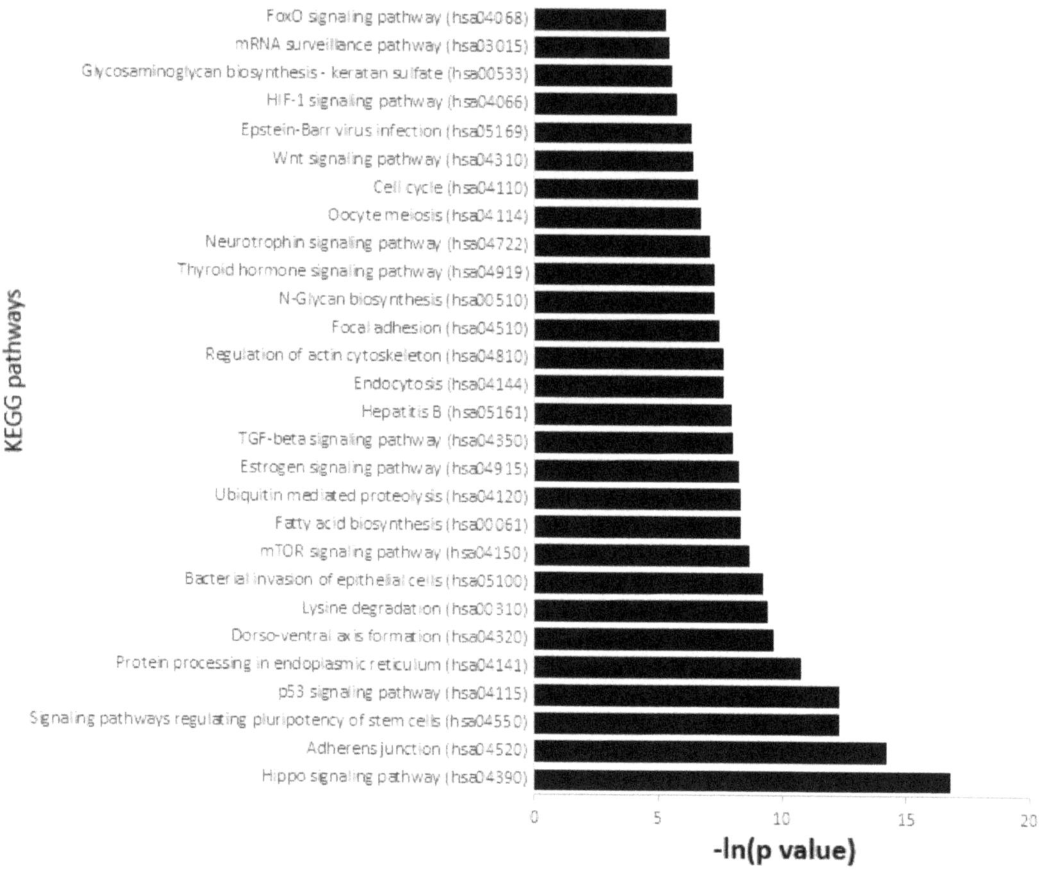


**Figure 12.** The KEGG pathway enrichment analysis of 11 microRNAs dysregulated in early pregnancies destinated to develop GDM. The analysis revealed a total of 62 various pathways, where a majority of pathways have been shown to play a role in physiological processes and besides to the pathogenesis of cancer. The results were expressed as –ln of the *p*-value (−ln(*p*-value)).

The microRNA/KEGG pathway heatmap and hierarchical clustering demonstrated the level of involvement of particular microRNAs in various biological pathways (Figure 13).

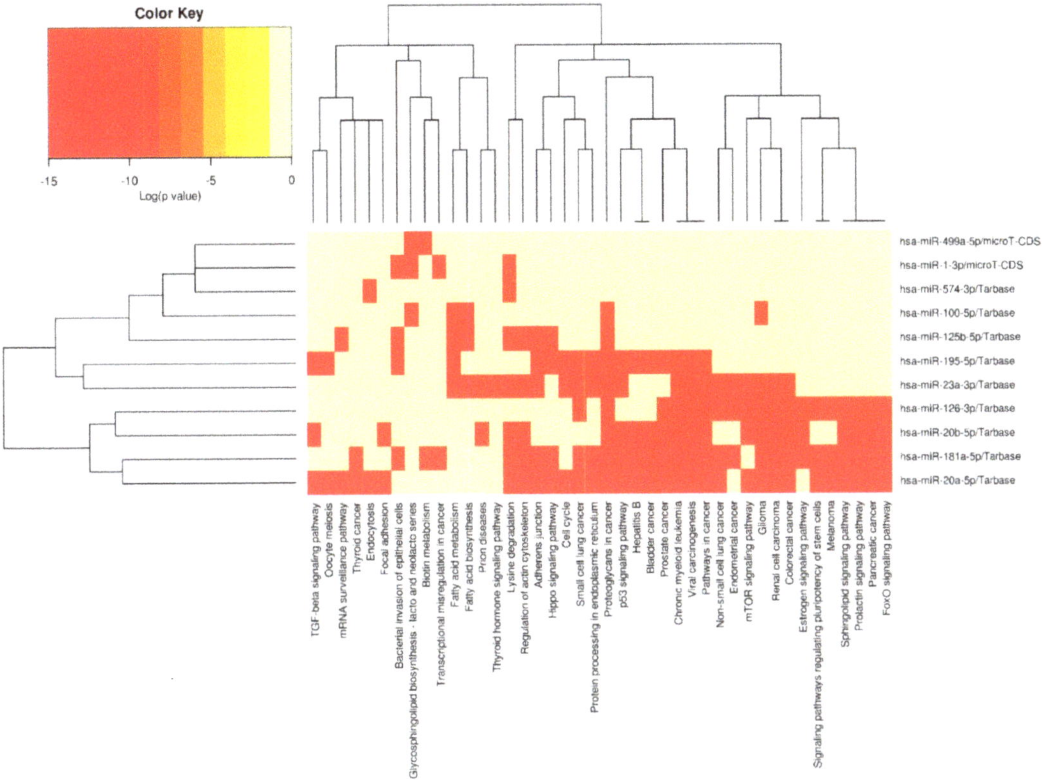


**Figure 13.** The microRNA/KEGG pathway heatmap and hierarchical clustering in early pregnancies destinated to develop GDM. The heatmap represents the level of involvement of particular microRNAs in various biological pathways. The results were expressed as log of the *p*-value (log(*p*-value)).

## 3. Discussion

Gene expression of 29 preselected cardiovascular disease-associated microRNAs was compared between pregnancies destinated to develop GDM and normal term pregnancies in the whole peripheral venous blood during the first trimester of gestation. The study was held within the framework of routine screening to assess the risk for a wide array of major fetal chromosomal and non-chromosomal defects as well as other pregnancy-related complications such as PE and/or FGR.

Upregulation of 11 cardiovascular disease-associated microRNAs (miR-1-3p, miR-20a-5p, miR-20b-5p, miR-23a-3p, miR-100-5p, miR-125b-5p, miR-126-3p, miR-181a-5p, miR-195-5p, miR-499a-5p, and miR-574-3p) was detected during the early stages of gestation in the entire group of pregnancies destinated to develop GDM.

To our knowledge, several studies have reported promising data on the early diagnosis of GDM during the first trimester of gestation via screening of circulating microRNAs in maternal plasma/serum or peripheral blood samples. Our study produced similar findings to Yoffe et al. [226], Lamadrid-Romero et al. [130], and Legare et al. [227].

Yoffe et al. validated two upregulated microRNAs (miR-23a and miR-223) as potential plasma biomarkers for early prediction of GDM (after the ninth gestational week and before completion of the 12th week of gestation) in women diagnosed with GDM via a 75 g OGTT performed at 22–24 weeks of gestation [226].

The study of Lamadrid-Romero et al. [130] reported higher miR-125b-5p expression levels in first-trimester serum samples in GDM pregnancies when compared with the control group. On the other hand, the study of Zhang et al. [229] reported downregulation of miR-125b in circulating plasma exosomes in patients with confirmed diagnosis of GDM within 26–40 weeks of pregnancy. Nevertheless, microRNA expression profile may differ between free circulating microRNAs and circulating exosomes; therefore, these findings are not necessarily contradictory results.

Our data may also support the data presented by Tagoma et al. [190], who observed upregulation of miR-100-5p and miR-195-5p in maternal plasma samples collected during the late second and early third pregnancy trimesters in patients who had a positive glucose tolerance test between 23 and 31 weeks of gestation, in which case miR-195-5p showed the highest fold upregulation, similar to our first-trimester study. Our data and the data of Tagoma et al. [190] are also consistent with the data of Wang et al. [230], who also observed increased expression levels of miR-195-5p in serum samples of GDM patients at 25 weeks of gestation.

Concerning miR-20a-5p, our first-trimester data may support the data of Zhu et al. [51] and Cao et al. [52]. Zhu et al. [51] observed upregulation of miR-20a-5p in peripheral blood samples of women at 16–19 weeks of pregnancy, whereas GDM was diagnosed via a 50 g glucose challenge test at 24–28 weeks of pregnancy. Cao et al. [52] observed upregulation of miR-20a-5p in plasma samples derived from patients at the time of diagnosis of GDM determined at 24–28 gestational weeks via performance of 50 g glucose challenge test and 75 g OGTT test.

Nevertheless, our data are inconsistent with the results of other researchers concerning miR-16-5p and miR-17-5p [51,52,228,231]. While in our study, first-trimester whole peripheral blood levels did not differ between pregnancies destinated to develop GDM and control groups, the expression levels of miR-16-5p and miR-17-5p have been reported to be significantly increased in patients with a diagnosis of GDM confirmed at 24–28 gestational weeks [52]. Similarly, Zhu et al. [51], Sorensen et al. [231], and Juchnicka et al. [228] presented similar findings to Cao et al. [52]. Zhu et al. [5] was able to observe upregulation of miR-16-5p and miR-17-5p in peripheral blood samples of women with subsequent onset of GDM at 16–19 weeks of pregnancy. Similarly, Sorensen et al. [231] observed elevated serum levels of miR-16-5p even in the earlier stages of gestation (mean 15th gestational week) in women destinated to develop GDM. Juchnicka et al. [228] showed upregulation of miR-16-5p in first-trimester serum samples of normoglycemic women that developed GDM within the 24–26 gestational weeks.

In addition, Zhao et al. [232] and Sorensen et al. [231] identified miR-29a and miR-29a-3p as other potentially predictive circulating GDM biomarkers. Unfortunately, they did not show any dysregulation when first-trimester expression levels were compared between pregnancies destinated to develop GDM and the control group in our study.

Parallelly, our data concerning miR-155-5p are inconsistent with the study of Wander et al. [112], who observed a positive association between early–mid-pregnancy plasma miR-155-5p levels and occurrence of GDM.

With regard to miR-1-3p, our study produced supportive findings to the study of Kennedy et al. [233], in which they reported increased levels of miR-1-3p in serum extracellular vesicles in patients with confirmed GDM diagnoses within 26–28 gestational weeks that subsequently delivered large-for-gestational-age new-borns (LGA) when compared with appropriately grown-for-gestational-age new-borns (AGA). Nevertheless, our data concerning miR-133a-3p and miR-145-5p are inconsistent with the study of Kennedy et al. [233]. While they observed reduced levels of miR-145-5p and increased levels of miR-133a-3p in GDM pregnancies delivering LGA new-borns, we did not detect any changes in the gene expression of miR-133a-3p and miR-145-5p during the early stages of gestation in pregnancies destinated to develop GDM.

Similarly, our data concerning miR-143-3p and miR-221-3p did not confirm the data of Legare et al. [227], that implemented these first-trimester dysregulated plasmatic microRNAs into the Lasso regression model for prediction of insulin sensitivity estimated by the Matsuda index at the end of the second trimester of pregnancy. However, our data concerning miR-100-5p concurred with Legare et al. [227], who also observed increased levels of miR-100-5p in plasma samples in the early stages of gestation in pregnancies that subsequently developed GDM.

In addition, other studies have introduced a whole range of other circulating microRNAs which were not subject of interest in our study as biomarkers with predictive or diagnostic potential for GDM. These are the following: let-7b-3p [227], miR-10b-5p [227], miR-16-1-3p [227] miR-19a and miR-19b [234], miR-21-3p [53,112], miR-33a-5p [235], miR-130a-3p [227], miR-132 [232], miR-134-5p [231], miR-141-3p [227], miR-142-3p [228], miR-144 [229], miR-144-3p [228], miR-200a-3p [227], miR-205-5p [227], miR-215-5p [227], miR-218-5p [227], miR-222 [232], miR-330-3p [236], miR-338-3p [227], miR-340 [237], miR-375 [227], miR-429 [227], miR-483-5p [227], miR-499a-3p [233], miR-503 [238], miR-512-3p [227], miR-515-5p [227], miR-516a-5p [227], miR-516b-5p [227], miR-517a-3p [227], miR-517b-3p [227], miR-518e-3p [227], miR-518e-5p [227], miR-519a-5p [227], miR-519b-5p [227], miR-519c-5p [227], miR-519d-5p [227], miR-520a-3p [227], miR-520d-3p [227], miR-522-5p [227], miR-523-5p [227], miR-524-3p [227], miR-582-5p [227], miR-873-5p [227], miR-877-5p [227], miR-1283 [227], miR-1323 [239], miR-2116-3p [227], miR-3183 [227], and miR-4772-5p [227].

The current study revealed that aberrant gene expression of miR-1-3p, miR-20a-5p, miR-20b-5p, miR-23a-3p, miR-100-5p, miR-125b-5p, miR-126-3p, miR-181a-5p, miR-195-5p, miR-499a-5p, and miR-574-3p expression is present during the early stages of gestation in pregnancies destinated to develop GDM.

During the first trimester of gestation, we have also recently observed an aberrant expression profile of these cardiovascular disease-associated microRNAs in pregnancies with chronic hypertension (miR-1-3p, miR-20a-5p, and miR-126-3p) and in normotensive pregnancies with subsequent onset of PE (miR-20a-5p, miR-126-3p, miR-181a-5p, and miR-574-3p), FGR (miR-20a-5p, miR-100-5p, miR-181a-5p, miR-195-5p, and miR-574-3p), SGA (miR-1-3p, miR-20a-5p, miR-20b-5p, miR-126-3p, miR-181a-5p, and miR-499a-5p), and/or preterm delivery (miR-20b-5p) [240–242].

Parallelly, not long ago we observed the upregulation of 11 microRNAs (miR-1-3p, miR-20a-5p, miR-20b-5p, miR-23a-3p, miR-100-5p, miR-125b-5p, miR-126-3p, miR-181a-5p, miR-195-5p, miR-499a-5p, and miR-574-3p) in the whole peripheral blood samples of mothers with a history of GDM [243]. At the same time, the upregulation of multiple other cardiovascular disease-associated microRNAs (miR-16-5p, miR-17-5p, miR-21-5p, miR-24-3p, miR-26a-5p, miR-29a-3p, miR-103a-3p, miR-130b-3p, miR-133a-3p, miR-143-3p, miR-145-5p, miR-146a-5p, miR-199a-5p, miR-221-3p, and miR-342-3p) was identified postpartum in mothers with a history of GDM [243], which had not yet been present in the early stages of gestation, and probably appeared later with the onset of GDM.

Existing data suggest that dysregulated microRNAs in early pregnancies destinated to develop GDM play a role, not only in the pathogenesis of cardiovascular and cerebrovascular diseases, but also in the pathogenesis of cancer. Since women with a history of GDM were reported to have a higher risk of developing both cardiovascular diseases [244–248] and cancer [249–256], cardiovascular risk assessment [243] together with cancer screening [249] should be implemented into the routine preventive programmes of women with a previous occurrence of GDM.

## 4. Materials and Methods

### 4.1. Patients Cohort

Within the framework of the retrospective case-control study held at the Institute for the Care of Mother and Child, Prague, Czech Republic, within the period 11/2012–5/2018, the whole peripheral venous blood samples were collected at 10–13 gestational weeks from a total of 4187 singleton pregnancies of Caucasian descent. Finally, 3028 out of 4187 pregnancies had complete medical records from the first trimester of gestation until the time of delivery. Out of these 3028 pregnancies, 121 women were consecutively confirmed to only have GDM, where 101 GDM pregnancies were managed by diet only and 20 GDM pregnancies were managed by the combination of diet and therapy (15 patients required insulin administration and metformin was prescribed for 5 patients). GDM was rarely diagnosed during the first trimester of gestation—only in four patients. Otherwise, the onset of GDM was confirmed in majority of patients ($n$ = 117) within 24–28 gestational weeks.

Gestational diabetes mellitus was defined as any degree of glucose intolerance with the first onset during gestation [2,3,257]. The International Association of Diabetes and Pregnancy Study Groups' (IADPSG) recommendations on the diagnosis and classification of hyperglycaemia in pregnancy were followed, and universal early testing was performed in all pregnancies [2]. The first screening phase, during the first trimester of gestation, detected patients with overt diabetes (fasting plasma glucose level $\geq$ 7.0 mmol/L) and patients with GDM (fasting plasma glucose level $\geq$ 5.1 mmol/L–<7.0 mmol/L). The second screening phase, 2 h 75 g OGTT at 24–28 weeks of gestation, was performed for all patients not previously found to have overt diabetes or GDM and identified GDM if fasting plasma glucose level was $\geq$5.1 mmol/L, 1 h plasma glucose was $\geq$10.0 mmol/L, or 2 h plasma glucose was $\geq$8.5 mmol/L [2].

Patients newly diagnosed with diabetes mellitus, patients with the occurrence of chronic hypertension, and those carrying growth-restricted or small-for-gestational-age fetuses, or fetuses with anomalies or chromosomal abnormalities were intentionally excluded from the study. Likewise, patients concurrently demonstrating other pregnancy-related complications such as gestational hypertension, preeclampsia, HELLP syndrome, in utero infections, spontaneous preterm birth, preterm prelabour rupture of membranes, fetal demise in utero, or stillbirth were also excluded from the study.

The control group was selected with regard to the uniformity of gestational age at sampling and storage times of biological samples, and included 80 women with normal courses of gestation that delivered healthy infants after the completion of 37 weeks of gestation with a weight above 2500 g.

No woman had a history of any cardiovascular disease (a positive anamnesis of cardiac remodelling, cardiac hypertrophy, heart failure, or acute myocardial infarction). All pregnant women had normal clinical findings (electrocardiography and echocardiography).

### 4.2. Processing of Samples

Homogenized leukocyte lysates were prepared from 200 µL maternal whole peripheral venous blood samples immediately after collection using a QIAamp RNA Blood Mini Kit (Qiagen, Hilden, Germany), according to the manufacturer's instructions. Firstly, lysis of erythrocytes was performed using EL buffer. Then, the pelleted leukocytes were stored in a mixture of RLT buffer and β-mercaptoethanol (β-ME) at −80 °C for several months until further processing.

Subsequently, a mirVana microRNA Isolation kit (Ambion, Austin, TX, USA) was used to isolate the RNA fraction highly enriched for small RNAs from whole peripheral blood leukocyte lysates.

Concentration and quality of RNA was assessed using a NanoDrop ND-1000 spectrophotometer (NanoDrop Technologies, Wilmington, DE, USA). The A(260/280) absorbance ratio of isolated RNA samples was 1.8–2.1, demonstrating that the RNA samples were pure and could be used for further analysis. The concentration of the isolated RNA ranged within 2.0–10.0 ng/µL.

Real-time RT-PCR analyses were performed regularly every six months to process the collection of frozen samples derived from GDM and normal term pregnancies. The gene expression levels of 29 cardiovascular disease-associated microRNAs (miR-1-3p, miR-16-5p, miR-17-5p, miR-20a-5p, miR-20b-5p, miR-21-5p, miR-23a-3p, miR-24-3p, miR-26a-5p, miR-29a-3p, miR-92a-3p, miR-100-5p, miR-103a-3p, miR-125b-5p, miR-126-3p, miR-130b-3p, miR-133a-3p, miR-143-3p, miR-145-5p, miR-146-5p, miR-155-5p, miR-181a-5p, miR-195-5p, miR-199a-5p, miR-210-3p, miR-221-3p, miR-342-3p, miR-499a-5p, and miR-574-3p) (Table 5) was determined.

**Table 5.** Characteristics of microRNAs involved in the study.

| Assay Name | ID | NCBI Location Chromosome | Sequence |
|---|---|---|---|
| hsa-miR-1 | hsa-miR-1-3p | Chr.20: 62554306–62554376 [+] | 5′-UGGAAUGUAAAGAAGUAUGUAU-3′ |
| hsa-miR-16 | hsa-miR-16-5p | Chr.13: 50048973–50049061 [−] | 5′-UAGCAGCACGUAAAUAUUGGCG- 3′ |
| hsa-miR-17 | hsa-miR-17-5p | Chr.13: 91350605–91350688 [+] | 5′-CAAAGUGCUUACAGUGCAGGUAG-3′ |
| hsa-miR-20a | hsa-miR-20a-5p | Chr.13: 91351065–91351135 [+] | 5′-UAAAGUGCUUAUAGUGCAGGUAG-3′ |
| hsa-miR-20b | hsa-miR-20b-5p | Chr.X: 134169809–134169877 [−] | 5′-CAAAGUGCUCAUAGUGCAGGUAG-3′ |
| hsa-miR-21 | hsa-miR-21-5p | Chr.17: 59841266–59841337 [+] | 5′-UAGCUUAUCAGACUGAUGUUGA-3′ |
| hsa-miR-23a | hsa-miR-23a-3p | Chr.19: 13836587–13836659 [−] | 5′-AUCACAUUGCCAGGGAUUUCC-3′ |
| hsa-miR-24 | hsa-miR-24-3p | Chr.9: 95086021–95086088 [+] | 5′-UGGCUCAGUUCAGCAGGAACAG-3′ |
| hsa-miR-26a | hsa-miR-26a-5p | Chr.3: 37969404–37969480 [+] | 5′-UUCAAGUAAUCCAGGAUAGGCU-3′ |
| hsa-miR-29a | hsa-miR-29a-3p | Chr.7: 130876747–130876810 [−] | 5′-UAGCACCAUCUGAAAUCGGUUA-3′ |
| hsa-miR-92a | hsa-miR-92a-3p | Chr.13: 91351314–91351391 [+] | 5′-UAUUGCACUUGUCCCGGCCUGU-3′ |
| hsa-miR-100 | hsa-miR-100-5p | Chr.11: 122152229–122152308 [−] | 5′-AACCCGUAGAUCCGAACUUGUG-3′ |
| hsa-miR-103 | hsa-miR-103a-3p | Chr.5: 168560896–168560973 [−] | 5′-AGCAGCAUUGUACAGGGCUAUGA-3′ |
| hsa-miR-125b | hsa-miR-125b-5p | Chr.11: 122099757–122099844 [−] | 5′-UCCCUGAGACCCUAACUUGUGA-3′ |
| hsa-miR-126 | hsa-miR-126-3p | Chr.9: 136670602–136670686 [+] | 5′-UCGUACCGUGAGUAAUAAUGCG-3′ |
| hsa-miR-130b | hsa-miR-130b-3p | Chr.22: 21653304–21653385 [+] | 5′-CAGUGCAAUGAUGAAAGGGCAU-3′ |
| hsa-miR-133a | hsa-miR-133a-3p | Chr.18: 21825698–21825785 [−] | 5′-UUUGGUCCCCUUCAACCAGCUG-3′ |
| hsa-miR-143 | hsa-miR-143-3p | Chr.5: 149428918–149429023 [+] | 5′-UGAGAUGAAGCACUGUAGCUC-3′ |
| hsa-miR-145 | hsa-miR-145-5p | Chr.5: 149430646–149430733 [+] | 5′-GUCCAGUUUUCCCAGGAAUCCCU-3′ |
| hsa-miR-146a | hsa-miR-146a-5p | Chr.5: 160485352–160485450 [+] | 5′-UGAGAACUGAAUUCCAUGGGUU-3′ |
| hsa-miR-155 | hsa-miR-155-5p | Chr.21: 25573980–25574044 [+] | 5′-UUAAUGCUAAUCGUGAUAGGGGU-3′ |
| hsa-miR-181a | hsa-miR-181a-5p | Chr.1: 198859044–198859153 [−] | 5′-AACAUUCAACGCUGUCGGUGAGU-3′ |
| hsa-miR-195 | hsa-miR-195-5p | Chr.17: 7017615–7017701 [−] | 5′-UAGCAGCACAGAAAUAUUGGC-3′ |
| hsa-miR-199a | hsa-miR-199a-5p | Chr.19: 10817426–10817496 [−] | 5′-CCCAGUGUUCAGACUACCUGUUC-3′ |
| hsa-miR-210 | hsa-miR-210-3p | Chr.11: 568089–568198 [−] | 5′-CUGUGCGUGUGACAGCGGCUGA-3′ |
| hsa-miR-221 | hsa-miR-221-3p | Chr.X: 45746157–45746266 [−] | 5′-AGCUACAUUGUCUGCUGGGUUUC-3′ |
| hsa-miR-342-3p | hsa-miR-342-3p | Chr.14: 100109655–100109753 [+] | 5′-UCUCACACAGAAAUCGCACCCGU-3′ |
| mmu-miR-499 | hsa-miR-499a-5p | Chr.20: 34990376–34990497 [+] | 5′-UUAAGACUUGCAGUGAUGUUU-3′ |
| hsa-miR-574-3p | hsa-miR-574-3p | Chr.4: 38868032–38868127 [+] | 5′-CACGCUCAUGCACACACCCACA-3′ |
| RNU58A | 664243 | Chr.18: 49491283–49491347 [−] | 5′-CTGCAGTGATGACTTTCTTGGGACACCTTTGGA TTTACCGTGAAAATTAATAAATTCTGAGCAGC-3′ |
| RNU38B | 568914 | Chr.1: 44778390–44778458 [+] | 5′-CCAGTTCTGCTACTGACAGTAAGTGAAGATAA AGTGTGTCTGAGGAGA-3′ |

mRNAs of the appropriate microRNAs were reverse transcribed into cDNA using a TaqMan MicroRNA assays containing miRNA-specific stem loop primers and a TaqMan MicroRNA Reverse Transcription Kit (Applied Biosystems, Branchburg, NJ, USA). The total reaction volumes were 10 µL. Furthermore, 3 µL of cDNA was mixed with the components of TaqMan MicroRNA assays (specific primers and the TaqMan MGB probes) and the components of the TaqMan Universal PCR Master Mix (Applied Biosystems, Branchburg, NJ, USA). The total reaction volumes were 15 µL. Reverse transcription and real-time qPCR were performed on a 7500 Real-Time PCR System using the TaqMan PCR conditions described in the TaqMan guidelines. The reverse transcription thermal cycling parameters were the following: 30 min at 16 °C, 30 min at 42 °C, 5 min at 85 °C, and then held at 4 °C.

The real-time qPCR thermal cycling parameters were the following: 2 min at 50 °C, 10 min at 95 °C, then 50 cycles at 95 °C for 15 s, and 60 °C for 1 min.

Assessment of microRNA gene expression was performed using the comparative Ct method [258]. The geometric mean of the Ct values of selected endogenous controls (RNU58A and RNU38B) was used as a normalization factor [259] to normalize microRNA gene expression. Selection and validation of endogenous controls for microRNA expression studies in whole peripheral blood samples affected by pregnancy-related complications has already been described in one of our previous studies [260]. In brief, the expression of 20 candidate endogenous controls (HY3, RNU6B, RNU19, RNU24, RNU38B, RNU43, RNU44, RNU48, RNU49, RNU58A, RNU58B, RNU66, RPL21, U6 snRNA, U18, U47, U54, U75, Z30, and cel-miR-39) was investigated using NormFinder (NormFinder v.5, Aarhus University Hospital, Aarhus, Denmark) [261]. RNU58A and RNU38B were identified as the most stable small nucleolar RNAs (ncRNA) and equally expressed in patients with normal and abnormal courses of gestation. Therefore, these ncRNA were selected as the most suitable endogenous controls for the normalization of microRNA qPCR expression studies performed on whole peripheral blood samples affected by pregnancy-related complications.

### 4.3. Statistical Analysis

Initially, power analysis was used to determine the sample size required to detect an effect of a given size with a given degree of confidence (G * Power Version 3.1.9.6, Franz Faul, University of Kiel, Kiel, Germany). A total of 51 cases and 51 controls were required to achieve a power of 0.805 and a total of 70 cases and 70 controls were required to achieve a power of 0.902.

With respect to non-normal data distribution, unpaired nonparametric tests were used for subsequent statistical analyses. Initially, microRNA gene expression was compared between GDM and normal term pregnancies using the Mann–Whitney test. Subsequently, microRNA gene expression was compared between particular groups with respect to the treatment strategies using the Kruskal–Wallis one-way analysis of variance. Afterwards, a post-hoc test for comparison between groups and the Benjamini–Hochberg correction were applied [262] (Tables 6 and 7).

**Table 6.** Benjamini–Hochberg correction: comparison of microRNA gene expression between GDM and normal term pregnancies.

| K | i | Alpha = 0.05 | Alpha = 0.01 | Alpha = 0.001 |
|---|---|---|---|---|
| **2** | | **0.05** | **0.01** | **0.001** |
| | 1 | 0.025 | 0.005 | 0.001 |

**Table 7.** Benjamini–Hochberg correction for multiple comparisons: comparison of microRNA gene expression between GDM and normal term pregnancies with respect to the treatment strategies (GDM pregnancies managed by diet only vs. GDM pregnancies managed by diet and therapy vs. normal term pregnancies).

| K | i | Alpha = 0.05 | Alpha = 0.01 | Alpha = 0.001 |
|---|---|---|---|---|
| **3** | | **0.05** | **0.01** | **0.001** |
| | 1 | 0.017 | 0.003 | 0.000 |
| | 2 | 0.033 | 0.007 | 0.001 |
| | 3 | 0.050 | 0.010 | 0.001 |

Boxplots display the median, the 75th and 25th percentiles (the upper and lower limits of the boxes), the maximum and minimum values that are no more than 1.5 times the span of the interquartile range (the upper and lower whiskers), outliers (circles), and extremes (asterisks). Statistica software (version 9.0; StatSoft, Inc., Tulsa, OK, USA) was used to produce the boxplots.

Receivers operating characteristic (ROC) curve analyses state the areas under the curves (AUC), $p$-values, the best cutoff point-related sensitivities, specificities, positive and negative likelihood ratios (LR+, LR−) together with the 95% CI (confidence interval). Furthermore, estimated specificities at fixed sensitivities and estimated sensitivities at fixed specificities are stated (MedCalc Software bvba, Ostend, Belgium). Sensitivities at a 90.0% specificity corresponding to a 10.0% false positive rate (FPR) were selected for data presentation. To select the optimal microRNA combinations, logistic regression with subsequent ROC curve analyses were applied (MedCalc Software bvba, Ostend, Belgium).

### *4.4. Information on MicroRNA-Gene-Biological Pathways Interactions*

The DIANA miRPath v.3 database (DIANA TOOLS-mirPath v.3 (uth.gr)) and genes union mode were used as an a priori analysis method to perform KEGG pathway enrichment analysis to investigate the regulatory mechanisms of the microRNAs dysregulated in the early stages of gestation in the whole peripheral blood of mothers destinated to develop GDM. The results of this enrichment analysis were expressed as –ln of the $p$-value (−ln($p$-value)). Preferentially, the database of experimentally verified microRNA targets (Tarbase v7.0) was used. In case that Tarbase v7.0 database did not provide a sufficient list of experimentally verified microRNA targets, the target prediction algorithm (microT-CDS v5.0) was used as an alternative.

In addition, the pathways/categories union mode, an a posteriori analysis method, was applied with the aim to identify merged $p$-values for each pathway significantly enriched with the gene targets of microRNAs dysregulated in early pregnancies destinated to develop GDM. Furthermore, the targeted pathway clusters/heatmap mode was applied to obtain the microRNA/KEGG pathway heatmap with hierarchical clustering.

## 5. Conclusions

Overall, we observed aberrant expression profiles of 11 microRNAs in the whole peripheral venous blood during the first trimester of gestation in pregnancies destinated to develop GDM. We confirmed the observations of other researchers that miR-23a-3p, miR-100-5p, and miR-125b-5p may serve as microRNA biomarkers with early predictive potential for GDM. In addition, novel microRNA biomarkers (miR-1-3p, miR-20a-5p, miR-20b-5p, miR-126-3p, miR-181a-5p, miR-195-5p, miR-499a-5p, and miR-574-3p) were identified, with the potential to predict GDM during the early stages of gestation.

Combined screening of all 11 dysregulated microRNA biomarkers (miR-1-3p, miR-20a-5p, miR-20b-5p, miR-23a-3p, miR-100-5p, miR-125b-5p, miR-126-3p, miR-181a-5p, miR-195-5p, miR-499a-5p, and miR-574-3p) showed the highest accuracy for the early identification of pregnancies destinated to develop GDM irrespective of the severity of the disease. This screening identified, in the early stages of gestation, 47.93% of pregnancies destinated to develop GDM at a 10.0% FPR.

The predictive model for GDM based on microRNA aberrant expression profile was further improved via the implementation of a minimal number of basic clinical characteristics (maternal age and BMI at early stages of gestation and an infertility treatment by assisted reproductive technology). Following this, 69.17% of pregnancies destinated to develop GDM were identified during the early stages of gestation at a 10.0% FPR.

The simplified prediction model for severe GDM (requiring management of diet and administration of appropriate therapy) using the combination of three basic clinical characteristics and three dysregulated microRNA biomarkers (miR-20a-5p, miR-20b-5p, and miR-195-5p) was able to identify 78.95% of cases at a 10.0% FPR during the early stages of gestation.

Parallelly, the simplified prediction model for GDM with a milder course (managed well by diet only) was more complex and required the involvement of three basic clinical characteristics and eight dysregulated microRNA biomarkers (miR-1-3p, miR-20a-5p, miR-

20b-5p, miR-100-5p, miR-125b-5p, miR-195-5p, miR-499a-5p, and miR-574-3p). Following this, the model was able to identify 50.50% of cases at a 10.0% FPR during the early stages of gestation.

The implementation of additional clinical variables into the final GDM predictive model is feasible; however, it depends on the availability of the clinical data, which differs between various health care providers.

The screening based on the combination of seven clinical characteristics (maternal age and BMI at early stages of gestation, an infertility treatment by assisted reproductive technology, history of miscarriage, the presence of trombophilic gene mutations, positive first-trimester screening for preeclampsia and/or FGR by FMF algorithm, and family history of diabetes mellitus in first-degree relatives) and microRNA biomarkers showed the highest possible accuracy for the early identification of pregnancies destinated to develop GDM either regardless or with regard to the severity of the disease. The screening was able to identify, in the early stages of gestation, 72.50% of GDM cases in total—89.47% of GDM cases requiring management by diet and administration of appropriate therapy and 56.44% GDM cases managed well by diet only—at a 10.0% FPR. Nevertheless, we prefer to leave the first-trimester GDM screening simplified as much as possible.

The implementation of a novel first-trimester GDM predictive model based on the combination of basic maternal clinical characteristics and aberrant microRNA expression profile into routine screening programmes may significantly improve the care of pregnancies at risk of the development of GDM. In pregnancies identified to be destinated to develop GDM, effective dietary counselling may be already provided during the early stages of gestation, and a healthy-eating plan naturally rich in nutrients and low in fat and calories may be developed to control blood glucose, manage weight, and control heart disease risk factors, such as a high blood pressure and high blood fats. This preventive measure may contribute to lowering the incidence of GDM overall and may also contribute to a reduction in the number of severe GDM cases that require the administration of an appropriate therapy. This may also contribute to a decrease in the occurrence of other pregnancy-related complications such as gestational hypertension, preeclampsia, and fetal growth restriction.

Since women with a history of GDM have an increased risk of developing diabetes (predominantly type 2 diabetes) and cardiovascular diseases later in life, the implementation of effective early screening programme for GDM alongside subsequent preventive measures into early prenatal care may contribute to a subsequent decrease in the occurrence of diabetes and cardiovascular diseases in young and middle-aged mothers. This would also have a large impact on the offspring descending from GDM-affected pregnancies. Accumulating data suggest that exposure to hyperglycemia in utero, as occurs in gestational diabetes mellitus, may expose the offspring to short-term and long-term adverse effects.

The cost of the implementation of the novel first-trimester GDM predictive model based on the combination of basic maternal clinical characteristics and aberrant microRNA expression profile into routine screening programmes is minimal when compared to the costs related to prenatal, peripartal, postpartal, neonatal, postnatal, and lifelong healthcare. In this manner, a significant reduction in healthcare cost can be achieved.

Large-scale follow-up studies need to be performed to verify diagnostic potential of cardiovascular disease-associated microRNA biomarkers to predict the subsequent occurrence of GDM.

Any changes to the epigenome, including the dysregulation of cardiovascular microRNAs induced during the early stages of gestation in pregnancies complicated by GDM, may predispose mothers to later development of diabetes mellitus and cardiovascular/cerebrovascular diseases. This hypothesis may also be supported by our previous finding that epigenetic changes (upregulation of serious cardiovascular microRNAs) appeared in a proportion of women with a history of GDM throughout postpartal life [243].

## 6. Patents

National patent application—Industrial Property Office, Czech Republic (Patent n. PV 2022-335).

**Supplementary Materials:** The following supporting information can be downloaded at: https://www.mdpi.com/article/10.3390/ijms231810635/s1.

**Author Contributions:** Conceptualization, I.H. and L.K.; methodology, I.H. and K.K.; software, I.H., K.K. and L.K.; validation, I.H., K.K. and L.K.; formal analysis, I.H., K.K.; investigation, K.K.; resources, I.H. and L.K.; data curation, I.H. and K.K.; writing—original draft preparation, I.H.; writing—review and editing, I.H. and K.K.; visualization, K.K.; supervision, I.H. and L.K.; project administration, I.H. and L.K.; funding acquisition, I.H. and L.K. All authors have read and agreed to the published version of the manuscript.

**Funding:** This work was supported by the Charles University research program Cooperatio—Mother and Childhood Care (no. 207035) and research grant SVV (no. 260529). All rights reserved.

**Institutional Review Board Statement:** The approval for the study was initially obtained from the Ethics Committee of the Third Faculty of Medicine, Charles University (Implication of placental specific microRNAs in maternal circulation for diagnosis and prediction of pregnancy-related complications, date of approval: 7 April 2011). Ongoing approvals for the study were obtained from the Ethics Committee of the Third Faculty of Medicine, Charles University (Long-term monitoring of complex cardiovascular profiles in mother, foetus, and offspring descending from pregnancy-related-complications; date of approval: 27 March 2014) and the Ethics Committee of the Institute for the Care of the Mother and Child, Charles University (Long-term monitoring of complex cardiovascular profiles in mother, foetus, and offspring descending from pregnancy-related-complications; date of approval: 28 May 2015; number of approval: 1/4/2015). This informed consent is very complex and involves consent for the collection of peripheral blood samples at the beginning of pregnancy. In addition, in case of the onset of pregnancy-related complications, it also involves consent for the collection of peripheral blood samples at the time of the onset of pregnancy-related complications and the collection of a piece of placenta sample during the childbirth. All procedures were in compliance with the Helsinki Declaration of 1975, as revised in 2000.

**Informed Consent Statement:** Informed consent for the study was obtained from patients during the first trimester of gestation when the collection of peripheral blood samples for the first-trimester screening was held. Informed consent was signed by all pregnant women involved in the study.

**Data Availability Statement:** The data presented in this study are available on request from the corresponding author. The data are not publicly available due to rights reserved by funding supporters.

**Acknowledgments:** We would like to thank the staff of the Institute for the Care of Mother and Child for assistance with the collection of the patients' biological samples.

**Conflicts of Interest:** The authors declare no conflict of interest.

## References

1. Committee on Practice Bulletins—Obstetrics. ACOG Practice Bulletin No. 190: Gestational Diabetes Mellitus. *Obstet. Gynecol.* **2018**, *131*, e49–e64. [CrossRef]
2. International Association of Diabetes and Pregnancy Study Groups Consensus Panel; Metzger, B.E.; Gabbe, S.G.; Persson, B.; Buchanan, T.A.; Catalano, P.A.; Damm, P.; Dyer, A.R.; Leiva, A.D.; Hod, M.; et al. International association of diabetes and pregnancy study groups recommendations on the diagnosis and classification of hyperglycemia in pregnancy. *Diabetes Care* **2010**, *33*, 676–682. [CrossRef]
3. American Diabetes Association. Diagnosis and classification of diabetes mellitus (Position Statement). *Diabetes Care* **2009**, *32*, S62–S67. [CrossRef]
4. England, L.J.; Dietz, P.M.; Njoroge, T.; Callaghan, W.M.; Bruce, C.; Buus, R.M.; Williamson, D.F. Preventing type 2 diabetes: Public health implications for women with a history of gestational diabetes mellitus. *Am. J. Obstet. Gynecol.* **2009**, *200*, 365.e1–365.e8. [CrossRef]
5. O'Sullivan, J.B. Body weight and subsequent diabetes mellitus. *JAMA* **1982**, *248*, 949–952. [CrossRef]
6. Kim, C.; Newton, K.M.; Knopp, R.H. Gestational diabetes and the incidence of type 2 diabetes: A systematic review. *Diabetes Care* **2002**, *25*, 1862–1868. [CrossRef]

7. International Diabetes Federation. IDF GDM Model of Care. Available online: https://www.idf.org/e-library/guidelines/77-idf-gdm-model-of-care-implementation-protocol-guidelines-for-healthcare-professionals.html (accessed on 16 May 2022).
8. National Institute for Health and Care Excellence (NICE). Diabetes in Pregnancy: Management from Preconception to the Postnatal Period. Available online: https://www.guidelines.co.uk/diabetes/nice-diabetes-in-pregnancy-guideline/252595.article (accessed on 16 May 2022).
9. Syngelaki, A.; Pastides, A.; Kotecha, R.; Wright, A.; Akolekar, R.; Nicolaides, K.H. First-Trimester Screening for Gestational Diabetes Mellitus Based on Maternal Characteristics and History. *Fetal Diagn. Ther.* **2015**, *38*, 14–21. [CrossRef]
10. Nanda, S.; Savvidou, M.; Syngelaki, A.; Akolekar, R.; Nicolaides, K.H. Prediction of gestational diabetes mellitus by maternal factors and biomarkers at 11 to 13 weeks. *Prenat. Diagn.* **2011**, *31*, 135–141. [CrossRef] [PubMed]
11. Sweeting, A.N.; Appelblom, H.; Ross, G.P.; Wong, J.; Kouru, H.; Williams, P.F.; Sairanen, M.; Hyett, J.A. First trimester prediction of gestational diabetes mellitus: A clinical model based on maternal demographic parameters. *Diabetes Res. Clin. Pract.* **2017**, *127*, 44–50. [CrossRef]
12. Sweeting, A.N.; Wong, J.; Appelblom, H.; Ross, G.P.; Kouru, H.; Williams, P.F.; Sairanen, M.; Hyett, J.A. A Novel Early Pregnancy Risk Prediction Model for Gestational Diabetes Mellitus. *Fetal Diagn. Ther.* **2019**, *45*, 76–84. [CrossRef]
13. Wang, P.; Ma, H.H.; Hou, X.Z.; Song, L.L.; Song, X.L.; Zhang, J.F. Reduced plasma level of irisin in first trimester as a risk factor for the development of gestational diabetes mellitus. *Diabetes Res. Clin. Pract.* **2018**, *142*, 130–138. [CrossRef]
14. Wu, Y.T.; Zhang, C.J.; Mol, B.W.; Kawai, A.; Li, C.; Chen, L.; Wang, Y.; Sheng, J.Z.; Fan, J.X.; Shi, Y.; et al. Early Prediction of Gestational Diabetes Mellitus in the Chinese Population via Advanced Machine Learning. *J. Clin. Endocrinol. Metab.* **2021**, *106*, e1191–e1205. [CrossRef]
15. Kumar, M.; Chen, L.; Tan, K.; Ang, L.T.; Ho, C.; Wong, G.; Soh, S.E.; Tan, K.H.; Chan, J.K.Y.; Godfrey, K.M.; et al. Population-centric risk prediction modeling for gestational diabetes mellitus: A machine learning approach. *Diabetes Res. Clin. Pract.* **2022**, *185*, 109237. [CrossRef]
16. Sakurai, K.; Eguchi, A.; Watanabe, M.; Yamamoto, M.; Ishikawa, K.; Mori, C. Exploration of predictive metabolic factors for gestational diabetes mellitus in Japanese women using metabolomic analysis. *J. Diabetes Investig.* **2019**, *10*, 513–520. [CrossRef]
17. Tenenbaum-Gavish, K.; Sharabi-Nov, A.; Binyamin, D.; Møller, H.J.; Danon, D.; Rothman, L.; Hadar, E.; Idelson, A.; Vogel, I.; Koren, O.; et al. First trimester biomarkers for prediction of gestational diabetes mellitus. *Placenta* **2020**, *101*, 80–89. [CrossRef]
18. Zheng, Y.; Hou, W.; Xiao, J.; Huang, H.; Quan, W.; Chen, Y. Application Value of Predictive Model Based on Maternal Coagulation Function and Glycolipid Metabolism Indicators in Early Diagnosis of Gestational Diabetes Mellitus. *Front. Public Health* **2022**, *10*, 850191. [CrossRef]
19. Xiong, Y.; Lin, L.; Chen, Y.; Salerno, S.; Li, Y.; Zeng, X.; Li, H. Prediction of gestational diabetes mellitus in the first 19 weeks of pregnancy using machine learning techniques. *J. Matern. Fetal Neonatal Med.* **2022**, *35*, 2457–2463. [CrossRef]
20. Zhang, Y.Z.; Zhou, L.; Tian, L.; Li, X.; Zhang, G.; Qin, J.Y.; Zhang, D.D.; Fang, H. A mid-pregnancy risk prediction model for gestational diabetes mellitus based on the maternal status in combination with ultrasound and serological findings. *Exp. Ther. Med.* **2020**, *20*, 293–300. [CrossRef]
21. Cremona, A.; O'Gorman, C.S.; Ismail, K.I.; Hayes, K.; Donnelly, A.E.; Hamilton, J.; Cotter, A. A risk-prediction model using parameters of maternal body composition to identify gestational diabetes mellitus in early pregnancy. *Clin. Nutr. ESPEN* **2021**, *45*, 312–321. [CrossRef]
22. Shaarbaf Eidgahi, E.; Nasiri, M.; Kariman, N.; Safavi Ardebili, N.; Salehi, M.; Kazemi, M.; Zayeri, F. Diagnostic accuracy of first and early second trimester multiple biomarkers for prediction of gestational diabetes mellitus: A multivariate longitudinal approach. *BMC Pregnancy Childbirth* **2022**, *22*, 13. [CrossRef]
23. Lappharat, S.; Rothmanee, P.; Jandee, K.; Suksai, M.; Liabsuetrakul, T. A model for predicting gestational diabetes mellitus in early pregnancy: A prospective study in Thailand. *Obstet. Gynecol. Sci.* **2022**, *65*, 156–165. [CrossRef]
24. Wang, X.; Zheng, X.; Yan, J.; Xu, R.; Xu, M.; Zheng, L.; Xu, L.; Lin, Z. The Clinical Values of Afamin, Triglyceride and PLR in Predicting Risk of Gestational Diabetes During Early Pregnancy. *Front. Endocrinol.* **2021**, *12*, 723650. [CrossRef]
25. Wang, J.; Lv, B.; Chen, X.; Pan, Y.; Chen, K.; Zhang, Y.; Li, Q.; Wei, L.; Liu, Y. An early model to predict the risk of gestational diabetes mellitus in the absence of blood examination indexes: Application in primary health care centres. *BMC Pregnancy Childbirth* **2021**, *21*, 814. [CrossRef]
26. Kang, M.; Zhang, H.; Zhang, J.; Huang, K.; Zhao, J.; Hu, J.; Lu, C.; Shao, J.; Weng, J.; Yang, Y.; et al. A Novel Nomogram for Predicting Gestational Diabetes Mellitus During Early Pregnancy. *Front. Endocrinol.* **2021**, *12*, 779210. [CrossRef]
27. Buerger, O.; Elger, T.; Varthaliti, A.; Syngelaki, A.; Wright, A.; Nicolaides, K.H. First-Trimester Screening for Gestational Diabetes Mellitus in Twin Pregnancies. *J. Clin. Med.* **2021**, *10*, 3814. [CrossRef]
28. Snyder, B.M.; Baer, R.J.; Oltman, S.P.; Robinson, J.G.; Breheny, P.J.; Saftlas, A.F.; Bao, W.; Greiner, A.L.; Carter, K.D.; Rand, L.; et al. Early pregnancy prediction of gestational diabetes mellitus risk using prenatal screening biomarkers in nulliparous women. *Diabetes Res. Clin. Pract.* **2020**, *163*, 108139. [CrossRef]
29. Zhang, X.; Zhao, X.; Huo, L.; Yuan, N.; Sun, J.; Du, J.; Nan, M.; Ji, L. Risk prediction model of gestational diabetes mellitus based on nomogram in a Chinese population cohort study. *Sci. Rep.* **2020**, *10*, 21223. [CrossRef]
30. van Hoorn, F.; Koster, M.; Naaktgeboren, C.A.; Groenendaal, F.; Kwee, A.; Lamain-de Ruiter, M.; Franx, A.; Bekker, M.N. Prognostic models versus single risk factor approach in first-trimester selective screening for gestational diabetes mellitus: A prospective population-based multicentre cohort study. *BJOG* **2021**, *128*, 645–654. [CrossRef]

31. Benhalima, K.; Van Crombrugge, P.; Moyson, C.; Verhaeghe, J.; Vandeginste, S.; Verlaenen, H.; Vercammen, C.; Maes, T.; Dufraimont, E.; De Block, C.; et al. Estimating the risk of gestational diabetes mellitus based on the 2013 WHO criteria: A prediction model based on clinical and biochemical variables in early pregnancy. *Acta Diabetol.* **2020**, *57*, 661–671. [CrossRef]
32. Meertens, L.; Smits, L.; van Kuijk, S.; Aardenburg, R.; van Dooren, I.; Langenveld, J.; Zwaan, I.M.; Spaanderman, M.; Scheepers, H. External validation and clinical usefulness of first-trimester prediction models for small- and large-for-gestational-age infants: A prospective cohort study. *BJOG* **2019**, *126*, 472–484. [CrossRef]
33. Pan, Y.; Hu, J.; Zhong, S. The joint prediction model of pBMI and eFBG in predicting gestational diabetes mellitus. *J. Int. Med. Res.* **2020**, *48*, 300060519889199. [CrossRef] [PubMed]
34. Zheng, T.; Ye, W.; Wang, X.; Li, X.; Zhang, J.; Little, J.; Zhou, L.; Zhang, L. A simple model to predict risk of gestational diabetes mellitus from 8 to 20 weeks of gestation in Chinese women. *BMC Pregnancy Childbirth* **2019**, *19*, 252. [CrossRef] [PubMed]
35. Lamain-de Ruiter, M.; Kwee, A.; Naaktgeboren, C.A.; de Groot, I.; Evers, I.M.; Groenendaal, F.; Hering, Y.R.; Huisjes, A.J.; Kirpestein, C.; Monincx, W.M.; et al. External validation of prognostic models to predict risk of gestational diabetes mellitus in one Dutch cohort: Prospective multicentre cohort study. *BMJ* **2016**, *354*, i4338. [CrossRef] [PubMed]
36. Ravnsborg, T.; Andersen, L.L.; Trabjerg, N.D.; Rasmussen, L.M.; Jensen, D.M.; Overgaard, M. First-trimester multimarker prediction of gestational diabetes mellitus using targeted mass spectrometry. *Diabetologia* **2016**, *59*, 970–979. [CrossRef]
37. Hassiakos, D.; Eleftheriades, M.; Papastefanou, I.; Lambrinoudaki, I.; Kappou, D.; Lavranos, D.; Akalestos, A.; Aravantinos, L.; Pervanidou, P.; Chrousos, G. Increased Maternal Serum Interleukin-6 Concentrations at 11 to 14 Weeks of Gestation in Low Risk Pregnancies Complicated with Gestational Diabetes Mellitus: Development of a Prediction Model. *Horm. Metab. Res.* **2016**, *48*, 35–41. [CrossRef]
38. Alptekin, H.; Çizmecioğlu, A.; Işık, H.; Cengiz, T.; Yildiz, M.; Iyisoy, M.S. Predicting gestational diabetes mellitus during the first trimester using anthropometric measurements and HOMA-IR. *J. Endocrinol. Investig.* **2016**, *39*, 577–583. [CrossRef]
39. Papastefanou, I.; Eleftheriades, M.; Kappou, D.; Lambrinoudaki, I.; Lavranos, D.; Pervanidou, P.; Sotiriadis, A.; Hassiakos, D.; Chrousos, G.P. Maternal serum osteocalcin at 11–14 weeks of gestation in gestational diabetes mellitus. *Eur. J. Clin. Investig.* **2015**, *45*, 1025–1031. [CrossRef]
40. Eleftheriades, M.; Papastefanou, I.; Lambrinoudaki, I.; Kappou, D.; Lavranos, D.; Akalestos, A.; Souka, A.P.; Pervanidou, P.; Hassiakos, D.; Chrousos, G.P. Elevated placental growth factor concentrations at 11–14 weeks of gestation to predict gestational diabetes mellitus. *Metabolism* **2014**, *63*, 1419–1425. [CrossRef]
41. Li, J.; Dong, X.; Wang, Z.; Wu, J. MicroRNA-1 in Cardiac Diseases and Cancers. *Korean J. Physiol. Pharmacol.* **2014**, *18*, 359–363. [CrossRef]
42. Li, Y.Q.; Zhang, M.F.; Wen, H.Y.; Hu, C.L.; Liu, R.; Wei, H.Y.; Ai, C.M.; Wang, G.; Liao, X.X.; Li, X. Comparing the diagnostic values of circulating microRNAs and cardiac troponin T in patients with acute myocardial infarction. *Clinics* **2013**, *68*, 75–80. [CrossRef]
43. Gasiulė, S.; Stankevičius, V.; Patamsytė, V.; Ražanskas, R.; Zukovas, G.; Kapustina, Z.; Zaliaduonytė, D.; Benetis, R.; Lesauskaitė, V.; Vilkaitis, G. Tissue-Specific miRNAs Regulate the Development of Thoracic Aortic Aneurysm: The Emerging Role of KLF4 Network. *J. Clin. Med.* **2019**, *8*, 1609. [CrossRef] [PubMed]
44. Gerlinger-Romero, F.; Yonamine, C.Y.; Junior, D.C.; Esteves, J.V.; Machado, U.F. Dysregulation between TRIM63/FBXO32 expression and soleus muscle wasting in diabetic rats: Potential role of miR-1-3p, -29a/b-3p, and -133a/b-3p. *Mol. Cell. Biochem.* **2017**, *427*, 187–199. [CrossRef] [PubMed]
45. Kokkinopoulou, I.; Maratou, E.; Mitrou, P.; Boutati, E.; Sideris, D.C.; Fragoulis, E.G.; Christodoulou, M.I. Decreased expression of microRNAs targeting type-2 diabetes susceptibility genes in peripheral blood of patients and predisposed individuals. *Endocrine* **2019**, *66*, 226–239. [CrossRef]
46. Hromadnikova, I.; Kotlabova, K.; Dvorakova, L.; Krofta, L. Evaluation of Vascular Endothelial Function in Young and Middle-Aged Women with Respect to a History of Pregnancy, Pregnancy-Related Complications, Classical Cardiovascular Risk Factors, and Epigenetics. *Int. J. Mol. Sci.* **2020**, *21*, 430. [CrossRef] [PubMed]
47. Wang, X.; Shang, Y.; Dai, S.; Wu, W.; Yi, F.; Cheng, L. MicroRNA-16-5p aggravates myocardial infarction injury by targeting expression of insulin receptor substrates 1 and mediating myocardial apoptosis and angiogenesis. *Curr. Neurovasc. Res.* **2019**. *Epub ahead of print*. [CrossRef]
48. O'Sullivan, J.F.; Neylon, A.; McGorrian, C.; Blake, G.J. miRNA-93-5p and other miRNAs as predictors of coronary artery disease and STEMI. *Int. J. Cardiol.* **2016**, *224*, 310–316. [CrossRef]
49. Vegter, E.L.; Schmitter, D.; Hagemeijer, Y.; Ovchinnikova, E.S.; van der Harst, P.; Teerlink, J.R.; O'Connor, C.M.; Metra, M.; Davison, B.A.; Bloomfield, D.; et al. Use of biomarkers to establish potential role and function of circulating microRNAs in acute heart failure. *Int. J. Cardiol.* **2016**, *224*, 231–239. [CrossRef]
50. Gacoń, J.; Badacz, R.; Stępień, E.; Karch, I.; Enguita, F.J.; Żmudka, K.; Przewłocki, T.; Kabłak-Ziembicka, A. Diagnostic and prognostic micro-RNAs in ischaemic stroke due to carotid artery stenosis and in acute coronary syndrome: A four-year prospective study. *Kardiol. Pol.* **2018**, *76*, 362–369. [CrossRef]
51. Zhu, Y.; Tian, F.; Li, H.; Zhou, Y.; Lu, J.; Ge, Q. Profiling maternal plasma microRNA expression in early pregnancy to predict gestational diabetes mellitus. *Int. J. Gynaecol. Obstet.* **2015**, *130*, 49–53. [CrossRef]
52. Cao, Y.L.; Jia, Y.J.; Xing, B.H.; Shi, D.D.; Dong, X.J. Plasma microRNA-16-5p, -17-5p and -20a-5p: Novel diagnostic biomarkers for gestational diabetes mellitus. *J. Obstet. Gynaecol. Res.* **2017**, *43*, 974–981. [CrossRef]

53. Hocaoglu, M.; Demirer, S.; Senturk, H.; Turgut, A.; Komurcu-Bayrak, E. Differential expression of candidate circulating microRNAs in maternal blood leukocytes of the patients with preeclampsia and gestational diabetes mellitus. *Pregnancy Hypertens.* **2019**, *17*, 5–11. [CrossRef] [PubMed]
54. Duan, Y.R.; Chen, B.P.; Chen, F.; Yang, S.X.; Zhu, C.Y.; Ma, Y.L.; Li, Y.; Shi, J. Exosomal microRNA-16-5p from human urine-derived stem cells ameliorates diabetic nephropathy through protection of podocyte. *J. Cell Mol. Med.* 2019, *Epub ahead of print*. [CrossRef] [PubMed]
55. Assmann, T.S.; Recamonde-Mendoza, M.; Costa, A.R.; Puñales, M.; Tschiedel, B.; Canani, L.H.; Bauer, A.C.; Crispim, D. Circulating miRNAs in diabetic kidney disease: Case-control study and in silico analyses. *Acta Diabetol.* **2019**, *56*, 55–65. [CrossRef]
56. Alicka, M.; Major, P.; Wysocki, M.; Marycz, K. Adipose-Derived Mesenchymal Stem Cells Isolated from Patients with Type 2 Diabetes Show Reduced "Stemness" through an Altered Secretome Profile, Impaired Anti-Oxidative Protection, and Mitochondrial Dynamics Deterioration. *J. Clin. Med.* **2019**, *8*, 765. [CrossRef] [PubMed]
57. Mogilyansky, E.; Rigoutsos, I. The miR-17/92 cluster: A comprehensive update on its genomics, genetics, functions and increasingly important and numerous roles in health and disease. *Cell Death Differ.* **2013**, *20*, 1603–1614. [CrossRef] [PubMed]
58. Zhou, L.; Qi, R.Q.; Liu, M.; Xu, Y.P.; Li, G.; Weiland, M.; Kaplan, D.H.; Mi, Q.S. microRNA miR-17-92 cluster is highly expressed in epidermal Langerhans cells but not required for its development. *Genes Immun.* **2014**, *15*, 57–61. [CrossRef]
59. Danielson, L.S.; Park, D.S.; Rotllan, N.; Chamorro-Jorganes, A.; Guijarro, M.V.; Fernandez-Hernando, C.; Fishman, G.I.; Phoon, C.K.; Hernando, E. Cardiovascular dysregulation of miR-17-92 causes a lethal hypertrophic cardiomyopathy and arrhythmogenesis. *FASEB J.* **2013**, *27*, 1460–1467. [CrossRef]
60. Du, W.; Pan, Z.; Chen, X.; Wang, L.; Zhang, Y.; Li, S.; Liang, H.; Xu, C.; Zhang, Y.; Wu, Y.; et al. By targeting Stat3 microRNA-17-5p promotes cardiomyocyte apoptosis in response to ischemia followed by reperfusion. *Cell. Physiol. Biochem.* **2014**, *34*, 955–965. [CrossRef]
61. Kaucsár, T.; Révész, C.; Godó, M.; Krenács, T.; Albert, M.; Szalay, C.I.; Rosivall, L.; Benyó, Z.; Bátkai, S.; Thum, T.; et al. Activation of the miR-17 family and miR-21 during murine kidney ischemia-reperfusion injury. *Nucleic Acid Ther.* **2013**, *23*, 344–354. [CrossRef]
62. Fang, L.; Ellims, A.H.; Moore, X.L.; White, D.A.; Taylor, A.J.; Chin-Dusting, J.; Dart, A.M. Circulating microRNAs as biomarkers for diffuse myocardial fibrosis in patients with hypertrophic cardiomyopathy. *J. Transl. Med.* **2015**, *13*, 314. [CrossRef]
63. Wu, J.; Du, K.; Lu, X. Elevated expressions of serum miR-15a, miR-16, and miR-17-5p are associated with acute ischemic stroke. *Int. J. Clin. Exp. Med.* **2015**, *8*, 21071–21079. [PubMed]
64. Chen, J.; Xu, L.; Hu, Q.; Yang, S.; Zhang, B.; Jiang, H. MiR-17-5p as circulating biomarkers for the severity of coronary atherosclerosis in coronary artery disease. *Int. J. Cardiol.* **2015**, *197*, 123–124. [CrossRef] [PubMed]
65. Tian, L.; Song, Z.; Shao, W.; Du, W.W.; Zhao, L.R.; Zeng, K.; Yang, B.B.; Jin, T. Curcumin represses mouse 3T3-L1 cell adipogenic differentiation via inhibiting miR-17-5p and stimulating the Wnt signalling pathway effector Tcf7l. *Cell Death Dis.* **2017**, *8*, e2559. [CrossRef] [PubMed]
66. Chen, T.C.; Sung, M.L.; Kuo, H.C.; Chien, S.J.; Yen, C.K.; Chen, C.N. Differential regulation of human aortic smooth muscle cell proliferation by monocyte-derived macrophages from diabetic patients. *PLoS ONE* **2014**, *9*, e113752.
67. Mendell, J.T. miRiad roles for the miR-17-92 cluster in development and disease. *Cell* **2008**, *133*, 217–222. [CrossRef]
68. Brock, M.; Samillan, V.J.; Trenkmann, M.; Schwarzwald, C.; Ulrich, S.; Gay, R.E.; Gassmann, M.; Ostergaard, L.; Gay, S.; Speich, R.; et al. AntagomiR directed against miR-20a restores functional BMPR2 signalling and prevents vascular remodelling in hypoxia-induced pulmonary hypertension. *Eur. Heart J.* **2014**, *35*, 3203–3211. [CrossRef]
69. Pheiffer, C.; Dias, S.; Rheeder, P.; Adam, S. Decreased Expression of Circulating miR-20a-5p in South African Women with Gestational Diabetes Mellitus. *Mol. Diagn. Ther.* **2018**, *22*, 345–352. [CrossRef]
70. Platania, C.B.M.; Maisto, R.; Trotta, M.C.; D'Amico, M.; Rossi, S.; Gesualdo, C.; D'Amico, G.; Balta, C.; Herman, H.; Hermenean, A.; et al. Retinal and circulating miRNA expression patterns in diabetic retinopathy: An in silico and in vivo approach. *Br. J. Pharmacol.* **2019**, *176*, 2179–2194.
71. Lareyre, F.; Clément, M.; Moratal, C.; Loyer, X.; Jean-Baptiste, E.; Hassen-Khodja, R.; Chinetti, G.; Mallat, Z.; Raffort, J. Differential micro-RNA expression in diabetic patients with abdominal aortic aneurysm. *Biochimie* **2019**, *162*, 1–7. [CrossRef]
72. Dickinson, B.A.; Semus, H.M.; Montgomery, R.L.; Stack, C.; Latimer, P.A.; Lewton, S.M.; Lynch, J.M.; Hullinger, T.G.; Seto, A.G.; van Rooij, E. Plasma microRNAs serve as biomarkers of therapeutic efficacy and disease progression in hypertension-induced heart failure. *Eur. J. Heart Fail.* **2013**, *15*, 650–659. [CrossRef]
73. Flowers, E.; Aouizerat, B.E.; Abbasi, F.; Lamendola, C.; Grove, K.M.; Fukuoka, Y.; Reaven, G.M. Circulating microRNA-320a and microRNA-486 predict thiazolidinedione response: Moving towards precision health for diabetes prevention. *Metabolism* **2015**, *64*, 1051–1059. [CrossRef] [PubMed]
74. Katayama, M.; Wiklander, O.P.B.; Fritz, T.; Caidahl, K.; El-Andaloussi, S.; Zierath, J.R.; Krook, A. Circulating Exosomal miR-20b-5p Is Elevated in Type 2 Diabetes and Could Impair Insulin Action in Human Skeletal Muscle. *Diabetes* **2019**, *68*, 515–526. [CrossRef] [PubMed]
75. Xiong, Y.; Chen, L.; Yan, C.; Zhou, W.; Endo, Y.; Liu, J.; Hu, L.; Hu, Y.; Mi, B.; Liu, G. Circulating Exosomal miR-20b-5p Inhibition Restores Wnt9b Signaling and Reverses Diabetes-Associated Impaired Wound Healing. *Small* **2020**, *16*, e1904044. [CrossRef] [PubMed]

76. Zhu, K.; Hu, X.; Chen, H.; Li, F.; Yin, N.; Liu, A.L.; Shan, K.; Qin, Y.W.; Huang, X.; Chang, Q.; et al. Downregulation of circRNA DMNT3B contributes to diabetic retinal vascular dysfunction through targeting miR-20b-5p and BAMBI. *EBioMedicine* **2019**, *49*, 341–353. [CrossRef]
77. Sekar, D.; Venugopal, B.; Sekar, P.; Ramalingam, K. Role of microRNA 21 in diabetes and associated/related diseases. *Gene* **2016**, *582*, 14–18. [CrossRef] [PubMed]
78. Suárez, Y.; Fernández-Hernando, C.; Pober, J.S.; Sessa, W.C. Dicer dependent microRNAs regulate gene expression and functions in human endothelial cells. *Circ. Res.* **2007**, *100*, 1164–1173. [CrossRef]
79. Dong, S.; Ma, W.; Hao, B.; Hu, F.; Yan, L.; Yan, X.; Wang, Y.; Chen, Z.; Wang, Z. microRNA-21 promotes cardiac fibrosis and development of heart failure with preserved left ventricular ejection fraction by up-regulating Bcl-2. *Int. J. Clin. Exp. Pathol.* **2014**, *7*, 565–574.
80. Zhang, J.; Xing, Q.; Zhou, X.; Li, J.; Li, Y.; Zhang, L.; Zhou, Q.; Tang, B. Circulating miRNA 21 is a promising biomarker for heart failure. *Mol. Med. Rep.* **2017**, *16*, 7766–7774. [CrossRef]
81. Licholai, S.; Blaż, M.; Kapelak, B.; Sanak, M. Unbiased Profile of MicroRNA Expression in Ascending Aortic Aneurysm Tissue Appoints Molecular Pathways Contributing to the Pathology. *Ann. Thorac. Surg* **2016**, *102*, 1245–1252. [CrossRef]
82. Kriegel, A.J.; Baker, M.A.; Liu, Y.; Liu, P.; Cowley, A.W., Jr.; Liang, M. Endogenous microRNAs in human microvascular endothelial cells regulate mRNAs encoded by hypertension-related genes. *Hypertension* **2015**, *66*, 793–799. [CrossRef]
83. Velle-Forbord, T.; Eidlaug, M.; Debik, J.; Sæther, J.C.; Follestad, T.; Nauman, J.; Gigante, B.; Røsjø, H.; Omland, T.; Langaas, M.; et al. Circulating microRNAs as predictive biomarkers of myocardial infarction: Evidence from the HUNT study. *Atherosclerosis* **2019**, *289*, 1–7. [CrossRef] [PubMed]
84. Demirsoy, İ.H.; Ertural, D.Y.; Balci, Ş.; Çınkır, Ü.; Sezer, K.; Tamer, L.; Aras, N. Profiles of Circulating MiRNAs Following Metformin Treatment in Patients with Type 2 Diabetes. *J. Med. Biochem.* **2018**, *37*, 499–506. [CrossRef] [PubMed]
85. Olivieri, F.; Spazzafumo, L.; Bonafè, M.; Recchioni, R.; Prattichizzo, F.; Marcheselli, F.; Micolucci, L.; Mensà, E.; Giuliani, A.; Santini, G.; et al. MiR-21-5p and miR-126a-3p levels in plasma and circulating angiogenic cells: Relationship with type 2 diabetes complications. *Oncotarget* **2015**, *6*, 35372–35382. [CrossRef] [PubMed]
86. Assmann, T.S.; Recamonde-Mendoza, M.; De Souza, B.M.; Crispim, D. MicroRNA expression profiles and type 1 diabetes mellitus: Systematic review and bioinformatic analysis. *Endocr. Connect.* **2017**, *6*, 773–790. [CrossRef] [PubMed]
87. Lakhter, A.J.; Pratt, R.E.; Moore, R.E.; Doucette, K.K.; Maier, B.F.; DiMeglio, L.A.; Sims, E.K. Beta cell extracellular vesicle miR-21-5p cargo is increased in response to inflammatory cytokines and serves as a biomarker of type 1 diabetes. *Diabetologia* **2018**, *61*, 1124–1134. [CrossRef]
88. Grieco, G.E.; Cataldo, D.; Ceccarelli, E.; Nigi, L.; Catalano, G.; Brusco, N.; Mancarella, F.; Ventriglia, G.; Fondelli, C.; Guarino, E.; et al. Serum Levels of miR-148a and miR-21-5p Are Increased in Type 1 Diabetic Patients and Correlated with Markers of Bone Strength and Metabolism. *Noncoding RNA* **2018**, *4*, 37. [CrossRef]
89. Gholaminejad, A.; Abdul Tehrani, H.; Gholami Fesharaki, M. Identification of candidate microRNA biomarkers in diabetic nephropathy: A meta-analysis of profiling studies. *J. Nephrol.* **2018**, *31*, 813–831. [CrossRef]
90. Long, B.; Gan, T.Y.; Zhang, R.C.; Zhang, Y.H. miR-23a Regulates Cardiomyocyte Apoptosis by Targeting Manganese Superoxide Dismutase. *Mol. Cells* **2017**, *40*, 542–549. [CrossRef]
91. Wang, S.; He, W.; Wang, C. MiR-23a Regulates the Vasculogenesis of Coronary Artery Disease by Targeting Epidermal Growth Factor Receptor. *Cardiovasc. Ther.* **2016**, *34*, 199–208. [CrossRef]
92. Cong, X.; Li, Y.; Lu, N.; Dai, Y.; Zhang, H.; Zhao, X.; Liu, Y. Resveratrol attenuates the inflammatory reaction induced by ischemia/reperfusion in the rat heart. *Mol. Med. Rep.* **2014**, *9*, 2528–2532. [CrossRef]
93. Černá, V.; Ostašov, P.; Pitule, P.; Moláček, J.; Třeška, V.; Pešta, M. The Expression Profile of MicroRNAs in Small and Large Abdominal Aortic Aneurysms. *Cardiol. Res. Pract.* **2019**, *2019*, 8645840. [CrossRef] [PubMed]
94. Lozano-Bartolomé, J.; Llauradó, G.; Portero-Otin, M.; Altuna-Coy, A.; Rojo-Martínez, G.; Vendrell, J.; Jorba, R.; Rodríguez-Gallego, E.; Chacón, M.R. Altered Expression of miR-181a-5p and miR-23a-3p Is Associated With Obesity and TNFα-Induced Insulin Resistance. *J. Clin. Endocrinol. Metab.* **2018**, *103*, 1447–1458. [CrossRef] [PubMed]
95. Dolz, S.; Górriz, D.; Tembl, J.I.; Sánchez, D.; Fortea, G.; Parkhutik, V.; Lago, A. Circulating MicroRNAs as Novel Biomarkers of Stenosis Progression in Asymptomatic Carotid Stenosis. *Stroke* **2017**, *48*, 10–16. [CrossRef] [PubMed]
96. De Gonzalo-Calvo, D.; Cenarro, A.; Garlaschelli, K.; Pellegatta, F.; Vilades, D.; Nasarre, L.; Camino-Lopez, S.; Crespo, J.; Carreras, F.; Leta, R.; et al. Translating the microRNA signature of microvesicles derived from human coronary artery smooth muscle cells in patients with familial hypercholesterolemia and coronary artery disease. *J. Mol. Cell. Cardiol.* **2017**, *106*, 55–67. [CrossRef]
97. Gecys, D.; Tatarunas, V.; Veikutiene, A.; Lesauskaite, V. New potential modulators of CYP4F2 enzyme activity in angina pectoris: Hsa-miR-24-3p and hsa-miR-34a-5p. *Biomarkers* **2020**, *25*, 40–47. [CrossRef] [PubMed]
98. Onrat, S.T.; Onrat, E.; Ercan Onay, E.; Yalım, Z.; Avşar, A. The Genetic Determination of the Differentiation between Ischemic Dilated Cardiomyopathy and Idiopathic Dilated Cardiomyopathy. *Genet. Test. Mol. Biomark.* **2018**, *22*, 644–651. [CrossRef] [PubMed]
99. Tan, H.; Qi, J.; Fan, B.Y.; Zhang, J.; Su, F.F.; Wang, H.T. MicroRNA-24-3p Attenuates Myocardial Ischemia/Reperfusion Injury by Suppressing RIPK1 Expression in Mice. *Cell. Physiol. Biochem.* **2018**, *51*, 46–62. [CrossRef]

100. Xiao, X.; Lu, Z.; Lin, V.; May, A.; Shaw, D.H.; Wang, Z.; Che, B.; Tran, K.; Du, H.; Shaw, P.X. MicroRNA miR-24-3p Reduces Apoptosis and Regulates Keap1-Nrf2 Pathway in Mouse Cardiomyocytes Responding to Ischemia/Reperfusion Injury. *Oxidative Med. Cell. Longev.* **2018**, *2018*, 7042105. [CrossRef]
101. Gao, J.; Liu, Q.G. The role of miR-26 in tumors and normal tissues. *Oncol. Lett.* **2011**, *2*, 1019–1023. [CrossRef]
102. Zheng, L.; Lin, S.; Lv, C. MiR-26a-5p regulates cardiac fibroblasts collagen expression by targeting ULK1. *Sci Rep.* **2018**, *8*, 2104. [CrossRef]
103. Bye, A.; Røsjø, H.; Nauman, J.; Silva, G.J.; Follestad, T.; Omland, T.; Wisløff, U. Circulating microRNAs predict future fatal myocardial infarction in healthy individuals—The HUNT study. *J. Mol. Cell. Cardiol.* **2016**, *97*, 162–168. [CrossRef]
104. Hsu, A.; Chen, S.J.; Chang, Y.S.; Chen, H.C.; Chu, P.H. Systemic approach to identify serum microRNAs as potential biomarkers for acute myocardial infarction. *BioMed Res. Int.* **2014**, *2014*, 418628. [CrossRef] [PubMed]
105. Xing, X.; Guo, S.; Zhang, G.; Liu, Y.; Bi, S.; Wang, X.; Lu, Q. miR-26a-5p protects against myocardial ischemia/reperfusion injury by regulating the PTEN/PI3K/AKT signaling pathway. *Braz. J. Med. Biol. Res.* **2020**, *53*, e9106. [CrossRef]
106. Chouvarine, P.; Geldner, J.; Giagnorio, R.; Legchenko, E.; Bertram, H.; Hansmann, G. Trans-Right-Ventricle and Transpulmonary MicroRNA Gradients in Human Pulmonary Arterial Hypertension. *Pediatr. Crit. Care Med.* **2019**. *Epub ahead of print*. [CrossRef]
107. Garavelli, S.; Bruzzaniti, S.; Tagliabue, E.; Prattichizzo, F.; Di Silvestre, D.; Perna, F.; La Sala, L.; Ceriello, A.; Mozzillo, E.; Fattorusso, V.; et al. Blood Co-Circulating Extracellular microRNAs and Immune Cell Subsets Associate with Type 1 Diabetes Severity. *Int. J. Mol. Sci.* **2020**, *21*, 477. [CrossRef]
108. Ye, Y.; Hu, Z.; Lin, Y.; Zhang, C.; Perez-Polo, J.R. Downregulation of microRNA-29 by antisense inhibitors and a PPAR-gamma agonist protects against myocardial ischaemia-reperfusion injury. *Cardiovasc. Res.* **2010**, *87*, 535–544. [CrossRef]
109. Moraes, L.N.; Fernandez, G.J.; Vechetti-Júnior, I.J.; Freire, P.P.; Souza, R.W.A.; Villacis, R.A.R.; Rogatto, S.R.; Reis, P.P.; Dal-Pai-Silva, M.; Carvalho, R.F. Integration of miRNA and mRNA expression profiles reveals microRNA-regulated networks during muscle wasting in cardiac cachexia. *Sci. Rep.* **2017**, *7*, 6998. [CrossRef]
110. Zhao, Y.; Yuan, Y.; Qiu, C. Underexpression of CACNA1C Caused by Overexpression of microRNA-29a Underlies the Pathogenesis of Atrial Fibrillation. *Med. Sci. Monit.* **2016**, *22*, 2175–2181. [CrossRef]
111. Zhang, L.; Zhang, Y.; Xue, S.; Ding, H.; Wang, Y.; Qi, H.; Wang, Y.; Zhu, W.; Li, P. Clinical significance of circulating microRNAs as diagnostic biomarkers for coronary artery disease. *J. Cell. Mol. Med.* **2020**, *24*, 1146–1150. [CrossRef] [PubMed]
112. Wander, P.L.; Boyko, E.J.; Hevner, K.; Parikh, V.J.; Tadesse, M.G.; Sorensen, T.K.; Williams, M.A.; Enquobahrie, D.A. Circulating early- and mid-pregnancy microRNAs and risk of gestational diabetes. *Diabetes Res. Clin. Pract.* **2017**, *132*, 1–9. [CrossRef]
113. Kong, L.; Zhu, J.; Han, W.; Jiang, X.; Xu, M.; Zhao, Y.; Dong, Q.; Pang, Z.; Guan, Q.; Gao, L.; et al. Significance of serum microRNAs in pre-diabetes and newly diagnosed type 2 diabetes: A clinical study. *Acta Diabetol.* **2011**, *48*, 61–69. [CrossRef] [PubMed]
114. Widlansky, M.E.; Jensen, D.M.; Wang, J.; Liu, Y.; Geurts, A.M.; Kriegel, A.J.; Liu, P.; Ying, R.; Zhang, G.; Casati, M.; et al. miR-29 contributes to normal endothelial function and can restore it in cardiometabolic disorders. *EMBO Mol. Med.* **2018**, *10*, e8046. [CrossRef]
115. Bulent Vatan, M.; Kalaycı Yigin, A.; Akdemir, R.; Tarik Agac, M.; Akif Cakar, M.; Aksoy, M.; Tatli, E.; Kilic, H.; Gunduz, H.; Guzel, D.; et al. Altered Plasma MicroRNA Expression in Patients with Mitral Chordae Tendineae Rupture. *J. Heart Valve Dis.* **2016**, *25*, 580–588. [PubMed]
116. Gumus, G.; Giray, D.; Bobusoglu, O.; Tamer, L.; Karpuz, D.; Hallioglu, O. MicroRNA values in children with rheumatic carditis: A preliminary study. *Rheumatol. Int.* **2018**, *38*, 1199–1205. [CrossRef] [PubMed]
117. Rogg, E.M.; Abplanalp, W.T.; Bischof, C.; John, D.; Schulz, M.H.; Krishnan, J.; Fischer, A.; Poluzzi, C.; Schaefer, L.; Bonauer, A.; et al. Analysis of Cell Type-Specific Effects of MicroRNA-92a Provides Novel Insights Into Target Regulation and Mechanism of Action. *Circulation* **2018**, *138*, 2545–2558. [CrossRef]
118. Marques, F.Z.; Vizi, D.; Khammy, O.; Mariani, J.A.; Kaye, D.M. The transcardiac gradient of cardio-microRNAs in the failing heart. *Eur. J. Heart Fail.* **2016**, *18*, 1000–1008. [CrossRef]
119. Liu, Y.; Li, Q.; Hosen, M.R.; Zietzer, A.; Flender, A.; Levermann, P.; Schmitz, T.; Frühwald, D.; Goody, P.; Nickenig, G.; et al. Atherosclerotic Conditions Promote the Packaging of Functional MicroRNA-92a-3p Into Endothelial Microvesicles. *Circ. Res.* **2019**, *124*, 575–587. [CrossRef]
120. Wiese, C.B.; Zhong, J.; Xu, Z.Q.; Zhang, Y.; Ramirez Solano, M.A.; Zhu, W.; Linton, M.F.; Sheng, Q.; Kon, V.; Vickers, K.C. Dual inhibition of endothelial miR-92a-3p and miR-489-3p reduces renal injury-associated atherosclerosis. *Atherosclerosis* **2019**, *282*, 121–131. [CrossRef]
121. Moncini, S.; Salvi, A.; Zuccotti, P.; Viero, G.; Quattrone, A.; Barlati, S.; De Petro, G.; Venturin, M.; Riva, P. The role of miR-103 and miR-107 in regulation of CDK5R1 expression and in cellular migration. *PLoS ONE* **2011**, *6*, e20038. [CrossRef]
122. Huang, L.; Li, L.; Chen, X.; Zhang, H.; Shi, Z. MiR-103a targeting Piezo1 is involved in acute myocardial infarction through regulating endothelium function. *Cardiol. J.* **2016**, *23*, 556–562. [CrossRef]
123. Deng, B.; Du, J.; Hu, R.; Wang, A.P.; Wu, W.H.; Hu, C.P.; Li, Y.J.; Li, X.H. MicroRNA-103/107 is involved in hypoxia-induced proliferation of pulmonary arterial smooth muscle cells by targeting HIF-1β. *Life Sci.* **2016**, *147*, 117–124. [CrossRef] [PubMed]
124. Trajkovski, M.; Hausser, J.; Soutschek, J.; Bhat, B.; Akin, A.; Zavolan, M.; Heim, M.H.; Stoffel, M. MicroRNAs 103 and 107 regulate insulin sensitivity. *Nature* **2011**, *474*, 649–653. [CrossRef]

125. Assmann, T.S.; Recamonde-Mendoza, M.; Puñales, M.; Tschiedel, B.; Canani, L.H.; Crispim, D. MicroRNA expression profile in plasma from type 1 diabetic patients: Case-control study and bioinformatic analysis. *Diabetes Res. Clin. Pract.* **2018**, *141*, 35–46. [CrossRef] [PubMed]
126. Shaham, L.; Binder, V.; Gefen, N.; Borkhardt, A.; Izraeli, S. MiR-125 in normal and malignant hematopoiesis. *Leukemia* **2012**, *26*, 2011–2018. [CrossRef]
127. Tiedt, S.; Prestel, M.; Malik, R.; Schieferdecker, N.; Duering, M.; Kautzky, V.; Stoycheva, I.; Böck, J.; Northoff, B.H.; Klein, M.; et al. RNA-Seq Identifies Circulating miR-125a-5p, miR-125b-5p, and miR-143-3p as Potential Biomarkers for Acute Ischemic Stroke. *Circ. Res.* **2017**, *121*, 970–980. [CrossRef]
128. Jia, K.; Shi, P.; Han, X.; Chen, T.; Tang, H.; Wang, J. Diagnostic value of miR-30d-5p and miR-125b-5p in acute myocardial infarction. *Mol. Med. Rep.* **2016**, *14*, 184–194. [CrossRef]
129. Bayoumi, A.S.; Park, K.M.; Wang, Y.; Teoh, J.P.; Aonuma, T.; Tang, Y.; Su, H.; Weintraub, N.L.; Kim, I.M. A carvedilol-responsive microRNA, miR-125b-5p protects the heart from acute myocardial infarction by repressing pro-apoptotic bak1 and klf13 in cardiomyocytes. *J. Mol. Cell. Cardiol* **2018**, *114*, 72–82. [CrossRef]
130. Lamadrid-Romero, M.; Solís, K.H.; Cruz-Reséndiz, M.S.; Pérez, J.E.; Díaz, N.F.; Flores-Herrera, H.; García-López, G.; Perichart, O.; Reyes-Muñoz, E.; Arenas-Huertero, F.; et al. Central nervous system development-related microRNAs levels increase in the serum of gestational diabetic women during the first trimester of pregnancy. *Neurosci. Res.* **2018**, *130*, 8–22. [CrossRef]
131. Satake, E.; Pezzolesi, M.G.; Md Dom, Z.I.; Smiles, A.M.; Niewczas, M.A.; Krolewski, A.S. Circulating miRNA Profiles Associated With Hyperglycemia in Patients With Type 1 Diabetes. *Diabetes* **2018**, *67*, 1013–1023. [CrossRef]
132. Samandari, N.; Mirza, A.H.; Kaur, S.; Hougaard, P.; Nielsen, L.B.; Fredheim, S.; Mortensen, H.B.; Pociot, F. Influence of Disease Duration on Circulating Levels of miRNAs in Children and Adolescents with New Onset Type 1 Diabetes. *Noncoding RNA* **2018**, *4*, 35. [CrossRef]
133. Yu, C.Y.; Yang, C.Y.; Rui, Z.L. MicroRNA-125b-5p improves pancreatic β-cell function through inhibiting JNK signaling pathway by targeting DACT1 in mice with type 2 diabetes mellitus. *Life Sci.* **2019**, *224*, 67–75. [CrossRef] [PubMed]
134. Wu, X.J.; Zhao, Z.F.; Kang, X.J.; Wang, H.J.; Zhao, J.; Pu, X.M. MicroRNA-126-3p suppresses cell proliferation by targeting PIK3R2 in Kaposi's sarcoma cells. *Oncotarget* **2016**, *7*, 36614–36621. [CrossRef] [PubMed]
135. Matsha, T.E.; Kengne, A.P.; Hector, S.; Mbu, D.L.; Yako, Y.Y.; Erasmus, R.T. MicroRNA profiling and their pathways in South African individuals with prediabetes and newly diagnosed type 2 diabetes mellitus. *Oncotarget* **2018**, *9*, 30485–30498. [CrossRef] [PubMed]
136. Lan, X.; Wu, L.; Wu, N.; Chen, Q.; Li, Y.; Du, X.; Wei, C.; Feng, L.; Li, Y.; Osoro, E.K.; et al. Long Noncoding RNA lnc-HC Regulates PPARγ-Mediated Hepatic Lipid Metabolism through miR-130b-3p. *Mol. Ther. Nucleic Acids* **2019**, *18*, 954–965. [CrossRef]
137. Tryggestad, J.B.; Vishwanath, A.; Jiang, S.; Mallappa, A.; Teague, A.M.; Takahashi, Y.; Thompson, D.M.; Chernausek, S.D. Influence of gestational diabetes mellitus on human umbilical vein endothelial cell miRNA. *Clin. Sci.* **2016**, *130*, 1955–1967. [CrossRef] [PubMed]
138. Zhang, J.; Jazii, F.R.; Haghighi, M.M.; Alvares, D.; Liu, L.; Khosraviani, N.; Adeli, K. miR-130b is a potent stimulator of hepatic very-low-density lipoprotein assembly and secretion via marked induction of microsomal triglyceride transfer protein. *Am. J. Physiol. Endocrinol. Metab.* **2020**, *318*, E262–E275. [CrossRef]
139. Li, P.; Zhang, Q.; Wu, X.; Yang, X.; Zhang, Y.; Li, Y.; Jiang, F. Circulating microRNAs serve as novel biological markers for intracranial aneurysms. *J. Am. Heart Assoc.* **2014**, *3*, e000972. [CrossRef]
140. Tian, C.; Li, Z.; Yang, Z.; Huang, Q.; Liu, J.; Hong, B. Plasma MicroRNA-16 Is a Biomarker for Diagnosis, Stratification, and Prognosis of Hyperacute Cerebral Infarction. *PLoS ONE* **2016**, *11*, e0166688. [CrossRef]
141. Prabu, P.; Rome, S.; Sathishkumar, C.; Aravind, S.; Mahalingam, B.; Shanthirani, C.S.; Gastebois, C.; Villard, A.; Mohan, V.; Balasubramanyam, M. Circulating MiRNAs of 'Asian Indian Phenotype' Identified in Subjects with Impaired Glucose Tolerance and Patients with Type 2 Diabetes. *PLoS ONE* **2015**, *10*, e0128372. [CrossRef]
142. Feng, T.; Li, K.; Zheng, P.; Wang, Y.; Lv, Y.; Shen, L.; Chen, Y.; Xue, Z.; Li, B.; Jin, L.; et al. Weighted Gene Coexpression Network Analysis Identified MicroRNA Coexpression Modules and Related Pathways in Type 2 Diabetes Mellitus. *Oxidative Med. Cell. Longev.* **2019**, *2019*, 9567641. [CrossRef]
143. Liang, H.W.; Yang, X.; Wen, D.Y.; Gao, L.; Zhang, X.Y.; Ye, Z.H.; Luo, J.; Li, Z.Y.; He, Y.; Pang, Y.Y.; et al. Utility of miR 133a 3p as a diagnostic indicator for hepatocellular carcinoma: An investigation combined with GEO, TCGA, meta analysis and bioinformatics. *Mol. Med. Rep.* **2018**, *17*, 1469–1484. [CrossRef] [PubMed]
144. van Rooij, E.; Olson, E.N. MicroRNAs: Powerful new regulators of heart disease and provocative therapeutic targets. *J. Clin. Investig.* **2007**, *117*, 2369–2376. [CrossRef] [PubMed]
145. Wang, J.; Xu, R.; Lin, F.; Zhang, S.; Zhang, G.; Hu, S.; Zheng, Z. MicroRNA: Novel regulators involved in the remodeling and reverse remodeling of the heart. *Cardiology* **2009**, *113*, 81–88. [CrossRef]
146. Kukreja, R.C.; Yin, C.; Salloum, F.N. MicroRNAs: New players in cardiac injury and protection. *Mol. Pharmacol.* **2011**, *80*, 558–564. [CrossRef]
147. Duisters, R.F.; Tijsen, A.J.; Schroen, B.; Leenders, J.J.; Lentink, V.; van der Made, I.; Herias, V.; van Leeuwen, R.E.; Schellings, M.W.; Barenbrug, P.; et al. miR-133 and miR-30 regulate connective tissue growth factor: Implications for a role of microRNAs in myocardial matrix remodeling. *Circ. Res.* **2009**, *104*, 170–178. [CrossRef]

148. Liu, W.; Ling, S.; Sun, W.; Liu, T.; Li, Y.; Zhong, G.; Zhao, D.; Zhang, P.; Song, J.; Jin, X.; et al. Circulating microRNAs correlated with the level of coronary artery calcification in symptomatic patients. *Sci. Rep.* **2015**, *5*, 16099. [CrossRef]
149. Jiang, Y.; Zhang, M.; He, H.; Chen, J.; Zeng, H.; Li, J.; Duan, R. MicroRNA/mRNA profiling and regulatory network of intracranial aneurysm. *BMC Med. Genom.* **2013**, *6*, 36. [CrossRef]
150. Liu, H.; Xiong, W.; Liu, F.; Lin, F.; He, J.; Liu, C.; Lin, Y.; Dong, S. Significant role and mechanism of microRNA-143-3p/KLLN axis in the development of coronary heart disease. *Am. J. Transl. Res.* **2019**, *11*, 3610–3619.
151. Li, C.; Li, J.; Xue, K.; Zhang, J.; Wang, C.; Zhang, Q.; Chen, X.; Gao, C.; Yu, X.; Sun, L. MicroRNA-143-3p promotes human cardiac fibrosis via targeting sprouty3 after myocardial infarction. *J. Mol. Cell. Cardiol.* **2019**, *129*, 281–292. [CrossRef]
152. Yu, B.; Zhao, Y.; Zhang, H.; Xie, D.; Nie, W.; Shi, K. Inhibition of microRNA-143-3p attenuates myocardial hypertrophy by inhibiting inflammatory response. *Cell Biol. Int.* **2018**, *42*, 1584–1593. [CrossRef]
153. Jiao, M.; You, H.Z.; Yang, X.Y.; Yuan, H.; Li, Y.L.; Liu, W.X.; Jin, M.; Du, J. Circulating microRNA signature for the diagnosis of childhood dilated cardiomyopathy. *Sci. Rep.* **2018**, *8*, 724. [CrossRef] [PubMed]
154. Deng, L.; Blanco, F.J.; Stevens, H.; Lu, R.; Caudrillier, A.; McBride, M.; McClure, J.D.; Grant, J.; Thomas, M.; Frid, M.; et al. MicroRNA-143 Activation Regulates Smooth Muscle and Endothelial Cell Crosstalk in Pulmonary Arterial Hypertension. *Circ. Res.* **2015**, *117*, 870–883. [CrossRef]
155. Shi, L.; Tian, C.; Sun, L.; Cao, F.; Meng, Z. The lncRNA TUG1/miR-145-5p/FGF10 regulates proliferation and migration in VSMCs of hypertension. *Biochem. Biophys. Res. Commun.* **2018**, *501*, 688–695. [CrossRef] [PubMed]
156. Yang, X.; Niu, X.; Xiao, Y.; Lin, K.; Chen, X. MiRNA expression profiles in healthy OSAHS and OSAHS with arterial hypertension: Potential diagnostic and early warning markers. *Respir. Res.* **2018**, *19*, 194. [CrossRef] [PubMed]
157. Toro, R.; Blasco-Turrión, S.; Morales-Ponce, F.J.; Gonzalez, P.; Martínez-Camblor, P.; López-Granados, A.; Brugada, R.; Campuzano, O.; Pérez-Serra, A.; Rosa Longobardo, F.; et al. Plasma microRNAs as biomarkers for Lamin A/C-related dilated cardiomyopathy. *J. Mol. Med.* **2018**, *96*, 845–856. [CrossRef]
158. Yuan, M.; Zhang, L.; You, F.; Zhou, J.; Ma, Y.; Yang, F.; Tao, L. MiR-145-5p regulates hypoxia-induced inflammatory response and apoptosis in cardiomyocytes by targeting CD40. *Mol. Cell. Biochem.* **2017**, *431*, 123–131. [CrossRef]
159. Wu, G.; Tan, J.; Li, J.; Sun, X.; Du, L.; Tao, S. miRNA-145-5p induces apoptosis after ischemia-reperfusion by targeting dual specificity phosphatase *J. Cell. Physiol.* **2019**, *234*, 16281–16289. [CrossRef]
160. Xie, X.; Peng, L.; Zhu, J.; Zhou, Y.; Li, L.; Chen, Y.; Yu, S.; Zhao, Y. miR-145-5p/Nurr1/TNF-α Signaling-Induced Microglia Activation Regulates Neuron Injury of Acute Cerebral Ischemic/Reperfusion in Rats. *Front. Mol. Neurosci.* **2017**, *10*, 383. [CrossRef]
161. Nunez Lopez, Y.O.; Retnakaran, R.; Zinman, B.; Pratley, R.E.; Seyhan, A.A. Predicting and understanding the response to short-term intensive insulin therapy in people with early type 2 diabetes. *Mol. Metab.* **2019**, *20*, 63–78. [CrossRef]
162. Zhang, J.; Cui, C.; Xu, H. Downregulation of miR-145-5p elevates retinal ganglion cell survival to delay diabetic retinopathy progress by targeting FGF5. *Biosci. Biotechnol. Biochem.* **2019**, *83*, 1655–1662. [CrossRef]
163. Zamanian Azodi, M.; Rezaei-Tavirani, M.; Rezaei-Tavirani, M.; Robati, R.M. Gestational Diabetes Mellitus Regulatory Network Identifies hsa-miR-145-5p and hsa-miR-875-5p as Potential Biomarkers. *Int. J. Endocrinol. Metab.* **2019**, *17*, e86640. [CrossRef] [PubMed]
164. Taganov, K.D.; Boldin, M.P.; Chang, K.J.; Baltimore, D. NF-kappaB-dependent induction of microRNA miR-146, an inhibitor targeted to signaling proteins of innate immune responses. *Proc. Natl. Acad. Sci. USA* **2006**, *103*, 12481–12486. [CrossRef] [PubMed]
165. Paterson, M.R.; Kriegel, A.J. MiR-146a/b: A family with shared seeds and different roots. *Physiol. Genom.* **2017**, *49*, 243–252. [CrossRef] [PubMed]
166. Zhang, X.; Ye, Z.H.; Liang, H.W.; Ren, F.H.; Li, P.; Dang, Y.W.; Chen, G. Down-regulation of miR-146a-5p and its potential targets in hepatocellular carcinoma validated by a TCGA- and GEO-based study. *FEBS Open Bio* **2017**, *7*, 504–521. [CrossRef]
167. Wang, X.; Ha, T.; Liu, L.; Zou, J.; Zhang, X.; Kalbfleisch, J.; Gao, X.; Williams, D.; Li, C. Increased expression of microRNA-146a decreases myocardial ischaemia/reperfusion injury. *Cardiovasc. Res.* **2013**, *97*, 432–442. [CrossRef]
168. Quan, X.; Ji, Y.; Zhang, C.; Guo, X.; Zhang, Y.; Jia, S.; Ma, W.; Fan, Y.; Wang, C. Circulating MiR-146a May be a Potential Biomarker of Coronary Heart Disease in Patients with Subclinical Hypothyroidism. *Cell. Physiol. Biochem.* **2018**, *45*, 226–236. [CrossRef]
169. Li, S.H.; Chen, L.; Pang, X.M.; Su, S.Y.; Zhou, X.; Chen, C.Y.; Huang, L.G.; Li, J.P.; Liu, J.L. Decreased miR-146a expression in acute ischemic stroke directly targets the Fbxl10 mRNA and is involved in modulating apoptosis. *Neurochem. Int.* **2017**, *107*, 156–167. [CrossRef]
170. Barberio, M.D.; Kasselman, L.J.; Playford, M.P.; Epstein, S.B.; Renna, H.A.; Goldberg, M.; DeLeon, J.; Voloshyna, I.; Barlev, A.; Salama, M.; et al. Cholesterol efflux alterations in adolescent obesity: Role of adipose-derived extracellular vesical microRNAs. *J. Transl. Med.* **2019**, *17*, 232. [CrossRef]
171. Gaudet, A.D.; Fonken, L.K.; Gushchina, L.V.; Aubrecht, T.G.; Maurya, S.K.; Periasamy, M.; Nelson, R.J.; Popovich, P.G. miR-155 Deletion in Female Mice Prevents Diet-Induced Obesity. *Sci. Rep.* **2016**, *6*, 22862. [CrossRef]
172. Chen, L.; Zheng, S.Y.; Yang, C.Q.; Ma, B.M.; Jiang, D. MiR-155-5p inhibits the proliferation and migration of VSMCs and HUVECs in atherosclerosis by targeting AKT1. *Eur. Rev. Med. Pharmacol. Sci* **2019**, *23*, 2223–2233.

173. Zhu, M.; Wei, Y.; Geißler, C.; Abschlag, K.; Corbalán Campos, J.; Hristov, M.; Möllmann, J.; Lehrke, M.; Karshovska, E.; Schober, A. Hyperlipidemia-Induced MicroRNA-155-5p Improves β-Cell Function by Targeting Mafb. *Diabetes* **2017**, *66*, 3072–3084. [CrossRef] [PubMed]
174. Li, S.; Lee, C.; Song, J.; Lu, C.; Liu, J.; Cui, Y.; Liang, H.; Cao, C.; Zhang, F.; Chen, H. Circulating microRNAs as potential biomarkers for coronary plaque rupture. *Oncotarget* **2017**, *8*, 48145–48156. [CrossRef]
175. Mukai, N.; Nakayama, Y.; Murakami, S.; Tanahashi, T.; Sessler, D.I.; Ishii, S.; Ogawa, S.; Tokuhira, N.; Mizobe, T.; Sawa, T.; et al. Potential contribution of erythrocyte microRNA to secondary erythrocytosis and thrombocytopenia in congenital heart disease. *Pediatr. Res.* **2018**, *83*, 866–873. [CrossRef]
176. Klimczak, D.; Kuch, M.; Pilecki, T.; Żochowska, D.; Wirkowska, A.; Pączek, L. Plasma microRNA-155-5p is increased among patients with chronic kidney disease and nocturnal hypertension. *J. Am. Soc. Hypertens.* **2017**, *11*, 831–841.e4. [CrossRef] [PubMed]
177. Wang, M.; Sun, L.; Ding, W.; Cai, S.; Zhao, Q. Ablation alleviates atrial fibrillation by regulating the signaling pathways of endothelial nitric oxide synthase/nitric oxide via miR-155-5p and miR-24-3p. *J. Cell. Biochem.* **2019**, *120*, 4451–4462. [CrossRef]
178. Sun, X.; Sit, A.; Feinberg, M.W. Role of miR-181 family in regulating vascular inflammation and immunity. *Trends Cardiovasc. Med.* **2014**, *24*, 105–112. [CrossRef] [PubMed]
179. Hulsmans, M.; Sinnaeve, P.; Van der Schueren, B.; Mathieu, C.; Janssens, S.; Holvoet, P. Decreased miR-181a expression in monocytes of obese patients is associated with the occurrence of metabolic syndrome and coronary artery disease. *J. Clin. Endocrinol. Metab.* **2012**, *97*, E1213–E1218. [CrossRef]
180. Du, X.; Yang, Y.; Xu, C.; Peng, Z.; Zhang, M.; Lei, L.; Gao, W.; Dong, Y.; Shi, Z.; Sun, X.; et al. Upregulation of miR-181a impairs hepatic glucose and lipid homeostasis. *Oncotarget* **2017**, *8*, 91362–91378. [CrossRef]
181. Wu, J.; Fan, C.L.; Ma, L.J.; Liu, T.; Wang, C.; Song, J.X.; Lv, Q.S.; Pan, H.; Zhang, C.N.; Wang, J.J. Distinctive expression signatures of serum microRNAs in ischaemic stroke and transient ischaemic attack patients. *Thromb. Haemost.* **2017**, *117*, 992–1001.
182. Zhu, J.; Yao, K.; Wang, Q.; Guo, J.; Shi, H.; Ma, L.; Liu, H.; Gao, W.; Zou, Y.; Ge, J. Circulating miR-181a as a Potential Novel Biomarker for Diagnosis of Acute Myocardial Infarction. *Cell. Physiol. Biochem.* **2016**, *40*, 1591–1602. [CrossRef]
183. Nabih, E.S.; Andrawes, N.G. The Association Between Circulating Levels of miRNA-181a and Pancreatic Beta Cells Dysfunction via SMAD7 in Type 1 Diabetic Children and Adolescents. *J. Clin. Lab. Anal.* **2016**, *30*, 727–731. [CrossRef] [PubMed]
184. He, J.F.; Luo, Y.M.; Wan, X.H.; Jiang, D. Biogenesis of MiRNA-195 and its role in biogenesis, the cell cycle, and apoptosis. *J. Biochem. Mol. Toxicol.* **2011**, *25*, 404–408. [CrossRef] [PubMed]
185. van Rooij, E.; Sutherland, L.B.; Liu, N.; Williams, A.H.; McAnally, J.; Gerard, R.D.; Richardson, J.A.; Olson, E.N. A signature pattern of stress-responsive microRNAs that can evoke cardiac hypertrophy and heart failure. *Proc. Natl. Acad. Sci. USA* **2006**, *103*, 18255–18260. [CrossRef] [PubMed]
186. You, X.Y.; Huang, J.H.; Liu, B.; Liu, S.J.; Zhong, Y.; Liu, S.M. HMGA1 is a new target of miR-195 involving isoprenaline-induced cardiomyocyte hypertrophy. *Biochemistry* **2014**, *79*, 538–544. [CrossRef]
187. Zampetaki, A.; Attia, R.; Mayr, U.; Gomes, R.S.; Phinikaridou, A.; Yin, X.; Langley, S.R.; Willeit, P.; Lu, R.; Fanshawe, B.; et al. Role of miR-195 in aortic aneurysmal disease. *Circ. Res.* **2014**, *115*, 857–866. [CrossRef] [PubMed]
188. Du, J.; Zheng, R.; Xiao, F.; Zhang, S.; He, K.; Zhang, J.; Shao, Y. Downregulated MicroRNA-195 in the Bicuspid Aortic Valve Promotes Calcification of Valve Interstitial Cells via Targeting SMAD7. *Cell. Physiol. Biochem.* **2017**, *44*, 884–896. [CrossRef] [PubMed]
189. Collares, C.V.; Evangelista, A.F.; Xavier, D.J.; Rassi, D.M.; Arns, T.; Foss-Freitas, M.C.; Foss, M.C.; Puthier, D.; Sakamoto-Hojo, E.T.; Passos, G.A.; et al. Identifying common and specific microRNAs expressed in peripheral blood mononuclear cell of type 1, type 2, and gestational diabetes mellitus patients. *BMC Res. Notes* **2013**, *6*, 491. [CrossRef]
190. Tagoma, A.; Alnek, K.; Kirss, A.; Uibo, R.; Haller-Kikkatalo, K. MicroRNA profiling of second trimester maternal plasma shows upregulation of miR-195-5p in patients with gestational diabetes. *Gene* **2018**, *672*, 137–142. [CrossRef]
191. Massaro, J.D.; Polli, C.D.; Costa e Silva, M.; Alves, C.C.; Passos, G.A.; Sakamoto-Hojo, E.T.; de Holanda Miranda, W.R.; Bispo Cezar, N.J.; Rassi, D.M.; Crispim, F.; et al. Post-transcriptional markers associated with clinical complications in Type 1 and Type 2 diabetes mellitus. *Mol. Cell. Endocrinol.* **2019**, *490*, 1–14. [CrossRef]
192. Li, M.; Luan, L.; Liu, Q.; Liu, Y.; Lan, X.; Li, Z.; Liu, W. MiRNA-199a-5p Protects Against Cerebral Ischemic Injury by Down-Regulating DDR1 in Rats. *World Neurosurg.* **2019**, *131*, e486–e494. [CrossRef]
193. Yan, M.; Yang, S.; Meng, F.; Zhao, Z.; Tian, Z.; Yang, P. MicroRNA 199a-5p induces apoptosis by targeting JunB. *Sci. Rep.* **2018**, *8*, 6699. [CrossRef] [PubMed]
194. Lynch, S.M.; Ward, M.; McNulty, H.; Angel, C.Z.; Horigan, G.; Strain, J.J.; Purvis, J.; Tackett, M.; McKenna, D.J. Serum levels of miR-199a-5p correlates with blood pressure in premature cardiovascular disease patients homozygous for the MTHFR 677C > T polymorphism. *Genomics* **2020**, *112*, 669–676. [CrossRef] [PubMed]
195. Tian, X.; Yu, C.; Shi, L.; Li, D.; Chen, X.; Xia, D.; Zhou, J.; Xu, W.; Ma, C.; Gu, L.; et al. MicroRNA-199a-5p aggravates primary hypertension by damaging vascular endothelial cells through inhibition of autophagy and promotion of apoptosis. *Exp. Ther. Med.* **2018**, *16*, 595–602. [CrossRef]
196. Zhou, Y.; Pang, B.; Xiao, Y.; Zhou, S.; He, B.; Zhang, F.; Liu, W.; Peng, H.; Li, P. The protective microRNA-199a-5p-mediated unfolded protein response in hypoxic cardiomyocytes is regulated by STAT3 pathway. *J. Physiol. Biochem.* **2019**, *75*, 73–81. [CrossRef] [PubMed]

197. Liu, Y.; Liu, G.; Zhang, H.; Wang, J. MiRNA-199a-5p influences pulmonary artery hypertension via downregulating Smad3. *Biochem. Biophys. Res. Commun.* **2016**, *473*, 859–866. [CrossRef] [PubMed]
198. Wang, J.; Yu, G. A Systems Biology Approach to Characterize Biomarkers for Blood Stasis Syndrome of Unstable Angina Patients by Integrating MicroRNA and Messenger RNA Expression Profiling. *Evid.-Based Complement. Altern. Med.* **2013**, *2013*, 510208. [CrossRef]
199. Yu, L.; Gu, T.; Shi, E.; Wang, Y.; Fang, Q.; Wang, C. Dysregulation of renal microRNA expression after deep hypothermic circulatory arrest in rats. *Eur. J. Cardiothorac Surg* **2016**, *49*, 1725–1731. [CrossRef]
200. Hirt, M.N.; Werner, T.; Indenbirken, D.; Alawi, M.; Demin, P.; Kunze, A.C.; Stenzig, J.; Starbatty, J.; Hansen, A.; Fiedler, J.; et al. Deciphering the microRNA signature of pathological cardiac hypertrophy by engineered heart tissue- and sequencing-technology. *J. Mol. Cell. Cardiol.* **2015**, *81*, 1–9. [CrossRef]
201. Aguado-Fraile, E.; Ramos, E.; Conde, E.; Rodríguez, M.; Martín-Gómez, L.; Lietor, A.; Candela, Á.; Ponte, B.; Liaño, F.; García-Bermejo, M.L. A Pilot Study Identifying a Set of microRNAs As Precise Diagnostic Biomarkers of Acute Kidney Injury. *PLoS ONE* **2015**, *10*, e0127175. [CrossRef]
202. Ma, H.; Chen, P.; Sang, C.; Huang, D.; Geng, Q.; Wang, L. Modulation of apoptosis-related microRNAs following myocardial infarction in fat-1 transgenic mice vs wild-type mice. *J. Cell. Mol. Med.* **2018**, *22*, 5698–5707. [CrossRef]
203. Qiao, X.R.; Wang, L.; Liu, M.; Tian, Y.; Chen, T. MiR-210-3p attenuates lipid accumulation and inflammation in atherosclerosis by repressing IGF2. *Biosci. Biotechnol. Biochem.* **2020**, *84*, 321–329. [CrossRef] [PubMed]
204. Derda, A.A.; Pfanne, A.; Bwangär, C.; Schimmel, K.; Kennel, P.J.; Xiao, K.; Schulze, P.C.; Bauersachs, J.; Thum, T. Blood-based microRNA profiling in patients with cardiac amyloidosis. *PLoS ONE* **2018**, *13*, e0204235.
205. Verjans, R.; Peters, T.; Beaumont, F.J.; van Leeuwen, R.; van Herwaarden, T.; Verhesen, W.; Munts, C.; Bijnen, M.; Henkens, M.; Diez, J.; et al. MicroRNA-221/222 Family Counteracts Myocardial Fibrosis in Pressure Overload-Induced Heart Failure. *Hypertension* **2018**, *71*, 280–288. [CrossRef] [PubMed]
206. Zhuang, X.; Li, R.; Maimaitijiang, A.; Liu, R.; Yan, F.; Hu, H.; Gao, X.; Shi, H. miR-221-3p inhibits oxidized low-density lipoprotein induced oxidative stress and apoptosis via targeting a disintegrin and metalloprotease-22. *J. Cell. Biochem.* **2019**, *120*, 6304–6314. [CrossRef] [PubMed]
207. Pereira-da-Silva, T.; Coutinho Cruz, M.; Carrusca, C.; Cruz Ferreira, R.; Napoleão, P.; Mota Carmo, M. Circulating microRNA profiles in different arterial territories of stable atherosclerotic disease: A systematic review. *Am. J. Cardiovasc. Dis.* **2018**, *8*, 1–13.
208. Coffey, S.; Williams, M.J.; Phillips, L.V.; Galvin, I.F.; Bunton, R.W.; Jones, G.T. Integrated microRNA and messenger RNA analysis in aortic stenosis. *Sci. Rep.* **2016**, *6*, 36904. [CrossRef]
209. Coskunpinar, E.; Cakmak, H.A.; Kalkan, A.K.; Tiryakioglu, N.O.; Erturk, M.; Ongen, Z. Circulating miR-221-3p as a novel marker for early prediction of acute myocardial infarction. *Gene* **2016**, *591*, 90–96. [CrossRef]
210. Sørensen, S.S.; Nygaard, A.B.; Nielsen, M.Y.; Jensen, K.; Christensen, T. miRNA expression profiles in cerebrospinal fluid and blood of patients with acute ischemic stroke. *Transl. Stroke Res.* **2014**, *5*, 711–718. [CrossRef]
211. Gusar, V.A.; Timofeeva, A.V.; Zhanin, I.S.; Shram, S.I.; Pinelis, V.G. Estimation of Time-Dependent microRNA Expression Patterns in Brain Tissue, Leukocytes, and Blood Plasma of Rats under Photochemically Induced Focal Cerebral Ischemia. *Mol. Biol.* **2017**, *51*, 683–695. [CrossRef]
212. Nie, X.; Chen, Y.; Tan, J.; Dai, Y.; Mao, W.; Qin, G.; Ye, S.; Sun, J.; Yang, Z.; Chen, J. MicroRNA-221-3p promotes pulmonary artery smooth muscle cells proliferation by targeting AXIN2 during pulmonary arterial hypertension. *Vascul. Pharmacol.* **2019**, *116*, 24–35. [CrossRef]
213. Villard, A.; Marchand, L.; Thivolet, C.; Rome, S. Diagnostic Value of Cell-free Circulating MicroRNAs for Obesity and Type 2 Diabetes: A Meta-analysis. *J. Mol. Biomark. Diagn.* **2015**, *6*, 251. [CrossRef] [PubMed]
214. Wang, L.; Xu, L.; Xu, M.; Liu, G.; Xing, J.; Sun, C.; Ding, H. Obesity-Associated MiR-342-3p Promotes Adipogenesis of Mesenchymal Stem Cells by Suppressing CtBP2 and Releasing C/EBPα from CtBP2 Binding. *Cell. Physiol. Biochem.* **2015**, *35*, 2285–2298. [CrossRef] [PubMed]
215. Hezova, R.; Slaby, O.; Faltejskova, P.; Mikulkova, Z.; Buresova, I.; Raja, K.R.; Hodek, J.; Ovesna, J.; Michalek, J. microRNA-342, microRNA-191 and microRNA-510 are differentially expressed in T regulatory cells of type 1 diabetic patients. *Cell. Immunol.* **2010**, *260*, 70–74. [CrossRef]
216. Eissa, S.; Matboli, M.; Bekhet, M.M. Clinical verification of a novel urinary microRNA panal: 133b, -342 and -30 as biomarkers for diabetic nephropathy identified by bioinformatics analysis. *Biomed. Pharmacother.* **2016**, *83*, 92–99. [CrossRef] [PubMed]
217. Cheng, S.; Cui, Y.; Fan, L.; Mu, X.; Hua, Y. T2DM inhibition of endothelial miR-342-3p facilitates angiogenic dysfunction via repression of FGF11 signaling. *Biochem. Biophys. Res. Commun.* **2018**, *503*, 71–78. [CrossRef]
218. Khalyfa, A.; Kheirandish-Gozal, L.; Bhattacharjee, R.; Khalyfa, A.A.; Gozal, D. Circulating microRNAs as Potential Biomarkers of Endothelial Dysfunction in Obese Children. *Chest* **2016**, *149*, 786–800. [CrossRef]
219. Hoekstra, M. MicroRNA-499-5p: A therapeutic target in the context of cardiovascular disease. *Ann. Transl. Med.* **2016**, *4*, 539. [CrossRef]
220. Zhao, L.; Wang, B.; Zhang, W.; Sun, L. Effect of miR-499a-5p on damage of cardiomyocyte induced by hypoxia-reoxygenation via downregulating CD38 protein. *J. Cell. Biochem.* **2020**, *121*, 996–1004. [CrossRef]

221. Neshati, V.; Mollazadeh, S.; Fazly Bazzaz, B.S.; de Vries, A.A.F.; Mojarrad, M.; Naderi-Meshkin, H.; Neshati, Z.; Mirahmadi, M.; Kerachian, M.A. MicroRNA-499a-5p Promotes Differentiation of Human Bone Marrow-Derived Mesenchymal Stem Cells to Cardiomyocytes. *Appl. Biochem. Biotechnol.* **2018**, *186*, 245–255. [CrossRef]
222. Boštjančič, E.; Zidar, N.; Glavač, D. MicroRNAs and cardiac sarcoplasmic reticulum calcium ATPase-2 in human myocardial infarction: Expression and bioinformatic analysis. *BMC Genom.* **2012**, *13*, 552. [CrossRef]
223. Salinas, J.; Lin, H.; Aparico, H.J.; Huan, T.; Liu, C.; Rong, J.; Beiser, A.; Himali, J.J.; Freedman, J.E.; Larson, M.G.; et al. Whole blood microRNA expression associated with stroke: Results from the Framingham Heart Study. *PLoS ONE* **2019**, *14*, e0219261. [CrossRef] [PubMed]
224. Baldeón Rojas, L.; Weigelt, K.; de Wit, H.; Ozcan, B.; van Oudenaren, A.; Sempértegui, F.; Sijbrands, E.; Grosse, L.; van Zonneveld, A.J.; Drexhage, H.A.; et al. Study on inflammation-related genes and microRNAs, with special emphasis on the vascular repair factor HGF and miR-574-3p, in monocytes and serum of patients with T2D. *Diabetol. Metab. Syndr.* **2016**, *8*, 6. [CrossRef]
225. Hromadnikova, I.; Kotlabova, K.; Dvorakova, L.; Krofta, L.; Sirc, J. Substantially Altered Expression Profile of Diabetes/Cardiovascular/Cerebrovascular Disease Associated microRNAs in Children Descending from Pregnancy Complicated by Gestational Diabetes Mellitus-One of Several Possible Reasons for an Increased Cardiovascular Risk. *Cells* **2020**, *9*, 1557.
226. Yoffe, L.; Polsky, A.; Gilam, A.; Raff, C.; Mecacci, F.; Ognibene, A.; Crispi, F.; Gratacós, E.; Kanety, H.; Mazaki-Tovi, S.; et al. Early diagnosis of gestational diabetes mellitus using circulating microRNAs. *Eur. J. Endocrinol.* **2019**, *181*, 565–577. [CrossRef] [PubMed]
227. Légaré, C.; Desgagné, V.; Poirier, C.; Thibeault, K.; White, F.; Clément, A.A.; Scott, M.S.; Jacques, P.É.; Perron, P.; Guérin, R.; et al. First trimester plasma microRNAs levels predict Matsuda Index-estimated insulin sensitivity between 24th and 29th week of pregnancy. *BMJ Open Diabetes Res. Care* **2022**, *10*, e002703. [CrossRef]
228. Juchnicka, I.; Kuźmicki, M.; Niemira, M.; Bielska, A.; Sidorkiewicz, I.; Zbucka-Krętowska, M.; Krętowski, A.J.; Szamatowicz, J. miRNAs as Predictive Factors in Early Diagnosis of Gestational Diabetes Mellitus. *Front. Endocrinol.* **2022**, *13*, 839344. [CrossRef]
229. Zhang, L.; Zhang, T.; Sun, D.; Cheng, G.; Ren, H.; Hong, H.; Chen, L.; Jiao, X.; Du, Y.; Zou, Y.; et al. Diagnostic value of dysregulated microribonucleic acids in the placenta and circulating exosomes in gestational diabetes mellitus. *J. Diabetes Investig.* **2021**, *12*, 1490–1500. [CrossRef]
230. Wang, J.; Pan, Y.; Dai, F.; Wang, F.; Qiu, H.; Huang, X. Serum miR-195-5p is upregulated in gestational diabetes mellitus. *J. Clin. Lab. Anal.* **2020**, *34*, e23325. [CrossRef]
231. Sørensen, A.E.; van Poppel, M.N.M.; Desoye, G.; Damm, P.; Simmons, D.; Jensen, D.M.; Dalgaard, L.T.; The DALI Core Investigator Group. The Predictive Value of miR-16, -29a and -134 for Early Identification of Gestational Diabetes: A Nested Analysis of the DALI Cohort. *Cells* **2021**, *10*, 170. [CrossRef]
232. Zhao, C.; Dong, J.; Jiang, T.; Shi, Z.; Yu, B.; Zhu, Y.; Chen, D.; Xu, J.; Huo, R.; Dai, J.; et al. Early second-trimester serum miRNA profiling predicts gestational diabetes mellitus. *PLoS ONE* **2011**, *6*, e23925. [CrossRef]
233. Kennedy, M.; Cartland, S.; Saravanan, P.; Simpson, N.; Scott, E.; Forbes, K. miR-1-3p and miR-133-3p are altered in maternal serum EVs and placenta in pregnancies complicated by gestational diabetes with large-for-gestational age babies. *Endocr. Abstr.* **2019**, *65*, 349. [CrossRef]
234. Wang, F.; Zhang, X.; Zhou, H. Role of cell free microRNA-19a and microRNA-19b in gestational diabetes mellitus patients. *3 Biotech.* **2019**, *9*, 406. [CrossRef]
235. Feng, Y.; Qu, X.; Chen, Y.; Feng, Q.; Zhang, Y.; Hu, J.; Li, X. MicroRNA-33a-5p sponges to inhibit pancreatic β-cell function in gestational diabetes mellitus LncRNA DANCR. *Reprod. Biol. Endocrinol.* **2020**, *18*, 61. [CrossRef] [PubMed]
236. Sebastiani, G.; Guarino, E.; Grieco, G.E.; Formichi, C.; Delli Poggi, C.; Ceccarelli, E.; Dotta, F. Circulating microRNA (miRNA) Expression Profiling in Plasma of Patients with Gestational Diabetes Mellitus Reveals Upregulation of miRNA miR-330-3p. *Front. Endocrinol.* **2017**, *8*, 345. [CrossRef] [PubMed]
237. Stirm, L.; Huypens, P.; Sass, S.; Batra, R.; Fritsche, L.; Brucker, S.; Abele, H.; Hennige, A.M.; Theis, F.; Beckers, J.; et al. Maternal whole blood cell miRNA-340 is elevated in gestational diabetes and inversely regulated by glucose and insulin. *Sci. Rep.* **2018**, *8*, 1366. [CrossRef] [PubMed]
238. Xu, K.; Bian, D.; Hao, L.; Huang, F.; Xu, M.; Qin, J.; Liu, Y. microRNA-503 contribute to pancreatic beta cell dysfunction by targeting the mTOR pathway in gestational diabetes mellitus. *EXCLI J.* **2017**, *16*, 1177–1187.
239. Liu, L.; Zhang, J.; Liu, Y. MicroRNA-1323 serves as a biomarker in gestational diabetes mellitus and aggravates high glucose-induced inhibition of trophoblast cell viability by suppressing TP53INP1. *Exp. Ther. Med.* **2021**, *21*, 230.
240. Hromadnikova, I.; Kotlabova, K.; Krofta, L. Cardiovascular Disease-Associated MicroRNA Dysregulation during the First Trimester of Gestation in Women with Chronic Hypertension and Normotensive Women Subsequently Developing Gestational Hypertension or Preeclampsia with or without Fetal Growth Restriction. *Biomedicines* **2022**, *10*, 256.
241. Hromadnikova, I.; Kotlabova, K.; Krofta, L. First-Trimester Screening for Fetal Growth Restriction and Small-for-Gestational-Age Pregnancies without Preeclampsia Using Cardiovascular Disease-Associated MicroRNA Biomarkers. *Biomedicines* **2022**, *10*, 718. [CrossRef]
242. Hromadnikova, I.; Kotlabova, K.; Krofta, L. First Trimester Prediction of Preterm Delivery in the Absence of Other Pregnancy-Related Complications Using Cardiovascular-Disease Associated MicroRNA Biomarkers. *Int. J. Mol. Sci.* **2022**, *23*, 3951. [CrossRef]

243. Hromadnikova, I.; Kotlabova, K.; Dvorakova, L.; Krofta, L. Diabetes Mellitus and Cardiovascular Risk Assessment in Mothers with a History of Gestational Diabetes Mellitus Based on Postpartal Expression Profile of MicroRNAs Associated with Diabetes Mellitus and Cardiovascular and Cerebrovascular Diseases. *Int. J. Mol. Sci.* **2020**, *21*, 2437. [CrossRef] [PubMed]
244. Shah, B.R.; Retnakaran, R.; Booth, G.L. Increased risk of cardiovascular disease in young women following gestational diabetes mellitus. *Diabetes Care* **2008**, *31*, 1668–1669. [CrossRef] [PubMed]
245. Kramer, C.K.; Campbell, S.; Retnakaran, R. Gestational diabetes and the risk of cardiovascular disease in women: A systematic review and meta-analysis. *Diabetologia* **2019**, *62*, 905–914. [CrossRef]
246. Yu, Y.; Soohoo, M.; Sørensen, H.T.; Li, J.; Arah, O.A. Gestational Diabetes Mellitus and the Risks of Overall and Type-Specific Cardiovascular Diseases: A Population- and Sibling-Matched Cohort Study. *Diabetes Care* **2022**, *45*, 151–159. [CrossRef] [PubMed]
247. Bentley-Lewis, R. Late cardiovascular consequences of gestational diabetes mellitus. *Semin. Reprod. Med.* **2009**, *27*, 322–329. [CrossRef]
248. Shostrom, D.C.V.; Sun, Y.; Oleson, J.J.; Snetselaar, L.G.; Bao, W. History of Gestational Diabetes Mellitus in Relation to Cardiovascular Disease and Cardiovascular Risk Factors in US Women. *Front. Endocrinol.* **2017**, *8*, 144. [CrossRef]
249. Peng, Y.S.; Lin, J.R.; Cheng, B.H.; Ho, C.; Lin, Y.H.; Shen, C.H.; Tsai, M.H. Incidence and relative risk for developing cancers in women with gestational diabetes mellitus: A nationwide cohort study in Taiwan. *BMJ Open* **2019**, *9*, e024583. [CrossRef]
250. Han, K.T.; Cho, G.J.; Kim, E.H. Evaluation of the Association between Gestational Diabetes Mellitus at First Pregnancy and Cancer within 10 Years Postpartum Using National Health Insurance Data in South Korea. *Int. J. Environ. Res. Public Health* **2018**, *15*, 2646. [CrossRef]
251. Liu, Y.; Chen, X.; Sheng, J.; Sun, X.; Chen, G.Q.; Zhao, M.; Chen, Q. Complications of Pregnancy and the Risk of Developing Endometrial or Ovarian Cancer: A Case-Control Study. *Front. Endocrinol.* **2021**, *12*, 642928. [CrossRef]
252. Simon, J.; Goueslard, K.; Arveux, P.; Bechraoui-Quantin, S.; Petit, J.M.; Quantin, C. Increased Risk of Hospitalization for Pancreatic Cancer in the First 8 Years after a Gestational Diabetes Mellitus regardless of Subsequent Type 2 Diabetes: A Nationwide Population-Based Study. *Cancers* **2021**, *13*, 308. [CrossRef]
253. Fuchs, O.; Sheiner, E.; Meirovitz, M.; Davidson, E.; Sergienko, R.; Kessous, R. The association between a history of gestational diabetes mellitus and future risk for female malignancies. *Arch. Gynecol. Obstet.* **2017**, *295*, 731–736. [PubMed]
254. Dawson, S.I. Long-term risk of malignant neoplasm associated with gestational glucose intolerance. *Cancer* **2004**, *100*, 149–155. [CrossRef] [PubMed]
255. Perrin, M.C.; Terry, M.B.; Kleinhaus, K.; Deutsch, L.; Yanetz, R.; Tiram, E.; Calderon-Margalit, R.; Friedlander, Y.; Paltiel, O.; Harlap, S. Gestational diabetes and the risk of breast cancer among women in the Jerusalem Perinatal Study. *Breast Cancer Res. Treat.* **2008**, *108*, 129–135. [PubMed]
256. Perrin, M.C.; Terry, M.B.; Kleinhaus, K.; Deutsch, L.; Yanetz, R.; Tiram, E.; Calderon, R.; Friedlander, Y.; Paltiel, O.; Harlap, S. Gestational diabetes as a risk factor for pancreatic cancer: A prospective cohort study. *BMC Med.* **2007**, *5*, 25. [CrossRef]
257. Metzger, B.E.; Coustan, D.R. Summary and recommendations of the Fourth International Workshop-Conference on Gestational Diabetes Mellitus. The Organizing Committee. *Diabetes Care* **1998**, *21*, B161–B167.
258. Livak, K.J.; Schmittgen, T.D. Analysis of relative gene expression data using real-time quantitative PCR and the 2(-Delta Delta C(T)) Method. *Methods* **2001**, *25*, 402–408.
259. Vandesompele, J.; De Preter, K.; Pattyn, F.; Poppe, B.; Van Roy, N.; De Paepe, A.; Speleman, F. Accurate normalization of re-al-time quantitative RT-PCR data by geometric averaging of multiple internal control genes. *Genome Biol.* **2002**, *3*, research0034.1.
260. Hromadnikova, I.; Kotlabova, K.; Hympanova, L.; Krofta, L. Gestational hypertension, preeclampsia and intrauterine growth restriction induce dysregulation of cardiovascular and cerebrovascular disease associated microRNAs in maternal whole peripheral blood. *Thromb. Res.* **2016**, *137*, 126–140.
261. Andersen, C.L.; Jensen, J.L.; Ørntoft, T.F. Normalization of real-time quantitative reverse transcription-PCR data: A model-based variance estimation approach to identify genes suited for normalization, applied to bladder and colon can-cer data sets. *Cancer Res.* **2004**, *64*, 5245–5250.
262. Benjamini, Y.; Hochberg, Y. Controlling the false discovery rate: A practical and powerful approach to multiple testing. *J. R. Stat. Soc. Ser. B* **1995**, *57*, 289–300.

Int. J. Mol. Sci. 2022, 23, 11448
International Journal of
*Molecular Sciences*
MDPI

*Article*

# Genome-Wide Copy Number Variant and High-Throughput Transcriptomics Analyses of Placental Tissues Underscore Persisting Child Susceptibility in At-Risk Pregnancies Cleared in Standard Genetic Testing

**Darina Czamara [1,†], Cristiana Cruceanu [1,2,†], Marius Lahti-Pulkkinen [3,4,5], Linda Dieckmann [1,6], Maik Ködel [1], Susann Sauer [1], Monika Rex-Haffner [1], Sara Sammallahti [7], Eero Kajantie [4,8,9,10], Hannele Laivuori [11,12,13], Jari Lahti [3], Katri Räikkönen [3] and Elisabeth B. Binder [1,14,*]**

1 Department of Translational Research in Psychiatry, Max-Planck-Institute of Psychiatry, 80804 Munich, Germany
2 Department of Physiology and Pharmacology, Karolinska Institutet, 171 77 Stockholm, Sweden
3 Department of Psychology and Logopedics, Faculty of Medicine, University of Helsinki, 00014 Helsinki, Finland
4 Finnish Institute for Health and Welfare, 00271 Helsinki, Finland
5 Centre for Cardiovascular Science, Queen's Medical Research Institute, University of Edinburgh, Edinburgh EH16 4TJ, UK
6 International Max Planck Research School for Translational Psychiatry, 80804 Munich, Germany
7 Department of Obstetrics and Gynaecology, Helsinki University Hospital and University of Helsinki, 00014 Helsinki, Finland
8 Children's Hospital, Helsinki University Hospital and University of Helsinki, 00014 Helsinki, Finland
9 Faculty of Medicine, PEDEGO Research Unit, MRC Oulu, Oulu University Hospital and University of Oulu, 90014 Oulu, Finland
10 Department of Clinical and Molecular Medicine, Norwegian University of Science and Technology, 7491 Trondheim, Norway
11 Institute for Molecular Medicine Finland, HiLIFE, University of Helsinki, 00014 Helsinki, Finland
12 Medical and Clinical Genetics, University of Helsinki and Helsinki University Hospital, 00014 Helsinki, Finland
13 Department of Obstetrics and Gynecology, Tampere University Hospital and Faculty of Medicine and Health Technology, Center for Child, Adolescent and Maternal Health Research, Tampere University, 33520 Tampere, Finland
14 Department of Psychiatry and Behavioral Sciences, School of Medicine, Emory University, Atlanta, GA 30322, USA
* Correspondence: binder@psych.mpg.de
† These authors contributed equally to this work.

**Citation:** Czamara, D.; Cruceanu, C.; Lahti-Pulkkinen, M.; Dieckmann, L.; Ködel, M.; Sauer, S.; Rex-Haffner, M.; Sammallahti, S.; Kajantie, E.; Laivuori, H.; et al. Genome-Wide Copy Number Variant and High-Throughput Transcriptomics Analyses of Placental Tissues Underscore Persisting Child Susceptibility in At-Risk Pregnancies Cleared in Standard Genetic Testing. *Int. J. Mol. Sci.* **2022**, *23*, 11448. https://doi.org/10.3390/ijms231911448

Academic Editor: Ilona Hromadnikova

Received: 9 August 2022
Accepted: 26 September 2022
Published: 28 September 2022



**Abstract:** Several studies have shown that children from pregnancies with estimated first-trimester risk based on fetal nuchal translucency thickness and abnormal maternal serum pregnancy protein and hormone levels maintain a higher likelihood of adverse outcomes, even if initial testing for known genetic conditions is negative. We used the Finnish InTraUterine cohort (ITU), which is a comprehensively characterized perinatal cohort consisting of 943 mothers and their babies followed throughout pregnancy and 18 months postnatally, including mothers shortlisted for prenatal genetic testing but cleared for major aneuploidies (cases: n = 544, 57.7%) and control pregnancies (n = 399, 42.3%). Using genome-wide genotyping and RNA sequencing of first-trimester and term placental tissue, combined with medical information from registry data and maternal self-report data, we investigated potential negative medical outcomes and genetic susceptibility to disease and their correlates in placenta gene expression. Case mothers did not present with higher levels of depression, perceived stress, or anxiety during pregnancy. Case children were significantly diagnosed more often with congenital malformations of the circulatory system (4.12 (95% CI [1.22–13.93]) higher hazard) and presented with significantly more copy number duplications as compared to controls (burden analysis, based on all copy number variants (CNVs) with at most 10% frequency, 823 called duplications in 297 cases versus 626 called duplications in 277 controls, $p = 0.01$). Fifteen genes showed differential gene expression (FDR < 0.1) in association with congenital malformations in first-trimester but not term placenta. These were significantly enriched for genes associated with placental dysfunction. In spite of normal routine follow-up prenatal testing results in early pregnancy, case children presented with an increased likelihood of negative outcomes, which should prompt vigilance in follow-up during pregnancy and after birth.

**Keywords:** chorionic villus sampling; congenital malformations; placenta; prenatal testing transcriptome sequencing

---

## 1. Introduction

Prenatal screening in early pregnancy is instrumental in monitoring maternal health and proper child development. Usually, women at risk for carrying babies with major aneuploidies are identified using an assessment of risk by a combined first-trimester screening battery that includes: fetal nuchal translucency (NT) thickness, maternal levels of serum pregnancy-associated plasma protein A (PAPP-A) and free β-human chorionic gonadotropin (β-hCG) [1] in addition to maternal age, maternal weight and gestational week. Although this screening test has proven to successfully identify risk for the major chromosomal aneuploidies, several studies have shown that fetuses with elevated risk at screening but cleared for the major aneuploidies in prenatal genetic testing remain at higher risk for negative outcomes diagnosed before or after birth [2–4]. Bardi et al. reported that 34% of congenital malformation cases would be missed if prenatal testing only relied on the standard screening for major aneuploidies [2]. Zhang et al. [4] showed that significantly more clinically relevant pathogenic copy number variants (CNVs) are present in fetuses with NT between 2.5 mm and 3.5 mm, which is below the commonly used threshold of 3.55 mm. Furthermore, low PAPP-A serum levels in the first trimester of pregnancy have been associated with short stature in offspring and de-novo development of maternal diabetes mellitus later in life [3]. This suggests that using these values on a continuum rather than as a threshold in the context of other indicators of risk might provide added value to the current clinical practices and detect individuals with underlying disease susceptibility otherwise not identified in the prenatal period.

Chorionic villus sampling (CVS), amniocentesis, or noninvasive prenatal testing (NIPT) of fetal DNA in combination with 11-to-13-week ultrasound examination, can detect major negative developmental outcomes with 70–100% detection accuracy [5,6]. Invasive approaches have been available since the 1970s [7], while noninvasive approaches have only become routine practice in developed countries in the last 10 years [6]. The latter have rapidly developed into the method of choice, given the much lower risk for mother and fetus compared to invasive techniques, and are being administered in more than half of pregnancies in developed countries [5]. However, NIPT is a screening test, and invasive techniques are still considered state-of-the-art in validating suspicious noninvasive results [6].

We set out to understand if children from pregnancies flagged for potential chromosomal aberrations but ultimately cleared for the major aneuploidies following CVS or amniocentesis differed from children of inconspicuous pregnancies and how detection of prevailing pathologies could be improved. To explore this question, we used a thoroughly characterized perinatal cohort, the Finnish InTraUterine cohort (ITU), consisting of 943 mothers and babies followed throughout pregnancy and postnatally until 18 months. We used genome-wide high-throughput analyses at the level of RNA and DNA methylation in cord blood, term placentas, and, to the best of our knowledge, in the largest sample to date of CVS biopsies in the subset of pregnancies in which chromosomal testing was performed. Our findings point to an enduring potential for concern for case children, with increased risk for congenital malformations.

## 2. Results

### 2.1. Case Mothers Do Not Differ from Controls in Psychological State across Pregnancy following Trisomy Screening Clearance

As shown by Kvist et al. [8], mothers shortlisted for prenatal screening (cases, n = 544) significantly differed from control mothers (n = 399): they were older, presented with higher body mass index (BMI), had had more pregnancies and smoked more often. By design, cases also significantly differed from controls with regards to NT, PAPP-A, and β-hCG levels as well as risk for Down syndrome or trisomy 18. The main characteristics of the cohort and of the screening variables are presented in Table 1.

**Table 1.** Demographic information of the ITU cohort.

| Maternal Characteristics | Cases (n = 544) | Controls (n = 399) | $p$-Value [5] |
|---|---|---|---|
| Age at delivery, years, mean (SD), CI | 35.54 (5.5) [35.52–35.56] | 33.71 (4.2) [33.69–33.73] | $\mathbf{7.16 \times 10^{-9}}$ |
| Antenatal corticosteroid treatment, n (%) | 23 (4.2%) | 9 (2.3%) | $1.41 \times 10^{-1}$ |
| Cesarean section, n (%) | 120 (22.1%) | 79 (19.8%) | $4.48 \times 10^{-1}$ |
| Diabetes Disorders in pregnancy, n (%) | 120 (22.06%) | 77 (19.3%) | $3.43 \times 10^{-1}$ |
| Early pregnancy BMI, median (IQR), CI | 23.43 (5.23) [23.05–23.74] | 22.65 (4.24) [22.27–23.03] | $\mathbf{1.42 \times 10^{-3}}$ |
| Hypertensive Disorders in pregnancy, n (%) | 46 (8.5%) | 20 (5.0%) | $5.50 \times 10^{-2}$ |
| Primiparous, n (%) | 221 (40.6%) | 243 (60.9%) | $\mathbf{1.15 \times 10^{-9}}$ |
| Smoking during pregnancy [1], n (%) | 36 (6.6%) | 4 (1.0%) | $\mathbf{1.83 \times 10^{-5}}$ |
| Thyroid disorders [2], n (%) | 9 (1.7%) | 8 (2.0%) | $8.79 \times 10^{-1}$ |
| **Child characteristics** | | | |
| Birth weight, g, median (IQR), CI | 3533 (675) [3470–3585] | 3562 (610) [3508–3620] | $3.83 \times 10^{-1}$ |
| Gestational age at birth, weeks, median (IQR), CI | 40.00 (2.0) [40.00–40.14] | 40.14 (1.7) [40.00–40.29] | $5.17 \times 10^{-1}$ |
| Preterm birth [3] (<37 weeks), n (%) | 33 (6.1%) | 13 (3.3%) | $6.80 \times 10^{-2}$ |
| Sex, girl, (%) | 256 (47.1%) | 204 (51.1%) | $2.42 \times 10^{-1}$ |
| **Screening variables** | | | |
| β-hCG level (MoM, ug/L), median (IQR), CI | 1.31 (1.17) [1.30–1.73] | 0.98 (0.74) [0.93–1.08] | $\mathbf{1.20 \times 10^{-11}}$ |
| NT, MoM mm, median (IQR), CI | 1.80 (1.15) [1.58–1.88] | 0.84 (0.32) [0.83–0.88] | $\mathbf{1.58 \times 10^{-36}}$ |
| PAPP-A level, MoM, mU/L, median (IQR), CI | 0.75 (0.80) [0.75–1.08] | 1.12 (0.73) [1.09–1.20] | $\mathbf{3.45 \times 10^{-18}}$ |
| Risk for Down syndrome [4], median (IQR), CI | 0.65 (1.01) [0.63–0.89] | 0.01 (0.2) [0.008–0.11] | $\mathbf{3.06 \times 10^{-101}}$ |
| Risk for trisomy 18 [4], median (IQR), CI | 0.003 (0.01) [0.003–0.014] | 0.001 (0.00) [0.001–0.001] | $\mathbf{1.15 \times 10^{-64}}$ |

BMI: body mass index, calculated as weight in kg divided by height in meters squared. CI: confidence interval, CI for median was calculated as suggested in [9]. IQR: interquartile range. MoM: multiple of median. NT: nuchal translucency. SD: standard deviation. [1] Women who quit smoking in the first trimester were counted as non-smokers. [2] Thyroid disorders are based on ICD-10 codes E00-E07. [3] Preterm birth was defined as birth at gestational weeks < 37. [4] Risk for Down syndrome and for trisomy 18 were calculated as risk ratios based on PAPP-A and β-hCG levels, NT as well as maternal age, BMI and gestational weeks (see also Kvist et al. [8]). [5] Nominal $p$-value of testing phenotype in cases versus phenotype in controls, based on $t$-tests/Wilcoxon- tests for quantitative traits and on chi-square tests for categorical traits, nominally significant $p$-values < 0.05 are depicted in **bold**. Quantitative variables were checked for normality. If the Kolmogorov-Smirnow-test provided no indication for deviation from normality, mean and SD are reported and $p$-values are based on $t$-tests. If the variable was not normally distributed, median and IQR are reported and $p$-values are based on Wilcoxon tests.

Furthermore, we assessed whether maternal psychological state differed in relation to case status, i.e., if mothers who had undergone prenatal testing possibly presented with higher levels of depression, anxiety or perceived stress due to the testing or its possible implications for the health of their child. Questionnaires on well-being throughout pregnancy were completed by 613 women, at three time points roughly representative of the three trimesters. We found no significant differences between cases and controls in any of the assessments during pregnancy (see Table 2) indicating that mothers did not differ in their psychological state across pregnancy.

**Table 2.** Self-reported maternal psychological health in cases and controls.

| Phenotype | Cases (n = 260) | Controls (n = 353) | *p*-Value [1] |
|---|---|---|---|
| CESD 19 gestational weeks, median (IQR), CI | 9 (9) [9–14] | 9 (8) [9–11] | 0.32 |
| CESD 26 gestational weeks, median (IQR), CI | 8 (10) [9–12] | 9 (8) [9–11] | 0.35 |
| CESD 38 gestational weeks, median (IQR), CI | 8 (9) [8–11] | 9 (9) [9–10] | 0.43 |
| Cohen's perceived stress scale 19 gestational weeks, median (IQR), CI | 5 (4.06) [6–8] | 6 (4) [6–7] | 0.48 |
| Cohen's 26 gestational weeks, median (IQR), CI | 5 (5) [5–7] | 5.5 (3) [6–6] | 0.30 |
| Cohen's perceived stress scale 38 gestational weeks, median (IQR), CI | 5 (4) [5–6] | 5 (4) [4.5–6] | 0.87 |
| STAI 19 gestational weeks, median (IQR). CI | 37 (11) [39–42] | 38 (9) [38.95–41] | 0.74 |
| STAI 26 gestational weeks, median (IQR), CI | 36 (9) [37–40] | 37 (10) [37–39] | 0.38 |
| STAI 38 gestational weeks, median (IQR), CI | 36 (10.42) [36–39] | 37 (10) [36–38] | 0.70 |

CESD: Center for Epidemiological Studies Depression. CI: confidence interval, CI for median were calculated as suggested in [9]. IQR: interquartile range. STAI: The State-Trait Anxiety Inventory. [1] *p*-value: Due to the non-normal distribution of the questionnaires, median and interquartile range (IQR) are presented, *p*-values are based on Wilcoxon tests.

## *2.2. Children from Pregnancies Recommended for Follow-Up Prenatal Genetic Testing Carry Higher Likelihood of Negative Developmental and Disease-Related Outcomes*

### 2.2.1. Congenital Malformations

We examined whether cases and controls differed with regards to negative medical outcomes and genetic susceptibility, starting with congenital malformations, as the first indication of underlying disease risk. Of 943 children, 98 (68 cases, 12.5% and 30 controls, 7.5%) presented with any congenital malformations diagnosed up until 2017. We ran Cox regression models to account for different diagnostic follow-up duration in the Finnish nationwide Care Register for Healthcare (HILMO) for children born in different years. Cox regression models showed no difference between groups in all malformation types ($p = 0.14$, hazard ratio = 1.39, 95% CI [0.47–1.11]). However, in circulatory system congenital malformations comprising 25 diagnoses (22 cases, 4.0%, and 3 controls, 0.8%), we found a 4.12 (95% CI [1.22–13.93]) higher hazard in the cases ($p = 0.02$, see Figure 1). The majority of affected children (n = 17, 65.4%) presented with ventricular septal defects. Beyond malformations of the circulatory system, there were some isolated cases of malformations of the respiratory, musculoskeletal, genital, digestive, or nervous systems that were too infrequent to permit statistical analysis.

In a next step, we investigated if congenital malformations of the circulatory system were specifically associated with any of the screening variables. While there were no significant associations with higher NT, higher β-hCG levels or higher risk for trisomy 18, children with congenital malformations of the circulatory system presented with significantly higher estimated ratio for Down syndrome and lower, albeit not significantly different, PAPP-A levels. These findings remained stable, even after correcting for case-control status, which was itself defined using the screening variables (risk for Down syndrome: $p = 0.021$, PAPP-A level: $p = 0.058$).

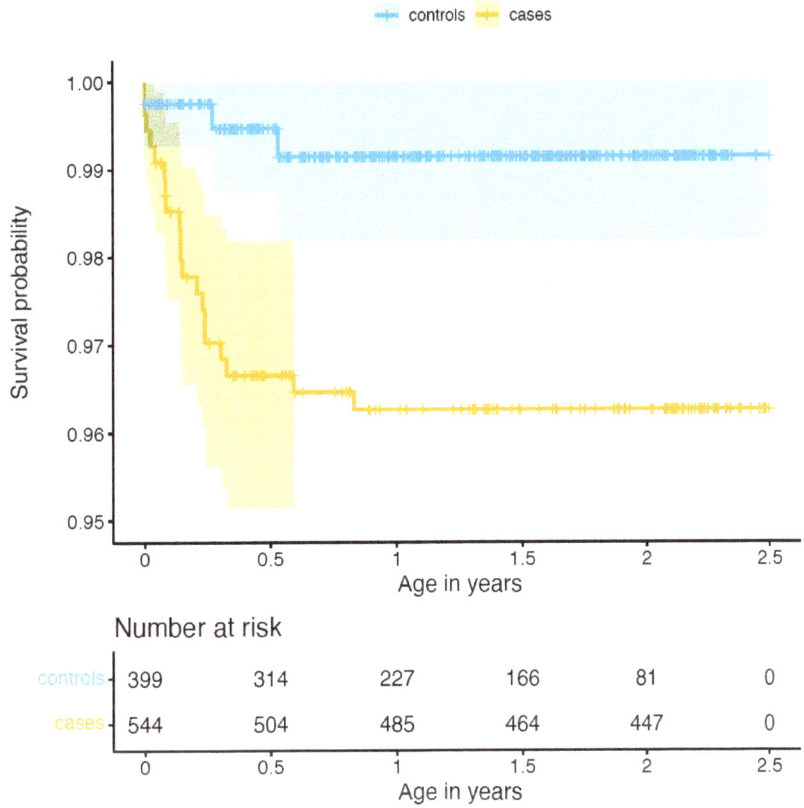


**Figure 1.** Survival curve for congenital malformations of the circulatory system compared between cases and controls. 'Number at risk' indicates the number of cases and controls available for analysis at 0–2.5 years of age. As cases presented with significantly longer follow-up time compared to controls, we limited the follow-up time of cases to the age of the eldest child in the control group (see also Methods).

2.2.2. Copy Number Variants

Given the importance of CNVs in disease susceptibility, including in association with congenital malformations [10], and the association of sub-threshold screening results with a higher number of CNVs [4], we next evaluated CNVs in our sample. We first called CNVs across all participants. The large majority of CNVs, over 92%, was classified as not pathogenic (see Methods). We found no significant differences between individuals with or without congenital malformations (any or circulatory system specific), neither on single CNV level nor based on burden analysis. This is possibly due to a reduced detection power given the low number of congenital malformations. Since significantly more clinically relevant pathogenic CNVs have been associated with risk-level prenatal screening variables such as maternal age and NT abnormalities [4,11], we also tested this association at the level of the entire cohort. In our cohort, cases presented with overall more duplications, but not more deletions, as compared to controls (burden analysis, based on all CNVs with at most 10% frequency, 823 called duplications in 297 cases versus 626 called duplications in 277 controls, $p = 0.01$, based on 10,000 permutations, see Figure 2). This range of CNV counts is in line with previous studies using similar technology and sample sizes [12,13].

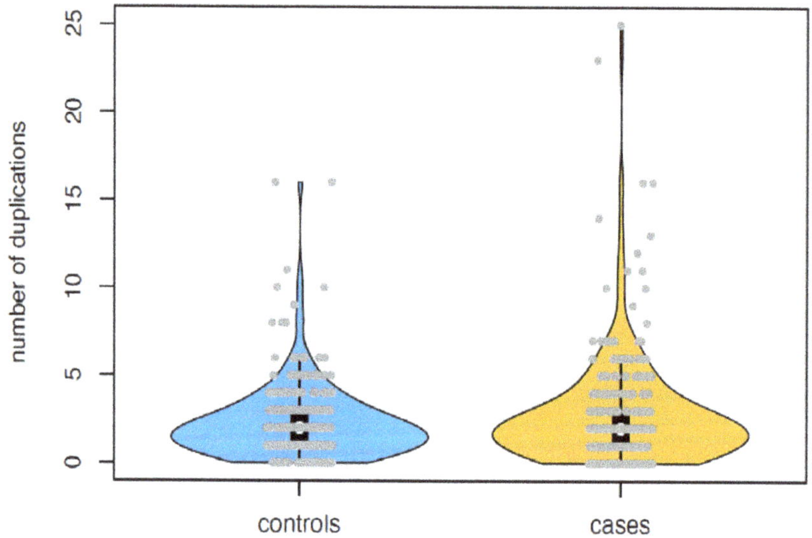


**Figure 2.** Differences in CNVs between cases and controls. Violin plot for numbers of called duplications between controls and cases. The difference remains significant after exclusion of two cases with more than 20 duplications ($p = 0.03$ based on 10,000 permutations).

### 2.3. *The Placental Transcriptome in Early Pregnancy Carries Signatures of Case-Control-Associated Negative Developmental Outcomes*

Given the observed differences in congenital malformations within the circulatory system and in CNVs, we next assessed if these were also associated with RNA expression in early and late placental tissue. Gene expression derived from CVS tissue was available for 266 case mothers who had undergone invasive prenatal testing via placental biopsies in the first trimester. We found no differentially expressed genes to associate with CNVs (number of duplications). However, 15 genes were significantly differentially upregulated in individuals with circulatory system congenital malformations (10 individuals with and 256 without diagnosed congenital malformations; see Figure 3 and Table 3).

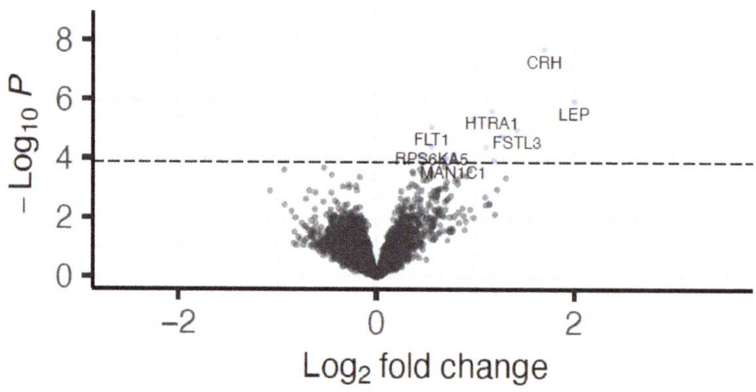


**Figure 3.** Volcano plot of differential gene expression in CVS with congenital malformations of the circulatory system. Genes differentially expressed at FDR of 0.10 are depicted in blue.

**Table 3.** Differentially expressed genes in CVS with congenital malformations within the circulatory system.

| Gene [1] | Position (hg19) [2] | Log2 (FC) [3] | *p*-Value [4] | Adjusted *p*-Value [5] | Correlation CVS and Placenta [6] |
|---|---|---|---|---|---|
| *LEP* | chr7: 127,881,331–127,897,682 | 1.99 | $1.16 \times 10^{-6}$ | $5.28 \times 10^{-3}$ | r = 0.27; $p = 0.01$ |
| *CRH* | chr8: 67,088,612–67,090,846 | 1.69 | $1.99 \times 10^{-8}$ | $1.81 \times 10^{-4}$ | r = 0.14; $p = 0.17$ |
| *FSTL3* | chr19: 676,389–683,392 | 1.42 | $1.11 \times 10^{-5}$ | $1.68 \times 10^{-2}$ | r = 0.26; $p = 0.01$ |
| *PAPPA2* | chr1: 176,432,307–176,811,970 | 1.26 | $1.79 \times 10^{-5}$ | $2.32 \times 10^{-2}$ | r = 0.30; $p < 0.01$ |
| *INHBA* | chr7: 41,733,514–41,818,976 | 1.18 | $1.18 \times 10^{-4}$ | $7.36 \times 10^{-2}$ | r = 0.23; $p = 0.03$ |
| *HTRA1* | chr10: 124,221,041–124,274,424 | 1.16 | $2.38 \times 10^{-6}$ | $7.22 \times 10^{-3}$ | r = 0.30; $p < 0.01$ |
| *DHRS2* | chr14: 24,105,573–24,114,848 | 1.11 | $4.16 \times 10^{-5}$ | $4.20 \times 10^{-2}$ | r < 0.01; $p = 0.97$ |
| *HS3ST3B1* | chr17: 14,204,506–14,249,492 | 0.98 | $4.27 \times 10^{-6}$ | $9.71 \times 10^{-3}$ | r = 0.13; $p = 0.23$ |
| *ANXA4* | chr2: 69,969,127–70,053,596 | 0.78 | $7.15 \times 10^{-5}$ | $6.47 \times 10^{-2}$ | r < 0.01; $p = 0.97$ |
| *MAN1C1* | chr1: 25,943,959–26,111,258 | 0.78 | $1.15 \times 10^{-4}$ | $7.36 \times 10^{-2}$ | r = 0.05; $p = 0.60$ |
| *MROH1* | chr8:145,202,919–145,316,843 | 0.70 | $9.64 \times 10^{-5}$ | $7.30 \times 10^{-2}$ | r = −0.14; $p = 0.17$ |
| *SEMA7A* | chr15: 74,701,630-74,726,299 | 0.68 | $1.21 \times 10^{-4}$ | $7.36 \times 10^{-2}$ | r = 0.07; $p = 0.53$ |
| *FLT1* | chr13: 28,874,483–29,069,265 | 0.56 | $8.27 \times 10^{-6}$ | $1.50 \times 10^{-2}$ | r = 0.32; $p < 0.01$ |
| *RPS6KA5* | chr14: 91,337,167–91,526,993 | 0.56 | $4.06 \times 10^{-5}$ | $4.20 \times 10^{-2}$ | r = 0.30; $p < 0.01$ |
| *HEXIM1* | chr17: 43,224,684–43,229,468 | 0.45 | $7.83 \times 10^{-5}$ | $6.47 \times 10^{-2}$ | r < 0.01; $p = 0.97$ |

[1] Gene: name of differentially expressed gene. [2] Position: gene position in hg19 coordinates. [3] Log2(FC): log2 (fold change) between cases and controls. [4] *p*-value: *p*-value for differential gene expression for congenital malformations within the circulatory system, adjusted for surrogate variable (SV), maternal age, maternal BMI in early pregnancy, smoking, parity, child's sex, and gestational age at sampling. [5] Adjusted *p*-value: *p*-value after Benjamini–Hochberg correction for multiple testing (FDR). [6] Correlation: Pearson's correlation coefficient and nominal *p*-value for correlation of gene expression between CVS and placental tissue (based on 93 individuals with gene expression available in both tissues and 7357 genes available in both tissues).

Correlation of gene expression for these 15 genes between CVS and term placenta (based on 93 individuals who had both tissues available) revealed significant, albeit weak to moderate, positive correlations for 7 of these transcripts (r = 0.23 to r = 0.30, see Table 3). Of these 15 genes, 11 had previously been shown to be dysregulated in pre-eclampsia [14–16]. This is significantly more than expected by chance ($p = 9.91 \times 10^{-11}$, see Methods). Importantly, it should be noted that only one case presenting with congenital malformations of the circulatory system also had a pre-eclampsia diagnosis. Removal of this case did not change the effect direction of the 15 genes.

In term placenta, we found no significantly differentialy expressed genes with circulatory system congenital malformations. However, it should be noted that only six individuals with this phenotype and term placenta gene expression were available hence this null result may be due to power issues.

Number of CNV duplications was associated with differential placental gene expression of *RFLNB* (based on 433 samples with information on CNVs and placental gene expression, log2(FC) = 0.07, adjusted *p*-value = $3.72 \times 10^{-2}$, correlation with CVS gene expression r = 0.16, *p*-value = 0.13). This result stayed significant after accounting for case-control status with regard to prenatal screening.

## 3. Discussion

Advances in medical prenatal screening and care during early pregnancy have led to life-saving effects, with infant mortality decreasing steadily over the past two decades, by as much as three-fold in high-income countries and by half globally [17]. However, there are still improvements to be made regarding access to state-of-the-art prenatal screening, as well as improved interpretation of available diagnostic tools. While sampling of CVS tissue or amniotic fluid to test the fetal DNA for major chromosomal rearrangements has been the gold standard for many decades, there is a significant risk for miscarriage linked with these invasive techniques [6]. In addition, they could lead to false positive or false negative outcomes in addition to sampling inaccuracy. Noninvasive prenatal screening techniques based on whole-genome DNA sequencing have become increasingly common and accurate and represent a suitable companion or replacement technique since they can detect cell-free fetal DNA in the mother's blood that is released from fetal placenta apoptotic trophoblasts [6]. However, a major limitation of this approach is sensitivity, given that circulating fetal DNA constitutes only a small fraction of the mother's blood [6]. Nonetheless, genetic screening, in addition to combined first-trimester testing for fetal NT and maternal PAPP-A and βhCG, has been very powerful in detecting individuals susceptible to congenital disease. With this work, we show that beyond indicating risk for major aneuploidies, common prenatal screening measurements can indicate the presence of additional risk for disease.

An interesting resource for improving prenatal or early-life diagnostics is the incorporation of perinatal tissues such as the placenta and cord blood into the interpretation of congenital and genetic risk of disease. In this manuscript, we investigated a case-control cohort of 943 mother-baby dyads whereby the case mothers (n = 544) were initially screened as having an increased risk for chromosomal abnormalities, but ultimately the children were deemed normal in regard to major aneuploidies following invasive genetic testing or NIPT. We compared these to a group of controls without any risk indicator at screening (n = 399). We found significant differences at the level of perinatal child outcomes regarding congenital malformations of the circulatory system, as well as CNV load. In addition, we found an association between child outcomes and first-trimester but not term placenta gene expression signatures. Interestingly, there was no difference in self-reported maternal stress or anxiety during pregnancy, in spite of the case mothers being exposed to the stressful experience of invasive prenatal testing and anticipation of results. However, it bares noting that there was a disproportionately lower adherence in the case mothers regarding the psychological self-assessment during pregnancy, which might influence the outcome of these analyses. Future studies should assess the impact of a similar exposure in additional cohorts.

We found an increased likelihood of a diagnosis of circulatory system congenital malformation for case children, which supports previous research linking abnormal NT results at screening with increased rates of congenital malformations and heart disease [2]. However, this study is the first to show that the congenital malformations phenotype is associated with gene expression in the first-trimester placenta from CVS sampling. A total of 15 genes showed differential expression association with circulatory system congenital malformations in first-trimester placenta obtained from CVS. Correlations with gene expression in placentas sampled at birth were weak to moderate, so we cannot directly translate our findings to late pregnancy. Of these 15 genes, 11 had previously been shown to be dysregulated in pre-eclampsia [14–16], significantly more than expected by chance.

A number of the genes differentially expressed in association with congenital malformations in the CVS tissue have previously been shown to be important mediators of placenta biology and associated with pregnancy diseases. The top-ranking gene by fold change, Leptin (*LEP*), is a circulating hormone involved in metabolism and energy homeostasis [18] and secreted primarily from adipose tissue but also from placenta cells such as cytotrophoblasts, syncytiotrophoblasts, and villous vascular endothelial cells. This hormone is upregulated in the placentas of mothers with pregnancy complications such as pre-eclampsia [19], which is the same direction of effect we found for early placentas from children with underlying congenital malformations. Secondly, corticotropin-releasing hormone (*CRH*) is a major mediator of hypothalamic–pituitary–adrenal axis (HPA axis) responses to stress and a key regulator of brain development. In addition to the paraventricular nucleus (PVN) of the hypothalamus, CRH is also synthesized by the placenta (pCRH) as early as post-conceptional week 7 and is found in both the maternal and the fetal compartments [20]. Down-regulation of this gene at the RNA level has been linked to placenta inflammation, while up-regulation, as in the case of our analyses, has been linked to motor dysfunctions [21] in children and pre-eclampsia in pregnancy [22]. Thirdly, elevated levels of the follistatin-like 3 protein (*FSTL3*) gene, which encodes a protein secreted by syncytiotrophoblast cells, have been linked to pre-eclampsia by several studies. In fact, placentally-derived FSTL3 detected in the maternal serum has been proposed as a third-trimester diagnostic biomarker of pre-eclampsia [23]. The next differentially expressed genes by fold change were pregnancy-associated plasma protein A2 (*PAPPA2*) [24], inhibin beta A subunit (*INHBA*) [25], and HtrA serine peptidase 1 (*HTRA1*) [26]. All these genes, in addition to several others identified through our analyses, were previously shown to be elevated at the mRNA and protein level in pre-eclampsia with severe features, early-onset pre-eclampsia, or fetal growth restriction. These findings may suggest that first-trimester placental tissue of fetuses with congenital anomalies presents with signatures of placental dysfunction with a possible additional impact on fetal health. Interestingly this signature disappeared with the maturation of the placenta, and there was no increase in rates of pre-eclampsia diagnoses.

Given the indication of existing susceptibility to disease in the early fetal placental tissue, we used child DNA extracted from cord blood to further investigate complex genetics that would not have been detected by prenatal genetic screening. We focused on CNVs, given their association with both congenital malformations [10] and general pregnancy risk factors such as advanced maternal age [11]. In addition to the previously suggested associations of increased numbers of pathogenic CNVs with positive prenatal screening [4], we demonstrated a novel connection whereby the cases presented with more duplications overall. Together these findings suggest an overall increase in potentially harmful genetic loading in the case children. Importantly, this detail of genetic variation could not be detected with currently available prenatal genetic screening methods due to limited fetal DNA amounts available at early stages.

The analyses described in this manuscript identified significant connections between first-trimester placenta gene expression (CVS) and child outcomes, specifically regarding congenital malformations. Conversely, associations to the number of CNV duplications were only identified in the term placenta RNA, related to the gene *RFLNB*, which is enriched in Hofbauer cells and previously associated with early signs of pre-eclampsia in the placental transcriptome [27]. This apparent discrepancy between first-trimester placentas from CVS and term placentas could be due to different reasons. Primarily, samples with CVS and term placenta tissue only partly overlap, and hence we cannot make robust conclusions on the individual developmental trajectories. In fact, only one individual with a circulatory system congenital malformation diagnosis also had both tissues available. However, having access to early pregnancy tissues, as in the ITU cohort, is extremely valuable in establishing the connections between placenta biology and underlying child disease risk. As mentioned above, our data may point to additional early placental dysfunction in children with congenital malformations of the circulatory system. Functional studies of early placental physiology could shed more light on the possible clinical relevance of this finding.

Some of the limitations of this study highlight the need for even larger, well-characterized longitudinal cohorts, where these findings can be independently replicated and extended with measures of placental function. Importantly, extended follow-up of the children will undoubtedly clarify and complete the picture of potential negative outcomes in this at-risk population. The ITU cohort, with the ongoing characterization of participants in addition to Finnish health registry data, will constitute a rich resource moving forward.

The findings identified in this study join other reports suggesting that children with elevated risk but not meeting formal diagnostic thresholds during prenatal screening maintain a higher likelihood of negative outcomes. Increased awareness is key at the level of medical professionals as well as the general public with regard to potential early indicators of underlying disease susceptibility and the prevention or intervention measures that can be taken to improve the outcome and quality of life of affected individuals. In our cohort, cases were significantly older as compared to control women, which is in line with what also Hayeems et al. reported [28]. Women of higher age are more likely to undergo prenatal testing, and higher maternal age has been associated with a higher risk for fetal aneuploidies [29]. Given that maternal age at first birth is shifting to older ages [30], this will likely be an issue of increasing importance. Our findings point to a need for redefining the risk classifications in relation to prenatal screening and more vigilance during prenatal and postnatal follow-up of at-risk children.

## 4. Materials and Methods

### 4.1. Sampling and Phenotypes

#### 4.1.1. Study Samples

The InTraUterine sampling in early pregnancy (ITU) cohort consists of 943 Finnish women and their children born between 2012 and 2017. The cohort, described by Kvist et al. [8], is a prospective, longitudinal pregnancy cohort study comprising a total of 943 women. Pregnant women were recruited at maternity clinics and through the Helsinki and Uusimaa Hospital District Fetomaternal Medical Center in Finland. Eligibility criteria included singleton pregnancy, no prenatal diagnosis of chromosomal abnormality, maternal age $\geq$ 18 years, and sufficient Finnish language ability to ensure informed consent. The ITU study comprises two study arms. Women in the chromosomal testing arm (cases, n = 544) had been referred to the Helsinki and Uusimaa Hospital District Fetomaternal Medical Center (FMC) because they had an increased risk of fetal chromosomal abnormalities based on routine serum and ultrasound screening, age, and patient characteristics. The screening is described in detail in Kvist et al. [8]. In brief, the screening program was a combination of serum screening and ultrasound examinations, including a nuchal translucency scan. Women who had a positive screening result (i.e., an estimated risk of fetal chromosomal abnormality >1:250, based on serum and ultrasound screening, age, maternal height and weight, and prior history) were then offered fetal chromosomal testing (CVS or amniocentesis followed by trisomy PCR, or noninvasive prenatal testing) at FMC. If the chromosomal test indicated no fetal chromosomal abnormalities, those who had expressed interest in participating were contacted for final recruitment. Those whose chromosomal test results suggested a fetal chromosomal abnormality were not recruited. Women in the no-chromosomal testing arm (controls, n = 399) were informed about ITU when attending the same routine serum and ultrasound screening at maternity clinics as the women in the chromosomal testing arm. Women who expressed interest in participating were contacted for final recruitment into this study arm if they had not been referred to FMC for fetal chromosomal testing. After careful inspection, one woman was excluded from the analyses post hoc as she did not meet the clear-cut case-control definition. One woman who had originally been accidentally coded as control was recoded as case. This resulted in 399 controls and 544 case women in the final analysis. Demographic information on the ITU cohort, as well as on the screening variables, is presented in Table 1.

#### 4.1.2. Phenotypes

Maternal Characteristics

Maternal characteristics are described in detail by Kvist et al. [8] and were extracted from the Finnish Medical Birth Register (FMBR) as well as through self-report questionnaires on their depressive, anxiety, and perceived stress symptoms up to three times during pregnancy. With regards to psychometric assessments, the mothers completed questionnaires on their depressive, anxiety, and perceived stress symptoms up to three times during pregnancy, on average (median) at 19.3 [Standard Deviation (SD) = 3.7], 26.1 (SD = 3.1), and 38.1 (SD = 8.5) weeks of gestation, respectively. Depressive symptoms were assessed with the Center for Epidemiologic Studies Depression Scale [31], perceived stress symptoms with a 5-item version of the Cohen's Perceived Stress Scale [32], and anxiety symptoms with the Spielberger State Anxiety Inventory (STAI) [33]. All three are validated questionnaires [33], and the CESD and the STAI have also been validated among pregnant women [34]. In our sample, the Cronbach's alphas indicating the internal consistencies of the scales ranged from 0.87 to 0.89 for CESD, from 0.94 to 0.95 for STAI, and from 0.69 to 0.71 for Cohen's Perceived Stress Scale.

Child Characteristics

Data on the child's sex (girl/boy), gestation length (weeks), and birth weight (grams) were extracted from the FMBR. We extracted data on diagnoses of congenital malformations and, more specifically, on malformations of the circulatory system from the Finnish nationwide Care Register for Healthcare (HILMO). The HILMO healthcare register includes primary and subsidiary diagnoses of all hospitalizations in Finland since 1969 and of all visits in specialized outpatient care since 1998. Diagnoses have been entered into the HILMO according to the International Statistical Classification of Diseases and Related Health Problems, Eighth Revision (ICD-8) until 1986, according to ICD-9 from 1987 to 1995, and according to ICD-10 since 1996. The HILMO is a validated tool for research [35]. We had lifetime data from HILMO available until 31 December 2017. Diagnosis of any congenital malformation, deformation, or chromosomal abnormalities was identified with ICD-10 diagnostic codes Q00-Q99. We also identified congenital malformations, specifically of the circulatory system, with diagnostic codes Q20-Q28. These diagnoses were identified from childbirth until 31 December 2017, when the children were between 2 days and 5.7 years old.

### 4.2. Biosampling, DNA/RNA Extractions

Placenta samples from the fetal side of the placenta, relatively close to the umbilical cord, were collected after birth by midwives who took nine-site biopsies within 120 min of delivery. The biopsies were stored in RNA storage solution (RNAprotect, Qiagen) until frozen by research staff at −80 °C (within 24 h of delivery) for long-term storage. Chorionic villus (CVS) biopsies were collected by experienced obstetricians in early pregnancy (weeks 8–12). After chromosomal analysis, surplus tissue was immediately stored at −80 °C. DNA and RNA were extracted from CVS and delivery placenta, and DNA was extracted from cord blood leukocytes using a bead-based method optimized by tissue type (Chemagic 360 Perkin Elmer). Total CVS biospecimens were homogenized and split 40–60% for RNA-DNA extraction. Delivery placenta samples preserved in RNAprotect reagent were thawed, and equal-sized aliquots were dissected, homogenized, and split 40–60% for RNA-DNA extraction. Quantification and quality assessments were performed using a TapeStation Automated Electrophoresis system (Agilent) and an Epoch Microplate Spectrophotometer (BioTek, Agilent). All extractions were performed at the BioPrep core unit, Max Planck Institute for Psychiatry.

### 4.3. Genotyping, Imputation, and MDS Components

Genotyping was performed on Illumina GSA-24v2-0_A1 arrays, according to the manufacturer's guidelines (Illumina Inc., San Diego, CA, USA). Quality control is described in detail in Dieckmann et al. [36]. After quality control, genotypes from 592 individuals and 338,132 SNPs were subjected to imputation. Imputation was performed using *shapeit2* [37] and *impute2* [38] based on the 1000 Genomes Phase III reference sample. After imputation, SNPs with an info score below 0.6, a minor allele frequency below 0.01, or deviating from Hardy–Weinberg equilibrium ($p$-value < $1 \times 10^{-5}$) were excluded from further analysis resulting in 9,826,011 SNPs. Multi-dimensional scaling (MDS) was performed in *PLINK* on the genotyped dataset after linkage disequilibrium pruning.

### 4.4. CNV Calling

Quality control on individuals' SNP genotyping was performed as described in Dieckmann et al. [36]. Briefly, individuals presenting with callrates < 98% or being outliers in the multi-dimensional scaling analysis were removed from the analysis. For calling of

CNVs, raw .idat files of these 592 IDs were converted into vcf files and tabular input files using *gtc2vcf* (https://github.com/freeseek/gtc2vcf, accessed on 9 July 2021), *VCF-simplify* (https://github.com/everestial/VCF-simplify, accessed on 9 July 2021), and R [39]. SNPs were excluded if any of these conditions were met: callrate < 98%, cluster separation < 0.3, AB R Mean <= 0.2, AB R Mean <= 0.2, BB R Mean <= 0.2, AB T Mean <= 0.1 or AB T Mean > 0.9, Het Excess < −0.9 or Het Excess > 0.9, minor allele frequency > 0 and AB Freq = 0, AA Freq = 1 and AA T Mean > 0.3, AA Freq = 1 and AA T Dev > 0.06, BB Freq = 1 and BB T Mean < 0.7, BB Freq = 1 and BB T Dev = 0.06. These values were chosen with regard to Illumina's recommendation (https://www.illumina.com/Documents/products/technotes/technote_infinium_genotyping_data_analysis.pdf, accessed on 9 July 2021). Afterward, CNVs were called using *PennCNV* [40], correcting for GC content as described in Diskin et al. [41]. Adjacent calls were merged and individuals with poor quality parameters were removed, according to the default settings (fraction < 0.2, LRR SD > 0.3, BAF drift > 0.01, WF > 0.05). Furthermore, CNV calls spanning less than 10 SNPs, individuals with more than 100 CNVs, and spurious CNVs, which are likely in centromeric and telomeric regions, were removed according to the recommendations of Lin et al. [42] and Li et al. [43] This resulted in a final sample size of 574 IDs (277 controls and 297 cases) and 9334 detected CNV calls including 3445 CNVs (2290 deletions and 1155 duplications). We used *ClassifyCNV* (https://github.com/Genotek/ClassifyCNV, accessed on 13 July 2021), to classify detected CNVs according to guidelines [44] of the American College of Medical Genetics and Genomics into the categories: benign, likely benign, uncertain significance, likely pathogenic, or pathogenic. Of the called CNVs, 3382 could be defined with *ClassifyCNV*, 239 CNVs were identified as likely pathogenic or pathogenic, 1177 as benign and 1966 with uncertain significance. A detailed list of classified CNVs is given in Supplementary Table S1.

### 4.5. RNA Sequencing

The QuantSeq 3′ mRNA-Seq Library Prep Kit (Lexogen) was used to generate messenger RNA (mRNA) sequencing libraries from both term placenta and CVS RNA samples. All libraries were multiplexed and sequenced on an Illumina HighSeq4000 system at a depth of 10 million reads per mRNA library. Adapter sequences were trimmed using cutadapt (https://cutadapt.readthedocs.io/en/stable/), and sequenced reads were aligned to the human genome reference using the STAR aligner [45]. We performed featureCounts [46] and filtered the datasets to genes presenting with a raw count of at least 10 in at least 90% of the individuals, resulting in a final dataset including 8245 transcripts and 493 individuals for the term placentas. For the CVS dataset, the same filtering led to 9089 transcripts quantified in 266 individuals.

### 4.6. Statistical Analysis

All statistical analyses were performed in R version 4.0.4 and SPSS 28.0. $p$-value thresholds are given separately for each sub-analysis.

#### 4.6.1. Differences in Congenital Malformations Using Cox Regression

We checked if cases and controls nominal significantly differed with regards to the hazard of congenital malformations using Cox regression models as implemented in the R-package *survival* 3.2.13. The follow-up time of cases was significantly longer as compared to controls ($p = 5.52 \times 10^{-86}$, Wilcoxon-test); hence, we limited the follow-up time of cases to the age of the eldest child in the control group and individuals who had been diagnosed with congenital malformations after that time, were set to not diagnosed.

First, we tested which covariates that were different between cases and controls were also significantly nominally associated with congenital malformations using Cox regression models. These covariates (none for any congenital malformation, maternal BMI in early pregnancy for congenital malformations within the circulatory system: $p = 0.004$, hazard ratio = 1.10 [1.03–1.18]) were then included in the final Cox regression model on case-control differences. Survival curves were plotted using the R-package *survminer* 0.4.9. All *p*-values < 0.05 were considered significant.

4.6.2. Association of Congenital Malformations with Screening Variables

Association was tested using linear regression models, with the screening variables as dependent, congenital malformations as independent variable, and case-control status as covariate. Due to high skewness, screening variables were first log-transformed and then Z-standardized, while risk percentages were transformed into normally distributed data using inverse rank transformation. All *p*-values < 0.05 were considered significant.

4.6.3. CNV Associations

Associations with CNVs were computed in *PLINK* [47]. We calculated burden tests, testing if the total number of detected CNVs differed between cases and controls or between carriers of congenital malformations and non-carriers. Due to the high number of tests, empirical *p*-values, which are then already corrected for multiple testing, based on 10,000 permutations, were computed. Empirical *p*-values < 0.05 were considered significant. Enrichment of CNV positions for GO terms and GWAS hits were tested using *FUMA* [48]. Positions of identified CNVs were not enriched for any GO terms or reported hits for neurodevelopmental disorders (at a false discovery rate (FDR) of 0.05).

4.6.4. Differential Gene Expression

Analysis of differential gene expression was conducted in R [39]. Raw gene counts were *voom* transformed [49], and afterward, differential gene expression was calculated using the R-package *eBayes* function in the R-package *limma* [50]. Surrogate variable (SV) analysis was used to correct for possible batch effects as well as cell type heterogeneity [51]. For both CVS and term placental tissue, the first SV was detected as significant (according to the permutation procedure implemented in the package) and used as a covariate in the subsequent analyses. Furthermore, those variables that were significantly different between cases and controls, i.e., maternal age, maternal BMI in early pregnancy, as well as smoking, and parity (uniparous vs. multiparous), were used as covariates. We also included the child's sex and gestational age (in weeks at sampling), as both have been associated with placental gene expression [52]. Differential gene expression was calculated for congenital malformations within the circulatory system and CNVs in both tissues. Within each analysis, multiple testing correction was applied based on the Benjamini–Hochberg [53] approach, and all *p*-values were considered significant at FDR of $\leq 0.1$.

4.6.5. *p*-Values

All reported *p*-values are two-sided. For the association of congenital malformations with screening variables, we hypothesized that individuals with congenital malformations were comparable to cases with regard to their outcome in screening variables. Hence, one-sided *p*-values are presented.

4.6.6. Enrichment for Pre-Eclampsia Genes

Of the 15 genes that were differentially expressed in CVS tissue with congenital malformations within the circulatory, 11 genes had been associated with pre-eclampsia before. To assess if this overlap was nominal significantly higher than expected by chance, we assumed that 9% of all investigated genes are associated with pre-eclampsia, based on the results for Saei et al. [14]. Using a binomial distribution, the chance to get at least 11 hits out of 15 genes is $9.91 \times 10^{-11}$.

**Supplementary Materials:** The supporting information can be downloaded at: https://www.mdpi.com/article/10.3390/ijms231911448/s1.

**Author Contributions:** Conceptualization, D.C., C.C., E.K., H.L., K.R. and E.B.B.; data curation, M.L.-P., M.K., S.S. (Susann Sauer), M.R.-H., S.S. (Sara Sammallahti), E.K. and J.L.; formal analysis, D.C., M.L.-P. and L.D.; funding acquisition, K.R.; investigation, D.C., C.C., M.K., S.S. (Susann Sauer), M.R.-H., S.S. (Sara Sammallahti), H.L., J.L., K.R. and E.B.B.; methodology, C.C.; project administration, S.S. (Sara Sammallahti), E.K. and H.L.; software, D.C.; supervision, K.R. and E.B.B.; validation, C.C.; visualization, D.C.; writing—original draft, D.C. and C.C.; writing—review and editing, D.C., C.C., M.L.-P., L.D., M.K., S.S. (Susann Sauer), M.R.-H., S.S. (Sara Sammallahti), E.K., H.L., J.L., K.R. and E.B.B. All authors have read and agreed to the published version of the manuscript.

**Funding:** The ITU is funded by the Academy of Finland (award numbers: 284859, 312670, 1324596). CC received funding from the Banting Postdoctoral Fellowship. ML-P receives funding from the Academy of Finland, University of Helsinki Funds. The funders had no role in the design of the study and collection, analysis, and interpretation of data and in writing the manuscript.

**Institutional Review Board Statement:** This study involves human participants. The ITU research protocol has been approved by the Coordinating Ethics Committee of the Helsinki and Uusimaa Hospital District (approval date: 18 May 2010, reference number: 269/13/03/00/09). The study protocol follows the Helsinki Declaration.

**Informed Consent Statement:** Participants provided written informed consent to participate in the study before taking part.

**Data Availability Statement:** Due to the sensitive nature of the patient data used in the current study and consent, the data sets are not and cannot be made publicly available. However, an interested researcher can obtain a de-identified data set after approval from ITU Study Board. Data requests may be subject to further review by the national register authority and by the ethical committees.

**Acknowledgments:** We thank the BioPrep biobank core unit at MPI Psychiatry in Munich, Germany, for their work on DNA and RNA extractions.

**Conflicts of Interest:** The authors declare no conflict of interest. The funders had no role in the design of the study, in the collection, analyses, or interpretation of data, in the writing of the manuscript, or in the decision to publish the results.

## References

1. Carlson, L.M.; Vora, N.L. Prenatal Diagnosis: Screening and Diagnostic Tools. *Obstet. Gynecol. Clin. N. Am.* **2017**, *44*, 245–256. [CrossRef] [PubMed]
2. Bardi, F.; Bosschieter, P.; Verheij, J.; Go, A.; Haak, M.; Bekker, M.; Sikkel, E.; Coumans, A.; Pajkrt, E.; Bilardo, C. Is there still a role for nuchal translucency measurement in the changing paradigm of first trimester screening? *Prenat Diagn.* **2020**, *40*, 197–205. [CrossRef] [PubMed]
3. Fruscalzo, A.; Cividino, A.; Rossetti, E.; Maurigh, A.; Londero, A.P.; Driul, L. First trimester PAPP-A serum levels and long-term metabolic outcome of mothers and their offspring. *Sci. Rep.* **2020**, *10*, 5131. [CrossRef] [PubMed]
4. Zhang, Z.; Hu, T.; Wang, J.; Li, Q.; Wang, H.; Liu, S. Prenatal Diagnostic Value of Chromosomal Microarray in Fetuses with Nuchal Translucency Greater than 2.5 mm. *BioMed Res. Int.* **2019**, *2019*, 6504159. [CrossRef]
5. Hui, L.; Hutchinson, B.; Poulton, A.; Halliday, J. Population-based impact of noninvasive prenatal screening on screening and diagnostic testing for fetal aneuploidy. *Genet. Med.* **2017**, *19*, 1338–1345. [CrossRef] [PubMed]
6. Pös, O.; Budis, J.; Szemes, T. Recent trends in prenatal genetic screening and testing. *F1000Research* **2019**, *8*, 764. [CrossRef] [PubMed]
7. Nadler, H.L.; Gerbie, A.B. Role of Amniocentesis in the Intrauterine Detection of Genetic Disorders. *N. Engl. J. Med.* **1970**, *282*, 596–599. [CrossRef] [PubMed]
8. Kvist, T.; Sammallahti, S.; Lahti-Pulkkinen, M.; Cruceanu, C.; Czamara, D.; Dieckmann, L.; Tontsch, A.; Röh, S.; Rex-Haffner, M.; Wolford, E.; et al. Cohort profile: InTraUterine sampling in early pregnancy (ITU), a prospective pregnancy cohort study in Finland: Study design and baseline characteristics. *BMJ Open* **2022**, *12*, e049231. [CrossRef] [PubMed]
9. Conover, W.J. *Pracitcal Nonparametric Statistics*, 3rd ed.; Wiley: Hoboken, NJ, USA, 1999.
10. Southard, A.E.; Edelmann, L.J.; Gelb, B.D. Role of Copy Number Variants in Structural Birth Defects. *Pediatrics* **2012**, *129*, 755–763. [CrossRef] [PubMed]
11. Zhang, R.; Chen, X.; Wang, D.; Chen, X.; Wang, C.; Zhang, Y.; Xu, M.; Yu, J. Prevalence of chromosomal abnormalities identified by copy number variation sequencing in high-risk pregnancies, spontaneous abortions, and suspected genetic disorders. *J. Int. Med. Res.* **2019**, *47*, 1169–1178. [CrossRef] [PubMed]
12. Heo, Y.; Heo, J.; Han, S.-S.; Kim, W.J.; Cheong, H.S.; Hong, Y. Difference of copy number variation in blood of patients with lung cancer. *Int. J. Biol. Mark.* **2020**, *36*, 3–9. [CrossRef] [PubMed]

13. Redon, R.; Ishikawa, S.; Fitch, K.R.; Feuk, L.; Perry, G.H.; Andrews, T.D.; Fiegler, H.; Shapero, M.H.; Carson, A.R.; Chen, W.; et al. Global variation in copy number in the human genome. *Nature* **2006**, *444*, 444–454. [CrossRef] [PubMed]
14. Saei, H.; Govahi, A.; Abiri, A.; Eghbali, M.; Abiri, M. Comprehensive transcriptome mining identified the gene expression signature and differentially regulated pathways of the late-onset preeclampsia. *Pregnancy Hypertens.* **2021**, *25*, 91–102. [CrossRef]
15. van Uitert, M.; Moerland, P.D.; Enquobahrie, D.A.; Laivuori, H.; van der Post, J.A.; Ris-Stalpers, C.; Afink, G.B. Meta-Analysis of Placental Transcriptome Data Identifies a Novel Molecular Pathway Related to Preeclampsia. *PLoS ONE* **2015**, *10*, e0132468. [CrossRef] [PubMed]
16. Vennou, K.E.; Kontou, P.I.; Braliou, G.G.; Bagos, P.G. Meta-analysis of gene expression profiles in preeclampsia. *Pregnancy Hypertens. Int. J. Women's Cardiovasc. Health* **2019**, *19*, 52–60. [CrossRef]
17. Paulson, K.R.; Kamath, A.M.; Alam, T.; Bienhoff, K.; Abady, G.G.; Abbas, J.; Abbasi-Kangevari, M.; Abbastabar, H.; Abd-Allah, F.; Abd-Elsalam, S.M.; et al. Global, regional, and national progress towards Sustainable Development Goal 3.2 for neonatal and child health: All-cause and cause-specific mortality findings from the Global Burden of Disease Study 2019. *Lancet* **2021**, *398*, 870–905. [CrossRef]
18. de Knegt, V.E.; Hedley, P.L.; Kanters, J.K.; Thagaard, I.N.; Krebs, L.; Christiansen, M.; Lausten-Thomsen, U. The Role of Leptin in Fetal Growth during Pre-Eclampsia. *Int. J. Mol. Sci.* **2021**, *22*, 4569. [CrossRef]
19. Sitras, V.; Paulssen, R.; Grønaas, H.; Leirvik, J.; Hanssen, T.; Vårtun, A.; Acharya, G. Differential Placental Gene Expression in Severe Preeclampsia. *Placenta* **2009**, *30*, 424–433. [CrossRef]
20. Kassotaki, I.; Valsamakis, G.; Mastorakos, G.; Grammatopoulos, D.K. Placental CRH as a Signal of Pregnancy Adversity and Impact on Fetal Neurodevelopment. *Front. Endocrinol.* **2021**, *12*, 714214. [CrossRef]
21. Leviton, A.; Allred, E.N.; Kuban, K.C.; O'Shea, T.M.; Paneth, N.; Majzoub, J. ELGAN study investigators Brain disorders associated with corticotropin-releasing hormone expression in the placenta among children born before the 28th week of gestation. *Acta Paediatr.* **2015**, *105*, e7–e11. [CrossRef]
22. Karteris, E.; Vatish, M.; Hillhouse, E.W.; Grammatopoulos, D.K. Preeclampsia Is Associated with Impaired Regulation of the Placental Nitric Oxide-Cyclic Guanosine Monophosphate Pathway by Corticotropin-Releasing Hormone (CRH) and CRH-Related Peptides. *J. Clin. Endocrinol. Metab.* **2005**, *90*, 3680–3687. [CrossRef]
23. Founds, S.A.; Ren, D.; Roberts, J.M.; Jeyabalan, A.; Powers, R.W. Follistatin-Like 3 Across Gestation in Preeclampsia and Uncomplicated Pregnancies Among Lean and Obese Women. *Reprod. Sci.* **2014**, *22*, 402–409. [CrossRef] [PubMed]
24. Macintire, K.; Tuohey, L.; Ye, L.; Palmer, K.; Gantier, M.; Tong, S.; Kaitu'U-Lino, T.J. PAPPA2 is increased in severe early onset pre-eclampsia and upregulated with hypoxia. *Reprod. Fertil. Dev.* **2014**, *26*, 351–357. [CrossRef] [PubMed]
25. Gurusinghe, S.; Wallace, E.M.; Lim, R. The relationship between Activin A and anti-angiogenic factors in the development of pre-eclampsia. *Pregnancy Hypertens. Int. J. Women's Cardiovasc. Health* **2014**, *4*, 3–6. [CrossRef] [PubMed]
26. Liu, C.; Xing, F.; He, Y.; Zong, S.; Luo, C.; Li, C.; Duan, T.; Wang, K.; Zhou, Q. Elevated HTRA1 and HTRA4 in severe preeclampsia and their roles in trophoblast functions. *Mol. Med. Rep.* **2018**, *18*, 2937–2944. [CrossRef]
27. Founds, S.; Conley, Y.; Lyons-Weiler, J.; Jeyabalan, A.; Hogge, W.A.; Conrad, K. Altered Global Gene Expression in First Trimester Placentas of Women Destined to Develop Preeclampsia. *Placenta* **2009**, *30*, 15–24. [CrossRef]
28. Hayeems, R.Z.; Campitelli, M.; Ma, X.; Huang, T.; Walker, M.; Guttmann, A. Rates of prenatal screening across health care regions in Ontario, Canada: A retrospective cohort study. *CMAJ Open* **2015**, *3*, E236–E243. [CrossRef]
29. Zhu, H.; Jin, X.; Xu, Y.; Zhang, W.; Liu, X.; Jin, J.; Qian, Y.; Dong, M. Efficiency of non-invasive prenatal screening in pregnant women at advanced maternal age. *BMC Pregnancy Childbirth* **2021**, *21*, 86. [CrossRef]
30. Kim, Y.-N.; Choi, D.-W.; Kim, D.S.; Park, E.-C.; Kwon, J.-Y. Maternal age and risk of early neonatal mortality: A national cohort study. *Sci. Rep.* **2021**, *11*, 814. [CrossRef]
31. Radloff, L.S. The CES-D Scale: A Self-Report Depression Scale for Research in the General Population. *Appl. Psychol. Meas.* **1977**, *1*, 385–401. [CrossRef]
32. Cohen, S.; Kamarck, T.; Mermelstein, R. A global measure of perceived stress. *J. Health Soc. Behav.* **1983**, *24*, 385–396. [CrossRef]
33. Spielberger, C.D. *State-Trait Anxiety Inventory: Bibliography, 2nd ed*; Consulting Psychologists Press: Washington, DC, USA, 1989.
34. Natamba, B.K.; Achan, J.; Arbach, A.; Oyok, T.O.; Ghosh, S.; Mehta, S.; Stoltzfus, R.J.; Griffiths, J.K.; Young, S.L. Reliability and validity of the center for epidemiologic studies-depression scale in screening for depression among HIV-infected and -uninfected pregnant women attending antenatal services in northern Uganda: A cross-sectional study. *BMC Psychiatry* **2014**, *14*, 303. [CrossRef] [PubMed]
35. Sund, R. Quality of the Finnish Hospital Discharge Register: A systematic review. *Scand. J. Public Health* **2012**, *40*, 505–515. [CrossRef]
36. Dieckmann, L.; Lahti-Pulkkinen, M.; Kvist, T.; Lahti, J.; DeWitt, P.E.; Cruceanu, C.; Laivuori, H.; Sammallahti, S.; Villa, P.M.; Suomalainen-König, S.; et al. Characteristics of epigenetic aging across gestational and perinatal tissues. *Clin. Epigenet.* **2021**, *13*, 97. [CrossRef]
37. Delaneau, O.; Marchini, J.; Zagury, J.-F. A linear complexity phasing method for thousands of genomes. *Nat. Methods* **2011**, *9*, 179–181. [CrossRef] [PubMed]
38. Marchini, J.; Howie, B.; Myers, S.; McVean, G.; Donnelly, P. A new multipoint method for genome-wide association studies by imputation of genotypes. *Nat. Genet.* **2007**, *39*, 906–913. [CrossRef]

39. R Development Core Team. *R: A Language and Environment for Statistical Computing*; R Foundation for Statistical Computing: Vienna, Austria, 2017.
40. Wang, K.; Li, M.; Hadley, D.; Liu, R.; Glessner, J.; Grant, S.F.; Hakonarson, H.; Bucan, M. PennCNV: An integrated hidden Markov model designed for high-resolution copy number variation detection in whole-genome SNP genotyping data. *Genome Res.* **2007**, *17*, 1665–1674. [CrossRef]
41. Diskin, S.J.; Li, M.; Hou, C.; Yang, S.; Glessner, J.; Hakonarson, H.; Bucan, M.; Maris, J.M.; Wang, K. Adjustment of genomic waves in signal intensities from whole-genome SNP genotyping platforms. *Nucleic Acids Res.* **2008**, *36*, e126. [CrossRef]
42. Lin, C.-F.; Naj, A.C.; Wang, L.-S. Analyzing Copy Number Variation Using SNP Array Data: Protocols for Calling CNV and Association Tests. *Curr. Protoc. Hum. Genet.* **2013**, *79*, 1–27. [CrossRef]
43. Li, Y.R.; Glessner, J.T.; Coe, B.P.; Li, J.; Mohebnasab, M.; Chang, X.; Connolly, J.; Kao, C.; Wei, Z.; Bradfield, J.; et al. Rare copy number variants in over 100,000 European ancestry subjects reveal multiple disease associations. *Nat. Commun.* **2020**, *11*, 255. [CrossRef]
44. Riggs, E.R.; Andersen, E.F.; Cherry, A.M.; Kantarci, S.; Kearney, H.; Patel, A.; Raca, G.; Ritter, D.I.; South, S.T.; Thorland, E.C.; et al. Technical standards for the interpretation and reporting of constitutional copy-number variants: A joint consensus recommendation of the American College of Medical Genetics and Genomics (ACMG) and the Clinical Genome Resource (ClinGen). *Genet. Med.* **2020**, *22*, 245–257, Correction in *Genet. Med.* **2021**, *23*, 2230. [CrossRef]
45. Dobin, A.; Davis, C.A.; Schlesinger, F.; Drenkow, J.; Zaleski, C.; Jha, S.; Batut, P.; Chaisson, M.; Gingeras, T.R. STAR: Ultrafast universal RNA-seq aligner. *Bioinformatics* **2013**, *29*, 15–21. [CrossRef]
46. Liao, Y.; Smyth, G.K.; Shi, W. feature Counts: An efficient general purpose program for assigning sequence reads to genomic features. *Bioinformatics* **2014**, *30*, 923–930. [CrossRef]
47. Purcell, S.; Neale, B.; Todd-Brown, K.; Thomas, L.; Ferreira, M.A.R.; Bender, D.; Maller, J.; Sklar, P.; de Bakker, P.I.W.; Daly, M.J.; et al. PLINK: A Tool Set for Whole-Genome Association and Population-Based Linkage Analyses. *Am. J. Hum. Genet.* **2007**, *81*, 559–575. [CrossRef]
48. Watanabe, K.; Taskesen, E.; van Bochoven, A.; Posthuma, D. Functional mapping and annotation of genetic associations with FUMA. *Nat. Commun.* **2017**, *8*, 1826. [CrossRef] [PubMed]
49. Law, C.W.; Chen, Y.; Shi, W.; Smyth, G.K. Voom: Precision weights unlock linear model analysis tools for RNA-seq read counts. *Genome Biol.* **2014**, *15*, R29. [CrossRef]
50. Ritchie, M.E.; Belinda, P.; Wu, D.; Hu, Y.; Law, C.W.; Shi, W.; Smyth, G.K. limma powers differential expression analyses for RNA-sequencing and microarray studies. *Nucleic Acids Res.* **2015**, *43*, e47. [CrossRef]
51. Leek, J.T.; Johnson, W.E.; Parker, H.S.; Jaffe, A.E.; Storey, J.D. The sva package for removing batch effects and other unwanted variation in high-throughput experiments. *Bioinformatics* **2012**, *28*, 882–883. [CrossRef]
52. Gong, S.; Gaccioli, F.; Dopierala, J.; Sovio, U.; Cook, E.; Volders, P.J.; Martens, L.; Kirk, P.D.W.; Richardson, S.; Smith, G.C.S.; et al. The RNA landscape of the human placenta in health and disease. *Nat. Commun.* **2021**, *12*, 2639. [CrossRef]
53. Benjamini, Y.; Hochberg, Y. Controlling the False Discovery Rate: A Practical and Powerful Approach to Multiple Testing. *J. R. Stat. Soc. Ser. B* **1995**, *57*, 289–300. [CrossRef]

International Journal of
*Molecular Sciences*

MDPI

*Article*

# PEG2-Induced Pyroptosis Regulates the Expression of HMGB1 and Promotes hEM15A Migration in Endometriosis

**Yi Huang [1], Ruiyun Li [1], Rui Hu [1], Jia Yao [2] and Yuan Yang [1,3,*]**

[1] The First Clinical Medical College, Lanzhou University, Lanzhou 730000, China
[2] Gansu Province Key Laboratory Biotherapy and Regenerative Medicine, Lanzhou 730000, China
[3] Reproductive Medicine Center, The First Hospital of Lanzhou University, Lanzhou 730000, China
[*] Correspondence: ldyy_yangy@lzu.edu.cn

**Abstract:** Endometriosis (EMS) is a common gynecological disease. Prostaglandin E2 (PGE2), which induces chronic pelvic inflammation and cell pyroptosis, a form of programmed cell death based on inflammasome activation, are involved in EMS, but the extent of their involvement and roles remain unclear. The present study aimed to evaluate PGE2-induced pyroptosis in EMS and the influence of PGE2 in EMS progression. Using western blotting, it was found that the expressions of PGE2 and pyroptosis-related proteins (NLRP3, cleaved caspase-1, interleukin (IL)-1β and IL-18) were higher in EMS tissues than in normal endometrial tissues. The levels of PGE2, IL-1β, and IL-18 in the serum of patients with EMS and cell culture fluids were also detected. Using the transwell assay, we verified that PGE2 promoted hEM15A migration via the NLRP3/caspase-1 pyroptotic pathway, and PGE2-induced pyroptosis upregulated the expressions of high mobility group box 1 (HMGB1), E-cadherin, and vimentin. Immunohistochemistry analysis confirmed that PGE2-induced pyroptosis contributed to EMS invasion. These results suggest that PGE2-induced pyroptosis affects the progression of EMS by changing the migration ability of pyroptotic cells and upregulating the expression of HMGB1, E-cadherin, and vimentin. Our findings provide crucial evidence for new treatment pathways and use of anti-inflammatory drugs in EMS.

**Keywords:** endometriosis; pyroptosis; PGE2; cell migration

**Citation:** Huang, Y.; Li, R.; Hu, R.; Yao, J.; Yang, Y. PEG2-Induced Pyroptosis Regulates the Expression of HMGB1 and Promotes hEM15A Migration in Endometriosis. *Int. J. Mol. Sci.* **2022**, *23*, 11707. https://doi.org/10.3390/ijms231911707

Academic Editor: Ilona Hromadnikova

Received: 11 September 2022
Accepted: 30 September 2022
Published: 3 October 2022



## 1. Introduction

Endometriosis (EMS) is a common gynecological disease that occurs in women of childbearing age, with an incidence of 10–15%. In patients with infertility, the incidence is approximately 40–50% [1,2]. The main clinical manifestations are dysmenorrhea, chronic pelvic pain, sexual pain, pelvic mass, and infertility. EMS presents malignant tumor characteristics, such as implantation, invasion, distant metastasis, and recurrence, and, hence, seriously affects the physical and mental health of women. Ectopic lesions themselves can lead to the occurrence of pelvic inflammation, which promotes further growth of ectopic endometrial lesions. These repeated inflammatory reactions induce an abnormal increase in inflammatory cytokines mediating the adhesion, proliferation, differentiation, and invasion of EMS lesions [3,4]. In addition, the malignant transformation of endometriosis is closely related to ovarian clear cell carcinoma and ovarian endometrioid carcinoma, and is even considered as the precancerous lesion of the latter [5–7]. Previous clinical studies have shown that patients with endometriosis associated ovarian carcinoma (EAOC) have some special clinicopathological features and a relatively good prognosis [8]. Basic research has found that specific genes are associated with abnormal expression [9]. However, studies on the role of endometriosis in the pathogenesis of ovarian cancer are still lacking.

Prostaglandin E2 (PGE2) is a common inflammatory factor that significantly increases in the serum of patients with EMS; therefore, it has been thoroughly studied in the pathogenesis of EMS [10]. In patients with EMS, PGE2 not only inhibits apoptosis of endometrial fragments, but also promotes cell proliferation in these fragments to form EMS lesions. In addition, elevated levels of PGE2 allow pain-causing substances and inflammatory mediators to leak out of blood vessels into local areas, causing dysmenorrhea [11]. Despite its recognized involvement in the progress of EMS, the inability of PGE2 inhibition as well as other targeted drugs in the treatment of EMS remains elusive. Moreover, previous studies have suggested that PGE2 may also be involved in other mechanisms for EMS progression.

Pyroptosis is a form of programmed cell death, based on activated inflammasome. In the cytoplasm, multiprotein complexes are formed to activate inflammatory NLR family pyrin domain-containing 3 (NLRP3), caspase-1, and caspase-4/5/11, which further lyse the perforated cell membrane of gasdermin D protein to promote pyroptosis [12–14]. Pyroptotic cells secrete a large number of inflammatory factors, such as interleukin (IL)-1β and IL-18. These inflammatory factors recruit inflammatory cells, induce the synthesis and release of IL-1α, IL-6, tumor necrosis factor (TNF)-α, and other substances, inducing an inflammatory response leading to various pathological processes, such as inflammatory exclusion, tissue destruction, and edema formation. It has been found that the pyroptosis-related protein NLRP3 is highly expressed at the early stage of breast cancer, increasing immune protection and inhibiting tumor progression [15]. However, when the tumor metastasizes, NLRP3 expression is significantly reduced and promotes tumor development. NLRP3 is also involved in innate immune response during cervical cancer. Reactive oxygen species activate the NLRP3 inflammasome to induce pyroptosis in cervical cancer cells, and then participate in tumor progression. Tripartite motif-containing 24 (TRIM24) may promote NLRP3/caspase-1/IL-1β-mediated pyroptosis during EMS through NLRP3 ubiquitination, revealing an important molecular mechanism [16]. Despite the crucial role of pyroptosis in EMS, no study has focused on the relationship between pyroptosis and inflammatory factors during EMS, or on whether the release of such factors triggers a series of secondary inflammatory reactions that participate in EMS progression.

The aim of the present study was to evaluate whether PGE2-induced pyroptosis is involved in EMS progression. Specifically, we examined the expression of PGE2 and pyroptosis-related proteins (NLRP3, cleaved caspase-1, IL-1β, and IL-18) in ovarian endometriotic cyst tissues. We further determined if PGE2-induced pyroptosis regulated the expression of HMGB1 and changed hEM15A migration both in vitro and in vivo. Cell migration signature proteins, E-cadherin and vimentin, were also detected. We hypothesized that PGE2 release due to pyroptosis was involved in the progression of EMS lesions by changing cell migration. The results of the present study advocate exploring new insights into the pathogenesis of EMS via inflammation and provide new treatment routes of EMS in clinical practice.

## 2. Results

### *2.1. PGE2 and Pyroptosis-Related Proteins Highly Expressed in Ectopic Ovarian Endometrium*

Inflammatory cytokine PGE2 production and pyroptosis are common phenomena in inflammatory diseases, but there is no study on their co-expression during EMS. Therefore, the expression of PGE2 and pyroptosis-related proteins (NLRP3, cleaved caspase-1, IL-1β, and IL-18) in ectopic ovarian endometriotic cyst tissues (n = 28) and normal endometrial tissues (n = 15) were investigated. Pyroptosis-related protein levels (NLRP3, cleaved caspase-1, IL-1β, and IL-18) were increased in ectopic ovarian endometriotic cyst tissues, compared to those in normal endometrial tissues (Figure 1a). Furthermore, PGE2, IL-1β, and IL-18 levels were elevated in the serum of patients with EMS, compared to those in normal controls (Figure 1b). These data suggested PGE2 may be related to pyroptosis during the occurrence and development of ovarian endometriotic cysts.

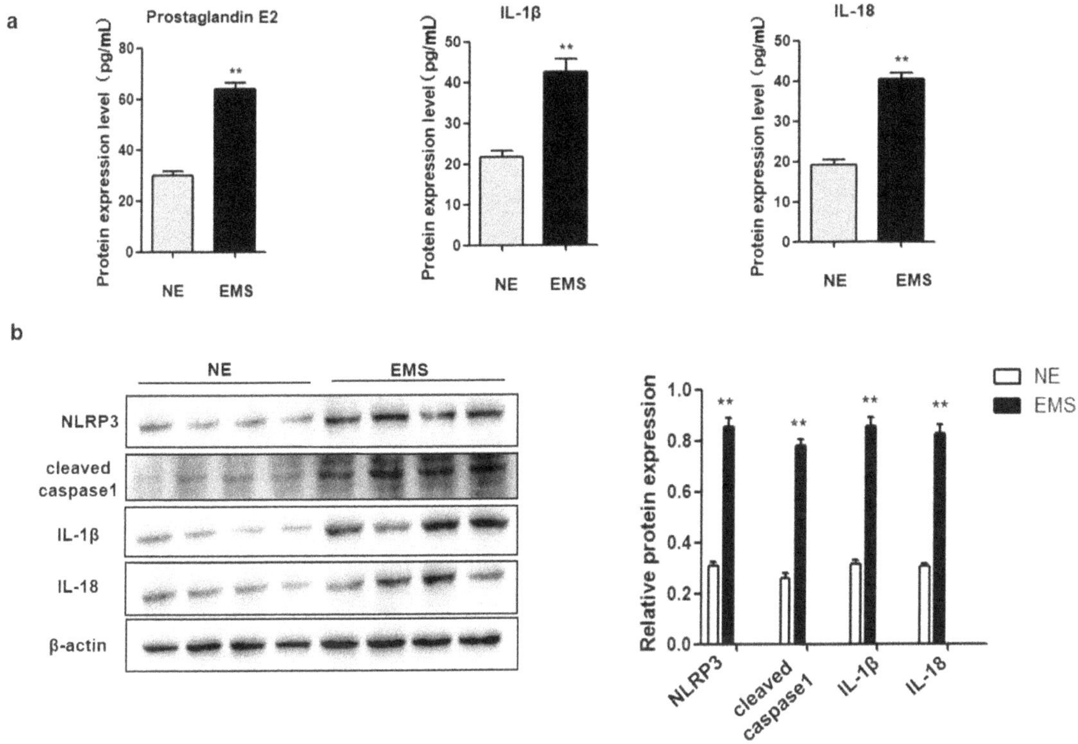


**Figure 1.** The expression of PGE2 and pyroptosis related proteins (NLRP3, cleaved caspase-1, IL-1β and IL-18) were increased in ectopic ovarian endometriosis cyst patients. (**a**) The expression of PEG2, IL-1β and IL-18 in serum of ectopic ovarian endometriosis cyst (EMS) and in the normal endometrium (NE) were detected by ELISA; (**b**) Pyroptosis related proteins (NLRP3, cleaved caspase-1, IL-1β and IL-18) protein expression were measured by Western blot analysis in NE and EMS. The data are expressed as means ± SEM. ** $p < 0.05$ by two-tailed Student's $t$ test.

### 2.2. PGE2 Induces hEM15A Pyroptosis

To determine whether PGE2-induced NLRP3/caspase-1/IL-1β-mediated pyroptosis is involved in EMS, hEM15A and hESCs were treated with 75 ng/mL PGE2. After 24 h, the expression of NLRP3 and cleaved caspase-1 was upregulated in hEM15A compared to that in hESCs (Figure 2b). The levels of IL-1β and IL-18 in the cell culture medium were significantly increased in hEM15A compared to those in hESCs (Figure 2c). These results indicated that PGE2 induced pyroptosis in hEM15A cells but not in hESCs.

### 2.3. PEG2 Promotes hEM15A Migration

It has been shown that EMS has similar properties to malignant tumors with enhanced cell migration and invasion. Therefore, we further evaluated the effect of PGE2 on hESCs and hEM15A migration using the transwell assay. After 24 h, the 75 ng/mL PGE2-treated hEM15A cells showed significantly higher cell migration than that by hESCs, suggesting that PGE2 is involved in hEM15A migration (Figure 3).

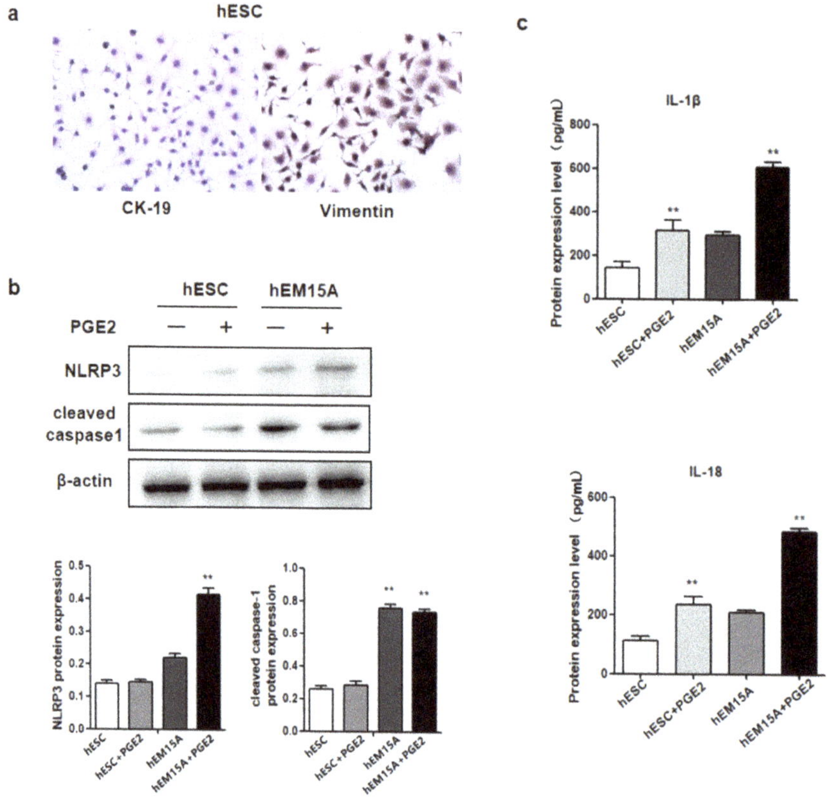


**Figure 2.** PGE2 induced hEM15A pyroptosis. (**a**) After added 75 ng/mL PGE2 for 24 h in hEM15A and hESC, the protein expression of NLRP3, cleaved caspase-1, IL-1β and IL-18 were detected by Western blot analysis. (**b**) The level of IL-1β and IL-18 in cell culture medium were detected by ELISA in hEM15A and hESC. The data are expressed as means ± SEM. ** $p < 0.05$ by one-way ANOVA. (**c**) ELISA was used to detect the expression of IL-1β and IL-18 in hESC and hEM15A without and after PGE2 treatment.

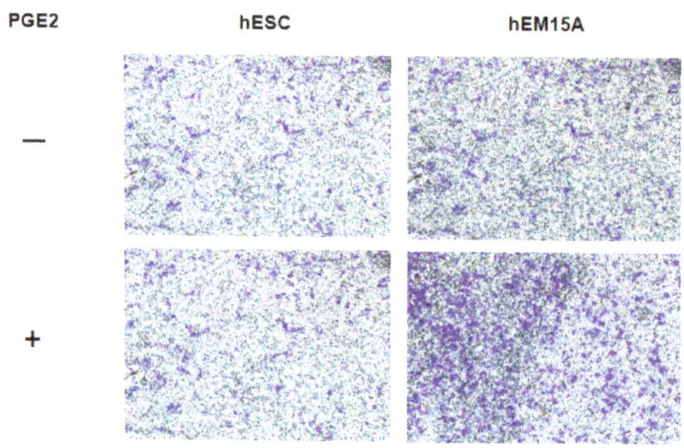


**Figure 3.** *Cont.*

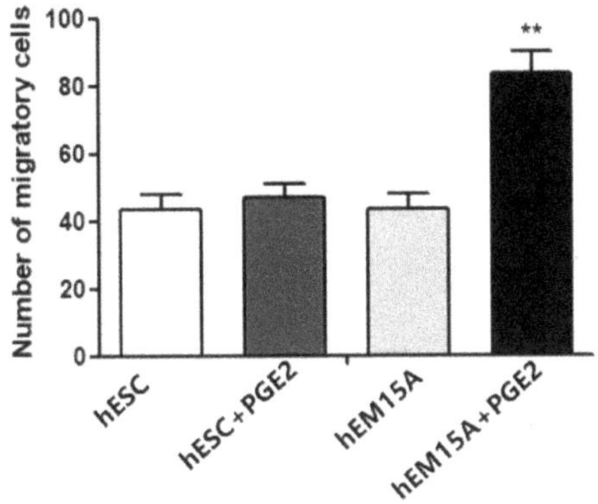


**Figure 3.** After added 75 ng/mL PGE2 for 24 h in hEM15A and hESC, Transwell migration assay was to evaluate the effect of PGE2 on the migration of hESC and hEM15A.The data are expressed as means ± SEM. ** $p < 0.05$ by one-way ANOVA.

*2.4. PEG2 Promotes hEM15A Migration through NLRP3/Caspase-1 Pyroptotic Pathways*

The relationship between cell migration and pyroptosis were then evaluated in hEM15A cells co-cultured in PGE2 and CY09 (the inhibitor of NLRP3). Pyroptosis-related proteins (NLRP3 and cleaved caspase-1) and the levels of IL-1β and IL-18 in the cell culture medium were decreased compared to that in PGE2-treated cells alone (Figure 4a,b). In addition, the migration ability of hEM15A cells under the effect of PGE2 and CY09 was significantly lower than that of PGE2-treated cells alone (Figure 4c). Similarly, in co-cultured PGE2 and VX-765 (the inhibitor of caspase-1) cells, cleaved caspase-1 and the levels of IL-1β and IL-18 in cell culture medium were decreased compared to that of PGE2-treated cells alone (Figure 4a,b). The migration ability of hEM15A cells under the effect of PGE2 and VX-765 was also significantly lower than that of PGE2-treated cells alone (Figure 4c). When both CY09 and VX-765 were added, PGE2 did not activate the expression of pyroptosis-related proteins (NLRP3, cleaved caspase-1, IL-1β, and IL-18), and the migration ability of hEM15A cells was only slightly activated. Thus, CY09 or/and VX-765 reduced the migration ability of hEM15A cells after PGE2 induction, and pyroptosis was involved in the process, in that PGE2 activated EMS invasion.

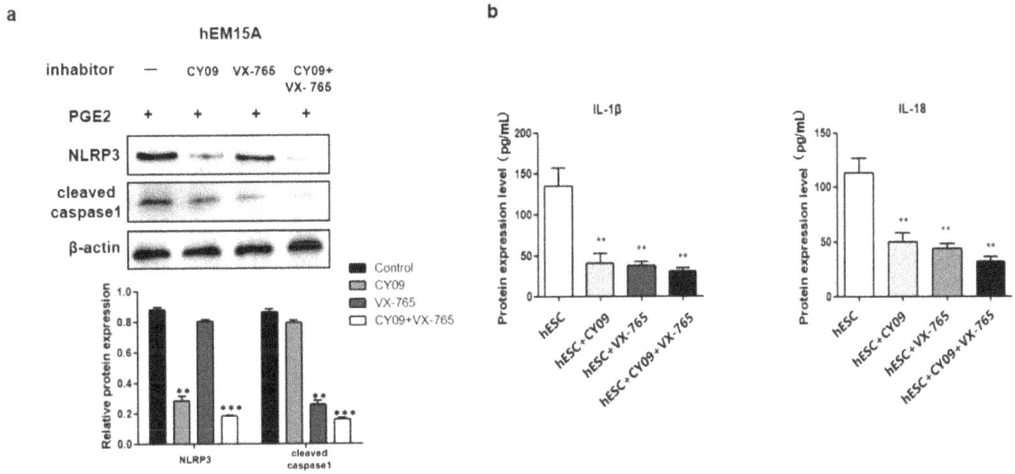


**Figure 4.** *Cont.*

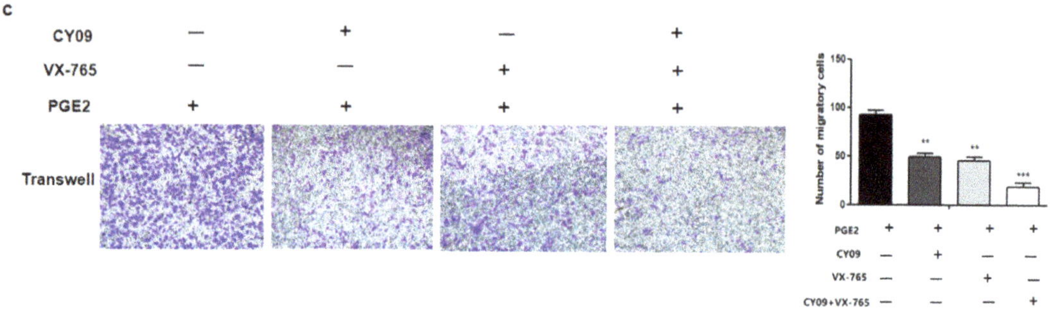


**Figure 4.** PEG2 promoted hEM15A migration through NLRP3/caspase-1 pyroptosis pathways. (**a**) In hEM15A, PGE2, co-cultured PGE2 + CY-09 (the inhibitor of NLRP3), co-cultured PGE2 + VX-765 (the inhibitor of caspase-1) and co-cultured PGE2 + CY-09 + VX-765, pyroptosis related proteins (NLRP3 and cleaved caspase-1) were detected by Western blot analysis. (**b**) In hEM15A, PGE2, co-cultured PGE2 + CY-09 (the inhibitor of NLRP3), co-cultured PGE2 + VX-765 (the inhibitor of caspase-1) and co-cultured PGE2 + CY-09+VX-765, the level of IL-1β and IL-18 in cell culture medium were detected by ELISA. (**c**) In hEM15A, PGE2, co-cultured PGE2 + CY-09 (the inhibitor of NLRP3), co-cultured PGE2 + VX-765 (the inhibitor of caspase-1) and co-cultured PGE2 + CY-09 + VX-765 were evaluated by Transwell migration assay in different groups.** $p < 0.05$, *** $p < 0.001$ by one-way ANOVA.

### *2.5. PGE2-Induced Pyroptosis Regulates the Expression of High Mobility Group Box 1 (HMGB1), E-Cadherin, and Vimentin*

In hEM15A cells co-cultured in PGE2 and CY09, HMGB1 and E-cadherin proteins had low expresssion while vimentin was highly expressed when compared to that in PGE2-treated cells alone. Similarly, co-cultured PGE2 and VX-765 cells showed low expressions of HMGB1 and E-cadherin proteins and high expression of vimentin when compared to PGE2-treated cells alone. When both CY09 and VX-765 were added, PGE2 did not influence the expressions of HMGB1, E-cadherin, and vimentin in hEM15A cells (Figure 5b).

### *2.6. PGE2-Induced Pyroptotic Pathways and Invasion of EMS Lesions In Vivo*

To further verify the induced pyroptosis effects of PGE2 on EMS in vivo, an EMS mouse model was built. As expected, PGE2 significantly induced the pyroptosis of EMS lesions. The weight and diameter (size) of EMS lesions in the PGE2 group were larger than that in the control group (Figure 6a). Additionally, pyroptosis-related proteins (NLRP3, cleaved caspase-1, IL-1β, and IL-18), HMGB1, E-cadherin, and vimentin were tested by immunohistochemistry analysis. The expression of NLRP3, cleaved caspase-1, IL-18, IL-1β, HMGB1, and vimentin was increased with increasing concentrations of PGE2 (Figure 6b). Moreover, the expression of HMGB1, E-cadherin, and vimentin was positively correlated with the expression of pyroptosis-related proteins, which was consistent with the results obtained in the in vitro experiments.

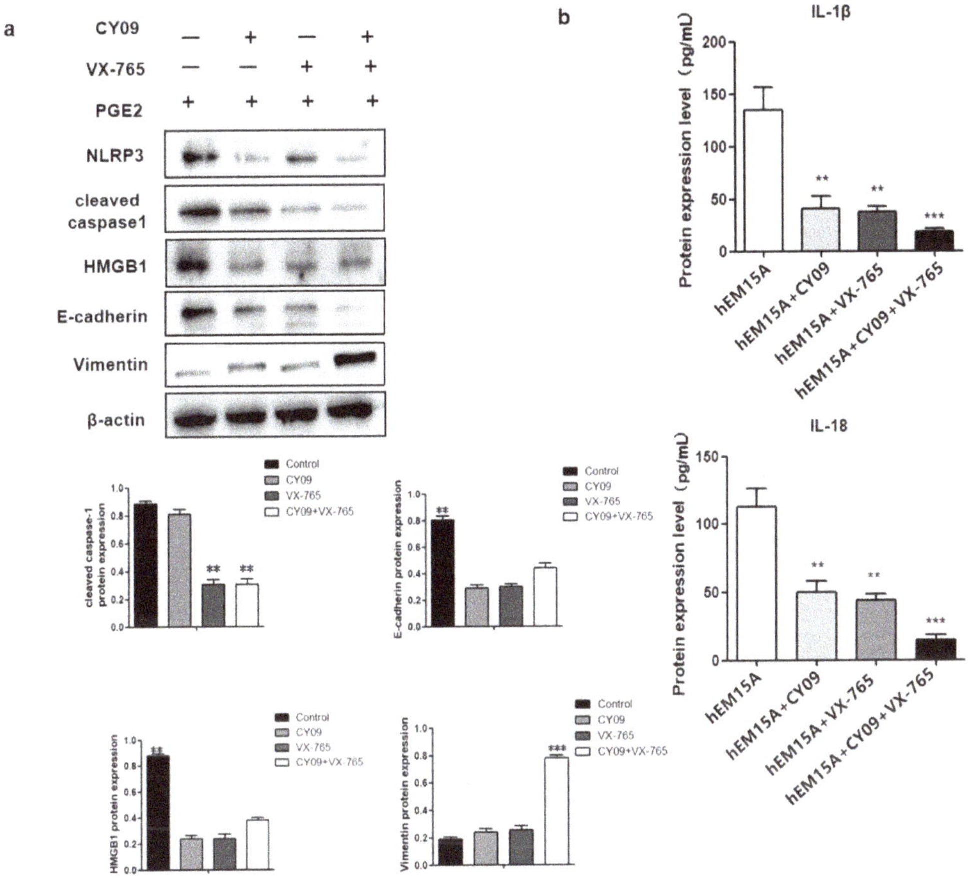


**Figure 5.** PEG2-induced pyroptosis regulates the expression of HMGB1 and promotes hEM15A migration ability. (**a**) In hEM15A, PGE2, co-cultured PGE2 + CY-09 (the inhibitor of NLRP3), co-cultured PGE2 + VX-765 (the inhibitor of caspase-1) and co-cultured PGE2 + CY-09 + VX-765, the protein of NLRP3, cleaved caspased1, HMGB1, E-cadherin and Vimentin were detected by Western blot analysis. (**b**) In hEM15A, PGE2, co-cultured PGE2 + CY-09 (the inhibitor of NLRP3), co-cultured PGE2 + VX-765 (the inhibitor of caspase-1) and co-cultured PGE2 + CY-09 + VX-765, the level of IL-1β and IL-18 in cell culture medium were detected by ELISA in hEM15A and hESC. The data are expressed as means ± SEM. ** $p < 0.05$, *** $p < 0.001$ by one-way ANOVA.

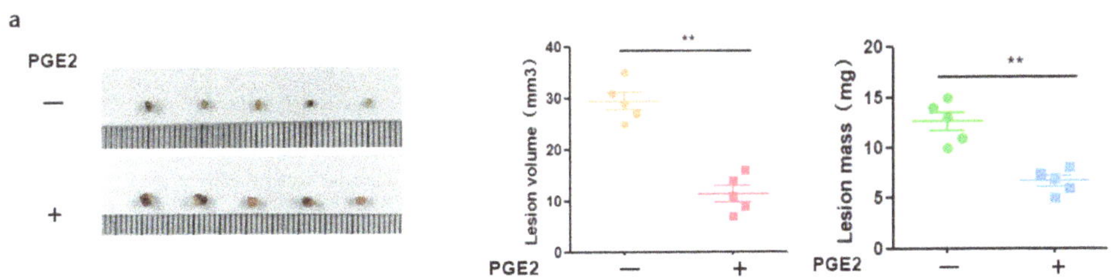


**Figure 6.** *Cont.*

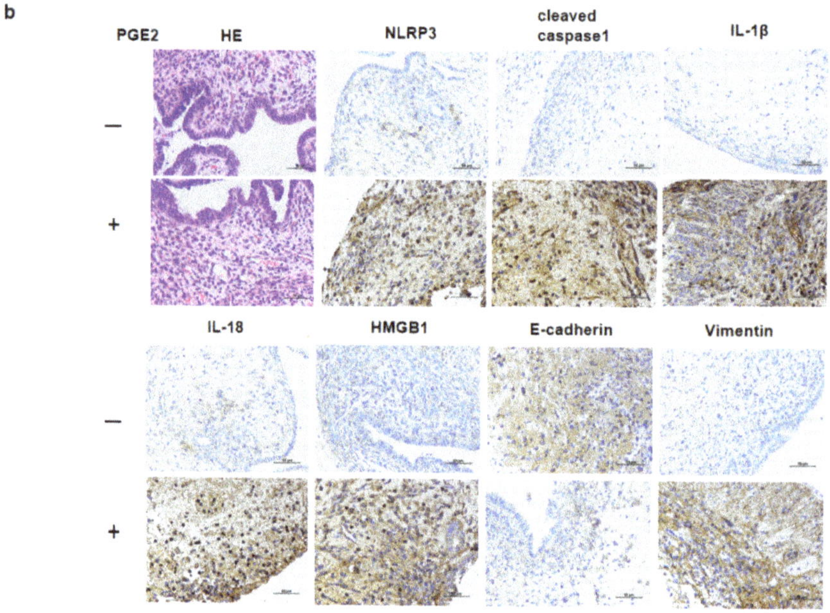


**Figure 6.** PEG2-induced pyroptosis pathways and invasion of EMS lesion in vivo. (**a**) After intervention with PGE2 in endometriosis mouse model, ectopic lesions were enlarged. Ectopic lesions obtained from endometriosis mice after continuous treatment with PGE2. The volume of ectopic lesions was statistically analyzed (** $p < 0.05$); (**b**) immunohistochemistry showed that NLRP3, cleaved caspase-1, Il-1β, Il-18, HMGB1 and Vimentin were highly expressed in the endometrial tissues of mice treated with PGE2.E-cadherin was down-regulated in the endometrial tissues of mice treated with PGE2 (** $p < 0.05$).

## 3. Discussion

EMS is a progressive inflammatory disease that affects women's health and has become a major threat to fertility and individual quality of life. Two important factors of EMS to human health are ectopic lesion-induced inflammation and lesion invasion/metastasis, which are not only a key factor in disease progression, but also a root cause of recurrence after surgical and medication treatments of EMS. Previous research has shown that persistent inflammation promotes local tissue fibrosis and disease progression [17]. Therefore, exploring the relationship between inflammation and disease aggressiveness is critical to understand the underlying molecular pathogenesis of EMS and to find novel therapeutic targets. The present study described the mechanism by which PEG2-induced pyroptosis regulates the migration of EMS cells (Figure 6). A notable increase in PGE2 and pyroptosis-related proteins (NLRP3, pro-caspase-1, caspase-1, IL-1β, and IL-18) was found in EMS tissue compared with that in normal tissue, and we demonstrated that PGE2 can induce pyroptosis in EMS cells, while PGE2-induced pyroptosis can affect cell migration and upregulate the expression of HMGB1 and vimentin. These results indicated that PEG2 could significantly promote EMS invasion, activated by PEG2, which was accomplished via pyroptosis induction.

PGE2 can activate multiple anti-apoptotic factors, such as Bcl-2, and prolong cell life and accelerate ectopic lesions growth. PGE2 stimulates protein kinase A and promotes the phosphorylation of Ras-like estrogen-regulated growth to induce the proliferation of ectopic lesions. In addition, PGE2 leads to the formation of large neovascularization in ectopic lesions and provides rich blood and nutrients for ectopic lesions, which is conducive to endometrial implantation and lesion growth. The present study is, to the best of our knowledge, the first to clarify the inducing effect of PGE2 on pyroptosis, promoting cell migration and the progress of EMS. It is, therefore, of great significance for further understanding PGE2 pathogenesis in EMS.

Inflammatory PGE2 induces the expression of large amounts of IL-1β and IL-18 inflammatory factors through the pyroptotic pathway, which is a positive cycle of inflammatory response in the EMS, leading to subsequent inflammation. Inflammation is a complex and huge network. In tumors, inflammatory cells widely receive proliferative signals by releasing inflammatory factors, leading to rapid cell proliferation and differentiation and increasing the risk of cell cancerization [18–20]. Additionally, inflammatory cells release chemokines and cytokines, which affect the cancerous organ and regulate the growth, migration, and differentiation of cells in the tumor microenvironment [21–23]. In the tumorigenesis process, tumor cells promote tumor spread and metastasis through inflammation by releasing TNF-α, IL-6, IL-1, and interferon factors, which in turn stimulate tumor cell growth, motility, and invasion [24,25].

In EMS, periodic bleeding occurs in the ectopic endometrium along with the menstrual cycle, which forms a wound with repeated injury and continuous healing. In the repair process, multiple systems, such as the inflammatory system, coagulation system, and immune system are often involved. The inflammatory response leads to the production of pro-inflammatory factors, and this inflammatory microenvironment promotes the proliferation, migration, metastasis, and angiogenesis of lesions through a series of pathways. It has been hypothesized that in PGE2-induced pyroptosis cells secrete a large number of IL-1β and IL-18 inflammatory factors. On the one hand, these inflammatory factors recruit more inflammatory cells, induce the synthesis and release of other factors such as IL-1α, IL-6, and TNF-α expanding the inflammatory response, and are widely involved in various pathological injury processes such as inflammatory exclusion, tissue destruction, and edema formation. In pneumonia, for instance, IL-1β and IL-18 can activate inflammatory cells, such as macrophages to release inflammatory mediators, causing inflammatory responses characterized by infiltration of macrophages and granulocytes in the lung [26]. On the other hand, inflammatory factors promote the proliferation and migration of cells. For example, IL-1β promotes the proliferation, migration, angiogenesis, and release of HMGB1 in smooth muscle cells, while IL-18 promotes the migration of breast cancer cells by down-regulating claudin-12 and inducing the P38/MAPK pathway [27]. In the present study, cleaved caspase 1, IL-1β, and IL-18 were activated after PGE2 stimulation. After PGE2-induced pyroptosis, hEM15A cell migration was increased, HMGB1 and vimentin expressions were upregulated, and E-cadherin expression was downregulated. Thus, in EMS, PGE2-induced pyroptosis seems to promote the expression of inflammatory factors that alter cell migration accompanied by abnormal expression of E-cadherin and vimentin (the markers of cell migration). To the best of our knowledge, this phenomenon was observed in EMS in this study for the first time. Although its specific regulatory mechanism needs to be further studied, the present findings contribute to revealing the underlying molecular pathogenesis of EMS and to the search for new therapeutic targets.

In general, inflammation is a double-edged sword. While protecting the body against disease, it also contributes to the progress of the disease. In the present study, the effect of PGE2-induced pyroptosis on the progression of EMS was mainly reflected in the changed migration ability of pyroptosis cells and upregulated expressions of HMGB1 and vimentin, as verified in in vivo experiments. The potential relationship found between cell pyroptosis and migration provides crucial evidence for new routes in the clinical development and use of anti-inflammatory drugs.

## 4. Materials and Methods

### *4.1. Patients and Tissues*

This study was approved by the Ethics Committee of the First Hospital of Lanzhou University (Lanzhou, China), and informed consent was obtained from each patient. Ectopic ovarian endometriotic cyst tissues were obtained from patients with EMS, who underwent laparoscopic treatment in Lanzhou University First Hospital from January 2020 to January 2021 (n = 28). Normal endometrial tissue was collected from women of childbearing age observed in clinical practice (n = 15). Collected tissues were rapidly placed in liquid nitrogen and then stored at −80 °C until analyses. The following were inclusion criteria: (1) patients diagnosed as EMs by laparoscopic surgery; (2) normal hormone level and regular menstrual cycle' (3) no history of hereditary or familial diseases, normal chromosomes, and infectious disease tests were negative. The following were exclusion criteria (1) receipt of hormone or immunosuppressive therapy in the last 3 months; (2) other endocrine diseases; (3) recurrent EMS; (4) other gynecological diseases, abnormal uterus, polycystic ovary syndrome and low ovarian response or malignant tumors. Control group: Fifteen patients with malformed mediastinal uterus who underwent hysteroscopic surgery during the same period were selected, and their normal endometrium was taken as the control group.

### 4.2. Cell Culture

The hEM15A cell line was purchased from China Center for Typical Cultures Preservation (Wuhan, China) and was cultured at 5% $CO_2$ and 37 °C in RPMI/1640 (HyClone, Logan, UT, USA) containing 15% fetal bovine serum (FBS) (Gibco, Thermo Fisher Scientific, Waltham, MA, USA). Primary human endometrial stromal cells (hESCs) were isolated and cultured as follows: the endometrial tissue was washed with phosphate-buffered saline (PBS) three times within 30 min to remove blood stains. Then, the tissue was cut into pieces using a sterile ophthalmic scissor, and 3–5 mL of 0.1% type I collagenase digestion solution (Solarbio, Beijing, China) was added to the tissue fragments. After incubation in a water bath at 37 °C for 60 min, Dulbecco's modified Eagle's medium/nutrient mixture F-12 (DMEM/F12) medium (HyClone) containing 10% FBS and 1% penicillomycin (Solarbio) was added. The tissue suspension was then filtered through a 40-mesh screen to discard the residue of tissue retained on the mesh. After centrifugation for 10 min at 1500 rpm, the obtained cell precipitates were suspended in the complete culture medium, inoculated in Petri dishes, and incubated at 37 °C and 5% $CO_2$. Cell adherence was observed after 12 h. The original medium was replaced with a fresh medium.

### 4.3. Transwell Assay

Different cell suspensions were added into the upper chamber. RPMI/1640 or DMEM/F12 containing 20% FBS was added to the lower chamber. Then, cells were cultured at 37 °C and 5% $CO_2$ for 24 h. The submembrane cells were fixed with 4% paraformaldehyde, stained with 0.1% crystal violet, and the inserts were cleaned with PBS. After drying, cells were selected and counted under a light microscope.

### 4.4. Immunocytochemistry

Cell suspensions ($1 \times 10^6$ cells/mL) were inoculated in a six-well plate, covered with sterile cover glass, and incubated for 24 h. When cell confluence reached 80%, cells were fixed in 4.0% paraformaldehyde and permeated using 0.1–0.2% polyethylene glycoloctylphenylether. After incubation for 15 min, 5.0% normal goat abandoned blood serum was added. Anti-CK19 (1:50; Abcam, Cambridge, UK) and anti-vimentin (1:50; Abcam) were added and a negative control (PBS instead of primary antibody) was set up simultaneously; all samples were placed in wet box at 4 °C overnight. After rewarming on the next day, diluted secondary antibody was added, incubated at 37 °C for 15 min, and dyed using the SP method (i.e., adding horseradish-labeled chain enzyme lactalbumin working solution). The neutral gum was then sealed and observed under a light microscope.Immunohistochemistry analysis showed that CK19 was negative and vimentin was positive in hESCs and hEM15A. CY09 and VX-765 were purchased from Selleck (Shanghai, China).

### 4.5. Enzyme-Linked Immunosorbent Assay (ELISA)

The concentrations of PGE2, IL-1β, and IL-18 in the serum of patients and cell culture medium were detected using ELISA kits (Solarbio), according to the manufacturer's protocol. The optical density (OD) at 450 nm was used to calculate the concentrations of PGE2, IL-1β, and IL-18.

### 4.6. Western Blot

Cells were collected and lysed. Total protein was extracted and quantified using the BCA Protein Assay Kit (Solarbio), following the manufacturer's instructions. Primary antibodies (all Abcam) used included anti-NLRP3 (1:1000), anti-cleaved caspase-1 (1:1000), anti-IL-1β (1:1000), anti-E-cadherin (1:1000), anti-vimentin (1:1000), and anti-β-actin (1:5000) and were incubated at 4 °C overnight. After washing with tris-buffered saline with Tween (TBS-T) three times, the secondary antibody (Solarbio) was added for 1 h at room temperature. The ECL substrate (Solarbio) was used to detect the expression of target proteins in the FUSION FX5 imaging system (Bio-Rad, Hercules, CA, USA).

### 4.7. Immunohistochemistry

Paraffin-embedded tissues were sectioned, placed for 20 min in a 60 °C constant temperature box after dewaxing hydration, and subjected to high-pressure repair for 2.5 min. After washing three times with PBS (5 min each), endogenous peroxidase blocking agent was added at room temperature for 10 min. The PBS washing step (5 min each) was performed for 1 h and conducted three times. Primary antibodies (anti-NLRP3, 1:100; anti-cleaved caspase-1, 1:100; anti-IL-1β, 1:100; anti-IL-1β, 1:100; anti-E-cadherin, 1:200; anti-vimentin, 1:200; and anti-β-actin, 1:500; all Abcam) were incubated overnight. After PBS washing for 5 min three times, secondary antibodies (Solarbio) were incubated before 3,3′-diaminobenzidine (DAB) was added as the chromogen. Tissue sections were then observed and photographed under the microscope.

### 4.8. Animals

Specific pathogen free female BALB/C mice (4–5 weeks old) were purchased from the Experimental Animal Center of Lanzhou University Medical College (Lanzhou, China). Animals were raised in the laboratory animal center for 1 week before the onset of the experiments. Recipient and donor mice were subcutaneously injected with estradiol (100 μg/kg body weight) once a week for modeling until the end of the experiment. On the day of mouse modeling, both uterine horns of donor mice were extracted and placed in a glass dish containing PBS to remove fat, blood, and other tissues. The ratio of donor mice to recipient mice was 1:2. The uteri of donor mice were cut into 1–2 mm pieces and then divided into several parts of equal weight, which were injected into the abdominal cavity of recipient mice. EMS recipient mice were randomly divided into two groups (n = 5 per group), (I) control group (0.9% sodium 0.01 mL/g/day) and (II) PGE2 group (0.01 mL/g/day), which were intraperitonially injected with the respective treatment. The mice weight was measured every 3 days until day 21. After the experiment, the mice were sacrificed. The weight and size of lesions were determined.

### 4.9. Statistical Analyses

All graphics were generated using GraphPad Prism 8.0. Data are expressed as mean ± standard error of the mean (SEM) and were analyzed in SPSS 22.0 software. Two-tailed unpaired t-test was used to analyze the differences between the two groups, and one-way analysis of variance (ANOVA) was used for statistical comparisons. $p < 0.05$ was considered statistically significant.

**Author Contributions:** Y.H. was mainly responsible for the conception and design, acquisition of data, drafting the manuscript. R.L. and R.H. provided acquisition of data, analysis and interpretation of data and drafting the manuscript. J.Y. and Y.H. contributed to analysis and interpretation of data and drafting the manuscript. Y.Y. supervised the whole project, who contributed to the conception and design, drafting the manuscript, critically revising the manuscript and approval of the final version submitted for publication, considered the corresponding author. All authors have read and agreed to the published version of the manuscript.

**Funding:** The study was supported by the Regional Scientists Fund of the National. Natural Science Foundation of China (No. 81960275); Gansu Provincial. Department of Science and Technology Foundation (21YF5FA119).

**Institutional Review Board Statement:** The study protocol for this retrospective cohort study was approved by The First Hospital of Lanzhou University Research Ethics Committee (LDYYLL2019-148).

**Informed Consent Statement:** Informed consent was obtained from all subjects involved in the study.

**Data Availability Statement:** Not applicable.

**Conflicts of Interest:** The authors have no commercial or other associations that might pose a conflict of interest.

## References

1. Peiris, A.N.; Chaljub, E.; Medlock, D. Endometriosis. *JAMA* **2018**, *320*, 2608. [CrossRef]
2. Zondervan, K.T.; Becker, C.M.; Missmer, S.A. Endometriosis. *N. Engl. J. Med.* **2020**, *382*, 1244–1256. [CrossRef] [PubMed]
3. Rolla, E. Endometriosis: Advances and controversies in classification, pathogenesis, diagnosis, and treatment. *F1000Research* **2019**, *8*, 529. [CrossRef] [PubMed]
4. Zhang, T.; De Carolis, C.; Man, G.C.W.; Wang, C.C. The link between immunity, autoimmunity and endometriosis: A literature update. *Autoimmun. Rev.* **2018**, *17*, 945–955. [CrossRef] [PubMed]
5. D'Alessandris, N.; Travaglino, A.; Santoro, A.; Arciuolo, D.; Scaglione, G.; Raffone, A.; Inzani, F.; Zannoni, G.F. TCGA molecular subgroups of endometrial carcinoma in ovarian endometrioid carcinoma: A quantitative systematic review. *Gynecol. Oncol.* **2021**, *163*, 427–432. [CrossRef]
6. Zhou, L.; Yao, L.; Dai, L.; Zhu, H.; Ye, X.; Wang, S.; Cheng, H.; Ma, R.; Liu, H.; Cui, H.; et al. Ovarian endometrioid carcinoma and clear cell carcinoma: A 21-year retrospective study. *J. Ovarian Res.* **2021**, *14*, 63. [CrossRef]
7. Woodbeck, R.; Kelemen, L.E.; Kobel, M. Ovarian Endometrioid Carcinoma Misdiagnosed as Mucinous Carcinoma: An Underrecognized Problem. *Int. J. Gynecol. Pathol.* **2019**, *38*, 568–575. [CrossRef]
8. Parra-Herran, C.; Bassiouny, D.; Vicus, D.; Olkhov-Mitsel, E.; Cesari, M.; Ismiil, N.; Nofech-Mozes, S. FIGO Versus Silverberg Grading Systems in Ovarian Endometrioid Carcinoma: A Comparative Prognostic Analysis. *Am. J. Surg. Pathol.* **2019**, *43*, 161–167. [CrossRef]
9. Zou, Y.; Liu, F.-Y.; Liu, H.; Wang, F.; Li, W.; Huang, M.-Z.; Huang, Y.; Yuan, X.-Q.; Xu, X.-Y.; Huang, O.-P.; et al. Frequent POLE1 p.S297F mutation in Chinese patients with ovarian endometrioid carcinoma. *Mutat. Res.* **2014**, *761*, 49–52.
10. Malik, K.; Dua, A. Prostaglandins. 15 May 2022. In *StatPearls [Internet]*; StatPearls Publishing: Treasure Island, FL, USA, 2022. [PubMed]
11. Pilbeam, C. Prostaglandins and Bone. *Handb. Exp. Pharmacol.* **2020**, *262*, 157–175.
12. Frank, D.; Vince, J.E. Pyroptosis versus necroptosis: Similarities, differences, and crosstalk. *Cell Death Differ.* **2019**, *26*, 99–114. [CrossRef]
13. Wang, J.; Shi, K.; An, N.; Li, S.; Bai, M.; Wu, X.; Shen, Y.; Du, R.; Cheng, J.; Wu, X.; et al. Direct Inhibition of GSDMD by PEITC Reduces Hepatocyte Pyroptosis and Alleviates Acute Liver Injury in Mice. *Front. Immunol.* **2022**, *13*, 825428. [CrossRef]
14. Sharma, B.R.; Kanneganti, T.D. NLRP3 inflammasome in cancer and metabolic diseases. *Nat. Immunol.* **2021**, *22*, 550–559. [CrossRef]
15. Zhang, Z.; Zhang, H.; Li, D.; Zhou, X.; Qin, Q.; Zhang, Q. Caspase-3-mediated GSDME induced Pyroptosis in breast cancer cells through the ROS/JNK signalling pathway. *J. Cell. Mol. Med.* **2021**, *25*, 8159–8168. [CrossRef]
16. Yu, S.; Zhao, N.; He, M.; Zhang, K.; Bi, X. MiRNA-214 promotes the pyroptosis and inhibits the proliferation of cervical cancer cells via regulating the expression of NLRP3. *Cell. Mol. Biol.* **2020**, *66*, 59–64. [CrossRef] [PubMed]
17. Giacomini, E.; Minetto, S.; Piani, L.L.; Pagliardini, L.; Somigliana, E.; Viganò, P. Genetics and Inflammation in Endometriosis: Improving Knowledge for Development of New Pharmacological Strategies. *Int. J. Mol. Sci.* **2021**, *22*, 9033. [CrossRef]
18. Zhao, L.; Hu, H.; Gustafsson, J.; Zhou, S. Nuclear Receptors in Cancer Inflammation and Immunity. *Trends Immunol.* **2020**, *41*, 172–185. [CrossRef]
19. Camilli, M.; Iannaccone, G.; La Vecchia, G.; Cappannoli, L.; Scacciavillani, R.; Minotti, G.; Massetti, M.; Crea, F.; Aspromonte, N. Platelets: The point of interconnection among cancer, inflammation and cardiovascular diseases. *Expert Rev. Hematol.* **2021**, *14*, 537–546. [CrossRef] [PubMed]
20. Maiorino, L.; Egeblad, M. Tumours pick the path to cancer inflammation. *Nat. Cell Biol.* **2019**, *21*, 1055–1057. [CrossRef] [PubMed]
21. Dey, M.; Ayan, B.; Yurieva, M.; Unutmaz, D.; Ozbolat, I.T. Studying Tumor Angiogenesis and Cancer Invasion in a Three-Dimensional Vascularized Breast Cancer Micro-Environment. *Adv. Biol.* **2021**, *5*, e2100090. [CrossRef]

22. Aprupe, L.; Litjens, G.; Brinker, T.J.; Van Der Laak, J.; Grabe, N. Robust and accurate quantification of biomarkers of immune cells in lung cancer micro-environment using deep convolutional neural networks. *PeerJ* **2019**, *7*, e6335. [CrossRef] [PubMed]
23. Wertel, I.; Barczynski, B.; Kotarski, J. The role of dendritic cells in cytotoxic immune response regulation in ovarian cancer micro-environment. *Front. Biosci.* **2008**, *13*, 2177–2190. [CrossRef] [PubMed]
24. Osipyan, A.; Chen, D.; Dekker, F.J. Epigenetic regulation in macrophage migration inhibitory factor (MIF)-mediated signaling in cancer and inflammation. *Drug Discov. Today* **2021**, *26*, 1728–1734. [CrossRef] [PubMed]
25. Wang, X.; Ye, X.; Zhang, Y.; Ji, F. Flurbiprofen suppresses the inflammation, proliferation, invasion and migration of colorectal cancer cells via COX2. *Oncol. Lett.* **2020**, *20*, 132. [CrossRef]
26. Sun, J.; LaRock, D.L.; Skowronski, E.A.; Kimmey, J.M.; Olson, J.; Jiang, Z.; O'Donoghue, A.J.; Nizet, V.; LaRock, C.N. The Pseudomonas aeruginosa protease LasB directly activates IL-1beta. *EBioMedicine* **2020**, *60*, 102984. [CrossRef]
27. Nisar, M.A.; Zheng, Q.; Saleem, M.Z.; Ahmmed, B.; Ramzan, M.N.; Ud Din, S.R.; Tahir, N.; Liu, S.; Yan, Q. IL-1beta Promotes Vasculogenic Mimicry of Breast Cancer Cells Through p38/MAPK and PI3K/Akt Signaling Pathways. *Front. Oncol.* **2021**, *11*, 618839. [CrossRef]

International Journal of
*Molecular Sciences*

MDPI

*Article*

# Impact of Placental SLC2A3 Deficiency during the First-Half of Gestation

**Cameron S. Lynch [1], Victoria C. Kennedy [1], Amelia R. Tanner [1], Asghar Ali [1], Quinton A. Winger [1], Paul J. Rozance [2] and Russell V. Anthony [1,*]**

[1] College of Veterinary Medicine, Colorado State University, Fort Collins, CO 80523, USA
[2] Anschutz Medical Campus, University of Colorado, Aurora, CO 80045, USA
* Correspondence: russ.anthony@colostate.edu

**Abstract:** In the ruminant placenta, glucose uptake and transfer are mediated by facilitative glucose transporters SLC2A1 (GLUT1) and SLC2A3 (GLUT3). SLC2A1 is located on the basolateral trophoblast membrane, whereas SLC2A3 is located solely on the maternal-facing, apical trophoblast membrane. While SLC2A3 is less abundant than SLC2A1, SLC2A3 has a five-fold greater affinity and transport capacity. Based on its location, SLC2A3 likely plays a significant role in the uptake of glucose into the trophoblast. Fetal hypoglycemia is a hallmark of fetal growth restriction (FGR), and as such, any deficiency in SLC2A3 could impact trophoblast glucose uptake and transfer to the fetus, thus potentially setting the stage for FGR. By utilizing in vivo placenta-specific lentiviral-mediated RNA interference (RNAi) in sheep, we were able to significantly diminish ($p \leq 0.05$) placental SLC2A3 concentration, and determine the impact at mid-gestation (75 dGA). In response to SLC2A3 RNAi ($n = 6$), the fetuses were hypoglycemic ($p \leq 0.05$), exhibited reduced fetal growth, including reduced fetal pancreas weight ($p \leq 0.05$), which was associated with reduced umbilical artery insulin and glucagon concentrations, when compared to the non-targeting sequence (NTS) RNAi controls ($n = 6$). By contrast, fetal liver weights were not impacted, nor were umbilical artery concentrations of IGF1, possibly resulting from a 70% increase ($p \leq 0.05$) in umbilical vein chorionic somatomammotropin (CSH) concentrations. Thus, during the first half of gestation, a deficiency in SLC2A3 results in fetal hypoglycemia, reduced fetal development, and altered metabolic hormone concentrations. These results suggest that SLC2A3 may be the rate-limiting placental glucose transporter during the first-half of gestation in sheep.

**Keywords:** placenta; glucose uptake; SLC2A3; SLC2A1; insulin; glucagon

**Citation:** Lynch, C.S.; Kennedy, V.C.; Tanner, A.R.; Ali, A.; Winger, Q.A.; Rozance, P.J.; Anthony, R.V. Impact of Placental SLC2A3 Deficiency during the First-Half of Gestation. *Int. J. Mol. Sci.* **2022**, *23*, 12530. https://doi.org/10.3390/ijms232012530

Academic Editor: Ilona Hromadnikova

Received: 20 September 2022
Accepted: 17 October 2022
Published: 19 October 2022

**Publisher's Note:** MDPI stays neutral with regard to jurisdictional claims in published maps and institutional affiliations.



## 1. Introduction

Glucose is the primary energy substrate for fetal oxidative processes and growth [1]. Due to a lack of endogenous fetal glucose production, until near term, the maternal circulation is the only source of glucose for the placenta and fetus [2,3]. Placental glucose uptake and transfer to the fetus requires a positive maternal-to-fetal glucose concentration gradient that is mediated by facilitative glucose transporter (GLUT) proteins on both the maternal-facing apical microvillus and fetal-facing basal trophoblast membranes. In sheep, SLC2A3 (GLUT3) is localized to the apical microvillus trophoblast membrane, while SLC2A1 (GLUT1) is localized to the basolateral trophoblast membrane, and as such, both transporters must be utilized sequentially for glucose uptake and transfer from maternal to fetal circulation [4]. While both SLC2A1 and SLC2A3 are present throughout gestation [5], SLC2A1 has been regarded as the primary placental glucose transporter as it is the most abundant glucose transporter in the mammalian placenta and increases in abundance as gestation progresses [6]. Conversely, SLC2A3 is less abundant than SLC2A1, but SLC2A3 has a five-fold greater affinity and transport capacity for glucose [7].

Functional placental insufficiency is a major cause of fetal growth restriction (FGR), however, the specific causes of placental insufficiency are not well characterized. A common hallmark of FGR is reduced placental transfer of glucose, resulting in the fetuses becoming hypoglycemic [8,9]. When assessed in pre-term or term FGR placentas, there is a lack of down-regulation of either SLC2A1 or SLC2A3 [10,11], indicating that the fetal hypoglycemia does not stem from a deficit in glucose transport mechanisms. However, these observations may not reflect the glucose transport capacity of the FGR placenta throughout gestation. Any deficit in placental glucose transport during the first-half of pregnancy could impact placental development and function, thus potentially resulting in functional placental insufficiency and setting the stage for fetal hypoglycemia and FGR. For obvious ethical reasons, the relative importance of SLC2A1 and SLC2A3 in placental glucose transport at different stages of gestation, and how a deficit in either glucose transporter may alter glucose transport, cannot be addressed in humans. As SLC2A1 is localized solely to the basolateral trophoblast membrane and SLC2A3 is localized solely to the maternal-facing apical microvillous membrane [4], the sheep placenta provides the opportunity to differentiate the relative importance of each placental glucose transporter as well as apical versus basolateral transport.

The development of in vivo lentiviral-mediated RNA interference (RNAi) [12–17], that specifically targets trophoblast cells, provides the opportunity to directly assess the relative importance of placental SLC2A3. While the abundance of SLC2A3 is less than SLC2A1 [5], SLC2A3 appears to play an important role in trophoblast uptake of glucose due to its location on the maternal-facing apical trophoblast membrane [4] and its five-fold greater affinity and transport capacity for glucose [7]. Thus, any deficit in SLC2A3 may have a major impact on placental uptake of glucose, placental development and function, and fetal development. Accordingly, we hypothesized that SLC2A3 deficiency would result in impaired placental development and significant FGR by mid-gestation (75 dGA). Therefore, it was our objective to use lentiviral-mediated RNAi to attenuate the expression of placental SLC2A3 to assess its relative importance in placental glucose transport, as assessed at mid-gestation.

## 2. Results

### 2.1. RNA Interference of SLC2A3 in iOTR Cells

To assess the effectiveness of the SLC2A3 RNAi construct, iOTR cells were infected at a MOI of 500 with either the NTS RNAi or SLC2A3 RNAi lentivirus, approximately the MOI used for blastocyst infection, and SLC2A3 concentrations were determined. SLC2A3 RNAi resulted in a 91% reduction in SLC2A3 concentration ($p \leq 0.05$; Figure 1) as compared to NTS RNAi-infected iOTR cells.

### 2.2. Fetal and Placental Measurements at Mid-Gestation

At 70 dGA, as assessed by ultrasonography, fetal binocular distance tended to be reduced ($p \leq 0.10$), and both femur length and tibia length were significantly shorter ($p \leq 0.05$), whereas there appeared to be no impact of SLC2A3 RNAi on fetal crown-rump length or abdominal circumference (Table 1). Doppler assessment of umbilical artery velocimetry revealed no differences in umbilical artery pulsatility indices, resistance indices, systolic/diastolic ratios, fetal heart rates, umbilical artery cross-sectional areas, or cross-sectional diameters between NTS RNAi and SLC2A3 RNAi fetuses ($p \geq 0.10$; Table 1). When assessed at the 75 dGA terminal surgery, head circumference, femur length, and tibia length were all significantly reduced in the SLC2A3 RNAi pregnancies ($p \leq 0.05$; Table 2), and fetal weight tended ($p \leq 0.10$) to be less (Table 2), whereas crown-rump length and abdominal circumference were not impacted by SLC2A3 RNAi. While fetal liver weight was not impacted (Table 2), fetal pancreas weight was significantly reduced ($p \leq 0.05$). Placental weight (total placentome weight) was less, but did not reach statistical significance ($p \geq 0.10$), and placentome number was not impacted by SLC2A3 RNAi (Table 2).

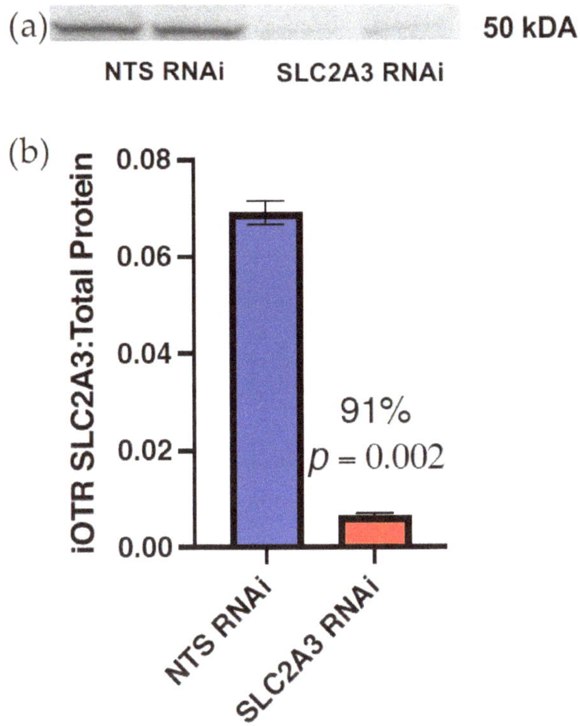


**Figure 1.** Efficiency of SLC2A3 RNAi in iOTR cells: (**a**) iOTR SLC2A3 detected by Western blot analysis following infection with either the NTS RNAi or SLC2A3 RNAi lentivirus; (**b**) concentration of SLC2A3, relative to total protein transferred, following infection with NTS RNAi or SLC2A3 RNAi lentivirus. Data are shown as means ± SEM. NTS, non-targeting sequence; RNAi, RNA interference.

**Table 1.** Fetal and Doppler velocimetry measurements assessed at 70 dGA.

| | NTS RNAi | SLC2A3 RNAi | *p*-Value | % Change |
|---|---|---|---|---|
| Binocular Distance, cm | 3.53 ± 0.10 | 3.29 ± 0.06 | 0.06 | 6.68 |
| Crown-rump length, cm | 14.88 ± 1.13 | 14.31 ± 0.40 | 0.58 | 3.83 |
| Abdominal circumference, cm | 13.09 ± 0.55 | 12.20 ± 0.46 | 0.24 | 6.82 |
| Femur Length, cm | 2.97 ± 0.12 | 2.34 ± 0.18 | 0.01 | 21.19 |
| Tibia Length, cm | 2.70 ± 0.20 | 1.99 ± 0.18 | 0.03 | 26.38 |
| Pulsatility Index | 2.80 ± 0.27 | 2.97 ± 0.15 | 0.59 | 6.15 |
| Resistance Index | 0.86 ± 0.05 | 0.85 ± 0.04 | 0.92 | 0.79 |
| Systolic: Diastolic | 16.18 ± 6.60 | 11.04 ± 3.96 | 0.52 | 31.76 |
| Fetal heart rate, bpm | 204.72 ± 1.91 | 200.03 ± 8.69 | 0.61 | 2.29 |
| Umbilical artery cross-sectional area, $cm^2$ | 0.09 ± 0.007 | 0.104 ± 0.01 | 0.44 | 11.52 |
| Umbilical artery cross-sectional diameter, cm | 0.34 ± 0.01 | 0.35 ± 0.02 | 0.74 | 2.37 |

Data are shown as means ± SEM for all ewes in each treatment group. NTS, non-targeting sequence; RNAi, RNA interference.

SLC2A3 RNAi resulted in a 37% reduction in placental SLC2A3 ($p \leq 0.05$; Figure 2), as well as a 38% increase in SLC2A1 concentration ($p \leq 0.10$; Figure 2) in the SLC2A3 RNAi pregnancies. Additionally, there was no effect of SLC2A3 RNAi on placental SLC2A8 ($p \geq 0.10$; Figure 2).

**Table 2.** Placental and fetal measurements obtained at 75 dGA.

| | NTS RNAi | SLC2A3 RNAi | p-Value | % Change |
|---|---|---|---|---|
| Fetal weight, g | 208.61 ± 9.93 | 179.89 ± 10.84 | 0.08 | 13.77 |
| Head circumference, cm | 13.90 ± 0.19 | 12.88 ± 0.43 | 0.05 | 7.37 |
| Crown-rump length, cm | 19.33 ± 0.4 | 19.17 ± 0.36 | 0.76 | 0.86 |
| Abdominal circumference, cm | 13.62 ± 0.51 | 12.92 ± 0.35 | 0.28 | 5.14 |
| Femur length, cm | 4.33 ± 0.20 | 3.67 ± 0.17 | 0.03 | 15.22 |
| Tibia length, cm | 3.42 ± 0.20 | 2.78 ± 0.19 | 0.05 | 18.54 |
| Liver weight, g | 13.21 ± 0.95 | 12.65 ± 1.06 | 0.71 | 4.20 |
| Pancreas weight, mg | 470.00 ± 34.35 | 363.33 ± 15.85 | 0.02 | 22.70 |
| Placentome weight, g | 536.10 ± 54.05 | 423.39 ± 34.81 | 0.13 | 21.02 |
| Placentome number | 81.83 ± 5.51 | 73.33 ± 7.17 | 0.37 | 10.39 |

Data are shown as mean values ± SEM for all ewes in each treatment group. NTS, non-targeting sequence; RNAi, RNA interference.

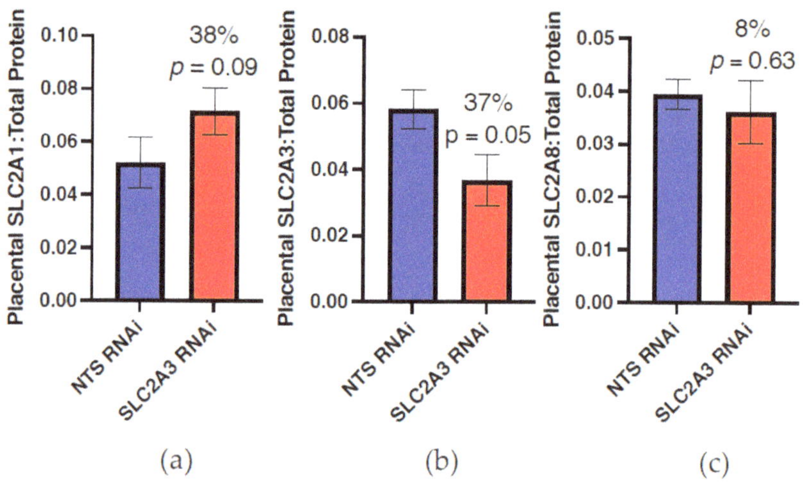


**Figure 2.** Impact of SLC2A3 RNAi on placental concentrations of (**a**) SLC2A1, (**b**) SLC2A3 and (**c**) SLC2A8 at mid-gestation in sheep. Data are shown as means ± SEM. NTS, non-targeting sequence; RNAi, RNA interference.

### *2.3. Maternal and Fetal Plasma Measurements at Mid-Gestation*

At 75 dGA, uterine artery and vein concentrations of glucose (Figure 3) and lactate (Figure 3) were not significantly impacted by SLC2A3 RNAi ($p \geq 0.10$), although SLC2A3 RNAi maternal glucose concentrations were 20–25% lower. In contrast, umbilical vein and artery glucose concentrations were significantly reduced by 42% and 46%, respectively, in the SLC2A3 RNAi pregnancies ($p \leq 0.05$; Figure 4). There were no statistical differences observed in lactate concentrations in either the umbilical vein or artery ($p \geq 0.10$; Figure 4). Individual amino acid concentrations in the uterine and umbilical vasculature are presented in Tables 3 and 4. There were few SLC2A3 RNAi-induced changes in maternal plasma amino acid concentrations, with the exceptions being greater ($p \leq 0.05$) concentrations of asparagine and lysine in the uterine artery, and tendencies ($p \leq 0.10$) for increased valine and ornithine in the uterine artery and citrulline in the uterine vein. Similarly, there were few changes in fetal plasma amino acids, other than significant reductions ($p \leq 0.05$) in the umbilical vein and artery concentrations of arginine, and increased ($p \leq 0.05$) asparagine in the umbilical artery.

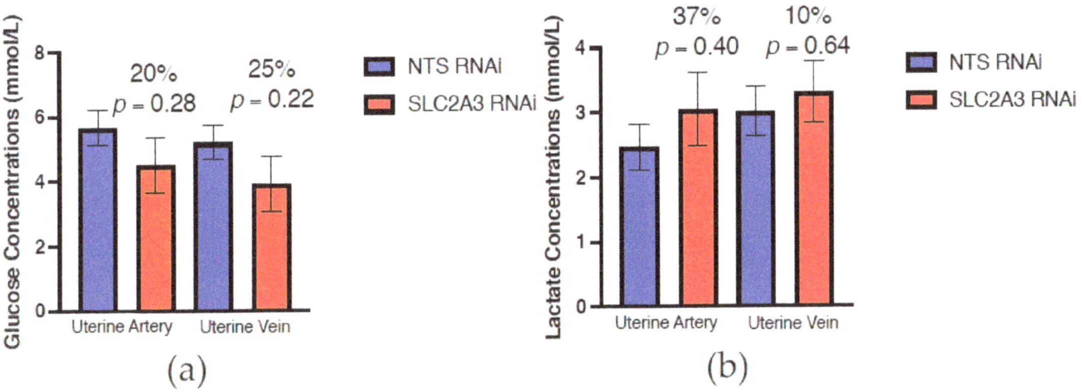


**Figure 3.** Impact of SLC2A3 RNAi on both uterine artery and vein concentrations of (**a**) glucose and (**b**) lactate in plasma samples harvested at 75 dGA. Data are shown as means ± SEM. NTS, non-targeting sequence; RNAi, RNA interference.

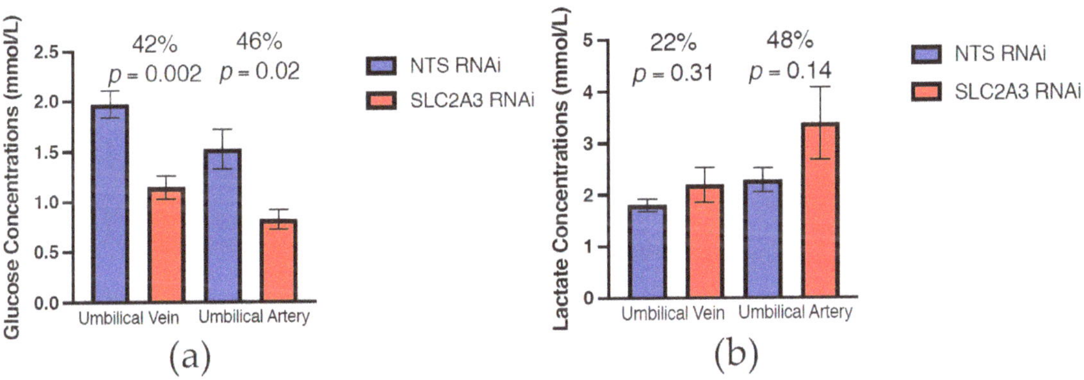


**Figure 4.** Impact of SLC2A3 RNAi on both umbilical vein and artery concentrations of (**a**) glucose and (**b**) lactate in plasma samples harvested at 75 dGA. Data are shown as means ± SEM. NTS, non-targeting sequence; RNAi, RNA interference.

As evidenced in Figure 5, SLC2A3 RNAi did not impact uterine artery insulin or uterine vein CSH concentrations, but did result in a significant reduction ($p \leq 0.05$) in the uterine artery concentrations of both glucagon and IGF1. In contrast, both umbilical artery insulin and glucagon (Figure 6) were reduced ($p \leq 0.10$) 44% and 53%, respectively, whereas umbilical artery IGF1 was not impacted by SLC2A3 RNAi, nor was fetal liver INSR concentration (data not presented). However, umbilical vein concentrations of CSH were increased 70% ($p \leq 0.05$; Figure 6) in SLC2A3 RNAi pregnancies.

**Table 3.** Maternal plasma amino acid concentrations (75 dGA).

| | NTS RNAi Uterine Artery | SLC2A3 RNAi Uterine Artery | *p*-Value | % Change | NTS RNAi Uterine Vein | SLC2A3 RNAi Uterine Vein | *p*-Value | % Change |
|---|---|---|---|---|---|---|---|---|
| TAU | 43.50 ± 8.70 | 49.55 ± 8.83 | 0.64 | 13.91 | 78.85 ± 17.57 | 74.16 ± 17.13 | 0.98 | 0.92 |
| ASP | 18.53 ± 2.66 | 17.65 ± 2.58 | 0.82 | 4.78 | 27.42 ± 3.50 | 34.10 ± 10.13 | 0.55 | 24.38 |
| THR | 108.92 ± 15.67 | 147.58 ± 17.81 | 0.13 | 35.50 | 153.15 ± 45.77 | 151.47 ± 19.38 | 0.97 | 1.10 |
| SER | 63.61 ± 6.28 | 75.04 ± 7.41 | 0.27 | 17.97 | 109.98 ± 50.80 | 71.67 ± 8.15 | 0.47 | 34.84 |
| ASN | 26.70 ± 3.15 | 42.09 ± 3.93 | 0.01 | 57.67 | 32.27 ± 5.42 | 42.23 ± 5.13 | 0.21 | 30.85 |
| GLU | 105.34 ± 11.72 | 106.04 ± 6.40 | 0.96 | 0.66 | 158.64 ± 12.88 | 172.02 ± 34.99 | 0.73 | 8.43 |
| GLN | 309.41 ± 17.70 | 317.86 ± 25.19 | 0.79 | 2.73 | 308.16 ± 30.39 | 328.82 ± 28.26 | 0.63 | 6.70 |
| PRO | 78.25 ± 9.15 | 81.99 ± 5.96 | 0.74 | 4.78 | 84.70 ± 6.03 | 89.16 ± 6.52 | 0.63 | 5.27 |
| GLY | 628.54 ± 53.55 | 578.48 ± 32.09 | 0.44 | 7.96 | 630.56 ± 63.84 | 687.23 ± 66.76 | 0.55 | 8.99 |
| ALA | 198.69 ± 14.19 | 219.90 ± 10.41 | 0.26 | 10.68 | 219.66 ± 19.76 | 245.43 ± 13.65 | 0.31 | 11.73 |
| CIT | 188.30 ± 23.08 | 256.7 ± 34.01 | 0.13 | 36.32 | 171.15 ± 19.28 | 253.27 ± 35.50 | 0.07 | 47.98 |
| VAL | 197.76 ± 13.18 | 231.82 ± 13.41 | 0.10 | 17.22 | 201.25 ± 24.21 | 219.92 ± 12.70 | 0.51 | 9.28 |
| CYS | 21.92 ± 2.88 | 17.99 ± 3.70 | 0.42 | 17.93 | 11.94 ± 4.15 | 21.28 ± 4.84 | 0.17 | 78.20 |
| MET | 22.15 ± 1.96 | 27.58 ± 2.98 | 0.16 | 24.51 | 33.38 ± 10.20 | 28.49 ± 3.39 | 0.66 | 14.65 |
| ILE | 102.26 ± 3.79 | 115.23 ± 6.93 | 0.13 | 12.69 | 94.24 ± 4.77 | 104.06 ± 3.18 | 0.12 | 10.42 |
| LEU | 120.45 ± 4.05 | 134.11 ± 9.62 | 0.22 | 11.33 | 114.62 ± 9.81 | 125.2 ± 6.66 | 0.39 | 9.23 |
| TYR | 39.20 ± 3.28 | 49.47 ± 5.61 | 0.14 | 26.22 | 52.23 ± 12.11 | 54.75 ± 7.68 | 0.86 | 4.82 |
| PHE | 34.06 ± 1.89 | 38.38 ± 3.34 | 0.29 | 12.66 | 44.80 ± 10.34 | 42.56 ± 4.15 | 0.84 | 4.99 |
| TRP | 34.27 ± 1.78 | 36.78 ± 2.54 | 0.44 | 7.30 | 35.72 ± 4.65 | 39.95 ± 1.99 | 0.96 | 0.65 |
| ORN | 94.49 ± 6.14 | 115.83 ± 8.20 | 0.06 | 22.58 | 97.07 ± 17.58 | 105.10 ± 8.02 | 0.69 | 8.27 |
| LYS | 103.41 ± 6.97 | 144.09 ± 17.23 | 0.05 | 39.33 | 117.46 ± 15.91 | 148.69 ± 19.41 | 0.24 | 26.59 |
| HIS | 59.89 ± 1.70 | 61.65 ± 3.16 | 0.63 | 2.94 | 57.60 ± 2.77 | 62.99 ± 3.74 | 0.27 | 9.38 |
| ARG | 210.83± 9.11 | 233.19 ± 20.86 | 0.35 | 10.61 | 219.04 ± 11.37 | 231.31 ± 23.04 | 0.64 | 5.60 |

Data are shown as mean values ± SEM for all ewes in each treatment group. TAU, taurine; ASP, aspartic acid; THR, threonine; SER, serine; ASN, asparagine; GLU, glutamic acid; GLN, glutamine; PRO, proline; GLY, glycine; ALA, alanine; CIT, citrulline; VAL, valine; CYS, cysteine; MET, methionine; ILE, isoleucine; LEU, leucine; TYR, tyrosine; PHE, phenylalanine; TRP, tryptophan; ORN, ornithine; LYS, lysine, HIS, histidine; ARG, arginine.

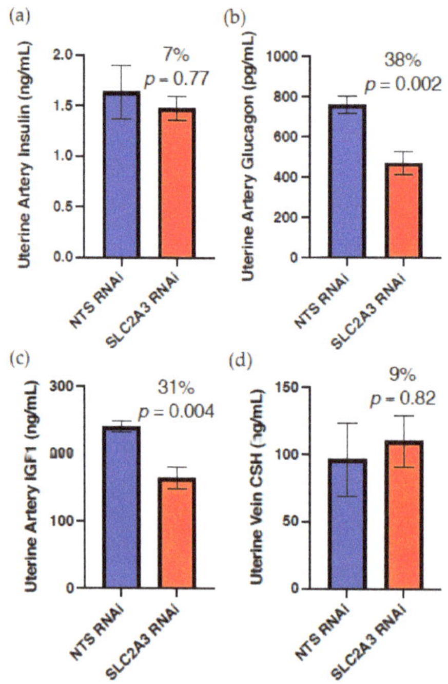


**Figure 5.** Impact of SLC2A3 RNAi on uterine artery concentrations of (**a**) insulin, (**b**) glucagon and (**c**) IGF1, and (**d**) uterine vein concentrations of CSH. Data are shown as means ± SEM. NTS, non-targeting sequence; RNAi, RNA interference.

**Table 4.** Fetal plasma amino acid concentrations (75 dGA).

| | NTS RNAi Umbilical Art. | SLC2A3 RNAi Umbilical Art. | p-Value | % Change | NTS RNAi Umbilical Vein | SLC2A3 RNAi Umbilical Vein | p-Value | % Change |
|---|---|---|---|---|---|---|---|---|
| TAU | 132.41 ± 16.08 | 141.54 ± 19.40 | 0.73 | 6.89 | 132.14 ± 13.98 | 144.70 ± 23.33 | 0.64 | 9.50 |
| ASP | 43.10 ± 4.27 | 41.54 ± 3.05 | 0.80 | 3.63 | 40.87 ± 3.80 | 32.67 ± 2.58 | 0.12 | 20.06 |
| THR | 453.57 ± 32.87 | 561.20 ± 66.67 | 0.15 | 23.73 | 497.44 ± 24.57 | 531.22 ± 57.58 | 0.58 | 6.79 |
| SER | 375.99 ± 25.84 | 420.26 ± 48.66 | 0.40 | 11.77 | 328.51 ± 15.49 | 338.54 ± 38.41 | 0.80 | 3.05 |
| ASN | 52.01 ± 2.62 | 66.88 ± 5.40 | 0.02 | 28.60 | 76.66 ± 2.62 | 86.04 ± 6.95 | 0.21 | 12.24 |
| GLU | 182.77 ± 23.49 | 181.57 ± 15.68 | 0.97 | 0.65 | 49.03 ± 3.88 | 40.59 ± 8.67 | 0.37 | 17.22 |
| GLN | 453.66 ± 36.48 | 495.05 ± 38.65 | 0.47 | 9.12 | 591.28 ± 28.78 | 589.07 ± 32.74 | 0.96 | 0.37 |
| PRO | 114.12 ± 14.32 | 136.71 ± 9.92 | 0.28 | 19.80 | 143.05 ± 12.36 | 163.25 ± 9.12 | 0.24 | 14.12 |
| GLY | 465.10 ± 46.10 | 458.02 ± 48.14 | 0.92 | 1.52 | 550.47 ± 30.18 | 477.85 ± 26.96 | 0.11 | 13.19 |
| ALA | 319.75 ± 28.75 | 349.43 ± 50.04 | 0.59 | 9.28 | 417.10 ± 18.71 | 384.69 ± 17.40 | 0.24 | 7.77 |
| CIT | 151.75 ± 16.05 | 184.56 ± 19.39 | 0.23 | 21.62 | 152.28 ± 16.16 | 168.38 ± 19.32 | 0.54 | 10.57 |
| VAL | 263.06 ± 19.26 | 313.77 ± 42.27 | 0.25 | 19.28 | 318.44 ± 26.91 | 332.76 ± 36.17 | 0.75 | 4.50 |
| CYS | 16.52 ± 1.99 | 19.53 ± 2.05 | 0.34 | 18.25 | 12.72 ± 2.86 | 12.39 ± 1.50 | 0.93 | 2.59 |
| MET | 79.89 ± 6.43 | 89.07 ± 11.02 | 0.46 | 11.48 | 98.43 ± 6.07 | 103.53 ± 4.21 | 0.53 | 5.18 |
| ILE | 75.66 ± 4.25 | 91.57 ± 11.79 | 0.18 | 21.03 | 106.30 ± 5.92 | 107.65 ± 10.54 | 0.91 | 1.27 |
| LEU | 130.62 ± 7.42 | 150.52 ± 17.26 | 0.26 | 15.24 | 188.21 ± 9.92 | 181.61 ± 14.04 | 0.70 | 3.51 |
| TYR | 104.88 ± 3.21 | 116.85 ± 14.31 | 0.35 | 11.42 | 134.05 ± 9.35 | 136.81 ± 14.15 | 0.87 | 2.05 |
| PHE | 92.42 ± 2.39 | 108.34 ± 11.15 | 0.13 | 17.22 | 123.35 ± 5.79 | 127.38 ± 7.21 | 0.67 | 3.27 |
| TRP | 50.73 ± 1.83 | 45.51 ± 4.21 | 0.23 | 10.28 | 59.73 ± 2.75 | 55.44 ± 2.72 | 0.30 | 7.18 |
| ORN | 160.01 ± 15.14 | 196.74 ± 43.89 | 0.38 | 22.95 | 165.92 ± 11.63 | 186.65 ± 39.12 | 0.59 | 12.49 |
| LYS | 202.84 ± 16.16 | 210.34 ± 19.88 | 0.78 | 3.70 | 260.43 ± 15.82 | 270.66 ± 30.56 | 0.76 | 3.93 |
| HIS | 50.08 ± 3.82 | 52.69 ± 4.12 | 0.66 | 5.20 | 68.09 ± 1.68 | 66.35 ± 7.37 | 0.81 | 2.55 |
| ARG | 251.54 ± 15.84 | 193.81 ± 9.20 | 0.03 | 22.95 | 313.01 ± 19.29 | 246.22 ± 15.08 | 0.03 | 21.34 |

Data are shown as mean values ± SEM for all ewes in each treatment group. TAU, taurine; ASP, aspartic acid; THR, threonine; SER, serine; ASN, asparagine; GLU, glutamic acid; GLN, glutamine; PRO, proline; GLY, glycine; ALA, alanine; CIT, citrulline; VAL, valine; CYS, cysteine; MET, methionine; ILE, isoleucine; LEU, leucine; TYR, tyrosine; PHE, phenylalanine; TRP, tryptophan; ORN, ornithine; LYS, lysine, HIS, histidine; ARG, arginine.

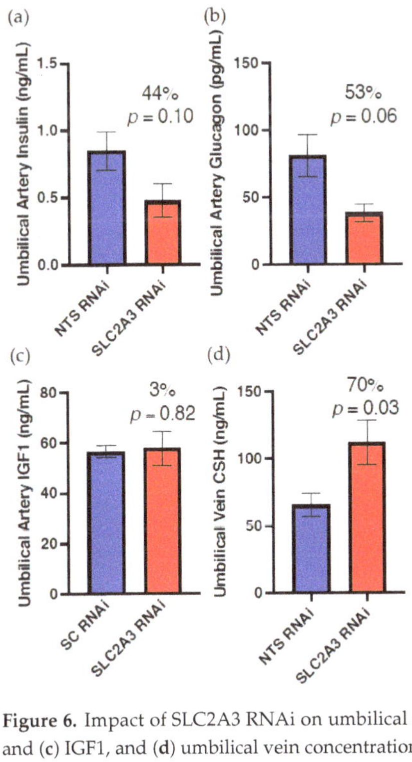


**Figure 6.** Impact of SLC2A3 RNAi on umbilical artery concentrations of (**a**) insulin, (**b**) glucagon, and (**c**) IGF1, and (**d**) umbilical vein concentrations of CSH. Data are shown as means ± SEM. NTS, non-targeting sequence; RNAi, RNA interference.

### 2.4. Placental mRNA Concentration of the Insulin-like Growth Factor Axis

Placental tissues harvested at 75 dGA were assessed for IGF, IGFBP, and IGFR (IGF receptor) mRNA concentrations. While there were no differences in placental IGF1 mRNA concentration ($p \geq 0.10$; Table 5), placental IGF2 mRNA concentration was increased by 71% ($p \leq 0.05$; Figure 7) in SLC2A3 RNAi pregnancies. Additionally, in SLC2A3 RNAi pregnancies, placental IGF1R and IGF2R mRNA concentrations were increased by 40% and 69% ($p \leq 0.05$; Figure 7), respectively. There were no differences in placental IGFBP1, IGFBP2, or IGFBP3 mRNA concentrations between treatments ($p \geq 0.10$; Table 5).

**Table 5.** Placental insulin-like growth factor mRNA concentrations (75 dGA).

| mRNA | NTS RNAi | SLC2A3 RNAi | *p*-Value | % Change |
|---|---|---|---|---|
| *IGF1*, pg/pg | 0.0011 ± 0.0002 | 0.0011 ± 0.00013 | 0.88 | 3.18 |
| *IGFBP1*, pg/pg | 0.00018 ± 0.000076 | 0.00013 ± 0.000066 | 0.64 | 26.81 |
| *IGFBP2*, pg/pg | 0.00051 ± 0.000082 | 0.00057 ± 0.000057 | 0.57 | 11.58 |
| *IGFBP3*, pg/pg | 0.025 ± 0.0056 | 0.028 ± 0.0049 | 0.69 | 12.19 |

Data are shown as mean values ± SEM for the starting quantity of the mRNA of interest (pg) divided by the starting quantity (pg) of the housekeeping mRNA (*RPS15*).

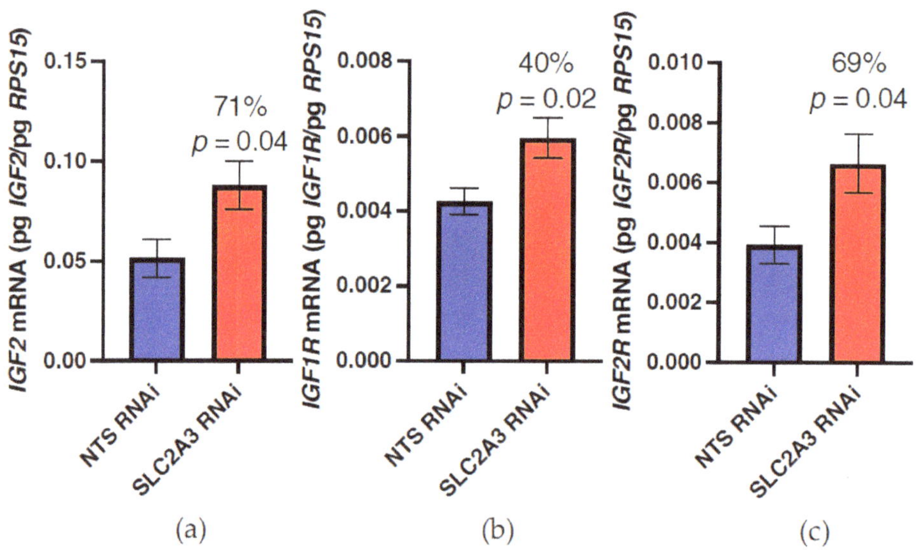


**Figure 7.** Impact of SLC2A3 RNAi on placental concentrations of (**a**) IGF2, (**b**) IGF1R, and (**c**) IGF2R mRNA. Data are shown as means ± SEM. NTS, non-targeting sequence; RNAi, RNA interference.

## 3. Discussion

As glucose is the primary energy substrate supporting fetal development [1], the fetus is reliant upon placental glucose uptake and transfer, mediated by facilitative glucose transporters. The importance of placental glucose transfer is exemplified in FGR pregnancies, with the magnitude of fetal hypoglycemia being correlated with the severity of FGR [8,9]. SLC2A1 (GLUT1) and SLC2A3 (GLUT3) are the primary transporters in human and ruminant placenta [4,18] believed responsible for uptake and transfer, yet the relative importance of each is debatable. In humans, SLC2A1 is found in both the microvillous (apical) and basal membranes of the syncytiotrophoblast [6,19], whereas SLC2A3 is localized to just the microvillous membrane [20]. By contrast, in sheep, SLC2A1 is localized to the basolateral trophoblast membrane and SLC2A3 is localized to the microvillous trophoblast membrane [4]. Beyond the individual transporters, the more important question is whether microvillous glucose transport is more or less important than basal membrane transport. In vivo maternal and fetal glucose clamp studies led to the conclusion that placental glucose transport capacity is greater on the fetal surface than the maternal surface [21]. However, the placenta is a highly metabolic organ in itself, and placental glucose utilization accounted for 80 and 72% of uterine glucose uptake at mid- and late-gestation, respectively [22,23], directly impacting the maternal-fetal glucose gradient, thereby requiring sufficient microvillous glucose uptake to maintain placental function.

As SLC2A3 is localized on the apical microvillous membrane in both sheep and humans [4,20], SLC2A3 likely plays an important role in trophoblast uptake of glucose in both species. Accordingly, we used our lentiviral-mediated in vivo RNAi methods [12–17] to diminish SLC2A3 in the sheep placenta to evaluate the relative importance of microvillous trophoblast glucose uptake during the first half of gestation. SLC2A3 RNAi resulted in a 37% reduction (Figure 2) in placental SLC2A3 concentration at 75 dGA. This reduction in SLC2A3 was sufficient to induce significant fetal hypoglycemia (Figure 4) and reduce fetal growth (Table 4), as assessed at mid-gestation. In contrast, SLC2A1 was increased (Figure 2), which we hypothesize is in an attempt to offset the deficit in SLC2A3 in terms of glucose uptake and transport. In mice, the homozygous *Slc2a3*$^{-/-}$ genotype results in embryonic lethality, whereas the heterozygous *Slc2a3*$^{+/-}$ genotype resulted in late-gestation FGR [24], further supporting the requirement of SLC2A3. In humans, placental SLC2A3 is more abundant in early gestation [20], and it has been suggested that its greater affinity and glucose transport capacity may be important during the early stages of gestation when glucose delivery to the developing placenta is low [20]. Interestingly, during late-gestation in FGR pregnancies, an upregulation of SLC2A3 is thought to be an adaptive response to increase placental glucose uptake and transfer [11,25,26]. Our results further support the important role of placental SLC2A3, and that a deficiency in microvillous trophoblast glucose uptake is crucial in supplying the fetus adequate glucose during the first half of gestation.

As noted above, SLC2A3 RNAi resulted in significant fetal hypoglycemia. Due to these pregnancies being studied at mid-gestation, they did not undergo maternal and fetal catheterization that would have allowed steady-state assessment of uterine and umbilical blood flows, as well as uterine and umbilical uptakes and uteroplacental utilization of nutrients [27]. A common hallmark of FGR pregnancies is fetal hypoxia and increased fetal lactate concentrations [28,29]. While we were unable to quantify oxygen content in the collected blood samples, there was a 48% increase in umbilical artery lactate concentrations in SLC2A3 pregnancies, although this was not statistically different ($p$ = 0.14). As the majority of fetal lactate is produced by the fetus [30,31], as evidenced by greater lactate concentrations in the umbilical artery as compared to the umbilical vein (Figure 4), these data indicate that the increase in umbilical artery lactate is the result of fetal hypoxia. However, as uterine and umbilical blood flows are the major determinants of fetal oxygen delivery [32], the lack of differences observed in the 70 dGA Doppler velocimetry assessment would suggest that umbilical blood flow in the SLC2A3 RNAi pregnancies, is not restricting fetal oxygen delivery. We did assess amino acid concentrations in the uterine and umbilical blood samples collected and there did not appear to be an overall impact of SLC2A3 RNAi on amino acid concentrations. This would indicate that there was not an increase in amino acid oxidation in response to fetal hypoglycemia, to maintain fetal oxidative metabolism [33,34]. Using maternal hyperinsulinemia clamps to induce fetal hypoglycemia, DiGiacomo and Hay [35], demonstrated that fetal oxygen consumption was reduced proportionally to fetal hypoglycemia and the rate of fetal growth reduction, diminishing the likelihood of increased amino acid oxidation. This was further demonstrated, following 8 weeks of fetal hypoglycemia, in which it was determined that fetal plasma leucine disposal, leucine flux into protein synthesis, and leucine oxidation were not impacted by fetal hypoglycemia [36].

There were impacts of SLC2A3 RNAi on fetal growth, which may have resulted solely from the hypoglycemia. Notably, fetal pancreas weight was reduced 23% ($p \leq 0.05$). As expected, umbilical artery insulin was reduced 44% ($p \leq 0.10$), essentially equivalent to the reduction in umbilical glucose concentrations. Surprisingly, umbilical artery glucagon concentrations were also reduced ($p \leq 0.10$) in the SLC2A3 RNAi pregnancies. The fact that both insulin and glucagon concentrations were reduced suggests that the fetal hypoglycemia induced by SLC2A3 RNAi during the first half of gestation had an overall effect on pancreas development and growth, rather than a β cell-specific effect. Glucose-stimulated insulin secretion at mid-gestation is 20% of the rate near term [37], indicating that during the first half of gestation the pancreas may not be responding to glucose concentrations in the specific fashion that occurs during the second half of gestation. Notably, arginine-induced insulin secretion increases with gestational age in a similar fashion to glucose-stimulated insulin secretion [37], and of all of the amino acids assessed, only arginine was significantly reduced ($p \leq 0.05$) in both the umbilical vein and artery.

In various experimental models that produce functional placental insufficiency that results in FGR, fetal hypoglycemia, fetal hypoinsulinemia, and decreased fetal liver growth are common characteristics [13,15,28,38,39]. Additionally, when fetal liver growth is decreased in FGR pregnancies, a decrease in IGF1 concentrations is often observed [13,15,40], as well as an upregulation of the fetal liver INSR in response to fetal hypoinsulinemia [41]. However, in response to SLC2A3 RNAi, fetal liver weight, INSR concentration, and umbilical artery IGF1 concentrations were unaffected. The lack of impact on the fetal liver may be due in part to the 70% increase in umbilical vein CSH (Figure 6) observed in the SLC2A3 RNAi pregnancies. A similar increase in umbilical CSH, in response to fetal hypoglycemia and hypoinsulinemia, was reported in fasted late-gestation pregnant ewes that was then subsequently reversed upon refeeding of the ewes [42]. The increase in circulating CSH concentrations may have preserved fetal liver weight and function, as CSH deficiency results in reduced fetal liver weights during early [17] and late-gestation sheep pregnancies [13,14], as well as significant reductions in umbilical artery IGF1 concentrations [13,14]. Collectively, these data may suggest that by enhancing umbilical concentrations of CSH, the placenta indirectly salvages fetal liver growth and function in the face of fetal hypoglycemia, at least during the first half of gestation.

With the SLC2A3 pregnancies, placental weight was reduced 21%, which did not reach statistical significance ($p = 0.13$). Within the placenta, *IGF2*, *IGF1R* and *IGF2R* mRNA concentrations were the only members of the insulin-like growth factor axis impacted, and all three mRNA were significantly ($p \leq 0.05$) elevated in SLC2A3 RNAi placenta (Figure 7). The upregulation of these mRNA may be a compensatory mechanism to stimulate placental growth, thus increasing the total nutrient exchange surface area to combat fetal hypoglycemia. Targeted mutagenesis studies have demonstrated that placental IGF2 plays a role in modulating placental growth as overexpression of *Igf2* results in placental overgrowth [43] and total ablation of *Igf2* results in placental growth restriction [44]. Placental IGF2 mediates its effects through IGF1R, as IGF2R has been demonstrated to be a clearance receptor for IGF2 [45,46] that prevents IGF2 from overstimulating IGF1R and producing placental overgrowth [46,47]. A similar increase in placental *IGF2* expression has been shown to be an adaptive response to decreased placental growth in other sheep FGR models [48,49].

Placenta-specific SLC2A3 RNAi did not have a statistically significant impact on uterine artery or vein glucose and lactate concentrations (Figure 3), but numerically there was a 20–25% reduction in uterine blood glucose concentrations in SLC2A3 RNAi pregnancies. In contrast to the significant changes observed in umbilical circulation, uterine artery insulin and uterine vein CSH concentrations were not impacted by SLC2A3 RNAi (Figure 5). However, to our surprise, the uterine artery concentrations of both IGF1 and glucagon were significantly ($p \leq 0.05$) reduced in SLC2A3 RNAi pregnancies. We did not hypothesize that there would be an impact on maternal hormone concentrations as the use of replication-deficient lentivirus to infect hatched blastocysts results in the RNAi being limited to the trophectoderm lineage of the placenta [12,50,51], such that the RNAi is placenta-specific. The non-significant reductions in maternal glucose may be tied to the reduction in maternal glucagon, but what is driving the reduction in maternal IGF1 and glucagon is not apparent. We can only speculate that this resulted from reductions in trophoblast uptake of glucose, altering placental secretory products that impact maternal IGF1 and glucagon.

Since SLC2A3 is limited to the microvillous membrane of the placental trophoblast in sheep [4], using lentiviral-mediated RNAi, we were able to assess the impact of SLC2A3 deficiency during the first half of gestation. Our results confirm that microvillous glucose uptake can be rate-limiting to fetal growth and development during early gestation, while not fully resolving whether microvillous or basal glucose transport is more important. Some of our results could be viewed as being predictable, in response to limiting glucose transfer to the fetus, while others were not. The "global" impact of fetal hypoglycemia on pancreas growth and function, with both insulin and glucagon concentrations being diminished, were not expected, especially since fetal liver growth and IGF1 secretion were not affected. The most surprising result was the diminished uterine artery IGF1 and glucagon concentrations. While considerable effort has been expended on examining maternal glucose, insulin and IGF1 in normal and compromised pregnancies, the role of maternal glucagon for the most part has been overlooked, although Qiao et al. [52] recently reported that pregnancy in mice induces an expansion of maternal $\alpha$-cell mass, and an increase in maternal glucagon concentrations during early pregnancy. Furthermore, the demonstration [53] that fetal hyperglucagonemia during late-gestation results in significant reductions in uterine artery blood flow and placental CSH production and secretion into maternal circulation, without impacting umbilical blood flow or CSH concentrations, highlights not only the importance of glucagon during pregnancy, but also that the three compartments (maternal, placental and fetal) are intimately integrated and need to be investigated together [54]. This research, therefore, also exemplifies the utility of integrating in vivo RNAi in an animal model that can allow steady-state assessment of altered maternal-placental-fetal physiology [27].

## 4. Materials and Methods

All procedures conducted with animals were approved by the Colorado State University Institutional Animal Care and Use Committee (Protocol 1483), as well as the Institutional Biosafety Committee (17-039B).

### *4.1. Lentiviral Generation*

Lentiviral infection was used to stably integrate and express shRNA targeting *SLC2A3* mRNA in the host cell. The shRNA sequences for hLL3.7 472 (SLC2A3 RNAi) and hLL3.7 NTS (non-targeting sequence; control RNAi) constructs are presented in Table 6. All subsequent virus generation and titering followed the procedures extensively described previously [13].

**Table 6.** Non-targeting sequence (NTS) RNAi and SLC2A3 RNAi shRNA sequences.

| Oligonucleotide | Sequence (5′–3′) |
|---|---|
| NTS shRNA sense strand | GAGTTAAAGGTTCGGCACGAATTCAAGAGATTCGTGCCG AACCTTTAACTC |
| SLC2A3 shRNA sense strand | GCGCAACTCAATGCTTATTGTTTCAAGAGAACAATAAGCA TTGAGTTGCGC |

### *4.2. Generation of SLC2A3 RNAi Pregnancies*

All ewes (Dorper breed composition) were group housed in pens at the Colorado State University Animal Reproduction and Biotechnology Laboratory, and were provided access to hay, trace minerals, and water to meet or slightly exceed their National Research Council [55] requirements. Animal management, estrus synchronization, and embryo transfers were done as previously described [13–15]. In summary, after synchronization and subsequent breeding, at 9 days post-conception, donor ewes were euthanized (88 mg/kg Euthasol; VetOne, Conshohocken, PA) and the uteri were harvested and flushed to collect hatched and fully expanded blastocysts. Each blastocyst was infected with

150,000 transducing units of either NTS RNAi or SLC2A3 RNAi virus as previously described [13–15]. Following 5 h of incubation with the virus, each blastocyst was washed and a single blastocyst was surgically transferred into the uterine horn ipsilateral to the corpus luteum of a synchronized recipient ewe. All recipient ewes (NTS RNAi $n$ = 10; SLC2A3 RNAi $n$ = 13) were then monitored daily for return to standing estrus and confirmed pregnant at 50 days of gestational age (dGA) by ultrasonography (Mindray Medical Equipment, Mahway, NJ, USA). At 70 dGA, all successful pregnancies (6 NTS RNAi and 6 SLC2A3 RNAi) underwent Doppler velocimetry assessment as described previously [14].

### 4.3. Tissue Collection

At 75 dGA, six NTS RNAi (5 males and 1 female) and six SLC2A3 RNAi (3 males and 3 females) pregnancies underwent a terminal surgery as previously described [13]. In summary, pregnant recipient ewes were food restricted for 18 h before surgery. The fetus and umbilical cord were exposed and fetal blood was collected from the umbilical artery and vein, while maternal blood was collected from the uterine artery and vein ipsilateral to the fetus, with the resulting serum stored in −80 °C until further analysis. The fetus was then euthanized (88 mg/kg, Euthasol; VetOne), excised, and fetal weight, head circumference, crown-rump length, abdominal circumference, femur and tibia length were recorded. The fetal liver and pancreas were harvested, weighed, and snap frozen in liquid nitrogen. The ewe was euthanized (88 mg/kg, Euthasol; VetOne) and a complete hysterectomy was performed and all placentomes were excised and recorded for total placentome number and weight. Thirty placentomes were randomly selected and snap frozen in liquid nitrogen. The resulting tissue was pulverized using a mortar and pestle and stored at −80 °C for later use.

### 4.4. Biochemical Analysis of Blood Samples

Plasma glucose and lactate were measured by Yellow Spring Instrument 2900 (YSI Incorporated, Yellow Springs, OH), and plasma amino acids were measured by HPLC as described previously [14,15]. Maternal and fetal plasma concentrations of insulin and IGF1 were assessed by enzyme-linked immunosorbent assay (ALPCO Immunoassays, Salem, NH; 80-IN-SOV-E01 and 22-IGFHU-E01, respectively) as described previously [14,15]. The concentration of plasma CSH was assessed by radioimmunoassay (RIA) as described previously [13]. Maternal and fetal plasma glucagon concentrations were assessed by enzyme-linked immunosorbent assay (ALPCO; 48-GLUHU-E01), which was validated for use with sheep plasma and exhibited an intra-assay coefficient of variation ranging from 0.2 to 11.6%, for the highest to lowest plasma concentrations, respectively.

### 4.5. Cell Lines

Immortalized ovine trophoblast (iOTR) cells [16] were used to test the degree of RNAi of the SLC2A3 shRNA construct. To infect the cells, a frozen aliquot of SLC2A3 RNAi or NTS RNAi lentivirus was resuspended in 500 µL of DMEM-F12 medium [16] (supplemented with 10% FBS, 1× penicillin-streptomycin-amphotericin B solution, 10 µg/mL insulin, 0.1 mM non-essential amino acids, 2 mM glutamine, and 1 mM sodium pyruvate) with 8 µg/mL polybrene (Sigma-Aldrich, St. Louis, MO, USA). The iOTR cells were incubated with lentiviral particles at a multiplicity of infection (MOI) of 500 for 8 h at 37 °C and 5% $CO_2$, after which the transfection media was replaced with fresh complete media. The subsequent cells were passaged up to a 150-mm tissue culture plate, pelleted and stored in −80 °C until further analysis.

### 4.6. Western Blot Analysis

Cellular protein from 75 dGA placentomes was assessed using Western immunoblot analysis. Protein isolation and analysis were done in accordance with the methods described previously [14,15]. Pulverized placentome tissue (75 mg) was lysed in 500 µL of lysis buffer and sonicated on ice. For placental SLC2A3 analysis, 25 µg of protein from each sample were electrophoresed through NuPAGE 4–12% Bis Tris gels (Life Technologies, Carlsbad, CA, USA), and transferred to a 0.45 μm pore nitrocellulose membrane. For iOTR cell SLC2A3 analysis, 10 μg of protein from each sample were electrophoresed through NuPAGE 4–12% Bis-Tris gels (Life Technologies), and transferred to a 0.45 μm pore nitrocellulose membrane. The resulting blots were stained with Ponceau S (Sigma Aldrich, St. Louis, MO, USA) to assess total protein per lane using the ChemiDoc XRS+ (BioRad, Hercules, CA, USA). To visualize SLC2A3, the blots were incubated in a 1:1000 dilution of CSU-α-SCL2A3-22 [15] for 24 h at 4 °C. After washing, the blots were incubated in a 1:5000 dilution of goat α-rabbit IgG conjugated to horse radish peroxidase (ab97051; Abcam, Cambridge, MA, USA). Membranes were developed using an ECL Western Blotting Detection Reagent chemiluminescent kit (Amersham, Pittsburgh, PA, USA) and imaged using the ChemiDoc XRS+ (BioRad). Densitometry of SLC2A3 was normalized on the total protein per lane. To account for technical error between immunoblots, a common sample was included in each immunoblot and densitometry measurements were adjusted based on the average densitometry measurements of the common sample.

For analysis of placental SLC2A1, 5 μg of protein from each sample were electrophoresed through a 4–15% Tris-Glycine stain-free gel (BioRad) and transferred to a 0.45 μm pore nitrocellulose membrane. After transfer, the nitrocellulose membrane was imaged using the ChemiDoc XRS+ chemiluminescence system (BioRad) to assess the total protein per lane to use for normalization. To visualize SLC2A1, the blots were incubated in a 1:40,000 dilution of rabbit α-SLC2A1 (07-1401; EMD Millipore, Burlington, MA, USA) for 24 h at 4 °C. After washing, the blots were incubated in a 1:80,000 dilution of goat α-rabbit IgG conjugated to horse radish peroxidase (ab205718; Abcam). As described above, densitometry analysis of SLC2A1 was performed using Image Lab software (version 6.1; BioRad) and normalized on total protein/lane.

For analysis of placental SLC2A8, 20 μg of each sample were electrophoresed through 4–15% Tris-Glycine stain-free gels (BioRad) and transferred and analyzed as described for SLC2A1. SLC2A8 was visualized using a 1:2000 dilution of rabbit α-SLC2A8 (LS-C757596; LifeSpan BioSciences, Seattle, WA) and a 1:10,000 dilution of goat α-rabbit IgG conjugated to horse radish peroxidase (ab97051; Abcam). INSR was visualized using a 1:1000 dilution of mouse α-INSR-β (ab69058; Abcam) and a 1:5000 dilution of goat alpha-mouse IgG conjugated to horse radish peroxidase (ab6789; Abcam).

### 4.7. RNA Isolation

Total cellular RNA was isolated from 75 dGA pulverized placentome samples using the RNeasy Mini Kit (Qiagen, Hilden, Germany) according to the manufacturer's protocol. RNA concentration was quantified using the BioTek Synergy 2 Microplate Reader (BioTek, Winooski, VT, USA), and RNA quality was measured by the 260/280 nm absorbance ratio. Samples were stored at −80 °C until use.

### 4.8. cDNA Synthesis and Quantitative Real-Time PCR

cDNA was generated from 2 μg of total cellular RNA using iScript Reverse Transcription Supermix (BioRad) according to the manufacturer's protocol. To control for variance in the efficiency of the reverse transcription reaction, cDNA was quantified using the Quant-iT OliGreen ssDNA Assay Kit (Invitrogen, Carlsbad, CA, USA) according to the manufacturer's protocol, and quality was measured by the 260/280 absorbance ratio. An equal mass of cDNA (10 ng/μL) was used for each sample in the quantitative real-time PCR (qRT-PCR) reaction. qRT-PCR was performed using the CFX384 Real-Time System (BioRad). Forward and reverse primers for qRT-PCR were designed using Oligo software (Molecular Biology Insights, Cascade, CO, USA) to amplify an intron-spanning product. Primer sequences and amplicon size are summarized in Table 7. Standard curves were generated as described previously [17]. Briefly, a PCR product for each gene was generated using cDNA from 135 dGA fetal placenta as a template and cloned into the StrataClone vector (Agilent Technologies), and each PCR product was sequenced to verify amplification of the correct cDNA. Using the PCR products amplified from the sequenced plasmids, standard curves were generated for each mRNA from $1 \times 10^2$ to $1 \times 10^{-5}$ pg, and were used to measure amplification efficiency. The starting quantity (pg) was normalized by dividing the starting quantity of mRNA of interest by the starting mRNA quantity (pg) of ribosomal protein S15 (RPS15) [17].

**Table 7.** Primers and product sizes for cDNA used in qRT-PCR.

| cDNA | Forward Primer (5′–3′) | Reverse Primer (5′–3′) | Product, bp |
|---|---|---|---|
| *RPS15* | ATCATTCTGCCCGAGATGGTG | TGCTTGACGGGCTTGTAGGTG | 134 |
| *IGF1* | TCGCATCTCTTCTATCTGGCCCT | ACAGTACATCTCCAGCCTCCTCA | 240 |
| *IGF2* | GACCGCGGCTTCTACTTCAG | AAGAACTTGCCCACGGGGTAT | 203 |
| *IGFBP1* | TGATGACCGACTCCAGTGAG | GTCCAGCGAAGTCTCACAC | 248 |
| *IGFBP2* | CAATGGCGAGGAGCACTCTG | TGGGGATGTGTAGGGAATAG | 330 |
| *IGFBP3* | CTCAGACGACAGACACCCA | GGCATATTTGAGCTCCAC | 336 |
| *IGF1R* | AACTGTCATCTCCAACCTC | CAAGCCTCCCACTATCAAC | 493 |
| *IGF2R* | GACTTGTGTCCAGACCAGATTC | GCCGTCGTCCTCACTCTCATC | 674 |

### 4.9. Statistical Analysis

Data were analyzed by two-way analysis of variance using GraphPad Prism (version 9) to analyze the main effects of treatment and fetal sex, as well as the treatment × sex interaction. The pregnancy success rate limited the final number, such that the study was not sufficiently powered to examine the effect of fetal sex. As there were no treatment by fetal sex interactions, the data are presented as the main effect of treatment only. Statistical significance was set at $p \leq 0.05$ and a statistical tendency at $p \leq 0.10$. Data are reported as the mean ± standard error of the mean (SEM).

## 5. Conclusions

Using lentiviral-mediated in vivo RNAi, we determined that a deficiency of SLC2A3, which is localized specifically to the microvillous apical membrane of placental trophoblasts in sheep, reduced the placental transfer of glucose to the fetus during the first-half of gestation resulting in impaired fetal growth and development. Furthermore, by impacting placental glucose uptake, placental function was altered in a fashion, which either directly or indirectly impacted maternal metabolic hormone secretion, highlighting the integration of the maternal, placental and fetal compartments of pregnancy.

**Author Contributions:** R.V.A. conception and design of research; C.S.L., V.C.K., A.R.T., A.A., Q.A.W., P.J.R. and R.V.A. performed experiments; C.S.L. and R.V.A. analyzed data; C.S.L. and R.V.A. interpreted results of experiments; C.S.L. and R.V.A. prepared figures; C.S.L. and R.V.A. drafted manuscript; C.S.L., V.C.K., A.R.T., A.A., Q.A.W., P.J.R. and R.V.A. edited and revised manuscript; C.S.L. and R.V.A. All authors have read and agreed to the published version of the manuscript.

**Funding:** This work was supported by National Institutes of Health grants HD094952 and HD093701.

**Institutional Review Board Statement:** All procedures conducted with animals were approved by the Colorado State University Institutional Animal Care and Use Committee (Protocol 1483), as well as the Institutional Biosafety Committee (17-039B).

**Informed Consent Statement:** Not applicable.

**Data Availability Statement:** Data available upon request to the corresponding author.

**Acknowledgments:** The authors wish to thank Richard Brandes, Bailyn Furrow and Gregory Harding at the Animal Reproduction and Biotechnology Laboratory for animal care, and Vince Abushaban, David Caprio and Karen Trembler at the Perinatal Research Center for additional technical support.

**Conflicts of Interest:** The authors declare no conflict of interest. The funders had no role in the design of the study; in the collection, analyses, or interpretation of data; in the writing of the manuscript; or in the decision to publish the results.

## References


1. Hay, W.W. Placental-fetal glucose exchange and fetal glucose metabolism. *Trans. Am. Clin. Climatol. Assoc.* **2006**, *117*, 321–340. [PubMed]
2. Anand, R.S.; Ganguli, S.; Sperling, M.A. Effect of insulin-induced maternal hypoglycemia on glucose turnover in maternal and fetal sheep. *Am. J. Physiol. Endocrinol. Metab.* **1980**, *238*, E524–E532. [CrossRef] [PubMed]
3. Marconi, A.M.; Cetin, I.; Davoli, E.; Baggiani, A.M.; Fanelli, R.; Fennessey, P.V.; Battaglia, F.C.; Pardi, G. An evaluation of fetal glucogenesis in intrauterine growth-retarded pregnancies. *Metabolism* **1993**, *42*, 860–864. [CrossRef]
4. Wooding, F.B.P.; Fowden, A.L.; Bell, A.W.; Ehrhardt, R.A.; Limesand, S.W.; Hay, W.W. Localisation of glucose transport in the ruminant placenta: Implications for sequential use of transporter isoforms. *Placenta* **2005**, *26*, 626–640. [CrossRef] [PubMed]
5. Ehrhardt, R.A.; Bell, A.W. Developmental increases in glucose transporter concentration in the sheep placenta. *Am. J. Physiol. Regul. Integr. Comp. Physiol.* **1997**, *273*, R1132–R1141. [CrossRef]
6. Jansson, T.; Wennergren, M.; Illsley, N.P. Glucose transporter protein expression in human placenta throughout gestation and in intrauterine growth retardation. *J. Clin. Endocrinol. Metab.* **1993**, *77*, 1554–1562.
7. Simpson, I.A.; Dwyer, D.; Malide, D.; Moley, K.H.; Travis, A.; Vannucci, S.J. The facilitative glucose transporter GLUT3: 20 years of distinction. *Am. J. Physiol. Endocrinol. Metab.* **2008**, *295*, E242–E253. [CrossRef]
8. Marconi, A.M.; Paolini, C.L. Nutrient transport across the intrauterine growth-restricted placenta. *Semin. Perinatol.* **2008**, *32*, 178–181. [CrossRef]
9. Economides, D.L.; Nicolaides, K.H. Blood glucose and oxygen tension levels in small-for-gestational-age fetuses. *Am. J. Obstet. Gynecol.* **1989**, *160*, 385–389. [CrossRef]
10. Jansson, T.; Ylvén, K.; Wennergren, M.; Powell, T.L. Glucose transport and system A activity in syncytiotrophoblast microvillous and basal plasma membranes in intrauterine growth restriction. *Placenta* **2002**, *23*, 392–399. [CrossRef]
11. Janzen, C.; Lei, M.Y.Y.; Cho, J.; Sullivan, P.; Shin, B.-C.; Devaskar, S.U. Placental glucose transporter 3 (GLUT3) is up-regulated in human pregnancies complicated by late-onset intrauterine growth restriction. *Placenta* **2013**, *34*, 1072–1078. [CrossRef] [PubMed]
12. Purcell, S.H.; Cantlon, J.D.; Wright, C.D.; Henkes, L.E.; Seidel, G.E.; Anthony, R.V. The involvement of proline-rich 15 in early conceptus development in sheep. *Biol. Reprod.* **2009**, *81*, 1112–1121. [CrossRef] [PubMed]
13. Baker, C.M.; Goetzmann, L.N.; Cantlon, J.D.; Jeckel, K.M.; Winger, Q.A.; Anthony, R.V. Development of ovine chorionic somatomammotropin hormone-deficient pregnancies. *Am. J. Physiol. Regul. Integr. Comp. Physiol.* **2016**, *310*, R837–R846. [CrossRef] [PubMed]
14. Tanner, A.R.; Lynch, C.S.; Kennedy, V.C.; Ali, A.; Winger, Q.A.; Rozance, P.J.; Anthony, R.V. CSH RNA interference reduces global nutrient uptake and umbilical blood flow resulting in intrauterine growth restriction. *Int. J. Mol. Sci.* **2021**, *22*, 8150. [CrossRef] [PubMed]
15. Tanner, A.R.; Lynch, C.S.; Ali, A.; Winger, Q.A.; Rozance, P.J.; Anthony, R.V. Impact of chorionic somatomammotropin RNA interference on uterine blood flow and placental glucose uptake in the absence of intrauterine growth restriction. *Am. J. Physiol. Regul. Integr. Comp. Physiol.* **2021**, *320*, R138–R148. [CrossRef]
16. Ali, A.; Stenglein, M.D.; Spencer, T.E.; Bouma, G.J.; Anthony, R.V.; Winger, Q.A. Trophectoderm-specific knockdown of LIN28 decreases expression of genes necessary for cell proliferation and reduces elongation of sheep conceptus. *Int. J. Mol. Sci.* **2020**, *21*, 2549. [CrossRef]
17. Jeckel, K.M.; Boyarko, A.C.; Bouma, G.J.; Winger, Q.A.; Anthony, R.V. Chorionic somatomammotropin impacts early fetal growth and placental gene expression. *J. Endocrinol.* **2018**, *237*, 301–310. [CrossRef]
18. Illsley, N.P. Glucose transporters in the human placenta. *Placenta* **2000**, *21*, 14–22. [CrossRef]
19. Barros, L.F.; Yudilevich, D.L.; Jarvis, S.M.; Beaumont, N.; Baldwin, S.A. Quantitation and immunolocalization of glucose transporters in the human placenta. *Placenta* **1995**, *16*, 623–633. [CrossRef]
20. Brown, K.; Heller, D.S.; Zamudio, S.; Illsley, N.P. Glucose transporter 3 (GLUT3) protein expression in human placenta across gestation. *Placenta* **2011**, *32*, 1041–1049. [CrossRef]
21. Hay, W.W.; Molina, R.A.; DiGiacomo, J.E.; Meschia, G. Model of placental glucose consumption and glucose transfer. *Am. J. Physiol. Regul. Integr. Comp. Physiol.* **1990**, *258*, R569–R577. [CrossRef] [PubMed]
22. Bell, A.W.; Kennaugh, J.M.; Battaglia, F.C.; Makowski, E.L.; Meschia, G. Metabolic and circulatory studies of fetal lamb at midgestation. *Am. J. Physiol. Endocrinol. Metab.* **1986**, *250*, E538–E544. [CrossRef] [PubMed]
23. Meschia, G.; Battaglia, F.C.; Hay, W.W.; Sparks, J.W. Utilization of substrates by the ovine placenta in vivo. *Fed. Proc.* **1980**, *39*, 245–249. [PubMed]
24. Ganguly, A.; McKnight, R.A.; Raychaudhuri, S.; Shin, B.-C.; Ma, Z.; Moley, K.; Devaskar, S.U. Glucose transporter isoform-3 mutations cause early pregnancy loss and fetal growth restriction. *Am. J. Physiol. Endocrinol. Metab.* **2007**, *292*, E1241–E1255. [CrossRef] [PubMed]
25. Stanirowski, P.J.; Szukiewicz, D.; Majewska, A.; Wątroba, M.; Pyzlak, M.; Bomba-Opoń, D.; Wielgoś, M. Differential expression of glucose transporter proteins GLUT-1, GLUT-3, GLUT-8 and GLUT-12 in the placenta of macrosomic, small-for-gestational-age and growth-restricted foetuses. *J. Clin. Med.* **2021**, *10*, 5833. [CrossRef]
26. Chang, Y.-L.; Chao, A.-S.; Chang, S.-D.; Cheng, P.-J. Placental glucose transporter 1 and 3 gene expression in monochorionic twin pregnancies with selective fetal growth restriction. *BMC Pregnancy Childbirth* **2021**, *21*, 260. [CrossRef]
27. Tanner, A.R.; Kennedy, V.C.; Lynch, C.S.; Hord, T.K.; Winger, Q.A.; Rozance, P.J.; Anthony, R.V. In vivo investigation of ruminant placental function and physiology—A review. *J. Anim. Sci.* **2022**, *100*, 1–14. [CrossRef]
28. Hay, W.W.; Brown, L.D.; Rozance, P.J.; Wesolowski, S.R.; Limesand, S.W. Challenges in nourishing the intrauterine growth-restricted foetus—Lessons learned from studies in the intrauterine growth-restricted foetal sheep. *Acta Paediatr.* **2016**, *105*, 881–889. [CrossRef]

29. Pardi, G.; Cetin, I.; Marconi, A.M.; Lanfranchi, A.; Bozzetti, P.; Ferrazzi, E.; Buscaglia, M.; Battaglia, F.C. Diagnostic value of blood sampling in fetuses with growth retardation. *N. Engl. J. Med.* **1993**, *328*, 692–696. [CrossRef]
30. Sparks, J.W.; Hay, W.W.; Bonds, D.; Meschia, G.; Battaglia, F.C. Simultaneous measurements of lactate turnover rate and umbilical lactate uptake in the fetal lamb. *J. Clin. Investig.* **1982**, *70*, 179–192. [CrossRef]
31. Hay, W.W. Regulation of placental metabolism by glucose supply. *Reprod. Fertil. Dev.* **1995**, *7*, 365–375. [CrossRef] [PubMed]
32. Carter, A.M. Factors affecting gas transfer across the placenta and the oxygen supply to the fetus. *J. Dev. Physiol.* **1989**, *12*, 305–322. [PubMed]
33. Limesand, S.W.; Rozance, P.J.; Brown, L.D.; Hay, W.W. Effects of chronic hypoglycemia and euglycemic correction on lysine metabolism in fetal sheep. *Am. J. Physiol. Endocrinol. Metab.* **2009**, *296*, E879–E887. [CrossRef] [PubMed]
34. Ganguly, A.; Collis, L.; Devaskar, S.U. Placental glucose and amino acid transport in calorie-restricted wild-type and Glut3 null heterozygous mice. *Endocrinology* **2012**, *153*, 3995–4007. [CrossRef] [PubMed]
35. DiGiacomo, J.E.; Hay, W.W. Fetal glucose metabolism and oxygen consumption during sustained hypoglycemia. *Metabolism* **1990**, *39*, 193–202. [CrossRef]
36. Carver, T.D.; Quick, A.A.; Teng, C.C.; Pike, A.W.; Fennessey, P.V.; Hay, W.W. Leucine metabolism in chronically hypoglycemic hypoinsulinemic growth restricted fetal sheep. *Am. J. Physiol. Endocrinol. Metab.* **1997**, *272*, E107–E117. [CrossRef]
37. Aldoretta, P.W.; Carver, T.D.; Hay, W.W. Maturation of glucose-stimulated insulin secretion in fetal sheep. *Biol. Neonate* **1998**, *73*, 375–386. [CrossRef]
38. Anthony, R.V.; Scheaffer, A.N.; Wright, C.D.; Regnault, T.R.H. Ruminant models of prenatal growth restriction. *Reprod. Suppl.* **2003**, *61*, 183–194. [CrossRef]
39. Limesand, S.W.; Rozance, P.J.; Zerbe, G.O.; Hutton, J.C.; Hay, W.W. Attenuated insulin release and storage in fetal sheep pancreatic islets with intrauterine growth restriction. *Endocrinology* **2006**, *147*, 1488–1497. [CrossRef]
40. Lassarre, C.; Hardouin, S.; Daffos, F.; Forestier, F.; Frankenne, F.; Binoux, M. Serum insulin-like growth factors and insulin-like growth factor binding proteins in the human fetus. Relationships with growth in normal subjects and in subjects with intrauterine growth retardation. *Pediatr. Res.* **1991**, *29*, 219–225. [CrossRef]
41. Thorn, S.R.; Rozance, P.J.; Brown, L.D.; Hay, W.W. The intrauterine growth restriction phenotype: Fetal adaptations and potential implications for later life insulin resistance and diabetes. *Semin. Reprod. Med.* **2011**, *29*, 225–236. [CrossRef] [PubMed]
42. Freemark, M.; Keen, A.; Fowlkes, J.; Mularoni, T.; Comer, M.; Grandis, A.; Kodack, L. The placental lactogen receptor in maternal and fetal sheep liver: Regulation by glucose and role in the pathogenesis of fasting during pregnancy. *Endocrinology* **1992**, *130*, 1063–1070. [PubMed]
43. Constância, M.; Hemberger, M.; Hughes, J.; Dean, W.; Ferguson-Smith, A.; Fundele, R.; Stewart, F.; Kelsey, G.; Fowden, A.; Sibley, C.; et al. Placental-specific IGF-II is a major modulator of placental and fetal growth. *Nature* **2002**, *417*, 945–948. [CrossRef] [PubMed]
44. Coan, P.M.; Fowden, A.L.; Constancia, M.; Ferguson-Smith, A.C.; Burton, G.J.; Sibley, C.P. Disproportional effects of Igf2 knockout on placental morphology and diffusional exchange characteristics in the mouse. *J. Physiol.* **2008**, *586*, 5023–5032. [CrossRef] [PubMed]
45. Baker, J.; Liu, J.-P.; Robertson, E.J.; Efstratiadis, A. Role of insulin-like growth factors in embryonic and postnatal growth. *Cell* **1993**, *75*, 73–82. [CrossRef]
46. Ludwig, T.; Eggenschwiler, J.; Fisher, P.; D'Ercole, A.J.; Davenport, M.L.; Efstratiadis, A. Mouse mutants lacking the type 2 IGF receptor (IGF2R) are rescued from perinatal lethality in Igf2 and Igf1r null backgrounds. *Dev. Biol.* **1996**, *177*, 517–535. [CrossRef]
47. Lau, M.M.; Stewart, C.E.; Liu, Z.; Bhatt, H.; Rotwein, P.; Stewart, C.L. Loss of the imprinted IGF2/cation-independent mannose 6-phosphate receptor results in fetal overgrowth and perinatal lethality. *Genes Dev.* **1994**, *8*, 2953–2963. [CrossRef]
48. Zhang, S.; Barker, P.; Botting, K.J.; Roberts, C.T.; McMillan, C.M.; McMillen, I.C.; Morrison, J.L. Early restriction of placental growth results in placental structural and gene expression changes in late gestation independent of fetal hypoxemia. *Physiol. Rep.* **2016**, *4*, e13049. [CrossRef]
49. de Vrijer, B.; Davidsen, M.L.; Wilkening, R.B.; Anthony, R.V.; Regnault, T.R.H. Altered placental and fetal expression of IGFs and IGF-binding proteins associated with intrauterine growth restriction in fetal sheep during early and mid-pregnancy. *Pediatr. Res.* **2006**, *60*, 507–512. [CrossRef]
50. Georgiades, P.; Cox, B.; Gertsenstein, M.; Chawengsaksophak, K.; Rossant, J. Trophoblast-specific gene manipulation using lentivirus-based vectors. *BioTechniques* **2007**, *42*, 317–324. [CrossRef]
51. Vaughn, O.R.; Maksym, K.; Silva, E.; Barentsen, K.; Anthony, R.V.; Brown, T.L.; Hillman, S.L.; Spencer, R.; David, A.L.; Rosario, F.J., et al. Placental-specific *Slc38a2*/SNAT2 knockdown causes fetal growth restriction in mice. *Clin. Sci.* **2021**, *135*, 2049–2066. [CrossRef] [PubMed]
52. Qiao, L.; Saget, S.; Lu, C.; Zang, T.; Dzyuba, B.; Hay, W.W.; Shao, J. The essential role of pancreatic α-cells in maternal metabolic adaptation to pregnancy. *Diabetes* **2022**, *71*, 978–988. [CrossRef] [PubMed]
53. Cilvik, S.N.; Wesolowski, S.R.; Anthony, R.V.; Brown, L.D.; Rozance, P.J. Late gestation fetal hyperglucagonaemia impairs placental function and results in diminished fetal protein accretion and decreased fetal growth. *J. Physiol.* **2021**, *599*, 3403–3427. [CrossRef] [PubMed]
54. Hay, W.W.; Sparks, J.W.; Battaglia, F.C.; Meschia, G. Maternal-fetal glucose exchange: Necessity of a three-pool model. *Am. J. Physiol. Endocrinol. Metab.* **1984**, *246*, E528–E534. [CrossRef] [PubMed]
55. National Research Council. *Nutrient Requirements of Small Ruminants: Sheep, Goats, Cervids, and New World Camelids*; The National Academies Press: Washington, DC, USA, 2007.

International Journal of
*Molecular Sciences*

MDPI

*Article*

# Molecular Indicators of Blood-Brain Barrier Breakdown and Neuronal Injury in Pregnancy Complicated by Fetal Growth Restriction

**Natalia Misan [1,*], Sławomir Michalak [2], Piotr Rzymski [3,4], Barbara Poniedziałek [3], Katarzyna Kapska [1], Krystyna Osztynowicz [2] and Mariola Ropacka-Lesiak [1]**

1 Department of Perinatology and Gynecology, Poznan University of Medical Sciences, 60-535 Poznan, Poland
2 Department of Neurochemistry and Neuropathology, Poznan University of Medical Sciences, 60-355 Poznan, Poland
3 Department of Environmental Medicine, Poznan University of Medical Sciences, 61-848 Poznan, Poland
4 Integrated Science Association (ISA), Universal Scientific Education and Research Network (USERN), 60-806 Poznań, Poland
* Correspondence: natalia.podkowa@wp.pl

**Citation:** Misan, N.; Michalak, S.; Rzymski, P.; Poniedziałek, B.; Kapska, K.; Osztynowicz, K.; Ropacka-Lesiak, M. Molecular Indicators of Blood-Brain Barrier Breakdown and Neuronal Injury in Pregnancy Complicated by Fetal Growth Restriction. *Int. J. Mol. Sci.* **2022**, *23*, 13798. https://doi.org/10.3390/ijms232213798

Academic Editor: Ilona Hromadnikova

Received: 22 September 2022
Accepted: 2 November 2022
Published: 9 November 2022



**Abstract:** This study evaluated the damage to the endothelial tight junctions (TJs) in pregnancies complicated by fetal growth restriction (FGR) and investigated whether FGR is related to blood–brain barrier disintegration and, subsequently, to the appearance of proteins indicative of neuronal injury in maternal blood. The studied group included 90 pregnant women diagnosed with FGR. The control group consisted of 70 women with an uncomplicated pregnancy. The biochemical measurements included serum neuronal proteins (subunit of the N-methyl-D-aspartate receptor—NR1, nucleoside diphosphate kinase A—NME1, and S100 calcium-binding protein B—S100B), serum TJ proteins (occludin—OCLN, claudin-5—CLN5, zonula occludens—zo-1, and OCLN/zo-1 and CLN5/zo-1 ratios), and placental expression of TJ proteins (OCLN, claudin-4 CLN4, CLN5, zo-1). The significantly higher serum S100B and CLN5 levels and serum CLN5/zo-1 ratio were observed in FGR compared to healthy pregnancies. Moreover, FGR was characterized by increased placental CLN5 expression. Both serum NME1 levels and placental CLN4 expression in FGR pregnancies were significantly related to the incidence of neurological disorders in newborns. Mothers of FGR neonates who developed neurological complications and intraventricular hemorrhage (IVH) had statistically higher NME1 concentrations during pregnancy and significantly lower placental CLN4 expression than mothers of FGR neonates without neurological abnormalities. The serum NME1 levels and placental CLN4 expression were predictive markers of IVH in the FGR group. The blood–brain barrier is destabilized in pregnancies complicated by FGR. Neurological disorders, including IVH, are associated with higher serum concentrations of NME1 and the decreased placental expression of CLN4. The serum NME1 levels and placental CLN4 expression may serve as biomarkers, helpful in predicting IVH in FGR. It may allow for more precise monitoring and influence decision-making on the optimal delivery time to avoid developing neurological complications.

**Keywords:** blood–brain barrier; endothelium; fetal growth restriction; fetal hypoxia; neuronal damage; tight junctions; tight junction proteins

## 1. Introduction

Fetal growth restriction (FGR) is a pregnancy complication characterized by an inability of the fetus to reach its genetically predicted growth potential [1–3]. According to the criteria offered by Figueras and Gratacos, FGR is diagnosed when the estimated fetus weight (EFW) is lower than the 3rd percentile or, if the EFW is less than the 10th percentile, and a Doppler blood flow test indicates abnormalities associated with a poorer perinatal outcome [4]. FGR affects approximately 5 to 10% of pregnancies and is the second most common cause of perinatal mortality [5]. It has a multifactorial etiology, which is still not fully understood [6]. Its leading cause is a uteroplacental unit failure resulting in dysregulated blood flow across the placenta [7]. It is likely to result from an impaired trophoblast invasion of the uterine vasculature in the early stage of pregnancy, leading to abnormal conversion of the spiral arteries into low-resistance vessels that are limited in their delivery of nutrients and oxygen to the fetus [8–10]. A low birth weight is related to long-term consequences, evident from the in utero nervous system programming (Barker's theory) [11–14]. FGR is associated not only with a higher risk of intrauterine fetal death, but also with poorer perinatal outcomes compared to those appropriate for gestational-age fetuses (AGA) [15–17].

The failure of invasive trophoblasts to sufficiently remodel the uterine arteries can lead to reduced blood flow, persistent placental hypoxia, and oxidative stress with consequences for fetal growth [18,19]. Initially, the fetus adapts to conditions of inadequate oxygen delivery, increasing cerebral, myocardial, and upper body blood flow while decreasing renal, gastrointestinal, and lower extremity perfusion. Circulatory centralization allows blood redistribution and preferential delivery of nutrients and oxygen to vital organs [20–23]. Eventually, as adaptive mechanisms are exhausted, the brain-sparing phenomenon does not provide sufficient protection against hypoxia for the developing fetal brain. It may lead to neurodevelopmental disorders on a heterogeneous spectrum [24–26].

Children with FGR show deficits in both brain function and structure [27]. FGR is associated with a reduction in total brain volume and gray matter, indicating its particular sensitivity to hypoxia and which is reflected in neurobehavioral impairments in children, including the ability to focus their attention [28–30]. In addition, postmortem studies of neonates with growth disorders have found a reduction in the total number of brain nerve cells [27,31]. Live-born FGR neonates show morphological differences in neurostructure, including abnormal corrugations of the cortex, and at 12 months of age, less structural complexity of the gray and white matter [27,32]. Additionally, prematurely born FGR neonates have lower global and local neural networks and reduced cortico-basal ganglia connectivity mainly in the prefrontal cortex and limbic system, compared to prematurely born neonates with AGA [27,33,34].

There are limited possibilities to study processes occurring in the developing fetal brain during pregnancy. Currently, the monitoring of a pregnancy complicated by FGR is based on repeated Doppler ultrasound and evaluation of cardiotocography records. There is a lack of sensitive and specific biomarkers of hypoxia in maternal blood that could be used in clinical practice for more precise monitoring of an FGR pregnancy to determine optimal delivery time and predict neonatal complications. Despite in vivo experimental studies and postmortem observations of fetuses with FGR, understanding the outcomes of placental insufficiency and in utero hypoxia on fetal neurodevelopmental processes is a challenge for modern perinatology. The sequence of events that occur in the brain during hypoxia is well understood, and these processes were observed in patients diagnosed with ischemic stroke [35,36].

The blood–brain barrier is a component of the neurovascular unit (NVU) that protects the nervous system from harmful agents and ensure the selective transport of substances from the blood to neurons. Under normal conditions, it is formed by tightly adherent endothelial cells, astrocytes, pericytes, the extracellular matrix, and the basal membrane [37–39]. The endothelial tight junctions (TJs) determine the paracellular permeability of molecules across the blood–brain barrier [40]. They consist of integral transmembrane proteins, i.e., occludin (OCLN), claudin (CLN), and junctional adhesion molecules (JAMs) [41], which are linked to the actin cytoskeleton by the zonula occludens proteins, zo-1, zo-2, and zo-3 [42]. Hypoxia-induced changes in blood–brain barrier permeability in pregnancies with FGR may result in the appearance of specific proteins in the blood that co-form the vascular or nonvascular (e.g., astroglial) part of the blood–brain barrier under normal conditions, the identification of which would allow monitoring of the pregnancy and a prognosis of neonatal complications. These markers could become a diagnostic tool to properly identify growth-restricted fetuses, among others, especially in those with mild abnormalities without severe placental insufficiency, which are more difficult both in terms of diagnosis and monitoring [43,44]. Given the hypothesis that abnormalities of TJs in FGR pregnancies are also reflected in processes occurring in the placenta, the searching for hypoxia indices in that maternal–fetal unit also seems justified.

The present study aimed to evaluate the usefulness of molecular indicators in identifying the damage to the endothelial TJs in pregnancies complicated by FGR and to investigate to what extend FGR is related to the release of blood–brain barrier proteins and molecules indicative of neuronal injury in maternal blood. The objectives were realized based on the evaluation of serum neuronal protein concentrations (subunit of the N-methyl-D-aspartate receptor—NR1, nucleoside diphosphate kinase A—NME1, and S100 calcium-binding protein B—S100B), serum levels (OCLN, claudin-5—CLN5, zonula occludens—zo-1, and OCLN/zo-1 and CLN5/zo-1 ratios), and the assessment of the placental expression of TJ proteins (OCLN, claudin-4—CLN4, CLN5, and zo-1). Moreover, the relationship between these biochemical parameters and neurological disorders in newborns was investigated. Furthermore, the usefulness of serum and placental proteins in predicting newborn neurological complications was evaluated.

## 2. Results

### 2.1. Group Characteristics

The basic characteristics of the studied and control groups, which did not differ in age, BMI at the first prenatal visit, and gestational age, are summarized in Table 1. The first, second, third, and fourth stage of FGR was diagnosed in 50.0, 12.2, 2.2, and 35.6% of patients, respectively. FGR newborns had significantly lower birth weights as compared to healthy neonates. They also revealed statistically lower values of all anthropometric measurements and lower body weight on the day of hospital discharge. FGR newborns had lower Apgar scores in the 1st and 5th minute and were hospitalized longer than healthy infants (Table 2).

**Table 1.** The characteristics of the group with fetal growth restriction (FGR) and physiological pregnancy.

| Characteristics | FGR ($n$ = 90) | Physiological Pregnancy ($n$ = 70) | $p$-Value |
|---|---|---|---|
| Age (years, mean ± SD) | 29 ± 5 | 31 ± 5 | 0.0662 |
| BMI at the first prenatal visit (kg/m$^2$) median, min–max) | 22.2 (15.2–42.0) | 21.9 (17.4–30.5) | 0.8443 |
| Gestational age at study eligibility (weeks, median, min–max) | 35 (24–41) | 36 (29–41) | 0.0677 |
| Estimated fetal weight (g, median, min–max) | 1932 (439–2920) | 3255 (756–4259) | **<0.0001** |
| Percentile of estimated fetal weight (median, min–max) | 1.0 (0.1–9.0) | 41.0 (4.0–98.0) | **<0.0001** |
| Mode of delivery (%) | | | |
| Spontaneous | 20.5 | 41.8 | **0.0039** |
| Cesarean section | 75.0 | 41.8 | **<0.0001** |
| Vacuum extractor | 4.5 | 16.4 | **0.0132** |
| Forceps | 0.0 | 0.0 | 1.0000 |

bold values means statistical significance at the $p < 0.05$ level.

**Table 2.** The perinatal outcomes of newborns in pregnancies complicated by fetal growth restriction (FGR) and physiological pregnancy. The statistically significant differences between groups are highlighted in bold.

| Parameters | FGR ($n$ = 90) | Physiological Pregnancy ($n$ = 70) | $p$-Value |
|---|---|---|---|
| Delivery week (median, min–max) | 37 (26–41) | 39 (37–42) | **<0.0001** |
| Premature delivery (%) | 42.2 | 0.0 | **<0.0001** |
| Fetal distress (%) | 50.0 | 20.0 | **0.0001** |
| Birth weight (g) (median, min–max) | 2260 (420–3080) | 3450 (2500–4600) | |
| <2500 overall (%) | 70.5 | 0.0 | |
| 1500–2500 (%) | 46.6 | 0.0 | **<0.0001** |
| 1000–1500 (%) | 8.0 | 0.0 | |
| <1000 (%) | 15.9 | 0.0 | |
| Anthropometric measurements (cm) (median, min–max) | | | |
| Head circumference | 31.5 (22.0–35.0) | 34.0 (31.0–38.0) | **<0.0001** |
| Thoracic circumference | 29.0 (17.5–33.0) | 34.0 (27.0–37.5) | **<0.0001** |
| Body length | 49.0 (29.5–55.0) | 55.0 (45.0–61.0) | **<0.0001** |
| Apgar score (points) (median, min–max) | | | |
| 1st minute | 10 (0–10) | 10 (4–10) | **0.0017** |
| 3rd minute | 8 (2–10) | 9 (6–9) | 0.6209 |
| 5th minute | 10 (4–10) | 10 (9–10) | **0.0001** |
| pH (median, min–max) | | | |
| Venous | 7.33 (7.01–7.46) | 7.33 (7.15–7.48) | 0.5835 |
| Arterial | 7.27 (6.95–7.45) | 7.27 (7.06–7.40) | 0.9758 |
| Base excess (mEq/L) (median, min–max) | | | |
| Venous | −2.6 (−11.3–3.2) | −2.7 (−10.3–2.7) | 0.2271 |
| Arterial | −2.3 (−13.4–3.4) | −2.6 (−11.8–3.4) | 0.1345 |
| Metabolic acidosis (%) | | | |
| Arterial pH < 7,30 overall | 56.7 | 42.9 | 0.1108 |
| 7.20–7.29 | 42.2 | 30.0 | 0.1376 |
| 7.10–7.19 | 12.2 | 10.0 | 0.8023 |
| 7.00–7.09 | 0.0 | 2.9 | 0.1899 |
| <7.00 | 2.2 | 0.0 | 0.5047 |
| Length of hospitalization (days) (median, min–max) | 6 (3–84) | 4 (2–22) | **<0.0001** |
| >5 days (%) | 50.6 | 14.5 | **<0.0001** |
| Neurological disorders overall (%) | 8.9 | 0.0 | **0.0098** |
| Intraventricular hemorrhage (%) | 7.8 | 0.0 | **0.0186** |
| First degree | 5.6 | 0.0 | 0.0683 |
| Second degree | 1.1 | 0.0 | 1.0000 |
| Third degree | 1.1 | 0.0 | 1.0000 |
| Fourth degree | 0.0 | 0.0 | - |
| Periventricular leucomalacia (%) | 2.2 | 0.0 | 0.5044 |

bold values means statistical significance at the $p < 0.05$ level.

## 2.2. Serum and Placental Markers

The significantly higher serum S100B and CLN5 levels and serum CLN5/zo-1 ratio were observed in FGR compared to the control pregnancy. Serum NR1, NME1, OCLN, and zo-1 levels and the serum OCLN/zo-1 index did not differ between groups (Table 3). The FGR group revealed significantly higher placental CLN5 expression. The placental expressions of OCLN, CLN4, and zo-1 were comparable in both groups (Table 4). A significant positive correlation between placental CLN5 expression and serum CLN5 levels was observed in the FGR group (Rs = 0.38, $p$ = 0.0252).

**Table 3.** Serum concentrations of biochemical parameters in pregnancies complicated by fetal growth restriction (FGR) and physiological pregnancy. The statistically significant differences between groups are highlighted in bold.

| Serum Concentrations | FGR ($n$ = 90) | Physiological Pregnancy ($n$ = 70) | $p$-Value |
|---|---|---|---|
| NR1 (pg/mL) mean ± SD (min–max) | 1295.2 ± 2852.6 (0.0–16,442.1) | 657.2 ± 1531.2 (0.0–9030.8) | 0.0799 |
| NME1 (pg/mL) mean ± SD (min–max) | 112.4 ± 650.7 (0.0–6085.6) | 77.1 ± 317.3 (0.0–2557.8) | 0.9246 |
| S100B (pg/mL) mean ± SD (min–max) | 29.6 ± 40.7 (0.0–165.9) | 14.9 ± 25.1 (0.0–72.8) | **0.0290** |
| OCLN (pg/mL) mean ± SD (min–max) | 32.8 ± 116.3 (0.0–676.0) | 8.4 ± 41.9 (0.0–284.0) | 0.0835 |
| CLN5 (pg/mL) mean ± SD (min–max) | 74.2 ± 167.6 (0.0–828.0) | 30.7 ± 150.6 (0.0–1115.0) | **0.0039** |
| zo-1 (RU/mL) mean ± SD (min–max) | 2.6 ± 4.6 (0.0–24.9) | 1.8 ± 4.2 (0.0–29.3) | 0.3122 |
| OCLN/zo-1 (pg/RU) mean ± SD (min–max) | 11.4 ± 50.5 (0.0–320.9) | 1.7 ± 8.9 (0.0–48.1) | 0.0612 |
| CLN5/zo-1 (pg/RU) mean ± SD (min–max) | 122.8 ± 6378.0 (0.0–4127.0) | 4.7 ± 12.2 (0.0–40.8) | **0.0181** |

bold values means statistical significance at the $p < 0.05$ level.

**Table 4.** Placental expression of tight junction proteins in pregnancies complicated by fetal growth restriction (FGR) and physiological pregnancy.

| Placental Expression | FGR ($n$ = 90) | Physiological Pregnancy ($n$ = 70) | $p$-Value |
|---|---|---|---|
| OCLN (ng/mg total protein) mean ± SD (min–max) | 0.18 ± 0.16 (0.00–0.53) | 0.12 ± 0.13 (0.00–0.43) | 0.0791 |
| CLN5 (ng/mg total protein) mean ± SD (min–max) | 0.02 ± 0.02 (0.00–0.11) | 0.01 ± 0.02 (0.00–0.09) | **0.0119** |
| CLN4 (ng/mg total protein) mean ± SD (min–max) | 0.16 ± 0.10 (0.00–0.42) | 0.13 ± 0.10 (0.02–0.43) | 0.1806 |
| zo-1 (RU/mL) mean ± SD (min–max) | 0.24 ± 0.16 (0.02–0.66) | 0.20 ± 0.16 (0.02–0.54) | 0.2615 |

bold values means statistical significance at the $p < 0.05$ level.

### 2.3. Association of Serum and Placental Markers and Neurological Outcomes

In FGR pregnancies, a significant relationship between serum NME1 levels and perinatal fetal distress was observed. Women from the FGR group, whose fetuses experienced life-threatening symptoms, had significantly higher serum NME1 concentrations compared to women with FGR without perilabour fetal distress (Figure 1A). Neurological disorders were observed in 8.9% of FGR newborns. They were not observed in children born from uncomplicated pregnancies. IVH and periventricular leukomalacia (PVL) were diagnosed in 7.8 and 2.2% of FGR neonates, respectively, with no cases in the control group. Mothers of FGR neonates who developed neurological complications had statistically higher NME1 concentrations during pregnancy (Figure 1B) and significantly lower placental CLN4 expression than mothers of FGR neonates without neurological abnormalities (Figure 2A). Moreover, mothers of FGR neonates diagnosed with IVH also showed significantly higher serum NME1 levels (Figure 1C) and lower placental CLN4 expression (Figure 2B) during pregnancy compared to mothers of FGR neonates without IVH.

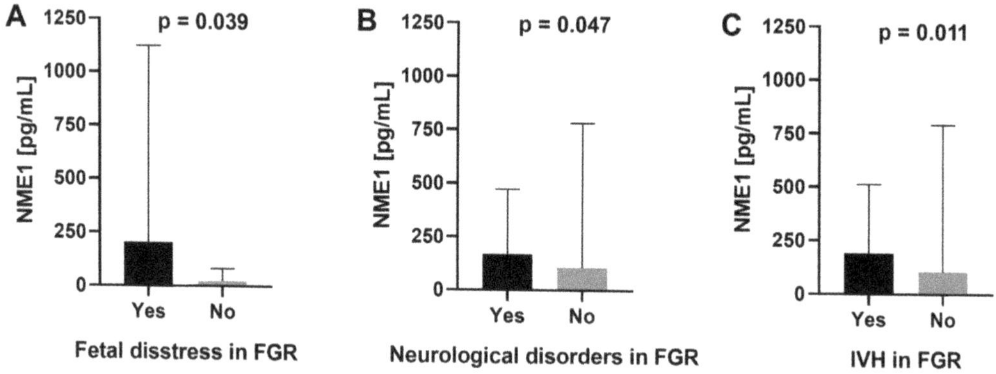


**Figure 1.** (**A**) Comparison of serum NME1 concentrations with reference to perinatal fetal distress in fetal growth restriction (FGR), (**B**) neurological disorders in FGR, and (**C**) intraventricular hemorrhage (IVH) in FGR newborns.

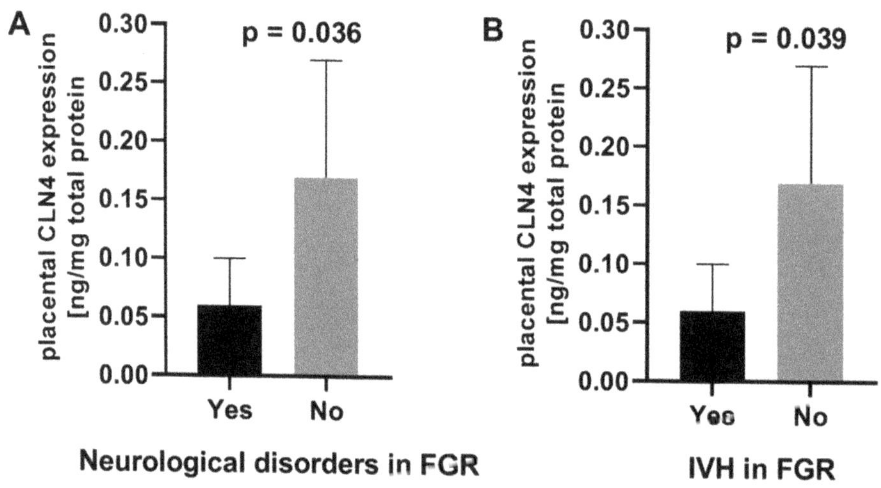


**Figure 2.** (**A**) Comparison of placental CLN4 expression with reference to neurological disorders in fetal growth restriction (FGR) and (**B**) intraventricular hemorrhage (IVH) in FGR newborns.

The serum NME1 levels showed a prognostic value for IVH in the FGR group (cut-off value for NME1: 40.45 pg/mL; sensitivity: 0.71; specificity: 0.76; PPV: 0.21; NPV: 0.97; LR+: 2.93; LR-: 0.38; ACC: 0.753; AUC: 0.745; $p = 0.0315$). The placental CLN4 expression was useful in the prediction of newborn IVH (cut-off value for CLN4: 20.82 ng/mg total protein; sensitivity: 1.00; specificity: 0.83; AUC: 0.879; $p < 0.0001$).

## 3. Discussion

The present study is the first to provide data on serum concentrations of NR1, NME1, OCLN, CLN5, and zo-1 in human FGR pregnancies. Previously, only one study, which included a small group of patients, reported placental expressions of OCLN and zo-1 in pre-eclampsia with coexisting FGR [45]. Developing programs of neonatal neurocritical care (NNCP), which include disorders related not only to neonatal, but also peripartum and parturition periods, focus on diagnostic and monitoring procedures. The identification of fetal or neonatal neurological disorders requires neuroimaging, neurophysiological, and laboratory tools. "The First Thousand Days" approach was used by Michael Scher to define the Fetal/Neonatal Neurology Program. This program focuses on trimester-specific mechanisms that influence the maternal/placental/fetal (MPF) triad. Disturbing the MPF triad triggers maternal immune responses, which modify pre-neuronal and pre-glial cell populations and their interactions. As a result, embryonic/fetal central nervous system structures are lesioned, leading to a spectrum of neurological complications, e.g., encephalopathy of prematurity, cortical dysplasia, seizures, etc. [46]. The understanding of the mechanisms involved in the development of placental vasculature is required to identify manifestations of trimester-specific MPF associated with great obstetrical syndromes (GOSs): pre-eclampsia, FGR, prematurity, fetal demise, placental abruption, and morbidly adherent placenta. Moreover, the development of the neurovascular unit relies on the interaction between neurogenesis and angiogenesis. The proper course of pathways that leads to the final destinations requires interactions between vascular and neural components. A motor for an attractive or repulsive force forms as a result of intersensing between endothelial tip cell caps, which represent the vascular component, and the axon growth cone, which represents the neural component. Thus, already at the developmental stage of the blood–brain barrier, the involvement of angiogenesis in this process also indicates the significant role played later in blood–brain barrier function and integrity, where TJ proteins (i.e., OCLN, CLN, e-cadherin, zo, JAMs, catenins, cingulin, and actin) are crucial [47].

Our study adds to the general understanding of the potential association between indicators of neuronal lesions (NR1, NME1, and S100B), TJ/blood–brain barrier stability (OCLN, CLN5, and zo-1) and FGR, and its neurological outcomes in neonates with potential implications for clinical practice, early diagnosis, and management.

Animal studies do not provide a clear answer on the changes in TJ proteins under hypoxic conditions. Warrington et al. observed no altered expression of zo-1 and OCLN in either the anterior or posterior cerebrum in placental ischemic rats [48]. Kuvacheva et al. reported an increased number of cells with CLN5 expression and a decreased number of zo-1-positive cells after perinatal hypoxia in postpartum rats [49]. Ma et al. showed a downregulated expression of zo-1 and CLN5 in mouse-brain microvascular endothelial cells under hypoxia/aglycemia conditions [50]. Zehender et al. observed the destruction of zo-1 and CLN5 in hypoxia with subsequent reoxygenation [51], whereas Mark and Davis found changes in OCLN, zo-1, and zo-2 localization and an increased expression of these proteins during reoxygenation [52]. Similar to Mark and Davis, Yamagata et al. reported increased OCLN expression at the mRNA level [53]. Animal studies indicate the unquestionable effect of hypoxia on TJ protein distribution and point out that early life stress causes an imbalance between TJ protein expression, but the observed changes may be opposite in direction. This may require further studies on larger and more homogeneous groups to determine how reduced oxygen delivery affects endothelial TJ breakdown.

A similar increase in maternal serum and umbilical artery S100B levels in the FGR was independently observed in two previous studies and is in line with our observations [54,55]. Additionally, Gazzolo et al. determined the usefulness of S100B concentration in FGR pregnancy, with a cut-off value of 720 pg/mL, a sensitivity of 100%, and a specificity of 99.3% as a single marker to predict IVH in newborns. Moreover, S100B was suggested to be a potential marker for the early detection of IVH in infants with perinatal asphyxia before clinical examination and transtemporal ultrasound display pathological changes [56].

The incidence of neonatal IVH in our study was three-fold lower (7.8%) than reported by Gazzolo et al. [54]. Moreover, most newborns in our study presented with the first degree of IVH (5.6%), with an equal frequency of the second and third degrees (1.1%), and no occurrence of the fourth stage. These differences may be caused by the update of the guidelines for the management and recommended delivery time in FGR pregnancies. One should note that S100B levels in neonates with perinatal hypoxia reported by Gazzolo et al.

are significantly higher than those observed in pregnancies with FGR. This suggests that the fetal blood–brain barrier breakdown, which results in the release of the S100B protein into the maternal bloodstream, is a dynamic process that progresses over time. Only its early detection, followed by decision-making about the optimal delivery time, may protect against neurological deficits in newborns. Contrary to this, Mazarico et al. and Boutsikou et al. did not report increased maternal S100B concentrations in FGR compared to physiological pregnancy [57,58]. This may have been because the researchers collected the maternal venous blood at the time of delivery. Moreover, they used the chemiluminescent immunoassay for the in vitro determination of serum S100B with an analytical range between 0.02 and 30 μg/l, whereas the limit of detection in our study was 5 pg/mL. Despite these discrepancies, S100B appears to be associated with FGR as its increased urine and serum levels were reported for FGR neonates, including those with neurological abnormalities detected one week after delivery [59,60]. All in all, S100B may be useful in the prediction of brain damage in neonates.

In addition, one study considered the possibility of a partial placental release of S100B in hypoxemic conditions because of its localization in intermediate villi and trophoblast cells. However, its expression does not change in pregnancies complicated by FGR and in physiological gestation [61]. Therefore, it is reasonable to assume that a significant increase in S100B in the fetus and/or in the placenta would lead to higher maternal blood levels, despite the increased total blood volume in pregnant women. It should be considered that S100B could be derived from maternal tissues [62], but all pregnant women included in the study were healthy and had no detectable neurological symptoms; thus, that possibility seems unlikely. We observed no relationship between the analyzed serum or placental proteins and perinatal outcomes in pregnant women. Thus, we can speculate that the changes in the placental expression of TJ proteins and their associations with neurological complications in the newborn reflect processes associated with fetal blood–brain barrier destabilization in pregnancies complicated by FGR.

The present study provides novel information on serum NME1 as a valuable marker in IVH prediction among FGR newborns. Furthermore, significantly higher NME1 concentrations were observed in FGR pregnancies when perinatal fetal distress occurred, indirectly suggesting the association of this protein with neuronal damage. Although serum NR1 concentrations showed no difference between FGR and physiological pregnancies, it oscillated close to the significance limit, with a predominance of higher values in pregnancies complicated by FGR. NR1 is one of the subunits of N-methyl-D-aspartate receptors (NMDAR), which is reported to decrease in response to prenatal hypoxia (to which the fetal brain is particularly vulnerable). That results from the high expression and activity of these receptors in fetuses, which are specific roles of NMDAR neurotransmission in the maturation and plasticity of developing neurons, and which change the NMDAR configuration or their affinity to neurotransmitters in response to noxious stimuli [63,64]. Therefore, we hypothesized that a decrease in the expression of these receptors in response to intrauterine hypoxia could occur with the persistence of adverse conditions during fetal development in utero. At the time of blood collection from pregnant women, the disorder may have been moderately severe, as indicated by the predominance of the first stage of FGR diagnoses. Therefore, the disintegration of the blood–brain barrier could already be identified (as evidenced by significantly higher serum CLN5 and S100B concentrations and the serum CLN5/zo-1 index) with a relative increase in NR1 in maternal blood. However, over time, as the Doppler blood flow worsens, a significant change in levels of NR1, as well as NR2 and NR3, could occur due to the degradation of NMDAR. These hypotheses were put forward by Schober et al. and Phillips et al., who observed the reduction in the NR1 subunit and NR2A to NR2B ratio, dendrite shortening, and the reduction in the density of immunostaining of NR1 due to hypoxia in rats with induced growth restriction because of placental insufficiency [65,66]. Understanding the role of the NMDAR and the balance between their different subunits in FGR pregnancy requires further research, perhaps including the role of their agonists and antagonists.

Our study did not observe any difference in OCLN expression, which is in line with some previous research on pre-eclampsia and laboratory-induced hypoxia [67,68]. Contrary to this, Lim et al. and Wang et al. reported a decrease in OCLN expression at the protein level, although there were no changes in mRNA expression in patients diagnosed with pre-eclampsia compared to physiological pregnancy [45,69]. Moreover, pre-eclampsia coexisted with a disorganized pattern of TJ proteins [69]. Similar to our study, Lim et al. and Liévano et al. did not observe changes in zo-1 expression [45,67], whereas Zhang et al. noticed a decrease in zo-1 and CLN4 expression, as well as an increase in CLN8 expression [68]. Contrary to the increase in CLN5 expression observed in the present study, Liévano et al. found a decrease in CLN5, CLN1, and CLN3 expression [67]. This may suggest the existence of further pathways which control the placental claudin expression and distinct mechanisms of "up-" and "down-" regulation of TJ proteins in response to the conditions in utero. Zhang et al. concluded that placental TJ dysfunction induced by reduced oxygen concentrations might be important in the pathogenesis of pre-eclampsia [68].

Similarly, in our study, along with increased placental CLN5 expression, significantly higher serum CLN5 levels were observed in the FGR group. This suggests that the disorganization of placental TJ proteins, contributing to reduced endothelial cell tightness, may influence further blood flow deterioration, leading to placental insufficiency and affecting fetal blood–brain barrier stability with the subsequent release of CLN5 into the maternal blood circulation. It remains significant that CLN5 is localized in the blood–brain barrier on the outer part of the endothelial cell membrane with the internal organization of OCLN and zo-1. Thus, the observed relationship may be derived from partial destabilization of TJ proteins in the developing fetal brain, resulting in a higher release of CLN5 into the maternal blood compared to other TJ proteins or indicating the involvement of additional molecules supporting these connections. The lower placental expression of CLN4 was found in FGR pregnancies with neurological disorders of newborns, including IVH. Since the multinucleated syncytiotrophoblast layer shows the presence of CLN4 in the basolateral part of the cell membrane [67], it can be speculated that the disorganization of the placental TJ proteins, leading to lower CLN4 expression, may be related to newborn neurological complications, including IVH. Moreover, it should be pointed out that the finding of significantly higher serum NME1 concentrations in pregnant women with FGR was also associated with neurological abnormalities and IVH. Because Lööv et al. reported the appearance of NME1 only in cultures of damaged neurons and, therefore, suggested the neuroprotective or regenerative function of extracellular NME1 as well as the lack of its secretion in normal conditions [70], the link between the change in NME1 levels in pregnant patients with FGR and the secondarily found reduced placental CLN4 expression seems reasonable. A limitation of the study was the lack of sequential analysis in the newborns' urine, which could show the changes in the studied parameters occurring during the delivery period and postnatal life. Perhaps the studied parameters could be used in the prognosis of the later abnormalities found through transcranial ultrasound. Moreover, the studied groups, although homogeneous, were relatively small. In addition, the long-term neurological consequences were not evaluated. The final limitation of our study is the inability to exclude the subtle lesions in the placenta, such as those from the maternal or fetal inflammatory response to an ascending intrauterine infection, the diagnosis of which may not be available through conventional histopathological detection [71].

## 4. Materials and Methods

### 4.1. The Studied and Control Groups

The study was conducted in collaboration between the Department of Perinatology and Gynecology and the Department of Neurochemistry and Neuropathology of the Poznan University of Medical Sciences between 2015 and 2019.

The studied group included 90 pregnant women between 24 and 41 weeks of gestation and who were diagnosed with FGR according to the Figueras and Gratacós criteria [4].

The control group comprised 70 women with an uncomplicated pregnancy between 29 and 41 gestational weeks. Before inclusion in the study, a detailed medical interview was conducted regarding obstetric history, course of current pregnancy, chronic diseases, and medications. Then, each pregnant woman underwent an obstetric examination and Doppler ultrasound velocimetry (Voluson E10 BT18, GE Healthcare, Chicago, IL, USA). All newborns underwent a routine neonatal examination after delivery, but in the cases of severe asphyxia, it was performed after the stabilization of the vital functions. The neurological examination included an evaluation of a neonate's level of alertness, cranial nerve function, sensory and motor system function, and the presence of primitive reflexes. In the case of an infant born with less than 32 weeks of gestation, the first transcranial ultrasound examination was not performed until the 3rd day of life, and the second examination between the 5th and 7th days after delivery. If IVH was diagnosed using Papile's classification [72,73], the frequency of subsequent examinations depended on identifying primary changes and the baby's clinical condition. If the ultrasound image was unclear, a magnetic resonance (MR) was performed. MR was performed at the postconceptional age between 38 and 42 weeks if no changes were observed. In the case of a newborn between 32 and 35 weeks of gestation, an ultrasound was performed at the same time intervals as mentioned above, and if an abnormal or unclear image was found, MR was performed. If the neonate, born after 35 weeks of gestation with an Apgar score between 0 and 3 or an umbilical cord blood pH below 7.0, received therapeutic hypothermia, a Doppler ultrasound followed by MR was performed between the 7th and 10th day after delivery. If therapeutic hypothermia was not applied, ultrasound examinations were repeated until the 3rd day of life and between the 5th and 7th day after delivery. MR was performed when neurological abnormalities were observed. PVL was diagnosed using MR [74,75]. In newborns from physiological pregnancies born at term, because of the absence of risk factors for central nervous system damage and no neurological abnormalities at birth, diagnostic imaging was not necessary.

The exclusion criteria were as follows: maternal malnutrition, nicotinism, alcohol consumption, drug abuse, taking medications such as warfarin, antiepileptic drugs, anticancer drugs, folic acid antagonists (trimethoprim-sulfamethoxazole, phenobarbital), cyanotic congenital heart defects, heart failure NYHA III/IV, uncontrolled asthma, chronic obstructive pulmonary disease, cystic fibrosis, pregestational and gestational diabetes mellitus, chronic renal failure, nephrotic syndrome, renal transplantation, continuous hemodialysis, systemic lupus erythematosus, antiphospholipid syndrome, Crohn's disease, ulcerative colitis, severe anemia, sickle cell anemia, beta-thalassemia, hemoglobin H disease, and uterine malformations. Fetal exclusion factors were chromosomal aberrations, autosome abnormalities, uniparental disomies, microdeletion syndromes, congenital malformations, and confirmed infection with cytomegalovirus, rubella virus, herpes simplex virus, varicella-zoster virus, human immunodeficiency virus, *Toxoplasma gondii*, *Treponema pallidum*, *Chlamydia* sp., *Mycoplasma* sp., *Listeria monocytogenes*, or *Mycobacterium tuberculosis*. The women with placenta previa, placenta accreta, placental infarcts, placental villous thrombosis, circumvallate placenta, hemangiomas, or other placental tumors were also excluded from the study.

### 4.2. *Collection of Blood and Placental Samples*

Three tubes per clot, each with 7.5 mL of whole blood, were collected from each patient and centrifuged for 10 min at 2750$\times$ *g*. The serum was transferred into Eppendorf tubes and frozen at $-80$ °C. The external section of the placental plate, 25 $cm^2$ in size, was taken from the opposite side of the umbilical cord. If the umbilical cord was centrally located, the external sample of the placental plate was cut, regardless of the location. Fresh tissue was frozen at $-80$ °C.

### 4.3. *Laboratory Serum Assays*

The commercial ELISA assays were applied to measure the levels of NR1 (Human Glutamate [NMDA] receptor subunit zeta-1, GRIN1 ELISA Kit, MyBioSource, San Diego, CA,

USA), NME1 (Human Nucleoside diphosphate kinase A, NME1 ELISA Kit, MyBioSource, San Diego, CA, USA) and S100B (S100B human ELISA kit, DRG MedTek, Warsaw, Poland). Because of the lack of commercial diagnostic tests for OCLN, zo-1, and CLN5, these concentrations were assessed using an in-house ELISA method. The Nunc MaxiSorpTM plates (Thermo Fisher, Waltham, MA, USA) were used for all measurements. The rabbit anti-human and mouse anti-human antibodies were used as capture and detection antibodies for the determination of serum OCLN levels (Occludin Polyclonal Antibody, Zymed, South San Francisco, CA, USA, RRID AB_2533977; Occludin Monoclonal Antibody (OC-3F10), Invitrogen, Waltham, MA, USA, RRID AB_2533101) and zo-1 levels (ZO-1 Polyclonal Antibody, Zymed, South San Francisco, CA, USA, RRID AB_2533938; ZO-1 Monoclonal Antibody (ZO1-1A12), Invitrogen, Waltham, MA, USA, AB_2533147), respectively. For CLN5, mouse anti-human antibodies (Claudin 5 Monoclonal Antibody (4C3C2), Zymed, South San Francisco, CA, USA, RRID AB_2533200) were used as capture antibodies, whereas rabbit anti-human antibodies (Claudin 5 Polyclonal Antibody, Abcam, Cambridge, UK, RRID AB_2533157) were used for detection. Goat anti-mouse IgG (Goat anti-Mouse IgG (H+L) Cross-Adsorbed Secondary Antibody, HRP, Invitrogen, Waltham, MA, USA, RRID AB_2536527) served as the secondary antibodies for OCLN and zo-1, whereas goat anti-rabbit IgG (H+L, HRP, Invitrogen, Waltham, MA, USA) was used for CLN5. The Substrate Reagent Pack (Substrate Reagent Pack, R&D Systems™, Minneapolis, MN, USA) was used in that reaction. Recombinant human OCLN (Recombinant Human Occludin GST (N-Term) Protein, Novus Biologicals, Littleton, CO, USA) and recombinant human CLN5 (Recombinant Human Claudin-5 GST (N-Term) Protein, Novus Biologicals, Littleton, CO, USA) served as standards. The concentrations of all tested factors, except zo-1, were expressed in pg/mL. Due to the lack of an acceptable standard for zo-1, relative units (RUs) were calculated from the optical density (OD), measured at 450 nm (OD450) as the quotient: OD of 10 samples per OD of the cut-off. The OD was statistically determined from zo-1 serum measurements of 48 healthy patients, whereas the 95th percentile was defined as the cut-off value. The serum zo-1 concentration was expressed in RU/mL. All ELISA steps were conducted using an automated RT-3100 microplate washer (Rayto Life and Analytical Sciences Co., Ltd., Shenzhen, China), and the final reading was performed using an EL×800 microplate reader (BioTek, Winooski, VT, USA). Since blood–brain barrier disintegration is associated with the release of TJ proteins, externally expressed CLN5 and internally expressed OCLN and zo-1, the ratios of OCLN/zo-1 and CLN5/zo-1 were calculated to estimate the degree of blood–brain barrier breakdown.

### 4.4. *Laboratory Placental Tests*

Placental tissues were homogenized in a buffer (1 L) consisting of 150 mM NaCl, 5 mM ethylenediaminetetraacetic acid (EDTA), and 50 mM Tris buffer solution (all Sigma-Aldrich, Saint Louis, MO, USA). The mixture of protease inhibitors and Triton X-100 (both Sigma-Aldrich, Saint Louis, MO, USA) was added to the buffer, giving a solution with a final concentration of 1%. The protease inhibitors included: fluorinated 4-(2-aminoethyl)benzenesulfonyl hydrochloride (AEBSF), aprotinin, bestatin hydrochloride, trans-epoxysuccinyl-L-leucylamido(4-guanidino)butane (E-64), EDTA, and leupeptin hemisulfate. The tissue homogenates were centrifuged for 15 min in Eppendorf tubes at 10,000 rpm. The obtained filtrate was used for TJ protein expression analyses. All ELISA steps were performed using an automated microplate washer (RT-3100 Microplate Washer, Rayto Life and Analytical Sciences Co., Ltd., Shenzhen, China), and the records were read out using an EL×800 microplate reader (BioTek, Winooski, VT, USA). The placental expression of TJ proteins was estimated by using the Lowry method and described as ng/mg of total protein [76]. The dependence of the absorbance of the comparison solutions on protein concentration was plotted, and linear regression was used to prepare a standard curve. The protein concentration was determined from the standard curve and the absorbance of the test solution [77]. The placental expression of OCLN, zo-1, and CLN5 was analyzed using an in-house ELISA (described in Section 2.3. Laboratory Serum Assays). The CLN4 expression was evaluated by using a commercial ELISA kit (ELISA Kit for Claudin 4 (CLDN4), USCN Life Science, Wuhan, China).

### *4.5. Statistical Analysis*

The statistical analysis was performed with Statistica StatSoft 13.1 (StatSoft, Krakóv Poland) and PQStat 1.8.0 (PQStat, Warsaw, Poland). The normality of the data distribution was checked using Kolmogorov–Smirnov, Lilliefors, and Shapiro–Wilk tests. If the assumption of the Gaussian distribution was met, Student's *t*-test was used for calculations; otherwise, the nonparametric U-Mann–Whitney test was performed. The chi-square test and Fisher's exact test were used to analyze the data expressed on a nominal scale. The correlations were assessed with Spearman's rank correlation coefficient (Rs). The usefulness of serum and placental measurements in the prognosis of neurological disorders was evaluated with a receiver operating curve (ROC). The prediction analysis included area under the curve (AUC), sensitivity, specificity, positive predictive value (PPV), negative predictive value (NPV), reliability quotient of the positive result (LR+), reliability quotient of the negative result (LR-), and accuracy (ACC), calculated using the DeLong's nonparametric method and the Clopper–Pearson method for a single proportion and cut-off value. A $p$-value < 0.05 was considered statistically significant.

## 5. Conclusions

The present study indicates that the blood–brain barrier is destabilized in pregnancies complicated by FGR and marked by increased serum levels of CLN5 and S100B and the CLN5/zo-1 ratio in maternal blood. The neurological complications in FGR, including IVH, are associated with the increased release of NME1 into the maternal blood and decreased placental CLN4 expression. The serum NME1 and placental CLN4 expression may be predictive markers of IVH in FGR. They may allow for more precise monitoring and facilitate decision-making about the optimal delivery date to avoid fetuses developing neurological complications. Further research is needed to consider the time of FGR diagnosis with groups divided into early- and late-onset FGR or regarding fetal circulatory centralization (brain sparing). Moreover, the follow-up of FGR newborns could provide valuable data on long-term neurological deficits.

**Author Contributions:** Conceptualization, N.M., S.M. and M.R.-L.; data curation, N.M.; investigation, N.M., K.K. and K.O.; methodology, S.M.; resources, M.R.-L.; supervision, S.M., P.R., B.P. and M.R.-L.; writing—original draft, N.M.; writing—review and editing, S.M., P.R., B.P. and M.R.-L. All authors have read and agreed to the published version of the manuscript.

**Funding:** This research received no external funding.

**Institutional Review Board Statement:** The study was performed in accordance with the Declaration of Helsinki and approved by the Local Bioethical Committee of the Poznan University of Medical Sciences, Poznan, Poland (667/15, annex 787/17).

**Informed Consent Statement:** Informed consent was obtained from all subjects involved in the study.

**Data Availability Statement:** The data presented in the study are available from the corresponding author upon reasonable request.

**Conflicts of Interest:** The authors declare no conflict of interest.

## References

1. Murki, S.; Sharma, D. Intrauterine growth retardation—A review article. *J. Neonatal Biol.* **2014**, *3*, 135. [CrossRef]
2. Sharma, D.; Shastri, S.; Sharma, P. Intrauterine growth restriction: Antenatal and postnatal aspects. *Clin. Med. Insights Pediatr.* **2016**, *10*, 67–83. [CrossRef] [PubMed]
3. Mamelle, N.; Cochet, V.; Claris, O. Definition of fetal growth restriction according to constitutional growth potential. *Biol. Neonate.* **2001**, *80*, 277–285. [CrossRef] [PubMed]
4. Figueras, F.; Gratacós, E. Update on the diagnosis and classification of fetal growth restriction and proposal of a stage-based management protocol. *Fetal Diagn. Ther.* **2014**, *36*, 86–98. [CrossRef] [PubMed]

5. Nardozza, L.M.M.; Caetano, A.C.R.; Zamarian, A.C.P.; Mazzola, J.B.; Silva, C.P.; Marçal, V.; Lobo, T.F.; Peixoto, A.B.; Júnior, E.A. Fetal growth restriction: Current knowledge. *Arch. Gynecol. Obstet.* **2017**, *295*, 1061–1077. [CrossRef]
6. Moh, W.; Graham, J.M.; Wadhawan, I.; Sanchez-Lara, P.A. Extrinsic factors influencing fetal deformations and intrauterine growth restriction. *J. Pregnancy* **2012**, *2012*, 750485. [CrossRef]
7. Salafia, C.M.; Minior, V.K.; Pezzullo, J.C.; Popek, E.J.; Rosenkrantz, T.S.; Vintzileos, A.M. Intrauterine growth restriction in infants of less than thirty-two weeks' gestation: Associated placental pathologic features. *Am. J. Obstet. Gynecol.* **1995**, *173*, 1049–1057. [CrossRef]
8. Brosens, I.; Dixon, H.G.; Robertson, W.B. Fetal growth retardation and the arteries of the placental bed. *Br. J. Obstet. Gynaecol.* **1977**, *84*, 656–663. [CrossRef]
9. Khong, T.Y.; De Wolf, F.; Robertson, W.B.; Brosens, I. Inadequate maternal vascular response to placentation in pregnancies complicated by pre-eclampsia and by small-for-gestational age infants. *Br. J. Obstet. Gynaecol.* **1986**, *93*, 1049–1059. [CrossRef]
10. Giles, W.B.; Trudinger, B.J.; Baird, P.J. Fetal umbilical artery flow velocity waveforms and placental resistance: Pathological correlation. *Br. J. Obstet. Gynaecol.* **1985**, *92*, 31–38. [CrossRef]
11. Barker, D.J.P. Fetal programming: Influences on development and disease in later life. In *NIH Monograph Series*; Marcel Dekker: New York, NY, USA, 2000.
12. Baker, J.L.; Olsen, L.W.; Sørensen, T.I. Childhood body-mass index and the risk of coronary heart disease in adulthood. *N. Engl. J. Med.* **2007**, *357*, 2329–2337. [CrossRef]
13. Barker, D.; Osmond, C.; Winter, P.; Margetts, B.; Simmonds, S. Weight in infancy and death from ischaemic heart disease. *Lancet* **1989**, *2*, 577–580. [CrossRef]
14. Ramírez-Vélez, R. In utero fetal programming and its impact on health in adulthood. *Endocrinol. Nutr.* **2012**, *59*, 383–393. [CrossRef]
15. Ego, A.; Zeitlin, J.; Batailler, P.; Cornec, S.; Fondeur, A.; Baran-Marszak, M.; Jouk, P.-S.; Debillon, T.; Cans, C. Stillbirth classification in population-based data and role of fetal growth restriction: The example of RECODE. *BMC Pregnancy Childbirth* **2013**, *13*, 182. [CrossRef]
16. Unterscheider, J.; O'Donoghue, K.; Daly, S.; Geary, M.P.; Kennelly, M.M.; McAuliffe, F.M.; Hunter, A.; Morrison, J.J.; Burke, G.; Dicker, P.; et al. Fetal growth restriction and the risk of perinatal mortality-case studies from the multicentre PORTO study. *BMC Pregnancy Childbirth* **2014**, *14*, 63. [CrossRef]
17. Im, D.H.; Kim, Y.N.; Cho, H.J.; Park, Y.H.; Kim, D.H.; Byun, J.M.; Jeong, D.H.; Lee, K.B.; Sung, M.S. Placental Pathologic Changes Associated with Fetal Growth Restriction and Consequent Neonatal Outcomes. *Fetal Pediatr. Pathol.* **2020**, *40*, 430–441. [CrossRef]
18. Colson, A.; Sonveaux, P.; Debiève, F.; Sferruzzi-Perri, A.N. Adaptations of the human placenta to hypoxia: Opportunities for interventions in fetal growth restriction. *Hum. Reprod. Update* **2021**, *27*, 531–569. [CrossRef]
19. Nowaczyk, J.; Poniedziałek, B.; Rzymski, P.; Sikora, D.; Ropacka-Lesiak, M. Platelets in Fetal Growth Restriction: Role of Reactive Oxygen Species, Oxygen Metabolism, and Aggregation. *Cells* **2022**, *11*, 724. [CrossRef]
20. Verburg, B.O.; Jaddoe, V.W.; Wladimiroff, J.W.; Hofman, A.; Witteman, J.C.; Steegers, E.A. Fetal hemodynamic adaptive changes related to intrauterine growth: The Generation R Study. *Circulation* **2008**, *117*, 649–659. [CrossRef]
21. Garcia-Canadilla, P.; Rudenick, P.A.; Crispi, F.; Cruz-Lemini, M.; Palau, G.; Camara, O.; Gratacos, E.; Bijens, B.H. A Computational Model of the Fetal Circulation to Quantify Blood Redistribution in Intrauterine Growth Restriction. *PLOS Comput. Biol.* **2014**, *10*, e1003667. [CrossRef]
22. MacDonald, T.M.; Hui, L.; Tong, S.; Robinson, A.J.; Dane, K.M.; Middleton, A.L.; Walker, S.P. Reduced growth velocity across the third trimester is associated with placental insufficiency in fetuses born at a normal birthweight: A prospective cohort study. *BMC Med.* **2017**, *15*, 164. [CrossRef] [PubMed]
23. Cohen, E.; Wong, F.Y.; Horne, R.S.; Yiallourou, S. Intrauterine growth restriction: Impact on cardiovascular development and function throughout infancy. *Pediatr. Res.* **2016**, *79*, 821–830. [CrossRef] [PubMed]
24. Thompson, L.; Crimmins, S.; Telugu, B.; Turan, S. Intrauterine hypoxia: Clinical consequences and therapeutic perspectives. *Res. Rep. Neonatol.* **2015**, *5*, 79–89. [CrossRef]
25. Malhotra, A.; Ditchfield, M.; Fahey, M.C.; Castillo-Melendez, M.; Allison, B.; Polglase, G.; Wallace, E.; Hodges, R.; Jenkin, G.; Miller, S. detection and assessment of brain injury in the growth-restricted fetus and neonate. *Pediatr. Res.* **2017**, *82*, 184–193. [CrossRef] [PubMed]
26. Rees, S.; Harding, R.; Walker, D. The biological basis of injury and neuroprotection in the fetal and neonatal brain. *Int. J. Dev. Neurosci.* **2011**, *29*, 551–563. [CrossRef] [PubMed]
27. Miller, S.L.; Huppi, P.S.; Mallard, C. The consequences of fetal growth restriction on brain structure and neurodevelopmental outcome. *J. Physiol.* **2016**, *594*, 807–823. [CrossRef]
28. Polat, A.; Barlow, S.; Ber, R.; Achiron, R.; Katorza, E. Volumetric MRI study of the intrauterine growth restriction fetal brain. *Eur. Radiol.* **2017**, *27*, 2110–2118. [CrossRef]
29. Padilla, N.; Junqué, C.; Figueras, F.; Sanz-Cortes, M.; Bargallo, N.; Arranz, A.; Donaire, A.; Figueras, J.; Gratacos, E. Differential vulnerability of gray matter and white matter to intrauterine growth restriction in preterm infants at 12 months corrected age. *Brain Res.* **2014**, *1545*, 1–11. [CrossRef]
30. Morsing, E.; Malova, M.; Kahn, A.; Lätt, J.; Björkman-Burtscher, I.; Maršál, K.; Ley, D. Brain Volumes and Developmental Outcome in Childhood Following Fetal Growth Restriction Leading to Very Preterm Birth. *Front. Physiol.* **2018**, *9*, 1583. [CrossRef]

31. Samuelsen, G.B.; Pakkenberg, B.; Bogdanović, N.; Gundersen, H.J.; Larsen, J.F.; Græm, N.; Laursen, H. Severe cell reduction in the future brain cortex in human growth-restricted fetuses and infants. *Am. J. Obstet. Gynecol.* **2007**, *197*, 56.e1–56.e7. [CrossRef]
32. Dubois, J.; Benders, M.; Borradori-Tolsa, C.; Cachia, A.; Lazeyras, F.; Leuchter, R.H.-V.; Sizonenko, S.V.; Warfield, S.K.; Mangin, J.F.; Hüppi, P.S. Primary cortical folding in the human newborn: An early marker of later functional development. *Brain* **2008**, *131 Pt 8*, 2028–2041. [CrossRef]
33. Batalle, D.; Eixarch, E.; Figueras, F.; Muñoz-Moreno, E.; Bargallo, N.; Illa, M.; Acosta-Rojas, R.; Amat-Roldan, I.; Gratacos, E. Altered small-world topology of structural brain networks in infants with intrauterine growth restriction and its association with later neurodevelopmental outcome. *Neuroimage* **2012**, *60*, 1352–1366. [CrossRef]
34. Fischi-Gómez, E.; Vasung, L.; Meskaldji, D.-E.; Lazeyras, F.; Borradori-Tolsa, C.; Hagmann, P.; Barisnikov, K.; Thiran, J.-P.; Hüppi, P.S. Structural Brain Connectivity in School-Age Preterm Infants Provides Evidence for Impaired Networks Relevant for Higher Order Cognitive Skills and Social Cognition. *Cereb. Cortex* **2015**, *25*, 2793–2805. [CrossRef]
35. Kazmierski, R.; Michalak, S.; Wencel-Warot, A.; Nowinski, W.L. Serum tight-junction proteins predict hemorrhagic transformation in ischemic stroke patients. *Neurology* **2012**, *79*, 1677–1685. [CrossRef]
36. Kazmierski, R.; Michalak, S.; Wencel-Warot, A.; Nowinski, W.L. Blood-brain barrier dysfunction and recovery after ischemic stroke. *Prog. Neurobiol.* **2018**, *163–164*, 144–171. [CrossRef]
37. Chow, B.W.; Gu, C. The molecular constituents of the blood-brain barrier. *Trends Neurosci.* **2015**, *38*, 598–608. [CrossRef]
38. Persidsky, Y.; Ramirez, S.; Haorah, J.; Kanmogne, G.D. Blood-brain barrier: Structural components and function under physiologic and pathologic conditions. *J. Neuroimmune Pharmacol.* **2006**, *1*, 223–236. [CrossRef]
39. Baeten, K.M.; Akassoglou, K. Extracellular matrix and matrix receptors in blood-brain barrier formation and stroke. *Dev. Neurobiol.* **2011**, *71*, 1018–1039. [CrossRef]
40. Günzel, D.; Yu, A.S. Claudins and the modulation of tight junction permeability. *Physiol. Rev.* **2013**, *93*, 525–569. [CrossRef]
41. Luissint, A.-C.; Artus, C.; Glacial, F.; Ganeshamoorthy, K.; Couraud, P.-O. Tight junctions at the blood brain barrier: Physiological architecture and disease-associated dysregulation. *Fluids Barriers CNS* **2012**, *9*, 23. [CrossRef]
42. Greene, C.; Campbell, M. Tight junction modulation of the blood brain barrier: CNS delivery of small molecules. *Tissue Barriers* **2016**, *4*, e1138017. [CrossRef] [PubMed]
43. Diaz-Arrastia, R.; Wang, K.; Papa, L.; Sorani, M.D.; Yue, J.; Puccio, A.M.; McMahon, P.J.; Inoue, T.; Yuh, E.L.; Lingsma, H.F.; et al. Acute biomarkers of traumatic brain injury: Relationship between plasma levels of ubiquitin C-terminal hydrolase-L1 and glial fibrillary acidic protein. *J. Neurotrauma* **2014**, *31*, 19–25. [CrossRef] [PubMed]
44. Graham, E.M.; Burd, I.; Everett, A.D.; Northington, F.J. Blood Biomarkers for Evaluation of Perinatal Encephalopathy. *Front. Pharmacol.* **2016**, *7*, 196. [CrossRef] [PubMed]
45. Lim, S.-C.; Kwon, J.-Y.; Hwang, H.-S.; Kim, Y.-H.; Park, Y.-W. 488: Comparison of placental junctional proteins expression in preeclapsia with intrauterine growth restriction. *Am. J. Obstet. Gynecol.* **2007**, *197*, S142. [CrossRef]
46. Scher, M.S. "The First Thousand Days" Define a Fetal/Neonatal Neurology Program. *Front. Pediatr.* **2021**, *9*, 683138. [CrossRef]
47. Tam, S.J.; Watts, R.J. Connecting vascular and nervous system development: Angiogenesis and the blood-brain barrier. *Annu. Rev. Neurosci.* **2010**, *33*, 379–408. [CrossRef]
48. Warrington, J.P.; Fan, F.; Murphy, S.R.; Roman, R.J.; Drummond, H.A.; Granger, J.P.; Ryan, M.J. Placental ischemia in pregnant rats impairs cerebral blood flow autoregulation and increases blood-brain barrier permeability. *Physiol. Rep.* **2014**, *2*, e12134. [CrossRef]
49. Kuvacheva, N.V.; Morgun, A.; Malinovskaya, N.A.; Gorina, Y.V.; Khilazheva, E.D.; Pozhilenkova, E.A.; Panina, Y.; Boytsova, E.B.; Ruzaeva, V.A.; Trufanova, L.V.; et al. Tight junction proteins of cerebral endothelial cells in early postnatal development. *Cell Tissue Biol.* **2016**, *10*, 372–377. [CrossRef]
50. Ma, X.; Zhang, H.; Pan, Q.; Zhao, Y.; Chen, J.; Zhao, B.; Chen, Y. Hypoxia/Aglycemia-induced endothelial barrier dysfunction and tight junction protein downregulation can be ameliorated by citicoline. *PLoS ONE* **2013**, *8*, e82604. [CrossRef]
51. Zehendner, C.M.; Librizzi, L.; Hedrich, J.; Bauer, N.M.; Angamo, E.A.; De Curtis, M.; Luhmann, H.J. Moderate hypoxia followed by reoxygenation results in blood-brain barrier breakdown via oxidative stress-dependent tight-junction protein disruption. *PLoS ONE* **2013**, *8*, e82823. [CrossRef]
52. Mark, K.S.; Davis, T.P. Cerebral microvascular changes in permeability and tight junctions induced by hypoxia-reoxygenation. *Am. J. Physiol. Heart Circ. Physiol.* **2002**, *282*, H1485–H1494. [CrossRef]
53. Yamagata, K.; Tagami, M.; Takenaga, F.; Yamori, Y.; Itoh, S. Hypoxia-induced changes in tight junction permeability of brain capillary endothelial cells are associated with IL-1beta and nitric oxide. *Neurobiol. Dis.* **2004**, *17*, 491–499. [CrossRef]
54. Gazzolo, D.; Marinoni, E.; Di Iorio, R.; Lituania, M.; Marras, M.; Bruschettini, M.; Bruschettini, P.; Frulio, R.; Michetti, F.; Petraglia, F.; et al. High maternal blood S100B concentrations in pregnancies complicated by intrauterine growth restriction and intraventricular hemorrhage. *Clin. Chem.* **2006**, *52*, 819–826. [CrossRef]
55. Velipaşaoğlu, M.; Yurdakök, M.; Özyüncü, Ö.; Portakal, O. Neural injury markers to predict neonatal complications in intrauterine growth restriction. *J. Obstet. Gynaecol.* **2015**, *35*, 555–560. [CrossRef]
56. Gazzolo, D.; Di Iorio, R.; Marinoni, E.; Masetti, P.; Serra, G.; Giovannini, L.; Michetti, F. S100B protein is increased in asphyxiated term infants developing intraventricular hemorrhage. *Crit. Care Med.* **2002**, *30*, 1356–1360. [CrossRef]

57. Mazarico, E.; Llurba, E.; Cumplido, R.; Valls, A.; Melchor, J.C.; Iglesias, M.; Cabero, L.; Gratacós, E.; Gómez-Roig, M.D. Neural injury markers in intrauterine growth restriction and their relation to perinatal outcomes. *Pediatr. Res.* **2017**, *82*, 452–457. [CrossRef]
58. Boutsikou, T.; Mastorakos, G.; Kyriakakou, M.; Margeli, A.; Hassiakos, D.; Papassotiriou, I.; Kanaka-Gantenbein, C.; Malamitsi-Puchner, A. Circulating levels of inflammatory markers in intrauterine growth restriction. *Mediat. Inflamm.* **2010**, *2010*, 790605. [CrossRef]
59. Kirmemiş, Ö.; Tatli, M.M.; Tayman, C.; Koca, C.; Karadağ, A.; Uras, N.; Dilmen, U.; Kafali, H. Subclinical hypoxia of infants with intrauterine growth retardation determined by increased serum S100B protein levels. *Turk. J. Med. Sci.* **2011**, *41*, 795–800. [CrossRef]
60. Florio, P.; Marinoni, E.; Di Iorio, R.; Bashir, M.; Ciotti, S.; Sacchi, R.; Bruschettini, M.; Lituania, M.; Serra, G.; Michetti, F.; et al. Urinary S100B protein concentrations are increased in intrauterine growth-retarded newborns. *Pediatrics* **2006**, *118*, e747–e754. [CrossRef]
61. Marinoni, E.; Di Iorio, R.; Gazzolo, D.; Lucchini, C.; Michetti, F.; Corvino, V.; Cosmi, E.V. Ontogenetic localization and distribution of S-100beta protein in human placental tissues. *Obstet. Gynecol.* **2002**, *99*, 1093–1099. [CrossRef]
62. Schmidt, A.P.; Tort, A.; Amaral, O.; Schmidt, A.P.; Walz, R.; Vettorazzi, J.; Martins-Costa, S.H.; Ramos, J.G.L.; Souza, D.; Portela, L.V. Serum S100B in pregnancy-related hypertensive disorders: A case-control study. *Clin. Chem.* **2004**, *50*, 435–438. [CrossRef] [PubMed]
63. Tingley, W.G.; Ehlers, M.D.; Kameyama, K.; Doherty, C.; Ptak, J.B.; Riley, C.T.; Huganir, R.L. Characterization of protein kinase A and protein kinase C phosphorylation of the N-methyl-D-aspartate receptor NR1 subunit using phosphorylation site-specific antibodies. *J. Biol. Chem.* **1997**, *272*, 5157–5166. [CrossRef] [PubMed]
64. Mishra, O.P.; Delivoria-Papadopoulos, M. Modification of modulatory sites of NMDA receptor in the fetal guinea pig brain during development. *Neurochem. Res.* **1992**, *17*, 1223–1228. [CrossRef] [PubMed]
65. Schober, M.E.; McKnight, R.A.; Yu, X.; Callaway, C.W.; Ke, X.; Lane, R.H. Intrauterine growth restriction due to uteroplacental insufficiency decreased white matter and altered NMDAR subunit composition in juvenile rat hippocampi. *Am. J. Physiol. Regul. Integr. Comp. Physiol.* **2009**, *296*, R681–R692. [CrossRef] [PubMed]
66. Phillips, T.J.; Scott, H.; Menassa, D.A.; Bignell, A.L.; Sood, A.; Morton, J.S.; Akagi, T.; Azuma, K.; Rogers, M.F.; Gilmore, C.E.; et al. Treating the placenta to prevent adverse effects of gestational hypoxia on fetal brain development. *Sci. Rep.* **2017**, *7*, 9079. [CrossRef] [PubMed]
67. Liévano, S.; Alarcón, L.; Chávez–Munguía, B.; González–Mariscal, L. Endothelia of term human placentae display diminished expression of tight junction proteins during pre-eclampsia. *Cell Tissue Res.* **2006**, *324*, 433–448. [CrossRef]
68. Zhang, Y.; Zhao, H.-J.; Xia, X.-R.; Diao, F.-Y.; Ma, X.; Wang, J.; Gao, L.; Liu, J.; Gao, C.; Cui, Y.-G.; et al. Hypoxia-induced and HIF1α-VEGF-mediated tight junction dysfunction in choriocarcinoma cells: Implications for pre-eclampsia. *Clin. Chim. Acta* **2019**, *489*, 203–211. [CrossRef]
69. Wang, Y.; Gu, Y.; Granger, D.; Roberts, J.M.; Alexander, J. Endothelial junctional protein redistribution and increased monolayer permeability in human umbilical vein endothelial cells isolated during pre-eclampsia. *Am. J. Obstet. Gynecol.* **2002**, *186*, 214–220. [CrossRef]
70. Lööv, C.; Shevchenko, G.; Nadadhur, A.G.; Clausen, F.; Hillered, L.; Wetterhall, M.; Erlandsson, A. Identification of injury specific proteins in a cell culture model of traumatic brain injury. *PLoS ONE* **2013**, *8*, e55983. [CrossRef]
71. Khong, T.Y.; Mooney, E.E.; Ariel, I.; Balmus, N.C.M.; Boyd, T.K.; Brundler, M.-A.; Derricott, H.; Evans, M.J.; Faye-Petersen, O.M.; Gillan, J.E.; et al. Sampling and Definitions of Placental Lesions: Amsterdam Placental Workshop Group Consensus Statement. *Arch. Pathol. Lab. Med.* **2016**, *140*, 698–713. [CrossRef]
72. Papile, L.A.; Burstein, J.; Burstein, R.; Koffler, H. Incidence and evolution of subependymal and intraventricular hemorrhage: A study of infants with birth weights less than 1500 gm. *J. Pediatr.* **1978**, *92*, 529–534. [CrossRef]
73. Bowerman, R.A.; Donn, S.M.; Silver, T.M.; Jaffe, M.H. Natural history of neonatal periventricular/intraventricular hemorrhage and its complications: Sonographic observations. *AJR Am. J. Roentgenol.* **1984**, *143*, 1041–1052. [CrossRef]
74. De Vries, L.S.; Eken, P.; Dubowitz, L. The spectrum of leukomalacia using cranial ultrasound. *Behav. Brain Res.* **1992**, *49*, 1–6. [CrossRef]
75. Ment, L.R.; Bada, H.S.; Barnes, P.; Grant, P.E.; Hirtz, D.; Papile, L.A.; Pinto-Martin, J.; Rivkin, M.; Slovis, T.L. Practice parameter: Neuroimaging of the neonate: Report of the quality standards subcommittee of the American Academy of Neurology and the Practice Committee of the Child Neurology Society. *Neurology* **2002**, *58*, 1726–1738. [CrossRef]
76. Lowry, O.H.; Rosebrough, N.J.; Farr, A.L.; Randall, R.J. Protein measurement with the Folin phenol reagent. *J. Biol. Chem.* **1951**, *193*, 265–275. [CrossRef]
77. Urząd Rejestracji Produktów Leczniczych, Wyrobów Medycznych i Produktów Biobójczych. *Farmakopea Polska VII*; t. 1, wyd. 7; Urząd Rejestracji Produktów Leczniczych, Wyrobów Medycznych i Produktów Biobójczych: Warszawa, Poland, 2006; 236p, ISBN 83-88157-37-X.

International Journal of
*Molecular Sciences*

MDPI

*Article*

# First Trimester Maternal Plasma Aberrant miRNA Expression Associated with Spontaneous Preterm Birth

**Danai Mavreli [1], Mariana Theodora [2], Margaritis Avgeris [3], Nikolas Papantoniou [2], Panagiotis Antsaklis [2], George Daskalakis [2] and Aggeliki Kolialexi [4,*]**

[1] Laboratory of Medical Genetics, School of Medicine, National and Kapodistrian University of Athens, 106 79 Athens, Greece
[2] 1st Department of Obstetrics and Gynecology, School of Medicine, National and Kapodistrian University of Athens, 106 79 Athens, Greece
[3] Laboratory of Clinical Biochemistry–Molecular Diagnostics, Second Department of Pediatrics, School of Medicine, National and Kapodistrian University of Athens, "P. & A. Kyriakou" Children's Hospital, 106 79 Athens, Greece
[4] Department of Genetics, Institute of Child Health, 106 79 Athens, Greece
* Correspondence: akolial@med.uoa.gr; Tel.: +30-210-7467462; Fax: +210-7795553

**Abstract:** Spontaneous Preterm Delivery (sPTD) is one of the leading causes of perinatal mortality and morbidity worldwide. The present case–control study aims to detect miRNAs differentially expressed in the first trimester maternal plasma with the view to identify predictive biomarkers for sPTD, between $32^{0/7}$ and $36^{6/7}$ weeks, that will allow for timely interventions for this serious pregnancy complication. Small RNA sequencing (small RNA-seq) of five samples from women with a subsequent sPTD and their matched controls revealed significant down-regulation of miR-23b-5p and miR-125a-3p in sPTD cases compared to controls, whereas miR-4732-5p was significantly overexpressed. Results were confirmed by qRT-PCR in an independent cohort of 29 sPTD cases and 29 controls. Statistical analysis demonstrated that miR-125a is a promising early predictor for sPTL (AUC: 0.895; 95% CI: 0.814-0.972; $p < 0.001$), independent of the confounding factors tested, providing a useful basis for the development of a novel non-invasive predictive test to assist clinicians in estimating patient-specific risk.

**Keywords:** microRNAs; miRNAs; small RNA sequencing; small RNA-seq; sPTD; miR-23b-5p; miR-125a-5p; miR-4732-5p

**Citation:** Mavreli, D.; Theodora, M.; Avgeris, M.; Papantoniou, N.; Antsaklis, P.; Daskalakis, G.; Kolialexi, A. First Trimester Maternal Plasma Aberrant miRNA Expression Associated with Spontaneous Preterm Birth. *Int. J. Mol. Sci.* **2022**, 23, 14972. https://doi.org/10.3390/ijms232314972

Academic Editor: Ilona Hromadnikova

Received: 31 October 2022
Accepted: 23 November 2022
Published: 29 November 2022



## 1. Introduction

Spontaneous Preterm Delivery (sPTD), defined as delivery that occurs before the completion of 37 weeks of gestation, is a complex pregnancy-related complication with multiple etiologies affecting 5–18% of all pregnancies worldwide [1].

sPTD is a leading cause of perinatal mortality and morbidity accounting for approximately 16% of all deaths and 35% of deaths among newborns. [2]. Premature infants that survive are at higher risk of developmental and neurological dysfunction later in life [3]. It is therefore widely accepted that the foresight in women at risk for sPTD early in pregnancy may enable, through earlier intervention, prolonged pregnancy and improved neonatal outcome [4].

Despite extensive efforts, screening strategies to accurately predict sPTD are still unsatisfactory. Currently, the most effective predictive tool combines well-known risk factors including maternal age, maternal BMI, racial origin, maternal behavior, spontaneous or assisted conception and prior history of sPTD, resulting in the identification of ~18% of nulliparous and 38% of parous sPTD cases, with a 10% false positive rate [5]. The prediction rate has been shown to increase by 54.8% by combining a priori risk factors with the measurement of cervical length [6].

MicroRNAs (miRNAs) are a class of non-coding RNAs of 19–25nt long responsible for regulating gene expression at the posttranscriptional level, without altering the genetic code. Concerning mode of action, miRNAs target the 3′ untranslated region (UTR) of mRNAs, resulting in translational inhibition and finally in mRNA degradation [7]. miRNAs have been documented to modulate most biological processes, such as cell proliferation, survival, differentiation and apoptosis. Interestingly, miRNAs can be detected within the cells, but they are also released in a stable form, due to their small size, into body fluids, such as plasma and serum [8].

Often, altered expression levels of circulating miRNAs have been involved with the initiation and progression of various diseases, including pregnancy complications, suggesting their role as potential biomarkers.

Regarding sPTD, several studies have investigated circulating miRNAs as potential biomarkers for its prediction. However, most studies were restricted to miRNAs associated to cardiovascular disease, clusters from chromosome 14 (C14MC), chromosome 19 (C19MC) and miR-371-3 cluster or were performed using samples collected later during pregnancy, in the second or even the third trimester [9–15]. However, an ideal biomarker would be detectable during the first trimester of pregnancy, when routine screening for fetal aneuploidies is also offered, to reduce anxiety throughout pregnancy in low-risk women and allow for the close monitoring and timely interventions in the high-risk group [16].

The aim of the present study is to identify differentially expressed miRNAs in first trimester maternal plasma and to evaluate their clinical value as novel biomarkers for the early prediction of sPTD. Analysis was performed using small RNA next-generation sequencing (small RNA-seq) followed by quantitative real-time polymerase chain reaction (qRT-PCR) to confirm the results. The sPTD group consisted of women who experienced premature delivery between $32^{0/7}$ and $36^{6/7}$ weeks of gestation (moderate/late sPTD), which represent more than 8% of all premature deliveries [17]. Born at this gestational age, premature neonates, are at a relatively lower risk of mortality and morbidity than early preterm births. Still, the impact on healthcare worldwide is significant due to their increased risks compared to full-term births [18].

## 2. Results

### *2.1. Demographic and Clinical Characteristics*

Table 1 describes the comparison between the demographic and the clinical characteristics of the cases and controls. Beside pre-pregnancy BMI, no significant differences were noted between the two groups.

**Table 1.** Maternal and neonatal characteristics of moderate/late sPTD cases and controls included in the study.

| Characteristic | Controls (n = 34) | sPTD Cases (n = 34) | *p*-Value |
|---|---|---|---|
| Maternal age (y)<br>Median (min–max) | 31.00<br>(24.80–38.80) | 30.65<br>(24.80–38.20) | 0.898 |
| Pre-pregnancy BMI (kg/m$^2$)<br>Median (min–max) | 25.95<br>(22.00–31.20) | 24.65<br>(20.90–31.90) | 0.028 |
| Cigarette smoker | | | |
| • Yes<br>• No | 12<br>22 | 8<br>26 | 0.40 |
| Mode of conception | | | |
| • Spontaneous<br>• In vitro fertilization | 28 (82.4%)<br>6 (17.6%) | 26 (76.4%)<br>8 (23.6%) | 0.750 |

**Table 1.** *Cont.*

| Characteristic | Controls (n = 34) | sPTD Cases (n = 34) | *p*-Value |
|---|---|---|---|
| Parity | | | |
| • Nulliparous | 17 (50%) | 14 (41.2%) | 0.616 |
| • Parous | 17 (50%) | 20 (58.8%) | |
| Previous PTD | | | |
| • No | 32 (94.1%) | 27 (79.4%) | 0.104 |
| • Yes | 2 (5.88%) | 7 (20.58%) | |
| Mode of Delivery | | | |
| • Vaginal | 29 (85.2%) | 22 (64.7%) | 0.075 |
| • Caesarean section | 5 (14.7%) | 12 (35.2%) | |
| Neonatal Gender | | | |
| • Male | 18 (52.95%) | 16 (47.05%) | 0.95 |
| • Female | 16 (47.05%) | 18 (52.95%) | |

### 2.2. *Small RNA-seq Analysis*

Differentially expressed miRNAs were screened using small RNA-seq in five pairs of plasma samples. A total of 1682 unique circulating miRNAs were identified across all first trimester maternal plasma samples analyzed. Among these, 387 miRNAs were up-regulated and 95 miRNAs were down-regulated in sPTD cases compared to controls. Three circulating miRNAs showed significant changes ($p$-Value < 0.05, FC >1.5) in the sPTD group as compared to the control group. Specifically, significantly decreased levels of miR-23b-5p ($p < 0.047$, FC = 0.60) and miR-125a-3p ($p = 0.014$, FC = 0.59) were detected during the first trimester of gestation in women who subsequently experienced sPTD between $32^{0/7}$ and $36^{6/7}$ weeks of gestation whereas miR-4732-5p was significantly overexpressed ($p = 0.048$, FC = 1.70) (Supplement S1).

A scatter plot is used to display differentially expressed circulating RNAs (Figure 1). Significantly dysregulated circulating miRNAs are presented in a volcano plot (Figure 2).

### 2.3. *GO Analysis*

The downstream targets of the significantly dysregulated miRNAs were identified commonly between Targetscan and miRDB databases and used for GO analysis. The significant biological processes associated with these miRNA target genes are presented in Table 2. KEGG pathway analysis revealed that the significantly underexpressed miRNAs target signaling pathways associated with the T cell receptor signaling pathway, Homologous recombination and Osteoclast differentiation (Table 3). No significant pathways were identified for the target genes of the up-regulated miRNAs.

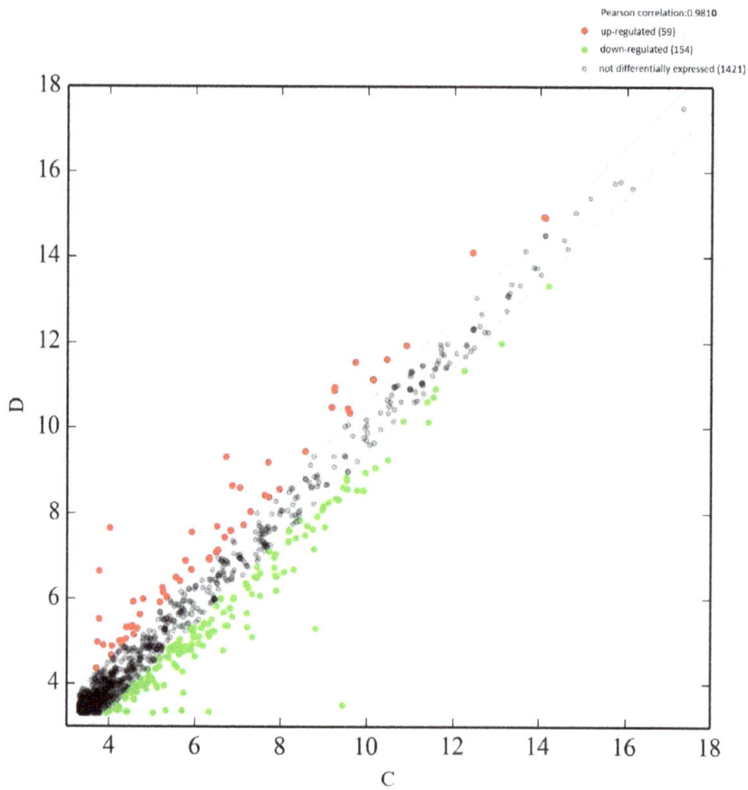


**Figure 1.** Scatter plot of circulating miRNA levels in the first trimester maternal plasma from women who subsequently experienced sPTD when compared to term ones. Red points show up-regulated miRNAs and green points show down-regulated miRNAs. Data from the sPTD group are plotted on the vertical axis ($y$-axis). Data from the term deliveries (controls) are plotted on the horizontal axis ($x$-axis).

**Table 2.** GO annotations of miRNA's target genes.

| Down-Regulated miRNAs | | | |
|---|---|---|---|
| **GO.ID** | **Term** | ***p*-Value** | **Genes** |
| GO:0048585 | Negative regulation of response to stimulus | 0.000 | RGS4//IL10//NDFIP1//GPRC5A//STAP1//MTMR4//TMEM127//BRCA1//GREM2//HGS//STK38//FAP//TPT1 |
| GO:0014854 | Response to inactivity | 0.001 | MTMR4//IL10 |
| GO:1900119 | Positive regulation of execution phase of apoptosis | 0.001 | FAP//BOK |
| GO:0009968 | Negative regulation of signal transduction | 0.001 | RGS4//GPRC5A//IL10//MTMR4//TMEM127//BRCA1//GREM2//HGS//STK38//STAP1//TPT1 |
| GO:0043032 | Positive regulation of macrophage activation | 0.001 | STAP1//IL10 |
| GO:0010324 | Membrane invagination | 0.001 | ABCA1//STAP1//HGS |
| GO:1900120 | Regulation of receptor binding | 0.002 | GREM2//IL10 |

**Table 2.** *Cont.*

| Down-Regulated miRNAs | | | |
|---|---|---|---|
| **GO.ID** | **Term** | ***p*-Value** | **Genes** |
| GO:0010648 | Negative regulation of cell communication | 0.002 | RGS4//GPRC5A//IL10//MTMR4//TMEM127//BRCA1//GREM2//HGS//STK38//STAP1//TPT1 |
| GO:0023057 | Negative regulation of signaling | 0.002 | RGS4//GPRC5A//IL10//MTMR4//TMEM127//BRCA1//GREM2//HGS//STK38//STAP1//TPT1 |
| GO:0070230 | Positive regulation of lymphocyte apoptotic process | 0.002 | IL10//PDCD1 |
| **Up-Regulated miRNAs** | | | |
| **GO.ID** | **Term** | ***p*-Value** | **GENES** |
| GO:0006397 | mRNA processing | 0.008 | RAVER2//LUC7L3//ALKBH5 |
| GO:0045625 | Regulation of T-helper 1 cell differentiation | 0.009 | SOCS5 |
| GO:0071071 | Regulation of phospholipid biosynthetic process | 0.009 | HTR2A |
| GO:0002829 | Negative regulation of type 2 immune response | 0.010 | SOCS5 |
| GO:0006054 | N-acetylneuraminate metabolic process | 0.010 | CMAS |
| GO:0007175 | Negative regulation of epidermal growth factor-activated receptor activity | 0.010 | SOCS5 |
| GO:0035970 | Peptidyl-threonine dephosphorylation | 0.010 | PPM1E |
| GO:0042118 | Endothelial cell activation | 0.010 | SOCS5 |
| GO:0045623 | Negative regulation of T-helper cell differentiation | 0.010 | SOCS5 |
| GO:0043371 | Negative regulation of CD4-positive, alpha-beta T cell differentiation | 0.011 | SOCS5 |

**Table 3.** Pathway enrichment analysis of the target genes of down-regulated miRNAs.

| Pathway ID | Definition | *p*-Value | FDR | Enricment Score | Gene Ratio | Genes |
|---|---|---|---|---|---|---|
| hsa04660 | T cell receptor signaling | 0.0021 | 0.658 | 2.671681 | 0.16 | IL10//PDCD1//TEC |
| hsa03440 | Homologous recombination | 0.0046 | 0.713 | 2.335566 | 0.11 | BRCA1//FAM175A |
| hsa04380 | Osteoclast differentiation | 0.0425 | 1 | 1.37106 | 0.11 | SIRPA//TEC |

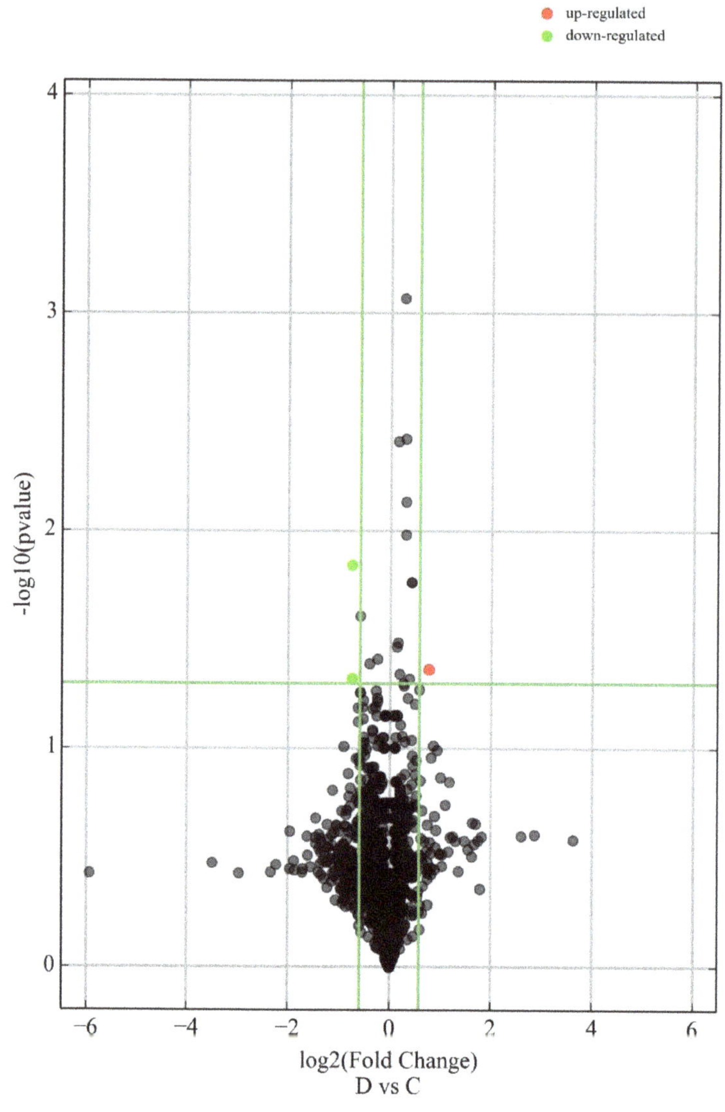


**Figure 2.** Volcano plot illustrates significantly differentially abundant circulating miRNAs. The -log10 (Benjamini–Hochberg corrected *p*-Value) is plotted against the log2 (fold change: sPTD cases/controls). The non-axial vertical lines denote ±1.5-fold change while the non-axial horizontal line denotes *p*-Value = 0.05, which is our significance threshold (prior to logarithmic transformation).

### 2.4. qRT-PCR Assays

The expression levels miR-23b-5p and miR-125a-3p were further determined in an independent cohort of 58 first trimester maternal plasma samples (29 sPTD cases and 29 controls) by qRT-PCR in order to confirm the reliability of the small RNA-seq analysis. The selection of miRNAs for verification by qRT-PCR was based on *p*-Values (Supplement S1). When compared with the control group, miR-125a-3p was significantly down-regulated in sPTD cases ($p < 0.001$; Figure 3A); where for miR-23b-5p, no statistically significant changes were observed ($p = 0.750$; Figure 3B). The qRT-PCR data were further used to assess the ability of miRNAs to discriminate between women at risk for sPTD and controls using ROC analysis. As expected, miR-125a-3p was highlighted to differentiate sPTD cases from uncomplicated pregnancies (AUC: 0.895; 95% CI: 0.814–0.972; $p < 0.001$; Figure 3C). Binomial logistic regression analysis demonstrated that miR-125a-3p represents a significant predictor of pregnancy outcome independent of the confounding factors tested (Table 4).

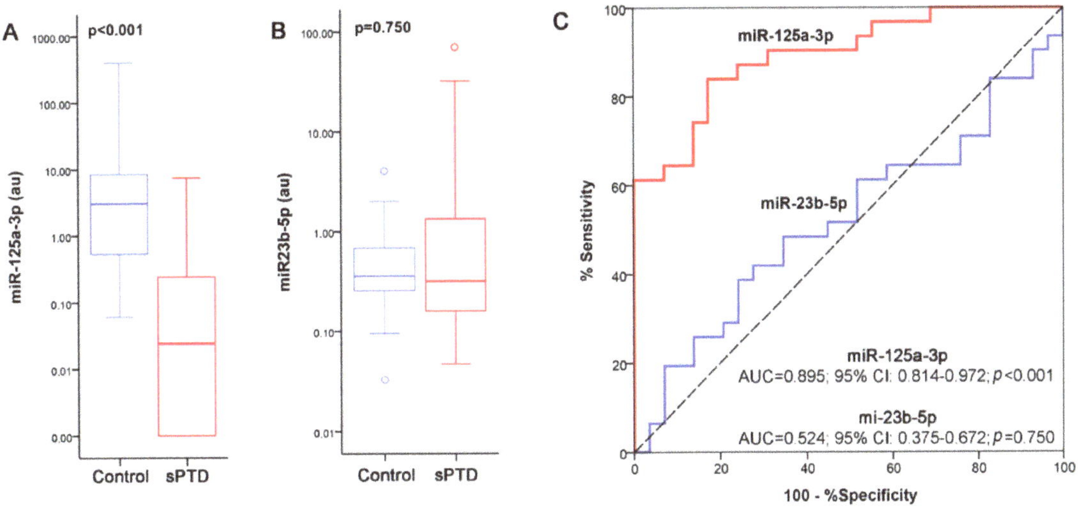


**Figure 3.** Statistical analysis of the data obtained using qRT-PCR. (**A**,**B**): Box-plots of the expression levels of miR-23b and miR-125a in the first trimester maternal plasma in women with sPTD versus uncomplicated pregnancies based on data derived from qRT-qPCR. *p*-Values calculated by Mann–Whitney U test. (**C**): ROC curve analysis for the prediction of sPTD based on miRNAs levels. The diagram is a plot of the % sensitivity (true-positive rate) vs. 100% specificity (false-positive rate).

**Table 4.** Logistic regression analysis for the prediction of sPTD patients according to maternal. plasma miRNAs expression levels.

| | | Univariate Analysis | |
|---|---|---|---|
| Covariant | OR [a] | 95% CI [b] | *p*-Value [c] |
| miR-23b-5p | 1.057 | 0.943–1.184 | 0.343 |
| miR-125a-3p | 0.654 | 0.490–0.872 | 0.004 |
| Previous PTD | 7.154 | 0.809–63.30 | 0.077 |
| Smoking | 0.535 | 0.176–1.624 | 0.269 |
| Maternal pre-pregnancy BMI | 0.808 | 0.650–1.003 | 0.053 |
| Mode of conception | | | |
| Natural | 1.00 | | |
| IVF | 1.512 | 0.425–5.384 | 0.523 |
| Fetal gender | | | |
| Male | 1.00 | | |
| Female | 1.319 | 0.470–3.703 | 0.600 |
| Maternal age | 1.016 | 0.885–1.167 | 0.819 |
| | | Multivariate analysis [d] | |
| Covariant | OR [a] | 95% CI [b] | *p*-Value [c] |
| miR125a-3p | 0.491 | 0.295–0.817 | 0.006 |

[a] Odds Ratio. [b] 95% Confidence Interval of the estimated OR. [c] calculated by test for trend. [d] Multivariate models adjusted for tested miRNAs, previous PTD, smoking, BMI, mode of conception, fetal gender and age.

## 3. Discussion

In the current study, we performed small RNA-seq in first trimester maternal plasma samples to investigate whether a differential miRNA expression profile is associated with subsequent sPTD and to provide a useful basis for the development of a novel non-invasive predictive test to assist clinicians in estimating patient-specific risks.

Compared to women who delivered at term, maternal plasma miRNA profiling of women who subsequently developed sPTD revealed significant alterations in the expression level of miR-23b-5p, mIR-125a-5p and mIR-4732-5p. It is noteworthy that none has previously been related to sPTD. Downstream targets, however, have been associated with biological processes with an important role in the pathogenesis of sPTD, e.g., immune response, inflammation and apoptosis.

More precisely, miR-4732-5p showed a significant over expression in the first trimester maternal plasma of women destined to deliver sPTD compared to those with uncomplicated pregnancies. miR4732-5p has a proven role in tumorigenesis [19,20]. However, our GO analysis revealed that its downstream targets include members of suppressors of the cytokine signaling (SOCS) family, namely SOCS1-SOCS7. The SOCS protein family is implicated in the negative regulation of cytokine signaling and therefore in the regulation of pregnancy and labor [21,22]. Blumenstein et al. reported overexpression of SOCS1, SOCS2 and SOCS3 proteins in human placenta in an attempt to either prevent the entrance of pathogens into the maternal circulation or suppress trophoblast cytokine production and inhibit their damaging effects on the fetus [23]. In the control group, i.e., women who delivered at term, a decrease in SOCS1 and SOCS3 was observed possibly reflecting the involvement of inflammatory cytokines in mechanisms leading to delivery. However, no alteration in SOCS protein levels was demonstrated in the sPTD group. Still, further investigation is required to uncover the relationship between circulating miR-4732-5p and SOCS proteins as the lack of change in SOCS levels in the sPTD group could be attributed to immaturity.

Moreover, 23b-5p, a pleiotropic miRNA demonstrating diverse effects on various biological processes, was significantly down-regulated in the first trimester maternal plasma in women who later delivered prematurely as compared to uncomplicated pregnancies. miR-23b is a member of the miR-23-27-24 family which consists of two paralogs with the miR-23a cluster (miR-23a-27a-24-2) found on chromosome 19 and the intragenic miR-23b cluster (miR-23b27b-24-1) located on chromosome 9 within the C9orf3 gene [24]. Previous research suggested that miR-23b plays vital roles in cancer development where it exerts either oncogenic or tumor suppressor activity [25,26]. It is also implicated in the regulation of angiogenesis and endothelial cells homeostasis and may serve as biomarkers for the early diagnosis of acute myocardial infarction [27,28]. Moreover, research findings suggest that miR-23b overexpression enhanced the expression of IL-10 which is vital for normal pregnancy, and low IL-10 levels are associated with pregnancy complications [29]. Hence, down-regulation of miR-23b has been reported in first trimester maternal plasma in women who later developed late onset pre-eclampsia as compared to uncomplicated pregnancies, indicating a possible involvement in the pathogenesis of the condition [24].

Finally, miR-125a-5p is involved in development and cell differentiation and therefore it is implicated in several malignancies including gastric, cervical, lung and ovarian cancers, retinoblastoma and neuroblastoma [30–32]. Recently, it was demonstrated that miR-125a suppresses cell proliferation and migration and inhibits angiogenesis by regulating its downstream target vascular endothelial growth factor A (VEGFA), indicating a potential role to the pathophysiology of pre-eclampsia [33].

Down-regulation of miR-125a was detected in the discovery cohort using small RNA-seq and was further confirmed by the verification cohort. Following statistical analysis, the AUC for miR-125a-5p was 89%, demonstrating promising diagnostic potential as a first trimester screening test for the prediction of subsequent sPTD at 33–36 weeks of gestation. Furthermore, logistic regression analysis revealed that miR-125a-5p is an early predictor of sPTD independently of well-known risk factors (previous sPTD, maternal smoking, maternal pre-pregnancy BMI, maternal age, fetal gender and mode of conception).

Hence, once validated in a large cohort, miR-125a-5p might be used to identify women at high risk for sPTD, allowing for close monitoring and/or clinical interventions such as cervical cerclage. They can also assist the development of novel therapeutics. To date, although the role of miRNA therapeutics has not yet been translated into clinical practice, miRNA mimics and miRNA suppressors have shown significant efficiency in various cancers [34].

To the best of our knowledge, this is the first study that aimed to identify novel biomarkers for the early prediction of women at risk for sPTD using a sensitive high-throughput method for miRNA plasma profiling followed by verification of the results obtained using RT-qPCR. We acknowledge, however, that the small sample size poses a significant limitation for our study.

Still, this pilot study aims to offer improved candidate biomarkers predictive of a subsequent sPTD to be validated in a larger series of pregnant women.

## 4. Materials and Methods

### 4.1. Study Population

Maternal samples for this retrospective study for sPTD were collected from pregnant women during first trimester prenatal screening for fetal aneuploidies between March 2018 and December 2020 as previously described [35]. At the same time, maternal demographic characteristics and medical history were obtained and recorded in an electronic database. Gestational age was determined based on the last menstrual period and confirmed by ultrasound measurement of the fetal crown rump length (CRL). Pregnancy outcomes were made known after the completion of pregnancies from the maternity hospital files and were also recorded in the database.

The selection of samples for analysis was carried out using a case–control design. Cases were pregnant women with a subsequent sPTD, between $32^{0/7}$ and $36^{6/7}$ weeks of gestation. The control group consisted of participants with termly delivered neonates. Cases and controls were matched with respect to maternal age and duration of storage at −80°. Only European women with singleton pregnancies who delivered a phenotypically normal live born neonate were included in the study. Women with missing information or insufficient plasma for analysis, those with chronic diseases or pregnancy complications other than sPTD (e.g., pre-eclampsia, gestational diabetes, fetal growth restriction) as well as women with PPROM or signs of intra-uterine infection/inflammation at the time of admission were excluded from the study.

Over the study period, a total of 1809 plasma samples were collected and stored at −80 °C. Through the database search, 63 women were identified to be diagnosed with sPTD. Of those, 28 were excluded from the study because of fetal chromosomal abnormality/major fetal malformations (n = 4), presence of a multiple pregnancy (n = 6), miscarriage or fetal death before 21 weeks or termination (n = 5), lost to follow-up (n = 4) or due to inadequate plasma sample (n = 9). Plasma sample of a pregnant woman who subsequently delivered preterm before the 32nd week of gestation (early sPTD) was excluded to ensure increased homogeneity among cohort. Different aliquots from the same cohort have been used previously in a proteomic study [35].

Finally, 34 plasma samples from women who subsequently delivered prematurely, between $32^{0/7}$ and $36^{6/7}$ weeks, fulfilled the inclusion criteria and were retrieved for analyses along with their matched controls. Selected samples were analyzed in two phases: samples from five women with a subsequent sPTD between $32^{0/7}$ and $36^{6/7}$ weeks of gestation and with their matched controls were analyzed using small RNA-seq (discovery cohort) to identify miRNAs with significantly aberrant expression levels between the compared groups. Based on discovery cohort data, the remaining 58 samples (29 sPTL cases and 29 controls) were used in the second phase of the study to confirm the differential expression of selected miRNAs. None of the samples were previously thawed and refrozen.

Written informed consent to collect and use the biological samples and clinical information for research purposes was obtained from all participants prior to inclusion in the study. The study was approved by Alexandra's Hospital ethics committee (P.N. 9/5-1-2018) and conducted according to the standards of the 1975 Declaration of Helsinki, as revised in 2008.

### 4.2. *Clinical Definitions*

sPTD is defined by the presence of regular uterine contractions (at least two uterine contractions every 10 min for 30 min, as confirmed by external tocometry) in combination with cervical changes occurring prior to $36^{6/7}$ weeks that required hospitalization [36]. A term pregnancy is defined as a delivery from 37 completed weeks to less than 42 and was used to describe the optimal timing for a good outcome for the mother and baby [37].

### 4.3. *Methods*

#### 4.3.1. Total RNA and miRN Isolation

Total RNA enriched in miRNAs was extracted from EDTA-preserved maternal plasma using the mirVana miRNA Isolation kit (Thermo Fisher, Waltham, MA, USA) according to the manufacturer's instructions. The RNA integrity was determined with polyacrylamide gel electrophoresis. The concentration and purity of RNA were checked by NanoDrop ND-1000 spectrophotometer (Thermo Fisher, Wilmington, DE, USA) and 2100 Bioanalyzer Instruments (Agilent, Santa Clara, CA, USA).

#### 4.3.2. Small RNA-seq

Profiling of circulating miRNA was performed by small RNA-seq in Illumina NextSeq 500 platform as previously described [24]. Briefly, small RNA libraries were generated using the TruSeq® miRNA Sample Prep kit v2 (Illumina, San Diego, CA, USA). RNA template of each sample was sequentially ligated to 3′ and 5′ adapters, reverse transcribed, PCR amplified and selected on agarose gels by size of ~130–150 bp (correspond to ~15–35nt small RNAs). Then, PCR amplified fragments were eluted from gel pieces, purified and quantified by Agilent 2100 Bioanalyzer (Agilent, Santa Clara, CA, USA). The completed libraries were diluted to a final concentration of 8 pM and processed for cluster generation on the Illumina cBOT using TruSeq Rapid SR cluster kit (Illumina, San Diego, CA, USA). Sequencing was performed on Illumina NextSeq 500 platform using TruSeq Rapid SRB kits (Illumina, San Diego, CA, USA).

#### 4.3.3. Sequencing Data Analysis

The complete raw sequences from Illumina NextSeq 500 were generated as clean reads by real-time base calling and quality filtering. The clean reads were recorded in FASTQ format, containing the read information, sequences and quality encoding. The 3′ prime adapter sequences were trimmed and reads with lengths <15 nucleotides were removed. Those with length >15nt were aligned to all human miRNAs (miRBase v21) using NovoAlign software (NovoCraft, Selangor, Malaysia) allowing maximum one mismatch per sequence [38]. The miRNA read counts were used to estimate the expression level of each miRNA. All counts were normalized by reads per million [39]. Differential expression and statistical analysis of sequencing data were performed by the DESeq2 package in R.

Circulating miRNAs having $p$-Value $\leq 0.05$ and FC $\geq 1.5$ were considered significant. $p$-Values were adjusted for multiple hypotheses testing (<0.01 compared to control) using the method of Benjamini–Hochberg to establish a false discovery rate (FDR).

#### 4.3.4. miRNAs Target Gene Prediction and Gene Ontology Analysis

Potential targets of the significantly dysregulated miRNAs were obtained using Targetscan7 (http://www.targetscan.org/vert_71/) and MirdbV5 (http://mirdb.org/miRDB/) algorithms accessed 21 January 2020. Only overlapping results between these databases were accepted as potential targets.

Targets gene were submitted to the WebGestalt web-tool http://bioinfo.vanderbilt.edu/webgestalt/ accessed on 21 January 2020 for Gene Ontology (GO) annotation and Kyoto Encyclopedia of Genes and Genomes (KEGG) signaling enrichment analyses [40,41]. GO terms and KEGG pathways having an adjusted $p$-Value $\leq 0.05$ were considered significant.

4.3.5. Quantitative Real-Time Polymerase Chain Reaction Verification

The expression levels of representative miRNAs, differentially expressed in the first trimester maternal plasma of women who later experienced sPTD as compared to the control group were verified in an independent cohort consisting of 58 plasma samples (n = 29 sPTL and n = 29 controls) using qRT-PCR. cDNA synthesis and qRT-PCR were performed using a TaqMan miRNA Reverse Transcription kit (Applied Biosystems, Inc., Foster City, CA 94404, USA) and individual TaqMan MiRNA assays (Applied Biosystems, Inc., USA), following the manufacturer's recommendations. All reactions were run in triplicate in a LC480 LightCycler system (Roche GmbH, Rotkreuz, Switzerland). The miRNA gene expression was determined using the $2^{-\Delta\Delta Ct}$ method [42]. RNU44 (Applied Biosystems, Inc., Foster City, CA 94404, USA) was used for normalization purposes.

4.3.6. Statistical Analysis

Statistical analyses were conducted in IBM SPSS Statistics 20 software (IBM Corp., Armonk, NY, USA). Comparisons of maternal demographic and clinical characteristics between the two groups were compared using Pearson chi-square test for the evaluation of categorical variables or the Mann–Whitney $U$ test for continuous variables. A two-tailed Fisher's exact test was applied to test GO and pathway enrichment of the target genes of the significant differentially expressed miRNAs. Receiver Operating Characteristic curves (ROC) were applied to evaluate the diagnostic value of each miRNA using the qRT-qPCR data. Binomial logistic regression analyses were performed using the occurrence of sPTD as the dependent variable and miRNA expression levels, previous sPTD, maternal smoking, maternal pre-pregnancy BMI, maternal age, fetal gender and mode of conception as independent variables. $p$-Values $< 0.05$ were considered significant.

## 5. Conclusions

The present study demonstrated that small RNA profiling allows for the identification of novel biomarkers in the first trimester maternal plasma that can potentially be used in clinical practice for early, minimally invasive, prediction of sPTD. Analysis revealed for the first time that the expression levels of circulating, miR-125a-5p, miR-23b and miR-4732-5p are significantly different in plasma samples collected from women who subsequently experience sPTD, as compared to uncomplicated pregnancies, implying their critical roles in the pathology of sPTD. Further studies should be performed for a more complete understanding of the topic. More importantly, the results obtained in our cohort demonstrated that miR-125a-5p may be used as reliable independent biomarkers to predict poor pregnancy outcomes.

**Supplementary Materials:** The following supporting information can be downloaded at: https://www.mdpi.com/article/10.3390/ijms232314972/s1. Supplement S1. MiRNAs with significant altered expression in the 1st trimester maternal plasma of women who later delivered preterm compared to term deliveries.

**Author Contributions:** D.M.: experimental work, data analysis, literature search and manuscript preparation, M.T.: sample and data collection, sample selection, literature search and manuscript preparation, M.A.: bioinformatics, statistical analyses and manuscript critical review, N.P.: sample selection, literature search and manuscript critical review, P.A.: sample selection, literature search and manuscript critical review, G.D.: sample and data collection, sample selection, and critical review A.K.: conceived and designed the experiments and provided critical review. All authors have read and agreed to the published version of the manuscript.

**Funding:** This research received no external funding.

**Institutional Review Board Statement:** The study was approved by the Alexandra's hospital ethics committees (P.N. 9/5-1-2018) and conducted according to the standards of the 1975 Declaration of Helsinki, as revised in 2008.

**Informed Consent Statement:** Written informed consent to collect and use the biological samples and clinical information for research purposes was obtained from all participants prior to inclusion in the study.

**Data Availability Statement:** The datasets used and/or analyzed during the current study are available from the corresponding author on reasonable request. Please also refer to the Section 4 "Materials and Methods".

**Conflicts of Interest:** The authors declare no conflict of interest.

## References


1. Goldenberg, R.L.; Culhane, J.F.; Iams, J.D.; Romero, R. Epidemiology and causes of preterm birth. Epidemiology and causes of preterm birth. *Lancet* **2008**, *371*, 75–84. [CrossRef]
2. Chawanpaiboon, S.; Vogel, J.P.; Moller, A.B.; Lumbiganon, P.; Petzold, M.; Hogan, D.; Landoulsi, S.; Jampathong, N.; Kongwattanakul, K.; Laopaiboon, M.; et al. Global, regional, and national estimates of levels of preterm birth in 2014: A systematic review and modelling analysis. *Lancet Glob. Health* **2019**, *7*, e37–e46. [CrossRef]
3. Natarajan, G.; Shankaran, S. Short- and Long-Term Outcomes of Moderate and Late Preterm Infants. *Am. J. Perinatol.* **2016**, *33*, 305–317. [PubMed]
4. Suff, N.; Story, L.; Shennan, A. The prediction of preterm delivery: What is new? *Semin Fetal Neonatal Med.* **2019**, *24*, 27–32. [CrossRef] [PubMed]
5. Beta, J.; Akolekar, R.; Ventura, W.; Syngelaki, A.; Nicolaides, K.H. Prediction of spontaneous preterm delivery from maternal factors, obstetric history and placental perfusion and function at 11–13 weeks. *Prenat. Diagn.* **2011**, *31*, 75–83. [CrossRef] [PubMed]
6. Greco, E.; Gupta, R.; Syngelaki, A.; Poon, L.C.; Nicolaides, K.H. First-trimester screening for spontaneous preterm delivery with maternal characteristics and cervical length. *Fetal Diagn. Ther.* **2012**, *31*, 154–161. [CrossRef]
7. Mellios, N.; Sur, M. The Emerging Role of microRNAs in Schizophrenia and Autism Spectrum. *Disord. Front. Psychiatry* **2012**, *25*, 39. [CrossRef]
8. Valadi, H.; Ekström, K.; Bossios, A.; Sjöstrand, M.; Lee, J.J.; Lötvall, J.O. Exosome-mediated transfer of mRNAs and microRNAs is a novel mechanism of genetic exchange between cells. *Nat. Cell Biol.* **2007**, *9*, 654–659. [CrossRef]
9. Hromadnikova, I.; Kotlabova, K.; Krofta, L. First Trimester Prediction of Preterm Delivery in the Absence of Other Pregnancy-Related Complications Using Cardiovascular-Disease Associated MicroRNA Biomarkers. *Int. J. Mol. Sci.* **2022**, *23*, 3951. [CrossRef]
10. Winger, E.E.; Reed, J.L.; Ji, X. Early first trimester peripheral blood cell microRNA predicts risk of preterm delivery in pregnant women: Proof of concept. *PLoS ONE* **2017**, *12*, e0180124. [CrossRef]
11. Winger, E.E.; Reed, J.L.; Ji, X.; Gomez-Lopez, N.; Pacora, P.; Romero, R. MicroRNAs isolated from peripheral blood in the first trimester predict spontaneous preterm birth. *PLoS ONE* **2020**, *15*, e0236805. [CrossRef] [PubMed]
12. Ran, Y.; Yin, N.; Huang, D.; Zhao, Y.; Yang, J.; Zhang, H.; Qi, H. Identification and Characterization of Circular RNA as a Novel Regulator and Biomarker in Preterm Birth. *Front. Bioeng. Biotechnol.* **2020**, *8*, 566984. [CrossRef] [PubMed]
13. Gray, C.; McCowan, L.M.; Patel, R.; Taylor, R.S.; Vickers, M.H. Maternal plasma miRNAs as biomarkers during mid-pregnancy to predict later spontaneous preterm birth: A pilot study. *Sci. Rep.* **2017**, *7*, 815. [CrossRef]
14. Menon, R.; Debnath, C.; Lai, A.; Guanzon, D.; Bhatnagar, S.; Kshetrapal, P.K.; Sheller-Miller, S.; Salomon, C. Garbhini Study Team. Circulating Exosomal miRNA Profile During Term and Preterm Birth Pregnancies: A Longitudinal Study. *Endocrinology* **2019**, *160*, 249–275. [CrossRef]
15. Enquobahrie, D.A.; Hensley, M.; Qiu, C.; Abetew, D.F.; Hevner, K.; Tadesse, M.G.; Williams, M.A. Candidate Gene and MicroRNA Expression in Fetal Membranes and Preterm Delivery Risk. *Reprod Sci.* **2016**, *23*, 731–737. [CrossRef]
16. Mavreli, D.; Theodora, M.; Kolialexi, A. Known biomarkers for monitoring pregnancy complications. *Expert Rev. Mol. Diagn.* **2021**, *21*, 1115–1117. [CrossRef]
17. Frey, H.A.; Klebanoff, M.A. The epidemiology, etiology, and costs of preterm birth. *Semin Fetal Neonatal Med.* **2016**, *21*, 68–73. [CrossRef]
18. Di Renzo, G.C.; Cabero Roura, L.; Facchinetti, F.; Helmer, H.; Hubinont, C.; Jacobsson, B.; Jørgensen, J.S.; Lamont, R.F.; Mikhailov, A.; Papantoniou, N.; et al. Preterm Labor and Birth Management: Recommendations from the European Association of Perinatal Medicine. *J. Matern Fetal Neonatal Med.* **2017**, *30*, 2011–2030. [CrossRef]
19. Qi, L.; Gao, C.; Feng, F.; Zhang, T.; Yao, Y.; Wang, X.; Liu, C.; Li, J.; Li, J.; Sun, C. MicroRNAs associated with lung squamous cell carcinoma: New prognostic biomarkers and therapeutic targets. *J. Cell Biochem.* **2019**, *120*, 18956–18966. [CrossRef] [PubMed]
20. Chen, X.; Yu, L.; Hao, K.; Yin, X.; Tu, M.; Cai, L.; Zhang, L.; Pan, X.; Gao, Q.; Huang, Y. Fucosylated exosomal miRNAs as promising biomarkers for the diagnosis of early lung adenocarcinoma. *Front. Oncol.* **2022**, *12*, 935184. [CrossRef]

21. Huang, S.; Liu, K.; Cheng, A.; Wang, M.; Cui, M.; Huang, J.; Zhu, D.; Chen, S.; Liu, M.; Zhao, X.; et al. SOCS Proteins Participate in the Regulation of Innate Immune Response Caused by Viruses. *Front. Immunol.* **2020**, *11*, 558341. [CrossRef] [PubMed]
22. Blumenstein, M.; Bowen-Shauver, J.M.; Keelan, J.A.; Mitchell, M.D. Identification of suppressors of cytokine signaling (SOCS) proteins in human gestational tissues: Differential regulation is associated with the onset of labor. *J. Clin. Endocrinol. Metab.* **2002**, *87*, 1094–1097. [CrossRef] [PubMed]
23. Blumenstein, M.; Keelan, J.A.; Bowen-Shauver, J.M.; Mitchell, M.D. Suppressors of cytokine signaling proteins in human preterm placental tissues. *J. Mol. Endocrinol.* **2005**, *35*, 165–175. [CrossRef] [PubMed]
24. Mavreli, D.; Lykoudi, A.; Lambrou, G.; Papaioannou, G.; Vrachnis, N.; Kalantaridou, S.; Papantoniou, N.; Kolialexi, A. Deep Sequencing Identified Dysregulated Circulating MicroRNAs in Late Onset Preeclampsia. *In Vivo* **2020**, *34*, 2317–2324. [CrossRef] [PubMed]
25. Ru, L.; Wang, X.M.; Niu, J.Q. The miR-23-27-24 cluster: An emerging target in NAFLD pathogenesis. *Acta Pharmacol. Sin.* **2022**, *43*, 1167–1179. [CrossRef] [PubMed]
26. Guo, Y.X.; Wang, N.; Wu, W.C.; Li, C.Q.; Chen, R.H.; Zhang, Y.; Li, X. The Role of miR-23b in Cancer and Autoimmune Disease. *J. Oncol.* **2021**, *2021*, 6473038. [CrossRef] [PubMed]
27. Bang, C.; Fiedler, J.; Thum, T. Cardiovascular importance of the microRNA-23/27/24 family. *Microcirculation* **2012**, *19*, 208–214. [CrossRef] [PubMed]
28. Wang, G.K.; Zhu, J.Q.; Zhang, J.T.; Li, Q.; Li, Y.; He, J.; Qin, Y.W.; Jing, Q. Circulating microRNA: A novel potential biomarker for early diagnosis of acute myocardial infarction in humans. *Eur. Heart J.* **2010**, *31*, 659–666. [CrossRef]
29. Azizieh, F.Y.; Raghupathy, R. L-10 and pregnancy complications. *Clin. Exp. Obstet. Gynecol.* **2017**, *44*, 252–258. [CrossRef]
30. Li, S.; Zheng, S. Down-Regulation of Circ_0032833 Sensitizes Colorectal Cancer to 5-Fluorouracil and Oxaliplatin Partly Depending on the Regulation of miR-125-5p and MSI1. *Cancer Manag. Res.* **2020**, *12*, 11257–11269. [CrossRef]
31. Fu, K.; Zhang, L.; Liu, R.; Shi, Q.; Li, X.; Wang, M. MiR-125 inhibited cervical cancer progression by regulating VEGF and PI3K/AKT signaling pathway. *World J. Surg. Oncol.* **2020**, *18*, 115. [CrossRef]
32. Szebeni, G.J.; Vizler, C.; Kitajka, K.; Puskas, L.G. Inflammation and Cancer: Extra- and Intracellular Determinants of Tumor-Associated Macrophages as Tumor Promoters. *Mediat. Inflamm.* **2017**, *2017*, 9294018. [CrossRef] [PubMed]
33. 33 Wade, S.M.; Ohnesorge, N.; McLoughlin, H.; Biniecka, M.; Carter, S.P.; Trenkman, M.; Cunningham, C.C.; McGarry, T.; Canavan, M.; Kennedy, B.N.; et al. Dysregulated miR-125a promotes angiogenesis through enhanced glycolysis. *EBioMedicine* **2019**, *47*, 402–413. [CrossRef]
34. Hanna, H.; Hossain, G.S.; Kocerha, J. The Potential for microRNA Therapeutics and Clinical Research. *Front. Genet.* **2019**, *16*, 478. [CrossRef]
35. Mavreli, D.; Theodora, M.; Lambrou, G.; Avgeris, M.; Papantoniou, N.; Treager-Synodinos, J.; Daskalakis, G.; Kolialexi, A. First trimester maternal plasma proteomic changes predictive of spontaneous moderate/late preterm delivery. *J. Matern Fetal Neonatal Med.* 2017, submitted.
36. Tsiartas, P.; Holst, R.M.; Wennerholm, U.B.; Hagberg, H.; Hougaard, D.M.; Skogstrand, K.; Pearce, B.D.; Thorsen, P.; Kacerovsky, M.; Jacobsson, B. Prediction of spontaneous preterm delivery in women with threatened preterm labour: A prospective cohort study of multiple proteins in maternal serum. *BJOG* **2012**, *119*, 866–873. [CrossRef] [PubMed]
37. ACOG Committee Opinion No 579: Definition of term pregnancy. *Obstet. Gynecol.* **2013**, *122*, 1139–1140. [CrossRef]
38. Kozomara, A.; Griffiths-Jones, S. miRBase: Annotating high confidence microRNAs using deep sequencing data. *Nucleic Acids Res.* **2014**, *42*, D68. [CrossRef] [PubMed]
39. Bullard, J.H.; Purdom, E.; Hansen, K.D.; Dudoit, S. Evaluation of statistical methods for normalization and differential expression in mRNA-Seq experiments. *BMC Bioinform.* **2010**, *11*, 94. [CrossRef] [PubMed]
40. Zhang, B.; Kirov, S.; Snoddy, J. WebGestalt: An integrated system for exploring gene sets in various biological contexts. *Nucleic Acids Res.* **2005**, *33*, W741–W748. [CrossRef]
41. Wang, J.; Duncan, D.; Shi, Z.; Zhang, B. WEB-based GEne SeT AnaLysis Toolkit (WebGestalt): Update 2013. *Nucleic Acids Res.* **2013**, *41*, W77–W83. [CrossRef] [PubMed]
42. Livak, K.J.; Schmittgen, T.D. Analysis of relative gene expression data using real-time quantitative PCR and the 2(-Delta Delta C(T) Method. *Methods* **2001**, *25*, 402–408. [CrossRef] [PubMed]

International Journal of
*Molecular Sciences*

MDPI

*Article*

# Association of the *FCN2* Gene Promoter Region Polymorphisms with Very Low Birthweight in Preterm Neonates

**Agnieszka Szala-Poździej [1,†], Anna S. Świerzko [1,†], Gabriela Gajek [1], Maja Kufelnicka-Babout [2], Karolina Chojnacka [3], Paulina Kobiela [4], Dariusz Jarych [1], Katarzyna Sobczuk [2], Jan Mazela [5], Iwona Domżalska-Popadiuk [4], Jarosław Kalinka [2], Hideharu Sekine [6], Misao Matsushita [7] and Maciej Cedzyński [1,*]**

1 Laboratory of Immunobiology of Infections, Institute of Medical Biology, Polish Academy of Sciences, Lodowa 106, 93-232 Łódź, Poland
2 Department of Perinatology, First Chair of Gynaecology and Obstetrics, Medical University of Łódź, Wileńska 37, 94-029 Łódź, Poland
3 II Department of Neonatology, Poznań University of Medical Sciences, Polna 33, 60-535 Poznań, Poland
4 Department of Neonatology, Medical University of Gdańsk, Smoluchowskiego 17, 80-214 Gdańsk, Poland
5 Department of Neonatology, Poznań University of Medical Sciences, Polna 33, 60-535 Poznań, Poland
6 Department of Immunology, Fukushima Medical University School of Medicine, 1 Hikariga-oka, Fukushima City 960-1295, Japan
7 Department of Applied Biochemistry, Tokai University, 4-1-1 Kitakaname, Hiratsuka 259-129, Kanagawa, Japan
* Correspondence: mcedzynski@cbm.pan.pl
† These authors contributed equally to this work and share the first authorship.

**Citation:** Szala-Poździej, A.; Świerzko, A.S.; Gajek, G.; Kufelnicka-Babout, M.; Chojnacka, K.; Kobiela, P.; Jarych, D.; Sobczuk, K.; Mazela, J.; Domżalska-Popadiuk, I.; et al. Association of the *FCN2* Gene Promoter Region Polymorphisms with Very Low Birthweight in Preterm Neonates. *Int. J. Mol. Sci.* **2022**, *23*, 15336. https://doi.org/10.3390/ijms232315336

Academic Editor: Ilona Hromadnikova

Received: 30 September 2022
Accepted: 3 December 2022
Published: 5 December 2022



**Abstract:** Single nucleotide polymorphisms (SNPs) localised to the promoter region of the *FCN2* gene are known to influence the concentration of ficolin-2 in human serum and therefore potentially have clinical associations. We investigated the relationships between SNPs at positions −986 (A > G), −602 (G > A), −64 (A > C) and −4 (A > G) and clinical complications in 501 preterms. Major alleles at positions −986 and −64 and A/A homozygosity for both polymorphisms were less frequent among babies with very low birthweight (VLBW, ≤1500 g) compared with the reference group (OR = 0.24, $p = 0.0029$; and OR = 0.49, $p = 0.024$, respectively for A/A genotypes). A lower frequency of G/G homozygosity at position −4 was associated with gestational age <33 weeks and VLBW (OR = 0.38, $p = 0.047$; and OR = 0.07, $p = 0.0034$, respectively). The AGAG haplotype was protective for VLBW (OR = 0.6, $p = 0.0369$), whilst the GGCA haplotype had the opposite effect (OR = 2.95, $p = 0.0249$). The latter association was independent of gestational age. The AGAG/GGAA diplotype favoured both shorter gestational age and VLBW (OR = 1.82, $p = 0.0234$ and OR = 1.95, $p = 0.0434$, respectively). In contrast, AGAG homozygosity was protective for lower body mass (OR = 0.09, $p = 0.0155$). Our data demonstrate that some *FCN2* variants associated with relatively low ficolin-2 increase the risk of VLBW and suggest that ficolin-2 is an important factor for fetal development/intrauterine growth.

**Keywords:** ficolin-2; *FCN2*; newborn; neonate; prematurity; single nucleotide polymorphism; very low birthweight

## 1. Introduction

The single nucleotide polymorphisms (SNPs) localised to the promoter region of the *FCN2* gene are known to influence the concentration of ficolin-2 (or L-ficolin) in human serum and therefore are considered to have clinical associations. Possession of variant alleles at positions −986 (rs3124952, A > G) and −64 (rs7865453, A > C) is related to lower ficolin-2 levels, while minor alleles at positions −602 (rs3124953, G > A) and −4 (rs17514136, A > G) have the opposite effect [1–3]. The above-mentioned SNPs were shown to form 2 haplotype blocks: one created by rs3124952 and rs3124953, and another, by rs7865453 and rs17514136 [4].

Like other ficolins (ficolin-1, ficolin-3) and some collectins [mannose-binding lectin (MBL), collectin-10 (CL-10), collectin-11 (CL-11)], ficolin-2 recognises pathogen-associated molecular patterns (PAMP) exposed on a variety of microorganisms or viruses that enable opsonisation and, by forming complexes with MBL-associated serine proteases (MASP), activation of the complement cascade via the lectin pathway (reviewed in [5–7]).

In 2021, in Poland, 24,523 (7.4%) babies were born prematurely, including 1369 delivered before completing 28 weeks of gestation, 2376 at gestational age 28–31 weeks and 20,778 aged 32–36 weeks [8]. Preterm newborns, especially those born at gestational age < 33 weeks and with very low birthweight ($\leq$1500 g), are prone to severe adverse effects, including respiratory distress syndrome (RDS) and perinatal infections, due to immaturity of organs, immune system and other congenital defects. Disorders related to short gestation and low birthweight are the most common causes of neonatal death [9]. Furthermore, very low birthweight ($\leq$1500 g) is considered to continue to influence health during childhood, adolescence and adulthood. It was reported to be associated, for example, with higher risk of coronary heart disease, chronic kidney disease, type 2 diabetes, stroke and hypertension (reviewed in [10]); gout development [11]; brain abnormalities [12]; stunting in preschool children [13] and low bone mineral density [14].

Our previous data suggested that low ficolin-2 concentration (determined in cord serum) is associated with prematurity, low birthweight and perinatal infections [15]. Recently, we reported for the first time a relationship between SNPs of the *FCN2* gene 3′-untranslated region (3′UTR) and very low ($\leq$1500 g) birthweight as well as early onset of infection and pneumonia in preterm newborns [4]. Furthermore, two of the 3′UTR polymorphisms (rs4521835 and rs73664188) influenced ficolin-2 concentration in cord sera [4]. Here, we report associations of the aforementioned promoter region polymorphism with short gestational age and very low birthweight in the same cohort.

## 2. Results

The frequencies of genotypes corresponding to SNPs at positions −986 (rs3124952), −602 (rs3124953), −64 (rs7865453) and −4 (rs17514136) of the *FCN2* gene as well as minor allele frequencies (MAF) in preterm neonates are listed in Table 1. Each SNP adhered to Hardy–Weinberg expectations ($p > 0.01$, details are given in Supplementary Table S1). As mentioned, they created two haplotype blocks (rs3124952 and rs3124953; rs7865453 and rs17514136) (Supplementary Figure S1). None appeared associated with incidence of RDS, early- or late-onset perinatal infections, sepsis or pneumonia (Supplementary Table S2). However, major (A) alleles at positions −986 and −64 as well as A/A homozygosity for both polymorphisms were significantly less frequent among babies born with very low birthweight (VLBW) ($\leq$1500 g) when compared with the corresponding reference group. Those relationships remained significant after multiple logistic regression analysis (Table 2). Moreover, lower frequency of G/G (minor allele) homozygosity at position −4 was found to be associated not only with low body mass but also with shorter gestation (Table 2). Those associations were however not confirmed by multiple logistic regression ($p > 0.05$). It should be stressed that minor variants at both −986 and −64 positions are associated with lower ficolin-2 concentration in serum compared to A alleles while the G variant at −4 is associated with a higher *FCN2* gene expression level. VLBW was not significantly related to the sex of newborns, although a trend towards its higher incidence in girls was observed (13.5% vs. 8.7%, $p = 0.087$). The numbers of females and males born at gestational age <33 weeks did not differ significantly (23.3% vs. 18.9%, $p = 0.22$).

**Table 1.** Distribution of genotypes associated with *FCN2* gene promoter polymorphisms in preterm newborns (n = 501).

| Polymorphism | Genotype | N | % | MAF |
|---|---|---|---|---|
| rs3124952 −986 A > G | A/A<br>A/G<br>G/G | 165<br>265<br>61 | 32.9<br>52.9<br>12.2 | 0.386 |
| rs3124953 −602 G > A | G/G<br>G/A<br>A/A | 314<br>170<br>17 | 62.7<br>33.9<br>3.4 | 0.204 |
| rs7865453 −64 A > C | A/A<br>A/C<br>C/C | 390<br>107<br>4 | 77.8<br>21.4<br>0.8 | 0.115 |
| rs17514136 −4 A > G | A/A<br>A/G<br>G/G | 204<br>243<br>51 | 40.7<br>48.5<br>10.2 | 0.347 |

**Table 2.** Distribution of genotypes associated with *FCN2* gene promoter polymorphisms in preterm newborns, depending on gestational age and birthweight.

| | | Gestational Age (Weeks) | | Birthweight (g) | |
|---|---|---|---|---|---|
| Polymorphism | Genotype | <33 | ≥33 | ≤1500 | >1500 |
| | | N (%) | N (%) | N (%) | N (%) |
| rs3124952 −986 A > G | A/A<br>A/G<br>G/G | 32 (30.5)<br>61 (58.1)<br>12 (11.4) | 143 (36.1)<br>204 (51.5)<br>49 (12.4) | 7 (12.7) [2]<br>39 (70.9)<br>9 (16.4) | 168 (37.9)<br>223 (50.3)<br>52 (11.7) |
| rs3124953 −602 G > A | G/G<br>G/A<br>A/A | 69 (65.7)<br>32 (30.5)<br>4 (5.8) | 245 (61.9)<br>138 (34.8)<br>13 (3.3) | 35 (63.6)<br>20 (36.4)<br>0 (0) | 276 (62.3)<br>150 (33.9)<br>17 (3.8) |
| rs7865453 −64 A > C | A/A<br>A/C<br>C/C | 83 (79)<br>22 (21)<br>0 (0) | 307 (77.5)<br>85 (21.5)<br>4 (1) | 36 (65.5) [3]<br>18 (32.7)<br>1 (1.8) | 352 (79.5)<br>88 (19.9)<br>3 (0.7) |
| rs17514136 −4 A > G | A/A<br>A/G<br>G/G | 38 (36.2)<br>62 (59)<br>5 (4.8) [1] | 166 (41.9)<br>184 (46.5)<br>46 (11.6) | 26 (47.3)<br>29 (52.7)<br>0 (0) [4] | 177 (40)<br>215 (48.5)<br>51 (11.5) |

[1]—OR = 0.38, 95% CI (0.15—0.98), $p = 0.0447$. [2]—OR = 0.24. 95% CI (0.11—0.54), $p = 0.0029$; OR = 0.25, 95% CI (0.11–0.61), $p = 0.002$, after multiple logistic regression; A allele frequency: 0.482 vs. 0.631; OR = 0.54. 95% CI (0.37—0.81), $p = 0.0035$. [3]—OR = 0.49. 95% CI (0.27—0.89), $p = 0.0244$; OR = 0.33, 95% CI (0.15–0.71), $p = 0.005$, after multiple logistic regression; A allele frequency: 0.818 vs. 0.894; OR = 0.53, 95% CI (0.31—0.91), $p = 0.0254$. [4]—OR = 0.07, 95% CI (0.004—1.13), $p = 0.0034$.

Eleven promoter haplotypes were identified with the help of Haploview software, including three with MAF > 0.1 and two with a marginally lower value (Table 3). Their frequencies, depending on gestational age and body mass at birth, are shown in Table 4. The most common haplotype (AGAG) may be considered protective from very low birthweight, whilst the fifth most frequent (GGCA) seemed to have the opposite effect, confirmed by multiple logistic regression analysis as well (Table 4). That effect appeared independent of gestational age: the frequency of the GGCA variant was significantly higher among babies with very low birthweight compared with the corresponding reference group, born at <33 [0.146 vs. 0.048, OR = 3.43, 95% CI (1.23–9.54), $p = 0.021$] as well as ≥33 weeks [0.214 vs. 0.096, OR = 2.57, 95% CI (1.01–6.53), $p = 0.04$], respectively. Again, no association of any haplotype with RDS, infections, sepsis, pneumonia (Supplementary Table S3) or shorter gestational age was noted (Table 4).

**Table 3.** Frequencies of haplotypes identified in preterm newborns (n = 501).

| Haplotype | N | Frequency |
|---|---|---|
| AGAG | 307 | 0.306 |
| GGAA | 244 | 0.244 |
| AAAA | 192 | 0.192 |
| AGAA | 99 | 0.099 |
| GGCA | 98 | 0.098 |
| GGAG | 41 | 0.041 |
| AGCA | 6 | 0.006 |
| AACA | 5 | 0.005 |
| GACA | 4 | 0.004 |
| GAAA | 4 | 0.004 |
| AGCG | 2 | 0.002 |

**Table 4.** Frequencies of the most common haplotypes in preterm newborns, depending on gestational age and birthweight.

| Haplotype | Gestational Age (Weeks) | | | | Birthweight (g) | | | |
|---|---|---|---|---|---|---|---|---|
| | <33 | | ≥33 | | ≤1500 | | >1500 | |
| | N | Frequency [1] | N | Frequency [1] | N | Frequency [1] | N | Frequency [1] |
| AGAG | 65 | 0.31 | 242 | 0.306 | 24 [2] | 0.218 | 281 | 0.317 |
| GGAA | 57 | 0.271 | 187 | 0.236 | 35 | 0.318 | 207 | 0.234 |
| AAAA | 38 | 0.181 | 154 | 0.194 | 18 | 0.164 | 174 | 0.196 |
| AGAA | 20 | 0.095 | 79 | 0.1 | 7 | 0.064 | 91 | 0.103 |
| GGCA | 19 | 0.09 | 79 | 0.1 | 18 [3] | 0.164 | 79 | 0.089 |

[1]—haplotype frequency among newborns born at GA <33/≥33 weeks; with birthweight ≤1500/>1500 g, respectively. [2]—OR = 0.6, 95% CI (0.37—0.97), $p = 0.0369$; OR = 0.5, 95% CI (0.25–0.99), $p = 0.047$, after multiple logistic regression. [3]—OR = 2.95, 95% CI (1.15—3.48), $p = 0.0249$; OR = 3.53, 95% CI (1.57–7.97), $p = 0.002$, after multiple logistic regression.

Further analysis using PHASE software revealed thirty diplotypes, although the frequency of half of them was less than 1% (Table 5). Interestingly, the most common one, AGAG/GGAA, was associated with adverse events (shorter gestational age and very low birthweight) (Table 6). However, after multiple logistic regression analysis, that relationship lost statistical significance ($p > 0.05$). Furthermore, diplotypes possessing the GGCA haplotype (5, 6, 11, 14, 20, 24, 30, see Table 5) were significantly more frequent among neonates with birthweight ≤1500 g [17/55 (30.9%)] compared with the corresponding reference group [76/443 (17.2%)] [OR = 2.16, $p = 0.0175$, 95% CI (1.16–4.03)]. In contrast, AGAG homozygosity was found to be protective from lower body mass (Table 6), although that association lost significance in multiple logistic regression analysis ($p > 0.05$). There was also a trend for more babies with AGAG homozygosity to be born after 33 weeks of gestation (Table 6). None of the ten most common diplotypes was associated with RDS, infections, sepsis or pneumonia (Supplemetary Table S4).

We investigated the relationship of the ten most common diplotypes with ficolin-2 concentration in cord serum. The Kruskal–Wallis ANOVA revealed significant differences among genotypes (Supplementary Figure S2). Diplotype 2 (AGAG/AAAA, 2748 ng/mL) had a higher median than the others, although the difference between it and diplotype 3 did not quite reach statistical significance ($p = 0.067$). Diplotype 5 (AGAG/GGCA) had the lowest median at 1415 ng/mL (Supplementary Figure S2). However, we found no clinical associations with either diplotype 2 or 5 (at least when analysed individually) (Table 6). It is worth noting that, in general, diplotypes including the GGCA haplotype were associated with relatively low serum ficolin-2. Nevertheless, wide ranges were found for most diplotypes (Table 5; Supplementary Figure S2).

**Table 5.** Frequencies of diplotypes identified in preterm newborns (n = 501).

| | Diplotype | N | % | Ficolin-2 Concentration (ng/mL) | |
|---|---|---|---|---|---|
| | | | | Median | Range (n) |
| 1 | AGAG/GGAA | 94 | 18.8 | 1761 | 237–5068 (80) |
| 2 | AGAG/AAAA | 60 | 12 | 2748 | 481–5235 (56) |
| 3 | AAAA/GGAA | 46 | 9.2 | 2327 | 803–5166 (39) |
| 4 | AGAG/AGAG | 41 | 8.2 | 1950 | 632–5299 (38) |
| 5 | AGAG/GGCA | 33 | 6.6 | 1415 | 430–4081 (30) |
| 6 | AAAA/GGCA | 29 | 5.8 | 1900 | 504–5644 (27) |
| 7 | GGAA/GGAA | 28 | 5.6 | 1743 | 153–4772 (27) |
| 8 | AGAA/GGAA | 27 | 5.4 | 2192 | 372–5408 (25) |
| 9 | AGAA/AGAG | 24 | 4.8 | 1785 | 479–4426 (22) |
| 10 | AAAA/GGAG | 21 | 4.2 | 2159 | 407–4199 (19) |
| 11 | GGAA/GGCA | 17 | 3.4 | 1098 | 242–2157 (16) |
| 12 | AGAA/AGAA | 16 | 3.2 | 2165 | 853–5481 (16) |
| 13 | AAAA/AAAA | 15 | 3 | 2323 | 690–4038 (15) |
| 14 | AGAA/GGCA | 8 | 1.6 | 1479 | 480–3063 (8) |
| 15 | GGAG/GGAG | 6 | 1.2 | 2134 | 706–4165 (5) |
| 16 | AGAA/AAAA | 4 | 0.8 | 3240 | 1455–4954 (4) |
| 17 | AGAA/AGCA | 4 | 0.8 | 2562 | 1737–2733 (3) |
| 18 | AGAG/AACA | 4 | 0.8 | 2495 | 1698–2919 (4) |
| 19 | AGAG/GACA | 4 | 0.8 | 1105 | 652–1563 (4) |
| 20 | GGCA/GGCA | 4 | 0.8 | 681 | 331–947 (3) |
| 21 | AGAG/GGAG | 3 | 0.6 | 2120 | 1387–4756 (3) |
| 22 | GGAA/GGAG | 3 | 0.6 | 1745 | 312–2195 (3) |
| 23 | AGAG/AGCG | 2 | 0.4 | 2002 | 520–3483 (2) |
| 24 | GGAG/GGCA | 2 | 0.4 | 907 | 387–1426 (2) |
| 25 | AAAA/AACA | 1 | 0.2 | 2210 | 2210 (1) |
| 26 | AGAG/AGCA | 1 | 0.2 | 2221 | 2221 (1) |
| 27 | AGCA/AAAA | 1 | 0.2 | 239 | 239 (1) |
| 28 | GAAA/GAAA | 1 | 0.2 | 3531 | 3531 (1) |
| 29 | GGAA/GAAA | 1 | 0.2 | 2194 | 2194 (1) |
| 30 | GGCA/GAAA | 1 | 0.2 | 937 | 937 (1) |

**Table 6.** Frequencies of the most common diplotypes in preterm newborns, depending on gestational age and birthweight.

| | Diplotype | Gestational Age (Weeks) | | | | Birthweight (g) | | | |
|---|---|---|---|---|---|---|---|---|---|
| | | <33 | | ≥33 | | ≤1500 | | >1500 | |
| | | N | % [1] | N | % [1] | N | % [1] | N | % [1] |
| 1 | AGAG/GGAA | 28 [2] | 26.7 | 66 | 16.7 | 16 [3] | 29.1 | 77 | 17.4 |
| 2 | AGAG/AAAA | 11 | 10.5 | 49 | 12.4 | 3 | 5.5 | 57 | 12.9 |
| 3 | AAAA/GGAA | 9 | 8.6 | 37 | 9.3 | 7 | 12.7 | 39 | 8.8 |
| 4 | AGAG/AGAG | 5 | 4.8 | 36 | 9.1 | 0 | 0 [4] | 41 | 9.3 |
| 5 | AGAG/GGCA | 6 | 5.7 | 27 | 6.8 | 3 | 5.5 | 30 | 6.8 |
| 6 | AAAA/GGCA | 5 | 4.8 | 24 | 6.1 | 5 | 9.1 | 24 | 5.4 |
| 7 | AGAA/GGAA | 7 | 6.7 | 20 | 5.1 | 2 | 3.6 | 25 | 5.6 |
| 8 | GGAA/GGAA | 4 | 3.8 | 24 | 6.1 | 3 | 5.5 | 25 | 5.6 |
| 9 | AGAA/AGAG | 9 | 8.6 | 15 | 3.8 | 1 | 1.8 | 22 | 5 |
| 10 | AAAA/GGAG | 4 | 3.8 | 17 | 4.3 | 2 | 3.6 | 19 | 4.3 |

[1]—percentages of diplotype 1–10 carriers among newborns born at GA <33/≥33 weeks; with birthweight ≤1500/>1500 g, respectively. [2]—OR = 1.82, 95% CI (1.1—3.02), $p = 0.0243$. [3]—OR = 1.95, 95% CI (1.04—3.67), $p = 0.0434$. [4]—OR = 0.087, 95% CI (0.005—1.44), $p = 0.0155$.

## 3. Discussion

Genome-wide association studies (GWAS) have enabled identification of a variety of loci/SNPs associated with gestational age and birthweight. The majority of reports concerns maternal genome analysis while data from newborns are relatively scarce.

Tiensuu et al. [16] found an association of rs116461311 polymorphism (*SLIT2* gene, encoding slit guidance ligand 2) with spontaneous preterm birth. Furthermore, they observed higher expression of the SLIT2 protein and its receptor ROBO1 in placentas from preterm deliveries compared with those from term births. The SLIT2-ROBO1 signaling pathway is involved, among others, in regulation of expression of genes associated with inflammation [16]. Rappoport et al. [17], based on the analysis of >2 million SNPs in five populations, reported only two loci to be significantly related to prematurity: rs17591250 and rs1979081 in African and American populations, respectively. Later, Huusko et al. [18], based on GWAS and other methods, identified genes encoding heat shock proteins and nuclear receptors (*SEC63, HSPA1L, SACS, RORA*, and *AR*) to be associated with spontaneous preterm birth. Using another approach, whole exome sequencing (WES), Modi et al. [19] proposed candidate genes in which mutations were found to be risk factors for preterm premature rupture of membranes (pPROM), one of the major causes of prematurity. Those genes (*CARD6, CARD8, DEFB1, FUT2, MBL2, NLP10, NLRP12* and *NOD2*) are involved in host defence. Interestingly, data concerning association of the *MBL2* gene [encoding mannose-binding lectin (MBL), structurally and functionally related to ficolins] polymorphisms with preterm birth are contradictory. Several reports suggested MBL deficiency to be a risk factor [20–22]. In contrast, Swierzko et al. [15] found high MBL concentration/activity-conferring genotypes to be associated with prematurity.

A GWAS analysis concerning birthweight performed by Luo et al. [23] identified a variety of loci in both maternal and fetal genomes potentially affecting this parameter in four populations (Afro-Caribbean, European, Hispanic and Thai). Extensive meta-analyses [24,25], identified a variety of loci where fetal genotype was associated with birthweight and found their associations with height, body-mass index and some metabolic diseases in adulthood.

The role of ficolin-2 in neonatal health and disease has not been studied extensively. Kilpatrick et al. [26] first reported lower concentrations in cord sera compared with sera from adult donors. Furthermore, ficolin-2 levels correlated positively with both gestational age and birthweight. Later, Swierzko et al. [15] confirmed those findings with a large (>1800) cohort of newborns. An association of low ficolin-2 with prematurity was further reported by Schlapbach et al. [27], Sallenbach et al. [28] and Kilpatrick et al. [3]. However, Briana et al. [29] observed no impact of ficolin-2 concentration on intrauterine growth restriction in full-term newborns.

Ficolin-2 is known to recognise a variety of pathogens, including group B streptococci, pneumococci and enteroaggregative *E. coli*, that can cause severe infections in newborns and/or infants [30–33]. Cord serum concentrations of this protein <1 μg/mL were found significantly more often among preterm babies with perinatal infections, compared with gestational-age-matched controls [15]. Later, we reported markedly lower ficolin-2 levels in neonates suffering from perinatal sepsis versus those without infections before hospital discharge [34]. On the other hand, Schlapbach et al. [27] did not find such an association.

Much less data concerning the role of *FCN2* gene polymorphisms, including those affecting ficolin-2 concentration, in neonates has been published to date. Our previous report [3] demonstrated that the genotype A/G-G/G-A/A-A/A-A/G-C/T-G/G (corresponding to SNPs at positions −986, −602, −557, −64, −4, +6369 and +6424) was the most common among Polish newborns. That genotype corresponds to the commonest AGAG/GGAA diplotype described in this paper on the basis of analysis with PHASE software (version 2.2.1.) (Table 5). Our current data from a large cohort of preterm babies found it to be associated with a relatively high risk of short (<33 weeks) gestational age and very low (≤1500 g) body mass at birth.

It should be stressed that AGAG/AGAG homozygosity seems protective from very low birthweight (Table 6). It differs from the AGAG/GGAA in one haplotype only, by possessing a major allele (A) at rs3124952 (−986) and a minor one (G) at rs17514136 (−4), both related to higher ficolin-2 concentration. The possible causal relationship between ficolin-2 concentration and outcome is strengthened by the association of the GGCA haplotype (generally associated with low serum ficolin-2) with very low birthweight (Table 4).

Furthermore, when each polymorphic site was analysed separately, homozygosity for major alleles at −986 (rs3124952) and −64 (rs7865453) and minor allele at −4 (rs17514136), associated with higher ficolin-2 levels, appears protective from very short gestational age or very low birthweight (Table 2). Although no impact of the SNPs, haplo- or diplotypes investigated here on such adverse effects of prematurity as perinatal infections, sepsis, pneumonia or RDS was found (Supplementary Tables S2–S4), it does not exclude a possible influence of low ficolin-2 on such complications. The *FCN2* gene is highly polymorphic and the concentration of its product depends on the interplay between SNPs localised to the promoter, exon 8 and 3′UTR regions and, possibly, epigenetic mechanisms. Furthermore, the ficolin-2 protein has several active sites and genetic changes can influence both concentration and activity, making for a very complex situation. Individuals may possess genetic variants that influence the ficolin-2 level in opposite directions [35]. Together with results published previously by ourselves and others, the data presented here are consistent with the view that ficolin-2 is an important factor for fetal development and neonatal immunity. The most important message from the data presented here is that an association of the *FCN2* gene promoter polymorphisms with very low birthweight may have potentially severe clinical consequences not only in the neonatal period but also during later life. Supplementary Figure S3 shows an interplay between four investigated SNPs, corresponding haplo- and diplotypes, ficolin-2 concentrations and the aforementioned adverse effects of prematurity.

## 4. Materials and Methods

### 4.1. Cohort

The study group comprised 501 Polish preterm newborns born in the Department of Newborns' Infectious Diseases (University of Medical Sciences, Poznań, Poland), Department of Neonatology (Medical University of Gdańsk, Gdańsk, Poland) and Department of Perinatology (Medical University of Łódź, Łódź, Poland) [4]. Among them, 105 were born at gestational age < 33 weeks (mean: 30.3 ± 1.9; range: 24–32) and 396 were born between the 33rd and 37th week of gestation (mean: 35 ± 1.1). Fifty-five had very low birthweight (≤1500 g, according to WHO International Classification of Diseases). A total of 323 newborns came from singleton pregnancies, 172 from 97 twin pregnancies (in 22, material from only one sibling was collected) and 6 from 2 triple pregnancies. Data concerning the *FCN2* gene 3′UTR polymorphisms, concentrations of ficolin-2 in cord sera and their clinical associations were published recently [4]. However, 3 subjects were excluded from current analyses due to incomplete results of promoter SNP analysis. As mentioned, promoter SNPs analysed here were previously reported to form 2 haplotype blocks: one created by rs3124952 and rs3124953, and another, by rs7865453 and rs17514136 [4]. The study was approved by the corresponding local ethics committees: Bioethics Committee of The Karol Marcinkowski Poznań University of Medical Sciences, Independent Bioethics Committee for Scientific Research at The Medical University of Gdańsk, Bioethics Committee of The Medical University of Łódź. Written informed parental consent was obtained. This work conforms to the provisions of the Declaration of Helsinki.

### 4.2. Blood Samples and DNA Isolation

Cord blood samples for genomic DNA isolation were taken consecutively into tubes with sodium citrate and stored at −80 °C. DNA was isolated using GeneMATRIX Quick Blood Purification Kit (EURx Ltd. Gdańsk, Poland), according to the manufacturer's protocol. Blood for serum isolation was placed in tubes containing clot activator. Samples were kept at −80 °C.

### 4.3. Determination of the FCN2 Gene Polymorphisms

Promoter polymorphisms at positions −986 (rs3124952, A > G) and −602 (rs3124953, G > A) were investigated by PCR-RFLP analysis, according to the procedures published by

Metzger et al. [36]. SNPs at positions −64 (rs7865453, A > C) and −4 (rs17514136, A > G) were determined using allele-specific PCR or PCR-RFLP, respectively, as described by Szala et al. [37], with minor modifications.

### 4.4. Determination of Ficolin-2 Concentration in Cord Sera

Ficolin-2 concentrations in cord serum samples were determined in TRIFMA as described by Świerzko et al. [38], using specific mAb (ABS 005-16, BioPorto Diagnostics, Denmark) for coating and another biotinylated mAb (GN4, Hycult Biotech, Uden, The Netherlands) and $Eu^{3+}$-labelled streptavidin (Perkin Elmer, Waltham, MA, USA) for detection.

### 4.5. Statistical Analysis

Linkage disequilibrium (LD) and haplotype block analysis were performed by Haploview 4.2 software (http://www.broad.mit.edu/mpg/haploview/, accessed on 30 June 2022). LD analysis was performed for each pair of polymorphisms using D′ and $r^2$, indicating the amount of LD between two genetic loci. Haplotype block identification was performed based on the Four Gamete Rule. The PHASE software (http://stephenslab.uchicago.edu/phase/download.html, accessed on 30 June 2022; version 2.1.1.) was used for diplotype reconstruction from genotype data. The frequencies of genotypes were compared by Fischer's exact (two-tailed) test. Ficolin-2 concentrations were compared with Kruskal–Wallis ANOVA and Mann–Whitney *U* tests. The Statistica (version 13.3, TIBCO Software) and SigmaPlot (version 12, Systat Software) software packages were used for data management and statistical calculations. Odds ratio was calculated using online MedCalc software (https://www.medcalc.org, accessed on 30 June 2022). $p$ values < 0.05 were considered statistically significant.

**Supplementary Materials:** The following supporting information can be downloaded at https://www.mdpi.com/article/10.3390/ijms232315336/s1. Table S1. Hardy–Weinberg expectation statistics for investigated *FCN2* gene promoter polymorphisms. SNPs were considered to adhere to Hardy–Weinberg expectations when $p > 0.01$. Table S2. Distribution of genotypes associated with *FCN2* gene promoter polymorphisms in preterm newborns, depending on incidence of respiratory distress syndrome, early-onset infection, pneumonia and sepsis. None of the associations analysed was significant ($p > 0.05$). Table S3. Frequencies of the most common haplotypes in preterm newborns, depending on incidence of respiratory distress syndrome, early-onset infection, pneumonia and sepsis. None of the associations analysed was significant ($p > 0.05$). Table S4. Frequencies of the most common diplotypes in preterm newborns, depending on incidence of respiratory distress syndrome, early-onset infection, pneumonia and sepsis. None of the associations analysed was significant ($p > 0.05$). Supplementary Figure S1: Linkage disequilibrium analysis of promoter rs3124952 (−986 A > G), rs3124953 (−602 G > A), rs7865453 (−64 A > C) and rs17514136 (−4 A > G) *FCN2* single nucleotide polymorphisms. The numbers in the grid refer to D′ (**A**) and $r^2$ (**B**) parameters of the given pairs of SNPs. Bolded triangles show haplotype blocks identified using the four gamete rule test. Supplementary Figure S2: Individual concentrations of ficolin-2 in cord sera from preterm newborns, corresponding to the ten most common *FCN2* gene promoter diplotypes. Blue bars represent median values (given below the graph in bold). Medians related to diplotypes 2 (the highest) and 5 (the lowest one) were compared with the remaining values using a Mann–Whitney U test. Corresponding *p*-values are given below the graph in red and blue, respectively. Diplotypes: 1—AGAG/GGAA; 2—AGAG/AAAA; 3—AAAA/GGAA; 4—AGAG/AGAG; 5—AGAG/GGCA; 6—AAAA/GGCA; 7—GGAA/GGAA; 8—AGAA/GGAA; 9—AGAA/AGAG; 10—AAAA/GGAG. Supplementary Figure S3: **A:** Scheme of the *FCN2* gene with investigated promoter polymorphic sites. Alleles associated with higher gene expression are marked in green and those with lower, in red. Exons 1–8 are shown as blue rectangles. **B:** Genotypes corresponding to polymorphic sites, most common haplotypes, diplotypes and median ficolin-2 concentrations in cord sera (ng/mL), related to demonstrated promoter diplotypes. Alleles corresponding to particular sites associated with higher gene expression are marked in green and those with lower, in red. Median ficolin-2 levels higher than the median for the whole cohort are marked in green and those lower, in red. Genotypes: A/A (−986), A/A (−64), G/G (−4), all corresponding to relatively high FCN2 gene expression; the related AGAG haplotype and AGAG/AGAG diplotype were associated with lower risk of very low birthweight (green boxes). Furthermore, the G/G variant at −4 corresponds to a lower risk of birth at gestational age <33 weeks (blue box). The GGCA haplotype (all alleles related to lower gene expression) was associated with a higher risk of VLBW (red box) while the AGAG/GGAA diplotype was associated with a higher risk of both VLBW and GA < 33 weeks (red and orange boxes).

**Author Contributions:** A.S.-P., A.S.Ś., G.G. and D.J. determined *FCN2* polymorphisms; A.S.Ś. and M.C. designed the study, analysed and interpreted data and prepared the draft manuscript; A.S.Ś. determined ficolin-2 concentrations in sera; H.S. produced anti-ficolin-2 antibodies; M.M. contributed to data analysis and manuscript preparation; M.K.-B., K.C., P.K. and K.S. were responsible for recruiting patients, collection of samples and clinical data; J.M., I.D.-P. and J.K. supervised collection of material and were responsible for clinical data analysis. All authors have read and agreed to the published version of the manuscript.

**Funding:** This work was funded by National Science Center, Poland, grant 2015/17/B/NZ6/04250.

**Institutional Review Board Statement:** This study was approved by the local ethics committees: Bioethics Committee of the Karol Marcinkowski Poznań University of Medical Sciences (595/14), Independent Bioethics Committee for Scientific Research at the Medical University of Gdańsk (NKBBN/527/2014), Bioethics Committee of the Medical University of Łódź (RNN/257/19/KE). This work conforms to the provisions of the Declaration of Helsinki.

**Informed Consent Statement:** Written informed parental consent was obtained for all subjects involved in the study.

**Data Availability Statement:** The data are available from the corresponding author on reasonable request.

**Acknowledgments:** Authors are very grateful to David C. Kilpatrick for critical reading of the manuscript and helpful discussion.

**Conflicts of Interest:** The authors declare no conflict of interest.

## References

1. Hummelshoj, T.; Munthe-Fog, L.; Madsen, H.O.; Fujita, T.; Matsushita, M.; Garred, P. Polymorphisms in the FCN2 gene determine serum variation and function of Ficolin-2. *Hum. Mol. Genet.* **2005**, *14*, 1651–1658. [CrossRef] [PubMed]
2. Cedzynski, M.; Nuytinck, L.; Atkinson, A.P.M.; Swierzko, A.S.; Zeman, K.; Szemraj, J.; Szala, A.; Turner, M.L.; Kilpatrick, D.C. Extremes of l-ficolin concentration in children with recurrent infections are associated with single nucleotide polymorphisms in the *FCN2* gene. *Clin. Exp. Immunol.* **2007**, *150*, 99–104. [CrossRef] [PubMed]
3. Kilpatrick, D.C.; Swierzko, A.S.; Matsushita, M.; Domzalska-Popadiuk, I.; Borkowska-Klos, M.; Szczapa, J.; Cedzynski, M. The relationship between FCN2 genotypes and serum ficolin-2 (L-ficolin) protein concentrations from a large cohort of neonates. *Hum. Immunol.* **2013**, *74*, 867–871. [CrossRef] [PubMed]
4. Świerzko, A.S.; Jarych, D.; Gajek, G.; Chojnacka, K.; Kobiela, P.; Kufelnicka-Babout, M.; Michalski, M.; Sobczuk, K.; Szala-Poździej, A.; Matsushita, M.; et al. Polymorphisms of the FCN2 Gene 3′UTR Region and Their Clinical Associations in Preterm Newborns. *Front. Immunol.* **2021**, *12*, 741140. [CrossRef] [PubMed]
5. Thiel, S.; Gadjeva, M. Humoral Pattern Recognition Molecules: Mannan-Binding Lectin and Ficolins. *Adv. Exp. Med. Biol.* **2009**, *653*, 58–73. [CrossRef]
6. Endo, Y.; Matsushita, M.; Fujita, T. New Insights into the Role of Ficolins in the Lectin Pathway of Innate Immunity. *Int. Rev. Cell. Mol. Biol.* **2015**, *316*, 49–110. [CrossRef]
7. Świerzko, A.S.; Cedzyński, M. The Influence of the Lectin Pathway of Complement Activation on Infections of the Respiratory System. *Front. Immunol.* **2020**, *11*, 585243. [CrossRef]
8. Statistics Poland. *Demographic Yearbook of Poland, 2021*; Statistics Poland: Warsaw, Poland, 2022. Available online: https://stat.gov.pl/en/topics/statistical-yearbooks/statistical-yearbooks/demographic-yearbook-of-poland-2021,3,15.html (accessed on 4 July 2022).
9. Heron, M. Deaths: Leading causes for 2019. *Natl. Vital Stat. Rep.* **2021**, *70*, 1–114.
10. Van de Pol, C.; Allegaert, K. Growth patterns and body composition in former extremely low birth weight (ELBW) neonates until adulthood: A systematic review. *Eur. J. Pediatr.* **2020**, *179*, 757–771. [CrossRef]
11. Dehlin, M.; Jacobsson, L.T.H. Association between perinatal factors and future risk for gout—A nested case-control study. *Arthritis Res. Ther.* **2022**, *24*, 60. [CrossRef]
12. Kuula, J.; Martola, J.; Hakkarainen, A.; Räikkönen, K.; Savolainen, S.; Salli, E.; Hovi, P.; Björkqvist, J.; Kajantie, E.; Lundbom, N. Brain Volumes and Abnormalities in Adults Born Preterm at Very Low Birth Weight. *J. Pediatr.* **2022**, *246*, 48–55.e7. [CrossRef] [PubMed]
13. Halli, S.S.; Biradar, R.A.; Prasad, J.B. Low Birth Weight, the Differentiating Risk Factor for Stunting among Preschool Children in India. *Int. J. Environ. Res. Public Health* **2022**, *19*, 3751. [CrossRef] [PubMed]

14. Sandboge, S.; Kuula, J.; Björkqvist, J.; Hovi, P.; Mäkitie, O.; Kajantie, E. Bone mineral density in very low birthweight adults—A sibling study. *Paediatr. Peérinat. Epidemiol.* **2022**, *in press*. [CrossRef] [PubMed]
15. Swierzko, A.S.; Atkinson, A.P.; Cedzynski, M.; MacDonald, S.L.; Szala, A.; Domzalska-Popadiuk, I.; Borkowska-Klos, M.; Jopek, A.; Szczapa, J.; Matsushita, M.; et al. Two factors of the lectin pathway of complement, l-ficolin and mannan-binding lectin, and their associations with prematurity, low birthweight and infections in a large cohort of Polish neonates. *Mol. Immunol.* **2009**, *46*, 551–558. [CrossRef] [PubMed]
16. Tiensuu, H.; Haapalainen, A.M.; Karjalainen, M.K.; Pasanen, A.; Huusko, J.M.; Marttila, R.; Ojaniemi, M.; Muglia, L.J.; Hallman, M.; Ramet, M. Risk of spontaneous preterm birth and fetal growth associates with fetal SLIT2. *PLoS Genet.* **2019**, *15*, e1008107. [CrossRef]
17. Rappoport, N.; Toung, J.; Hadley, D.; Wong, R.J.; Fujioka, K.; Reuter, J.; Abbott, C.W.; Oh, S.; Hu, D.; Eng, C.; et al. A genome-wide association study identifies only two ancestry specific variants associated with spontaneous preterm birth. *Sci. Rep.* **2018**, *8*, 226. [CrossRef]
18. Huusko, J.M.; Karjalainen, M.K.; Graham, B.E.; Zhang, G.; Farrow, E.G.; Miller, N.A.; Jacobsson, B.; Eidem, H.R.; Murray, J.C.; Bedell, B.; et al. Whole exome sequencing reveals HSPA1L as a genetic risk factor for spontaneous preterm birth. *PLoS Genet.* **2018**, *14*, e1007394. [CrossRef]
19. Modi, B.P.; Teves, M.E.; Pearson, L.N.; Parikh, H.I.; Haymond-Thornburg, H.; Tucker, J.L.; Chaemsaithong, P.; Gomez-Lopez, N.; York, T.P.; Romero, R.; et al. Mutations in fetal genes involved in innate immunity and host defense against microbes increase risk of preterm premature rupture of membranes (PPROM). *Mol. Genet. Genom. Med.* **2017**, *5*, 720–729. [CrossRef] [PubMed]
20. Frakking, F.N.J.; Brouwer, N.; Zweers, D.; Merkus, M.P.; Kuijpers, T.W.; Offringa, M.; Dolman, K.M. High prevalence of mannose-binding lectin (MBL) deficiency in premature neonates. *Clin. Exp. Immunol.* **2006**, *145*, 5–12. [CrossRef]
21. Bodamer, O.A.; Mitterer, G.; Maurer, W.; Pollak, A.; Mueller, M.W.; Schmidt, W.M. Evidence for an association between mannose-binding lectin 2 (MBL2) gene polymorphisms and pre-term birth. *Genet. Med.* **2006**, *8*, 518–524. [CrossRef] [PubMed]
22. da Silva, L.V.C.; Javorski, N.; Brandão, L.A.C.; Lima, M.D.C.; Crovella, S.; Eickmann, S.H. Influence of *MBL2* and *NOS3* polymorphisms on spontaneous preterm birth in North East Brazil: Genetics and preterm birth. *J. Matern. Neonatal Med.* **2018**, *33*, 127–135. [CrossRef]
23. Liu, X.; Cui, Y. A Genome-wide Association Analysis in Four Populations Reveals Strong Genetic Heterogeneity for Birth Weight. *Curr. Genom.* **2016**, *17*, 416–426. [CrossRef]
24. Horikoshi, M.; Yaghootkar, H.; Mook-Kanamori, D.O.; Sovio, U.; Taal, H.R.; Hennig, B.J.; Bradfield, J.P.; Pourcain, B.S.; Evans, D.M.; Charoen, P.; et al. New loci associated with birth weight identify genetic links between intrauterine growth and adult height and metabolism. *Nat. Genet.* **2012**, *45*, 76–82. [CrossRef]
25. Thompson, W.D.; Beaumont, R.N.; Kuang, A.; Warrington, N.M.; Ji, Y.; Tyrrell, J.; Wood, A.R.; Scholtens, D.M.; Knight, B.A.; Evans, D.M.; et al. Fetal alleles predisposing to metabolically favorable adiposity are associated with higher birth weight. *Hum. Mol. Genet.* **2021**, *31*, 1762–1775. [CrossRef]
26. Kilpatrick, D.C.; Fujita, T.; Matsushita, M. P35, an opsonic lectin of the ficolin family, in human blood from neonates, normal adults, and recurrent miscarriage patients. *Immunol. Lett.* **1999**, *67*, 109–112. [CrossRef]
27. Schlapbach, L.J.; Mattmann, M.; Thiel, S.; Boillat, C.; Otth, M.; Nelle, M.; Wagner, B.; Jensenius, J.C.; Aebi, C.; Christoph, A. Differential Role of the Lectin Pathway of Complement Activation in Susceptibility to Neonatal Sepsis. *Clin. Infect. Dis.* **2010**, *51*, 153–162. [CrossRef]
28. Sallenbach, S.; Thiel, S.; Aebi, C.; Otth, M.; Bigler, S.; Jensenius, J.C.; Schlapbach, L.; Ammann, R.A. Serum concentrations of lectin-pathway components in healthy neonates, children and adults: Mannan-binding lectin (MBL), M-, L-, and H-ficolin, and MBL-associated serine protease-2 (MASP-2). *Pediatr. Allergy Immunol.* **2011**, *22*, 424–430. [CrossRef] [PubMed]
29. Briana, D.D.; Liosi, S.; Gourgiotis, D.; Boutsikou, M.; Baka, S.; Marmarinos, A.; Hassiakos, D.; Malamitsi-Puchner, A. The potential role of the lectin pathway of complement in the host defence of full-term intrauterine growth restricted neonates at birth. *J. Matern. Neonatal Med.* **2012**, *25*, 531–534. [CrossRef]
30. Aoyagi, Y.; Adderson, E.E.; Rubens, C.E.; Bohnsack, J.F.; Min, J.G.; Matsushita, M.; Fujita, T.; Okuwaki, Y.; Takahashi, S. L-Ficolin/Mannose-Binding Lectin-Associated Serine Protease Complexes Bind to Group B Streptococci Primarily through *N*-Acetylneuraminic Acid of Capsular Polysaccharide and Activate the Complement Pathway. *Infect. Immun.* **2008**, *76*, 179–188. [CrossRef] [PubMed]
31. Fujieda, M.; Aoyagi, Y.; Matsubara, K.; Takeuchi, Y.; Fujimaki, W.; Matsushita, M.; Bohnsack, J.F.; Takahashi, S. L-Ficolin and Capsular Polysaccharide-Specific IgG in Cord Serum Contribute Synergistically to Opsonophagocytic Killing of Serotype III and V Group B Streptococci. *Infect. Immun.* **2012**, *80*, 2053–2060. [CrossRef] [PubMed]
32. Brady, A.M.; Calix, J.J.; Yu, J.; Geno, K.A.; Cutter, G.R.; Nahm, M.H. Low Invasiveness of Pneumococcal Serotype 11A Is Linked to Ficolin-2 Recognition of O-acetylated Capsule Epitopes and Lectin Complement Pathway Activation. *J. Infect. Dis.* **2014**, *210*, 1155–1165. [CrossRef]
33. Sørensen, C.A.; Rosbjerg, A.; Jensen, B.H.; Krogfelt, K.A.; Garred, P. The Lectin Complement Pathway Is Involved in Protection Against Enteroaggregative Escherichia coli Infection. *Front. Immunol.* **2018**, *9*, 1153. [CrossRef] [PubMed]
34. Świerzko, A.S.; Szala-Poździej, A.; Kilpatrick, D.C.; Sobociński, M.; Chojnacka, K.; Sokołowska, A.; Michalski, M.; Mazerant, K.; Jensenius, J.C.; Matsushita, M.; et al. Components of the lectin pathway of complement activation in paediatric patients of intensive care units. *Immunobiology* **2016**, *221*, 657–669. [CrossRef] [PubMed]

35. Kilpatrick, D.C.; Świerzko, A.S.; Sobociński, M.; Krajewski, W.; Chojnacka, K.; Szczapa, J.; Cedzyński, M. Can ficolin-2 (L-ficolin) insufficiency be established by a single serum protein measurement? *Int. J. Immunogenet.* **2015**, *42*, 453–456. [CrossRef]
36. Metzger, M.-L.; Michelfelder, I.; Goldacker, S.; Melkaoui, K.; Litzman, J.; Guzman, D.; Grimbacher, B.; Salzer, U. Low ficolin-2 levels in common variable immunodeficiency patients with bronchiectasis. *Clin. Exp. Immunol.* **2015**, *179*, 256–264. [CrossRef]
37. Szala, A.; Swierzko, A.S.; Cedzynski, M. Cost-effective procedures for genotyping of human FCN2 gene single nucleotide polymorphisms. *Immunogenetics* **2013**, *65*, 439–446. [CrossRef]
38. Swierzko, A.S.; Michalski, M.; Sokołowska, A.; Nowicki, M.; Szala-Poździej, A.; Eppa, Ł.; Mitrus, I.; Szmigielska-Kapłon, A.; Sobczyk-Kruszelnicka, M.; Michalak, K.; et al. Associations of Ficolins With Hematological Malignancies in Patients Receiving High-Dose Chemotherapy and Autologous Hematopoietic Stem Cell Transplantations. *Front. Immunol.* **2020**, *10*, 3097. [CrossRef]

International Journal of *Molecular Sciences*

MDPI

*Article*

# Placental Transcriptome Profiling in Subtypes of Diabetic Pregnancies Is Strongly Confounded by Fetal Sex

**Sarah M. Kedziora [1,2,3,4,†], Benedikt Obermayer [5,†], Meryam Sugulle [6,7], Florian Herse [1,2,3], Kristin Kräker [1,2,3,4], Nadine Haase [1,2,3,4], Immaculate M. Langmia [1,2,3], Dominik N. Müller [1,2,3,4], Anne Cathrine Staff [6,7], Dieter Beule [5] and Ralf Dechend [1,4,8,*]**

1 Experimental and Clinical Research Center (ECRC), a Joint Cooperation between the Charité—Universitätsmedizin Berlin and Max-Delbrück-Center for Molecular Medicine, 13125 Berlin, Germany
2 Max-Delbrück-Center for Molecular Medicine in the Helmholtz Association (MDC), 13125 Berlin, Germany
3 Charité—Universitätsmedizin Berlin, Corporate Member of Freie Universität Berlin, Humboldt Universität zu Berlin, Berlin Institute of Health, 10178 Berlin, Germany
4 DZHK (German Centre for Cardiovascular Research), Partner Site, 10785 Berlin, Germany
5 Berlin Institute of Health, Charité—Universitätsmedizin Berlin, Core Unit Bioinformatics, 10117 Berlin, Germany
6 Faculty of Medicine, University of Oslo, 0372 Oslo, Norway
7 Division of Obstetrics and Gynaecology, Oslo University Hospital, 0424 Oslo, Norway
8 HELIOS Clinic, Department of Cardiology and Nephrology, 13125 Berlin, Germany
* Correspondence: ralf.dechend@charite.de; Tel.: +49-30-4505-40301
† These authors contributed equally to this work.

**Citation:** Kedziora, S.M.; Obermayer, B.; Sugulle, M.; Herse, F.; Kräker, K.; Haase, N.; Langmia, I.M.; Müller, D.N.; Staff, A.C.; Beule, D.; et al. Placental Transcriptome Profiling in Subtypes of Diabetic Pregnancies Is Strongly Confounded by Fetal Sex. *Int. J. Mol. Sci.* **2022**, *23*, 15388. https://doi.org/10.3390/ijms232315388

Academic Editor: Ilona Hromadnikova

Received: 25 October 2022
Accepted: 1 December 2022
Published: 6 December 2022



**Abstract:** The placenta is a temporary organ with a unique structure and function to ensure healthy fetal development. Placental dysfunction is involved in pre-eclampsia (PE), fetal growth restriction, preterm birth, and gestational diabetes mellitus (GDM). A diabetic state affects maternal and fetal health and may lead to functional alterations of placental metabolism, inflammation, hypoxia, and weight, amplifying the fetal stress. The placental molecular adaptations to the diabetic environment and the adaptive spatio–temporal consequences to elevated glucose or insulin are largely unknown (2). We aimed to identify gene expression signatures related to the diabetic placental pathology of placentas from women with diabetes mellitus. Human placenta samples ($n$ = 77) consisting of healthy controls, women with either gestational diabetes mellitus (GDM), type 1 or type 2 diabetes, and women with GDM, type 1 or type 2 diabetes and superimposed PE were collected. Interestingly, gene expression differences quantified by total RNA sequencing were mainly driven by fetal sex rather than clinical diagnosis. Association of the principal components with a full set of clinical patient data identified fetal sex as the single main explanatory variable. Accordingly, placentas complicated by type 1 and type 2 diabetes showed only few differentially expressed genes, while possible effects of GDM and diabetic pregnancy complicated by PE were not identifiable in this cohort. We conclude that fetal sex has a prominent effect on the placental transcriptome, dominating and confounding gene expression signatures resulting from diabetes mellitus in settings of well-controlled diabetic disease. Our results support the notion of placenta as a sexual dimorphic organ.

**Keywords:** diabetes mellitus; human; placenta; pregnancy; RNA sequencing

## 1. Introduction

The placenta is a temporary mammalian organ with the vital function of fetal nutrient supply and waste removal. Its unique structure, with tissue emerging from both embryonic and maternal origin, is necessary for healthy fetal development [1]. The rapid placental growth includes proliferation, differentiation and invasion of trophoblasts, implantation, remodeling, and angiogenesis, all of which are decisive factors for healthy pregnancies and outcomes [2]. Placental maldevelopment and dysfunction are involved in the major, often co-occurring obstetric syndromes, including preeclampsia (PE), fetal growth restriction, preterm birth and gestational diabetes mellitus (GDM) [3–5].

In a diabetic environment, a variety of structural and functional changes take place in the placenta. Besides heavier placentas, inflammation and DNA methylation, altered expression of genes regulating for growth, glucose metabolism, cytoskeletal structure, oxidative stress and apoptosis have been described [6,7]. The extent of structural and functional changes depends on multiple variables, including the type of diabetes mellitus (DM) and glycemic control [6].

DM can be present before, and throughout, gestation as in DM type 1 (DM1) or type 2 (DM2). In both cases, placentation takes places under the complex diabetic milieu, although the pathophysiology of DM1 and DM2 differs. During the first trimester, the placental growth rate is highest and the placenta is therefore extremely sensitive to environmental influences [6]. Secondly, DM can occur during gestation (gestational diabetes mellitus; GDM) and resolve postpartum in most cases [8]. In women with GDM, impaired glucose tolerance develops and leads to more short-term molecular alterations [2].

Both pregestational and gestational diabetes imply a well-described and significantly increased risk of adverse maternal and offspring pregnancy outcomes, as well as long-term diseases such as obesity and cardiovascular disease [5,9]. Central to short-term association is the placental pathology, most likely due to maternal metabolic and inflammatory changes in a diabetic milieu. The long-term consequences are likely explained by in utero programming impacting on health or disease later in life, independent of the DNA sequences that are inherited in a person's genetic code (i.e., developmental origins of health and disease) [10].

One potential complication to pregestational and gestational diabetes during pregnancy is PE [11], which is also closely linked to a dysfunctional placenta [6]. PE superimposed on a diabetic condition is a major risk factor for preterm delivery, maternal and offspring morbidity and future health [3,12].

The precise mechanisms by which pregestational or gestational diabetes contribute to placental dysfunction are unknown [1,6]. Besides the effects of glucose and insulin on the placenta, reactive oxygen species production in first trimester trophoblasts, insulin resistance of trophoblasts, and altered oxygen tension in the intervillous space have been addressed [13]. It is likely that pregestational and gestational diabetes may have different adaptation mechanisms, including differential effects on placentation, as GDM is usually not present during the most important placentation period [14].

The aim of this study is to unravel novel pathways related to the contribution of diabetes to pregnancy pathologies, including diabetic pregnancies complicated by PE, by transcriptome analysis of placenta tissues. Various studies suggest that placental inflammatory pathways, stress response and gene expression patterns are related to maternal pregnancy complications. Understanding the contribution of the placental transcriptome profile to placental differences responsible for fetal and maternal health—such as fetal growth, preterm birth and survival—is essential. By gene expression analysis, we aimed to identify differentially expressed genes (DEGs) and associated signaling pathways among GDM, pregestational DM, or diabetic pregnancies complicated with PE. Here we present, to our knowledge, the first RNA sequencing dataset of the placental transcriptome from women with GDM, DM1, or DM2, and in DM pregnancies with superimposed PE, to gain a profound understanding of transcriptional placental profiles during the various forms of a diabetic pregnancy.

## 2. Results

### 2.1. Clinical Characteristics of the Study Cohort

The study cohort consisted of healthy and diabetic women, while other diabetic patients developed superimposed PE during pregnancy. Clinical characteristics of these healthy and diabetic pregnant women (subgroups) are shown in Table 1. Characteristics of diabetic women with a pregnancy complicated with PE are shown in Table 2. Further, the characteristics of all women with the different types of DM or DM superimposed with PE are shown in Table 3. The BMI of women with GDM and DM2 was higher at the beginning of pregnancy and at delivery (GDM: 33.2 ± 6.77; DM2: 38.57 ± 2.89; Table 1) compared to healthy CTL (CTL: 28.79 ± 3.98), even when complicated with PE (Table 2). The diabetic pregnant women superimposed with PE not only developed hypertension (Table 2), but also proteinuria (Table 3) during pregnancy. Early-onset PE appeared in 37.5% of pregnant women. HbA1c was higher during pregnancy in pregestational diabetic women compared to women who developed GDM, independent of PE. All subgroups of women with DM without superimposed PE in pregnancy had, as expected, heavier babies and higher birthweight percentiles compared to CTL (Table 1). Pregnancies with DM superimposed with PE had, as expected, an earlier delivery than CTL (Table 3). Women with GDM + PE had consequently lower birth weight and smaller newborn weight percentiles than CTL, a proxy of placental dysfunction. While DM1 + PE had higher birthweight percentiles but similar birthweights as CTL, the birthweight and newborn weight percentile were not different in pregnancies of DM2 + PE compared to CTL (Table 2).

### 2.2. *Placentas of Patients with Diabetes Reveal Differentially Expressed Genes*

The gene expression between diabetic and healthy CTL placentas using DESeq2 [15] with fetal sex as covariate did not show any DEGs at adj. $p$-value < 0.05 (see MA-plot in Figure 1).

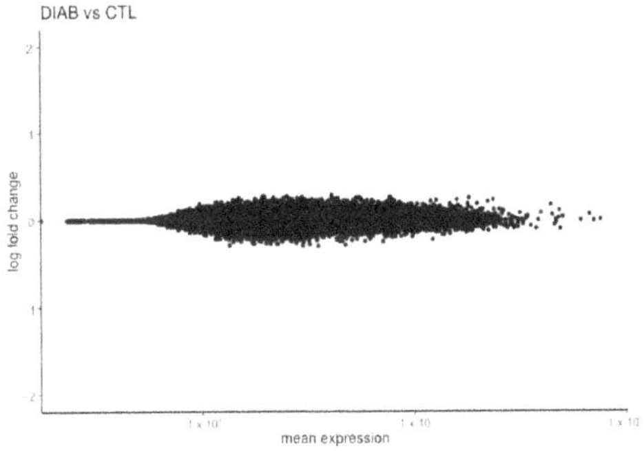


**Figure 1.** The effect of diabetes on the placental transcriptome in comparison to healthy controls. MA-plot showing the log2-fold change over mean expression for all genes for the diabetes group (DIAB) without PE ($n$ = 32) versus CTL ($n$ = 29). DIAB = patients with DM1, DM2 and GDM; CTL = healthy controls.

Subsequently we analyzed the diabetic subgroups and found relatively few significant gene expression differences between placentas from DM1 ($n$ = 17), DM2 ($n$ = 3), GDM ($n$ = 12) and CTL ($n$ = 29) patients (Figure 2). Most DEGs (75 up-regulated and 18 down-regulated at adj. $p$-value < 0.05) were observed when comparing DM2 to CTL (Figure 2a), summarized in Supplementary Table S1. Figure 2b shows the comparison between DM2 and GDM, where we identified four up- and 23 down-regulated genes (Supplementary Table S1). Pathway analysis showed that several DEGs in the comparison of DM2 with either CTL or GDM were associated with metabolism (such as *SELENBP1* [16], *CNNM* [17], and *SOXS3* [18]) and placental metabolism (such as *ACSL6* [19] and *ARID5A* [20]). DM1 versus DM2 (Figure 2c) showed 4 up-regulated genes (Supplementary Table S1), and GDM versus DM1 showed only two up-regulated genes (*SMCO3* and *NQO1*), both linked to DM [21,22] (Figure 2d). In contrast, DM1 versus CTL (Figure 2e) and GDM versus CTL (Figure 2f) did not show significant DEGs. Our results show that gene expression differences in DM1 and DM2 showed similar patterns to each other (Pearson's R = 0.47 between log2-fold changes), whereas GDM was more similar to CTL.

**Table 1.** Clinical characteristics of study group participants for diabetic subgroups without PE ($n$ = 61, including 29 CTL and 32 DM).

| Characteristics | CTL ($n$ = 29) | GDM ($n$ = 12) | $p$-Value CTL vs. GDM | DM1 ($n$ = 17) | $p$-Value CTL vs. DM1 | DM2 ($n$ = 3) | $p$-Value CTL vs. DM2 |
|---|---|---|---|---|---|---|---|
| **maternal characteristics <20 weeks gestation** | | | | | | | |
| BMI | 23.68 ± 4.06 | 28.03 ± 5.65 | 0.02 | 25.22 ± 3.95 | 0.81 | 34.58 ± 1.20 | 0.00 |
| SBP, mmHg | 111.03 ± 9.41 | 113.08 ± 9.27 | 0.99 | 113.81 ± 8.42 (16) | 0.98 | 127.67 ± 19.14 | 0.15 |
| DBP, mmHg | 66.21 ± 7.04 | 70.42 ± 7.35 | 0.48 | 70.06 ± 5.92 (16) | 0.48 | 75 ± 1.73 | 0.29 |
| HbA1c 2. Trimester, % | NA | 5.2 ± 0.62 (4) | NA | 6.17 ± 0.62 (15) | NA | 5.87 ± 0.09 | NA |
| smoking, % ($n$) | 13.79 (4) | 16.67 (2) | NA | 5.88 (1) | NA | 33.3 (1) | NA |
| height, m | 1.70 ± 0.06 | 1.67 ± 0.06 | 0.87 | 1.67 ± 0.05 | 0.75 | 1.68 ± 0.05 | 0.99 |
| weight, kg | 68.17 ± 11.96 | 78.34 ± 14.77 | 0.13 | 70.53 ± 11.48 | 0.99 | 98 ± 5.29 | 0.00 |
| **maternal characteristics at delivery** | | | | | | | |
| BMI at delivery | 28.79 ± 3.98 | 33.2 ± 6.77 (11) | 0.05 | 29.7 ± 4.03 | 0.99 | 38.57 ± 2.89 | 0.00 |
| pre-operative SBP, mmHg | 124.42 ± 12.22 (26) | 132 ± 10.89 (9) | 0.72 | 138.93 ± 14.35 (15) | 0.02 | 144.5 ± 0.70 (2) | 0.35 |
| pre-operative DBP, mmHg | 72.96 ± 10.78 (26) | 80.44 ± 15.48 (9) | 0.42 | 79.13 ± 9.95 (15) | 0.44 | 73.0 ± 2.83 (2) | >0.99 |
| proteinuria, % ($n$) | 4 (1) | 10 (1) | NA | 13,32 (1) | NA | 0 (0) | NA |
| HbA1c 3. Trimester, % | NA | 5.79 ± 0.55 (11) | NA | 6.27 ± 0.57 (15) | NA | 6.2 ± 0.22 | NA |
| medication, % ($n$) [insulin, metformin] | 0 (0) | 41.67 (5) | NA | 88.24 (15) | NA | 100 (3) | NA |
| gestational age, days | 271.93 ± 7.94 | 267.67 ± 13.03 | 0.94 | 260.65 ± 17.19 | 0.06 | 269.67 ± 4.16 | >0.99 |
| age, years | 32.66± 4.50 | 34.58 ± 3.61 | 0.77 | 33.59 ± 3.76 | 0.98 | 31.0 ± 6.56 | 0.99 |
| parity, count | 0.93 ± 0.88 | 1.08 ± 0.90 | 0.99 | 0.71 ± 0.59 | 0.93 | 0.33 ± 0.58 | 0.78 |
| gravidity, count | 2.72 ± 1.25 | 3.25 ± 1.42 | 0.77 | 2.29 ± 1.16 | 0.83 | 2.00 ± 1.00 | 0.91 |
| blood sugar, mmol/l | 3.98 ± 0.48 (9) | 4.78 ± 1.05 (9) | 0.72 | 6.57 ± 1.56 (13) | 0.00 | 6.53 ± 1.45 | 0.03 |
| **birth outcome** | | | | | | | |
| fetal sex, female/male, count | 15/14 | 7/5 | NA | 8/9 | NA | 1/2 | NA |
| birth weight, g | 3342 ± 450 | 3697 ± 844 | 0.72 | 3611 ± 864 (16) | 0.85 | 4515 ± 788 | 0.09 |
| birth length, cm | 50.05 ± 1.49 (21) | 50.55 ± 3.55 (10) | 0.99 | 49.57 ± 4.09 (14) | 0.99 | 53.67 ± 2.89 | 0.34 |
| percentile birthweight | 51.04 ± 27.68 | 73.18 ± 33.06 | 0.22 | 78.87 ± 30.69 (16) | 0.03 | 97.19 ± 2.99 | 0.09 |
| placental + umbilical cord weight, g | 584.23 ± 107.9 (26) | 680.11 ± 114.4 (9) | 0.51 | 653.53 ± 198.7 (15) | 0.67 | 906.33 ± 164.5 | 0.00 |

Data are shown as mean ± standard deviation (SD) or percentage (absolute numbers). Absolute numbers are shown in parentheses if the characteristic was not available for all participants of the group. CTL = healthy controls; GDM = gestational diabetes mellitus; DM1 = type 1 diabetes mellitus; DM2 = type 2 diabetes mellitus; NA = not available; BMI = body mass index; SBP = systolic blood pressure; DBP = diastolic blood pressure; HbA1c = hemoglobin A1c. Comparison of a single group to CTL was assessed by one-way ANOVA with multiple comparisons.

**Table 2.** Clinical characteristics of study group participants, including 29 CTLs and DM superimposed by PE ($n$ = 16).

| Characteristics | CTL ($n$ = 29) | GDM + PE ($n$ = 4) | *p*-Value CTL vs. GDM + PE | DM1 + PE ($n$ = 8) | *p*-Value CTL vs. DM1 + PE | DM2 + PE ($n$ = 4) | *p*-Value CTL vs. DM2 + PE |
|---|---|---|---|---|---|---|---|
| **maternal characteristics <20 weeks gestation** | | | | | | | |
| BMI | 23.68 ± 4.06 | 31.91 ± 4.47 | 0.00 | 24.68 ± 2.85 | 0.99 | 28.07 ± 5.69 | 0.30 |
| SBP, mmHg | 111.03 ± 9.41 | 118.5 ± 11.21 | 0.82 | 125.88 ± 16.3 | 0.02 | 124 ± 29.02 | 0.25 |
| DBP, mmHg | 66 21 ± 7.04 | 76.25 ± 8.96 | 0.08 | 72.25 ± 6.5 | 0.25 | 69.25 ± 16.4 | 0.97 |
| HbA1c 2. Trimester, % | NA | 5.5 ± 0.1 (2) | NA | 6.63 ± 0.66 | NA | 5.67 ± 0.12 (2) | NA |
| smoking, % (n) | 13.79 (4) | 25 (1) | NA | 25 (2) | NA | 0 (0) | NA |
| height, m | 1.70 ± 0.06 | 1.61 ± 0.11 | 0.06 | 1.69 ± 0.04 | >0.99 | 1.57 ± 0.14 | 0.00 |
| weight, kg | 68.17 ± 11.96 | 83.5 ± 20.76 | 0.16 | 68.75 ± 7.15 | >0.99 | 70.25 ± 20.76 | 0.99 |
| **maternal characteristics at delivery** | | | | | | | |
| BMI at delivery | 28.79 ± 3.98 | 38.05 ± 3.55 | 0.00 | 32.07 ± 3.65 | 0.39 | 31.54 ± 7.39 | 0.85 |
| pre-operative SBP, mmHg | 124.42 ± 12.22 (26) | 153 ± 16.92 (3) | 0.03 | 169.38 ± 25.05 | 0.00 | 152.67 ± 13.65 (3) | 0.02 |
| pre-operative DBP, mmHg | 72.96 ± 10.78 (26) | 90.33 ± 8.51 (3) | 0.08 | 99.13 ± 10.91 | 0.00 | 88.0 ± 8.66 (3) | 0.17 |
| proteinuria, % (n) | 4 (1) | 100 (4) | NA | 100 (8) | NA | 100 (4) | NA |
| HbA1c 3. Trimester, % | NA | 6.25 ± 0.05 (2) | NA | 6.6 ± 0.56 | NA | 6.65 ± 0.25 (2) | NA |
| medication, % (n) [insulin, metformin] | 0 (0) | 25 (1) | NA | 100 (8) | NA | 100 (4) | NA |
| gestational age, days | 271.93 ± 7.94 | 247.25 ± 18.77 | 0.00 | 248.38 ± 13.75 | 0.00 | 249.25 ± 31.10 | 0.02 |
| age, years | 32.66± 4.50 | 29.0 ± 4.08 | 0.57 | 32.13 ± 6.11 | 0.98 | 34.25 ± 5.38 | 0.98 |
| parity, count | 0.93 ± 0.88 | 0.75 ± 0.50 | 0.99 | 0.5 ± 0.76 | 0.70 | 0.5 ± 1.00 | 0.90 |
| gravidity, count | 2.72 ± 1.25 | 1.75 ± 0.50 | 0.60 | 2.0 ± 1.07 | 0.60 | 1.75 ± 1.50 | 0.60 |
| blood sugar, mmol/l | 3.98 ± 0.48 (9) | 4.28 ± 0.50 | 0.99 | 6.32 ± 1.8 (7) | 0.00 | 3.9 ± 0.30 (2) | >0.99 |
| **birth outcome** | | | | | | | |
| fetal sex, female/male, count | 15/14 | 3/1 | NA | 4/4 | NA | 3/1 | NA |
| birth weight, g | 3342 ± 450 | 2422 ± 671 | 0.17 | 3305 ± 830 | >0.99 | 2989 ± 1845 | 0.95 |
| birth length, cm | 50.05 ± 1.49 (21) | 46.75 ± 2.06 | 0.33 | 48.8 ± 0.84 (5) | 0.96 | 38 ± 11.30 (2) | 0.00 |
| percentile birthweight | 51.04 ± 27.68 | 28.29 ± 26.60 | 0.68 | 76.97 ± 31.06 | 0.21 | 51.6 ± 55.17 | >0.99 |
| placental + umbilical cord weight, g | 584.23 ± 107.9 (26) | 494.25 ± 111.7 | 0.86 | 615.75 ± 201.3 | 0.99 | 577.75 ± 218.0 | >0.99 |

Data are shown as mean ± standard deviation (SD) or percentage (absolute number). Total numbers are shown in parentheses if the characteristic was not available for all participants of the group. CTL = healthy controls; GDM + PE = gestational diabetes mellitus superimposed with preeclampsia; DM1 + PE = type 1 diabetes mellitus superimposed with preeclampsia; DM2 + PE = type 2 diabetes mellitus superimposed with preeclampsia; NA = not available; BMI = body mass index; SBP = systolic blood pressure; DBP = diastolic blood pressure; HbA1c = haemoglobin A1c. Comparison of a single group to CTL was assessed by one-way ANOVA with multiple comparisons.

**Table 3.** Clinical characteristics of study groups with all diabetes subgroups combined ($n$ = 77).

| | CTL ($n$ = 29) | Diabetes ($n$ = 32) | $p$-Value CTL vs. Diabetes | Diabetes + PE ($n$ = 16) | $p$-Value CTL vs. Diabetes + PE |
|---|---|---|---|---|---|
| **maternal characteristics <20 weeks gestation** | | | | | |
| BMI | 23.68 ± 4.06 | 27.15 ± 5.21 | 0.01 | 27.33 ± 4.87 | 0.03 |
| SBP, mmHg ($n$) | 111.03 ± 9.41 | 114.87 ± 10.47 (31) | 0.40 | 123.56 ± 18.09 | 0.00 |
| DBP, mmHg ($n$) | 66.21 ± 7.04 | 70.68 ± 6.28 (31) | 0.04 | 72.50 ± 9.81 | 0.02 |
| HbA1c 2. Trimester, % ($n$) | NA | 5.95 ± 0.69 (22) | NA | 6.23 ± 0.73 (13) | NA |
| smoking, % ($n$) | 13.79 (4) | 12.50 (4) | NA | 18.75 (3) | NA |
| height, m | 1.70 ± 0.06 | 1.76 ± 0.05 | 0.33 | 1.64 ± 0.10 | 0.01 |
| weight, kg | 68.17 ± 11.96 | 76.03 ± 14.58 | 0.06 | 72.81 ± 15.40 | 0.49 |
| **maternal characteristics at delivery** | | | | | |
| BMI at delivery | 28.79 ± 3.98 | 31.80 ± 5.69 (31) | 0.04 | 33.43 ± 5.22 | 0.00 |
| pre-operative SBP, mmHg | 124.42 ± 12.22 (26) | 136.96 ± 13.02 (26) | 0.00 | 161.93 ± 22.15 (14) | 0.00 |
| pre-operative DBP, mmHg | 72.96 ± 10.78 (26) | 79.12 ± 11.66 (26) | 0.09 | 94.86 ± 10.65 (14) | 0.00 |
| Proteinuria, % ($n$) | 4 (1) | 6.25 (2) | NA | 100 (16) | NA |
| HbA1c 3. Trimester, % | NA | 6.08 ± 0.58 (29) | NA | 6.55 ± 0.49 (12) | NA |
| medication metformin, % ($n$) | 0 (0) | 9.38 (3) | NA | 18.75 (3) | NA |
| insulin, % ($n$) | 0 (0) | 68.75 (22) | NA | 68.75 (11) | NA |
| early-onset PE % ($n$) | 0 (0) | 0 (0) | NA | 37.5 (6) | NA |
| age, years | 32.66± 4.50 | 33.72 ± 3.96 | 0.59 | 31.88 ± 5.50 | 0.82 |
| parity, count | 0.93 ± 0.88 | 0.81 ± 0.74 | 0.81 | 0.56 ± 0.72 | 0.26 |
| gravidity, count | 2.72 ± 1.25 | 2.63 ± 1.31 | 0.94 | 1.88 ± 1.03 | 0.06 |
| blood sugar, mmol/l | 3.98 ± 0.48 (9) | 5.92 ± 1.59 (25) | 0.00 | 5.32 ± 1.74 (13) | 0.08 |
| **birth outcome** | | | | | |
| fetal sex, female/male, count | 15/14 | 16/16 | NA | 10/6 | NA |
| birth weight, g | 3342 ± 450 | 3732 ± 863 (31) | 0.12 | 3005 ± 1110 | 0.32 |
| birth length, cm | 50.05 ± 1.49 (21) | 50.39 ± 3.87 (27) | 0.94 | 46.09 ± 5.59 (11) | 0.01 |
| percentile birthweight | 51.04 ± 27.68 | 78.44 ± 30.31 (31) | 0.00 | 58.46 ± 40.48 | 0.70 |
| placental + umbilical cord weight, g | 584.23 ± 107.9 (26) | 690.48 ± 183.2 (27) | 0.03 | 575.88 ± 183.1 | 0.98 |

Data are shown as mean ± standard deviation or percentage (total number). Absolute numbers are shown in parentheses if the characteristic was not available for all participants of the group. CTL = healthy controls; Diabetes = includes patients with DM1, DM2 and GDM; Diabetes + PE = includes patients with diabetes superimposed with preeclampsia (DM1 + PE, DM2 + PE, GDM + PE); NA = not available; BMI = body mass index; SBP = systolic blood pressure; DBP = diastolic blood pressure; HbA1c = hemoglobin A1c. Comparison of one group to healthy controls was performed by one-way ANOVA with multiple comparisons.

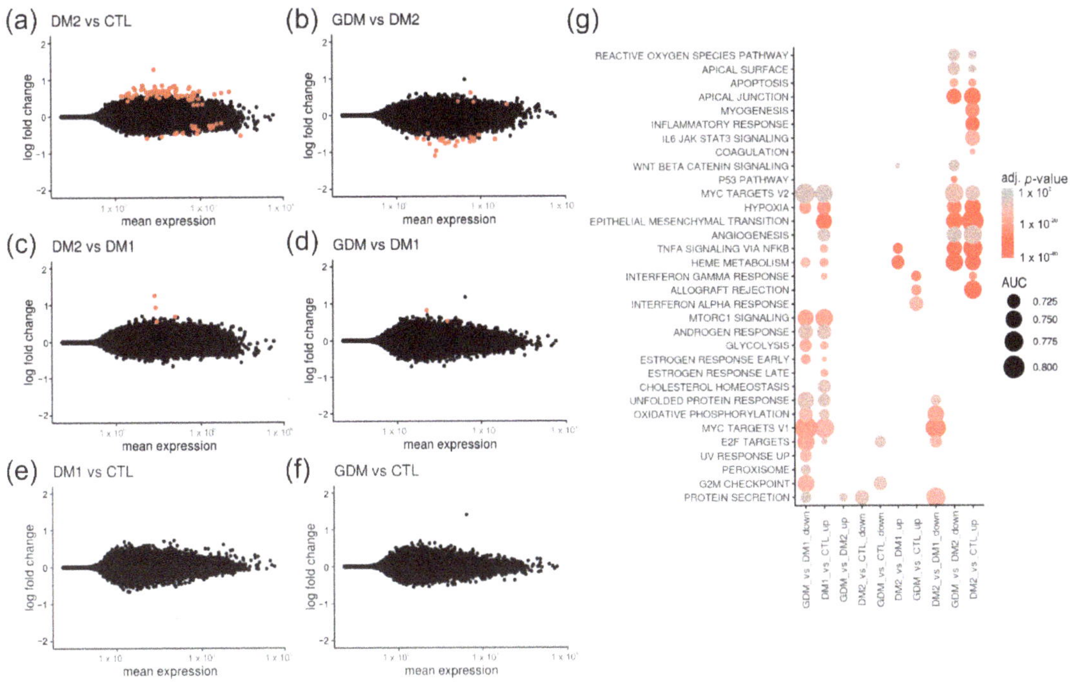


**Figure 2.** Diabetes subgroups show an altered placental transcriptome in comparison to healthy controls. (**a–f**) MA-plots showing log2-fold change as a function of mean expression for the diabetes subgroups DM1 ($n$ = 17), DM2 ($n$ = 3), GDM ($n$ = 12) versus CTL ($n$ = 29). DEGs (adj. $p$-value < 0.05) are marked in red. (**g**) Gene-set enrichment analysis with tmod for up- and down-regulated genes in these contrasts. The adjusted $p$-value is color-coded and the AUC statistic is displayed as the dot size. GDM = gestational diabetes mellitus; DM1 = type I diabetes mellitus; DM2 = type II diabetes mellitus; CTL = healthy controls.

Since only few informative DEGs were observed in the analysis of 77 placental transcriptomes, we next conducted gene-set enrichment analysis using HALLMARK transcriptional gene sets on the estimated log2-fold changes for all genes in the comparisons of different diabetic conditions in pregnancy (Figure 2g). Several gene sets were altered but no uniform dysregulated pathway between the different subtypes of diabetic pregnancy and CTL was observed. Genes involved in the epithelial–mesenchymal transition, hypoxia, angiogenesis or inflammation were consistently altered when comparing the placental transcriptome of DM2 versus GDM or CTL. DM1 versus GDM or CTL showed similar patterns, including in pathways involved in unfolded protein response, oxidative phosphorylation, and proliferation, but the differences were less pronounced.

When investigating the data of placentas from diabetic pregnancies complicated by PE ($n$ = 16) in comparison to CTL ($n$ = 29), we observed a similar pattern. The comparison of PE against CTL gave only one DEG (Figure 3a, Supplementary Table S1) and gene-set enrichment analysis mainly showed alterations of genes involved in the epithelial–mesenchymal transition and inflammatory response (Figure 3b).

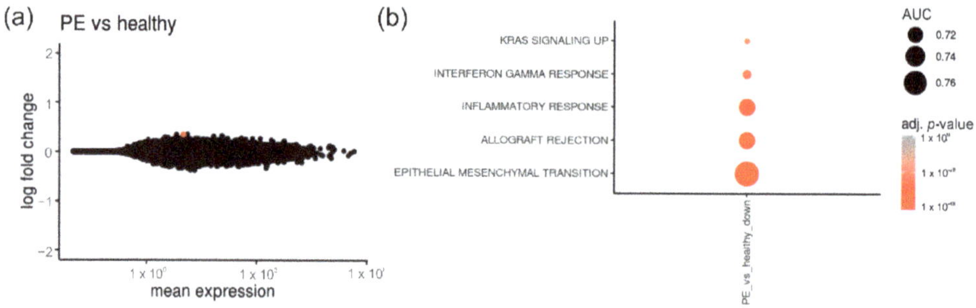


**Figure 3.** The effect of diabetes superimposed with preeclampsia on the placental transcriptome in comparison to healthy controls. (**a**) MA-plot shows log2-fold change over mean expression for all diabetic placenta samples superimposed with PE ($n$ = 16) versus CTL ($n$ = 29). DEGs are highlighted in red. (**b**) The gene-set enrichment analysis displays significantly altered gene sets in this comparison. PE = includes patients with DM1 + PE, DM2 + PE and GDM + PE; CTL = healthy controls.

### *2.3. Placental RNA Sequencing Samples Group Mainly According to Fetal Sex*

We next performed principal component analysis (PCA) to identify what factors could confound the differences between the clinical subgroups. In the first principal component with almost 11% explained variance, the data set clustered into two groups according to the fetal sex (Figure 4a), driven by sex-specific genes such as *XIST*, *UTY*, *USP9Y*, *DDX3Y* and *KDM5D*. The other PCA components were not clearly related to clinical parameters or driven by systematic gene groups. The second PC (7.76% explained variance, driven by *TAC3*, *AADACL3*, *DIO2*, *NOTUM* and *HTRA4*) showed an even distribution of samples without obvious clustering according to the clinical diagnosis (Figure 4a). Similarly, other components contributed little to the explained variance (Figure 4b) and did not induce a clustering of samples.

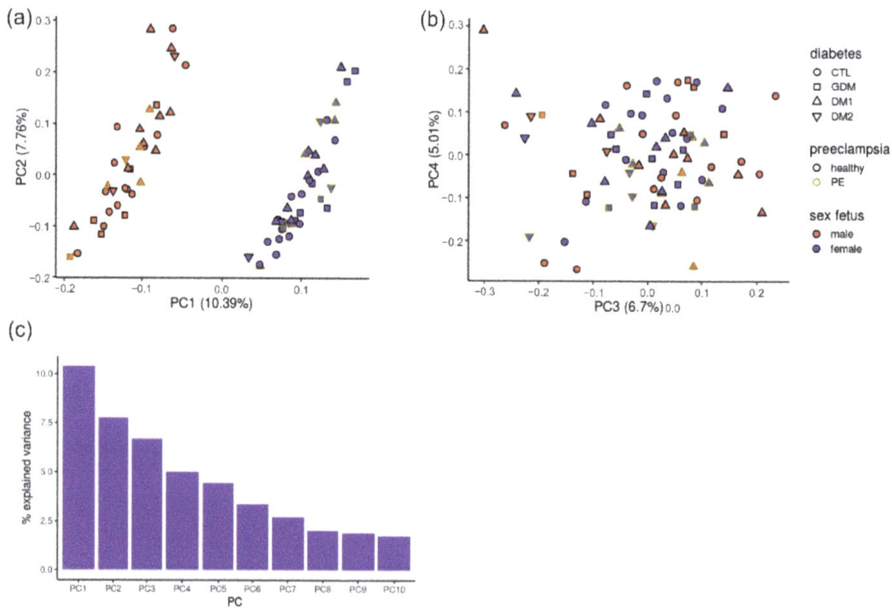


**Figure 4.** Placental samples group mainly due to fetal sex. (**a**) Principal component analysis (PCA) displays the highest amount of explained variance (10.39%) between samples in PC1 and the second

highest (7.76%) in PC2. Placental samples cluster according to the fetal sex. (**b**) PC4 and PC3 do not cluster subjects into groups in the PCA. Black border: CTL ($n$ = 29, circle); GDM ($n$ = 12, square); DM1 ($n$ = 17, triangle pointed top); DM2 ($n$ = 3, triangle pointed bottom). Green border: GDM + PE ($n$ = 4, square); DM1 + PE ($n$ = 8, triangle pointed top); DM2 + PE ($n$ = 4, triangle pointed bottom). Fetal sex is indicated in blue = male and red = female. (**c**) Percentage of explained variance by each PC from PC1 to PC10. Diabetes types are indicated with symbols. CTL = healthy control; GDM = gestational diabetes mellitus; DM1 = Diabetes mellitus type 1; DM2 = diabetes mellitus type 2.

### 2.4. *Of All Gene Sets and Clinical Data, the Fetal Sex and Diagnosis Contribute Most to Principal Components*

We next analyzed which gene sets contributed most to the PCA and used gene-set enrichment analysis of the gene loadings. PC 1 and, to a lesser extent, PC 2 were strongly associated with Y-chromosomal genes. Higher components showed much weaker association with chromosomal locations or functions such as epithelial–mesenchymal transition, mitochondria, cell cycle or heme-metabolism (Figure 5a).

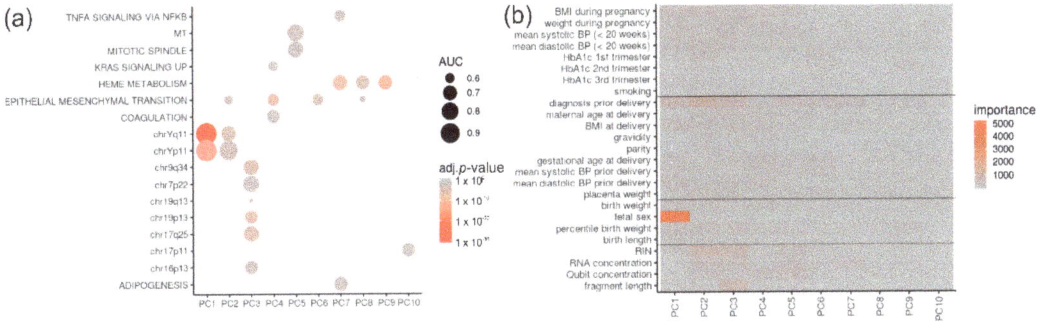


**Figure 5.** Association of principal components to Y-chromosome linked genes and fetal sex. (**a**) Gene-set enrichment analysis shows how principal components (PC) are associated with particular gene sets. PC1 and PC2 are strongly influenced by Y-chromosome genes. The effect size (AUC) is shown as dot size and the color indicates significance (adjusted $p$-value). (**b**) The heat map shows the contribution of clinical (meta data) and quality control parameters to the PCs. The impact ("importance") is color-coded with low contribution in grey and high contribution in red. Fetal sex clearly contributes to PC1. The figure is based on all CTL ($n$ = 29) and diabetic placenta samples without PE ($n$ = 32).

As none of the gene sets convincingly explained the variation in PC2 and below, we next investigated whether PCs were associated with clinical data (meta data) or technical quality-control parameters. We used random forest regression to determine the contribution of each clinical parameter to a particular PC. Again, fetal sex contributed overwhelmingly to the first PC, while the diagnosis prior to delivery and other parameters, such as BMI, Hba1c, blood pressure during pregnancy as well as RNA quality (RIN value), contributed much less to the PCs (Figure 5b).

### 2.5. *Comparison between Placentas of Male or Female Fetus Display Several DEGs*

We next performed differential expression analysis between placentas from female fetus ($n$ = 41) and male fetus ($n$ = 36) pregnancies with clinical diagnoses (diabetes subtypes and PE) as covariates. The comparison revealed 78 up-regulated and 76 down-regulated genes (Supplementary Table S2) that are highlighted in the MA-plot (Figure 6). Genes with increased expression in placentas with male fetuses (e.g., *DDX3Y*, *ZFY*, *KDM5D* and *UTY*) were mainly located on the Y-chromosome, but other DEGs such as *CTFR*, *SPP1* and *ZNF711* were located on the X-chromosome or autosomal chromosomes.

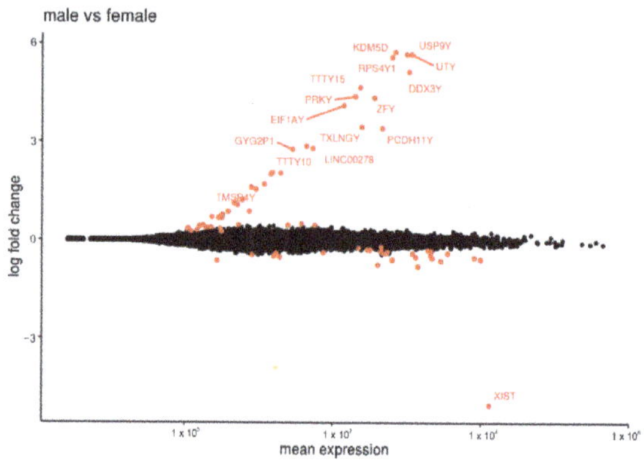


**Figure 6.** Fetal sex effect on gene expression in placenta. The MA-plot displays the effect of fetal sex on gene expression in placentas with male ($n$ = 36) or female fetuses ($n$ = 41). Differential genes (adj. $p$-value < 0.05) are marked in red. Positive log2-fold change indicates male-specific expression; negative log2-fold changes indicate female-specific expression.

The 76 genes with higher expression in placentas with female fetuses were mostly located on the X-chromosome; among them *XIST*, *FTX*, *ZFX*, *SMC1A*, *STS* and *FMR1*. Some DEGs were not located on the X-chromosome but were associated with placentation. For example, *PLAU* has been associated with trophoblast invasion [23] and *SWAP70* has been associated with placentation [24]. We performed analyses of up-regulated and down-regulated genes from the comparison of male and female placenta samples. There were no KEGG terms associated with the gene sets. In addition, no relevant gene ontology terms were present.

Finally, we compared our results to a study by Gonzales et al. on sex differences in the late first trimester in the human placenta transcriptome (GSE109120) [25] and found a good overlap: 33 male and 15 female genes were significantly different in both studies; six male and two female genes were discovered only in our data; and 13 male and 11 female genes were detected only in their study.

## 3. Discussion

In the present study, we show that the placental transcriptome signature from healthy CTL is similar to pregnancies complicated by DM. When diabetic pregnancy was complicated with PE, again only mild differences in the placental transcriptome were observed compared to healthy CTL. At the transcriptome level, the placentas from women with GDM were more similar to CTL than DM1 or DM2, while the latter two were relatively similar to each other. Genes linked to pregnancy complications or metabolic diseases contributed little to the observed variance, and no defined gene set or pathway could be directly attributed as the main contributor to the observed differences. The analysis of clinical data showed only minor contribution of the clinical diagnosis to the variances between diabetic subgroups. Remarkably, we identified fetal sex to be the strongest contributor to differences in the transcriptome profile. Our study has two unexpected results, which warrant further studies and follow up. Firstly, even though the placenta is heavily affected by the various forms of DM and by superimposing PE, the transcriptome profile appears to be only marginally altered. However, the transcriptome does neither fully reflect the profound metabolic and pro-inflammatory alterations in the pregnant mother, nor does it fully explain the adverse maternal or fetal morbidity and mortality. Our second remarkable finding is that fetal sex has a profound influence on the placental transcriptome, indicating that sex-specific alterations in placental function are more important than previously expected and supports that the sexual dimorphism of the placenta should not be ignored in scientific practice.

RNAseq is a powerful method to quantify transcriptomes [26]. It allows identification of pathologic alterations that are linked to clinical diagnosis, which can help to develop new biomarkers for future prediction, diagnosis, and therapy.

Several pregnancy complications have been linked to altered gene expression via studies on specific genes and transcriptome analysis using microarray or RNAseq [27–29]. Sõber et al. highlighted differences in gene expression pattern with RNAseq analysis ($n = 8$/group) of placentas from women with PE, while GDM or small- and large-for-gestational age showed less intensive expression differences [30]. Although the sample number was relatively small, the researchers found that the GDM placenta transcriptome differed the least from healthy CTL placentas, while placentas from preeclamptic women showed the strongest differences in gene expression pattern, followed by those small- or large-for-gestational-age. Lekva et al. did not find altered genes in the transcriptome profile of placentas from women with GDM, and only five DEGs in term placentas from women with PE [31]. The observations on GDM placental transcriptome of Sõber et al. and Lekva et al., along with our findings, suggest that gene expression in the placenta is not sufficiently altered during a diabetic pregnancy to produce observable effects beyond inter-individual variability, possibly due to well-treated disease. In our study cohort, 35.7% of women with GDM were treated with insulin and/or metformin (Tables 1 and 2). The lack of an effect on the placental transcriptome could result from relatively mild BMI at delivery (GMD: 33.2; GMD + PE: 38.5) in our study, which serves as an indicator for morbid obesity, and is related to inflammation and dysregulated placental function [32].

Multiple other studies highlight the transcriptome profile of placentas from women with GDM compared to controls. While two microarray studies identified seven [33] and 66 [34] DEGs, respectively, which were associated with apoptosis and inflammation, another RNAseq study found 281 DEGs [35]. Since GDM is a time-restricted disease during pregnancy (although the risk of DM2 is increased long-term), the question arises as to whether the placental transcriptome reflects the metabolic changes of that period. This question is so far unresolved, as some studies found no or only minor changes in placental gene expression [30,31] when comparing GDM to healthy controls, while other studies identified DEG patterns in placentas of women with GDM and assigned dysregulated genes to pathways of glucose metabolism and immunology [36–38].

Only few studies have considered pregestational diabetes subtypes. The published analyses include small sample numbers ($n = 3$ and $n = 6$, respectively), which makes expression differences difficult to be identified [39,40].

Limiting to our study design is the relatively small sample size in some subgroups, which is attributed to the limited number of women suffering from multiple pregnancy-related complications. Therefore, careful interpretation of the data is necessary. Nonetheless, the variety of subgroups characterized gives a rare insight into the placenta transcriptome and results should be judged wisely. These limitations go along with a low number of studies focusing on transcriptome analysis of placentas from women with DM2 because of limited numbers of pregnant women suffering from DM2 [41]. In one study, placentas from women with GDM ($n = 14$) and DM2 ($n = 3$) were analyzed by RNAseq and DNA methylation [39]. The authors report differences in methylation and transcriptome level in placentas from male and female offspring to be more pronounced than the difference between clinical diagnosis. In our transcriptome analysis, placentas from women with DM2 showed the most distinct pattern compared to both healthy CTL placentas and placentas from women with GDM, however the number was also small. These findings are reasonable as the placenta is exposed to a diabetic surrounding. While some studies focus on the analysis of pre-selected genes in placenta tissues from women with DM1 [42,43], global transcriptome analysis is lacking.

In contrast to our results, other studies have identified multiple DEGs in placentas from women with PE, yet these women did not have additionally diagnosed DM [44]. Buckberry et al. identified gene sets in the human placenta that were preserved at different timepoints of gestation, with altered expression patterns in placenta samples from women with PE [45]. In another study with previously identified gene expression-based PE subtypes, the severity of histopathological placental lesions matched the PE subtypes [46].

Although various studies suggest that placental inflammatory pathways, stress response and gene expression patterns are related to maternal pregnancy complications, we could not strengthen these observations with our study [38,47,48]. In our RNAseq data, fetal sex contributed most to the observed transcriptome pattern. The impact of fetal sex on the placental transcriptome had previously been observed in a microarray analysis study. Placentas from women with PE complicated with either HELLP (= hemolysis, elevated liver enzymes, low platelet), IUGR (= intrauterine growth restriction) or SGR (= small for gestational age), showed DEG patterns that varied due to fetal sex [49]. A RNAseq analysis of first trimester human placentas from healthy women highlighted early differences in the transcriptome with 58 DEGs between placentas from female and male fetuses [25]. Gonzales et al. identified genes located on Chromosome 19 contributing most to DEGs, followed by genes on the Y-chromosome [25]. Also, in our data set, some DEGs were located on Chromosome 19, while a larger number were located on gonosomes. Besides X-chromosomal-linked genes, Sood et al. detected some autosomal genes, suggesting that the difference in expression might be due to underlying differences in male and female physiology [50]. Another study analyzed placental transcriptomes of women with PE or fetal growth restriction using RNAseq; this found strong placental transcriptome clustering according to fetal sex and identified sex-biased pathways [51]. The meta-analysis of microarray data highlighted 88 autosomal genes that were differentially expressed between placentas bearing a male or female fetus [45]. Altogether, these observations are in agreement with our findings and verify that the fetal sex strongly contributes to the placental transcriptome profile.

Our findings comply with other specific features related to fetal sex. Sex-specific placental differences are relevant for fetal growth, preterm birth, and survival [52]. In addition, sex-specific alterations of gene expression have not only been reported in genes located on either the X- or Y-chromosome, but also on autosomal genes that encode immune and hormonal pathways [52].

Our data contribute to the concept of the placenta as a sexual dimorphic organ. It also suggests that the transcriptional signature of the placenta is not very informative for understanding maternal–placental–fetal health in the context of well-treated diabetic pregnancies and superimposed PE. Our data is in line with previous findings that emphasize the important influence of fetal sex on the placental transcriptome. In our analysis, the effect on variance by fetal sex is stronger than the clinical diagnosis. Thus, we feel confident to advise all researchers who aim to investigate the placental transcriptome profiles not only to adjust for fetal sex, but to consider fetal sex in their experimental planning, including for sample size calculation.

In summary, our data underlines the concept that the placenta is a sexual dimorphic organ with gonosomal genes strongly contributing to the transcriptome signature of the placenta. Future studies are needed to clarify which adjustments to a pathological pregnancy are sex-specific and which are not. Furthermore, the role of epigenetic alterations in the placenta as the result of exposure to the diabetic milieu in pregnancy should be explored.

## 4. Materials and Methods

### *4.1. Study Population and Sample Collection*

The placenta samples were collected between 2001–2013 at the Oslo University Hospital, Norway as a part of the Oslo Pregnancy Biobank. The study was approved by the Regional Committee of Medical Research Ethics in South East Norway (Oslo Pregnancy Biobank REK: 2010/1850/REK South East C). The population used in this study includes 77 placenta samples from women with either healthy or complicated pregnancy. The set consists of 29 healthy controls (CTL), 12 women with gestational diabetes (GDM), 17 women with type I diabetes mellitus (DM1) and three women with type II diabetes mellitus (DM2). The DM patients were grouped according to the World Health Organization criteria [53,54]. PE was diagnosed on the basis of new-onset hypertension (>140/90 mmHg) and proteinuria during pregnancy [55]. A written informed consent was provided by all patients.

The placenta was delivered following caesarean section as previously described [56,57]. Briefly, following the delivery of the baby, 3–5 IU oxytocin was given to the mother intravenously. The placenta was separated spontaneously from the uterine wall and gently removed. Placental plus umbilical cord weight was noted. The placental villous biopsies were taken from macroscopically normal-appearing cotyledons, avoiding the decidual layer as previously described [54,58]. After collection, tissue samples were immediately frozen and stored at −80 °C until further analyzed. Blood sample biochemistry, blood pressure and BMI were analyzed as previously described [59]. The newborn weight percentiles were calculated according to Norwegian fetal growth curves as previously described [60].

*4.2. RNA Isolation*

RNA was isolated from 77 placenta samples using Qiagen RNeasy Kit (Qiagen). After homogenization of the placental tissue sample, the RNA extraction was performed following the manufacturer's protocol. Only 160 µL of the watery phase was combined with an equal amount of 70% ethanol and loaded onto the RNeasy Mini Column. The RNA was eluted with 40 µL RNase-free water. The RNA concentration, size range and quality were measured using Agilent Bioanalyser 2100, Eukaryote Total RNA Nano Series II according to the manufacturer's protocol (Agilent RNA 6000 Nano Kit Guide). A Qubit Fluorometric Quantitation Assay was used to validate RNA concentration. Samples used for RNA sequencing data analysis had a mean RIN of 5.1 (±1.09 SD).

*4.3. RNA Sequencing*

The Illumina TruSeq stranded total RNA Library Prep Kit was used for library preparation and RNA samples were diluted in water to 1000 ng/µL. The sequencing was done on the Illumina HiSeq4000 system at the Scientific Genomics Platforms at the Max Delbrück Centre for Molecular Medicine, Berlin. A loading concentration of 200 pM, paired-end run-type mode and a read length of 75 bp was used.

*4.4. Sequencing Data Processing*

Sequencing reads were aligned to the human genome (GRCh38) using STAR (v2.6.1a, Dobin et al., USA) [61]. Gene expression was quantified using featureCounts (v1.6.3, online available at www.bioconductor.org, Liao et al., USA) [62] and the Gencode v25 reference, including non-coding genes. We then used DESeq2 (v1.18.1, online available at www.bioconductor.org, Love et al., USA, DE) [15] to detect differentially expressed genes for comparisons between groups (DM/CTL/PE) and subgroups (DM1/DM2/GDM with and without PE), using fetal sex as a covariate. Gene-set enrichment on estimated log2-fold change values was performed using tmod [63] and Hallmark gene sets from MSigDB (version 7, Broad institute, San Diego, CA, USA). For the principal component analysis (PCA) we used regularized log2-transformed counts for the top 5000 variable genes. Gene-set enrichment on PC gene scores was performed using tmod [63] and Hallmark and positional gene sets from MSigDB (version 7, Broad institute, San Diego, CA, USA). We used random forest regression with the randomForest package v4.7-14 [64] to infer the contribution of each clinical parameter to the principal components of the analysis, imputing missing values with the roughfix method.

*4.5. Statistics*

The sequencing data was statistically analyzed using R software (v3.4.4, R core team, online available www.r-project.org), SPSS (v1.2.0, IBM, USA), GraphPad Prism (v6, GraphPad Software, US) and Microsoft Excel (v2211, Microsoft 365, USA). Clinical parameters are displayed as mean ± standard deviation or a percentage. Group differences were tested with one-way ANOVA with Sidak's multiple comparisons, adjusted $p$-values are indicated and significant when $p < 0.05$.

**Supplementary Materials:** The following supporting information can be downloaded at: https://www.mdpi.com/article/10.3390/ijms232315388/s1; Supplementary Table S1, Supplementary Table S2.

**Author Contributions:** Conceptualization, M.S., F.H., A.C.S. and R.D.; methodology, S.M.K., B.O., M.S., K.K. and A.C.S.; software, S.M.K. and B.O.; formal analysis, S.M.K., B.O., F.H. and R.D.; writing—original draft preparation, S.M.K., B.O. and R.D.; writing—review and editing, S.M.K., B.O., M.S., F.H., K.K., N.H., I.M.L., D.N.M., A.C.S., D.B. and R.D.; visualization, S.M.K. and B.O.; supervision, F.H., D.B. and R.D.; project administration, M.S., F.H., N.H., D.N.M., A.C.S., D.B. and R.D.; funding acquisition, M.S., F.H., N.H., D.N.M., A.C.S., D.B. and R.D. All authors have read and agreed to the published version of the manuscript.

**Funding:** The Deutsche Forschungsgemeinschaft supported K. Kräker (DE 631/15-1). Meryam Sugulle receives a grant from the Research Council of Norway (297333) via ERA-CVD-NET.

**Institutional Review Board Statement:** The study was conducted in accordance with the Declaration of Helsinki, and approved by the Regional Committee of Medical Research Ethics in South East Norway (Oslo Pregnancy Biobank REK: 2010/1850/REK South East C).

**Informed Consent Statement:** Informed consent was obtained from all subjects involved in the study.

**Data Availability Statement:** The data sets generated during and/or analyzed during the current study are available from the corresponding authors on reasonable request.

**Conflicts of Interest:** The authors declare no conflict of interest.

## References


1. Aplin, J.D.; Myers, J.E.; Timms, K.; Westwood, M. Tracking placental development in health and disease. *Nat. Rev. Endocrinol.* **2020**, *16*, 479–494. [CrossRef]
2. Desoye, G.; Hauguel-De Mouzon, S. The human placenta in gestational diabetes mellitus. The insulin and cytokine network. *Diabetes Care* **2007**, *30* (Suppl. S2), S120–S126. [CrossRef]
3. Brown, M.A.; Magee, L.A.; Kenny, L.C.; Karumanchi, S.A.; McCarthy, F.; Saito, S.; Hall, D.R.; Warren, C.E.; Adoyi, G.; Ishaku, S. Hypertensive Disorders of Pregnancy: ISSHP Classification, Diagnosis, and Management Recommendations for International Practice. *Hypertension* **2018**, *72*, 24–43. [CrossRef] [PubMed]
4. Benirschke, K.; Burton, G.J.; Baergen, R.N. (Eds.) *Pathology of the Human Placenta*, 6th ed.; Springer: Berlin/Heidelberg, Germany, 2012.
5. American Diabetes Association. 14. Management of Diabetes in Pregnancy: Standards of Medical Care in Diabetes—2020. *Diabetes Care* **2020**, *43* (Suppl. S1), S183–S192. [CrossRef] [PubMed]
6. Desoye, G.; Cervar-Zivkovic, M. Diabetes Mellitus, Obesity, and the Placenta. *Obstet. Gynecol. Clin. N. Am.* **2020**, *47*, 65–79. [CrossRef] [PubMed]
7. Huynh, J.; Dawson, D.; Roberts, D.; Bentley-Lewis, R. A systematic review of placental pathology in maternal diabetes mellitus. *Placenta* **2015**, *36*, 101–114. [CrossRef]
8. Ferrara, A. Increasing prevalence of gestational diabetes mellitus: A public health perspective. *Diabetes Care* **2007**, *30* (Suppl. S2), S141–S146. [CrossRef]
9. Yessoufou, A.; Moutairou, K. Maternal diabetes in pregnancy: Early and long-term outcomes on the offspring and the concept of "metabolic memory". *Exp. Diabetes Res.* **2011**, *2011*, 218598. [CrossRef]
10. Nelson, D.M. How the placenta affects your life, from womb to tomb. *Am. J. Obstet. Gynecol.* **2015**, *213* (Suppl. S4), S12–S13. [CrossRef] [PubMed]
11. Duckitt, K.; Harrington, D. Risk factors for pre-eclampsia at antenatal booking: Systematic review of controlled studies. *BMJ* **2005**, *330*, 565. [CrossRef]
12. Rana, S.; Lemoine, E.; Granger, J.P.; Karumanchi, S.A. Preeclampsia: Pathophysiology, Challenges, and Perspectives. *Circ. Res.* **2019**, *124*, 1094–1112. [CrossRef] [PubMed]
13. Roberts, C.T. IFPA Award in Placentology Lecture: Complicated interactions between genes and the environment in placentation, pregnancy outcome and long term health. *Placenta* **2010**, *31*, S47–S53. [CrossRef] [PubMed]
14. Gauster, M.; Desoye, G.; Tötsch, M.; Hiden, U. The placenta and gestational diabetes mellitus. *Curr. Diab. Rep.* **2012**, *12*, 16–23. [CrossRef] [PubMed]
15. Love, M.I.; Huber, W.; Anders, S. Moderated estimation of fold change and dispersion for RNA-seq data with DESeq2. *Genome Biol.* **2014**, *15*, 550. [CrossRef] [PubMed]

16. Steinbrenner, H.; Micoogullari, M.; Hoang, N.A.; Bergheim, I.; Klotz, L.-O.; Sies, H. Selenium-binding protein 1 (SELENBP1) is a marker of mature adipocytes. *Redox Biol.* **2019**, *20*, 489–495. [CrossRef]
17. Huang, H.-H.; Condello, G.; Chen, C.-Y. Towards a link between magnesium, exercise, and risk of type 2 diabetes mellitus. *J Chin. Med. Assoc.* **2019**, *82*, 527–528. [CrossRef]
18. Zampieri, T.T.; Ramos-Lobo, A.M.; Furigo, I.C.; Pedroso, J.A.B.; Buonfiglio, D.C.; Donato, J. SOCS3 deficiency in leptin receptor-expressing cells mitigates the development of pregnancy-induced metabolic changes. *Mol. Metab.* **2015**, *4*, 237–245. [CrossRef]
19. Duttaroy, A.K. Transport of fatty acids across the human placenta: A review. *Prog. Lipid Res.* **2009**, *48*, 52–61. [CrossRef]
20. Chalise, J.P.; Hashimoto, S.; Parajuli, G.; Kang, S.; Singh, S.K.; Gemechu, Y.; Metwally, H.; Nyati, K.K.; Dubey, P.K.; Zaman, M.M.-U.; et al. Feedback regulation of Arid5a and Ppar-γ2 maintains adipose tissue homeostasis. *Proc. Natl. Acad. Sci. USA* **2019**, *116*, 15128–15133. [CrossRef]
21. Liu, G.-M.; Zeng, H.-D.; Zhang, C.-Y.; Xu, J.-W. Key genes associated with diabetes mellitus and hepatocellular carcinoma. *Pathol. Res. Pract.* **2019**, *215*, 152510. [CrossRef]
22. Yeo, S.-H.; Noh, J.-R.; Kim, Y.-H.; Gang, G.-T.; Kim, S.-W.; Kim, K.-S.; Hwang, J.H.; Shong, M.; Lee, C.-H. Increased vulnerability to β-cell destruction and diabetes in mice lacking NAD(P)H:quinone oxidoreductase 1. *Toxicol. Lett.* **2013**, *219*, 35–41. [CrossRef]
23. Lash, G.E.; Otun, H.A.; Innes, B.A.; Bulmer, J.N.; Searle, R.F.; Robson, S.C. Low oxygen concentrations inhibit trophoblast cell invasion from early gestation placental explants via alterations in levels of the urokinase plasminogen activator system. *Biol. Reprod.* **2006**, *74*, 403–409. [CrossRef] [PubMed]
24. Liu, J.; Li, D.; Cao, B.; Li, Y.-X.; Herva, R.; Piao, Y.-S.; Wang, Y.-L. Expression and localization of SWAP-70 in human fetomaternal interface and placenta during tubal pregnancy and normal placentation. *J. Histochem. Cytochem.* **2007**, *55*, 701–708. [CrossRef] [PubMed]
25. Gonzalez, T.L.; Sun, T.; Koeppel, A.F.; Lee, B.; Wang, E.T.; Farber, C.R.; Rich, S.S.; Sundheimer, L.W.; Buttle, R.A.; Chen, Y.-D.I.; et al. Sex differences in the late first trimester human placenta transcriptome. *Biol. Sex Differ.* **2018**, *9*, 4. [CrossRef] [PubMed]
26. Wang, Z.; Gerstein, M.; Snyder, M. RNA-Seq: A revolutionary tool for transcriptomics. *Nat. Rev. Genet.* **2009**, *10*, 57–63. [CrossRef]
27. Yong, H.E.J.; Chan, S.-Y. Current approaches and developments in transcript profiling of the human placenta. *Hum. Reprod. Update* **2020**, *26*, 799–840. [CrossRef]
28. Herse, F.; Dechend, R.; Harsem, N.K.; Wallukat, G.; Janke, J.; Qadri, F.; Hering, L.; Muller, D.N.; Luft, F.C.; Staff, A.C. Dysregulation of the circulating and tissue-based renin-angiotensin system in preeclampsia. *Hypertension* **2007**, *49*, 604–611. [CrossRef]
29. Gong, S.; Gaccioli, F.; Dopierala, J.; Sovio, U.; Cook, E.; Volders, P.-J.; Martens, L.; Kirk, P.D.W.; Richardson, S.; Smith, G.C.S.; et al. The RNA landscape of the human placenta in health and disease. *Nat. Commun.* **2021**, *12*, 2639. [CrossRef]
30. Sõber, S.; Reiman, M.; Kikas, T.; Rull, K.; Inno, R.; Vaas, P.; Teesalu, P.; Marti, J.M.L.; Mattila, P.; Laan, M. Extensive shift in placental transcriptome profile in preeclampsia and placental origin of adverse pregnancy outcomes. *Sci. Rep.* **2015**, *5*, 13336. [CrossRef]
31. Lekva, T.; Lyle, R.; Roland, M.C.P.; Friis, C.; Bianchi, D.W.; Jaffe, I.Z.; Norwitz, E.R.; Bollerslev, J.; Henriksen, T.; Ueland, T. Gene expression in term placentas is regulated more by spinal or epidural anesthesia than by late-onset preeclampsia or gestational diabetes mellitus. *Sci. Rep.* **2016**, *6*, 29715. [CrossRef]
32. Aye, I.L.; Lager, S.; Ramirez, V.I.; Gaccioli, F.; Dudley, D.J.; Jansson, T.; Powell, T. Increasing maternal body mass index is associated with systemic inflammation in the mother and the activation of distinct placental inflammatory pathways. *Biol. Reprod.* **2014**, *90*, 129. [CrossRef] [PubMed]
33. Magee, T.R.; Ross, M.G.; Wedekind, L.; Desai, M.; Kjos, S.; Belkacemi, L. Gestational diabetes mellitus alters apoptotic and inflammatory gene expression of trophobasts from human term placenta. *J. Diabetes Complicat.* **2014**, *28*, 448–459. [CrossRef] [PubMed]
34. Enquobahrie, D.A.; Williams, M.A.; Qiu, C.; Meller, M.; Sorensen, T.K. Global placental gene expression in gestational diabetes mellitus. *Am. J. Obstet. Gynecol.* **2009**, *200*, 206.e1–206.e13. [CrossRef] [PubMed]
35. Ding, R.; Guo, F.; Zhang, Y.; Liu, X.-M.; Xiang, Y.-Q.; Zhang, C.; Liu, Z.-W.; Sheng, J.-Z.; Huang, H.-F.; Zhang, J.; et al. Integrated Transcriptome Sequencing Analysis Reveals Role of miR-138-5p/ TBL1X in Placenta from Gestational Diabetes Mellitus. *Cell Physiol. Biochem.* **2018**, *51*, 630–646. [CrossRef]
36. Binder, A.M.; LaRocca, J.; Lesseur, C.; Marsit, C.J.; Michels, K.B. Epigenome-wide and transcriptome-wide analyses reveal gestational diabetes is associated with alterations in the human leukocyte antigen complex. *Clin. Epigenetics* **2015**, *7*, 79. [CrossRef]
37. Tang, L.; Li, P.; Li, L. Whole transcriptome expression profiles in placenta samples from women with gestational diabetes mellitus. *J. Diabetes Investig.* **2020**, *11*, 1307–1317. [CrossRef]
38. Radaelli, T.; Varastehpour, A.; Catalano, P.; Hauguel-De Mouzon, S. Gestational diabetes induces placental genes for chronic stress and inflammatory pathways. *Diabetes* **2003**, *52*, 2951–2958. [CrossRef]
39. Alexander, J.; Teague, A.M.; Chen, J.; Aston, C.E.; Leung, Y.-K.; Chernausek, S.; Simmons, R.; Pinney, S.E. Offspring sex impacts DNA methylation and gene expression in placentae from women with diabetes during pregnancy. *PLoS ONE* **2018**, *13*, e0190698. [CrossRef]
40. Radaelli, T.; Lepercq, J.; Varastehpour, A.; Basu, S.; Catalano, P.M.; Hauguel-De Mouzon, S. Differential regulation of genes for fetoplacental lipid pathways in pregnancy with gestational and type 1 diabetes mellitus. *Am. J. Obstet. Gynecol.* **2009**, *201*, 209.e1–209.e10. [CrossRef]

41. Coton, S.J.; Nazareth, I.; Petersen, I. A cohort study of trends in the prevalence of pregestational diabetes in pregnancy recorded in UK general practice between 1995 and 2012. *BMJ Open* **2016**, *6*, e009494. [CrossRef] [PubMed]
42. Lindegaard, M.L.S.; Damm, P.; Mathiesen, E.R.; Nielsen, L.B. Placental triglyceride accumulation in maternal type 1 diabetes is associated with increased lipase gene expression. *J. Lipid Res.* **2006**, *47*, 2581–2588. [CrossRef] [PubMed]
43. Iciek, R.; Wender-Ozegowska, E.; Zawiejska, A.; Mikolajczak, P.; Mrozikiewicz, P.M.; Pietryga, M.; Brazert, J. Placental leptin and its receptor genes expression in pregnancies complicated by type 1 diabetes. *J. Physiol. Pharmacol.* **2013**, *64*, 579–585. [PubMed]
44. Kaartokallio, T.; Cervera, A.; Kyllönen, A.; Laivuori, K.; Kere, J.; Laivuori, H. Gene expression profiling of pre-eclamptic placentae by RNA sequencing. *Sci. Rep.* **2015**, *5*, 14107. [CrossRef] [PubMed]
45. Buckberry, S.; Bianco-Miotto, T.; Bent, S.J.; Clifton, V.; Shoubridge, C.; Shankar, K.; Roberts, C.T. Placental transcriptome co-expression analysis reveals conserved regulatory programs across gestation. *BMC Genom.* **2017**, *18*, 10. [CrossRef] [PubMed]
46. Benton, S.J.; Leavey, K.; Grynspan, D.; Cox, B.J.; Bainbridge, S.A. The clinical heterogeneity of preeclampsia is related to both placental gene expression and placental histopathology. *Am. J. Obstet. Gynecol.* **2018**, *219*, 604.e1–604.e25. [CrossRef] [PubMed]
47. Li, H.-P.; Chen, X.; Li, M.-Q. Gestational diabetes induces chronic hypoxia stress and excessive inflammatory response in murine placenta. *Int. J. Clin. Exp. Pathol.* **2013**, *6*, 650–659.
48. Gauster, M.; Majali-Martinez, A.; Maninger, S.; Gutschi, E.; Greimel, P.H.; Ivanisevic, M.; Djelmis, J.; Desoye, G.; Hiden, U. Maternal Type 1 diabetes activates stress response in early placenta. *Placenta* **2017**, *50*, 110–116. [CrossRef]
49. Kang, Q.; Li, W.; Xiao, J.; Yu, N.; Fan, L.; Sha, M.; Ma, S.; Wu, J.; Chen, S. Integrated analysis of multiple microarray studies to identify novel gene signatures in preeclampsia. *Placenta* **2021**, *105*, 104–118. [CrossRef]
50. Sood, R.; Zehnder, J.L.; Druzin, M.L.; Brown, P.O. Gene expression patterns in human placenta. *Proc. Natl. Acad. Sci. USA* **2006**, *103*, 5478–5483. [CrossRef]
51. Gong, S.; Sovio, U.; Aye, I.L.; Gaccioli, F.; Dopierala, J.; Johnson, M.D.; Wood, A.M.; Cook, E.; Jenkins, B.J.; Koulman, A.; et al. Placental polyamine metabolism differs by fetal sex, fetal growth restriction, and preeclampsia. *JCI Insight* **2018**, *3*, e120723. [CrossRef]
52. Clifton, V.L. Review: Sex and the human placenta: Mediating differential strategies of fetal growth and survival. *Placenta* **2010**, *31*, S33–S39. [CrossRef] [PubMed]
53. Alberti, K.; Zimmet, P.Z. for the WHO Consultation. Definition, diagnosis and classification of diabetes mellitus and its complications. Part 1: Diagnosis and classification of diabetes mellitus. Provisional report of a WHO Consultation. *Diabet. Med.* **1998**, *15*, 539–553. [CrossRef]
54. Sugulle, M.; Dechend, R.; Herse, F.; Weedon-Fekjaer, M.S.; Johnsen, G.M.; Brosnihan, K.B.; Anton, L.; Luft, F.; Wollert, K.C.; Kempf, T.; et al. Circulating and placental growth-differentiation factor 15 in preeclampsia and in pregnancy complicated by diabetes mellitus. *Hypertension* **2009**, *54*, 106–112. [CrossRef] [PubMed]
55. Brown, M.A.; Lindheimer, M.D.; de Swiet, M.; Van Assche, A.; Moutquin, J.M. The classification and diagnosis of the hypertensive disorders of pregnancy: Statement from the International Society for the Study of Hypertension in Pregnancy (ISSHP). *Hypertens Pregnancy* **2001**, *20*, IX–XIV. [CrossRef]
56. Harsem, N.K.; Staff, A.C.; He, L.; Roald, B. The decidual suction method: A new way of collecting decidual tissue for functional and morphological studies. *Acta Obstet. Gynecol. Scand.* **2004**, *83*, 724–730. [CrossRef]
57. Staff, A.C.; Ranheim, T.; Khoury, J.; Henriksen, T. Increased contents of phospholipids, cholesterol, and lipid peroxides in decidua basalis in women with preeclampsia. *Am. J. Obstet. Gynecol.* **1999**, *180*, 587–592. [CrossRef]
58. Weedon-Fekjær, M.; Sheng, Y.; Sugulle, M.; Johnsen, G.; Herse, F.; Redman, C.; Lyle, R.; Dechend, R.; Staff, A. Placental miR-1301 is dysregulated in early-onset preeclampsia and inversely correlated with maternal circulating leptin. *Placenta* **2014**, *35*, 709–717. [CrossRef]
59. Moe, K.; Alnaes-Katjavivi, P.; Størvold, G.L.; Sugulle, M.; Johnsen, G.M.; Redman, C.W.; Dechend, R.; Staff, A.C. Classical Cardiovascular Risk Markers in Pregnancy and Associations to Uteroplacental Acute Atherosis. *Hypertension* **2018**, *72*, 695–702. [CrossRef]
60. Rieber-Mohn, A.B.; Sugulle, M.; Wallukat, G.; Alnæs-Katjavivi, P.; Størvold, G.L.; Bolstad, N.; Redman, C.W.; Dechend, R.; Staff, A.C. Auto-antibodies against the angiotensin II type I receptor in women with uteroplacental acute atherosis and preeclampsia at delivery and several years postpartum. *J. Reprod. Immunol.* **2018**, *128*, 23–29. [CrossRef]
61. Dobin, A.; Davis, C.A.; Schlesinger, F.; Drenkow, J.; Zaleski, C.; Jha, S.; Batut, P.; Chaisson, M.; Gingeras, T.R. STAR: Ultrafast universal RNA-seq aligner. *Bioinformatics* **2013**, *29*, 15–21. [CrossRef]
62. Liao, Y.; Smyth, G.K.; Shi, W. featureCounts: An efficient general purpose program for assigning sequence reads to genomic features. *Bioinformatics* **2014**, *30*, 923–930. [CrossRef] [PubMed]
63. Zyla, J.; Marczyk, M.; Domaszewska, T.; Kaufmann, S.H.E.; Polanska, J.; Weiner, J. Gene set enrichment for reproducible science: Comparison of CERNO and eight other algorithms. *Bioinformatics* **2019**, *35*, 5146–5154. [CrossRef] [PubMed]
64. Breiman, L. Random Forests. *Mach. Learn.* **2001**, *45*, 5–32. [CrossRef]

International Journal of *Molecular Sciences*

MDPI

*Article*

# SLC38A4 Amino Acid Transporter Expression Is Significantly Lower in Early Preterm Intrauterine Growth Restriction Complicated Placentas

**Elif Kadife [1,2], Alesia Harper [1,2], Natasha De Alwis [2,3], Keegan Chien [1,2], Natalie Hannan [2,3] and Fiona C. Brownfoot [1,2,*]**

1 Obstetric Diagnostics and Therapeutics Group, Department of Obstetrics and Gynaecology, University of Melbourne, 163 Studley Road, Heidelberg, VIC 3084, Australia
2 Mercy Perinatal, 163 Studley Road, Heidelberg, VIC 3084, Australia
3 Therapeutics Discovery and Vascular Function Group, Department of Obstetrics and Gynaecology, University of Melbourne, 163 Studley Road, Heidelberg, VIC 3084, Australia
* Correspondence: fiona.brownfoot@unimelb.edu.au

**Abstract:** Intrauterine growth restriction (IUGR), predominantly caused by placental insufficiency, affects partitioning of nutrients to the fetus. The system A sodium-coupled transporters (SNAT or SLC38), of types A1, A2, and A4, control non-essential amino acid uptake and supply. Here, we aimed to investigate the expression of these transporters across different placental disease cohorts and cells. To determine disease impact, transporter expressions at the gene (qPCR) and protein (western blots) level were assessed in gestationally matched placental tissues. Early (<34 weeks), and late (34–36 weeks) onset IUGR cases with/out preeclampsia were compared to preterm controls. We also investigated level of transporter expression in primary trophoblasts under glucose deprivation (n = 6) and hypoxia conditions (n = 7). SLC38A4 protein was significantly downregulated in early preterm pregnancies complicated with IUGR with/out preeclampsia. There were no differences in late preterm IUGR cohorts. Furthermore, we demonstrate for the first time in primary trophoblast cells, that gene expression of the transporters was sensitive to and induced by glucose starvation. SLC38A4 mRNA expression was also significantly upregulated in response to hypoxia. Thus, SLC38A4 expression was persistently low in early preterm IUGR pregnancies, regardless of disease aetiology. This suggests that gestational age at delivery, and consequently IUGR severity, may influence loss of its expression.

**Keywords:** fetal growth restriction; human; placenta; amino acid transporters; SLC38A1; SLC38A2; SLC38A4

**Citation:** Kadife, E.; Harper, A.; De Alwis, N.; Chien, K.; Hannan, N.; Brownfoot, F.C. SLC38A4 Amino Acid Transporter Expression Is Significantly Lower in Early Preterm Intrauterine Growth Restriction Complicated Placentas. *Int. J. Mol. Sci.* **2023**, *24*, 403. https://doi.org/10.3390/ijms24010403

Academic Editor: Ilona Hromadnikova

Received: 21 November 2022
Revised: 14 December 2022
Accepted: 17 December 2022
Published: 26 December 2022



## 1. Introduction

Intrauterine growth restriction (IUGR) is a leading risk factor for stillbirth [1]. It is predominantly caused by placental insufficiency. Placental insufficiency, with reduced blood flow, also affects the transport and partitioning of nutrients across the maternal microvillous membranes to the fetal facing basal membranes of the syncytiotrophoblasts [2]. Nutrient deprivation further restricts the growth and healthy development of the fetus. Various factors control nutrient exchange across the placenta and, of these, system A transporters coordinate non-essential neutral amino acid uptake and supply [3]. System A transporters (SNAT or SLC38), types A1, A2, and A4, are sodium-dependent and pH-sensitive transporters. Of these, SLC38A1 and SLC38A2 are abundantly expressed in a range of organs, predominantly found in the heart and brain, while SLC38A4 was thought to be exclusively expressed by the liver. However, over the years, all SLC38 subtypes were found to be indispensable for the healthy development of the placenta and the fetus. Recently, SLC38A4 was found to be critical for placental development in mice, with the knockout leading to significant placental hypoplasia and a reduction in placental and fetal weights [4].

However, the expression and activity of these transporters in the dysfunctional placenta and related complications have not been fully established. In some human placentas, the level of reduction in system A transporter expression correlates with the severity of fetal growth restriction and the degree of fetal compromise [5,6]. In contrast, other studies involving animal models and in vitro trophoblast cultures propose enhanced expression and activity of these transporters to compensate for growth-restricted placentas [7–9]. However, compensation in response to maternal undernutrition appears to be specific to the mouse. In primates and rats, calorie, and protein restriction (respectively) cause a net decrease in transporter systems in the placenta, which precedes fetal growth restriction in these models [10,11]. In humans, different disease aetiologies and molecular factors also add further complexities to these findings. Uptake of amino acids is reduced in small for gestational age placentas, but not if they were complicated by preeclampsia [12,13]. Others show that transporter expression in the placenta is increased with fetal macrosomia cases, but not necessarily reduced with low birthweight. The discrepancy in findings may be because of the different clinical features and measures that define IUGR vs small for gestational age vs low birthweight, which can overlap and are sometimes used incorrectly or interchangeably [14,15]. Other limitations that exist in the current literature are the focus and study of one type of transporter, instead of all, as well as the use of inaccurately matched controls.

In this study, we have considered the expression of all three system A transporters, SLC38A1, SLC38A2, and SLC38A4. To address the impact of gestational age at delivery and disease severity, we separated our cohorts as early (>34 weeks) and late (34–36 weeks) preterm cases. To investigate the effect of IUGR and its different aetiologies, we considered cases with IUGR alone or combined with preeclampsia (PE), comparing them to appropriately matched gestational controls. To determine the impact of IUGR relevant stress conditions on expression, we isolated cytotrophoblast cells from term placentas and subjected these cells to glucose deprivation and hypoxia conditions in vitro.

## 2. Results

### 2.1. System A Sodium-Coupled Transporter Expression in Placentas from Pregnancies Complicated by Early Preterm IUGR and Preeclampsia

We assessed gene and protein expressions of SLC38A1, SLC38A2, and SLC38A4 in IUGR alone or preeclampsia + IUGR placental samples collected from early preterm (<34 weeks) gestation and compared these to gestationally matched uncomplicated preterm controls. In the early preterm cohort, there were no significant differences in *SLC38A1, SLC38A2,* and *SLC38A4* mRNA expression in the pathological placentas compared to the controls (Figure 1A–C).

To investigate whether gene expression directly correlates to protein levels, we conducted western blot analysis (Figure 1D). While there were no significant differences in SLC38A1 (Figure 1E), SLC38A4 protein (Figure 1F) expression was significantly lower in both IUGR cohorts compared to early preterm controls ($p < 0.01$ for IUGR and $p < 0.001$ for PE + IUGR). SLC38A2 proteins could not be reliably detected in these samples and were excluded from analysis in all experiments.

### 2.2. System A Sodium-Coupled Transporter Expression in Placentas from Pregnancies Complicated by Late Preterm IUGR and Preeclampsia

We next assessed gene and protein expressions of SLC38A1, SLC38A2, and SLC38A4 in IUGR alone or in preeclampsia + IUGR placental samples collected from late preterm (34–36 weeks) gestation and compared these to gestationally matched controls. Like in early preterm tissues, gene expressions of the transporters were uniform across the late preterm cohorts (34–36 weeks average gestation) (Figure 2A–C). Western blot analysis (Figure 2D) revealed large variations at the protein level (Figure 2E,F). In contrast to the early preterm cases, the average protein expression of SLC38A1 and SLC38A4 in IUGR complicated placentas were not different than the controls (Figure 2E,F). These results highlight that gestational age may be a determining factor for transporter expression.

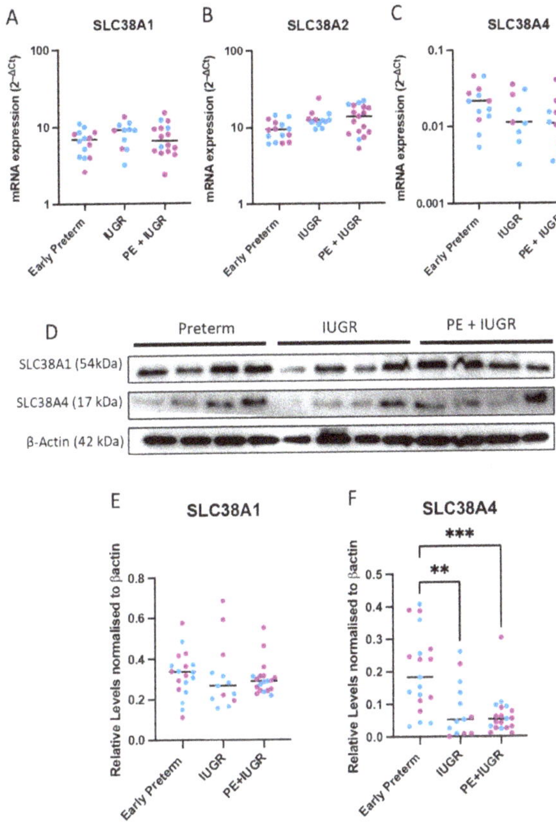


**Figure 1. In early preterm gestation patients, SLC38A4 protein levels were significantly reduced in IUGR complicated placentas.** In IUGR and PE + IUGR placentas, compared to preterm controls, mRNA expressions (assessed via qPCR) of *SLC38A1* (A), *SLC38A2* (B), and *SLC38A4* (C) were not different. Western blots (D) showed that, at the protein level, SLC38A1 (E) was unchanged between groups. SLC38A4 (F) was significantly decreased in both IUGR and PE + IUGR cohorts (n = 11–19 samples/cohort). Results are displayed as medians. ** $p < 0.01$, *** $p < 0.001$ (Kruskal Wallis non-parametric test for qPCR results and Western blot analysis). Blue dots (male pregnancies), purple dots (female pregnancies).

### *2.3. Expression of System A Sodium-Coupled Transporters in Trophoblast Cells under IUGR Relevant Stress Conditions*

To determine the effect of IUGR relevant stress conditions on the expression of these transporters, we isolated trophoblasts from term placenta and exposed them to glucose deprivation and hypoxic conditions in vitro.

In the glucose deprivation experiments, cells were exposed to normal 'high glucose', 'low glucose', and 'no glucose' culture media conditions for 48 h. We demonstrate that glucose deprivation significantly alters gene expression of the transporters (Figure 3A–C). Of these, *SLC38A1* mRNA was significantly upregulated under low glucose conditions ($p < 0.01$), while complete starvation did not significantly alter its expression (Figure 3A). *SLC38A2* mRNA expression was induced in a stepwise manner in response to lowering glucose concentration (Figure 3B) ($p < 0.001$). Finally, *SLC38A4* mRNA expression was significantly induced in glucose starved cells (Figure 3C) ($p < 0.05$). In western blots (Figure 3D), the protein levels of SLC38A1 (Figure 3E) and SLC38A4 (Figure 3F) were unaltered.

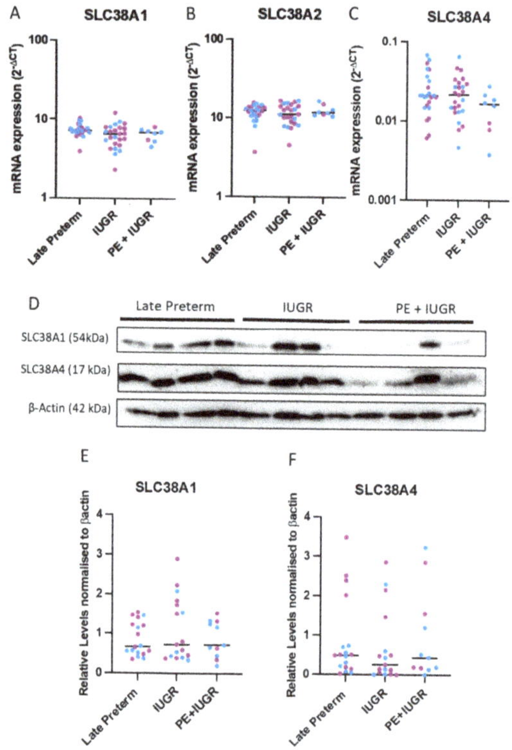


**Figure 2. In late gestation patients, the expression of transporters were indifferent between IUGR complicated placentas and preterm controls.** The mRNA levels (assessed via qPCR) of the transporters (**A–C**) and protein levels (assessed via Western blots) (**D**) when quantified (**E**,**F**) did not reveal changes to the expression of system A amino acid transporters. (n = 8–25 samples/cohort). Results are displayed as medians (Kruskal Wallis non-parametric test). Blue dots (male pregnancies), purple dots (female pregnancies).

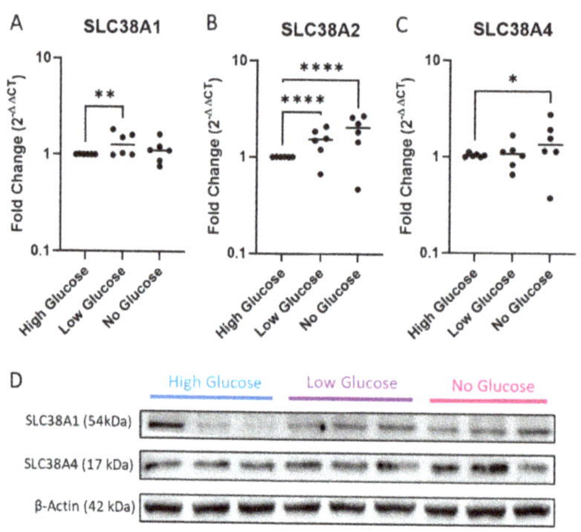


**Figure 3.** *Cont.*

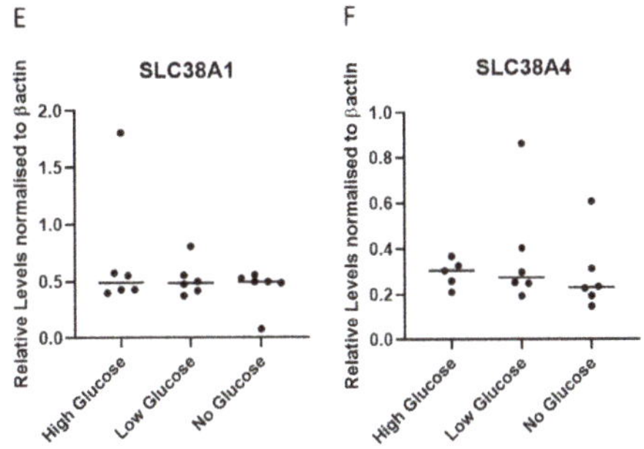


**Figure 3. Glucose deprivation significantly upregulated transporter expression in vitro trophoblast cultures.** Compared to normal media controls, glucose deprived cultures significantly upregulated *SLC38A1* (**A**), *SLC38A2* (**B**), and *SLC38A4* (**C**) expression (n = 6/condition). Additionally, in western blots (**D**), the expressions of SLC38A1 (**E**) and SLC38A4 (**F**) were not significantly altered (n = 6/condition). Results are displayed as medians. * $p < 0.05$, ** $p < 0.01$, **** $p < 0.0001$ (repeated measures are for one-way ANOVA for qPCR analysis and for Western blot analysis).

On average, hypoxic conditions (1% $O_2$) tended to reduce *SLC38A1* (Figure 4A) and *SLC38A2* (Figure 4B) expression while significantly promoting *SLC38A4* expression (Figure 4C) ($p < 0.05$). Western blots (Figure 4D) showed similar expression of transporters between cells incubated under hypoxia and normoxia (8% $O_2$) conditions in the 48-h experimental period (Figure 4E,F).

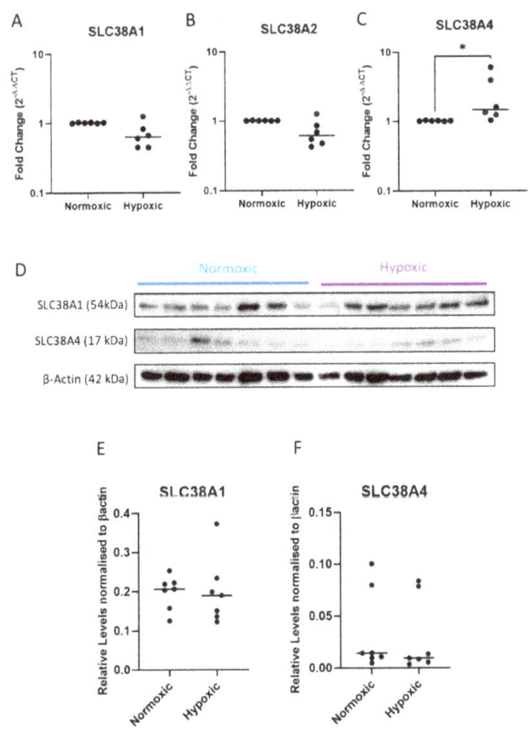


**Figure 4. IUGR relevant hypoxic stress conditions induced SLC38A4 expression at the gene level, however, protein expression remains unaffected in vitro trophoblast cultures.** Compared to normoxic

controls, hypoxic trophoblasts showed insignificant changes to the expressions of *SLC38A1* (**A**) and *SLC38A2* (**B**). *SLC38A4* expression was significantly upregulated under hypoxic conditions (**C**) (n = 6/condition) (assessed via qPCR). Western blots (**D**) showed patient derived variations in response to hypoxia, which did not consistently alter the expressions of SLC38A1 (**E**) or SLC38A4 (**F**) (n = 7/condition). Results are displayed as medians * $p < 0.05$ (the paired *t*-test was used for qPCR analysis, and the Wilcoxon nonparametric test was used for Western blot analysis).

These results suggest that short-term changes in nutrient composition in media may be a more dominant factor affecting gene expression of these transporters than hypoxia. Nonetheless, *SLC38A4* expression was more sensitive to and significantly upregulated in response to glucose and oxygen deprivation stresses. However, gene changes do not reflect protein levels, which may be more tightly regulated or compensated in trophoblasts.

## 3. Discussion

Amino acid transport across the placenta and supply to the fetus are crucial for healthy development. System A sodium-coupled transporters are key to partitioning of neutral amino acids to the fetus. However, genetic targeting of these transporters has also revealed their role in supporting proliferation and cellular composition of the developing rodent placentas [4]. In human preeclampsia and IUGR, altered transporter expression and function reduces cord plasma amino acid concentrations and delivery to the fetus [6,16]. Yet, it is unclear whether the changes in transporters precede placental dysfunction and contribute to the condition or if they are impacted because of suboptimal cellular environments during pathogenesis [5,11]. To elucidate the influence of early vs. late disease onset, we investigated the expression of all three subtypes of system A transporters in matched gestation cohorts.

Our early preterm IUGR cohorts (<34 weeks) exhibited significant reduction in SLC38A4 protein expression compared to gestationally matched controls. Recently, SLC38A4 knock-out in a mouse model caused placental hypoplasia and reduction in fetal weight [4]. Our investigation supports this finding, as the significant reduction in SLC38A4 appear to be an IUGR specific feature, regardless of comorbidities such as preeclampsia. Interestingly though, we show alternative regulation of gene and protein levels, as *SLC38A4* mRNA expression was not different than the early preterm controls. This type of disparity between gene and protein levels appears to be a feature of these transporters in healthy pregnancies as well. In normal placentas, SLC38A1 and SLC38A2 were found to be steady across first and third trimester samples. In contrast, while SLC38A4 mRNA significantly decreased, its protein expression increased across these gestations [17].

We did not observe any significant changes in transporter expression in the late gestation (>34 week) cohorts. Another study showed placentas from late rather than early onset IUGR had decreased SLC38A2 gene expression [18]. However, as acknowledged by the authors, there were small number of samples used for analysis in the early-onset cohort, and control samples were not appropriately matched for gestation. Furthermore, as highlighted by our findings, gene expression may not reflect protein or activity levels of these transporters. A strength of our study is the use of highly prized clinical samples from early and late preterm gestations, as well as appropriately matched controls. Therefore, gestational age, disease onset, and severity may be influencing factors for transporter expression.

Potential inter-species differences in transporter regulation have also been noted. In mice, studies suggest a compensatory mechanism where nutrient transporters are upregulated in the event of maternal undernutrition. While in non-human primates and rats, nutrient restriction causes a net decrease in transporter systems [10,11]. Reduced nutrient sensing and supply are important aspects of human IUGR [19,20]. Even though we did not take nutritional status into account when classifying our cohorts, there is existing evidence of amino acid deprivation positively regulating transporter expression in other systems [7,8]. However, little evidence exists about whether these transporters respond to other forms of nutritional inadequacies observed in IUGR. Low glucose transfer due to placental insufficiency causes premature activation of hepatic glucose production in

IUGR fetuses [21]. This form of glucose is insulin resistant and predisposes the infants to metabolic disease. This has also been demonstrated in sheep models, where a state of prolonged fasting (>2 weeks) and hypoglycemia instigates growth restriction in sheep and alters glucose tolerance in the fetus [21]. In muscle cells, molecular regulators of glucose uptake concomitantly stimulate system A amino acid transport, crucial for protein synthesis [22,23]. Similarly, in rat livers, glucagon drives SLC38A2 expression [24,25]. To the best of our knowledge, the effect of glucose levels on transporter expression has not been previously reported in placenta cells. Here, we observe that glucose deprivation in primary trophoblasts significantly upregulates gene expression of all transporters, while the protein levels remain unchanged.

We went on to investigate transporter expression under another IUGR specific stress condition, hypoxia. Hypoxia induced HIF1$\alpha$ positively regulates SLC38A1 and SLC38A2 expression in adipose cells during obesity and in breast cancer cells, respectively [26,27]. In contrast to other systems, a placental study suggested that hypoxia reduced the expressions of a system A transporter in cytotrophoblast cultures in an oxygen level dependent manner and impeded amino acid uptake [28]. Our results showed that low oxygen significantly upregulated *SLC38A4* mRNA levels. While, *SLC38A1* and *SLC28A2* mRNA expression tended to be lower than the controls, this was not significant. Despite the significant changes to SLC38A4 protein in IUGR and preeclampsia + IUGR placentas, trophoblast hypoxia models did not alter the protein expressions of any of the transporters in vitro. This emphasizes the complexity of the human condition, which cannot be fully replicated in culture. A point of difference in findings between our study and those reported in cytotrophoblasts [28] may be that, by and large at the 48 h timepoint, our primary cytotrophoblasts fuse to form syncytiotrophoblasts. Our study is limited by sample number and the investigation of transporter expression, which may not be an accurate depiction of function. Hence, amino acid transport will be an important measure in determining biological consequences of stress conditions. This was shown in BeWo choriocarcinoma cell lines, where amino acid starvation had opposing effects on SLC38A1 and SLC38A2 expression, yet the overall activity of system A transporters was increased [29]. Therefore, short-lived stress may be compensated by adaptation through gene, protein, or functional upregulation, however, this may not be adequate in overcoming persistent pregnancy complications. Thus, even though the transporters serve a similar purpose of maintaining amino acid balance, they may be regulated by distinct molecular factors that can overlap depending on the cellular context.

## 4. Materials and Methods

### 4.1. Patient Samples

Placental tissues were collected with informed, written consent from patients at the University of Melbourne, Department of Obstetrics and Gynaecology at the Mercy Hospital for Women Ethics (#R11/34). Placental samples were collected from pregnancies complicated by severe early-onset preeclampsia and fetal growth restriction (requiring delivery $\leq$ 34 + 0 weeks gestation) at caesarean section. Fetal growth was defined as birthweight <10th centile, according to Australian population charts [30]. Preeclampsia was defined according to the American College of Obstetricians and Gynecologists guidelines [31]. Control preterm placental samples were collected from women with normotensive preterm pregnancies ($\leq$ 34 + 0 weeks gestation) with fetal growth >10th birthweight centile at caesarean section. Preterm deliveries were predominantly for iatrogenic reasons other than fetal growth restriction (such as vasa previa) or premature rupture of membranes. Cases with evidence of chorioamnionitis (confirmed by placental histopathology) were excluded. At the time of collection, tissue samples were fixed in 4% paraformaldehyde and RNALater for 48 h, after which they were snap frozen and stored at −80 °C.

Early preterm (<34 weeks gestation) placental samples from cases of clinically diagnosed preeclampsia + intrauterine growth restriction (IUGR) (n = 22), IUGR alone (n = 13), and preterm controls (n = 20) were included in the analysis of genes and histological images. For late preterm (34–36 weeks), we utilised 24 preterm, 29 IUGR, and 11 preeclampsia + IUGR samples. All samples from the cohorts were included in the analysis and only excluded when there were experimental or technical errors within the experiments. Patient characteristics are presented in Tables 1 and 2.

**Table 1.** Clinical characteristics of <34 week patient cohort.

| Patient Characteristics | Preterm (n = 20) | IUGR (n = 13) | IUGR + PE (n = 22) |
|---|---|---|---|
| Maternal age (years) | 32 (25–37) | 30.5 (24.3–36) | 31 (25–32) |
| Body–mass index (kg/m$^2$) | 29.2 (25.3–35.2) | 23.6 (18.6–30.5) | 28 (25–35.5) |
| Smoking during pregnancy (%) | 4 (27%) | 5 (42%) | 1 (5.9%) |
| Diabetes during pregnancy (%) | 3 (21%) | 3 (25%) | 1 (5.9%) |
| Gestational age at delivery (weeks) | 30 (29–32) | 31.5 (30.1–32.8) | 30.1 (26.9–31.3) |
| Birthweight (grams) | 1585 (1237–2000) | 993.5 (861–1257.3) | 911 (557.5–1173) |
| Fetal Sex (Female/Male) | 9/11 | 8/5 | 14/8 |

**Table 2.** Clinical characteristics of >34 Week Patient Cohort.

| Patient Characteristics | Preterm (n = 24) | IUGR (n = 25) | IUGR + PE (n = 11) |
|---|---|---|---|
| Maternal age (years) | 30.5 (28–34.8) | 32.24 (30–35.5) | 35.5 (27.3–36) |
| Body-mass index (kg/m$^2$) | 24 (21–27) | 26.05 (22.2–28) | 23.5 (20.5–25.7) |
| Smoking during pregnancy (%) | 1 (4.2%) | 2 (6.9%) | 0 (0%) |
| Diabetes during pregnancy (%) | 2 (8.3%) | 6 (21%) | 3 (38%) |
| Gestational age at delivery, weeks | 34.9 (34.6–35.9) | 35.52 (34.4–37.4) | 36.1 (34.5–36.3) |
| Birthweight, grams | 2605 (2239.8–2810) | 1932.68 (1672.5–2240) | 1958 (1774.5–2085) |
| Fetal Sex (Female/Male) | 11/13 | 16/9 | 4/7 |

### 4.2. *Isolating and Treating Primary Human Cytotrophoblast Cells and Placental Explants*

Term placentas, from patients having elective caesarean sections, were sampled at four sites (clockwise direction), and the fetal and maternal membranes were removed. Then, the chorionic villi were used to isolate cytotrophoblast, as previously described [32]. Cells were seeded ($5 \times 10^5$ for RNA and $1 \times 10^6$ for protein/well in technical triplicates/condition) and were incubated under 8% (normoxic) or 1% (hypoxic) $O_2$ conditions at 37 °C, 5% $CO_2$, or with DMEM (#11966025) culture media, supplemented with 1000 mg/L D-glucose of 'high glucose', 500mg/L of 'low glucose' and glucose-free media, or 'no glucose' conditions for 48 h under standard culture conditions (8% $O_2$, 5% $CO_2$, 37 °C). The RNA was collected for qPCR analysis (n = 6 for hypoxia/normoxia experiments and n = 6 for glucose deprivation) and protein for Western blot analysis (n = 7 for hypoxia/normoxia experiments and n = 6 for glucose deprivation). The samples for RNA and Westerns are not matched (from the same patients) for normoxia/hypoxia experiments, but are matched for glucose deprivation studies.

### 4.3. *Quantitative Polymerase Chain Reaction (qPCR)*

qPCR analysis was conducted on mRNA extracted from preterm control, preeclampsia + IUGR and IUGR placentas. Extraction of RNA from placental cytotrophoblasts and explants were performed with the RNAeasy mini kit (Qiagen, Valencia, CA, USA), according to the manufacturer's instructions and quantified using the Nanodrop ND 1000 spectrophotometer (NanoDrop technologies Inc., Wilmington, DE, USA). RNA (0.2 μg) was converted to cDNA using the Applied Biosystems high-capacity cDNA reverse transcriptase kit (Life Technologies, Carlsbad, CA, USA) in line with the manufacturer's guidelines. We assessed gene expressions of SLC38A1 (Hs01562175_m1), SLC38A2 (Hs01089954_m1), and SLC38A4 (Hs00394339_m1) (Taqman probes, Life Technologies) by real time PCR (RT-PCR) on the CFX 384 (Bio-Rad, Hercules, CA) using FAM-labeled Taqman universal PCR mastermix and its specific primer/probe set (Life Technologies) with the following run conditions: 50 °C for 2 min, 95 °C for 10 min, 95 °C for 15 s, and 60 °C for 1 min (40 cycles).

Quantification was performed using the $2^{-\Delta Ct}$ method, normalising expression to the average expression of housekeeper genes CYC1 (Hs00357717_m1) and TOP1 (Hs00243257_m1). Results are presented as mRNA expression ($2^{-\Delta Ct}$).

### 4.4. Western Blots

Protein was extracted in RIPA buffer and quantified using Pierce BCA kit (Thermo Fisher, Waltham, MA, USA), according to the manufacturer's instructions. Protein (20 ug for patient samples and 7 µg for trophoblast) was loaded onto 12.5% gels and ran at 100 V before being transferred to a PVDF membrane at 100V for 1 h at 4 °C. The membranes were blocked in 5% skim milk and probed with primary antibodies SLC38A1 (NOVNBP259311, monoclonal) and SLC38A4 (NOVNBP155228, polyclonal) (Novus Biologicals) at 4 °C overnight. Secondary anti-rabbit-HRP antibodies were applied for 1 h RT, and then the membranes were imaged using chemiluminescence on the Bio-Rad ChemiDoc machine. The target bands were normalised against β-actin levels (#3700, cell signaling, Danvers, MA, US), and the band intensities were plotted a percentage change from normoxic controls for normoxia/hypoxia experiments and high glucose controls for glucose deprivation experiments.

### 4.5. Statistical Analysis

All experiments were performed with a minimum of three technical triplicates for each biological replicate, and there were at least three patients for each experiment. Statistical analysis was conducted using Kruskal-Wallis test for qPCR and Western blots of patient samples. Two-way ANOVA for glucose qPCR and Western blot experiments was used. Wilcoxon test for hypoxia qPCR and Western blots for hypoxia experiments were used. We used the GraphPad Prism 6 (GraphPad Software, La Jolla, CA, USA) for statistical analysis. All data were expressed as medians; $p$ values < 0.05 were considered significant.

## 5. Conclusions

In conclusion, we have investigated gene and protein expressions of system A sodium-coupled transporters in both idiopathic IUGR and IUGR, with preeclampsia, to determine if disease aetiology has distinct impact on outcome. An important aspect of our study was also the categorisation of early (<34 weeks) and late (>34 weeks) preterm diseases and comparing these to appropriately matched preterm controls. As such, we demonstrated that early gestation cohort is marked by dysregulation of SLC38A4, which is significantly downregulated in the IUGR cohort. Furthermore, in exploring different forms of nutrient deprivations, we demonstrated, for the first time in primary trophoblast cells, that the gene expression of the transporters is sensitive to and is upregulated by glucose starvation. Moreover, changes to expression and sensitivities to cellular stresses are not uniform between transporter subtypes. There are likely different factors influencing the expression and function of these transporters in placental pathologies.

**Author Contributions:** E.K.—experiments, data analysis, writing the original manuscript, and editing. A.H.—conducting the experiments and collecting the data. N.D.A.—writing, reviewing, and editing. K.C.—writing, reviewing, and editing. N.H.—writing, review, and editing, F.C.B.—conceptualizing, writing, reviewing, and editing. All authors have read and agreed to the published version of the manuscript.

**Funding:** This research was funded by Norman Beischer Innovation Grants.

**Institutional Review Board Statement:** Placental tissues were collected with informed, written consent from patients at the University of Melbourne, Department of Obstetrics and Gynaecology at the Mercy Hospital for Women Ethics (#R11/34).

**Informed Consent Statement:** Informed consent was obtained from all subjects involved in the study.

**Data Availability Statement:** The data presented in this study are available on request from the corresponding author.

**Conflicts of Interest:** The authors declare no conflict of interest.

## References


1. Malhotra, A.; Allison, B.J.; Castillo-Melendez, M.; Jenkin, G.; Polglase, G.R.; Miller, S.L. Neonatal Morbidities of Fetal Growth Restriction: Pathophysiology and Impact. *Front. Endocrinol.* **2019**, *10*, 55. [CrossRef] [PubMed]
2. Gaccioliy, F.; Huang, C.C.; Wang, C.; Bevilacqua, E.; Franchi-Gazzola, R.; Gazzola, G.C.; Bussolati, O.; Snider, M.D.; Hatzoglou, M. Amino Acid Starvation Induces the SNAT2 Neutral Amino Acid Transporter by a Mechanism That Involves Eukaryotic Initiation Factor 2$\alpha$ Phosphorylation and cap-independent Translation. *J. Biol. Chem.* **2006**, *281*, 17929–17940. [CrossRef] [PubMed]
3. MacKenzie, B.; Erickson, J.D. Sodium-coupled neutral amino acid (System N/A) transporters of the SLC38 gene family. *Pflügers Archiv.* **2004**, *447*, 784–795. [CrossRef] [PubMed]
4. Matoba, S.; Nakamuta, S.; Miura, K.; Hirose, M.; Shiura, H.; Kohda, T.; Nakamuta, N.; Ogura, A. Paternal knockout of *Slc38a4*/SNAT4 causes placental hypoplasia associated with intrauterine growth restriction in mice. *Proc. Natl. Acad. Sci. USA* **2019**, *116*, 21047–21053. [CrossRef] [PubMed]
5. Glazier, J.D.; Cetin, I.; Perugino, G.; Ronzoni, S.; Grey, A.M.; Mahendran, D.; Marconi, A.M.; Pardi, G.; Sibley, C.P. Association between the Activity of the System A Amino Acid Transporter in the Microvillous Plasma Membrane of the Human Placenta and Severity of Fetal Compromise in Intrauterine Growth Restriction. *Pediatr. Res.* **1997**, *42*, 514–519. [CrossRef]
6. Cetin, I.; Ronzoni, S.; Marconi, A.M.; Perugino, G.; Corbetta, C.; Battaglia, F.C.; Pardi, G. Maternal concentrations and fetal-maternal concentration differences of plasma amino acids in normal and intrauterine growth-restricted pregnancies. *Am. J. Obstet. Gynecol.* **1996**, *174*, 1575–1583. [CrossRef]
7. Hoffmann, T.M.; Cwiklinski, E.; Shah, D.S.; Stretton, C.; Hyde, R.; Taylor, P.M.; Hundal, H.S. Effects of Sodium and Amino Acid Substrate Availability upon the Expression and Stability of the SNAT2 (SLC38A2) Amino Acid Transporter. *Front. Pharmacol.* **2018**, *9*, 63. [CrossRef]
8. Gaccioli, F.; Lager, S. Placental Nutrient Transport and Intrauterine Growth Restriction. *Front. Physiol.* **2016**, *7*, 40. [CrossRef]
9. Coan, P.M.; Angiolini, E.; Sandovici, I.; Burton, G.J.; Constancia, M.; Fowden, A.L. Adaptations in placental nutrient transfer capacity to meet fetal growth demands depend on placental size in mice. *J. Physiol.* **2008**, *586*, 4567–4576. [CrossRef]
10. Rosario, F.J.; Kramer, A.; Li, C.; Galan, H.L.; Powell, T.L.; Nathanielsz, P.W.; Jansson, T. Reduction of In Vivo Placental Amino Acid Transport Precedes the Development of Intrauterine Growth Restriction in the Non-Human Primate. *Nutrients* **2021**, *13*, 2892. [CrossRef]
11. Jansson, N.; Pettersson, J.; Haafiz, A.; Ericsson, A.; Palmberg, I.; Tranberg, M.; Ganapathy, V.; Powell, T.L.; Jansson, T. Down-regulation of placental transport of amino acids precedes the development of intrauterine growth restriction in rats fed a low protein diet. *J. Physiol.* **2006**, *576*, 935–946. [PubMed]
12. Roos, S.; Jansson, N.; Palmberg, I.; Säljö, K.; Powell, T.; Jansson, T. Mammalian target of rapamycin in the human placenta regulates leucine transport and is down-regulated in restricted fetal growth. *J. Physiol.* **2007**, *582*, 449–459. [CrossRef] [PubMed]
13. Shibata, E.; Hubel, C.; Powers, R.; von Versen-Hoeynck, F.; Gammill, H.; Rajakumar, A.; Roberts, J. Placental System A Amino Acid Transport is Reduced in Pregnancies With Small For Gestational Age (SGA) Infants but Not in Preeclampsia with SGA Infants. *Placenta* **2008**, *29*, 879–882. [CrossRef] [PubMed]
14. Cutland, C.L.; Lackritz, E.M.; Mallett-Moore, T.; Bardají, A.; Chandrasekaran, R.; Lahariya, C.; Nisar, M.I.; Tapia, M.D.; Pathirana, J.; Kochhar, S.; et al. Low birth weight: Case definition & guidelines for data collection, analysis, and presentation of maternal immunization safety data. *Vaccine* **2017**, *35*, 6492–6500. [CrossRef]
15. Sharma, D.; Shastri, S.; Sharma, P. Intrauterine Growth Restriction: Antenatal and Postnatal Aspects. *Clin. Med. Insights Pediatr.* **2016**, *10*, 67–83. [CrossRef]
16. Norberg, S.; Powell, T.L.; Jansson, T. Intrauterine Growth Restriction Is Associated with a Reduced Activity of Placental Taurine Transporters. *Pediatr. Res.* **1998**, *44*, 233–238. [CrossRef]
17. Desforges, M.; Lacey, H.A.; Glazier, J.D.; Greenwood, S.L.; Mynett, K.J.; Speake, P.F.; Sibley, C.P. SNAT4 isoform of system A amino acid transporter is expressed in human placenta. *Am. J. Physiol. Physiol.* **2006**, *290*, C305–C312. [CrossRef]
18. Vaughan, O.R.; Maksym, K.; Silva, E.; Barentsen, K.; Anthony, R.V.; Brown, T.L.; Hillman, S.L.; Spencer, R.; David, A.L.; Rosario, F.J.; et al. Placenta-specific Slc38a2/SNAT2 knockdown causes fetal growth restriction in mice. *Clin. Sci.* **2021**, *135*, 2049–2066. [CrossRef]
19. Marconi, A.M.; Paolini, C.; Buscaglia, M.; Zerbe, G.; Battaglia, F.C.; Pardi, G. The impact of gestational age and fetal growth on the maternal-fetal glucose concentration difference. *Obstet. Gynecol.* **1996**, *87*, 937–942. [CrossRef] [PubMed]
20. Economides, D.; Nicolaides, K. Blood glucose and oxygen tension levels in small-for-gestational-age fetuses. *Am. J. Obstet. Gynecol.* **1989**, *160*, 385–389. [CrossRef]
21. Wesolowski, S.R.; Hay, W.W., Jr. Role of placental insufficiency and intrauterine growth restriction on the activation of fetal hepatic glucose production. *Mol. Cell Endocrinol.* **2016**, *435*, 61–68. [CrossRef] [PubMed]
22. Hundal, H.S.; Bilan, P.J.; Tsakiridis, T.; Marette, A.; Klip, A. Structural disruption of the trans-Golgi network does not interfere with the acute stimulation of glucose and amino acid uptake by insulin-like growth factor I in muscle cells. *Biochem. J.* **1994**, *297*, 289–295. [CrossRef] [PubMed]

23. Hajduch, E.; Alessi, D.R.; Hemmings, B.A.; Hundal, H.S. Constitutive activation of protein kinase B alpha by membrane targeting promotes glucose and system A amino acid transport, protein synthesis, and inactivation of glycogen synthase kinase 3 in L6 muscle cells. *Diabetes* **1998**, *47*, 1006–1013. [CrossRef] [PubMed]
24. Menchini, R.J.; Chaudhry, F.A. Multifaceted regulation of the system A transporter Slc38a2 suggests nanoscale regulation of amino acid metabolism and cellular signaling. *Neuropharmacology* **2019**, *161*, 107789. [CrossRef]
25. Ortiz, V.; Alemán, G.; Escamilla-Del-Arenal, M.; Recillas-Targa, F.; Torres, N.; Tovar, A.R. Promoter characterization and role of CRE in the basal transcription of the rat SNAT2 gene. *Am. J. Physiol. Metab.* **2011**, *300*, E1092–E1102. [CrossRef]
26. Horie, T.; Fukasawa, K.; Iezaki, T.; Park, G.; Onishi, Y.; Ozaki, K.; Kanayama, T.; Hiraiwa, M.; Kitaguchi, Y.; Kaneda, K.; et al. Hypoxic Stress Upregulates the Expression of Slc38a1 in Brown Adipocytes via Hypoxia-Inducible Factor-1$\alpha$. *Pharmacology* **2017**, *101*, 64–71. [CrossRef]
27. Morotti, M.; Bridges, E.; Valli, A.; Choudhry, H.; Sheldon, H.; Wigfield, S.; Gray, N.; Zois, C.E.; Grimm, F.; Jones, D.; et al. Hypoxia-induced switch in SNAT2/SLC38A2 regulation generates endocrine resistance in breast cancer. *Proc. Natl. Acad. Sci. USA* **2019**, *116*, 12452–12461. [CrossRef]
28. Nelson, D.M.; Smith, S.D.; Furesz, T.C.; Sadovsky, Y.; Ganapathy, V.; Parvin, C.A.; Smith, C.H. Hypoxia reduces expression and function of system A amino acid transporters in cultured term human trophoblasts. *Am. J. Physiol. Physiol.* **2003**, *284*, C310–C315. [CrossRef]
29. Jones, H.N.; Ashworth, C.J.; Page, K.R.; McArdle, H.J. Expression and adaptive regulation of amino acid transport system A in a placental cell line under amino acid restriction. *J. Reprod.* **2006**, *131*, 951–960. [CrossRef]
30. Dobbins, T.A.; Sullivan, E.A.; Roberts, C.L.; Simpson, J.M. Australian national birthweight percentiles by sex and gestational age, 1998–2007. *Med. J. Aust.* **2012**, *197*, 291–294. [CrossRef]
31. Roberts, J.M.; August, P.A.; Bakris, G.; Barton, J.R.; Bernstein, I.M.; Druzin, M.; Gaiser, R.R.; Granger, J.P.; Jeyabalan, A.; Johnson, D.D.; et al. Hypertension in Pregnancy: Executive Summary. *Obstet. Gynecol.* **2013**, *122*, 1122–1131.
32. Brownfoot, F.; Hannan, N.; Onda, K.; Tong, S.; Kaitu'U-Lino, T. Soluble endoglin production is upregulated by oxysterols but not quenched by pravastatin in primary placental and endothelial cells. *Placenta* **2014**, *35*, 724–731. [CrossRef] [PubMed]

International Journal of
*Molecular Sciences*

MDPI

*Article*

# Similar Pro- and Antiangiogenic Profiles Close to Delivery in Different Clinical Presentations of Two Pregnancy Syndromes: Preeclampsia and Fetal Growth Restriction

**Weronika Dymara-Konopka [1,*], Marzena Laskowska [1,*], Ewelina Grywalska [2], Anna Hymos [2], Anna Błażewicz [3] and Bożena Leszczyńska-Gorzelak [1]**

1 Department of Obstetrics and Perinatology, Medical University of Lublin, 8 Jaczewskiego Street, 20-095 Lublin, Poland
2 Department of Experimental Immunology, Medical University of Lublin, 4a Chodźki Street, 20-093 Lublin, Poland
3 Department of Pathobiochemistry and Interdisciplinary Applications of Ion Chromatography, Biomedical Sciences, Medical University of Lublin, 1 Chodźki Street, 20-093 Lublin, Poland
* Correspondence: weronika.dymara@gmail.com or weronikadymara@umlub.pl (W.D.-K.); marzena.laskowska@umlub.pl or melaskowska@go2.pl (M.L.); Tel.: +48-81-7244778 (W.D.-K. & M.L.)

**Abstract:** The purpose of this study was to evaluate serum levels of anti- and pro-angiogenic substances measured using enzyme-linked immunosorbent assays and their ratios in pregnancies complicated by different clinical subsets of placental ischemic syndrome: preeclampsia and/or fetal growth restriction. A prospective case-control study was performed consisting of 77 singleton pregnancies complicated by preeclampsia, preeclampsia with concurrent fetal growth restriction (FGR), and isolated normotensive FGR pairwise matched by gestational age with healthy pregnancies. The entire study cohort was analyzed with respect to adverse pregnancy outcomes that occurred. In all investigated subgroups, placental growth factor (PlGF) was lower and soluble endoglin (sEng), the soluble fms-like tyrosine kinase-1—sFlt-1/PlGF and sFlt-1*sEng/PlGF ratios were higher than in the control group. The differences were most strongly pronounced in the PE with concurrent FGR group and in the sFlt-1/PlGF ratio. The highest sFlt-1 values in preeclamptic patients suggest that this substance may be responsible for reaching the threshold needed for PE to develop as a maternal manifestation of ischemic placental disease. The FGR is characterized by an elevated maternal sFlt-1/PlGF ratio, which boosts at the moment of indicated delivery due to fetal risk. We concluded that angiogenic imbalance is reflective of placental disease regardless of its clinical manifestation in the mother, and may be used as support for the diagnosis and prognosis of FGR.

**Keywords:** ischemic placental syndrome; preeclampsia; fetal growth restriction; soluble endoglin (sEng); soluble fms-like tyrosine kinase-1 (sFlt-1); placental growth factor (PlGF); sFlt-1/PlGF and sFlt-1*sEng/PlGF ratios; pregnancy adverse outcomes

**Citation:** Dymara-Konopka, W.; Laskowska, M.; Grywalska, E.; Hymos, A.; Błażewicz, A.; Leszczyńska-Gorzelak, B. Similar Pro- and Antiangiogenic Profiles Close to Delivery in Different Clinical Presentations of Two Pregnancy Syndromes: Preeclampsia and Fetal Growth Restriction. *Int. J. Mol. Sci.* **2023**, *24*, 972. https://doi.org/10.3390/ijms24020972

Academic Editors: Berthold Huppertz and Ilona Hromadnikova

Received: 3 November 2022
Revised: 29 December 2022
Accepted: 30 December 2022
Published: 4 January 2023



## 1. Introduction

Preeclampsia and fetal growth restriction are considered major pregnancy syndromes that significantly contribute to perinatal morbidity and mortality [1–5]. In many cases, FGR occurs together with PE. They share the challenge that currently there is no effective treatment to stop the progress of clinical deterioration, and it is still delivery that definitely eliminates the direct threat for the mother and fetus during pregnancy, resulting in prematurity-associated consequences for the newborn.

Preeclampsia affects 3–6% of pregnant women worldwide, whereas fetal growth restriction (FGR) is a condition affecting approximately 8% of all pregnancies and contributing to 30% of stillbirths [1,2]. There is a significant overlap in the incidence of PE and FGR, but the association is particularly strong in early-onset disease, diagnosed before 34 weeks, with evidence suggesting that up to 75% of preeclamptic pregnancies will be further complicated by the presence of fetal growth restriction. In contrast, FGR presents simultaneously with late-onset PE in only 10% of preeclamptic cases [3–5]. However, concurrent FGR in the course of PE does not lead to a more severe maternal disease phenotype [5,6]. The maternal symptoms of PE do not show any differences between women, regardless of the growth percentiles of the fetus.

Mothers of babies with impaired growth are at increased risk of developing PE [7,8]. However, it has been shown that only approximately 15% of women with FGR will be diagnosed with superimposed preeclampsia later in pregnancy [9]. The management of FGR with concurrent PE should be combined with preeclamptic women's surveillance, which also takes into consideration the maternal condition and the possibility of rapid fetal deterioration in this disease [10].

In PE, an antiangiogenic state has been implicated as a mechanism of disease. Elevated levels of sFlt-1 and sEng and a lower level of PlGF are features of PE and have already established a significant role in the pathogenesis of the condition. They are also associated with maternal endothelium dysfunction and impaired nitric oxide production. The current knowledge in the area of angiogenic balance and dynamics implicates its possible use in clinical practice. Additionally, recent studies have reported similarities in the mechanisms that underlie the development of PE and FGR. They are both defined by some level of placental insufficiency and thus are considered placenta-originating disorders that share many pathological features [11]. Further, similar changes in maternal levels of angiogenic biomarkers are supposed to be found among them [11–14]. Disordered angiogenesis is one of the processes that is inherently connected with the development of placental insufficiency. The alterations in sFlt-1, PlGF, and sEng levels taken together are believed to describe placental insufficiency better than each marker separately.

In addition, it is now the subject of research to determine whether the significance of angiogenic factors in PE may be extrapolated to FGR as a part of the clinical picture of placental ischemic disease [15,16]. Although their role has been widely studied in the context of the prediction [17–30] and diagnosis [31–37] of PE, there is growing evidence that the angiogenic profile is reflective of placental disease and can also be observed in FGR [17,18,38–46], especially in patients with abnormal Doppler studies in uterine and umbilical arteries [43]. In PE, maternal and fetal conditions should be kept under surveillance. Tight control of blood pressure in PE results mostly in benefits for the mother [47]. At the same time, it does not improve the rate of adverse perinatal outcomes but may lead to a negative effect on fetal growth without increasing illness or death in the infant. As of yet, in FGR, there is no therapy that improves fetal growth in utero, and current management is to deliver the fetus before intrauterine death or irreversible organ damage occurs [48]. This is particularly challenging in early-onset FGR, where delivery entails additional risks to the baby from extremely preterm birth, with its own attendant short- and long-term complications [49–51]. In recent years, better monitoring of PE and FGR has been a subject of growing interest, and the range of diagnostic tools at our disposal involving angiogenic markers is expanding. The evaluation of adverse outcome risk in FGR and PE patients is of particular interest.

Further, a comparison between FGR alone and FGR associated with PE as well as isolated PE regarding levels of maternal sFlt-1, sEng, and PlGF and their ratios could provide valuable information for understanding these different manifestations of placental syndrome. In the definitions of small for gestational age (SGA) as fetal growth below the 10th percentile, which may comprise both small but normally grown babies and intrauterine/fetal growth restriction (IUGR, FGR) with placenta-mediated restricted growth, the terms are often confused, and limited studies in this area have led to inconsistent results. The ongoing advances in terms of the definitions of these two entities—FGR and PE—enable us to precisely select a study group of patients who meet the newest, very strict criteria for both, which were recently formed by international experts and societies to explain differences and similarities in these great obstetrical syndromes [52].

In this study, we hypothesized that FGR and PE demonstrate the same profile in the most widely analyzed pro- and antiangiogenic substances: PlGF, sEng, and sFlt-1, with possible differences in the degree of severity of imbalances. In putting our work into a clinical, practical perspective, the most widely analyzed values of the sFlt-1 were compared to the PlGF ratio in investigated subgroups according to gestational age with preliminary cutoffs and current evidence of the clinical application for the diagnosis, prognosis, and assessment of PE. Finally, we also investigated angiogenic imbalance in the context of adverse outcomes that occured across the entire study population.

## 2. Results

### 2.1. Patient Demographic Data and Clinical Characteristics

There were no statistically significant differences with regard to gravidity and parity, maternal age, weight, or height in patient profiles between groups. In the PE with FGR group, the mean gestational age at delivery was significantly lower than in the PE alone group (median value, 32 vs. 35 weeks), but there were no significant differences in gestational age between other subgroups or any of the subgroups and the control group. The systolic blood pressure (SBP) and mean arterial blood pressure (MAP) values were significantly higher in all investigated subgroups in comparison with control groups, and diastolic blood pressure (BDP) was higher than controls in the two PE groups (PE + FGR and iPE). Aspartate and alanine transferases (AST and ALT), uric acid (UA), and urea values were higher in preeclamptic patients in both groups (PE + FGR and iPE) than in healthy controls.

The patient demographic data and clinical characteristics of pregnant women in all investigated groups are shown in Tables 1 and 2. Sonographic parameters with Doppler flow evaluation are presented in Table 3.

In our study, 83% (30/36) of patients with FGR had abnormal Doppler study results (UtA or UA above 95 pc or MCA or CPR under 5 pc), and the remaining 17% (6/36) had exclusively EFW under 3 pc. The doppler study analysis revealed a statistically higher uterine artery mean pulsatility index in both FGR groups and a higher umbilical artery pulsatility index in all investigated subgroups compared to controls. It was observed that a positive correlation existed between the sFlt-1/PlGF ratio values and uterine artery PI ($R = 0.8$, $p < 0.00005$), umbilical artery PI ($R = 0.52$, $p < 0.00005$), and RI ($R = 0.47$, $p < 0.00005$) in the entire study population using Spearman's correlation analysis. These correlations were also observed separately for iFGR and both PE groups together (including iPE with PE + FGR).

Furthermore, with respect to perinatal results, neonatal birth weight was lower than controls in the PE + FGR, iPE, and iFGR groups, which is mostly due to lower gestational age at delivery as all the control patients delivered at term, but birth weight percentiles were significantly lower only in the FGR groups: iFGR and PE + FGR. The patients in the PE + FGR and iPE groups had significantly lower APGAR scores than controls. Additionally, the detailed perinatal outcomes are presented in Table 4.

**Table 1.** Basic characteristics and statistical analysis of the study groups (based on results of the Kruskal-Wallis test and analysis of variance ANOVA with a post hoc RIR Tukey test).

| | PE + FGR | | iPE | | iFGR | | Control | | | |
|---|---|---|---|---|---|---|---|---|---|---|
| | I | | II | | III | | IV | | | |
| Parameter | Median | Q1–Q3 | Median | Q1–Q3 | Median | Q1–Q3 | Median | Q1–Q3 | p-Value | Differences |
| Gravidity | 1 | 1–2 | 1 | 1–2 | 2 | 2–3 | 2 | 1.5–3 | ns | - |
| Parity | 1 | 1–2 | 1 | 1–2 | 2 | 1–2 | 2 | 1–2 | ns | - |
| Gestation age (weeks *) | 32 | 28–34 | 35 | 33–37 | 35 | 33–37 | 34 | 31–37 | ns | I < II |
| Age (years) | 29 | 27–35 | 30 | 27–34 | 33 | 30–37 | 29 | 28–37 | ns | - |

**Table 1.** *Cont.*

| | PE + FGR | | iPE | | iFGR | | Control | | | |
|---|---|---|---|---|---|---|---|---|---|---|
| | I | | II | | III | | IV | | | |
| **Parameter** | **Median** | **Q1–Q3** | **Median** | **Q1–Q3** | **Median** | **Q1–Q3** | **Median** | **Q1–Q3** | ***p*-Value** | **Differences** |
| Height (cm) | 167 | 160–170 | 164 | 160–168 | 167 | 164–171 | 165 | 164–168 | ns | - |
| Weight (kg) | 72 | 66–89 | 80 | 72–92 | 70 | 67–79 | 78 | 68–89 | ns | - |
| | | | | | **I trimester** | | | | | |
| Weight (kg) | 63 | 56–76 | 67 | 61–75 | 60 | 58–69 | 71 | 65–75 | ns | - |
| BMI | 24 | 20–26 | 25 | 23–28 | 23 | 21–26 | 28 | 24–28 | ns | - |
| MAP | 95 | 88–97 | 93 | 83–96 | 88 | 87–92 | 79 | 75–91 | <0.05 | I, I + II > IV |
| DBP (mmHg) | 80 | 76–82 | 77 | 68–80 | 75 | 70–78 | 67 | 62–76 | <0.05 | I, I + II > IV |
| SBP (mmHg) | 120 | 110–130 | 126 | 115–129 | 119 | 116–124 | 111 | 100–120 | <0.05 | I + II > IV |

*p*-value reflecting statistically significant differences between the study groups and the control group. ns refers to the nonsignificant differences between the control and investigated groups. Q1–Q3 refers to the interquartile range; BMI refers to the body mass index; MAP is the arterial pressure; DBP refers to the diastolic blood pressure; SBP refers to the systolic blood pressure; Gestation age (weeks *) refers to the gestational age at the moment of venipuncture.

**Table 2.** Clinical results at the time of evaluation before delivery and statistical analysis of the study groups (based on the results of the Kruskal-Wallis test and analysis of variance ANOVA with a post hoc RIR Tukey test).

| | PE + FGR | | iPE | | iFGR | | Control | | | |
|---|---|---|---|---|---|---|---|---|---|---|
| | I | | II | | III | | IV | | | |
| **Parameter** | **Median** | **Q1–Q3** | **Median** | **Q1–Q3** | **Median** | **Q1–Q3** | **Median** | **Q1–Q3** | ***p*-Value** | **Groups** |
| SBP max | 170 | 156–178 | 156 | 150–165 | 127 | 115–134 | 114 | 104–122 | <0.005 | I > II > III > IV |
| DBP max | 104 | 102–111 | 98 | 95–105 | 81.5 | 76–84 | 66 | 62–75 | <0.00005 | I,II > IV |
| MAP | 128 | 121–131 | 117.3 | 114–127 | 95.7 | 92–98.7 | 84 | 76–89 | <0.00005 | I > II > III > IV |
| Proteinuria (mg/24 h) | 1438 | 547–3483 | 668 | 295–1981 | 170 | 138–192 | 0 | 0 | <0.00005 | I,II > IV |
| Total protein (g/dL) | 6.05 | 5.8–6.3 | 5.8 | 5.6–6.3 | 6.35 | 6–6.7 | 6.0 | 5.5–6.2 | ns | - |
| Fibrinogen (g/L) | 4.4 | 3.7–5 | 5.2 | 4.3–5.6 | 4.7 | 4–5.2 | 3.85 | 3.5–4.2 | <0.05 | II > IV |
| INR | 0.9 | 0.86–0.93 | 0.9 | 0.9–0.98 | 0.92 | 0.9–0.97 | 1.0 | 1–1 | <0.00005 | I,II,III < IV |
| PT Index(%) | 109 | 105–114 | 106 | 100–110 | 105.3 | 102–110 | 98 | 96–100 | <0.005 | I,II,III > IV |
| PT (s) | 10 | 10.4–9.6 | 10.3 | 10.9–9.9 | 10.4 | 10.7–9.9 | 11.1 | 1.1–10.9 | <0.005 | I,II,III < IV |
| APTT (s) | 28.8 | 26.6–29.6 | 26.5 | 25.6–27.3 | 27.4 | 26.2–29.6 | 26.9 | 25.9–29.4 | ns | - |
| D-dimers (ng/mL) | 1277 | 1050–1800 | 1362 | 1140–1731 | 1306 | 951–1670 | 1359 | 923–2389 | ns | - |
| WBC ($\times 10^9$/L) | 10.3 | 8.8–11.4 | 9.8 | 9–12.4 | 9.4 | 7.5–11.4 | 9.05 | 8.4–10.9 | ns | - |
| RBC ($\times 10^{12}$/L) | 4.16 | 3.95–4.34 | 4.0 | 3.8–4.3 | 4.2 | 4–4.3 | 4.0 | 3.7–4.2 | ns | - |
| Hb | 12.5 | 11.8–13.5 | 12.2 | 11.4–13.2 | 12.6 | 12–13.1 | 12.25 | 11.5–12.7 | ns | - |
| HCT (%) | 37 | 34.3–38.8 | 34.9 | 33.7–38.5 | 36.8 | 35.6–37.9 | 35.3 | 33.9–37.2 | ns | - |
| PLT ($\times 10^9$/L) | 192 | 147–220 | 189 | 144–213 | 207 | 178–253 | 220 | 179–263 | ns | - |
| ALT (U/L) | 28.5 | 21–76 | 24.5 | 18–46 | 16 | 14–28 | 17 | 12–18 | <0.005 | I,II > IV |
| AST (U/L) | 42 | 30–69 | 30 | 25–41 | 23 | 20–28 | 20 | 15–21 | <0.00005 | I,II > IV |
| Creatinine (mg/dL) | 0.7 | 0.6–0.8 | 0.7 | 0.6–0.8 | 0.65 | 0.5–0.7 | 0.5 | 0.5–0.6 | <0.05 | I, I + II > IV |
| UA (mg/dL) | 7.2 | 6.7–8.2 | 6.5 | 5.5–7.2 | 5.5 | 3.95–6.6 | 4.2 | 3.2–4.6 | <0.00005 | I,II > IV |
| Urea | 31 | 26.3–42.1 | 24.5 | 20–35 | 19 | 19–26 | 15.6 | 13.7–17 | <0.00005 | I,II > IV |

*p*-value reflecting statistically significant differences between the study groups and the control group. ns refers to the nonsignificant differences between the control and investigated groups. Q1–Q3 refers to the interquartile range; MAP refers to the mean arterial pressure; DBP max refers to the maximum value of diastolic blood pressure; SBP max refers to the maximum value of systolic blood pressure; INR refers to the international normalized ratio; PT refers to the prothrombin time; APTT refers to the activated partial thromboplastin time; WBC refers to the white blood cell count; RBC refers to the red blood cell count; Hb refers to the hemoglobin concentration; HCT refers to the hematocrit; PLT refers to the platelet count; ALT refers to the alanine transaminase; AST refers to the aspartate transaminase; UA refers to uric acid.

**Table 3.** Characteristics and statistical analysis of ultrasound, including Doppler parameters, in the study groups (based on results of the Kruskal-Wallis test and analysis of variance ANOVA with a post hoc RIR Tukey test).

| | PE + FGR | | iPE | | iFGR | | Control | | | |
|---|---|---|---|---|---|---|---|---|---|---|
| | I | | II | | III | | IV | | | |
| Parameter | Median | Q1–Q3 | Median | Q1–Q3 | Median | Q1–Q3 | Median | Q1–Q3 | p | Differences |
| UtPI mean | 1.6 | 1.4–2.0 | 1.0 | 1–1 | 1.6 | 1.3–2 | 0.7 | 0.65–0.75 | <0.0005 | I,III > IV |
| Ut PI pc | 100 | 99–100 | 91 | 85–96 | 100 | 100–100 | 52 | 42–62 | <0.0005 | I,III > IV |
| UA PI | 1.3 | 1–2.42 | 1.0 | 0.9–1 | 1.2 | 1–1.7 | 0.8 | 0.65–0.87 | <0.00005 | I,II,III > IV |
| UA PI pc | 84 | 53–100 | 73 | 59–82 | 96 | 70–100 | 11 | 3–46 | <0.00005 | I,II,III > IV |
| UA RI | 0.8 | 0.6–1.0 | 0.6 | 0.6–0.7 | 0.7 | 0.6–0.8 | 0.6 | 0.49–0.58 | <0.0001 | I,III > IV |
| MCA PI | 1.2 | 1.1–1.7 | 1.6 | 1.5–2 | 1.3 | 1.3–1.5 | 1.6 | 1.2–1.7 | <0.05 | I < II |
| MCA PI pc | 1 | 1–23 | 32 | 12–46 | 3 | 1–15 | 21 | 8–35 | <0.05 | I < II |
| CPR | 1.2 | 0.6–1.6 | 1.6 | 1.5–1.8 | 1 | 0.7–1.6 | 2.0 | 1.7–2.2 | <0.0005 | I,III < IV |
| CPR pc | 1 | 1–16 | 19 | 6–35 | 1 | 1–13 | 51 | 30–83 | <0.0005 | I,III < IV |
| AFI | 6.5 | 3–10 | 10 | 8–14 | 10 | 7.5–11 | 11 | 9–14 | <0.05 | I < II |
| EFW | 1326 | 708–1714 | 2760 | 2167–3173 | 1915 | 1464–2255 | 2607 | 1773–3351 | <0.00005 | I < IV |
| EFW pc | 1 | 1–2 | 56 | 33–83 | 2 | 1–5 | 64 | 43–87 | <0.00005 | I,III < IV |
| AC | 242 | 217–260 | 323 | 299–336 | 281 | 245–286 | 304 | 260–342 | <0.0005 | I < IV |
| AC pc | 1 | 1–5 | 58 | 48–74 | 4 | 1–7 | 57 | 40–81 | <0.00005 | I,III < IV |

p-value reflecting statistically significant differences between the study groups and the control group. ns—nonsignificant differences between the control and investigated groups. Q1–Q3—interquartile range. Ut PI (uterine artery pulsatility index); Ut PI pc (centile of uterine artery pulsatility index); UA PI (umbilical artery pulsatility index); UA PI pc (centile of umbilical artery pulsatility index); UA RI (umbilical artery resistance index); MCA PI (middle cerebral artery pulsatility index); CPR (cerebroplacental ratio); CPR pc (centile of cerebroplacental ratio); AFI (amniotic fluid index); EFW (estimated fetal weight); EFW pc (centile of estimated fetal weight); AC (abdominal circumference); AC pc (centile of abdominal circumference).

**Table 4.** Characteristics and statistical analysis of neonatal outcomes of the study groups (based on results of the Kruskal-Wallis test and analysis of variance ANOVA with a post hoc RIR Tukey test).

| | PE + FGR | | iPE | | iFGR | | Control | | | |
|---|---|---|---|---|---|---|---|---|---|---|
| | I | | II | | III | | IV | | | |
| Parameter | Median | Q1–Q3 | Median | Q1–Q3 | Median | Q1–Q3 | Median | Q1–Q3 | p | Differences |
| Gestational age at birth (weeks) | 32 | 28–34 | 35 | 33–37 | 35 | 33–37 | 34 | 37–39 | ns | - |
| Birth weight (g) | 1370 | 680–1700 | 2500 | 1980–2980 | 1985 | 1480–2320 | 3340 | 3170–3520 | <0.00005 | I,II,III < IV |
| Birth percentile Hadlock * 1991 | 1 | 1–2 | 33 | 18–81 | 3.5 | 1–8 | 61 | 37–77 | <0.00005 | I,III < IV |
| Birth percentile Akolekar ** 2018 | 1 | 1–1 | 46 | 22–86 | 3 | 1–7 | 70 | 47–87 | <0.00005 | I,III < IV |
| Apgar 1 minute | 7 | 6–8 | 8 | 7–10 | 8 | 7–10 | 10 | 9–10 | <0.0005 | I,II < IV |
| Apgar 5 min | 7.5 | 6–9 | 9 | 8–10 | 8 | 8–10 | 10 | 9.5–10 | <0.00005 | I < IV |

p-value reflecting statistically significant differences between the study groups and the control group. Ns refers to the nonsignificant differences between controls and investigated groups. Q1–Q3 refers to interquartile range. * Hadlock FP, et al., In utero analysis of fetal growth: a sonographic weight standard. Radiology. 1991 Oct;181(1):129–33 [53]. ** Nicolaides KH, Wright D, Syngelaki A, Wright A, Akolekar R. Fetal Medicine Foundation fetal and neonatal population weight charts. Ultrasound ObstetGynecol 2018; doi:10.1002/uog.19073 [54].

### 2.2. Maternal Serum PlGF

The entire study group had significantly lower PlGF values ($p < 0.001$) than the control group. The PlGF serum levels were significantly lower in the iFGR (median 154 pg/mL, quartile range 117–221, $p < 0.002$), iPE (median 142 pg/mL, quartile range 27–227, $p < 0.001$) and PE + FGR groups (median 42 pg/mL, quartile range 22–113, $p < 0.001$) than in the normal pregnancy group (median 769 pg/mL, quartile range 444–1248). The differences between subgroups were not significantly different.

The maternal serum PlGF levels were significantly higher in healthy pregnant control patients at <34 weeks of gestational age (median 1247 pg/mL, quartile range 1078–1480) than in healthy pregnant women at more advanced gestation (median 500 pg/mL, quartile range 338–769, $p < 0.005$).

### 2.3. Maternal Serum sFlt-1

The entire study group had significantly higher sFlt-1 values ($p < 0.001$) than the control group.The sFlt-1 serum levels were significantly higher only in the preeclamptic subgroups: iPE (median 76,345 pg/mL, quartile range 8614–133,888 pg/mL, $p < 0.05$) and PE + FGR group (median 115,702 pg/mL, quartile range 14,981–221,278 pg/mL, $p < 0.001$) when compared to normal pregnancies (median 8878 pg/mL, quartile range 5574–10,809 pg/mL). The iFGR group presented clearly higher sFlt-1 concentrations than the control group, but the differences were not statistically significant (median 33,590 pg/mL, quartile range 13,871–66,994 pg/mL). The differences between all subgroups were not statistically significant.

The maternal serum sFlt-1 levels were significantly lower in healthy pregnant control patients at <34 weeks of gestational age (median 6567 pg/mL, quartile range 2647–9105 pg/mL) than in healthy pregnant women at more advanced gestation (median 10,170 pg/mL, quartile range 9501–11,407 pg/mL, $p < 0.05$).

### 2.4. Maternal Serum sEng

The entire study group had significantly higher sEng values ($p < 0.001$) than the control group. The sEng serum levels were significantly higher in the iFGR (median 11.7 ng/mL, quartile range 9.1–11.9 ng/mL, $p < 0.05$), iPE (median 11.5 ng/mL, quartile range 10.2–11.9 ng/mL, $p < 0.003$) and PE + FGR groups (median 12.0 ng/mL, quartile range 11.8–12.2 ng/mL, $p < 0.001$) when compared to normal pregnancy group (median 5.8 ng/mL, quartile range 4–8.3 ng/mL). The differences between the studied subgroups of patients with pregnancy complicated by PE and/or FGR were not significantly different.

There was a strong positive correlation between sEng and sFlt-1 levels in the entire study population ($R = 0.69$, $p < 0.001$).

The maternal serum sEng levels were significantly lower in healthy pregnant control patients at <34 weeks of gestational age (median 4.1 ng/mL, quartile range 3.4–6.0 ng/mL) than in healthy pregnant women at more advanced gestation (median 7.4 ng/mL, quartile range 6.0–9.9 ng/mL, $p < 0.05$).

### 2.5. sFlt-1/PlGF and sFlt-1*sEng/PlGF Ratios

The entire study group had significantly higher ratios ($p < 0.001$) than the control group. All of the investigated subgroups had significantly higher values of both ratios than the control group ($p < 0.005$). The sFlt-1/PlGF ratio was the highest and most pronounced in the iPE and PE with concurrent FGR groups (median values 30 and 100 times higher than controls, respectively). In all three investigated subsets the ratio was significantly higher than that in control subjects. None of the subgroups differed significantly from any other subgroup. At delivery, elevation of the sFlt-1 to PLGF ratio in iFGR, iPE and PE + FGR patients reached extreme, very high values exceeding the preliminary cutoff of 655 in most women (64%, 65% and 70% respectively), with similar occurrence in all investigated subgroups. The distributions of the values of sFlt-1, sEng, PlGF and the sFlt-1/PlGF and sFlt-1*sEng/PlGF ratios in women with isolated PE, isolated FGR, combined PE and FGR and pairwise gestational age-matched healthy controls are represented in Table 5 and Figure 1.

**Table 5.** Distributions of the values of sFlt-1, sEng, PlGF and the sFlt-1/PlGF and sFlt-1*sEng/PlGF ratios in women with isolated PE, isolated FGR, combined PE and FGR and in the control group (based on the results of the Kruskal-Wallis test with Bonferroni adjustment).

| | Group | M | Me | Q1 | Q3 | SD | H, *p* |
|---|---|---|---|---|---|---|---|
| sEng [ng/mL] | (I) PE + FGR | 11.9 | 12.1 | 11.8 | 12.2 | 1.0 | H = 34.598 |
| | (II) iPE | 10.5 | 11.5 | 10.2 | 11.9 | 2.4 | $p < 0.001$ |
| | (III) iFGR | 9.9 | 11.7 | 9.2 | 11.9 | 3.0 | I > IV, II > IV, III > IV |

**Table 5.** *Cont.*

| | Group | M | Me | Q1 | Q3 | SD | H, *p* |
|---|---|---|---|---|---|---|---|
| | (IV) Control | 6.4 | 5.8 | 4.1 | 8.3 | 2.9 | |
| PlGF [pg/mL] | (I) PE + FGR | 72 | 42 | 22 | 113 | 62 | H = 42.603 |
| | (II) iPE | 149 | 142 | 27 | 227 | 118 | $p < 0.001$ |
| | (III) iFGR | 216 | 154 | 117 | 221 | 261 | I > IV, II > IV, III > IV |
| | (IV) Control | 851 | 769 | 444 | 1248 | 480 | |
| sFlt-1 [pg/mL] | (I) PE + FGR | 129,263 | 115,702 | 14,981 | 221,278 | 123,234 | H = 14.510 |
| | (II) iPE | 87,234 | 76,345 | 8614 | 133,888 | 99,327 | $p = 0.002$ |
| | (III) iFGR | 51,193 | 33,590 | 13,871 | 66,994 | 49,647 | I > IV, II > IV |
| | (IV) Control | 9787 | 8878 | 5574 | 10,809 | 6416 | |
| RATIO sFlt-1/PlGF | (I) PE + FGR | 2577 | 1072 | 250 | 2833 | 4638 | H = 46.100 |
| | (II) iPE | 1181 | 314 | 143 | 547 | 3567 | $p < 0.001$ |
| | (III) iFGR | 408 | 219 | 81 | 846 | 438 | I > IV, II > IV, III > IV |
| | (IV) Control | 18 | 10 | 5 | 24 | 17 | |
| RATIO sFlt-1*sEng/PlGF | (I) PE + FGR | 31,549 | 12,859 | 3013 | 33,345 | 56,617 | H = 44.965 |
| | (II) iPE | 13,832 | 3251 | 1492 | 6418 | 41,999 | $p < 0.001$ |
| | (III) iFGR | 4727 | 2587 | 646 | 10,109 | 5350 | I > IV, II > IV, III > IV |
| | (IV) Control | 142 | 53 | 24 | 187 | 191 | |

*p*-value statistically significant differences between the study groups and the control group. M—mean value, Me—median value, Q1–Q3—interquartile range.

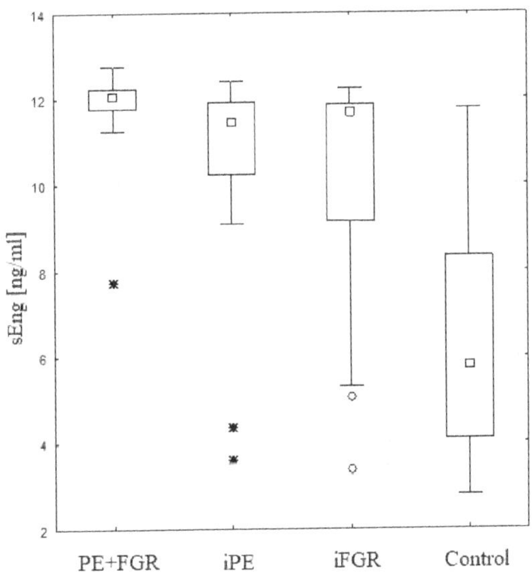


**(A)**

**Figure 1.** *Cont.*

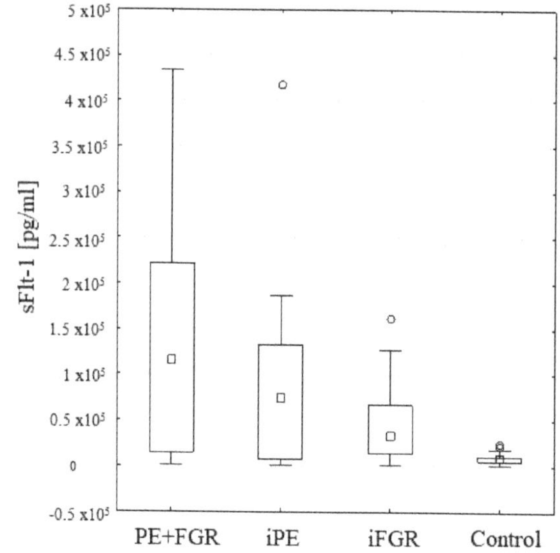


(**B**)

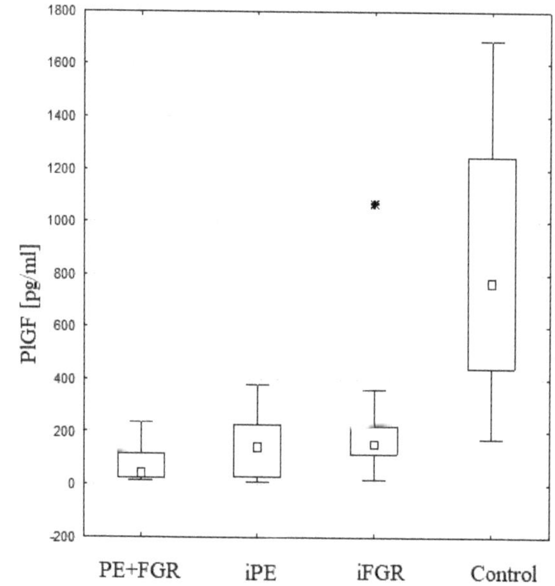


(**C**)

**Figure 1.** *Cont.*

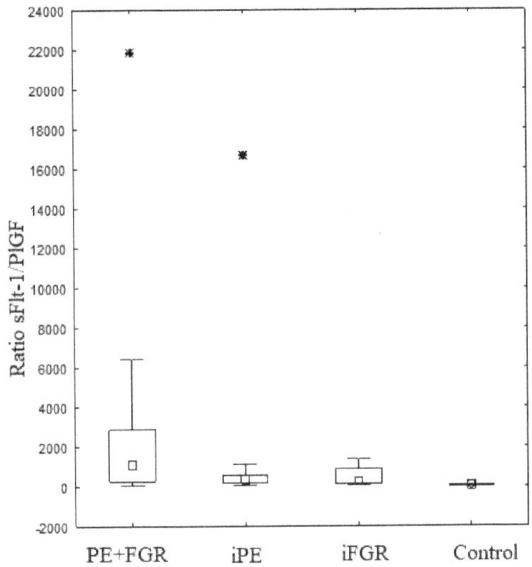


**(D)**

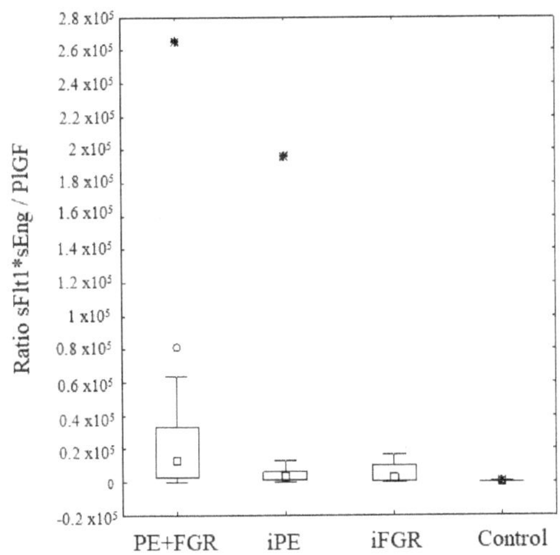


**(E)**

**Figure 1.** (**A–E**) Box and whisker plots for serum concentrations of sEng, sFlt-1, PlGF and sFlt-1/PlGF and sFlt-1*sEng/PlGF ratios in study groups: PE + FGR (preeclampsia with concurrent FGR), iPE (isolated PE), iFGR (isolated, normotensive FGR) and healthy control patients. Boxes indicate interquartile range; whiskers indicate range; square bars indicate median, asterisks indicate extreme values.

Forty-eight hours before delivery, elevation of the sFlt-1 to PlGF ratio in iFGR, iPE and PE + FGR patients reached extreme values exceeding the cutoff of 655/201 in most women (64%, 65% and 70% respectively) with similar occurrence in all investigated subgroups. The results are presented in Table 6.

**Table 6.** Distribution of the sFlt-1/PlGF ratio in each study subgroup in the low, intermediate, high or very high risk groups according to the currently recomended cutoff values.

| Ratio sFlt-1/PlGF | | PE + FGR (n = 22) | iPE (n = 21) | iFGR (n = 14) | Control (n = 20) |
|---|---|---|---|---|---|
| **Mean/Median** | | **2577/1072** | **1181/314** | **408/218** | **18/10** |
| Low | | 5% | 0 | 0 | 85% |
| <38 | | 1 (5%) | 0 | 1 (7%) | 17 (85%) |
| Intermediate | | 9% | 24% | 29% | 15% |
| 38–85 | <34 w | 2 (9%) | 1 (5%) | 0 | 0 |
| 38–110 | >34 w | 0 | 4 (19%) | 4 (29%) | 3 (15%) |
| High | | 14% | 14% | 0 | 0 |
| >85 | <34 w | 3 (14%) | 3 (14%) | 0 | 0 |
| >110 | >34 w | 0 | 0 | 0 | 0 |
| Very high | | 16 (73%) | 13 (65%) | 9 (64%) | 0 |
| >655 | <34 w | 10 (45%) | 3 (14%) | 4 (29%) | 0 |
| >201 | >34 w | 6 (27%) | 10 (48%) | 5 (36%) | 0 |

Cutoff values are based on the following: (1) Stepan H, Herraiz I, Schlembach D, Verlohren S, Brennecke S, Chantraine F, Klein E, Lapaire O, Llurba E, Ramoni A, Vatish M, Wertaschnigg D, Galindo A: Implementation of the sFlt-1/PlGF ratio for prediction and diagnosis of preeclampsia in singleton pregnancy: implications for clinical practice. Ultrasound ObstetGynecol 2015;45:241–246 [55]. (2) Herraiz I, Llurba E, Verlohren S, Galindo A. Spanish Group for the study of angiogenic markers in preeclampsia. Update on the diagnosis and prognosis of preeclampsia with the aid of the sFlt-1/PlGF ratio in singleton pregnancies. FetalDiagnTher 2018;43:81–89 [56].

### 2.6. Adverse Pregnancy Outcomes

Finally, the angiogenic disturbances were checked in the group with adverse (N = 23) and no adverse outcomes (N = 34) were observed among the study cohort. There were 41 adverse events in total, which occurred in 23 patients from the entire study group (40%, 23/57). Most of the events occurred in PE + FGR subgroup (59%, 24/41) which is also the group with the statistically lowest gestational age. A detailed list of adverse outcomes and their occurrence in each study subgroup as well as their relation to the proposed cutoff value of 655 for adverse outcomes increased risk are listed in Table 7. At the moment of prompt delivery, there were no significant differences in terms of angiogenic factor levels or their ratios between the adverse and no-adverse groups.

**Table 7.** Adverse outcomes, their occurrence in the PE + FGR, iPE and iFGR groups and relation to the sFlt/PlGF ratio cutoff values.

| | Adverse Outcome (AO) | Total | PE + FGR | iPE | iFGR |
|---|---|---|---|---|---|
| | PATIENTS WITH AO | 23/57 | 10/22 | 11/21 | 2/14 |
| | Ratio > 655 in the AO group | 10/23 | 6/10 | 3/11 | 1/2 |
| | % with the ratio > 655 in AO group | 44% | 60% | 27% | 50% |
| | sFlT-1/PlGF median value in AO group | 476 | 840 | 446 | 537 |
| | sFlT-1 /PlGF mean value in AO group | 2610 | 3776 | 1926 | 537 |
| | ANY ADVERSE EVENT | 41 (100%) | 24 (59%) | 15 (37%) | 2 (4%) |
| | **Maternal** | | | | |
| 1 | HELLP syndrome | 8 | 5 | 3 | 0 |
| 2 | Placental abruption | 3 | 2 | 0 | 1 |
| 3 | Blood product transfusion | 5 | 3 | 2 | 0 |

**Table 7.** *Cont.*

| | Adverse Outcome (AO) | Total | PE + FGR | iPE | iFGR |
|---|---|---|---|---|---|
| | **Maternal** | | | | |
| 4 | Neurological symptoms (eclamptic fits, visual disturbances, intense headache) | 9 | 3 | 6 | 0 |
| 5 | Renal oliguria or AKI | 3 | 1 | 2 | 0 |
| 6 | Intravenous antihypertensive therapy with 3rd drug | 9 | 7 | 2 | 0 |
| | **Fetal** | | | | |
| 5 | Not viable fetus/ IUD | 4 | 3 | 0 | 1 |

HELLP—hemolysis, elevated liver enzymes, low platelet syndrome; AKI—acute kidney injury; IUD—intrauterine demise. The results of sEng, PlGF, and sFlt-1 measurements and corresponding ratios in the adverse and non-adverse group are displayed in Table 8.

**Table 8.** Distributions of the values of sFlt-1, sEng, PlGF and the sFlt-1/PlGF and sFlt-1*sEng/PlGF ratios in pregnancies with and without adverse outcomes.

| | | Mean | Median | Q1 | Q3 | SD |
|---|---|---|---|---|---|---|
| sEng [ng/mL] | adverse | 11.6 | 11.8 | 11.3 | 12.2 | 1 |
| | non adverse | 10.4 | 11.8 | 9.8 | 12.1 | 2.8 |
| PlGF [pg/mL] | adverse | 107 | 58 | 25 | 186 | 100 |
| | non adverse | 156 | 123 | 26 | 190 | 189 |
| sFlt-1 [pg/mL] | adverse | 108,727 | 77,932 | 6201 | 156,292 | 124,261 |
| | non adverse | 82,512 | 48,260 | 13,052 | 135,395 | 87,355 |
| Ratio sFlt-1/PlGF | adverse | 2610 | 476 | 130 | 1032 | 5544 |
| | non adverse | 800 | 322 | 143 | 1090 | 982 |
| Ratio sFlt*sEng/PlGF | adverse | 31,599 | 5876 | 1560 | 12,714 | 66,852 |
| | non adverse | 9527 | 3423 | 1493 | 12,996 | 12,084 |
| Gestational age [weeks] | adverse | 31 | 32 | 27 | 35 | 4.8 |
| | non adverse | 35 | 35 | 33 | 37 | 2.7 |

## 3. Discussion

The pathophysiologic processes underlying preeclampsia and fetal growth restriction are complicated, multifactorial, and still unclear. In addition, the PE and FGR point to a common placental disorder and a common antiangiogenic state, but the development of maternal syndrome may require interaction with other factors to explain the disease in preeclampsia or perhaps the existence of some kind of protective agent against its appearance [11–14]. In reality, the balance between placental and maternal causations most likely varies among individuals across the spectrum of gestational age at clinical presentation. To date, it has not been sufficiently explained why the ischemic placenta in one disorder leads to growth restriction of the fetus and in the other to a maladaptation of systemic vasculature and inflammatory state, which constitute PE in the mother [15,16]. It was postulated that maternal PE arises from soluble factors released into the systemic circulation from the stressed placenta. Fetoplacental hypoxia is one of the consequences of placental impairment, which is a threat to the survival of the fetus.

The important findings of our study are that pregnancies with normotensive fetal growth restriction, isolated preeclampsia and preeclampsia with concurrent fetal growth restriction (iFGR, iPE and PE + FGR) share similar angiogenic profiles expressed in decreased

PlGF and increased sFlt-1 and sEng maternal serum concentrations. However, differences between controls and investigated groups were most strongly pronounced in the PE with concurrent FGR group and by the ratio sFlt-1/PlGF (its median values in iPE and PE + FGR groups were 30 and 100 times higher than controls, respectively). The absolute highest values of the sFlt-1/PlGF ratio in preeclamptic cases with concurrent growth restriction of the fetus suggest accumulation of pathological changes reflected in extreme angiogenic disturbances in this group. These results support the hypothesis that although FGR and PE differ in their association with maternal disease, they share similar placental pathology responsible for changes in the angiogenic profile and thus, they share a common pathogenesis. Despite the heterogeneity in the phenotypic classification, disordered function of the maternal-fetal unit circulation is its hallmark feature, leading to the idea of considering PE and FGR as a single pathology named "ischemic placental syndrome" [15,16].

Placental ischemia is a key to their pathogenesis, but in preeclampsia it is the maternal systemic vasculature that is predominantly affected, while in FGR the major defect is limited to the placental and uterine vasculature presenting at risk fetal development and well-being.

Nanjo et al. showed results consistent with our work, where in the PE complicated by FGR group, similar changes in sEng, sFlt-1 and PlGF serum levels were more markedly shown compared with PE alone [57]. Additionally, significant differences were observed between iFGR and controls with respect to sFlt-1 serum concentrations. sFlt-1 was the only substance analyzed that reached significantly higher values in both PE groups of patients, irrespective of concurrent FGR, when compared to healthy pregnant patients but did not differ between the iFGR group and controls.

Shibata et al. however found PlGF levels to be decreased in both PE and SGA, but sFlt-1 was increased only in the PE group [58]. However, authors included in the study isolated SGA, not precisely FGR patients.

This study is partially consistent with Alahakoon et al. who demonstrated similar, decreased PlGF profile in PE, PE + IUGR and isolated IUGR [59]. In their study sFlt-1 presented a significantly increased profile in all of these subgroups when compared to healthy controls.

Furthermore, significantly higher sEng serum values were observed compared to controls were identified in all investigated subgroups—iFGR, iPE and PE with simultaneous FGR. However, again (as in the case of PlGF and sFlt-1), the differences in sEng concentrations between the investigated subgroups were not significant. In addition, similar findings in pregnancies complicated by PE and SGA were presented by Levine et al. from the Karumanchi group, who demonstrated increased maternal levels of soluble endoglin in SGA patients from the second half of pregnancy onwards when compared to healthy controls, resulting in concentrations of this protein in the SGA group being as high as in term PE [26]. This study, however, did not involve the precise, updated definition of FGR based on placental insufficiency manifestations in Doppler studies but only fetuses with birth weight below the 10th percentile, which constitutes the definition of SGA. It was observed that sEng increased as sFlt-1 did in all investigated subgroups with either PE or FGR ($p < 0.005$, R = 0.57, 0.77 and 0.54 for the iFGR, iPE and PE + FGR groups, respectively), but there was no correlation between sFlt-1 and sEng in the control group. The sFlt-1*sEng/PlGF ratio was suggested to be predictive of PE [59,60]. There was a strong positive correlation in our study between sEng and sFlt-1 levels in the entire population ($R = 0.69$, $p < 0.001$).

It was suggested that both entities, PE and FGR, share a common biological pathway, but they differ in the sequence of symptoms with attenuated sEng activity in isolated IUGR [60,61]. Thus, the maternal reaction in IUGR is supposed to be delayed. In this study, sEng did not present significant differences between groups, but we measured its concentration at the moment of prompt delivery when clinical decompensation of the mother or fetus was overt. However, the trend of sEng being elevated in normotensive FGR to a lesser extent than in PE, similar to sFlt-1, was visible (Figure 1A,B), and the PE with

FGR group reached the highest levels of sEng, which is consistent with our previous work (mean values for PE + FGR, iPE and iFGR were 11.9, 10.5 and 9.9 ng/mL, respectively) [62].

It was found that pregnancies with fetal growth restriction, preeclampsia or HELLP, and preeclampsia or HELLP and fetal growth restriction showed significantly lower values of PlGF and a higher sFlt-1/PlGF ratio than control pregnancies in both groups with different gestational age: before and after 34 weeks (Tables 9 and 10). The iFGR group presented clearly higher sFlt-1 concentrations than the control group, but the differences were not statistically significant (median values 33,590 vs. 8878 pg/mL, respectively).

**Table 9.** Comparison of the results in early and late PE.

| | Early PE <34 Weeks | | Late PE ≥34 Weeks | | *p* Value |
|---|---|---|---|---|---|
| | *n* = 22 | | *n* = 21 | | |
| | **Median** | **Q1–Q3** | **Median** | **Q1–Q3** | |
| Gestational age [weeks] | 31.5 | 27–32 | 36 | 35–37 | 0.0000001 |
| PlGF pg/mL [pg/mL] | 33 | 21–125 | 124 | 36–238 | 0.01 |
| sFlt-1 [pg/mL] | 104,310 | 3152–172,024 | 74,759 | 14,981–135,395 | NS |
| sEng [ng/mL] | 12 | 11.6–12.2 | 11.7 | 10.2–12.1 | NS |
| sFlt-1/PlGF | 973 | 228–2247 | 364 | 228–960 | NS |

NS stands for Not Statistically Significant.

**Table 10.** Comparison of the results in early and late FGR.

| | Early FGR <34 Weeks | | Late FGR ≥34 Weeks | | *p* Value |
|---|---|---|---|---|---|
| | *n* = 5 | | *n* = 9 | | |
| | **Median** | **Q1–Q3** | **Median** | **Q1–Q3** | |
| Gestational age [weeks] | 31 | 27–32 | 36 | 35–37 | 0.0000001 |
| PlGF [pg/mL] | 119 | 117–150 | 172 | 143–221 | 0.01 |
| sFlt-1 [pg/mL] | 115,490 | 56,238–127,053 | 29,183 | 13,051–36,878 | <0.05 |
| sEng [ng/mL] | 11.7 | 11.6–12 | 10.9 | 9.2–11.8 | NS |
| sFlt-1/PlGF | 972 | 845–984 | 212 | 80–222 | 0.01 |

NS stands for Not Statistically Significant.

Crispi et al. found that in patients with early-onset PE and FGR, maternal PlGF levels were significantly lower and sFlt-1 levels were higher in all cases compared with healthy control subjects of similar gestational age and that all changes were more pronounced in PE than in FGR, but in the late-onset group, these differences were only seen in PE but not in the FGR group [39].

In our study, both PE groups with or without concomitant FGR presented significantly higher sFlt-1 in comparison to healthy controls. The median sFlt-1 values in isolated FGR were definitely lower than in iPE and PE + FGR, but higher than controls. The increased sFlt-1 values were more pronounced not only in PE with concomitant FGR, but also in early PE compared to the late form. Our interpretation is that it is sFlt-1 that is clearly directly linked to maternal disease and, considering generally lower values of sFlt-1 in late-onset disease when compared to early-onset, may be the factor responsible for never reaching the barrier needed to be crossed to trigger maternal response for the antiangiogenic state in iFGR patients. This is in line with previous studies by Levine, Chaiworapongsa and Koga that claimed that circulating sFlt-1 concentrations are increased in women with an established diagnosis of PE and may begin to increase weeks before the onset of clinical

symptoms [23,24,34]. Thus, sFlt-1 appears to be the central mediator of preecla
is not is not indifferent to the development of FGR.

Although sFlt-1 plays an important role in the pathogenesis of PE, it is
sFlt-1 alone governs disease onset. In our comparison, sFlt-1 or any of its ra
differ significantly between iFGR and PE pregnancies with or without concurre
led us to conclude that other circumstances and multiple factor interactions a
to progress from angiogenic imbalance caused by increased sFlt-1 to maternal

It has been hypothesized that the development of maternal endothelial
in preeclampsia occurs after a certain threshold of imbalance is reached.
women with predisposing conditions may develop PE earlier, with milder
imbalance. The cardiovascular predisposition of some pregnant women to PE is
investigated hypothesis on its etiology [63]. The hypothesis assumes that any
dysfunction during pregnancy is secondary to underlying subclinical defects in
cardiovascular system [64]. It is becoming increasingly evident that pregnancy
a significant strain on the maternal cardiovascular system and metabolism, and in
with evidence of significant maladaptation, preeclampsia is the clinical phenotype.
reason, pregnancy has been described as a 'stress test' that unmasks women who hav
cardiovascular reserve or dysfunction [65]. Individual women with prepregnancy r
developing cardiovascular syndromes, such as obesity, under this pregnancy stress
will develop PE. FGR develops in the absence of maternal constitutional predisposi
whereas PE develops when mediators of placental stress, such as sFlt-1, provoke sufficie
severe and prolonged endothelial insult and interact with chronic hypertension, re
disease, thrombophilia, insulin resistance/diabetes and obesity [66]. In women with lon
term cardiovascular risk factors, even physiological changes at the end of pregnancy an
especially late-onset placental disease, which is more commonly associated with less or no placental damage, may induce endothelial dysfunction leading to PE [56,67].

Thus, in some women the degree of placental dysfunction may be high, but with low predisposition to maternal inflammation the final result is FGR without maternal complications. Some women may also have a resistant-type of vasculature that may therefore take longer to develop clinical signs and symptoms in response to placental-shed inflammatory factors. Clinically, these pregnancies can be rescued by iatrogenic delivery when severe FGR is detected on time, before they develop the overt stage of PE. The removal of dysfunctional placenta protects them from symptomatic PE. In these women the threshold needed to develop maternal disease is never reached.

Therefore, it is possible that very high maternal soluble Flt-1 levels are required for preeclampsia to develop. The highest sFlt-1 values in both PE groups, with or without FGR, suggest that this substance is of essential significance for the development of PE. It may be one of the key substances needed for reaching the threshold for PE to develop as a maternal manifestation of ischemic placental disease. Nevertheless, sFlt-1 serum values overlap in many cases in PE and FGR, which Figure 1C illustrates very well. The threshold to be crossed in mothers might be wide range, as women developing PE are a heterogeneous group with different health backgrounds, predispositions and tolerance to physiological and pathological changes taking place in their organisms during pregnancy.

Herraiz et al. proposed a very interesting term of "antiangiogenic continuum" as a proxy of deteriorating placental function, which reflects an observed tendency of the sFlt-1/PlGF ratio to increase from iFGR, toward iPE and finally PE with FGR [14]. This gradual, significant growth is also visible in our study. It is mostly discernible for median concentration of sFlt-1 in each subgroup: starts with the lowest values in iFGR, reaches median results in iPE and finishes with the highest median value in PE + FGR, which is visualized on previous Figure 1B. It seems that this might be interpreted as a more severe placental involvement during FGR in the course of PE as well as at early-onset disease, as these two mechanisms are synergistic: these two entities, FGR and PE converge more frequently in pregnancies at younger gestational age which is attributed to our PE + FGR subgroup.

The division of our relatively small group of patients with iFGR into the early and late subgroups showed significant differences in the level of PlGF concentration, as well as sFlt-1, and thus also their ratio. There was no significant difference in sEng concentration between the early and late FGR groups. For early and late PE, we noted significant differences only in terms of serum concentration of PlGF (data presented in Tables 9 and 10).

The change in the equilibrium in serum between pro- (PlGF) and antiangiogenic (sFlt-1) factors as well as the level of the resulting sFlt-1/PlGF ratio seems to play a role not only in the diagnosis of PE but also as a predictor of adverse outcomes and gestational time remaining until delivery. Strongly elevated sFlt-1/PlGF ratios in early and late-onset PE have been observed to be correlated with the necessity of delivering the fetus within 48 h and the need for immediate and careful fetal monitoring. The ratio could be helpful in decision-making when inpatient monitoring and fetal lung maturation are being considered.

FGR is also an obstetric complication that presents huge challenges in terms of carrying out diagnostic tests and monitoring the fetus. The complexity of diagnostic criteria is a representative of how complicated the syndrome is. The ultrasound is of primary usefulness here, where apart from assessing the fetal weight, it is used to measure Doppler flows in the uterine, umbilical and middle cerebral arteries [68]. However, the results of Doppler studies may be somewhat delayed in relation to the onset of insufficiency, as some studies have shown quite severe ischemic placental lesions despite the diagnosis of normal UA flows [69]. It seems that combined with ultrasound (fetal biometry, feto-maternal Doppler studies), the sFlt-1/PlGF ratio appears to be useful as a supplementary criterion not only for the detection of FGR but also for the prediction of the time-to-delivery interval and associated adverse outcomes in isolated FGR cases [70–72]. In our research, we observed a positive correlation between sFlt-1/PlGF ratio values and uterine artery PI ($R = 0.8$, $p < 0.00005$), umbilical artery PI ($R = 0.52$, $p < 0.00005$) and RI ($R = 0.47$, $p < 0.00005$) in the entire study population using Spearman's correlation analysis. These correlations were also observed separately for iFGR and both PE groups together (including iPE with PE + FGR).

In putting our work into a clinical, practical perspective, we compared the most widely analyzed ratio of sFlt-1 to PlGF in investigated subgroups according to gestational age with preliminary cutoffs with suggested clinical application in PE syndrome. These preestablished cutoffs of ratio values were: 38, 85, 110, 201 and 655 and their recommended interpretation and clinical application in PE diagnosis and prognosis are shown in Table 11 [55,56].

**Table 11.** Preeclampsia diagnostic criteria formed by International Society for the Study of Hypertension in Pregnancy (ISSHP) in 2018.

| Preeclampsia |
| --- |
| Preeclampsia is gestational hypertension accompanied by ≥1 of the following new-onset conditions at or after 20 weeks' gestation:<br>Proteinuria<br>Other maternal organ dysfunction, including:<br>AKI (creatinine ≥ 90 umol/L; 1 mg/dL)<br>Liver involvement (elevated transaminases, e.g., alanine aminotransferase or aspartate aminotransferase >40 IU/L) with or without right upper quadrant or epigastric abdominal pain<br>Neurological complications (examples include eclampsia, altered mental status, blindness, stroke, clonus, severe headaches, and persistent visual scotomata)<br>Hematological complications (thrombocytopenia–platelet count <150,000/μL, disseminated intravascular coagulation, hemolysis)<br>Uteroplacental dysfunction (such as fetal growth restriction, abnormal umbilical artery [UA] Doppler wave form analysis, or stillbirth) |

In the control group, the median and mean values of the sFlt-1/PlGF ratio were 10 and 18, respectively, and the vast majority of patients (17/20, 85%) belonged to the low ratio group. This is quite expected, as these patients were totally healthy during the entire pregnancy, with no underlying medical conditions. Currently, the cutoff level of ≤38 is widely accepted for ruling out PE in patients with suspicion of the disease, and has already been implemented in some European countries (such as Spain, Italy, the UK, France and Germany) [56,73]. There was a small representative of the control group (3/20, 15%) in the intermediate ratio group but none of these patients exceeded the gestational-wide threshold of 85 for PE diagnosis [74]. Notably, all of these patients were at advanced gestational age (mean ratio value is 50 and mean gestational age is 37.3 weeks). The explanation for the finding that all patients from the control group that crossed the threshold of 38 are at term is the fact that advanced, close to term but eventually normal pregnancy exhibit changes in biomarkers sFlt-1 and PlGF similar to those in PE—the ratio increases. In early healthy pregnancy, proangiogenic substances prevail, whereas PE is considered to be an antiangiogenic state. However, the trend that aims to reach this antiangiogenic state can be observed in advancing physiological pregnancies as well. In our study we noted higher concentrations of sFlt-1 and sEng, and lower PlGF values in patients > 34 weeks in the control group in comparison to those at lower gestational age, <34 weeks ($p < 0.005$ for PlGF and <0.05 for sEng and sFlt-1). As the syncytiotrophoblast (STB) is the main placental source of soluble fms-like tyrosine kinase 1, PlGF and sEng and secretes these factors predominantly into the maternal circulation where they impact maternal vascular adaptation to pregnancy, we can interpret the increasing sFlt-1/PlGF ratio at term as an indicator of accumulating STB stress, expanding with advancing gestation [31,75]. STB pathology, at or after term (for example focal STB necrosis or syncytial knots) demonstrates this stress with or without the occurrence of PE [76], which is supported in our study although the number of women is very low. It should be emphasized that that postterm pregnancies are also complicated by increased rates of PE and more strikingly, eclampsia [77]. It also implies that all pregnant women may be somehow destined to suffer from PE, but spontaneous or induced delivery averts this outcome in most instances [78].

In each study cohort, 87%, 79% and 64% for PE + FGR, iPE and iFGR respectively, met the PE diagnostic threshold of 85. Furthermore, most of the patients from study subgroups reach extremely high, defined above 655 or 201 cutoffs for early and late onset PE [74,77]. These patients in our study delivered within the following 48 h. When the preliminary cutoff value of 85 [75] for PE diagnosis for gestation-wide was applied, all investigated subgroups—iFGR, iPE, PE + FGR had a mean sFlt-1/PlGF ratio above that cutoff meeting the diagnostic criteria of PE even with isolated FGR and no overt maternal disease. The mean/median values for the PE + FGR, iPE, and iFGR groups were 2577/1072, 1181/314, and 408/218, respectively. These patients in our study delivered within the following 48 h. At delivery, most patients with isolated FGR without maternal disease reach the diagnostic criteria for PE of a ratio of more than or equal to 85. In our cohort, all patients in this high-ratio group had only early-onset PE.

It was observed that a sFlt-1/PlGF ratio > 655 at diagnosis was associated with a 5-fold increased risk of delivery in ≤48 h [79]. In our study, all of the patients delivered within the next 48 h, and 67% of the entire study group was allocated to the very high sFlt-1/PlGF ratio cohort.

The blood samples were collected within the last 48 h prior to delivery, when disturbances in angiogenic substances are maximized at the most critical point of pregnancy, when aggravating the fetal state in utero is no longer acceptable and evacuation is necessary, disturbances in the angiogenic profile reach levels as high as those in most serious PE cases. It was expected that the angiogenic and antiangiogenic factors would be dysregulated at this point, but to a less dramatic extent.

In both preeclampsia and fetal growth restriction, placental dysfunction with fetoplacental hypoxia has been suggested as a common underlying condition that may provide an explanation for their strong associations: therefore, the sFlt-1/PlGF ratio may be useful for predicting adverse outcomes not only in PE, but also in women with FGR [80,81].

Table 7 lists the occurrence of adverse outcomes by mortality and morbidity events that occurred in our patient group. In our study, most events occurred in the PE + FGR group of patients, which was also the group with the statistically lowest gestational age. In the FGR group, one preterm abruptio placentae and one intrauterine death occurred.

It was discovered that the combination of both entities, fetal growth restriction age and preeclampsia are exceptionally strongly associated with the elevated sFlt-1 to PlGF ratio.

Additionally, these findings and data from literature seem support both entities e.g FGR and preeclampsia should be considered as different manifestations of the same placental disease. It may be also suggested that an extreme angiogenic imbalance reflects a status of severe placental disease, in which there is a high risk of developing a broad spectrum of maternal and fetal complications that must be assessed comprehensively, regardless of whether a reliable diagnosis of preeclampsia has been reached.

## 4. Materials and Methods

### 4.1. Patients

A prospective cross-sectional case control study was conducted on 77 patients aged 20–41 years, between 24 and 41 weeks of gestation. Eligible cases were live singleton pregnancies with a diagnosis of new-onset PE and/or FGR hospitalized in a Polish tertiary referral hospital. In the entire study group, 75% (43/57) were PE patients (with or without concurrent FGR) and 25% (14/57) were isolated FGR cases. The patients were recruited into three subgroups: PE with concurrent FGR ($n$ = 22), isolated PE ($n$ = 21), and isolated normotensive FGR ($n$ = 14). The entire study group was pairwise matched by gestational age with healthy control pregnancies ($n$ = 20). The angiogenic disturbances were also checked in the group of adverse ($n$ = 23) and no adverse outcomes ($n$ = 34) among the entire cohort levels were observed at the most critical moment of pregnancy, right before delivery. The samples were collected immediately before delivery due to maternal or fetal clinical aggravation where the differences in terms of angiogenic imbalance are maximized.

Additionally, very precise inclusion criteria were used for the study groups, according to the newest, updated definitions for the diagnosis of both entities formed by ISSHP for PE in 2018 [82] and by FIGO (based on a Delphi consensus by Gordijn, reached in 2016) for FGR in 2021 [83,84]. The aim was to assess whether disordered angiogenesis markers and their ratios differed between the study groups and the control group, and among study subgroups (isolated PE, isolated FGR, and PE with simultaneous FGR) at the moment of indicated prompt delivery due to maternal or fetal aggravation.

The PE was defined according to the criteria applied in 2018 by the International Society for the Study of Hypertension in Pregnancy (ISSHP) [82] Group: the new onset of hypertension (BP $\geq$ 140 mm Hg systolic or $\geq$90 mm Hg diastolic) on two or more consecutive occasions accompanied by new-onset proteinuria (>0.3 g/24-h in 24-h urine collection) or, in the absence of proteinuria, another maternal organ or uteroplacental dysfunction. The detailed signs and symptoms required for PE diagnosis by ISSHP are listed in Table 9. All the PE patients in our study met the diagnostic criteria independently of developing uteroplacental dysfunction manifested by growth restriction, i.e., apart from gestational hypertension they presented signs of maternal organ dysfunction or proteinuria, despite the concurrent FGR in the PE plus FGR group. We included only preeclampsia patients diagnosed after 20 weeks of gestation and excluded patients with multiple gestations, chronic hypertension, nephrotic syndrome diagnosed postpartum, diabetes, other metabolic disorders, or autoimmune disease. The absence of fetal structural or chromosome abnormalities was also required.

Fetal growth restriction was diagnosed according to the Delphi consensus-based definition for placenta-mediated FGR published by Gordijn et al. in 2016 [83], recognized recently in 2021 by the FIGO initiative on fetal growth, which uses a combination of measures of fetal size percentile and Doppler abnormalities for early and late FGR.

Data presented in Table 12. According to the consensus definition of FGR, we defined FGR as an estimated fetal weight (EFW) < 3rd percentile or EFW < 10th percentile in combination with at least one of the following Doppler abnormalities: umbilical artery (UA) pulsatility index(PI) > 95th percentile, middle cerebral artery (MCA) PI < 5th percentile, cerebroplacental ratio (CPR) < 5th percentile and/or a mean uterine artery (mUtA) PI > 95th percentile. The cases with a birth weight exceeding the 10th percentile for gestational age, and those with placental or umbilical cord abnormalities, anatomical malformations or suspicion of any genetic defect were excluded from the FGR group.

**Table 12.** Consensus-based definitions for early and late fetal growth restriction (FGR).

| | |
|---|---|
| Early FGR:<br>GA <32 weeks, in absence of congenital anomalies<br>AC/EFW < 3rd centile or UA-AEDF<br>Or<br>1. AC/EFW < 10th centile combined with<br>2. UtA-PI > 95th centile and/or<br>3. UA-PI >95th centile | Late FGR:<br>GA ≥32 weeks, in absence of congenital anomalies<br>AC/EFW < 3rd centile<br>Or at least two out of three of the following<br>1. AC/EFW < 10th centile<br>2. AC/EFW crossing centiles >2 quartiles on growth centiles *<br>3. CPR < 5th centile or UA-PI > 95th centile |

AC, fetal abdominal circumference; AEDF, absent end-diastolic flow; CPR, cerebroplacental ratio; EFW, estimated fetal weight; GA, gestational age; PI, pulsatility index; UA, umbilical artery; UtA, uterine artery. Adapted from: Gordijn SJ, Beune IM, Thilaganathan B, Papageorghiou A, Baschat AA, Baker PN, Silver RM, Wynia K, Ganzevoort W. Consensus definition of fetal growth restriction: a Delphi procedure. Ultrasound Obstet Gynecol. 2016;48:333–9. AC, fetal abdominal circumference; AEDF, absent end-diastolic flow; CPR, cerebroplacental ratio; EFW, estimated fetal weight; GA, gestational age; PI, pulsatility index; UA, umbilical artery; UtA, uterine artery [83]. * Hadlock FP, et al., In utero analysis of fetal growth: a sonographic weight standard. Radiology. 1991 Oct;181(1):129–33 [53].

The healthy normal pregnancies were recruited in our outpatient department to gestationally match the pathologic cases. Inclusion criteria were noncomplicated singleton pregnancy with absence of labor at the time of venipuncture. The patients were verified postdelivery, and the control group included only women with delivery of a term (>37 weeks) infant whose birth weight was between the 10th and 90th percentile for gestational age and no medical, obstetrical nor surgical complications during the entire gestation.

Additionally, in giving this work clinical relevance, the sFlt-1 to PlGF ratio at delivery was then classified according to current recommendations for its practical use in risk stratification in suspected or diagnosed PE. These preestablished cutoffs of ratio values were: 38, 85, 110, 201 and 655 and their recommended interpretation and clinical application in PE diagnosis and prognosis are shown in Table 11. The values of the sFlt-1 to PlGF ratio in each cohort were then allocated to low, intermediate, high or very high-risk groups according to up-to-date expert opinion on the practical use of angiogenic markers.

Finally, there was a comparison between angiogenic biomarker concentrations and the adverse and non-adverse pregnancy outcome groups. The maternal and fetal adverse outcomes were based on the CHIPS (Control of Hypertension in Pregnancy Study), PIERS (Preeclampsia Integrated Estimate of RiSk) and PREP studies, where components of the outcome were derived through an iterative Delphi consensus process [85–87]. An independent panel of experts ranked the outcomes for their importance to clinical practice [88]. Data presented in Table 13.

The table is based on work of: Herraiz I, Llurba E, Verlohren S, Galindo A. Spanish Group for the study of angiogenic markers in preeclampsia. The update on the diagnosis and prognosis of preeclampsia with the aid of the sFlt-1/PlGF ratio in singleton pregnancies. FetalDiagnTher 2018;43:81–89 [56] and Stepan H, Herraiz I, Schlembach D, Verlohren S, Brennecke S, Chantraine F, Klein E, Lapaire O, Llurba E, Ramoni A, Vatish M, Wertaschnigg D, Galindo A: Implementation of the sFlt-1/PlGF ratio for prediction and diagnosis of preeclampsia in singleton pregnancy: implications for clinical practice. Ultrasound Obstet Gynecol 2015;45:241–246 [55].

**Table 13.** Summary of the recommendations for the use of sFlt-1/PlGF ratio in women with signs and symptoms of PE based on the opinion of experts in the use of angiogenic markers.

| Ratio sFlt-1/PlGF (EP/LP) | Interpretation | Time to Delivery (ep) | What Should Be Done? |
|---|---|---|---|
| LOW: <38 | Rule out PE: 1 week: NPV ≈ 99% 4 weeks: NPV ≈ 95% | Unmodified | Reassuring the patient. No further determinations are needed unless new suspicion arises. |
| Intermediate: 38–85/38–110 | Rule in PE: 4 weeks PPV ≈ 40% | 20% remain pregnant after 1 month | Follow up visit and retest in 1–2 weeks. Maternal education about signs and symptoms of PE |
| High: >85/>110 | Diagnosis of PE (or PD-related disorder) is highly likely | 15% remain pregnant after 2 weeks | Follow up visit and retest in 2–4 days. EP: consider referral to higher-level center. LP: consider lowering the threshold for labour induction. |
| Very high: >655/>201 | Short-term complications and need to deliver are highly likely | 30% remain pregnant after 2 days | Close surveillance. EP: corticosteroids to the mother for fetal lung maturation |

NPV, ang. negative predictive value; PD, placental dysfunction; PE, preeclampsia; PlGF, placental growth factor; PPV, positive predictive value; sFlt-1, soluble form of fms-like tyrosine kinase-1; EP, early phase (<34 weeks); LP, late phase (≥34 weeks).

### 4.2. Methods

The patients who had been referred to the tertiary-level academic unit for further treatment because of preeclampsia and/or fetal growth restriction without any signs of labor were offered participation in this study.

In all pathologic pregnancies, samples were collected at the moment of clinical detrioration and indication for prompt delivery due to maternal or fetal clinical decompensation, when the differences in terms of angiogenic imbalance were expected to be maximized. As a result, all the samples were collected within the last 48 h of pregnancy duration. The angiogenic substance levels were not known at this point, and they were not included as part of patient management.

The serum samples were collected according to a common standard operating procedure at our center. Further, the venous blood was drawn by venipuncture in tubes without anticoagulant directly from the patient. In the case of immediate cesarean section, blood was collected for routine blood tests and the remaining serum samples were obtained from the hospital laboratory. Once collected, the blood samples were centrifuged at 2000× $g$, and the serum was separated, aliquoted and stored at −80 °C for later analysis. The levels of PlGF, sEng and sFlt-1 in maternal serum were measured by the corresponding sandwich enzyme immunoassay technique kits (R&D Systems Europe Ltd., Abingdon, UK) according to the manufacturer's instructions. The sEng, sFlt-1 and PIGF concentrations of each sample were determined in parallel.

The fetal measurements and Doppler studies were performed at the Clinical Department of Obstetrics and Perinatology using curvilinear transabdominal probe and a Voluson E10 device (GE Medical Systems). The ultrasound was carried out by senior obstetricians upon routine conditions and guidelines. Biometry was performed by measuring the abdominal circumference (AC), biparietal diameter (BPD), head circumference (HC) and femur length (FL). The fetal weight and the weight percentile were calculated using the Hadlock curves [53]. The following Doppler parameters were measured: PI of the uterine arteries (UtA), PI of the umbilical artery (UA), PI of the middle cerebral artery (MCA), PI and the cerebroplacental ratio (CPR) as the ratio between MCA PI and UA PI and calculations were performed according to up-to-date reference ranges [89,90].

The clinical data were ascertained prospectively and included age, height, weight, verified gestational age, past medical and obstetrical history, as well as clinical information including blood pressure, medication, and laboratory test values.

All data were entered into a database in a real time.

Early perinatal outcomes including birthweight and Apgar score were recorded. A total of 77 pregnant women out of initially involved 88 fulfilled the criteria for their inclusion in the study.

The study protocol was approved by the Medical University of Lublin Ethics Committee (KE-0254/258/2016). Written consent was obtained from all participants in the study.

### 4.3. Statistical Analysis

The statistical comparisons between healthy control participants and the entire study group as well as comparisons between study subgroups: PE + FGR, iPE, iFGR were performed using Statistica 13.1 by StatSoft. The analysis was conducted with Kruskal-Wallis test with Bonferroni's adjustment and analysis of variance ANOVA with post hoc RIR Tukey test when possible. Box plots were generated to represent the values of analytes and their ratios in different subgroups. The previously described gestational phase-specific cutoff values of the sFlt-1/PlGF ratio for PE diagnosis and short-term outcome prediction were applied for different outcome groups [55,56]. The criterion for statistical significance was $p < 0.05$.

## 5. Conclusions

A strong subsequent antiangiogenic activity may be interpreted as a response to hypoxia, and high levels of soluble Flt-1 and soluble endoglin may be markers of this activity. In addition, a positive correlation between sFlt-1 and sEng suggests that antiangiogenic pattern in VEGFs and TGF-beta pathways provides a coordinated reaction to fetoplacental hypoxia.

The fact that PlGF, sFlt-1 and sEng concentrations at delivery as well as their ratios overlap in different manifestations of placental ischemia—FGR and PE, and in their early and late forms constitute proof that mechanisms behind their development finally lead to similar placental responses and cannot be separated. Further, multiple factors must interact and this additive mechanism needs time to give manifestation sooner or later. The complex interactions described above highlight, that molecules such as PlGF, sFlt-1 and sEng cannot be considered in isolation but in combination with different substances that reflect maternal systemic vasculature function. The potential application of biomarkers of angiogenesis in both FGR and PE to detect or predict patients at risk of adverse outcomes would be of value to improve risk stratification strategies to avoid serious complications of ischemic placental disease.

Additionally, from a practical point of view, the fact that at the end of pathological pregnancy complicated by FGR and PE, disturbances in angiogenic profile in maternal blood reach similar levels, points to their possible use in FGR as they ultimately start to be applicable in PE management.

Finally, this work supports the hypothesis that the determination of the sFlt-1/PlGF ratio in patients with FGR might also have value for counseling, clinical supervision, and risk anticipation as it has just started to be applied in PE management. This necessitates further research and a better integration of biomarkers in the diagnostic work-up.

## 6. Strengths and Limitations

The cohort was constructed with a very precise, up-to-date criteria for PE and FGR. The FGR diagnosis was confirmed for each newborn after delivery and patients not meeting the criteria were excluded. In an attempt to decrease the likelihood of false-positive and false-negative diagnosis of FGR, the consensus definition was based on a combination of measures of fetal size (fetal weight estimation and abdominal circumference) and abnormal Doppler findings in the umbilical, uterine, and middle cerebral arteries, as described in

Table 10. In our study blood samples were collected directly before the delivery, which was performed within the next 48 h from the moment of venipuncture, due to maternal or fetal clinical decompensation, when the differences in terms of angiogenic imbalance are maximized, while many studies concentrate on prediction or moment of diagnosis in suspected PE cases, when changes in angiogenic profile are less pronounced in late- onset disease. It was only found in one work that was focused on angiogenic factor levels within one week prior to delivery in similar cohorts—iPE, iFGR and those entities combined together by Nanjo et al. The results regarding concentrations of PlGF, sFlt-1 and sEng in investigated subgroups were exactly the same as those presented by Japanese authors. This study has certain limitations. It was done in a single center and it is of relatively small size. This is why we did not separate early-onset from late-onset patients in study cohorts for deeper analysis. Although there was an interested in all preeclampsia-related adverse outcomes, rare but meaningful adverse outcomes such as pulmonary edema, DIC or maternal death, were not observed in the patients.

**Author Contributions:** Conceptualization, W.D.-K. and M.L.; methodology, W.D.-K. and M.L.; software, W.D.-K.; validation, W.D.-K. and M.L.; formal analysis, W.D.-K. and A.H.; investigation, W.D.-K.; resources, W.D.-K.; data curation, W.D.-K. and M.L.; writing—original draft preparation, W.D.-K.; writing—review and editing, M.L.; visualization, W.D.-K.; supervision, M.L., E.G., A.B., and B.L.-G.; project administration, W.D.-K. and M.L.; funding acquisition, M.L. All authors have read and agreed to the published version of the manuscript.

**Funding:** This research was supported by research Grant No. DS334/2019-21 and DS640 of the Medical University of Lublin.

**Institutional Review Board Statement:** The study was conducted in accordance with the Declaration of Helsinki, and approved by the Medical University of Lublin Ethics Committee (protocol code KE-0254/51/2010) for studies involving humans. Written consent was obtained from all participants in the study.

**Informed Consent Statement:** Informed consent for peripheral blood sampling was obtained from all subjects involved in the study.

**Data Availability Statement:** Data available on request due to restrictions e.g., privacy or ethical. The data presented in this study are available on request from the corresponding author.

**Conflicts of Interest:** The authors declare no conflict of interest. The funders had no role in the design of the study; in the collection, analyses, or interpretation of data; in the writing of the manuscript; or in the decision to publish the results.

## References

1. Abalos, E.; Cuesta, C.; Grosso, A.L.; Chou, D.; Say, L. Global and regional estimates of preeclampsia and eclampsia: A systematic review. *Eur. J. Obstet. Gynecol. Reprod. Biol.* **2013**, *170*, 1–7. [CrossRef]
2. Lawn, J.E.; Blencowe, H.; Pattinson, R.; Cousens, S.; Kumar, R.; Ibiebele, I.; Gardosi, J.; Day, L.T.; Stanton, C. Stillbirths: Where? When? Why? How to make the data count? *Lancet* **2011**, *377*, 1448–1463. [CrossRef]
3. Crovetto, F.; Crispi, F.; Scazzocchio, E.; Mercade, I.; Meler, E.; Figueras, F.; Gratacos, E. First-trimester screening for early and late small-for-gestational-age neonates using maternal serum biochemistry, blood pressure and uterine artery Doppler. *Ultrasound Obstet. Gynecol.* **2013**, *43*, 34–40. [CrossRef]
4. Lees, C.C.; Marlow, N.; van Wassenaer-Leemhuis, A.; Arabin, B.; Bilardo, C.M.; Brezinka, C.; Calvert, S.; Derks, J.B.; Diemert, A.; Duvekot, J.J.; et al. 2 year neurodevelopmental and intermediate perinatal outcomes in infants with very preterm fetal growth restriction (TRUFFLE): A randomised trial. *Lancet* **2015**, *385*, 2162–2172. [CrossRef]
5. Shear, R.M.; Rinfret, D.; Leduc, L. Should we offer expectant management in cases of severe preterm preeclampsia with fetal growth restriction? *Am. J. Obstet. Gynecol.* **2005**, *192*, 1119–1125. [CrossRef]
6. Weiler, J.; Tong, S.; Palmer, K.R. Is Fetal Growth Restriction Associated with a More Severe Maternal Phenotype in the Setting of Early Onset Pre-Eclampsia? A Retrospective Study. *PLoS ONE* **2011**, *6*, e26937. [CrossRef]
7. Churchill, D.; Perry, I.J.; Beevers, D. Ambulatory blood pressure in pregnancy and fetal growth. *Lancet* **1997**, *349*, 7–10. [CrossRef]
8. McCowan, L.M.; North, R.A.; Harding, J.E. Abnormal uterine artery Doppler in small-for-gestational-age pregnancies is associated with later hypertension. *Aust. N. Z. J. Obstet. Gynaecol.* **2001**, *41*, 56–60. [CrossRef]
9. Mitani, M.; Matsuda, Y.; Makino, Y.; Akizawa, Y.; Ohta, H. Clinical features of fetal growth restriction complicated later by preeclampsia. *J. Obstet. Gynaecol. Res.* **2009**, *35*, 882–887. [CrossRef]

10. Figueras, F.; Gratacos, E. An integrated approach to fetal growth restriction. *Best Pract. Res. Clin. Obstet. Gynaecol.* **2017**, *38*, 48–58. [CrossRef]
11. Kwiatkowski, S.; Dołęgowska, B.; Kwiatkowska, E.; Rzepka, R.; Torbè, A.; Bednarek-Jędrzejek, M. A Common Profile of Disordered Angiogenic Factor Production and the Exacerbation of Inflammation in Early Preeclampsia, Late Preeclampsia, and Intrauterine Growth Restriction. *PLoS ONE* **2016**, *11*, e0165060. [CrossRef]
12. Chaiworapongsa, T.; Romero, R.; Whitten, A.E.; Korzeniewski, S.J.; Chaemsaithong, P.; Hernandez-Andrade, E.; Yeo, L.; Hassan, S.S. The use of angiogenic biomarkers in maternal blood to identify which SGA fetuses will require a preterm delivery and mothers who will develop pre-eclampsia. *J. Matern. Neonatal Med.* **2015**, *29*, 1214–1228. [CrossRef]
13. Triunfo, S.; Parra-Saavedra, M.; Rodriguez-Sureda, V.; Crovetto, F.; Dominguez, C.; Gratacós, E.; Figueras, F. Angiogenic Factors and Doppler Evaluation in Normally Growing Fetuses at Routine Third-Trimester Scan: Prediction of Subsequent Low Birth Weight. *Fetal Diagn. Ther.* **2015**, *40*, 13–20. [CrossRef]
14. Herraiz, I.; Droge, L.; Gomez-Montes, E.; Henrich, W.; Galindo, A.; Verlohren, S. Characterization of the soluble fms-like Tyro-sine Kinase-1 to Placental Growth Factor Ratio in Pregnancies Compliated by Fetal Growth Restriction. *Obstet. Gynecol.* **2014**, *124*, 265–273. [CrossRef]
15. Ananth, C.V. Ischemic placental disease: A unifying concept for preeclampsia, intrauterine growth restriction, and placental abruption. *Semin. Perinatol.* **2014**, *38*, 131–132. [CrossRef]
16. Kwiatkowski, S.; Kwiatkowska, E.; Rzepka, R.; Torbe, A.; Dołęgowska, B. Ischemic placental syndrome-Prediction and new disease monitoring. *J. Matern. Fetal Neonatal. Med.* **2016**, *29*, 2033–2039. [CrossRef]
17. Romero, R.; Nien, J.K.; Espinoza, J.; Todem, D.; Fu, W.; Chung, H.; Kusanovic, J.P.; Gotsch, F.; Erez, O.; Mazaki-Tovi, S.; et al. A longitudinal study of angiogenic (placental growth factor) and anti-angiogenic (soluble endoglin and soluble vascular endothelial growth factor receptor-1) factors in normal pregnancy and patients destined to develop preeclampsia and deliver a small for gestational age neonate. *J. Matern. Neonatal Med.* **2008**, *21*, 9–23. [CrossRef]
18. Erez, O.; Romero, R.; Espinoza, J.; Fu, W.; Todem, D.; Kusanovic, J.P.; Gotsch, F.; Edwin, S.; Nien, J.K.; Chaiworapongsa, T.; et al. The change in concentrations of angiogenic and anti-angiogenic factors in maternal plasma between the first and second trimesters in risk assessment for the subsequent development of preeclampsia and small-for-gestational age. *J. Matern. Neonatal Med.* **2008**, *21*, 279–287. [CrossRef]
19. Torry, D.S.; Wang, H.-S.; Wang, T.-H.; Caudle, M.R.; Torry, R.J. Preeclampsia is associated with reduced serum levels of placenta growth factor. *Am. J. Obstet. Gynecol.* **1998**, *179*, 1539–1544. [CrossRef]
20. Reuvekamp, A.; Velsing-Aarts, F.V.; Poulina, I.E.; Capello, J.J.; Duits, A.J. Selectivedeficit of angiogenic growth factors characterises pregnancies complicated by pre-eclampsia. *Br. J. Obstet Gynaecol.* **1999**, *106*, 1019–1022. [CrossRef]
21. Tidwell, S.; Ho, H.-N.; Chiu, W.-H.; Torry, R.J.; Torry, D.S. Low maternal serum levels of placenta growth factor as an antecedent of clinical preeclampsia. *Am. J. Obstet. Gynecol.* **2001**, *184*, 1267–1272. [CrossRef]
22. Taylor, R.N.; Grimwood, J.; Taylor, R.S.; McMaster, M.T.; Fisher, S.J.; North, R.A. Longitudinal serum concentrations of pla-cental growth factor: Evidence for abnormal placental angiogenesis in pathologic pregnancies. *Am. J. Obstet. Gynecol.* **2003**, *188*, 177–182. [CrossRef]
23. Levine, R.J.; Maynard, S.E.; Qian, C.; Lim, K.-H.; England, L.J.; Yu, K.F.; Schisterman, E.F.; Thadhani, R.; Sachs, B.P.; Epstein, F.H.; et al. Circulating Angiogenic Factors and the Risk of Preeclampsia. *N. Engl. J. Med.* **2004**, *350*, 672–683. [CrossRef]
24. Chaiworapongsa, T.; Romero, R.; Kim, Y.M.; Kim, G.J.; Kim, M.R.; Espinoza, J.; Bujold, E.; Goncalves, L.; Gómez, R.; Edwin, S.; et al. Plasma soluble vascular endothelial growth factor receptor-1 concentration is elevated prior to the clinical diagnosis of pre-eclampsia. *J. Matern. Neonatal Med.* **2005**, *17*, 3–18. [CrossRef]
25. Park, C.-W.; Park, J.S.; Shim, S.-S.; Jun, J.K.; Yoon, B.H.; Romero, R. An elevated maternal plasma, but not amniotic fluid, soluble fms-like tyrosine kinase-1 (sFlt-1) at the time of mid-trimester genetic amniocentesis is a risk factor for preeclampsia. *Am. J. Obstet. Gynecol.* **2005**, *193*, 984–989. [CrossRef]
26. Levine, R.J.; Lam, C.; Qian, C.; Yu, K.F.; Maynard, S.E.; Sachs, B.P.; Sibai, B.M.; Epstein, F.H.; Romero, R.; Thadhani, R.; et al. Soluble Endoglin and Other Circulating Antiangiogenic Factors in Preeclampsia. *N. Engl. J. Med.* **2006**, *355*, 992–1005. [CrossRef]
27. Unal, E.R.; Robinson, C.J.; Johnson, D.D.; Chang, E.Y. Second-trimester angiogenic factors as biomarkers for future-onset preeclampsia. *Am. J. Obstet. Gynecol.* **2007**, *197*, 211.e1–211.e4. [CrossRef]
28. Kusanovic, J.P.; Romero, R.; Chaiworapongsa, T.; Erez, O.; Mittal, P.; Vaisbuch, E.; Mazaki-Tovi, S.; Gotsch, F.; Edwin, S.S.; Gomez, R.; et al. A prospective cohort study of the value of maternal plasma concentrations of angiogenic and anti-angiogenic factors in early pregnancy and midtrimester in the identification of patients destined to develop preeclampsia. *J. Matern. Neonatal Med.* **2009**, *22*, 1021–1038. [CrossRef]
29. Chaiworapongsa, T.; Romero, R.; Tarca, A.L.; Kusanovic, J.P.; Gotsch, F.; Mittal, P.; Kim, S.K.; Vaisbuch, E.; Mazaki-Tovi, S.; Erez, O.; et al. A decrease in maternal plasma concentrations of sVEGFR-2 precedes the clinical diagnosis of preeclampsia. *Am. J. Obstet. Gynecol.* **2010**, *202*, 550.e1–550.e10. [CrossRef]
30. Verlohren, S.; Stepan, H.; Dechend, R. Angiogenic growth factors in the diagnosis and prediction of pre-eclampsia. *Clin. Sci.* **2011**, *122*, 43–52. [CrossRef]
31. Maynard, S.E.; Min, J.-Y.; Merchan, J.; Lim, K.-H.; Li, J.; Mondal, S.; Libermann, T.A.; Morgan, J.P.; Sellke, F.W.; Stillman, I.E.; et al. Excess placental soluble fms-like tyrosine kinase 1 (sFlt1) may contribute to endothelial dysfunction, hypertension, and proteinuria in preeclampsia. *J. Clin. Investig.* **2003**, *111*, 649–658. [CrossRef] [PubMed]

32. Tsatsaris, V.; Goffin, F.; Munaut, C.; Brichant, J.-F.; Pignon, M.-R.; Noel, A.; Schaaps, J.-P.; Cabrol, D.; Frankenne, F.; Foidart, J.-M. Overexpression of the Soluble Vascular Endothelial Growth Factor Receptor in Preeclamptic Patients: Pathophysiological Consequences. *J. Clin. Endocrinol. Metab.* **2003**, *88*, 5555–5563. [CrossRef] [PubMed]
33. Luttun, A.; Carmeliet, P. Soluble VEGF receptor Flt1: The elusive preeclampsia factor discovered? *J. Clin. Investig.* **2003**, *111*, 600–602. [CrossRef]
34. Koga, K.; Osuga, Y.; Yoshino, O.; Hirota, Y.; Ruimeng, X.; Hirata, T.; Takeda, S.; Yano, T.; Tsutsumi, O.; Taketani, Y. Elevated Serum Soluble Vascular Endothelial Growth Factor Receptor 1 (sVEGFR-1) Levels in Women with Preeclampsia. *J. Clin. Endocrinol. Metab.* **2003**, *88*, 2348–2351. [CrossRef]
35. Chaiworapongsa, T.; Romero, R.; Espinoza, J.; Bujold, E.; Kim, Y.M.; Gonçalves, L.F.; Gomez, R.; Edwin, S. Evidence supporting a role for blockade of the vascular endothelial growth factor system in the pathophysiology of preeclampsia: Young Investigator Award. *Am. J. Obstet. Gynecol.* **2004**, *190*, 1541–1547. [CrossRef]
36. Bujold, E.; Romero, R.; Chaiworapongsa, T.; Kim, Y.M.; Kim, G.J.; Kim, M.R.; Espinoza, J.; Gonçalves, L.F.; Edwin, S.; Mazor, M. Evidence supporting that the excess of the sVEGFR-1 concentration in maternal plasma in preeclampsia has a uterine origin. *J. Matern. Neonatal Med.* **2005**, *18*, 9–16. [CrossRef]
37. Venkatesha, S.; Toporsian, M.; Lam, C.; Hanai, J.-I.; Mammoto, T.; Kim, Y.M.; Bdolah, Y.; Lim, K.-H.; Yuan, H.-T.; Libermann, T.A.; et al. Soluble endoglin contributes to the pathogenesis of preeclampsia. *Nat. Med.* **2006**, *12*, 642–649. [CrossRef]
38. Boutsikou, T.; Malamitsi-Puchner, A.; Economou, E.; Boutsikou, M.; Puchner, K.-P.; Hassiakos, D. Soluble vascular endothelial growth factor receptor-1 in intrauterine growth restricted fetuses and neonates. *Early Hum. Dev.* **2006**, *82*, 235–239. [CrossRef] [PubMed]
39. Crispi, F.; Domínguez, C.; Llurba, E.; Martín-Gallán, P.; Cabero, L.; Gratacos, E. Placental angiogenic growth factors and uterine artery Doppler findings for characterization of different subsets in preeclampsia and in isolated intrauterine growth restriction. *Am. J. Obstet. Gynecol.* **2006**, *195*, 201–207. [CrossRef]
40. Savvidou, M.D.; Yu, C.K.; Harland, L.C.; Hingorani, A.D.; Nicolaides, K.H. Maternal serum concentration of soluble fms-like tyrosine kinase 1 and vascular endothelial growth factor in women with abnormal uterine artery Doppler and in those with fetal growth restriction. *Am. J. Obstet. Gynecol.* **2006**, *195*, 1668–1673. [CrossRef]
41. Schlembach, D.; Wallner, W.; Sengenberger, R.; Stiegler, E.; Mörtl, M.; Beckmann, M.W.; Lang, U. Angiogenic growth factor levels in maternal and fetal blood: Correlation with Doppler ultrasound parameters in pregnancies complicated by pre-eclampsia and intrauterine growth restriction. *Ultrasound Obstet. Gynecol.* **2007**, *29*, 407–413. [CrossRef]
42. Wallner, W.; Sengenberger, R.; Strick, R.; Strissel, P.L.; Meurer, B.; Beckmann, M.W.; Schlembach, D. Angiogenic growth factors in maternal and fetal serum in pregnancies complicated by intrauterine growth restriction. *Clin. Sci.* **2006**, *112*, 51–57. [CrossRef]
43. Chaiworapongsa, T.; Espinoza, J.; Gotsch, F.; Kim, Y.M.; Kim, G.J.; Goncalves, L.F.; Edwin, S.; Kusanovic, J.P.; Erez, O.; Than, N.G.; et al. The maternal plasma soluble vascular endothelial growth factor receptor-1 concentration is elevated in SGA and the magnitude of the increase relates to Doppler abnormalities in the maternal and fetal circulation. *J. Matern. Neonatal Med.* **2008**, *21*, 25–40. [CrossRef] [PubMed]
44. Chaiworapongsa, T.; Romero, R.; Gotsch, F.; Espinoza, J.; Nien, J.K.; Goncalves, L.; Edwin, S.; Kim, Y.M.; Erez, O.; Kusanovic, J.P.; et al. Low maternal concentrations of soluble vascular endothelial growth factor receptor-2 in preeclampsia and small for gestational age. *J. Matern. Neonatal Med.* **2008**, *21*, 41–52. [CrossRef]
45. Åsvold, B.O.; Vatten, L.J.; Romundstad, P.R.; Jenum, P.A.; Karumanchi, S.A.; Eskild, A. Angiogenic Factors in Maternal Circulation and the Risk of Severe Fetal Growth Restriction. *Am. J. Epidemiol.* **2011**, *173*, 630–639. [CrossRef]
46. Sibiude, J.; Guibourdenche, J.; Dionne, M.-D.; Le Ray, C.; Anselem, O.; Serreau, R.; Goffinet, F.; Tsatsaris, V. Placental Growth Factor for the Prediction of Adverse Outcomes in Patients with Suspected Preeclampsia or Intrauterine Growth Restriction. *PLoS ONE* **2012**, *7*, e50208. [CrossRef]
47. Magee, L.A.; von Dadelszen, O.; Rey, E. and the CHIPS Study Group. Less-tight versus tight control of hypertension in pregnancy. *N. Eng. J. Med.* **2015**, *372*, 407–417. [CrossRef]
48. Spencer, R.N.; Carr, D.J.; David, A.L. Treatment of poor placentation and the prevention of associated adverse outcomes—What does the future hold? *Prenat Diagn.* **2014**, *34*, 677–684. [CrossRef]
49. Marlow, N.; Wolke, D.; Bracewell, M.A.; Samara, M. Neurologic and developmental disability at six years of age after extremely preterm birth. *N. Engl. J. Med.* **2005**, *352*, 9–19. [CrossRef]
50. Brodszki, J.; Morsing, E.; Malcus, P.; Thuring, A.; Ley, D.; Maršál, K. Early intervention in management of very preterm growth-restricted fetuses: 2-year outcome of infants delivered on fetal indication before 30 gestational weeks. *Ultrasound Obstet. Gynecol.* **2009**, *34*, 288–296. [CrossRef]
51. Morsing, E.; Åsard, M.; Ley, D.; Stjernqvist, K.; Maršál, K. Cognitive Function After Intrauterine Growth Restriction and Very Preterm Birth. *Pediatrics* **2011**, *127*, e874–e882. [CrossRef] [PubMed]
52. Brosens, I.; Pijnenborg, R.; Vercruysse, L.; Romero, R. The "Great Obstetrical Syndromes" are associated with disorders of deep placentation. *Am. J. Obstet. Gynecol.* **2011**, *204*, 193–201. [CrossRef]
53. Hadlock, F.P.; Harrist, R.B.; Martinez-Poyer, J. In utero analysis of fetal growth: A sonographic weight standard. *Radiology* **1991**, *181*, 129–133. [CrossRef] [PubMed]
54. Nicolaides, K.; Wright, D.; Syngelaki, A.; Wright, A.; Akolekar, R. Fetal Medicine Foundation fetal and neonatal population weight charts. *Ultrasound Obstet. Gynecol.* **2018**, *52*, 44–51. [CrossRef] [PubMed]

55. Stepan, H.; Herraiz, I.; Schlembach, D.; Verlohren, S.; Brennecke, S.; Chantraine, F.; Klein, E.; Lapaire, O.; Llurba, E.; Ramoni, A.; et al. Implementation of the sFlt-1/PlGF ratio for prediction and diagnosis of pre-eclampsia in singleton pregnancy: Implications for clinical practice. *Ultrasound Obstet. Gynecol.* **2015**, *45*, 241–246. [CrossRef] [PubMed]
56. Herraiz, I.; Llurba, E.; Verlohren, S.; Galindo, A.; on behalf of the Spanish Group for the Study of Angiogenic Markers in Preeclampsia. Update on the Diagnosis and Prognosis of Preeclampsia with the Aid of the sFlt-1/PlGF Ratio in Singleton Pregnancies. *Fetal Diagn. Ther.* **2017**, *43*, 81–89. [CrossRef]
57. Nanjo, S.; Minami, S.; Mizoguchi, M.; Yamamoto, M.; Yahata, T.; Toujima, S.; Shiro, M.; Kobayashi, A.; Muragaki, Y.; Ino, K. Levels of serum-circulating angiogenic factors within 1 week prior to delivery are closely related to conditions of pregnant women with pre-eclampsia, gestational hypertension, and/or fetal growth restriction. *J. Obstet. Gynaecol. Res.* **2017**, *43*, 1805–1814. [CrossRef]
58. Shibata, E.; Rajakumar, A.; Powers, R.W.; Larkin, R.W.; Gilmour, C.; Bodnar, L.M.; Crombleholme, W.R.; Ness, R.B.; Roberts, J.M.; Hubel, C.A. Soluble fms-Like Tyrosine Kinase 1 Is Increased in Preeclampsia But Not in Normotensive Pregnancies with Small-for-Gestational-Age Neonates: Relationship to Circulating Placental Growth Factor. *J. Clin. Endocrinol. Metab.* **2005**, *90*, 4895–4903. [CrossRef]
59. Alahakoon, T.I.; Zhang, W.; Trudinger, B.J.; Lee, V.W. Discordant clinical presentations of preeclampsia and intrauterine fetal growth restriction with similar pro- and anti-angiogenic profiles. *J. Matern. Neonatal Med.* **2014**, *27*, 1854–1859. [CrossRef]
60. Stepan, H.; Krämer, T.; Faber, R. Maternal Plasma Concentrations of Soluble Endoglin in Pregnancies with Intrauterine Growth Restriction. *J. Clin. Endocrinol. Metab.* **2007**, *92*, 2831–2834. [CrossRef]
61. Stepan, H.; Jank, A. Angiogenic factors and their role in pathogenesis and prediction of preeclampsia. *Z. Geburtshilfe Neonatol.* **2009**, *213*, 101–105. [CrossRef] [PubMed]
62. Laskowska, M.; Laskowska, K.; Oleszczuk, J. Endoglin in pregnancy complicated by fetal IUGR in normotensive and preeclamptic pregnant women: A comparison between preeclamptic patients with appropriate-for-gestational-age weight in-fants and healthy pregnant women. *J. Matern. Fetal Neonatal. Med.* **2012**, *25*, 806–811. [CrossRef] [PubMed]
63. Perry, H.; Khalil, A.; Thilaganathan, B. Preeclampsia and the cardiovascular system: An update. *Trends Cardiovasc. Med.* **2018**, *28*, 505–513. [CrossRef] [PubMed]
64. Thilaganathan, B. Pre-eclampsia and the cardiovascular-placental axis. *Ultrasound Obstet. Gynecol.* **2018**, *51*, 714–717. [CrossRef] [PubMed]
65. Craici, I.; Wagner, S.; Garovic, V.D. Review: Preeclampsia and future cardiovascular risk: Formal risk factor or failed stress test? *Ther. Adv. Cardiovasc. Dis.* **2008**, *2*, 249–259. [CrossRef] [PubMed]
66. Duckitt, K.; Harrington, D. Risk factors for pre-eclampsia at antenatal booking: Systematic review of controlled studies. *BMJ* **2005**, *330*, 565. [CrossRef]
67. Roberts, J.M.; Catov, J.M. Preeclampsia more than 1 disease: Or is it? *Hypertension* **2008**, *51*, 989–990. [CrossRef]
68. Friedman, A.M.; Cleary, K.L. Prediction and prevention of ischemic placental disease. *Semin. Perinatol.* **2014**, *38*, 177–182. [CrossRef]
69. Figueras, F.; Gratacos, E. Stage-based approach to the management of fetal growth restriction. *Prenat. Diagn.* **2014**, *34*, 655–659. [CrossRef]
70. Triunfo, S.; Lonmaier, S.; Parra-Saavedra, M.; Crovetto, F.; Nadal, A.; Gratscos, E.; Figureas, F. Angiogenic factors at late-onset small for gestational age and histological placental under perfusion. *Placenta* **2014**, *35*, 398e40. [CrossRef]
71. Gaccioli, F.; Sovio, U.; Cook, E. Screening for fetal growth restriction using ultrasound and the sFLT1/PlGF ratio in nulliparous women: A prospective cohort study. *Lancet Child Adolesc. Health* **2018**, *2*, 569–581. [CrossRef] [PubMed]
72. Quezada, M.S.; Rodríguez-Calvo, J.; Villalaín, C. sFlt-1/PlGF ratio and timing of delivery in early-onset fetal growth restriction with antegrade umbilical artery flow. *Ultrasound Obstet. Gynecol.* **2020**, *56*, 549–556. [CrossRef]
73. Mendoza, M.; Hurtado, I.; Bonacina, E.; Garcia-Manau, P.; Serrano, B.; Tur, H.; Rodo, C.; Maiz, N.; Carreras, E. Individual risk assessment for prenatal counseling in early-onset growth-restricted and small-for-gestational-age fetuses. *Acta Obstet. et Gynecol. Scand.* **2020**, *100*, 504–512. [CrossRef] [PubMed]
74. Di Martino, D.; Cetin, I.; Frusca, T.; Ferrazzi, E.; Fuse, F.; Gervasi, M.T.; Plerbani, M.; Todros, T. Italian advisory board: sFlt-1/PlGF ratio and preeclampsia, state of the art and developments in diagnostic, therapeutic and clinical management. *Eur. J. Obstet. Gynecol. Reprod. Biol.* **2016**, *206*, 70–73. [CrossRef] [PubMed]
75. Verlohren, S.; Galindo, A.; Schlembach, D.; Zeisler, H.; Herraiz, I.; Moertl, M.G.; Pape, J.; Dudenhausen, J.W.; Denk, B.; Stepan, H. An automated method for the determination of the sFlt-1/PIGF ratio in the assessment of preeclampsia. *Am. J. Obstet. Gynecol.* **2010**, *202*, 161. [CrossRef]
76. Brownbill, P.; McKeeman, G.C.; Brockelsby, J.C.; Crocker, I.P.; Sibley, C.P. Vasoactive and Permeability Effects of Vascular Endothelial Growth Factor-165 in the Term in Vitro Dually Perfused Human Placental Lobule. *Endocrinology* **2007**, *148*, 4734–4744. [CrossRef]
77. Jones, C.J.P.; Fox, H. Ultrastructure of the placenta in prolonged pregnancy. *J. Pathol.* **1978**, *126*, 173–179. [CrossRef]
78. Caughey, A.B.; Stotland, N.E.; Escobar, G.J. What is the best measure of maternal complications of term pregnancy: Ongoing pregnancies or pregnancies delivered? *Am. J. Obstet. Gynecol.* **2003**, *189*, 1047e52. [CrossRef]
79. Verlohren, S.; Herraiz, I.; Lapaire, O.; Schlembach, D.; Moertl, M.; Zeisler, H.; Calda, P.; Holzgreve, W.; Galindo, A.; Engels, T.; et al. The sFlt-1/PlGF ratio in different types of hypertensive pregnancy disorders and its prognostic potential in preeclamptic patients. *Am. J. Obstet. Gynecol.* **2012**, *206*, 58. [CrossRef]

80. Kenneth, L.; Hall, D.R.; Gebhardt, S.; Grové, D. Late Onset Preeclampsia is not an Innocuous Condition. *Hypertens. Pregnancy* **2010**, *29*, 262–270. [CrossRef]
81. Schoofs, K.; Grittner, U.; Engels, T.; Pape, J.; Denk, B.; Henrich, W.; Verlohren, S. The importance of repeated measurements of the sFlt-1/PlGF ratio for the prediction of preeclampsia and intrauterine growth restriction. *J. Perinat. Med.* **2014**, *42*, 61–68. [CrossRef] [PubMed]
82. Brown, M.A.; Magee, L.A.; Kenny, L.C.; Karumanchi, S.A.; McCarthy, F.P.; Saito, S.; Hall, D.R.; Warren, C.E.; Adoyi, G.; Ishaku, S. The hypertensive disorders of pregnancy: ISSHP classification, diagnosis & management recommendations for international practice. *Pregnancy Hypertens.* **2018**, *13*, 291–310. [PubMed]
83. Gordijn, S.J.; Beune, I.M.; Thilaganathan, B.; Papageorghiou, A.; Baschat, A.A.; Baker, P.N.; Silver, R.M.; Wynia, K.; Ganzevoort, W. Consensus definition of fetal growth restriction: A Delphi procedure. *Ultrasound Obstet. Gynecol.* **2016**, *48*, 333–339. [CrossRef]
84. Melamed, N.; Baschat, A.; Yinon, Y.; Athanasiadis, A.; Mecacci, F.; Figueras, F.; Berghella, V.; Nazareth, A.; Tahlak, M.; McIntyre, H.D.; et al. FIGO (International Federation of Gynecology and Obstetrics) initiative on fetal growth: Best practice advice for screening, diagnosis, and management of fetal growth restriction. *Int. J. Gynecol. Obstet.* **2021**, *152*, 3–57. [CrossRef]
85. Magee, L.A.; von Dadelszen, P.; Singer, J.; Lee, T.; Rey, E.; Ross, S.; Asztalos, E.; Murphy, K.E.; Menzies, J.; Sanchez, J.; et al. The CHIPS Randomized Controlled Trial (Control of Hypertension in Pregnancy Study): Is Severe Hypertension Just an Elevated Blood Pressure? *Hypertension* **2016**, *68*, 1153–1159. [CrossRef] [PubMed]
86. von Dadelszen, P.; Menzies, J.M.; Payne, B.; Magee, L.A.; PIERS (Pre-eclampsia Integrated Estimate of RiSk) Study Group. Predicting adverse outcomes in women with severe pre-eclampsia. *Semin. Perinatol.* **2009**, *33*, 152–157. [CrossRef] [PubMed]
87. Thangaratinam, S.; Allotey, J.; Marlin, N.; Dodds, J.; Cheong-See, F.; von Dadelszen, P.; Ganzevoort, W.; Akkermans, J.; Kerry, S.; Mol, B.W.; et al. PREP Collaborative Network. Prediction of complications in early-onset pre-eclampsia (PREP): Development and external multinational validation of prognostic models. *BMC Med.* **2017**, *15*, 68. [CrossRef]
88. Cheong-See, F.; Allotey, J.; Marlin, N.; Mol, B.; Schuit, E.; Riet, G.; Riley, R.; Moons, K.; Khan, K.; Thangaratinam, S. Prediction models in obstetrics: Understanding the treatment paradox and potential solutions to the threat it poses. *BJOG Int. J. Obstet. Gynaecol.* **2015**, *123*, 1060–1064. [CrossRef]
89. Gómez, O.; Figueras, F.; Fernández, S.; Bennasar, M.; Martínez, J.M.; Puerto, B.; Gratacos, E. Reference ranges for uterine artery mean pulsatility index at 11-41 weeks of gestation. *Ultrasound Obstet. Gynecol.* **2008**, *32*, 128–132. [CrossRef]
90. Ciobanu, A.; Wright, A.; Syngelaki, A.; Wright, D.; Akolekar, R.; Nicolaides, K.H. Fetal Medicine Foundation reference ranges for umbilical artery and middle cerebral artery pulsatility index and cerebroplacental ratio. *Ultrasound Obstet. Gynecol.* **2019**, *53*, 465–472. [CrossRef]

International Journal of
*Molecular Sciences*

MDPI

*Case Report*

# A Fatal Case of Metastatic Pulmonary Calcification during the Puerperium

**Alberto Chighine [1,*], Andrea Corona [2], Gualtiero Catani [1], Celeste Conte [3], Roberto Demontis [1] and Matteo Nioi [1]**

1 Department of Medical Sciences and Public Health, Section of Legal Medicine, University of Cagliari, 09042 Monserrato, Italy
2 Kantonsspital Graubünden, 7000 Chur, Switzerland
3 Department of Health Surveillance and Bioethics, Università Cattolica del Sacro Cuore, Fondazione Policlinico Universitario Agostino Gemelli IRCCS, 00168 Rome, Italy
* Correspondence: alberto.chighine@unica.it

**Abstract:** We present an unusual case of a fatal respiratory failure in a young woman developed two weeks after she gave birth at home. Circumstantial and clinical features of the case were strongly suggestive for a 'classical' septic origin of the respiratory symptoms. Autopsy, together with histopathological and immunohistochemical analyses allowed demonstrating a massive calcium redistribution consisting of an important osteolysis, especially from cranial bones and abnormal accumulation in lungs and other organs. Such physiopathology was driven by a primary hyperparathyroidism secondary to a parathyroid carcinoma as demonstrated by immunohistochemistry. This very rare case is furthermore characterised by a regular pregnancy course, ended with the birth of a healthy new-born. A complex interaction between pregnancy physiology and hyperparathyroidism might be hypothesised, determining the discrepancy between the relative long period of wellness and the tumultuous cascade occurred in the puerperium.

**Keywords:** primary hyperparathyroidism; forensic; pulmonary metastatic calcification

**Citation:** Chighine, A.; Corona, A.; Catani, G.; Conte, C.; Demontis, R.; Nioi, M. A Fatal Case of Metastatic Pulmonary Calcification during the Puerperium. *Int. J. Mol. Sci.* **2022**, *23*, 15131. https://doi.org/10.3390/ijms232315131

Academic Editor: Ilona Hromadnikova

Received: 28 October 2022
Accepted: 29 November 2022
Published: 1 December 2022



## 1. Introduction

Primary hyperparathyroidism (pHPT) is one of the most common endocrine disorders in the adult population, being significantly higher in women. A solitary parathyroid adenoma (80%) generally sustains pHPT, whereas only a residual number of cases (1%) are caused by malignancies [1]. Hyperparathyroidism results in high parathormone and serum calcium levels, the latter being responsible for symptoms classically addressed as '*bones, stones, abdominal moans, and psychic groans*', consistent in fractures, nephrolithiasis, gastritis, and psychiatric symptoms. Although rare, hypertension, rhythm disorders, and pancreatitis may occur as well [2].

This disorder rarely strikes during pregnancy and several issues make the diagnosis in pregnant women extremely challenging. *Hyperemesis gravidarum* may indeed mask hypercalcemia symptoms while calcium levels are seldom elevated because of pregnancy physiological modifications (haemodilution, hypoalbuminemia, maternal hypercalciuria and high foetal calcium demand, amongst others) [3]. In pregnancy, symptoms may even be more occult with the possibility of severe events such as pre-eclampsia and hypercalcaemic crisis, not to mention foetal complications [4].

What is more, in cases of women asymptomatic or with unspecific symptoms, calcium levels are not routinely screened during pregnancy [5]. Once diagnosed, as pHPT presents significant complication rates on both mother and foetus, surgical approach is the treatment of choice, although it should be ideally performed during the second trimester. The risk of complications seems even higher after delivery, as pregnancy has a 'buffer' effect on hypercalcemia cessation [6].

Only a few reports have described a metastatic pulmonary calcification (MPC) secondary to pHPT [7–9].

We present the case of a 35-year-old woman, experiencing severe symptoms during the second week of puerperium following regular pregnancy and childbirth. To the best of the authors' knowledge, this is the first report of a pHPT occurring in the puerperium complicated by a severe MPC, which eventually led to a fatal respiratory failure.

## 2. Case Report

### 2.1. Medical History and CT Findings

We present the case of a 35-year-old tertigravida (160 cm × 55 kg) with past medical history positive for aspirin allergy; no endocrine disorders were revealed in the familiar anamnesis. The patient underwent routine check-ups during pregnancy showing slight anaemia, whereas electrolytes were never investigated. Echography check-ups ruled out foetal abnormalities. A home-delivery was planned and conducted by professional midwives. Gestation and delivery occurred without clinical complications and mother and new-born underwent routine gynaecologist and paediatric postpartum home check-ups.

After the first week of puerperium, the patient experienced low back and leg pain, responsive to painkillers. During the second puerperal week a worsening tachypnoea settled in, shifting in severe dyspnoea on the 14th day postpartum. On the next day the patient was admitted to the emergency department, showing clinical signs of shock (blood pressure = 100/60 mmHg, heart rate = 130 bpm, high respiratory frequency with face mask 10 L/min and $SpO_2$ = 80%, weak peripheral pulses, cold and mottled limbs) and decreased breath sounds on both sides. Blood and arterial gas tests at admission are reported in Supplementary Materials (Table S1).

Arterial blood gas test (Table S1b) was consistent with a primary hypoxemic respiratory insufficiency and metabolic acidosis with inadequate respiratory compensation. A contrast CT scan ruled out pulmonary embolism and cardiac tamponade whereas provided evidence of bilateral diffused parenchymal consolidation was consistent with an inflammatory aetiology. A 20 mm inhomogeneous thyroid nodule was also spotted on the left lobe.

Because of respiratory and haemodynamic instability, the patient underwent a mechanical ventilation trial and was admitted to the intensive care unit (ICU) with the suspicious of pneumonia with septic shock and multi organ failure (MOF). The patient was conscious but anxious, objective examination highlighted hypothermia (34.7 °C), tachypnoea (respiratory rate > 25/min), severe bilateral reduction of breath sounds, eyes and mouth dryness, diffused skin mottling, and perioral cyanosis. A wide spectrum antibiotic therapy was initiated, together with fluid supplementation and inotropic support. Soon after admission she was intubated and ventilated with $O_2$ 100%; arterial gas test parameters at ICU admission are shown in Table S1b.

Less than 2 h after ICU admission the patient experienced a cardiac arrest, successfully treated with resuscitation manoeuvrers. After return of spontaneous circulation, a chest X-ray showed a massive consolidation on both lungs' parenchyma. A large amount of secretion was aspirated from endotracheal tube with mixed content of enteric fluid and pulmonary oedema. A second cardiac arrest did not answer to resuscitation manoeuvres and, six hours after the emergency department admission, the patient died.

### 2.2. Autopsy Findings

A judicial autopsy was ordered by the local prosecutor's office. A team of forensic and clinical pathologists performed the autopsy. Organ weights are reported in Table 1.

**Table 1.** Organ weights in grams (g).

| | |
|---|---|
| Brain | 1300 |
| Heart | 300 |
| Right lung | 1430 |
| Left lung | 1300 |
| Liver | 850 |
| Spleen | 250 |
| Right Kidney | 200 |
| Left Kidney | 180 |

Most remarkable macroscopic findings are reported as follows. Skullcap (Figure 1) appeared diffusely eroded in a grainy reddish pattern resulting in a global bone thinning more accentuated in temporal and parietal regions.

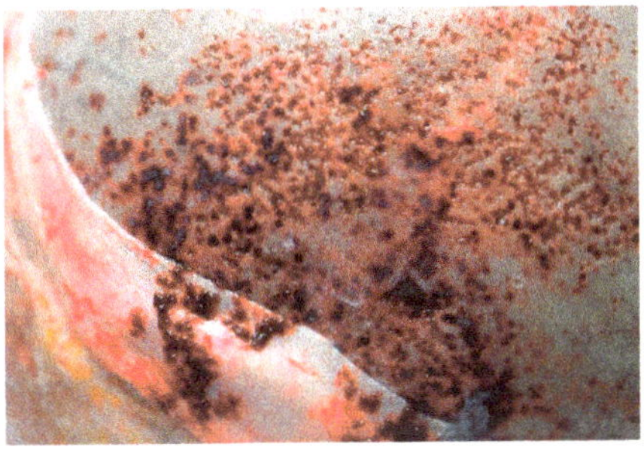

**Figure 1.** Skullcap erosion, parietal bone. Macroscopic finding.

The right thyroid lobe was enlarged due to a 2.5 cm diameter mass, compact and greyish on section. Lungs (Figure 2a) were utterly increased in weight, hyper-inflated, and consolidated. Diffused subpleural petechiae were present bilaterally. On section (Figure 2b), parenchyma appeared diffusely grey pinkish, extremely compact with porous appearance, and, when squeezed, gave a crunchy sensation and released abundant, slightly foamy, liquid.

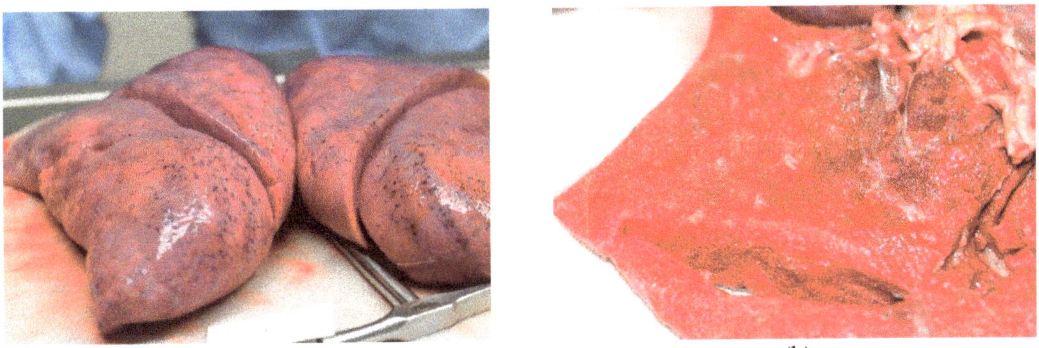

(a) (b)

**Figure 2.** Lungs, macroscopic findings. (**a**) Hyper-inflated and consolidated lungs with thick margins; (**b**) on section, pulmonary tissue appears compact and porous.

Internal genital findings were consistent with puerperium. Due to personal beliefs of the family, placenta was unconventionally preserved (under salt) while macroscopic evaluation did not highlight any noticeable pathological finding.

### 2.3. *Histopathological and Immunohistochemical Findings*

Histopathological analysis was performed on all the specimens collected. Skullcap bone (Figure 3) showed a diffused demineralization with multiple foci of hyperplastic multinucleate osteoclasts and osteofibrosis.

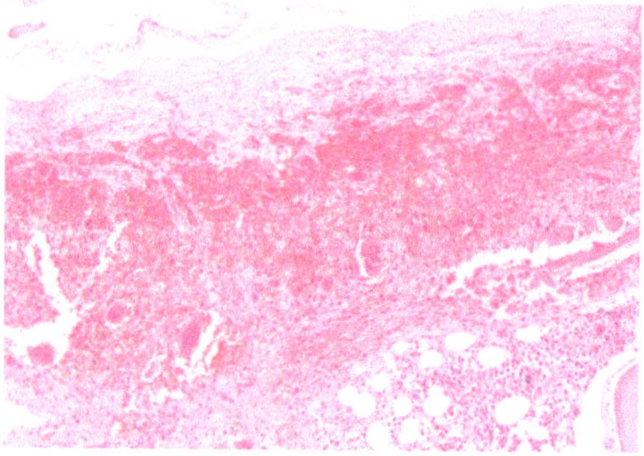

**Figure 3.** Skullcap erosion, parietal bone. H&E staining—magnification 40×. Immediately below the external theca of the bone cap is present a large osteolytic hotbed, bone tissue with massive demineralization replaced by granulation tissue, and massive red blood cell infiltration.

Thyroid mass detected at CT scan was better identified as a parathyroid carcinoma consisting of a capsulated lesion with proliferation of round cells with slight nuclear atypia, anisokaryosis, and scarce cytoplasm, infiltrating the capsule and the surrounding thyroid parenchyma (Figure 4a). Immunohistochemical analyses of such lesion revealed positivity for CK8/CK18, chromogranin (Figure 4b), and NSE, and negativity for CK19, CD31, calcitonin and thyroglobulin. Cellular proliferation index (Ki67) was <1%.

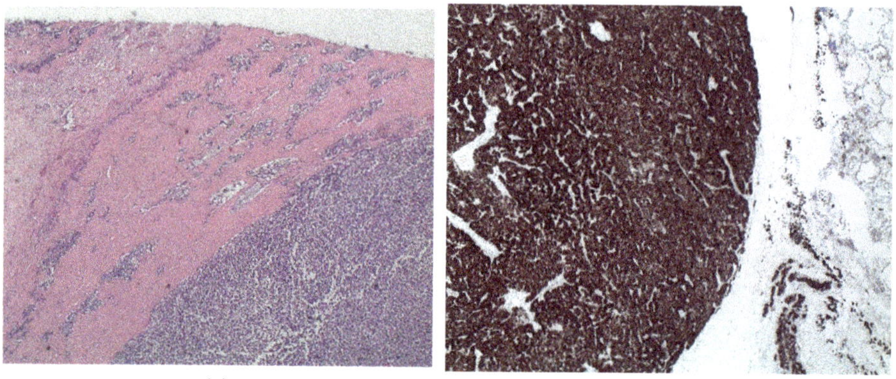

(a) (b)

**Figure 4.** Parathyroid. (**a**) H&E staining—magnification 120×. The nodule tightly stuck to the thyroid disclosed a uniform patternless solid proliferation of round cells with a scant amount of cytoplasm, seldom arranged in a lobular pattern, surrounded by a dense fibrous capsule that adheres to adjacent tissues. Nuclei are irregular and hyperchromic. There is evidence of capsular and vascular invasion. (**b**) IHC chromogranin—magnification 120×. Cells are widely positive compared to negative thyroid tissue (right side).

Lungs were characterised by a massive, diffuse calcium deposition regarding alveolar septa and bronchi walls (especially in basal membrane and submucosa). Analogous deposits were found in arterial and venous walls. Alveoli were filled with amorphous material, slightly eosinophils, mixed with red blood cells and epithelial cells (Figure 5a,b). Some sections provided evidence of neutrophile exudation in both bronchi lumen and alveoli. A diffuse calcium phosphate deposition was confirmed with von Kossa stain (Figure 5c).

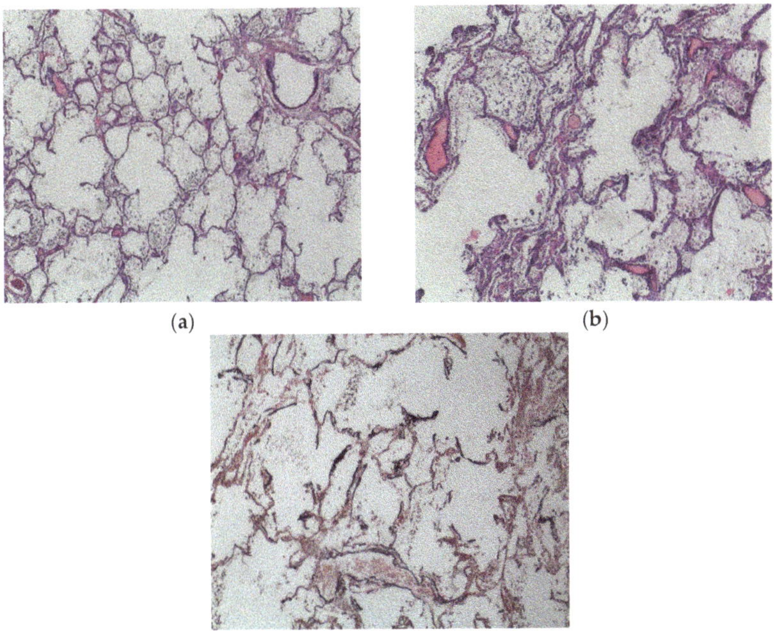

(a) (b) (c)

**Figure 5.** Lungs. (**a**) H&E staining—magnification 120×. Alveolar pattern and bronchial architecture are preserved although the septal walls are slightly thickened and basophilic. Calcium deposition can be spotted in a small bronchus basal membrane. (**b**) H&E staining—magnification 240×. The walls of small blood vessels and alveoli are particularly affected by metastatic calcification with a heavy diffuse deposition of calcium salts resulting in thickened, basophilic, and vaguely hyaline alveolar walls. The alveolar cavities contain few leukocytes and scarce amount of amorphous material. (**c**) IHC von Kossa stain—magnification 120×. Calcium salts deposition appears as dark stain because of its affinity for silver nitrate.

A diffuse nephrocalcinosis was bilaterally spotted; renal arterial walls were also involved by calcium deposition (Figure 6b). Diffuse calcifications were also observed in the gastroenteric tract, particularly in the gastric mucosa lamina propria and small vessels' walls (Figure 6a).

Taken together, histopathological and immunohistochemical findings were consistent with a solid-trabecular parathyroid carcinoma and a widespread calcium deposition, particularly severe on lungs, kidneys, and gastric wall. Alveolar septa, bronchi, and vessels walls were massively affected by calcium deposition; lungs also presented multiple foci of acute alveolar pneumonia.

To the best of the authors' knowledge, the new-born was reported in good health and, to date, has never experienced any complication.

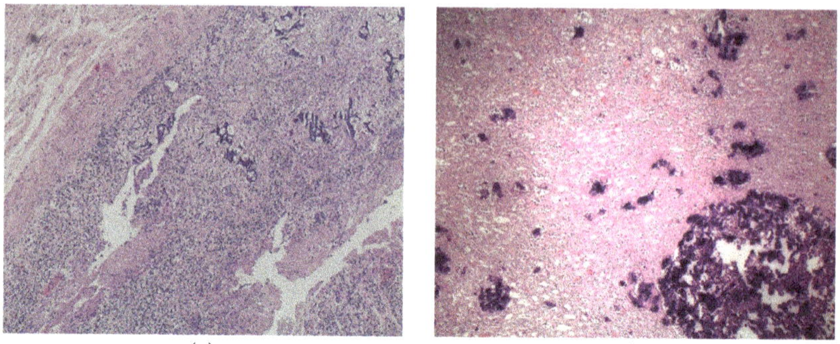

(a) (b)

**Figure 6.** Metastatic calcification involving viscera—calcium is recognisable in routine sections due to its affinity for haematoxylin. (**a**) Stomach, H&E staining—magnification 60×. Granular depositions of amorphous calcium salts in the lamina propria of the gastric mucosa. (**b**) Kidney, H&E staining—magnification 60×. Granular depositions of amorphous calcium salts in the kidney medullar interstitium.

## 3. Discussion

The case presented here was a harsh challenge for clinicians due to the rapid onset of symptoms that led to death within just a few hours after the admission into hospital. The difficult differential diagnosis reflected a peculiar physiopathology based on the complex interaction of a rare pathology and pregnancy.

The patient died before a defined diagnosis was made. Clinical course, CT scan, lab parameters, and circumstances (14 days following home delivery) were highly suggestive of a severe septic shock with respiratory failure resulting in MOF. On the other hand, the clinical course did not allow confirmation, nor excluding, of an infection.

Autopsy and histopathological analyses were essential to unravel the clinical conundrum. Post-mortem investigations ascertained the aetiology of the respiratory failure and its physio-pathological basis. Firstly, the thyroid mass described on CT scan was found instead to be a parathyroid carcinoma. Secondly, autopsy provided evidence of a massive calcium redistribution in the body due to the gross finding in the skullcap and the microscopic analyses of lungs and, to a lesser extent, kidneys, and stomach.

Combined with clinical biochemistry parameters, these findings indicate a condition of pHPT. Clinical suspicion was diverted from HPT since parameters' modification could have supported terminal pneumonia. With the benefit of the hindsight, others may be owing to possible paraneoplastic activity.

A few cases of respiratory symptoms sustained by calcium deposition were reported in literature, often linked to pHPT due to parathyroid adenoma [7–9], none of which regarded pregnant women. In other reports, MPC was linked to another medical conditions, such as multiple myeloma [10] and chronic renal failure [11].

Macroscopic, histopathological and immunohistochemical lungs findings suggested an etiological relationship between MPC and the developing of respiratory failure. Massive calcium deposition can indeed alter, thickening, the alveolar capillary membrane leading to an alteration of the oxygen and carbon dioxide exchange.

It is hard to exclude the influence of the recent pregnancy in such peculiar physiopathology. Although the understanding of the complex interaction between these biological phenomena is way beyond that of the aim of this report, a possible explanation may be addressed.

The parathyroid carcinoma was most likely already present during the pregnancy, which happened without clinical complications or specific symptoms. Physiological changes linked to pregnancy, such as haemodilution, hypoalbuminemia, maternal hypercalciuria, and high foetal calcium demand may have buffered the hypercalcemia, preventing the symptoms. We are not able to prove the occurrence of calcium redistribution yet during pregnancy, since no specific screenings were conducted. However, this hypothesis may be supported in its entirety due to the massive erosions spotted on the skullcap. The resulting physio-pathological and clinical stability may have dramatically ended in the puerperium as the 'buffer' mechanism ceased, with the unspecific symptoms' onset at the end of the first week and the tragical course on the 14th day with dyspnoea rapidly evolving into a massive, fatal respiratory insufficiency.

The case presented here has several limitations. As pregnancy was reported to occur without complications, only routine screenings were performed, regarding the calcium/phosphate balance. Due to unconventional preservation of placenta, pathological findings may have been lost. The CT report without images was only available to the authors.

## 4. Conclusions

The case presented provides evidence of lung complications in the case of a pHPT, with a silent clinical course during pregnancy, resulting in a fatal MPC occurred in the puerperium. This report hence suggests a potential implementation of calcium metabolism screening during pregnancy even when specific symptoms are missing. The authors believe that sharing this unique case with the scientific audience stresses the importance of the clinical autopsy either in completing a diagnosis or preventing potential litigations.

**Supplementary Materials:** The following supporting information can be downloaded at: https://www.mdpi.com/article/10.3390/ijms232315131/s1.

**Author Contributions:** Conceptualization A.C. (Alberto Chighine) and M.N.; C.C., G.C. and A.C. (Andrea Corona) contributed to the final version of the manuscript; supervision, R.D.; R.D. and M.N. contributed equally. All authors have read and agreed to the published version of the manuscript.

**Funding:** This research received no external funding.

**Institutional Review Board Statement:** Since only normal clinical practice is described in this case report, formal ethical approval by the Independent Review Board was not required in accordance with the policy of our institution.

**Informed Consent Statement:** Informed consent has been obtained from the Local Prosecutor as the case was the object of forensic investigation.

**Data Availability Statement:** Data, after adequate anonymization, will be available on request from the corresponding author (A.C.).

**Conflicts of Interest:** The authors declare no conflict of interest.

## References

1. Walker, M.D.; Silverberg, S.J. Primary hyperparathyroidism. *Nat. Rev. Endocrinol.* **2018**, *14*, 115–125. [CrossRef] [PubMed]
2. Komarowska, H.; Bromińska, B.; Luftmann, B.; Ruchała, M. Primary hyperparathyroidism in pregnancy—A review of literature. *Ginekol. Pol.* **2017**, *88*, 270–275. [CrossRef] [PubMed]
3. Baretić, M.; Tomić Brzac, H.; Dobrenić, M.; Jakovčević, A. Parathyroid carcinoma in pregnancy. *World J. Clin. Cases* **2014**, 2, 151–156. [CrossRef] [PubMed]
4. Ali, D.S.; Dandurand, K.; Khan, A.A. Primary Hyperparathyroidism in Pregnancy: Literature Review of the Diagnosis and Management. *J. Clin. Med.* **2021**, *10*, 2956. [CrossRef] [PubMed]
5. DiMarco, A.N.; Meeran, K.; Christakis, I.; Sodhi, V.; Nelson-Piercy, C.; Tolley, N.S.; Palazzo, F.F. Seventeen Cases of Primary Hyperparathyroidism in Pregnancy: A Call for Management Guidelines. *J. Endocr. Soc.* **2019**, *3*, 1009–1021. [CrossRef] [PubMed]
6. Dochez, V.; Ducarme, G. Primary hyperparathyroidism during pregnancy. *Arch. Gynecol. Obstet.* **2015**, *291*, 259–263. [CrossRef] [PubMed]
7. Cohen, A.M.; Maxon, H.R.; Goldsmith, R.E.; Schneider, H.J.; Wiot, J.F.; Loudon, R.G.; Altemeier, W.A. Metastatic pulmonary calcification in primary hyperparathyroidism. *Arch. Intern. Med.* **1977**, *137*, 520–522. [CrossRef] [PubMed]
8. Ikegaki, S.; Otoshi, T.; Hirai, T.; Hirabayashi, M. A rare case of parathyroid crisis with respiratory failure successfully treated using extracorporeal membrane oxygenation. *Resp. Med. Case Rep.* **2020**, *30*, 101088. [CrossRef] [PubMed]
9. Xu, W.; Hu, T.; Lv, S.; Dong, M.; Yang, J. Metastatic microcalcification in the lungs and stomach due to hyperparathyroidism: A case report. *J. Int. Med. Res.* **2021**, *49*, 300060521994420. [CrossRef] [PubMed]

10. Yamamoto, K.; Ando, K.; Tanaka, M.; Yura, H.; Sakamoto, N.; Zaizen, Y.; Ashizawa, K.; Fukuoka, J.; Miyazaki, Y.; Mukae, H. A 71-Year-Old Man With Dyspnea and Cough During Chemotherapy. *Chest* **2020**, *157*, e165–e171. [CrossRef] [PubMed]
11. Georges, S.; Srour, N. Metastatic pulmonary calcification in end-stage renal failure. *CMAJ Can. Med. Assoc. J.=J. De L'association Med. Can.* **2016**, *188*, E394. [CrossRef] [PubMed]

International Journal of *Molecular Sciences*

MDPI

*Communication*

# Modeling Preeclampsia In Vitro: Polymorphic Variants of STOX1-A/B Genes Can Downregulate CD24 in Trophoblast Cell Lines

**Marei Sammar [1,*], Clara Apicella [2], Peter Altevogt [3], Hamutal Meiri [4] and Daniel Vaiman [2]**

1 Prof. Ephraim Katzir's Department of Biotechnology Engineering, Braude College of Engineering, 51 Snunit St., Karmiel 2161002, Israel
2 Institute Cochin, U1016, INSERM, UMR 8504 CNRS, Paris-Descartes Université, 75014 Paris, France
3 Skin Cancer Unit, DKFZ and Department of Dermatology, Venereology and Allergology, University Medical Center Mannheim, Ruprecht-Karl University of Heidelberg, Theodor-Kutzer-Ufer 1–3, 68167 Mannheim, Germany
4 Hylabs, Rehovot and TeleMarpe, 21 Beit El St., Tel Aviv 6908742, Israel
* Correspondence: sammar@braude.ac.il; Tel.: +972-(04)-9901769; Fax: +972-(04)-99017

**Abstract:** CD24 is a mucin-like immunosuppressing glycoprotein whose levels increase during pregnancy and decrease in the syncytio- and cytotrophoblasts in early and preterm preeclampsia. We used two modified cell lines that mimic in vitro features of preeclampsia to identify if this phenomenon could be reproduced. Our model was the immortalized placental-derived BeWo and JEG-3 cell lines that overexpress the STOX1 A/B transcription factor gene that was discovered in familial forms of preeclampsia. BeWo and JEG-3 cells stably transduced with the two major isoforms of STOX1-A/B or by an empty vector (control), were propagated, harvested, and analyzed. CD24 mRNA expression was determined by quantitative real-time polymerase nuclear chain reaction (qRT-PCR). CD24 protein levels were determined by Western blots. In STOX1-A/B overexpressing in BeWo cells, CD24 mRNA was downregulated by 91 and 85%, respectively, compared to the control, and by 30% and 74%, respectively in JEG-3 cells. A 67% and 82% decrease in CD24 protein level was determined by immunoblot in BeWo overexpressing STOX1-A/B, respectively, while the reduction in JEG-3 cells was between 47 and 62%. The immortalized BeWo and JEG-3 cell lines overexpressing STOX1-A/B had reduced CD24. Although both cell lines were affected, BeWo appears to be more susceptible to downregulation by STOX-1 than JEG-3, potentially because of their different cell origin and properties. These results strengthen the in vivo results of reduced CD24 levels found in early and preterm preeclampsia. Accordingly, it implies the importance of the reduced immune tolerance in preeclampsia, which was already demonstrated in vivo in the STOX1-A/B model of preeclampsia, and is now implied in the in vitro STOX-1 model, a subject that warrants further investigations.

**Keywords:** CD24; preeclampsia; placental-derived immortalized cells; JEG-3 cells; BeWo cells; STOX1-A/B; in vitro preeclampsia models; immune tolerance

**Citation:** Sammar, M.; Apicella, C.; Altevogt, P.; Meiri, H.; Vaiman, D. Modeling Preeclampsia In Vitro: Polymorphic Variants of STOX1-A/B Genes Can Downregulate CD24 in Trophoblast Cell Lines. *Int. J. Mol. Sci.* **2022**, 23, 15927. https://doi.org/10.3390/ijms232415927

Academic Editors: Ilona Hromadnikova and Udo Jeschke

Received: 22 November 2022
Accepted: 12 December 2022
Published: 14 December 2022



## 1. Introduction

The pathogenesis of preeclampsia and other hypertensive pregnancy disorders are poorly understood, despite a substantial worldwide research effort and the high burden of maternal and neonatal morbidity associated with this condition [1–3]. In particular, the role of genetic variants as determinants of disease susceptibility has not been discovered [4], although family history [5] and ethnic origin are major prior risk factors for the syndrome [6].

In 2005, van Dijk et al. [7] showed that storkhead box 1 (STOX1), a transcription factor belonging to the enlarged FOX family, has a strong association with genetic polymorphisms located inside the open reading frame of this gene in patients with familial forms of preeclampsia. Clinical studies have shown that in subsets of patients with the Y153H variant of STOX1 there was systemic endothelial dysfunction, hypertension, and proteinuria. Despite some later studies that raised questions on the composition of the originally used cohort, subsequent studies by Tyberghein et al. [8] and van Dijk et al. [9] showed the STOX1 impact on placental cell migration/invasion mechanisms. Doridot et al. [10] have demonstrated the role of STOX1 in balancing oxidative/nitrosative stress.

Studies have shown that STOX1 has two major isoforms: STOX1-A (the most complete, encompassing, in particular, a DNA-binding domain and a transactivator domain, 989 amino acids), and STOX1-B, coding for a 227 amino acid polypeptide, which does not encompass the transactivator domain [7,11]. Vaiman and Miralles [12] hypothesized that the two isoforms could compete for the same DNA binding site(s) thus triggering different responses and that their imbalance could result in placental damage and hypertensive pregnancy disorders. Recently, it was revealed by exome sequencing that there are rare variants of STOX1 associated with HELLP syndrome, a preeclampsia subtype with serious hepatic complications [13].

In 2013, Doridot et al. [14] developed a transgenic mouse model expressing STOX1 at the fetoplacental level. During pregnancy, these mice developed hypertension and proteinuria, often combined with fetal growth restriction, which is reversed by aspirin and also by alpha-1 microglobulin [14,15]. These preeclampsia features found in the STOX1 mouse model were also accompanied by changes in the placental vascular and extracellular matrix, and are linked to impaired electron transfer in the mitochondria of placental cells [10,14,16]. In some cases, STOX1 expression, particularly in extravillous trophoblasts, was linked to the activation of the uteroplacental renin–angiotensin system [17,18]. Using in vitro and ex vivo approaches, van Dijk et al. [7,9] showed that the risk allele (Y153H) of the preeclampsia susceptibility gene STOX1 negatively regulates trophoblast invasion by upregulating the cell–cell adhesion protein alpha-T-catenin (CTNNA3) [19].

In vitro models were also developed in immortalized placental cell lines transfected with STOX1-A and STOX1-B [20] using BeWo and JEG-3 cell lines; in these models, there is a >20–30-fold overexpression of STOX1-A and a >6–10-fold overexpression for STOX1B. Transcriptome analysis showed that in these cells, STOX1 elicits the down or upregulation of 12.5% of the genes [11,20], mimicking the modification of gene expression observed in the preeclamptic placenta including the genes for Endoglin, Syncytins-1 and 2, human chorionic gonadotrophin (hCG), Progesterone, LGALS13 (PP13), and LGALS14 (PP14), among others. This gene list includes preeclampsia's main risk prediction markers or modulators [5,16,20–22].

The potential impact of STOX1 on immune suppression (pivotal for immunotolerance during pregnancy) has not yet been specifically explored. Among the suppressor proteins involved in these pathways is CD24 which has been shown to be an important factor in the immune suppression of these pathways. CD24 is a marker of regulatory B cells [23] that are involved in increasing immunotolerance, especially in the context of grafts [24].

CD24 is a small (27 amino acids) protein attached to the membrane via a glycosylphosphatidylinositol (GP-I) anchor [25,26]. It has many potential glycosylation sites for N- and O-linked carbohydrate binding, rendering the molecule structurally similar to mucins [27]. CD24 binds to Siglec-10, and together they form a strong immunosuppressive axis [28]. The binding of the Siglec10–CD24 axis was demonstrated to be an important immune checkpoint for immune tolerance in mouse autoimmune models [29,30]. Recently, CD24 was identified as an immune modulator in cancer cells inhibiting the phagocytic potential of macrophages, as was shown by Barkal et al. [31].

Recombinant CD24, in the form of CD24 fusion protein bound to the fragment crystalized (Fc) part of immunoglobulins of the antibodies (CD24-Fc), was recently demonstrated to be a promising drug for blocking over-shooting immune reactions ("the cytokine storm") in SARS-CoV-2 infections [32]. CD24-containing exosomes were also found in a clinical trial to reduce the symptoms and the severity of COVID-19 [33], indicating a pharmacological link between immune suppression and CD24 in different diseases.

In previous publications, our group has studied placental CD24 and found that its first-trimester expression is linked to the glandular epithelial cells of the uterine glands and other decidual cells [34]. As was previously reported [27,28], it was also found that the protein is co-expressed with Siglec-10, mainly in the close vicinity of the invasive extravillous trophoblasts [34]. In another study, we used qRT-PCR analysis to show a significant increase in CD24 expression from the first and early second trimester to the third trimester and term delivery [35]. In contrast to the normal course of pregnancy, in cases of early (before 34 weeks gestation), and preterm (before 37 weeks gestation) preeclampsia, the level of CD24 mRNA is reduced [35] compared to cases of term delivery and preterm delivery (before 37 weeks gestation). Such a reduction in CD24 mRNA was also found before [36]. Moreover, in addition to mRNA reduction, it was found that in cases of early and preterm preeclampsia, immunohistochemistry labeling by CD24 was reduced in the syncytio-and-cytotrophoblasts, compared to preterm and term controls [35]. In contrast, a higher expression of CD24 protein was found in term preeclampsia cases compared to term and preterm controls [35].

In the present study, we evaluated the impact of STOX1A/B overexpression in BeWo and JEG-3 cells on CD24 expression. We found a reduced expression level of CD24 mRNA and protein in BeWo cells, and to a lower extent in JEG-3 cells.

## 2. Results

### *2.1. CD24 Expression in Cell Line Transfected by the Polymorphic Variants of STOX1-A and B*

We applied qRT-PCR analyses to study the CD24 expression in BeWo and JEG-3 cells overexpressing STOX1-A or STOX1-B compared to those encompassing the empty vector (control). Measured against two reference genes (HPRT and WYHAZ), a reduced CD24 expression was found for both STOX1 isoforms in BeWo and JEG-3 (Figure 1A,B). The expression of CD24 ($2^{-\Delta\Delta CT}$) was massively and significantly reduced in the BeWo cell line for STOX1-B [85%] and STOX1-A [91%] (Figure 1A) compared to the control (** $p < 0.01$). The downregulation of CD24 in the JEG-3 cell line was more moderate [74%] in STOX1-B and [30%] in STOX1-A-overexpressing cells (Figure 1B).

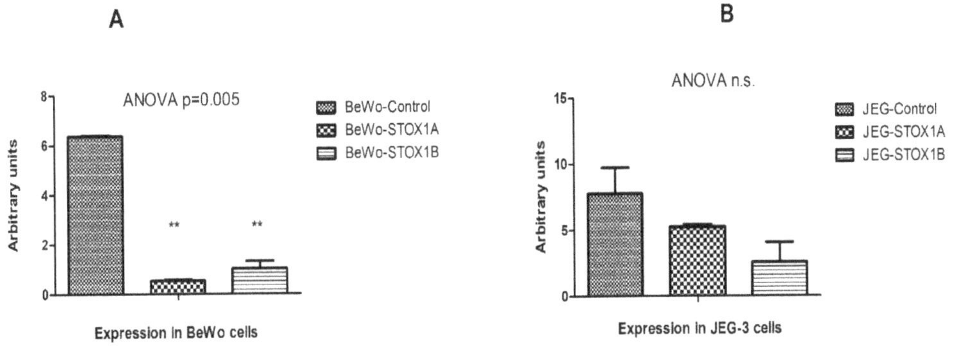


**Figure 1.** Downregulation of CD24 expression in (**A**) BeWo and (**B**) JEG-3 cells overexpressing STOX1-A and STOX1-B variants. The qPCR experiments were conducted in duplicates using, as a normalizer, the geometric average of the ΔCT of the two reference genes (HPRT and YWHAZ). The error bars are SEM. The two asterisks in A represent the individual one-way ANOVA significant level of each STOX1 variant compared to the control (** $p < 0.01$). The lower degree of downregulation of CD24 in the JEG-3 vs. BeWo cells could be due to the different origins and features of these two placental-derived cell lines.

The $2^{-\Delta\Delta CT}$ of CD24 expression is presented for the cell line overexpressing STOX1-A/B relative to the mock empty control vector (100%), all standardized against the housekeeping genes HPRT and YWHAZ; A—BeWo cells, B—JEG-3 cells. The analysis used one-way ANOVA, followed by a post-hoc test: Student–Neumann–Keuls. The stars correspond to these post hoc tests, i.e., comparing controls vs. STOX1-A and vs. STOX1-B, separately.

### 2.2. Bioinformatic Analysis of the CD24 Promoter

A FIMO analysis [36] was carried out on the position −2000 to +500 of the CD24 promoter obtained from the EPD database [37] for STRE1 (CATTTCACGG) and STRE2 (GGTGYGGAMA), as identified in Ducat et al. (2020) [11]. A unique hit was found, but with an FDR (q-value) of 0.204 (Table 1).

**Table 1.** A FIMO analysis on the position of CD24 promotor on the position −200 to +500 was extracted from the EPD database for the STRE1 sequence (CATTTCACGG) and STRE2 sequence (GGTGYGGAMA).

| Motif ID | Alt ID | Sequence Name | Strand | Start | End | *p*-Value | q-Value | Matched Sequence |
|---|---|---|---|---|---|---|---|---|
| 1 | GGTGYGGAMA | CD24Promoter | - | 2466 | 2475 | $4.35 \times 10^{-5}$ | 0.204 | GGTGTGGAAT |

### 2.3. Western Blot Analysis

Western Blot analysis of CD24 showed that the protein level was decreased in both cell lines overexpressing STOX1-A and B.

The total amount of CD24 in BeWo was significantly decreased by 67% ($p = 0.015$) when they overexpressed STOX1 A, and by 82% ($p = 0.008$) when they overexpressed STOX1-B compared to the control BeWo-C cells (Figure 2B).

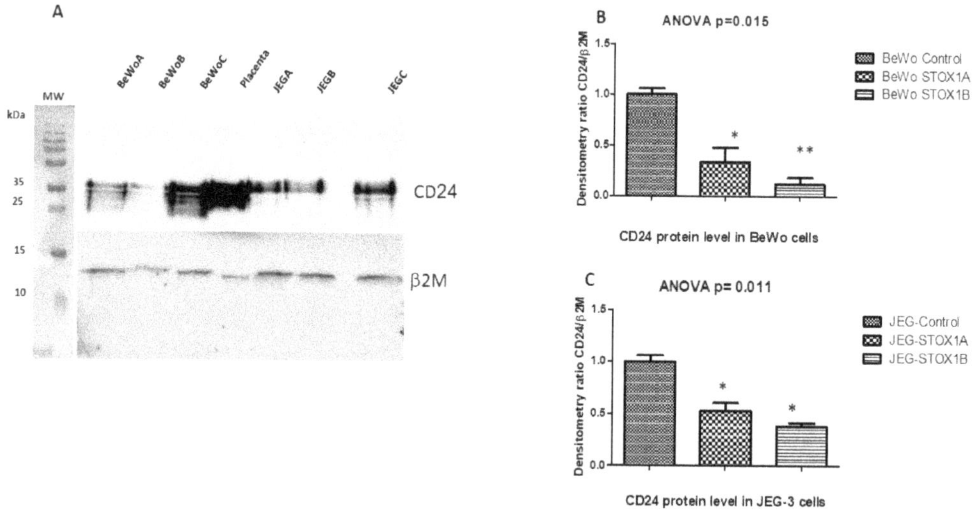


**Figure 2.** Western blot analyses of the CD24 expression in STOX1-A/B-stably transfected BeWo and JEG-3 cells. (**A**) Western blot analysis of the CD24 protein expression in trophoblast cell transfectants by STOX1-A/B and mock control of BeWo cells and of JEG-3 cells. CD24 was analyzed by Western blotting with a specific anti-CD24 monoclonal antibody (upper panel). Beta2-microglobulin was used as a control for loading equal amounts of proteins on the gel (lower panel). (**B**,**C**) Semiquantitative densitometric assessment of CD24 and B2-microglobulin bands was performed using ImageJ software. The data represents density at two exposure time points. The intensity of the CD24 bands was normalized to β2-microglobulin as a control for the equal loading of proteins. The results have a similar pattern in three different repeats. The statistic is based on one-way ANOVA followed by a post hoc test: Student–Neumann–Keuls. The stars were obtained by comparing controls vs. STOX1-A and STOX1-B, separately (* $p < 0.05$ and ** $p < 0.01$). The figure is representative of the Western blot experiments that were conducted three times and showed a similar pattern of CD24 reduction. The histogram error bars are SEM.

In the JEG-3 cells, the reduced CD24 level was more moderate when compared to the reduction in the mRNA levels. In JEG-3 cells, CD24 was significantly reduced by 47–62% ($p = 0.011$) in both STOX1-A and B compared to the control (Figure 2C).

The lower degree of the downregulation of CD24 in JEG-3 vs. BeWo could be due to the capability of BeWo cells to go to the end of their differentiation program, since they can fuse, while JEG-3 may be blocked in the status of nondifferentiated trophoblasts, and are possibly less sensitive to fusion signaling modulation.

The downregulation of CD24 was found in early and preterm preeclampsia [35]; the higher impact of STOX1-B over STOX1-A may be related to their differential features as transcription factors in gene expression, as it was demonstrated earlier for other proteins [11].

## 3. Discussion

CD24 appears to be one of the major actors in immune tolerance. It is a cell-surface protein marking B-reg cells, a subset of B cells that suppress immune reactions [38], as was already found in organ transplantation [38,39]. Since gestation can be defined as a semi-allograft, it would be desirable to identify whether B-reg-cell-mediated immune modulation is involved in a successful pregnancy.

The importance of STOX1 variants in preeclampsia was initially shown among Dutch familial cases [7,9], and also recently in the Turkish population [40]. On the contrary, no clear link to preeclampsia was found among Korean women [41] or in a number of European ancestries. Familial aspects of the disease are debatable [42], although a family history of preeclampsia is a worldwide known prior risk factor to develop preeclampsia [6].

Here, we attempt to make a connection between the overexpression of STOX1 and its link to reduced immune tolerance in pregnancy. For this purpose, we used JEG-3 and BeWo cells that overexpressed STOX1. BeWo cells have similarities with the villous trophoblasts (they can be induced to generate syncytia, but do not express HLA-G) [21,24]. JEG-3 cells are more similar to extravillous trophoblasts since they are highly proliferative but do not present syncytial fusion [21,24,43–45]. Therefore, these two cell types could serve as models, albeit imperfect ones, for two major types of placental cells. In vivo, the overexpression of STOX1 reproduced a few preeclamptic features [10,11], such as oxidative/nitrosative stress, and in previous in vitro studies, members of our team have shown that it is associated with membrane damage and impaired syncytialization.

Our main findings in this study are (1) reduced CD24 mRNA expression in BeWo and JEG-3 cells overexpressing STOX1-A/B, and (2) decreased level of CD24 protein in BeWo and to a lower extent in JEG-3 cells overexpressing STOX1-A/B. Altogether, our data demonstrate that there is a downregulation of CD24 in the overexpression of STOX1-A/B in trophoblast cell lines and that this effect is stronger in BeWo than in JEG-3 cells. Regarding the mechanism of deregulation, we attempted to identify STOX1 binding sites in the CD24 promoter. A putative STRE2 is found, however, STRE2 is not bound directly by STOX1 [11], but probably by a partner, since in most cases where a direct regulation occurs, both STRE1 and STRE2 are present together and STOX1 binds exclusively to STRE1 [11]. In summary, we cannot exclude an indirect interaction of STOX1 on the CD24 promoter that could explain the deregulation.

In early pregnancy, villous trophoblasts come into close contact with various types of maternal blood vessels, and with immune cells from the decidua, they are exposed to different oxygen concentrations throughout pregnancy (hypoxia, then normoxia) [21,43]. In this study, we found that the overexpression of STOX1 leads to the reduced expression of CD24, which is anticipated to negate immune tolerance. This suggests a novel link between the STOX1 gene in pregnancy and CD24 in a direction that may suppress the immune tolerance offered by CD24 (i.e., the link between overexpression of the STOX1 transcription factor could reduce the expression of CD24 and impair the CD24-Siglice10 axis which may impair immune tolerance the in the trophoblast cells). The physiological importance of CD24 expression by nonimmunological cells has not been clarified. However, we noticed that the expression level of CD24 in JEG-3 cells is similar to that which has already been found for the most expressed genes in this model. It implies that there is physiological importance of CD24 in trophoblasts, as was already shown for B cells. The verification of the importance of CD24 in JEG-3 cells, however, warrants further investigation.

The extravillous trophoblasts (modelized here by JEG-3) are formed early in pregnancy. They come into contact with a number of cells from the uterine wall, including decidual stromal cells and some immune cells, e.g., uNK cells. Since HLA-G is expressed in those cells, their need for control by the STOX1–CD24 axis is less essential than for villous trophoblasts [21,24,43–45]. In addition, during pregnancy, the extravillous trophoblasts do not encounter varying oxygen concentrations. When the extravillous trophoblasts invade the uterine wall, it occurs under standard oxygen concentrations. Thus, the electron transfer, which involves STOX1, may not be essential to the trophoblasts' function in comparison to the function of villous trophoblasts. If so, the overexpression of STOX1-A/B may not be so damaging to the JEG-3 cell line that resembles the extravillous trophoblast layer. This is consistent with the differential efficacy of STOX1 overexpression in decreasing CD24 protein levels, as we have found for JEG-3 in this study. Interestingly, another protein related to immune tolerance is PP13 or LGALS13. This protein is also expressed in the syncytiotrophoblasts and its level was increased in BeWo cells stimulated to syncytialization [5]. The expression of PP13 was not found in extravillous trophoblasts [45], and consistently, its level in JEG-3 cells is also very low [46].

We overexpressed the STOX1 A/B transcription factor gene that was discovered in familial forms of preeclampsia in immortalized placental cell lines. The results presented here are solid, despite being limited in scope, and may drive more extensive research by us and by other groups.

In accordance with our results, we propose an additional aspect to the role of STOX1 in increasing the risk of preeclampsia through the regulation of immune suppression. Since STOX1 suppresses CD24 which confers immune tolerance, this study indicates that there is an additional pathway for STOX1 involvement in the development of preeclampsia. Accordingly, the following process is delineated: In normal pregnancy, the blastocyst is implanted into the uterine wall and the invading trophoblasts enter the decidua. These cells express cell-surface proteins of a paternal origin; hence, a normal pregnancy is dependent on immune tolerance to enable the maintenance of the gestation. The development of immune tolerance involves certain sets of proteins including HLA-G and others; we propose that CD24 is one of them. Although our culture model of overexpression of STOX1 in BeWo and JEG-3 cells does not mimic major aspects of preeclampsia such as hypertension and proteinuria, it could still serve in the study of certain other molecular aspects of the disorder such as the reduction in the expression and in the level of proteins that are involved in immune tolerance. Since, in the past, we showed that in preeclampsia cases there is a reduction in the expression of CD24, our results in this in-vitro model enable us to consider the reduced level of CD24 as an additional aspect of reduced immune tolerance in preeclampsia. This direction may lead to exploring the supplement of CD24 as a therapeutic agent to fight preeclampsia.

A model of CD24 action in normal and preeclamptic pregnancy is proposed in Figure 3. Part A shows the interaction of Siglec-10 on immune cells with the expression of CD24 on the placental trophoblasts that were invading the decidua (or in our model, with BeWo cells). Such activity will result in an active immunosuppressive process during normal pregnancy (Figure 3A). When CD24 expression is limited such as in preeclampsia or BeWo trophoblasts transfected with STOX1A/B, the immunosuppression and immunotolerance fail to protect the pregnancy (Figure 3B).

Therefore, in this hypothetic model (Figure 3C), the replenishment of CD24 deems necessary for maintaining immune suppression during pregnancy. Therefore, a putative treatment with either CD24-Fc molecule or extracellular vesicles enriched with CD24 [32,33] could serve as a potential therapy to cure preeclampsia.

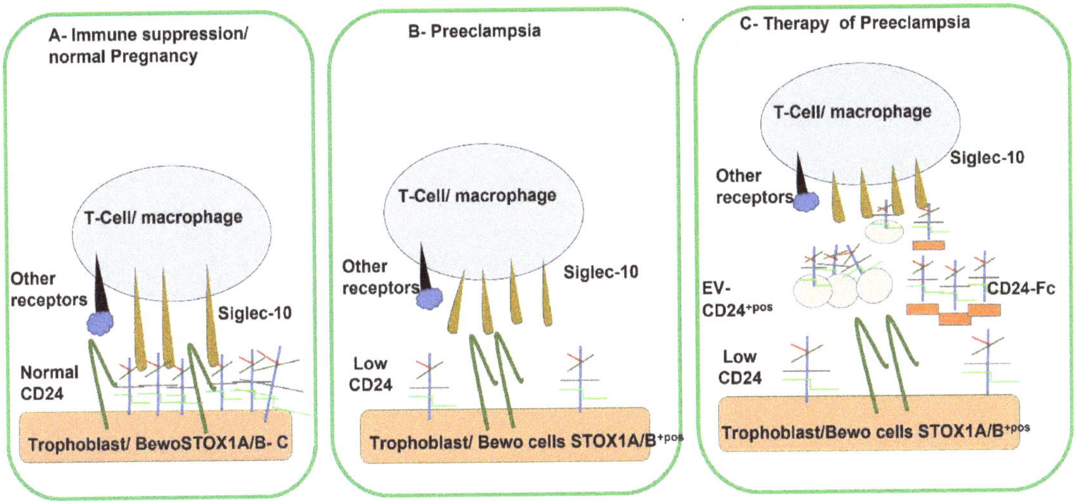


**Figure 3.** A schematic model of the potential role of CD24 in a normal and in preeclampsia is depicted to indicate how CD24 replenishment could help fighting preeclampsia. (**A**) In normal pregnancy, the immunosuppressive process involves the formation of a complex made of CD24, Siglec-10 and the control STOX1 A/B (STOX1-A/B—C) on the surface of BeWo/trophoblast cells. The complex triggers the inhibitory signals and promotes immunosuppression. (**B**) in preeclampsia there is a reduced amount of CD24-Siglec 10 due to the over expression of the mutated variants of sTOX1 A/B on the surface of the BeWo cells, leading to the loss of immune suppression. (**C**) When replenishment the cultured cells with extracellular vesicles (EV) that have a rich expression of CD24 on their surface, (EV-CD24$^{+pos}$) or with CD24-Fc soluble molecule, the BeWo/trophoblast regain immune tolerance and recover from preeclampsia.

## 4. Materials and Methods

### 4.1. Antibodies

The monoclonal antibody (mAb) clone SWA11 was used for the detection of CD24 [47,48]. This mAb is specific for CD24 and reacts with the leucine–alanine–proline (LAP) motif in the protein core, as shown by peptide inhibition studies [48]. In addition, SWA11 exhibits specific binding to CD24-transfected cells but not to a vector control [47].

### 4.2. Plasmid Preparation

The STOX1-A and STOX1-B constructs were a generous gift of Dr. C.B. Oudejans (VU University Medical Center, De Boelelaan 1117, 1081 HV, Amsterdam, The Netherlands). The ORF was isolated and subcloned in the pCMX expression vector and re-sequenced. Three mutations were found (ARC in 1658 (GAA = GluRGCA = Ala), a deleted T in 2948 (which creates an early stop codon), and the initiation codon (ACC (THR) instead of ATG (MET)) and corrected by site-directed mutagenesis. Hence, we obtained a pCMX vector containing the coding region of STOX1-A. STOX1-B was also re-sequenced from the relevant plasmid.

### 4.3. Transfection

Cell culture and stable transfection JEG-3 choriocarcinoma cells were seeded in DMEM medium (Gibco) supplemented with 10% FBS and 1% penicillin/streptomycin at a concentration of 106 cells per T25 culture flask. Passages of the cells were made between 5 to 10 min before processing.

At the time of transfection, the cells were at 60% confluence. The cells were transfected with the PGK-neo expression vector [25] using Lipofectamine 2000 Reagent (Invitrogen, Waltham, MA, USA), according to the manufacturer's recommendations. Four (4) µg of pCMX-STOX1-A or 4 μg empty pCMX together with 0.4 μg of PGK-neo per T 25 $cm^2$ culture dish with Opti-MEMH I Reduced Serum Medium. This plasmid ratio is known to ensure a co-transfection by both plasmids when the cells become resistant. The cells were passaged at 1:10 dilution into a selective medium 72 h post-transfection. The selection was continuously applied using Geneticin (G-418) (Invitrogen) at 500 mg/mL concentration for approximately 3 weeks. Resistant clones (nine cell lines transfected with pCMX-STOX1-A and three cell lines transfected with empty pCMX) were grown individually in the continual selection and used for further analysis or frozen in DMSO. Similarly, BeWo cells were stably transfected except that the cells were cultivated in an F12 medium (Life Technologies, Carlsbad, CA, USA) supplemented with FBS and antibiotics. JEG-3 cells were supplemented with 5-to-15 passages before processing.

Samples of mRNA were prepared in triplicates from two independent cell culture experiments and RT-PCR was performed using the MMLV kit from Invitrogen. The expression of STOX1-A or B was then assessed by qRT-PCR. The cells and their detailed preparation and selection were published previously [11,19]. They were all cultivated in an adequate medium complemented with Geneticin G-418 at 500 μg/mL [11]. BeWo-A and BeWo-B overexpressed STOX1-A and STOX1-B (20 and 6-fold, respectively), and BeWo-C was their related control (with the empty G-418 resistance plasmid). For JEG-3 cells, AA6 (JEG-3A) overexpressed STOX1-A (~30 fold), B10 (JEG-3B) overexpressed STOX1-B (~4–6 fold), and BD3 (JEG-3C) was the related control cell line.

### 4.4. RT-PCR Amplification of CD24 mRNA

RNA preparation was carried out in cells by direct lysis using the Trizol–Chloroform method, following the manufacturer's protocol.

Reverse transcription was carried out using the MMLTV reverse transcriptase kit from Invitrogen (Thermo-Fisher, Bd Sébastien Brant, Parc d'Innovation, France).

The expression of the CD24 gene was quantified by TaqMan RT-PCR utilizing the Applied Biosystem StepOne Plus cycler (Applied Biosystems, Austin, TX, USA) and TaqMan Gene Expression Assay with primer and probe sets (Applied Biosystems) for CD24 (Hs02379687_s1).

HPRT and WYHAZ (ABI, Branchburg, NJ, USA) were used to standardize the expression level. The relative amount of CD24 was calculated by employing the comparative CT method ($2^{-\Delta\Delta CT}$) [49]. Amplification was performed for the JEG-3 and BeWo cells transfected by STOX1-A or STOX1-B polymorphic variants compared with the empty expression vector used as a mock control.

### 4.5. Western Blot

BeWo and JEG-3 stable transfectant cells were trypsinized, pelleted, washed twice in PBS, and resuspended for 1 h at 4 °C in RIPA (5 mM Tris HCl pH 7.6, 150 mM NaCl, 1% NP-40, 1% sodium deoxycholate, 0.1% SDS) containing a cocktail of protease inhibitors (100× Thermo-Fisher, DTT 50 mM, and PMSF 50 mM). Insoluble material was removed by centrifugation (20,000× *g*, 4 °C), and the protein concentration of the soluble fraction was determined by a BCA reagent (Pierce, Rockford, IL, USA). Placental lysate samples were aliquoted and stored at −70 °C until use.

For Western blot analysis, 50 μg of total protein lysates were separated on 12.5% SDS-PAGE at 130 volts for 2 h. in Tris-Glycine, 2% SDS buffer, at room temperature; proteins were then electro-transferred to nitrocellulose membranes at 70 volts in Tris-Glycine buffer at 4 °C for 2 h. The transfer was evaluated by Ponceau red staining. Membranes were blocked in PBS with 0.1% Tween 20 and 5% defatted milk (Regilait) for 1 h, then rinsed thrice in PBS-Tween 0.1%. After blocking free binding sites, membranes were probed with the anti-CD24 mAb SWA11 and anti-β-globin antibodies (0.1 μg/mL) overnight at 4 °C. Bound immune complexes were detected by horseradish peroxidase-conjugated rabbit anti-mouse IgG and developed using an ECL detection kit (Biological Industries, Beit Haemek, Israel). The β2M was used as an internal control for the equal loading of proteins.

Signals were developed using chemiluminescence and were captured using an imager (Bio-Rad, Hercules, CA, USA).

### 4.6. Statistics

In the different experiments, statistics were based on one-way ANOVA, followed by a post hoc test: the Student–Neumann–Keuls test [50].

## 5. Conclusions

The downregulation of CD24 mRNA expression and its total protein levels were determined in placental-derived immortalized cell lines BeWo and JEG-3 cells that were overexpressing STOX1 A/B. STOX1-A appears more potent than STOX1-B in suppressing CD24 expression, and although both BeWo and JEG-3 cells were affected, BeWo appears more susceptible to downregulation by STOX-1 than JEG-3, potentially due to the differential cell origin. CD24 has already been found to convey immune suppression in cancer. Here, its reduced expression and protein levels were determined in the in vitro model of preeclampsia which is similar to the reduced CD24 found in early and preterm preeclampsia [35]. This may be another indication of the link between preeclampsia, a major pregnancy complication, and immune rejection, which has yet to be further explored.

**Author Contributions:** This study was initiated by M.S. who conceptualized and constructed the study on all fronts and designed its flow. C.A. and M.S. conducted all the experiments under the supervision, tools, and means of D.V. including cell culture, transfection, qRT-PCR, and immunoblots in BeWo and JEG-3 transfected cells. P.A., M.S. and H.M. organized the database and conducted the statistical analysis. All authors participated in data requisition, management, analysis, and the preparation of the manuscript. M.S. and H.M. All authors have read and agreed to the published version of the manuscript.

**Funding:** This work was supported by a travel grant from Braude College Research Collaboration Program Grant to M.S. The work in the laboratory of D.V. was funded by the recurrent dotation of Inserm/CNRS/University attributed to the team of Daniel Vaiman and by the European Union's Horizon 2020 Research and Innovation Programme, Grant No. 765274, acronym iPLACENTA (http://www.iplacenta.eu, accessed on 14 November 2022) to Danial Vaiman.

**Institutional Review Board Statement:** Not applicable.

**Informed Consent Statement:** Not applicable.

**Data Availability Statement:** Not applicable.

**Conflicts of Interest:** The authors declare no conflict of interest and there has been no significant financial support for this work that could have influenced its outcome.

## References

1. Magee, L.A.; Brown, M.A.; Hall, D.R.; Gupte, S.; Hennessy, A.; Ananth Karumanchi, S.; Kenny, L.C.; McCarthy, F.; Myers, J.; Poon, L.C.; et al. The Hypertensive Disorders of Pregnancy: The 2021 International Society for the Study of Hypertension in Pregnancy Classification, Diagnosis & Management Recommendations for International Practice. *Pregnancy Hypertens.* **2021**, *27*, 148–169. [CrossRef]
2. American College of Obstetricians and Gynecologists. Gestational Hypertension and Preeclampsia: ACOG Practice Bulletin, Number 222. *Obstet. Gynecol.* **2020**, *135*, e237–e260. [CrossRef]
3. World Health Organization. *Managing Complications in Pregnancy and Childbirth: A Guide for Midwives and Doctors*, 2nd ed.; World Health Organization: Geneva, Switzerland, 2017. Available online: http://apps.who.int/iris/betstream/handle/10665/255760/9789241565493-eng.pdf;sequence=1 (accessed on 14 November 2022).
4. Kaminski, V.L.; Chies, J.A.B. Genetic variants in preeclampsia: Lessons from studies in Latin-American populations. *Front. Physiol.* **2018**, *9*, 1771. [CrossRef]
5. Madar-Shapiro, L.; Karady, I.; Trahtenherts, A.; Syngelaki, A.; Akolekar, R.; Poon, L.; Cohen, R.; Sharabi-Nov, A.; Huppertz, B.; Sammar, M.; et al. Predicting the Risk to Develop Preeclampsia in the First Trimester Combining Promoter Variant-98A/C of LGALS13 (Placental Protein 13), Black Ethnicity, Previous Preeclampsia, Obesity, and Maternal Age. *Fetal Diagn. Ther.* **2018**, *43*, 250–265. [CrossRef]

6. Poon, L.C.; Kametas, N.A.; Chelemen, T.; Leal, A.; Nicolaides, K.H. Maternal risk factors for hypertensive disorders in pregnancy: A multivariate approach. *J. Hum. Hypertens.* **2010**, *24*, 104–110. [CrossRef]
7. van Dijk, M.; Mulders, J.; Poutsma, A.; Konst, A.A.; Lachmeijer, A.M.; Dekker, G.A.; Blankenstein, M.A.; Oudejans, C.B. Maternal segregation of the Dutch preeclampsia locus at 10q22 with a new member of the winged helix gene family. *Nat. Genet.* **2005**, *37*, 514–519. [CrossRef]
8. Tyberghein, K.; Goossens, S.; Haigh, J.J.; van Roy, F.; van Hengel, J. Tissue-wide overexpression of alpha-T-catenin results in aberrant trophoblast invasion but does not cause embryonic mortality in mice. *Placenta* **2012**, *33*, 554–560. [CrossRef]
9. van Dijk, M.; van Bezu, J.; van Abel, D.; Dunk, C.; Blankenstein, M.A.; Oudejans, C.B.; Lye, S.J. The STOX1 genotype associated with pre-eclampsia leads to a reduction of trophoblast invasion by alpha-T-catenin upregulation. *Hum. Mol. Genet.* **2010**, *19*, 2658–2667. [CrossRef]
10. Doridot, L.; Châtre, L.; Ducat, A.; Vilotte, J.L.; Lombès, A.; Méhats, C.; Barbaux, S.; Calicchio, R.; Ricchetti, M.; Vaiman, D. Nitroso-redox balance and mitochondrial homeostasis are regulated by STOX1, a pre-eclampsia-associated gene. *Antioxid. Redox Signal.* **2014**, *21*, 819–834. [CrossRef]
11. Ducat, A.; Couderc, B.; Bouter, A.; Biquard, L.; Aouache, R.; Passet, B.; Doridot, L.; Cohen, M.B.; Ribaux, P.; Apicella, C.; et al. Molecular Mechanisms of Trophoblast Dysfunction Mediated by Imbalance between STOX1 Isoforms. *iScience* **2020**, *23*, 101086. [CrossRef]
12. Vaiman, D.; Miralles, F. Targeting STOX1 in the therapy of preeclampsia. *Expert. Opin. Targets* **2016**, *20*, 1433–1443. [CrossRef] [PubMed]
13. Jiménez, K.M.; Morel, A.; Parada-Niño, L.; Alejandra González-Rodriguez, M.; Flórez, S.; Bolívar-Salazar, D.; Becerra-Bayona, S.; Aguirre-García, A.; Gómez-Murcia, T.; Fernanda Castillo, L.; et al. Identifying new potential genetic biomarkers for HELLP syndrome using a massive parallel sequencing. *Pregnancy Hypertens.* **2020**, *22*, 181–190. [CrossRef] [PubMed]
14. Doridot, L.; Passet, B.; Méhats, C.; Rigourd, V.; Barbaux, S.; Ducat, A.; Mondon, F.; Vilotte, M.; Castille, J.; Breuiller-Fouché, M.; et al. Preeclampsia-like symptoms induced in mice by fetoplacental expression of STOX1 are reversed by aspirin treatment. *Hypertension* **2013**, *61*, 662–668. [CrossRef] [PubMed]
15. Erlandsson, L.; Ducat, A.; Castille, J.; Zia, I.; Kalapotharakos, G.; Hedström, E.; Vilotte, J.L.; Vaiman, D.; Hansson, S.R. Alpha-1 microglobulin as a potential therapeutic candidate for treatment of hypertension and oxidative stress in the STOX1 preeclampsia mouse model. *Sci. Rep.* **2019**, *9*, 8561. [CrossRef] [PubMed]
16. Ducat, A.; Doridot, L.; Calicchio, R.; Méhats, C.; Vilotte, J.L.; Castille, J.; Barbaux, S.; Couderc, B.; Jacques, S.; Letourneur, F.; et al. Endothelial cell dysfunction and cardiac hypertrophy in the STOX1 model of preeclampsia. *Sci. Rep.* **2016**, *6*, 19196. [CrossRef]
17. Dunk, C.E.; van Dijk, M.; Choudhury, R.; Wright, T.J.; Cox, B.; Leavey, K.; Harris, L.K.; Jones, R.L.; Lye, S.J. Functional Evaluation of STOX1 (STORKHEAD-BOX PROTEIN 1) in Placentation, Preeclampsia, and Preterm Birth. *Hypertension* **2021**, *77*, 475–490. [CrossRef]
18. Parchem, J.G.; Kanasaki, K.; Lee, S.B.; Kanasaki, M.; Yang, J.L.; Xu, Y.; Earl, K.M.; Keuls, R.A.; Gattone, V.H., II; Kalluri, R. STOX1 deficiency is associated with renin-mediated gestational hypertension and placental defects. *JCI Insight* **2021**, *6*, e141588. [CrossRef]
19. van Dijk, M.; Oudejans, C.B. STOX1: Key player in trophoblast dysfunction underlying early onset preeclampsia with growth retardation. *J Pregnancy.* **2011**, *2011*, 521826. [CrossRef]
20. Rigourd, V.; Chauvet, C.; Chelbi, S.T.; Rebourcet, R.; Mondon, F.; Letourneur, F.; Mignot, T.M.; Barbaux, S.; Vaiman, D. STOX1 overexpression in choriocarcinoma cells mimics transcriptional alterations observed in preeclamptic placentas. *PLoS ONE* **2008**, *3*, e3905. [CrossRef]
21. Rothbauer, M.; Patel, N.; Gondola, H.; Siwetz, M.; Huppertz, B.; Ertl, P. A comparative study of five physiological key parameters between four different human trophoblast-derived cell lines. *Sci. Rep.* **2017**, *7*, 5892. [CrossRef]
22. van de Veen, W.; Stanic, B.; Wirz, O.F.; Jansen, K.; Globinska, A.; Akdis, M. Role of regulatory B cells in immune tolerance to allergens and beyond. *J. Allergy Clin. Immunol.* **2016**, *138*, 654–665. [CrossRef]
23. Piloni, D.; Morosini, M.; Magni, S.; Balderacchi, A.; Inghilleri, S.; Cova, E.; Oggionni, T.; Frangipane, V.; Pandolfi, L.; Scudeller, L.; et al. Peripheral CD19+CD24 high CD38 high B-regulatory cells in lung transplant recipients. *Transpl. Immunol.* **2019**, *57*, 101245. [CrossRef]
24. Hannan, N.J.; Paiva, P.; Dimitriadis, E.; Salamonsen, L.A. Models for study of human embryo implantation: Choice of cell lines? *Biol. Reprod.* **2010**, *82*, 235–245. [CrossRef]
25. Kay, R.; Rosten, P.M.; Humphries, R.K. CD24, a signal transducer modulating B cell activation responses, is a very short peptide with a glycosyl phosphatidylinositol membrane anchor. *J. Immunol.* **1991**, *147*, 1412–1416.
26. Rougon, G.; Alterman, L.A.; Dennis, K.; Guo, X.J.; Kinnon, C. The murine heat-stable antigen: A differentiation antigen expressed in both the hematolymphoid and neural cell lineages. *Eur. J. Immunol.* **1991**, *21*, 1397–1402. [CrossRef]
27. Aigner, S.; Sthoeger, Z.M.; Fogel, M.; Weber, E.; Zarn, J.; Ruppert, M.; Zeller, Y.; Vestweber, D.; Stahel, R.; Sammar, M.; et al. CD24, a mucin-type glycoprotein, is a ligand for P-selectin on human tumor cells. *Blood* **1997**, *89*, 3385–3395. [CrossRef]
28. Liu, Y.; Zheng, P. CD24: A genetic checkpoint in T cell homeostasis and autoimmune diseases. *Trends Immunol.* **2007**, *28*, 315–320. [CrossRef]
29. Crocker, P.R.; Paulson, J.C.; Varki, A. Siglecs and their roles in the immune system. *Nat. Rev. Immunol.* **2007**, *7*, 255–266. [CrossRef]

30. Pillai, S.; Netravali, I.A.; Cariappa, A.; Mattoo, H. Siglecs and immune regulation. *Ann. Rev. Immunol.* **2012**, *30*, 357–392. [CrossRef]
31. Barkal, A.A.; Brewer, R.E.; Markovic, M.; Kowarsky, M.; Barkal, S.A.; Zaro, B.W.; Krishnan, V.; Hatakeyama, J.; Dorigo, O.; Barkal, L.J.; et al. CD24 signalling through macrophage Siglec-10 is a target for cancer immunotherapy. *Nature* **2019**, *572*, 392–396. [CrossRef]
32. Song, N.J.; Allen, C.; Vilgelm, A.E.; Riesenberg, B.P.; Weller, K.P.; Reynolds, K.; Chakravarthy, K.B.; Kumar, A.; Khatiwada, A.; Sun, Z.; et al. Treatment with soluble CD24 attenuates COVID-19-associated systemic immunopathology. *J. Hematol. Oncol.* **2022**, *15*, 5. [CrossRef] [PubMed]
33. Shapira, S.; Ben Shimon, M.; Hay-Levi, M.; Shenberg, G.; Choshen, G.; Bannon, L.; Tepper, M.; Kazanov, D.; Seni, J.; Lev-Ari, S.; et al. A novel platform for attenuating immune hyperactivity using EXO-CD24 in COVID-19 and beyond. *EMBO Mol. Med.* **2022**, *14*, e15997. [CrossRef] [PubMed]
34. Sammar, M.; Siwetz, M.; Meiri, H.; Fleming, V.; Altevogt, P.; Huppertz, B. Expression of CD24 and Siglec-10 in first trimester placenta: Implications for immune tolerance at the fetal-maternal interface. *Histochem. Cell Biol.* **2017**, *147*, 565–574. [CrossRef] [PubMed]
35. Sammar, M.; Siwetz, M.; Meiri, H.; Sharabi-Nov, A.; Altevogt, P.; Huppertz, B. Reduced Placental CD24 in Preterm Preeclampsia Is an Indicator for a Failure of Immune Tolerance. *Int. J. Mol. Sci.* **2021**, *22*, 8045. [CrossRef] [PubMed]
36. Grant, C.E.; Bailey, T.L.; Noble, W.S. FIMO: Scanning for occurrences of a given motif. *Bioinformatics* **2011**, *27*, 1017–1018. [CrossRef]
37. Meylan, P.; Dreos, R.; Ambrosini, G.; Groux, R.; Bucher, P. EPD in 2020: Enhanced data visualization and extension to ncRNA promoters. *Nucleic. Acids Res.* **2020**, *48*, D65–D69. [CrossRef]
38. Katz, S.I.; Parker, D.; Turk, J.L. B-cell suppression of delayed hypersensitivity reactions. *Nature* **1974**, *251*, 550–551. [CrossRef]
39. Chong, A.S.; Sage, P.T.; Alegre, M.L. Regulation of Alloantibody Responses. *Front. Cell Dev. Biol.* **2021**, *9*, 706171. [CrossRef]
40. Pinarbasi, E.; Cekin, N.; Bildirici, A.E.; Akin, S.; Yanik, A. STOX1 gene Y153H polymorphism is associated with early-onset preeclampsia in Turkish population. *Gene* **2020**, *754*, 144894. [CrossRef]
41. Shin-Young, K.; Soyeon, P.; Ji Hyae, L.; Jae Hyug, Y.; Moon-Young, K.; Hyun-Young, P.; d Kwang-Soo, L.; Young Ju, K.; Hyun Mee, R. The Y153H Variant of the STOX1 Gene in Korean Patients with Preeclampsia. 2009. *J. Genet. Med.* **2009**, *6*, 56–61.
42. Kivinen, K.; Peterson, H.; Hiltunen, L.; Laivuori, H.; Heino; Tiala, I.; Knuutila, S.; Rasi, V.; Kere, J. Evaluation of STOX1 as a preeclampsia candidate gene in a population-wide sample. *Eur. J. Hum. Genet.* **2007**, *15*, 494–497. [CrossRef] [PubMed]
43. Huppertz, B. Placental origins of preeclampsia: Challenging the current hypothesis. *Hypertension* **2008**, *51*, 970–975. [CrossRef]
44. Orendi, K.; Gauster, M.; Moser, G.; Meiri, H.; Huppertz, B. Effects of vitamins C and E, acetylsalicylic acid and heparin on fusion, beta-hCG and PP13 expression in BeWo cells. *Placenta* **2010**, *31*, 431–438. [CrossRef]
45. Orendi, K.; Gauster, M.; Moser, G.; Meiri, H.; Huppertz, B. The choriocarcinoma cell line BeWo: Syncytial fusion and expression of syncytium-specific proteins. *Reproduction* **2010**, *140*, 759–766. [CrossRef]
46. Nagy, B.; Berkes, E.; Rigo, B.; Ban, Z.; Papp, Z.; Hupuczi, P. Under-expression of CD24 in pre-eclamptic placental tissues determined by quantitative real-time RT-PCR. *Fetal Diagn.* **2008**, *23*, 263–266. [CrossRef]
47. Weber, E.; Lehmann, H.P.; Beck-Sickinger, A.G.; Wawrzynczak, E.J.; Waibel, R.; Folkers, G.; Stahel, R.A. Antibodies to the protein core of the small cell lung cancer workshop antigen cluster-w4 and to the leucocyte workshop antigen CD24 recognize the same short protein sequence leucine-alanine-proline. *Clin. Exp. Immunol.* **1993**, *93*, 279–285. [CrossRef]
48. Kristiansen, G.; Machado, E.; Bretz, N.; Rupp, C.; Winzer, K.J.; Konig, A.K.; Moldenhauer, G.; Marme, F.; Costa, J.; Altevogt, P. Molecular and clinical dissection of CD24 antibody specificity by a comprehensive comparative analysis. *Lab. Investig.* **2010**, *90*, 1102–1116. [CrossRef]
49. Rao, X.; Huang, X.; Zhou, Z.; Lin, X. An improvement of the 2^(-delta delta CT) method for quantitative real-time polymerase chain reaction data analysis. *Biostat. Bioinforma. Biomath.* **2013**, *3*, 71–85.
50. Williams, L.J.; Abdi, H.; French, R.; Orange, J.B. A tutorial on multi-block discriminant correspondence analysis (MUDICA): A new method for analyzing discourse data from clinical populations. *J. Speech Lang Hear Res.* **2010**, *53*, 1372–1393. [CrossRef]

MDPI
St. Alban-Anlage 66
4052 Basel
Switzerland
Tel. +41 61 683 77 34
Fax +41 61 302 89 18
www.mdpi.com

*International Journal of Molecular Sciences* Editorial Office
E-mail: ijms@mdpi.com
www.mdpi.com/journal/ijms

www.ingramcontent.com/pod-product-compliance
Lightning Source LLC
Chambersburg PA
CBHW041016270926
51850CB00024B/427